PROGRESS IN BRAIN RESEARCH

VOLUME 116

# THE GLUTAMATE SYNAPSE AS A THERAPEUTICAL TARGET: MOLECULAR ORGANIZATION AND PATHOLOGY OF THE GLUTAMATE SYNAPSE

Other volumes in PROGRESS IN BRAIN RESEARCH


Volume 93: The Human Hypothalamus in Health and Disease, by D.F. Swaab, M.A. Hofman, M. Mirmiran, R. Ravid and F.W. van Leeuwen (Eds.) – 1992, ISBN 0-444-89538-8.

Volume 94: Neuronal-Astrocytic Interactions: Implications for Normal and Pathological CNS Function, by A.C.H. Yu, L. Hertz, M.D. Norenberg, E. Syková and S.G. Waxman (Eds.) – 1992, ISBN 0-444-89537-X.

Volume 95: The Visually Responsive Neuron: From Basic Neurophysiology to Behavior, by T.P. Hicks, S. Molotchnikoff and T. Ono (Eds.) – 1993, ISBN 0-444-89492-6.

Volume 96: Neurobiology of Ischemic Brain Damage, by K. Kogure, K.-A. Hossman and B.K. Siesjö (Eds.) – 1993, ISBN 0-444-89603-1.

Volume 97: Natural and Artificial Control of Hearing and Balance, by J.H.J. Allum, D.J. Allum-Mecklenburg, F.P. Harris and R. Probst (Eds.) – 1993, ISBN 0-444-81252-0.

Volume 98: Cholinergic Function and Dysfunction, by A.C. Cuello (Ed.) – 1993, ISBN 0-444-89717-8.

Volume 99: Chemical Signalling in the Basal Ganglia, by G.W. Arbuthnott and P.C. Emson (Eds.) – 1993, ISBN 0-444-81562-7.

Volume 100: Neuroscience: From the Molecular to the Cognitive, by F.E. Bloom (Ed.) – 1994, ISBN 0-444-81678-X.

Volume 101: Biological Function of Gangliosides, by L. Svennerholm et al. (Eds.) – 1994, ISBN 0-444-81658-5.

Volume 102: The Self-Organizing Brain: From Growth Cones to Functional Networks, by J. van Pelt, M.A. Corner, H.B.M. Uylings and F.H. Lopes da Silva (Eds.) – 1994, ISBN 0-444-81819-7.

Volume 103: Neural Regeneration, by F.J. Seil (Ed.) – 1994, ISBN 0-444-81727-1.

Volume 104: Neuropeptides in the Spinal Cord, by F. Nyberg, H.S. Sharma and Z. Wiesenfeld-Hallin (Eds.) – 1995, ISBN 0-444-81719-0.

Volume 105: Gene Expression in the Central Nervous System, by A.C.H. Yu et al. (Eds.) – 1995, ISBN 0-444-81852-9.

Volume 106: Current Neurochemical and Pharmacological Aspects of Biogenic Amines, by P.M. Yu, K.F. Tipton and A.A. Boulton (Eds.) – 1995, ISBN 0-444-81938-X.

Volume 107: The Emotional Motor System, by G. Holstege, R. Bandler and C.B. Saper (Eds.) – 1996, ISBN 0-444-81962-2.

Volume 108: Neural Development and Plasticity, by R.R. Mize and R.S. Erzurumlu (Eds.) – 1996, ISBN 0-444-82433-2.

Volume 109: Cholinergic Mechanisms: From Molecular Biology to Clinical Significance, by J. Klein and K. Löffelholz (Eds.) – 1996, ISBN 0-444-82166-X.

Volume 110: Towards the Neurobiology of Chronic Pain, by G. Carli and M. Zimmermann (Eds.) – 1996, ISBN 0-444-82149-X.

Volume 111: Hypothalamic Integration of Circadian Rhythms, by R.M. Buijs, A. Kalsbeek, H.J. Romijn, C.M.A. Pennartz and M. Mirmiran (Eds.) – 1996, ISBN 0-444-82443-X.

Volume 112: Extrageniculostriate Mechanisms Underlying Visually-Guided Orientation Behavior, by M. Norita, T. Bando and B.E. Stein (Eds.) – 1996, ISBN 0-444-82347-6.

Volume 113: The Polymodal Receptor: A Gateway to Pathological Pain, by T. Kumazawa, L. Kruger and K. Mizumura (Eds.) – 1996, ISBN 0-444-82473-1.

Volume 114: The Cerebellum: From Structure to Control, by C.I. de Zeeuw, P. Strata and J. Voogd (Eds.) – 1997, ISBN 0-444-82313-1.

Volume 115: Brain Function in Hot Environment, by H.S. Sharma and J. Westman (Eds.) – 1998, ISBN 0-444-82377-8.

PROGRESS IN BRAIN RESEARCH

VOLUME 116

# THE GLUTAMATE SYNAPSE AS A THERAPEUTICAL TARGET: MOLECULAR ORGANIZATION AND PATHOLOGY OF THE GLUTAMATE SYNAPSE

EDITED BY

O.P. OTTERSEN

*Department of Anatomy, Institute of Basic Medical Sciences, University of Oslo, N-0317 Oslo, Norway*

I.A. LANGMOEN

*Department of Neurosurgery, Karolinska Hospital, S-171 76 Stockholm, Sweden*

L. GJERSTAD

*Department of Neurology, The National Hospital, N-0027 Oslo, Norway*

ELSEVIER
AMSTERDAM – LAUSANNE – NEW YORK – OXFORD – SHANNON – SINGAPORE – TOKYO
1998

ISBN 0-444-82754-4 (volume)
ISBN 0-444-80104-9 (series)

Published by:
Elsevier Science B.V.
P.O. Box 211
1000 AE Amsterdam
The Netherlands

Library of Congress Cataloging in Publication Data
A catalog record from the Library of Congress has been applied for.



Transferred to digital print 2008
Printed and bound by CPI Antony Rowe, Eastbourne

# List of Contributors

B. Alessandri, Medical College of Virginia, Department of Neurosurgery, P.O. Station Box 693, 1225 E. Marshall St., Richmond, VA 23298, USA, e-mail: balessandro@gems.vcu.edu.
J. Andaloro, Department of Pharmacology, University of Nebraska Medical Centre, 600 S 42nd Street, Omaha, NE 68198-6260, USA.
L. Aniksztejn, INSERM U 29, 123 Boulevard de Port-Royal, 75674 Paris Cedex 14, France.
M. Ankarcrona, Karolinska Institutet, Department of Clin. Neuroscience & Family Medicine, Division of Geriatric Medicine, KFC, Novum, S-141 86 Huddinge, Sweden, e-mail: maria.ankarcrona@kfcmail.hs.sll.se.
D. Attwell, Department of Physiology, University College London, Gower Street, London WC1E 6BT, UK, e-mail: D.Attwell@ucl.ac.uk.
G. Battaglia, I.N.M. Neuromed, Pozzilli, University of Catania, Catania, Italy.
M.F. Beal, Department of Neurology, Warren 408, Massachusetts General Hospital, 32 Fruit Street, Boston, MA 02114, USA, e-mail: beal@helix.mgh.harvard.edu.
Y. Ben-Ari, INSERM U 29, 123 Boulevard de Port-Royal, 75674 Paris Cedex 14, France.
L. Bergersen, Department of Anatomy, Institute of Basic Medical Sciences, University of Oslo, POB 1105 Blindern, N-0317 Oslo, Norway, e-mail: linda.bergersen@basalmed.uio.no.
J. Berg-Johnsen, Department of Neurosurgery, The National Hospital, N-0027 Oslo, Norway.
B. Billups, Department of Physiology, University College London, Gower Street, London WC1E 6BT, UK.
K. Borges, Department of Pharmacology, Emory University, 1510 Clifton Road, Atlanta, GA 30322, USA, e-mail: kborges@bimcore.emory.edu.
E. Brodtkorb, Department of Neurology, University of Trondheim, University Hospital, Olav Kyrresgt. 17, N-7006 Trondheim, Norway.
P. Brundin, Section for Neuronal Survival, Wallenberg Neuroscience Center, Department of Physiology and Neuroscience, Lund University, Sölvegatan 17, S-223 62 Lund, Sweden, e-mail: patrik.brundin@neurol.lu.se.
V. Bruno, I.N.M, Neuromed, Pozzilli, University of Catania, Catania, Italy.
R. Bullock, Medical College of Virginia, Department of Neurosurgery, P.O. Station Box 693, 1225 E. Marshal St., Richmond, VA 23298, USA.
G. Casabona, I.N.M, Neuromed, Pozzilli, University of Catania, Catania, Italy.
A. Chapman, Department of Clinical Neuroscience, Institute of Psychiatry, De Crespigny Park, Denmark Hill, London SE5 8AF, UK, e-mail: Astrid.Chapman@iop.bpmf. ac.uk.
F.A. Chaudhry, Department of Anatomy, Institute of Basic Medical Sciences, University of Oslo, POB 1105 Blindern, N-0317 Oslo, Norway, e-mail: f.a.chaudhry@basalmed.uio.no.
P. Congar, INSERM U 29, 123 Boulevard de Port-Royal, 75674 Paris Cedex 14, France.
A. Copani, Institute of Pharmacology, School of Pharmacy, University of Catania, Catania, Italy.
V Crépel, INSERM U 29, 123 Boulevard de Port Royal, 75674 Paris Cedex 14, France.
N.C. Danbolt, Department of Anatomy, Institute of Basic Medical Sciences, University of Oslo, P.O. Box 1105 Blindern, N-0317 Oslo, Norway, e-mail: ncd@pons.uio.no.

Y. Dehnes, Department of Anatomy, Institute of Basic Medical Sciences, University of Oslo, POB 1105 Blindern, N-0317 Oslo, Norway, e-mail: yvette.dehnes@basalmed.uio.no.

R. Dingledine, Department of Pharmacology, Emory University, 1510 Clifton Road, Atlanta, GA 30322, USA.

N.B. Farber, Department of Psychiatry, Washington University, 4940 Children's Place, St. Louis, MO 63110-1093, USA, e-mail: farbern@psychiatry.wustl.edu.

F. Fonnum, Norwegian Defence Research Establishment, Division for Environmental Toxicology, P.O. Box 25, N-2007 Kjeller, Norway, e-mail: frode.fonnum@ffi.no.

E.M. Fykse, Norwegian Defence Research Establishment, Division for Environmental Toxicology, P.O. Box 25, N-2007 Kjeller, Norway.

G.F. Gebhart, Department of Pharmacology, The University of Iowa College of Medicine, 2-471 Bowen Science Building, Iowa City, IA 52242-1109, USA, e-mail: gf.gebhart@-uiowa.edu.

V.D.G. Gerevine, Institue of Pharmacology, School of Pharmacy, University of Catania, Catania, Italy.

L. Gjerstad, Department of Neurology, The National Hospital, N-0027 Oslo, Norway, e-mail: leif.gjerstad@klinmed.uio.no.

H. Gozlan, INSERM U 29, 123 Boulevard de Port-Royal, 75674 Paris Cedex 14, France.

C. Hammond, INSERM U 159, Centre Paul Broca, 2 ter rue d'Alésia, 75014, Paris, France.

T.S. Haugstad, Department of Neurosurgery, The National Hospital, N-0027 Oslo, Norway.

Y-P. Hsueh, Howard Hughes Medical Institute, Massachusetts General Hospital and Harvard Medical School, Boston, MA 02114, USA.

J.G.R. Jefferys, Neuroscience Unit, Department of Physiology, The Medical School, University of Birmingham, Vincent Drive, Edgbaston, Birmingham B15 2TT, UK, e-mail: j.g.r.jefferys@bham.ac.uk.

E. Kvamme, Neurochemical Laboratory, Institute of Basic Medical Sciences, University of Oslo, POB 1115, Blindern, N-0317 Oslo, Norway.

E. Lancelot, Department of Neurology, Warren 408, Massachusetts General Hospital, 32 Fruit Street, Boston MA 02114, USA.

A.S. Landsend, Department of Anatomy, Institute of Basic Medical Sciences, University of Oslo, POB 1105 Blindern, N-0317 Oslo, Norway.

I.A. Langmoen, Department of Neurosurgery, The National Hospital, N-0027 Oslo, Norway, e-mail: a.langmoen@klinmed.uio.no.

K.P. Lehre, Department of Anatomy, Institute of Basic Medical Sciences, University of Oslo, POB 1105 Blindern, N-0317 Oslo, Norway, e-mail: k.p.lehre@basalmed.uio.no.

L.M. Levy, Department of Anatomy, Institute of Basic Medical Sciences, University of Oslo, POB 1105 Blindern, N-0317 Oslo, Norway, e-mail: l.m.levy@basalmed.uio.no.

W-Y. Lu, Department of Physiology, Medical Science Building, University of Toronto, 1 King's College Circle, Toronto, ON M5S 1A8, Canada.

J.F. MacDonald, Department of Physiology, Medical Science Building, University of Toronto, 1 Kings's College Circle, Toronto, ON M5S 1A8, Canada, e-mail: j.mac-donald@utoronto.ca.

B.S. Meldrum, Department of Clinical Neurosciences, Institute of Psychiatry, De Crespigny Park, London SE5 8AF, UK, e-mail: b.meldrum@iop.bpmf.ac.uk.

D.T. Monaghan, Department of Pharmacology, University of Nebraska Medical Centre, 600 S 42nd Street, Omaha, NE 68198-6260, USA, e-mail: dtmonagh@mail.unmc.edu.

N. Nakao, Department of Neurological Surgery, Wakayama Medical College, 7-27 Wakayama, Japan.

J.W. Newcomer, Department of Psychiatry, Washington University, 4940 Children's Place, St. Louis, MO 63110-1093, USA.
R. Ngomba, I.N.M, Neuromed, Pozzilli, University of Catania, Catania, Italy.
D.G. Nicholls, Neurosciences Institute, Department of Pharmacology, University of Dundee, Dundee DD1 9SY, Scotland, UK, e-mail: d.g.nicholls@dundee.ac.uk.
F. Nicoletti, Institute of Pharmacology, School of Pharmacy, University of Catania, Catania, Italy.
J.W. Olney, Department of Psychiatry, Washington University, 4940 Children's Place, St. Louis, MO 63110-1093, USA.
B.A. Orser, Department of Physiology, Medical Science Building, University of Toronto, 1 King's College Circle, Toronto, ON M5S 1A8, Canada.
T. Oshima, Department of Physiology, University College London, Gower Street, London WC1E 6BT, UK.
O.P. Ottersen, Department of Anatomy, Institute of Basic Medical Sciences, University of Oslo, POB 1105 Blindern, N-0317 Oslo, Norway, e-mail: o.p.ottersen@basalmed.uio.no.
R. Raouf, Department of Physiology, Medical Science Building, University of Toronto, 1 King's College Circle, Toronto, ON M5S 1A8, Canada.
I.J. Reynolds, Department of Pharmacology, University of Pittsburgh, E1354 Biomedical Science Tower, Pittsburgh, PA 15261, USA, e-mail: ijr@prophet.pharm.pitt.edu.
E. Rinvik, Department of Anatomy, Institute of Basic Medical Sciences, University of Oslo, POB 1105 Blindern, N-0317 Oslo, Norway, e-mail: eric.rinvik@basalmed.uio.no.
K.W. Roche, Laboratory of Neurochemistry, National Institute on Deafness and Other Communication Disorders, NIH, Bethesda, MD, USA.
S. Roseth, Norwegian Defence Research Establishment, Division for Environmental Toxicology, P.O. Box 25, N-2007 Kjeller, Norway.
D. Rossi, Department of Physiology, University College London, Gower Street, London WC1E 6BT, UK.
M. Sarantis, Department of Physiology, University College London, Gower Street, London WC1E 6BT UK.
R.D. Scheyer, Global Clinical Pharmacology, Hoechst Marion Roussel, Inc., P.O. Box 6800, Bridgewater, NJ 08807-0800, USA, e-mail: Richard.Scheyer@hmrag.com.
M. Sheng, Howard Hughes Medical Institute, Massachusetts General Hospital and Harvard Medical School, Boston, MA 02114, USA, e-mail: sheng@helix.mgh.harvard.edu.
D.A. Skifter, Department of Pharmacology, University of Nebraska Medical Centre, 600 S 42nd Street, Omaha, NE 68198-6260, USA.
J. Storm-Mathisen, Department of Anatomy, Institute of Basic Medical Sciences, University of Oslo, POB 1105 Blindern, N-0317 Oslo, Norway, e-mail: jon.storm-mathisen@basalmed.uio.no.
M. Storto, I.N.M. Neuromed, Pozzilli, University of Catania, Catania, Italy.
M. Szatkowski, Department of Physiology, University College London, Gower Street, London WC1E 6BT, UK.
M. Takahashi, Department of Physiology, University College London, Gower Street, London WC1E 6BT, UK.
Y. Takumi, Department of Anatomy, Institute of Basic Medical Sciences, University of Oslo, POB 1105 Blindern, N-0317 Oslo, Norway, e-mail: yutaka.takumi@basalmed.uio.no.
E. Taubøll, Department of Neurology, The National Hospital, N-0027 Oslo, Norway.
R.D. Traub, Neuroscience Unit, Department of Physiology, The Medical School, University of Birmingham, Vincent Drive, Edgbaston, Birmingham B15 2TT, UK.

L. Trussell, Department of Physiology, University of Wisconsin, 1300 University Avenue, Madison, WI 53706, USA, e-mail: trussell@physiology.wisc.edu.

T.J. Turner, Department of Physiology, Tufts University School of Medicine, Boston, MA 02111-1800, USA, e-mail: tturner@opal.tufts.edu.

K. Ullensvang, Department of Anatomy, Institute of Basic Medical Sciences, University of Oslo, POB 1105 Blindern, N-0317 Oslo, Norway, e-mail: kyrre.ullensvang@studmed.uio.no.

M.O. Urban, Department of Pharmacology, The University of Iowa College of Medicine, 2-471 Bowen Science Building, Iowa City, IA 52242-1109, USA.

O. Warr, Department of Physiology, University College London, Gower Street, London WC1E 6BT, UK.

R.J. Wenthold, Laboratory of Neurochemistry, National Institute on Deafness and Other Communication Disorders, NIH, Bethesda, MD, USA, e-mail: wenthold@pop.nidcd.nih.gov.

X-G. Xiong, Department of Physiology, Medical Science Building, University of Toronto, 1 King's College Circle, Toronto, ON M5S 1A8, Canada.

# Preface

The purpose of this volume is twofold: to provide an update on basic research on glutamate neurotransmission and to identify promising directions for therapeutic strategies aimed at the glutamate synapse. While the book covers the entire sequence of events in signal transfer – from synthesis and vesicular accumulation of glutamate to its clearance following receptor activation – emphasis is placed on those steps that are currently held to be of particular importance from a pathophysiological and therapeutic point of view.

In keeping with the clinical perspective, several of the chapters deal with mechanisms of acute and chronic degeneration in the central nervous system, and with epilepsy. The link between glutamate and degeneration, discovered by Lucas and Newhouse in the late 1950s and explored in the subsequent studies by Olney and colleagues, is now widely recognized and commonly referred to by the term "excitotoxicity". Excitotoxicity may be involved in the pathogenesis of a number of neurological diseases and considerable efforts are made to find ways to mitigate it; at the glutamate receptor level, or downstream or upstream of this. As to epilepsy, the relation to glutamate is obvious from the fact that this transmitter provides the major excitatory drive of neurons in the central nervous system. Further, some of the antiepileptic drugs in current use may exert their effect at least in part through their interaction with glutamate transmission. For this reason the field of epilepsy is used as a backdrop for a discussion (Chapter 26) of some of the clinical concerns relating to the use of drugs that act on neuronal transmission and excitability.

The signal transduction at the glutamate synapse is influenced by a number of factors including presynaptic receptors, synaptic geometry, postsynaptic receptor profile, and the properties and density of glutamate transporters. How such factors interact to produce the postsynaptic response is now beginning to be resolved and will be considered in this book. In a fast synapse it is also a matter of importance how key molecules like receptors and transporters are arranged in relation to the release site and each other. Also in this area rapid advances are being made, based on the combined application of molecular biology and high resolution immunocytochemistry. Therefore, the molecular organization of the synapse will be a recurrent theme in the present volume.

More than ever is it clear that glutamatergic synapses are not merely stereotypical contacts for point-to-point transmission. Rather they are dynamic functional interfaces with substantial heterogeneity at the molecular as well as the structural level. In this complexity lies a challenge for future research, and promising prospects for the development of drugs targeting specific subpopulations of synapses.

The idea to prepare a volume of the present scope arose during a symposium that was arranged in Oslo, Norway, in May 1996 under the sponsorship of Glaxo Wellcome.

Glaxo Wellcome generously agreed to cover expenses incurred during the planning and editing of this book. Special thanks are extended to Gottfred Ansteensen for his continuous support.

We are also grateful to Ms. Kirsten Haga for her skillful and enthusiastic contribution toward the completion of this project.

Ole P. Ottersen
Iver A. Langmoen
Leif Gjerstad

# Dedication to Frode Fonnum

It is a pleasure, and highly appropriate, to dedicate the present volume on GLUTAMATE to Professor Dr. Philos., Frode Fonnum. Frode, who recently celebrated his 60th birthday, is a pioneer in the study of glutamate, and in establishing its role as transmitter in the brain. His 1984 review paper on this subject (*J. Neurochem.*, 42: 1–11) continues to be cited very frequently – and its author continues to do important work on glutamate.

Born in Oslo 3rd February 1937, Frode went to Scotland to study biochemistry, graduating in 1960 as B.Sc. in Applied Chemistry from Herriot Watt Technical University, Edinburgh. He trained in neurochemistry in the laboratory of Victor P. Whittaker at Babraham, Cambridge, UK, from 1965 to 1966. In 1970 he obtained his Doctor Philosophiae degree at the University of Oslo. He has worked since 1961 at the Norwegian Defence Research Establishment, Division for Environmental Toxicology, serving as Research Director from 1972 to 1996, when he relinquished his administrative duties, but continues to work in his laboratory there as VISTA Professor (granted by Statoil and the Norwegian Academy of Science and Letters). Since 1984 he has served in addition as Professor of Toxicology at the Faculty of Mathematics and Natural Sciences, University of Oslo.

Frode Fonnum has served on many national and international scientific councils (including the European Training Programme in Brain and Behaviour, Max-Planck-Institut für Biophysikalische Chemie, Göttingen, and the Norwegian Research Council), and on government-appointed boards. He has worked zealously for the control of environmental pollution, and has participated as expert consultant under the Disarmament Negotiations in Geneva. He continues to serve on the editorial board of several journals and was Deputy Chief Editor in the Journal of Neurochemistry from 1975 to 1985. From 1991 to 1993 he was chairman of the International Society for Neurochemistry (ISN). He was elected member of the Norwegian Academy of Science and letters in 1992.

Working at an institution that is principally dedicated to applied research, Frode Fonnum has nonetheless managed to establish a strong group doing basic research in neuroscience. Many of his former collaborators and students continue to be successful scientists in Norway and abroad. Among Frode's 200 published papers are significant studies on toxicology (e.g., organophosphates, trimethyltin, algal toxins, excitotoxicity) and endocrine changes in stress, but he is above all known for his studies on neurotransmitter biochemistry. His important early work on the cholinergic system, where he introduced new methodology (*Biochem. J.*, 115: 465–472; *J. Neurochem.*, 24: 407–409), is still very highly cited. At the 2nd ISN congress in Milan 1969 he presented the first demonstration that GABA is synthesised specifically in the terminals of an inhibitory neuron in the vertebrate brain (*Brain Res.*, 20: 259–275). He also provided early evidence that this amino acid has functions other than as a transmitter (*Nature*, 228: 1001). In several studies from 1977 onwards Fonnum and collaborators showed that glutamate uptake is selectively associated with excitatory nerve pathways (e.g., *Nature*, 266: 377–378; *Neuroscience*, 5: 1691–1698, 6: 863–873). They followed this up with studies on amino acid contents and evoked release, contributing important evidence for the transmitter role of glutamate. Behavioural studies provided evidence for the involvement of glutamatergic mechanisms in cognitive function (*Exp. Brain Res.*, 77: 499–506; *Brain Res. Dev. Brain Res.*, 99: 126–130). Fonnum and colleagues were the first to show uptake of GABA in synaptic vesicles isolated from brain (*J. Neurochem.*, 50: 1237–1242) and that glycine shares a synaptic vesicular carrier with GABA (*J. Neurochem.*, 54: 1142–1147). They have introduced potent inhibitors of the synaptic vesicular glutamate transporter (*J. Neurochem.*, 65: 96–103), and recently shown that arachidonic acid can modulate vesicular uptake of glutamate and GABA (*Eur. J. Pharmacol.*, 341: 281–288). In an impressive series of papers over three decades, using a wide range of methods, Fonnum with various collaborators has characterised the pathways of synthesis and metabolism of glutamate and related amino acids (e.g. *J. Cereb. Blood Flow Metab.*, 17: 1230–1238).

Apart from being a highly successful and respected scientist, Frode is a very honest and frank person, with a positive attitude and a good sense of humour, and with the ability to show concern for his fellow human beings. These too are features appreciated by friends and colleagues.

Leif Gjerstad
Iver A. Langmoen
Ole Petter Ottersen
Jon Storm-Mathisen

# Contents

## III. Glutamate receptors



## IV. Glutamate toxicity and cell death in the central nervous system

**VI Perspectives**

SECTION I

# Glutamate release and transport

O.P. Ottersen, I.A. Langmoen and L. Gjerstad (Eds.)
*Progress in Brain Research*, Vol 116


CHAPTER 1

# Calcium channels coupled to glutamate release

Timothy J. Turner

*Departments of Physiology and Neuroscience, Tufts University School of Medicine, Boston, MA 02111, USA*

Voltage-gated calcium channels stand as the pivotal regulatory element of excitation-secretion coupling at nerve endings in brain. In light of the critical role that calcium channels play in intercellular communication, a thorough understanding of the properties of these presynaptic calcium channels is a prerequisite to understanding the molecular interactions that are responsible for neurosecretion of glutamate and most other neurotransmitters.

The relationship between presynaptic calcium channels and the neurosecretory apparatus has been firmly established by work spanning the last 40 years. Recent advances have resulted in an explosive increase in information about the identity of both the calcium channels and the proteins that comprise the neurosecretory machinery. This information has helped to guide our search for a complete description of how calcium channel activity promotes vesicle fusion and exocytosis. The focus of this chapter is to review the current understanding of the properties of presynaptic calcium channels, and consider how defects in the channels or their interactions might result in pathological abnormalities. Mechanisms for presynaptic modulation of glutamate release are reviewed by Nicholls in this volume.

## Properties of neuronal calcium channels

While it has been appreciated for many years that multiple types of calcium channels are expressed in many different tissues, an accurate picture of the extent of the diversity of calcium channels has only begun to surface in the past five years. Analysis of calcium channel gene expression has revealed numerous related genes, each of which with splice variants that in principle could amount to impressive combinatorial possibilities. During the same period, studies of calcium currents in neurons as well as heterologous expression systems has led to a fuller understanding of the biophysical and pharmacological properties of the various channel types. A remaining challenge is to correlate these two lines of investigation, i.e., to develop a complete description of the molecular components of calcium channels that underlie the diverse current phenotypes observed in native neurons.

### *Pharmacological and biophysical characteristics of native neuronal calcium channels*

The development of tight-seal gigaohm recording techniques during the 1980's greatly facilitated the ability to record neuronal calcium currents, and it quickly became apparent that $Ca^{2+}$ currents in most neurons are heterogeneous with regard to activation voltages, inactivation rates, and toxin sensitivities. At that time, it was known that calcium currents in many types of muscle were blocked by the 1,4-dihydropyridine class of drugs widely used clinically as antihypertensive and antianginal drugs (Fleckenstein, 1977). These agents blocked a small fraction of somatic calcium currents, generally less than 50%, but had no effect on $^{45}Ca^{2+}$ flux or neurotransmitter release at the nerve terminal (Nachshen and Blaustein, 1979; Suszkiw et al., 1986). Around that time, Olivera and colleagues (Olivera et al., 1985; 1990; 1994) developed a number of peptide neurotoxins from various *Conus* species that were high-affinity inhibitors of neuronal but not muscle (Cruz et al., 1987) calcium currents. This was a crucial development as it provided a set of tools to distinguish different calcium channels. A toxin from *Conus geographus*, ω-conotoxin GVIA (ωCTX GVIA) referred to as an ω-toxin (since it blocked nerve "endings") was of particular interest, since it blocked somatic calcium currents as well as neurotransmitter release in a number of preparations.

Using the toxin in combination with a 1,4-dihydropyridine, Nowycky et al. (1985) demonstrated that chick sensory neurons had three types of currents that could be distinguished on biophysical and pharmacological criteria; T-type ("tiny", "transient"), the dihydropyridine-sensitive L type ("large", "long-lasting"), and N-type ("neither" L nor T) that was subsequently shown to be blocked by ω-CTX GVIA. The currents could be separated on biophysical grounds; the T current was inactivated at modestly depolarized potentials, and the N current often showed more prominent inactivation kinetics than L current, particularly at negative holding potentials.

Another prominent calcium channel type was identified by Llinas and colleagues. They recorded calcium currents in cerebellar Purkinje neurons that were distinct from L- or N-type currents in that they carried a sustained current (designated P-type for Purkinje) that showed little inactivation, and was not blocked by either 1,4-dihydropyridines or ω-CTX GVIA (Llinas et al., 1989). Specific inhibitors of P channels were soon discovered in spider venom of *Agelenopsis aperta* by Adams and colleagues (Olivera et al., 1994). One of the peptides purified from that venom, designated ω-Aga IVA, was shown to be a specific and high-affinity inhibitor of P channels in rat Purkinje neurons (Mintz et al., 1992a), with an IC50 of 1–3 nM (Mintz and Bean, 1993).

Two other peptide toxins, obtained from *Conus magus* and designated ω-CTX MVIIA and ω-CTX MVIIC, are valuable tools to study neuronal calcium channels. ω-CTX MVIIA, like ω-CTX GVIA, is a high-affinity specific blocker of N-type currents (Olivera et al., 1987). Unlike ω-CTX GVIA which is nearly irreversible in mammalian systems, ω-CTX MVIIA has a reasonable dissociation rate constant, making this a suitable ligand for equilibrium experiments. ω-CTX MVIIC was discovered by a cloning strategy, probing cDNA libraries of *Conus magus* venom ducts using sequences that capitalized on the high degree of conservation of the six cysteine residues in ω-conotoxins (Hillyard et al., 1992). The sequence obtained by this strategy was converted into a synthetic peptide that blocks both N-type and P-type currents in many neurons.

Many subsequent studies have demonstrated that the four classes of neuronal calcium channel are widely distributed throughout the CNS, and that diverse populations of neurons can express any combination of channel types in various proportions. At this time it is not clear why neurons require multiple channel types, what tasks each type subserves, and significantly, whether the properties of calcium channels at the nerve terminal is identical to those in the soma. Some of these issues will be taken on below.

*Multiple genes encode neuronal calcium channels*

Intense work over the past ten years has provided substantial information regarding the proteins that constitute neuronal calcium channels, and the genes that encode them (for reviews, see McCleskey, 1994; Dunlap et al., 1995). Initially, conventional strategies were used to obtain protein sequence information that was in turn used to clone neuronal calcium channels (Campbell et al., 1988). The skeletal muscle T-tubule preparation is highly enriched in L-type calcium channels that bind 1,4-dihydropyridines with nanomolar affinity. The high-affinity binding of radiolabeled 1,4-

dihydropyridines provided a tag that allowed several groups to purify the skeletal muscle calcium channel from the T-tubule preparation. The calcium channel complex consists of at least three subunits; the pore-forming subunit $\alpha_1$ (~180 kDa), a glycoprotein $\alpha_2$ subunit (~140 kDa) disulfide-linked to the $\delta$ subunit (~35 kDa), and a peripheral $\beta$ subunit (~54 kDa). The amino acid sequence of $\alpha_1$ subunit of the skeletal channel was used in the molecular cloning of its gene, and provided a strategy to clone homologous calcium channel genes from brain using low-stringency hybridization. Thus far, six distinct calcium channel $\alpha_1$ subunits have been cloned (each class designated by the subscript A, B, C, D, E, and S [Birnbaumer et al., 1994]), and all except the S (or skeletal) class are expressed in brain . At least one other subunit (the "95 kDa" subunit in neurons, the $\gamma$ subunit in muscle) co-purifies with calcium channel protein complexes purified from brain, but these have yet to be cloned.

*Functional expression of calcium channel gene products*

The use of heterologous expression systems has been instrumental to the provisional assignment of functional roles to the various subunits, as well as gross characterization of the broad current classes. The two most successful expression systems for calcium channels have been the *Xenopus* oocyte and human embryonic kidney cell line HEK293. Complementary DNA or RNA are injected or transfected into the cells in question, and channel expression can be assayed by standard electrophysiological recording methods under the appropriate ionic conditions designed to isolate calcium currents. In general, the $\alpha_1$ subunit (which by sequence analysis is thought to contain the ion-permeation pore as well as the voltage sensor) is necessary but not sufficient for robust expression of functional calcium currents. Co-expression of the $\beta$ subunit dramatically enhances the amplitude of calcium currents. In addition, co-expression of $\beta$ subunits can affect gating properties of the channel complex, thereby shifting the voltage sensitivity and inactivation rates (Castellano and Perez-Reyes, 1994). The remaining subunits, $\alpha_2$-$\delta$ and the 95 kDa subunit, do not have obvious functional roles, since co-expression of $\alpha_2$-$\delta$ has little consequence for currents recorded in the heterologous systems, and it is not yet possible to express the 95 kDa subunit due to the lack of a cloned sequence.

The complex properties of neuronal calcium currents are complemented by the broad array of calcium channel genes (as well as their splice variants) identified thus far. A remaining challenge is to relate the functional phenotype of neuronal calcium currents to discrete calcium channel gene products. Heterologous expression of channel will be expressed, facilitating the comparison between native channels and calcium channel gene products. However, a lingering concern over this strategy is the uncertain fidelity of expression of calcium channel complexes in injected or transfected, non-neuronal cells. Differences in post-translational processing of the channel could result in subtle differences in its physiological or pharmacological properties, confounding the efforts to dissect the contributions of unique channel complexes to the heterogeneous macroscopic currents.

At present, there is reasonable concordance between native N- and L-type currents and currents resulting from expression of class B and class C or D genes, respectively. In each case, the macroscopic biophysical (activation voltage range, inactivation rates, etc.) and pharmacological properties (e.g., sensitivity to $\omega$-CTX GVIA) of the expressed proteins are similar to expectations based on neuronal currents. Impressive studies by Yue and colleagues (Patil et al., 1996) have demonstrated that single channel properties of class B channels expressed in HEK 293 cells are consistent with those of N-type channels in the native environment, and have shown this reconstituted model system to be an important means of studying the modulation of calcium channel activity.

The situation is less clear for the class A channel. On the basis of expression patterns, where $\alpha_{1A}$ is enriched in cerebellum, it was proposed that the class A channel mediates the P-type current prominent in cerebellar Purkinje neurons. P currents are characterized by high-affinity blockade by ω-Aga-IVA and by slow kinetics of inactivation in response to prolonged depolarizing steps (Mintz et al., 1992b). By contrast, when $\alpha_{1A}$ channels are expressed in *Xenopus* oocytes, resulting calcium currents show considerable inactivation and are much less sensitive to ω-Aga-IVA (Sather et al., 1993). Tsien and colleagues have proposed that this current be designated "Q-type" to distinguish it from P-type currents described in cerebellum (Zhang et al., 1993). As discussed below, it remains to be determined whether the differences between P-type and Q-type currents are substantive, but nonetheless there is general agreement that the P/Q channel is encoded by the $\alpha_{1A}$ gene.

The remaining neuronal calcium channel gene, $\alpha_{1E}$, has remained something of an enigma. Initial reports suggested that it may encode the T-type calcium current, but expression studies indicate that its properties are rather divergent from those expected of a T channel (Soong et al., 1993). In particular, the T current is distinguished by a low-voltage activation range; currents mediated by $\alpha_{1E}$ fall into a high-voltage activation range like the other calcium channel classes. However, the ion permeation properties, especially with respect to block by $Ni^{2+}$, are in line with those expected of T currents (Williams et al., 1994). Because of the imperfect agreement between native and expression currents, it is premature to conclude that $\alpha_{1E}$ encodes T channels. It is expected that further research will reveal the nature of these differences and allow conclusions to be made with regard to the identity of the $\alpha_{1E}$ channel.

The conclusion drawn from these studies is that expression of defined combinations of genes encoding calcium channels has been of great value to our understanding of the identity of calcium channel proteins. The properties of the class B channel, and its modulation by G-protein coupled receptors, is well-replicated by heterologous expression, and this approach promises to be invaluable for study of the mechanisms that underlie channel regulation by cellular signaling systems. However, the results of class A and class E expression have been less clear-cut, reminding us that epigenetic factors involved in calcium channel expression (such as post-translational modifications as well as cellular environments) are important determinants of channel physiology and pharmacology.

**Identifying presynaptic calcium channels**

The nerve terminal is of particular importance to neuroscientists, since it is the cellular locus where electrical information is transduced to intercellular chemical messages, and calcium channels play a pivotal role in this transduction. Thus, it is of obvious significance to understand which calcium channels participate, how they are integrated into the secretory apparatus, and how their activities are regulated. Unfortunately, the nerve terminal presents a special technical problem due to the small size. This has made conventional electrophysiological approaches impossible in most cases with the exception of large calical nerve terminals (Stanley, 1993). In order to surmount this obstacle, a number of creative solutions have been employed in our effort to gain a more complete description of the working of calcium channels in mediating excitation-secretion coupling at CNS synapses.

*Synaptosomes as a biochemical preparation to study excitation-secretion coupling*

One approach to studying nerve terminals is to disrupt the structure of the brain by grinding (i.e., homogenizing) the tissue in isotonic sucrose. The shearing forces produced cause the presynaptic bouton to be severed from the axon and the postsynaptic dendrite, and many such boutons will reseal, resulting in a population of dissociated nerve terminals. These terminals can be separated from the remaining debris via differential centrifugation, resulting in a preparation enriched in resealed nerve terminals. This preparation has been characterized extensively over a period exceeding thirty years, and it has been shown that synaptosomes retain many of the functional properties expected for intact nerve terminals, including the ability to store and release neurotransmitters (McMahon and Nicholls, 1991; Nicholls, this volume). Our strategy for study of excitation-secretion coupling in synaptosomes emphasizes the use of as many rigorous functional criteria as possible to focus on high-level activities that could only be supported by relatively intact nerve terminals. Furthermore, we have improved on manual biochemical methods by concentrating on the subsecond kinetics of transmitter release, approaching the time scale on which physiological events are thought to occur (Turner and Dunlap, 1995). This results in a fruitful means of obtaining reproducible data on properties of neurosecretion common to most synaptic endings, since the results reflect the averaged activity of billions of terminals. Further, these experiments have proven to be of predictive value, guiding more delicate experiments such as brain slice recordings (Luebke et al., 1993).

The characterization of synaptosomal calcium channels has been primarily pharmacological. We have used the ability of calcium channel blockers to inhibit [$^3$H]glutamate release as an assay to investigate the relative contributions of neuronal calcium channels to neurosecretion. The most efficacious pharmacological agent is ω-Aga IVA, the selective P channel blocker for somatic calcium currents. In the synaptosomal system, the properties of ω-Aga IVA vary according to the stimulation conditions employed. Both the efficacy and potency of ω-Aga IVA are enhanced when low-intensity stimulation (modest depolarization, short stimulus duration, low external calcium concentrations) is used; at high levels of stimulation, little or no block is produced (Turner et al., 1992; Luebke et al., 1993; Turner and Dunlap, 1995). Likewise, the selective N channel blocker ω-CTX GVIA is ineffective except at weak stimulation conditions, and is generally much less efficacious than ω-Aga IVA. Finally, the 1,4-dihydropyridines are ineffective in altering glutamate release, suggesting that L channels are not present in significant numbers at glutamatergic presynaptic elements; non-selective peptide toxins (such as ω-Aga III) that block L channels corroborate this conclusion (Turner et al., 1993).

The perturbation of neurosecretion by selective calcium channel blockers reveals important details regarding the relationship between calcium channels and the neurosecretory apparatus. The observation that both ω-Aga IVA and ω-CTX GVIA block release indicate that both P- and N-type channels regulate glutamate release in mammalian CNS. The way in which these calcium channel types interact was demonstrated by combining the specific blockers (Turner and Dunlap, 1995). By examining the concentration-response relationship for ω-Aga IVA in the presence of a saturating concentration of ω-CTX GVIA, it is observed that the "apparent" efficacy and potency of ω-Aga IVA is enhanced when compared with the same experiment in the absence of ω-CTX GVIA. Under standard assay conditions, ω-Aga IVA blocks ~50% of synaptosomal glutamate release with an IC50 of ~200 nM. Using the same stimulus in the presence of saturating (100 nM) ω-CTX GVIA, ω-Aga IVA blocks ~90% of glutamate release with an IC50 of ~12 nM. The same effect is observed when ω-CTX GVIA is varied in the presence of saturating ω-Aga IVA. Thus, these peptides demonstrate synergy of action, and this synergy has an important ramification for the structure of the presynaptic machinery. In order for the reciprocal synergistic interaction to occur, both N- and P-type channels must each contribute to the neurosecretory process. When either N- or P-channels alone are blocked, the other unblocked type can support secretion, albeit at a reduced level. Blockade of both classes of channel is necessary to abolish secretion. This observation has important implications for the therapeutic use of calcium channel blockers.

### *Synaptic potentials in the brain slice as an assay of presynaptic calcium channels*

A sophisticated means of measuring synaptic events in a more intact system is the classical method of hippocampal slice recording. The hippocampus is a superb preparation to study synaptic properties due to the anatomy that delineates a well-defined circuitry, a uniform population of pyramidal neurons to record from, and the ability to pharmacologically isolate excitatory postsynaptic potentials (or currents) mediated by glutamate from inhibitory ones mediated by GABA. Further, synaptic events evoked by physiological stimuli (i.e., presynaptic action potentials) can be studied with a high degree of sensitivity. When calcium channel blockers are applied to hippocampal slices, both ω-Aga IVA and ω-CTX GVIA block glutamatergic transmission, but the efficacy of these blockers is substantially higher than seen in synaptosomes (Takahashi and Momiyama, 1993; Luebke et al., 1993; Castillo et al., 1994; Wheeler et al., 1994). Typically, ω-Aga IVA blocks 90% of synaptic release of glutamate, and ω-CTX GVIA blocks ~50% of the response. The greater-than-additive effects of the two inhibitors is another manifestation of the synergy of the P/Q and N type channels in regulating glutamate release.

A more informative technique for characterizing presynaptic calcium channels in brain slice preparations combines traditional electrophysiological recording of postsynaptic responses with optical recording of presynaptic calcium transients using calcium indicator dyes (Wu and Saggau, 1994; Regehr and Atluri, 1995). Anatomically-defined afferent tracts innervating discrete regions (e.g., the Schaffer collateral tract innervating CA3 pyramidal neurons in hippocampus or the parallel fibers innervating the Purkinje neurons of cerebellum) are loaded with calcium indicator dyes (such as fura-2 and related derivatives) by perfusing the afferent axons with the membrane-permeant form of the dye. The ester is then hydrolyzed by enzymes in the axoplasm, trapping the indicator and allowing it to diffuse intracellularly. Over a period of hours, the dye arrives in the terminals synapsing on the target neuron, providing a fluorescent signal proportional to the spatially-averaged presynaptic calcium concentrations, which can be used as a measure of calcium channel activity. By relating the presynaptic calcium channel activity to the postsynaptic response, investigators have been able to reaffirm the highly non-linear relationship between calcium entry and neurosecretion. As originally reported by Dodge and Rahamimoff (1967), this approximates a fourth-order (i.e., strongly sigmoidal) relationship between calcium and exocytosis. Thus, the effect of calcium channel blockers such as ω-Aga IVA is to reduce the presynaptic calcium transient by ~50%, but to diminish the postsynaptic response by ~95%. This relationship provides a mechanistic basis for the observed synergy between the toxins and release, since a 50% reduction would produce a 93.75% inhibition for a fourth power relationship ($[0.5]^4 = 0.0625$).

In summary, results obtained over a wide range of conditions indicate that glutamate release is controlled by multiple calcium channels that are co-localized at most active zones. ω-Aga IVA sensitive P-type channels provide most of the calcium to trigger release. ω-CTX GVIA sensitive N-type channels provide less calcium entry, but available evidence suggests that the N-type channel is more sensitive to modulation by G-protein coupled receptors than the P-type channel. Finally, a small amount of transmission remains after saturating concentrations of both selective antagonists are applied to the synapse, suggesting that a "resistant" type channel may contribute a small amount of calcium to the release process.

### *A more detailed pharmacology of neuronal calcium channels*

As a preface to consideration of the therapeutic potential of calcium channel blockers, it is germane to consider the detailed pharmacological properties of some of the more prominent agents, since there appears to be considerable confusion in the literature. Much of the confusion has arisen due to misunderstanding of the kinetics of peptide association with calcium channels. There are two fairly selective inhibitors of the N-type channel (as mentioned above), ω-CTX GVIA and ω-CTX MVIIA, differing primarily in the dissociation rate constant. The dissociation rate of ω-CTX GVIA is so slow as to make this peptide essentially irreversible, whereas ω-CTX MVIIA has a dissociation rate constant in the $10^{-3}$ $s^{-1}$ range, that is, it is fairly reversible. However, in lower species (e.g., amphibia), this distinction is less clear (Boland et al., 1994); the dissociation rate constant is higher and therefore ω-CTX GVIA is much less potent ($K_d = k_{-1}/k_1$).

The properties of inhibitors of ω-Aga IVA sensitive calcium channels are much more difficult to decode from the literature accounts of their effects on neuronal calcium currents. Most of the uncertainty revolves around whether the pharmacological properties of ω-Aga IVA on calcium currents in cerebellar Purkinje neurons are typical of all ω-Aga IVA sensitive currents. The potency of block of Purkinje cell calcium currents by this peptide is reported to be near 3 nM (Mintz et al., 1992; Mintz and Bean, 1993). By contrast, most quantitative studies of ω-Aga IVA action on neuronal calcium channels find that its potency is somewhat less, between 10 and 100 nM (Dunlap et al., 1995; Randall and Tsien, 1995). Conversely, the first studies of the action of ω-CTX MVIIC on Purkinje cell currents showed a weak but nearly complete inhibition when micromolar concentrations of peptide were used (Hillyard et al., 1992), while ω-CTX MVIIC is a fairly potent inhibitor of the ω-Aga IVA sensitive calcium channel in synaptosomal (Guar et al., 1994; Keith et al., 1995; Malva et al., 1995; Turner and Dunlap, 1995; Turner et al., 1995) and brain slice preparations (Wheeler et al., 1994). Further, when $\alpha_{1A}$ cDNA is expressed in oocytes, the currents are rather insensitive to ω-Aga IVA, but are blocked by ω-CTX MVIIC with a distinctly prolonged time course (Sather et al., 1993). On the basis of these initial observations, Tsien and colleagues proposed that a novel channel type (termed Q) was relatively insensitive to ω-Aga IVA and selectively inhibited by ω-CTX MVIIC (Zhang et al., 1993; Randall and Tsien, 1995).

More detailed investigations of the pharmacological properties of the toxins both in native neurons and in various expression systems indicate that the distinctions described above are not as clear as they initially appeared. First, the early experiments using ω-CTX MVIIC to block Purkinje cell currents did not take into account the unusually slow kinetics of association of this peptide with their targets. When sufficient time is allowed for the toxin to bind, it is consistently observed that the toxin blocks Purkinje cell currents with an IC50 of 10–50 nM (McDonough et al., 1996), similar to its potency in inhibiting ω-Aga IVA-sensitive glutamate release from nerve terminals. Secondly, the potency of ω-Aga IVA has certainly been understated in many experiments due to slow kinetics of association, especially at low nanomolar concentrations that have been used to define P-type channels. Consequently, arbitrary conditons (e.g., current or secretion blocked by 20 nM ω-Aga IVA) have been established as a litmus test to define P-type channels (Randall and Tsien, 1995). In many instances, this test has been wrongly applied, using experimental protocols that do not allow sufficient time for the toxin to reach equilibrium. Based on published values for kinetic constants for ω-Aga IVA block of Purkinje cell currents, 95% block would be achieved in ~40 minutes, provided there are no diffusion barriers to retard peptide access to the channel, such as in brain slice preparations (Dunlap et al., 1994). Third, ω-CTX MVIIC is not a selective blocker; it also reversibly blocks ω-CTX GVIA sensitive N current with a similar IC50 (~10 nM) values, albeit with faster kinetics than the block of Purkinje cell currents (Grantham et al., 1994; McDonough et al., 1996). It is not uncommon to see reports where 1 μM ω-CTX MVIIC is applied to a preparation, and the current blocked is identified as Q-type, when in fact it could be N, P, or Q.

The most satisfactory means to resolve these issues is to refine our nomenclature. There is broad agreement that the class B channel (encoded by $\alpha_{1B}$) is uniquely sensitive to ω-CTX GVIA in mammalian tissues, and thus the N current is mediated by class B channels. Likewise, while there may be some disagreement about the precise properties of P-type and Q-type currents, it is also generally agreed that they are both mediated by class A (encoded by $\alpha_{1A}$) channels. It will be of particular interest to determine whether perceived differences in the properties of the class A channel are due to different molecular properties of the class A channel complex (i.e., different $\beta$ subunits, splice variants, etc.). Finally, one must exercise caution when interpreting the results of toxin action. These experiments are best done when toxins are used in a quantitative, equilibrium manner rather than as a simple qualitative one. Caveat emptor.

## Therapeutic potential for inhibitors of presynaptic calcium channels

Calcium channels, by virtue of regulating excitation-secretion coupling, provide an attractive target for pharmacological management of neuronal activity. Drugs that selectively act at calcium channels in smooth and cardiac muscle (e.g., nifedipine) have been in wide use for almost two decades. Much research has gone towards developing inhibitors of presynaptic calcium channels, but there progress has been slowed by several confounding problems. As a means of illustrating the therapeutic potential of such drugs, we will consider three different scenarios that may have promise for clinical applications of inhibitors of presynaptic calcium channels.

### *Ischemic stroke – loss of regulated glutamate exocytosis*

It is well-established that ischemic episodes, such as those produced by stroke or trauma, can lead to massive release of glutamate that eventually leads to excitotoxicity and neuronal death (Crépel et al., this volume). Anoxia causes ATP levels to fall, resulting in diminished Na-K ATPase activity. Thus, intracellular $K^+$ levels fall and extracellular $K^+$ levels increase, resulting in a wide-spread and long lasting depolarization of neurons and glia. Such ionic conditions can lead to glutamate accumulation in the extracellular space due to activation of calcium channels followed by persistent calcium-dependent exocytosis accompanied by impaired reuptake by $Na^+$- and $K^+$-dependent glutamate transporters. In theory, inhibitors of presynaptic calcium channels could blunt the unregulated release process.

The pathologic scenario in ischemic conditions is an important factor in considering pharmacological applications for calcium channel blockers. Recall that the potency and efficacy of these compounds is diminished under conditions of strong stimulation akin to the conditions found in ischemic regions of brain due to the depolarized level of the $K^+$ equilibrium potential. Accordingly, more efficacious inhibitors might be desired, such as the class A channel inhibitors $\omega$-Aga IVA and $\omega$-CTX MVIIC. Two problems exist with the therapeutic use of such inhibitors. First, both peptides are large (5203 and 2749 Da, respectively), charged molecules that would not be predicted to cross the blood-brain barrier. However, some evidence points to a diminished integrity of the blood-brain barrier during ischemic episodes. Second, both of these class A channel inhibitors have been shown to block calcium channels at motor nerve endings, suggesting that the problematic side effect of neuromuscular paralysis might occur during administration. While such paralysis could be managed, it would be desirable to develop an inhibitor that crosses into the brain and that is selective for CNS terminals rather than peripheral ones.

In studies to date, the class B calcium channel blockers have shown the greastest promise as neuroprotective agents in models of global stroke and transient ischemia. In particular, $\omega$-CTX MVIIA (designated SNX-111) has been shown to have substantial neuroprotective actions as a post-ischemic treatment modality in several preclinical studies (Valentino et al., 1993; Buchan et al., 1994), whereas in some cases $\omega$-CTX GVIA does not appear to have these actions (Madden et al., 1994). An intriguing aspect of the action of this peptide is that its efficacy is enhanced when administered several hours after infarction, suggesting that the pathological mechanisms leading to neuronal death continue long after the ischemic event and that block of calcium entry during the postischemic period can ameliorate the pathogenesis.

*Neuropathic pain – long-lasting analgesia provided by class B inhibitors*

The class B inhibitor ω-CTX MVIIC has also shown promise as a potent, long-lasting analgesic, presumably acting by disrupting pain transmission pathways both at the site of injury (Xiao and Bennett, 1995) as well as in the dorsal horn of the spinal cord (Malmberg and Yaksh, 1995; Bowersox et al., 1996). The distinct advantage of the conopeptide in analgesia is that, in contrast to opiates, tolerance does not appear to develop, and potential for physical dependence would seem to be low. Clearly, the disadvantage to the peptide as an analgesic is that it must be administered intrathecally. Nonetheless, in a hospital setting as an adjunct to surgery, the intrathecal administration diminishes concerns over potential sympatholytic actions occuring from circulating peptide.

*Tottering mouse as a model for absence epilepsy – A role for calcium channels in seizure disorders?*

First described in 1962 (Green & Sidman, 1962), tottering (*tg*) is a mutant mouse strain with a neurological defect arising from a recessive mutation in a single gene on chromosome 8. Its phenotype is similar to absence epilepsy in humans, and has been used as one of the few animal models of this form of epilepsy. It is characterized by a fixed-gaze behavioral arrest (i.e., "absence") that coincides with abnormal bursts of bilateral synchronous spike waves in EEG, and infrequent, stereotyped focal motor seizures (Noebels & Sidman, 1979). These activities respond uniquely to the anticonvulsant drug ethosuximide, which has been reported to block T-type calcium currents in neurons (Gross et al., 1997). Neuroanatomical data reveal hyperinnervation of cortical areas by noradrenergic projections from locus ceruleus, but no gross abnormalities are observed (Noebels, 1984).

Recently, the genetic basis for this mutation has been identified. (Fletcher et al., 1996). Using a positional cloning strategy in yeast artificial chromosome libraries, it was found that the *tg* mutation maps to a region homologous to an exon of the rat $\alpha_{1A}$ class A calcium channel. In order to confirm this gene as the locus of *tg*, the mouse $\alpha_{1A}$ gene was cloned by RT-PCR. Sequence from the *tg* gene identified a C-to-T mutation leading to a predicted proline-to-leucine substitution in the putative pore region in the second transmembrane repeat of $\alpha_{1A}$. In situ hybridization with riboprobes specific to $\alpha_{1A}$ in *tg* brains show widespread expression of this gene.

The phenotype of tottering at the level of calcium currents is currently unknown. In the absence of information on the properties of the mutated channel, it is difficult to predict how its biophysical properties might differ from the wild type channels. However, this observation does suggest that disregulation of neuronal calcium channels could participate in some forms of generalized seizure disorder in humans. If the disregulation is a compensatory increase in the relative contribution of class B channels to control neurosecretion, it is conceivable that the pathology arises as a consequence of the more profound modulation of class B channels by G protein coupled receptors. Subtle alterations in the regulation of excitation-secretion coupling might be amenable to treatment by class B inhibitors.

**Prospects**

In a relatively short period, we have witnessed great strides in the detailed understanding of the structure and function of neuronal calcium channels (also see Reynolds, this volume). Nonetheless, many obstacles remain before calcium channel inhibitors will reach their maximal potential. From a theoretical point of view, we can consider the actions of these inhibitors on neurosecretion in a number of in vitro systems studied. One of the more interesting aspects of that research is that one could tailor the efficacy of block of glutamate release on the basis of which calcium channel is targeted. Recall that class A inhibitors would produce profound block of glutamatergic transmission due to the steep relationship between calcium entry and release. Likewise, class B channel inhibitors produce a modest degree of block (~50%) that would preserve transmission at a diminished level. Block of the putative resistant channel would be expected to produce a mild degree of inhibition, perhaps 5 to 10%. In principle, the desired level of inhibiton could be selected by administering saturating concentrations of a blocker that provides strong, moderate or mild inhibition. At the present rate of progress, we should not be surprised to witness the development of these and even more sophisticated pharmacological agents during the next ten years.

## References


Birnbaumer, L., Campbell, K.P., Catterall, W.A., Harpold, M.M., Hofmann, F., Horne, W.A., Mori, Y., Schwartz, A., Snutch, T.P., Tanabe, T. and Tsien, R.W. (1994) The naming of voltage-gated calcium channels. *Neuron*, 13: 505–506.

Boland, L.M., Morrill, J.A. and Bean, B.P. (1994) ω-Conotoxin block of N-type calcium channels in frog and rat sympathetic neurons. *J. Neurosci.*, 14: 5011–5027.

Bowersox, S.S., Gadbois, T., Singh, T., Pettus, M., Wang, Y.X. and Luther, R.R. (1996) Selective N-type neuronal voltage-sensitive calcium channel blocker, SNX-111, produces spinal antinociception in rat models of acute, persistent and neuropathic pain. *J. Pharmacol. Exp. Ther.*, 279: 1243–1249.

Bowersox, S.S., Miljanich, G.P., Sugiura, Y., Li, C., Nadasdi, L. and Hoffman, B.B. (1995) Differential blockade of voltage-sensitive calcium channels at the mouse neuromuscular junction by novel ω-conopeptides and ω-agatoxin-IVA. *J. Pharmacol. Exp. Ther.*, 273: 248–256.

Campbell, K.P., Leung, A.T. and Sharp, A.H. (1988) The biochemistry and molecular biology of the dihydropyridine-sensitive calcium channel. *Trends Neurosci.*, 11: 425–430.

Castellano, A. and Perez-Reyes, E. (1994) Molecular diversity of $Ca^{2+}$ channel β subunits. *Biochem. Soc. Trans.*, 22: 483–488.

Castillo, P.E., Weisskopf, M.G. and Nicoll, R.A. (1994) The role of $Ca^{2+}$ channels in hippocampal mossy fiber synaptic transmission and long-term potentiation. *Neuron*, 12: 261–269.

Cruz, L.J., Johnson, D.S. and Olivera, B.M. (1987) Characterization of the ω-conotoxin target. Evidence for tissue-specific heterogeneity in calcium channel types. *Biochemistry*, 26: 820–824.

Dodge, F.A. and Rahamimoff, R. (1967) Cooperative action of calcium ions in transmitter release at the neuromuscular junction. *J. Physiol. (Lond.)*, 193: 419–432.

Dunlap, K., Luebke, J. and Turner, T.J. (1994) Identification of calcium channels that control neurosecretion. *Science*, 266: 828–831.

Dunlap, K., Luebke, J. and Turner, T. (1995) Exocytotic $Ca^{2+}$ channels in mammalian central neurons. *Trends Neurosci.*, 18: 89–98.

Fleckenstein, A. (1977) Specific pharmacology of calcium in myocardium, cardiac pacemakers, and vascular smooth muscle. *Annu. Rev. Pharmacol. Toxicol.*, 17: 149–166.

Fletcher, C.F., Lutz, C.M., O'Sullivan, T.N., Shaughnessy, J.D., Hawkes R., Frankel, W.N., Copeland, N.G. and Jenkins, N.A. (1996) Absence epilepsy in tottering mutant mice is associated with calcium channel defects. *Cell*, 78: 607–617.

Gaur, S., Newcomb, R., Rivnay, B., Bell, J.R., Yamashiro, D., Ramachandran, J. and Miljanich, G.P. (1994) Calcium channel antagonist peptides define several components of transmitter release in the hippocampus. *Neuropharmacol.*, 33: 1211–1219.

Grantham, C.J., Bowman, D., Bath, C.P., Bell, D.C. and Bleakman, D. (1994) ω-Conotoxin MVIIC reversibly inhibits a human N-type calcium channel and calcium influx into chick synaptosomes. *Neuropharmacol.*, 33: 255–258.

Green, M.C. and Sidman, R.L. (1962) Tottering-a neuromuscular mutation in mouse. *J. Hered.*, 53: 79–94.

Gross, R.A., Covey, D.F. and Ferrendelli, J.A. (1997) Voltage-dependent calcium channels as targets for convulsant and anticonvulsant alkyl-substituted thiobutyrolactones. *J. Pharmacol. Exp. Ther.*, 280: 686–694.

Hillyard, D.R., Monje, V.D., Mintz, I.M., Bean, B.P., Nadasdi, L., Ramachandran, J., Miljanich, G., Azimi-Zoonooz, A., McIntosh, J.M., Cruz, L.J., Imperial, J.S. and Olivera, B.M. (1992) A new conus peptide ligand for mammalian presynaptic $Ca^{2+}$ channels. *Neuron*, 9: 69–77.

Keith, R.A, Mangano, T.J., Lampe, R.A., DeFeo, P.A., Hyde, M.J. and Donzanti, B.A. (1995) Comparative actions of synthetic ω-grammotoxin SIA and synthetic ω-Aga-IVA on neuronal calcium entry and evoked release of neurotransmitters in vitro and in vivo. *Neuropharmacol.*, 34: 1515–28.

Llinás, R., Sugimori, M., Lin, J.-W. and Cherksey, B. (1989) Blocking and isolation of a calcium channel from neurons in mammals and cephalopods utilizing a toxin fraction (FTX) from funne-web spider poison. *Proc. Natl. Acad. Sci.*, 86: 1689–1693.

Luebke, J.I., Dunlap, K. and Turner, T.J. (1993) Multiple calcium channel types control glutamatergic synaptic transmission in the hippocampus. *Neuron*, 11: 895–902.

Madden, K.P., Clark, W.M., Marcoux, F.W., Probert, A.W., Weber, M.L., Rivier J. and Zivin, J.A. (1990) Treatment with conotoxin, an 'N-type' calcium channel blocker, in neuronal hypoxic-ischemic injury. *Brain Res.*, 537: 256–262.

Malmberg, A.B. and Yaksh, T.L. (1995) Effect of continuous intrathecal infusion of ω-conopeptides, N-type calcium-channel blockers, on behavior and antinociception in the formalin and hot-plate tests in rats. *Pain*, 60: 83–90.

Malva, J.O., Ambrosio, A.F., Carvalho, A.P., Duarte, C.B. and Carvalho, C.M. (1995) Involvement of class A calcium channels in the KCl induced $Ca^{2+}$ influx in hippocampal synaptosomes. *Brain Res.*, 696: 242–245.

McCleskey, E.W. (1994) Calcium channels: Cellular roles and molecular mechanisms. *Curr Opin. Neurobiol.*, 4: 304–312.

McDonough, S.I., Swartz, K.J., Mintz, I.M., Boland, L.M. and Bean, B.P. (1996) Inhibition of calcium channels in rat central and peripheral neurons by ω-conotoxin MVIIC. *J. Neurosci.*, 16: 2612–2623.

McMahon, H.T. and Nicholls, D.G. (1991) The bioenergetics of neurotransmitter release. *Biochim. Biophys. Acta.*, 1059: 243–264.

Miljanich, G.P. and Ramachandran, J. (1995) Antagonists of neuronal calcium channels: Sturcture, function, and therapeutic implications. *Annu. Rev. Pharmacol. Toxicol.*, 35: 707–734.

Mintz, I.M., Adams, M.E. and Bean, B.P. (1992a) P-type calcium channels in rat central and peripheral neurons. *Neuron*, 9: 85–95.

Mintz, I.M., Venema, V.J., Swiderek, K.M., Lee, T.D., Bean, B.P. and Adams, M.E. (1992b). P-type calcium channels blocked by the spider toxin ω-Aga-IVA. *Nature*, 355: 827–829.

Mintz, I.M. and Bean, B.P. (1993) Block of calcium channels in rat neurons by synthetic ω-Aga-IVA *Neuropharmacol.*, 32: 1161–1169.

Nachshen, D.A. and Blaustein, M.P. (1979) Effects of some organic "calcium antagonists" on calcium influx in presynaptic nerve terminals. *Mol. Pharmacol.*, 16: 576–584.

Noebels, J.L and Sidman, R.L. (1979) Inherited epilepsy: Spike wave and focal motor seizures in the mutant mouse tottering. *Science*, 204: 1334–1336.

Noebels, J.L. (1984) A single gene error of noradrenergic axon growth synchronizes central neurons. *Nature*, 310: 409–411.

Nowycky, M.C., Fox, A.P. and Tsien, R.W. (1985). Three classes of neuronal calcium channels with different calcium agonist sensitivity. *Nature*, 316, 440–443.

Olivera, B.M., Miljanich, G., Ramachandran, J. and Adams, M.E. (1994) Calcium channel diversity and neurotransmitter release: The ω-conotoxins and ω-agatoxins. *Annu. Rev. Biochem.*, 63: 823–867.

Olivera, B.M., Cruz, L.J., de Santos, V., LeCheminant, G.W., Griffin, D., Zeikus, R., McIntosh, J.M., Galyean, R., Varga, J., Gray, W.R. and Rivier, J. (1987) Neuronal calcium channel antagonists. Discrimination between calcium channel subtypes using ω-conotoxin from *Conus magus*, venom. *Biochemistry*, 26: 2086–2090.

Olivera, B.M., Gray, W.R., Zeikus, R., McIntosh, J.M., Varga, J., Rivier, J., de Santos, V. and Cruz, L.J. (1985). Peptide neurotoxins from fish-hunting cone snails. *Science*, 230: 1338–1343.

Olivera, B.M., Rivier, J., Clark, C., Ramilo, C.A., Corpuz, G.P., Abogadie, F.C., Mena, E.E., Woodward, S.R., Hillyard, D.R. and Cruz, L.J. (1990) Diversity of *Conus*, neuropeptides. *Science*, 249: 257–263.

Patil, P.G., de Leon, M., Reed, R.R., Dubel, S., Snutch, T.P. and Yue, D.T. (1996) Elementary events underlying voltage-dependent G-protein inhibition of N-type calcium channels. *Biophys. J.*, 71: 2509–2521.

Randall, A. and Tsien, R.W. (1995) Pharmacological dissection of multiple types of $Ca^{2+}$ channel currents in rat cerebellar granule neurons. *J. Neurosci.*, 15: 2995–3012.

Regehr, W.G. and Atluri, P.P. (1995) Calcium transients in cerebellar granule cell presynaptic terminals. *Biophys. J.*, 68: 2156–2170.

Sather, W.A., Tanabe, T., Zhang, Z-F., Mori, Y., Adams, M. E. and Tsien, R.W. (1993) Distinctive biophysical and pharmacological properties of class A (BI) calcium channel $a_1$ subunits. *Neuron*, 11: 291–303.

Soong, T.W., Stea, A., Hodson, C.D., Dubel, S.J., Vincent, S.R. and Snutch, T.P. (1993) Structure and functional expression of a member of the low voltage-activated calcium channel family. *Science*, 260: 1133–1136.

Stanley, E.F. (1993) Single calcium channels and acetylcholine release at a presynaptic nerve terminal. *Neuron*, 11: 1007–1011.

Suszkiw, J.B., O'Leary, M.E., Murawsky, M.M. and Wang, T. (1986) Presynaptic calcium channels in rat cortical synaptosomes: Fast-kinetics of phasic calciuminflux, channel inactivation, and relationship to nitrendipine receptors. *J. Neurosci.*, 6: 1349–1357.

Takahashi, T. and Momiyama, M (1993) Different types of calcium channels mediate central synaptic transmission. *Nature*, 366: 156–158.

Turner, T.J., Adams, M.E. and Dunlap, K. (1993) Multiple $Ca^{2+}$ channel types coexist to regulate synaptosomal neurotransmitter release. *Proc. Natn. Acad. Sci. U.S.A.*, 90: 9518–9522.

Turner, T.J., Lampe, R. and Dunlap, K. (1995) Characterization of presynaptic calcium channels with ω-conotoxin MVIIC and ω-grammotoxin SIA: A role for a resistant calcium channel type in neurosecretion. *Mol. Pharmacol.*, 47: 348–353.

Turner, T.J. and Dunalp, K. (1995) Pharmacological characterization of presynaptic calcium channels using subsecond biochemical measurements of synaptosomal neurosecretion. *Neuropharmacol.*, 34: 1469–1478.

Wheeler, D.B., Randall, A. and Tsien, R.W. (1994) Roles of N-type and Q-type $Ca^{2+}$ channels in supporting synaptic transmission. *Science*, 264: 107–111.

Williams, M.E., Marubio, L.M., Deal, C.R., Hans, M., Brust, P.F., Philipson, L.H., Miller, R.J., Johnson, E.C., Harpold, M.M. and Ellis, S.B. (1994) Structure and functional characterization of neuronal α 1E calcium channel subtypes. *J. Biol. Chem.*, 269: 22347–22357.

Wu, L.G. and Saggau, P. (1994) Presynaptic calcium is increased during normal synaptic transmission and paired-pulse facilitation, but not in long-term potentiation in area CA1 of hippocampus. *J. Neurosci.*, 14: 645–654.

Xiao, W.H. and Bennett, G.J. (1995) Synthetic ω-conopeptides applied to the site of nerve injury suppress neuropathic pains in rats. *J. Pharmacol. Exp. Ther.*, 274: 666–672.

Zhang, J.F., Randall, A.D., Ellinor, P.T., Horne, W.A., Sather, W.A., Tanabe, T., Schwarz, T.L. and Tsien, R.W. (1993) Distinctive pharmacology and kinetics of cloned neuronal $Ca^{2+}$ channels and their possible counterparts in mammalian CNS neurons. *Neuropharm.*, 32: 1975–1988.

O.P. Ottersen, I.A. Langmoen and L. Gjerstad (Eds.)
*Progress in Brain Research*, Vol 116


CHAPTER 2

# Presynaptic modulation of glutamate release

David G. Nicholls

*Neurosciences Institute, Department of Pharmacology, University of Dundee, Dundee DD1 9SY, Scotland, UK*

## Introduction

The arrival of an action potential at a CNS glutamatergic varicosity does not automatically result in exocytosis of the transmitter. Instead glutamatergic varicosities in brain regions capable of synaptic plasticity show a low quantal release probability (Hessler et al., 1993), such that only a proportion of the action potentials passing through the varicosity lead to the release of a synaptic vesicle at a given release site. Since an action potential must pass through a varicosity to continue its propagation, this implies that there is an inherent unreliability in the coupling between the action potential and exocytosis. The high power dependency upon external $Ca^{2+}$ concentration of exocytosis (Augustine et al., 1985) suggests that $Ca^{2+}$ entry may be the prime probability-determining step in action potential/exocytosis coupling. Presynaptic receptors may either facilitate or inhibit release by directly or indirectly modulating this $Ca^{2+}$ entry and their existence enormously enhances the potential complexity of synaptic transmission by allowing release to be controlled at each individual varicosity along the axon.

The exocytosis of a single synaptic vesicle can elevate the concentration of the amino acid in the synaptic cleft to around 1mM (Clements et al., 1992), sufficient to saturate the ionotropic receptors at the post-synaptic membrane. Presynaptic receptors could modulate the probability of vesicle release in response to an invading action potential by a number of mechanisms: firstly by altering the kinetics of $K^+$ or $Na^+$ channels and influencing the waveform of the depolarization; secondly by activating or inhibiting the release-coupled $Ca^{2+}$ channel and directly controlling the access of $Ca^{2+}$ to the exocytotic trigger, or thirdly by acting at an intracellular locus to control some aspect of the exocytotic/endocytic cycle. The evidence, obtained from studies with isolated nerve terminals (synaptosomes) will now be reviewed that the two classes of receptor may modulate exocytosis by respectively enhancing presynaptic action potentials and inhibiting $Ca^{2+}$ channels. (see Turner, this volume, for a description of $Ca^{2+}$ channels coupled to glutamate release.)

## Synaptosomes and the study of glutamate exocytosis

Synaptosomes have proved a valuable preparation for the investigation of these presynaptic regulatory pathways (reviewed in Nicholls and Sánchez-Prieto, 1997; Nicholls, 1995a,b). Glutamate exocytosis greatly outweighs the release of other transmitters in preparations from cerebral cortex, hippocampus or cerebellum and is the simplest preparation which retains all the pathways for the synthesis, storage, release and reuptake of transmitter glutamate, together with a range of presynaptic receptors capable of modulating the release of the transmitter. In recent years this preparation has provided much information on the signal transduction pathways present in the glutamatergic terminal, and these provide the topic of this review.

The release of glutamate can be monitored continuously by including glutamate dehydrogenase and NADP+ in the incubation medium (Nicholls and Sihra, 1986; Sánchez-Prieto et al., 1987), the released glutamate being coupled to the generation of fluorescent NADPH. The $Ca^{2+}$-dependent component of KCl-evoked glutamate release amounts to some 15% of the total glutamate content of the preparation (McMahon and Nicholls, 1991) and is exocytotic judged by its ATP requirement (Sánchez-Prieto et al., 1987), inhibition by tetanus and botulinum toxins (McMahon et al., 1992) and by compartmental analysis (Wilkinson and Nicholls, 1989).

In addition to the conventional KCl depolarization technique for evoking transmitter release, a second and potentially more versatile approach for evoking glutamate exocytosis is by $K^+$-A type channel inhibition (Tibbs et al., 1989). A-type $K^+$ channels activate very rapidly in response to slight depolarizations and serve as negative feedback mechanisms limiting statistical fluctuations in membrane potential from triggering spontaneous action potentials. Such channels appear to exist in nerve terminals, sensitive to $\alpha$-dendrotoxin or to low concentrations of 4-aminopyridine (4AP) (Tapia and Sitges, 1982; Agoston et al., 1983; Dolezal and Tucek, 1983; Tibbs et al., 1989). Either agent causes a $Ca^{2+}$-dependent, tetrodotoxin-sensitive exocytosis of glutamate from the same pool as for KCl, and utilizes the same non-L non-N type $Ca^{2+}$ channels (Tibbs et al., 1989), triggering release via a highly localized pool of $Ca^{2+}$ in the immediate vicinity of the $Ca^{2+}$ channel (Verhage et al., 1991). The selective use of these two means of synaptosomal excitation have been central to studies dissecting the presynaptic signal transduction pathways controlling glutamatergic transmission.

## PKC and the receptor-mediated facilitation of transmitter release

It has been established for some time that the application of phorbol esters leads to an enhancement of glutamate release, which can be observed with preparations from the brain slice to the isolated synaptosome (Huang et al., 1989; Segal, 1989; Barrie et al., 1991; Herrero et al., 1992; Coffey et al., 1993). In our studies, a key finding was that the ability of phorbol esters to enhance glutamate exocytosis from cerebrocortical synaptosomes was dependent on the mode by which the preparation was depolarized: exocytosis evoked by a clamped depolarization by elevated KCl was independent of PKC activity – either phorbol ester mediated activation, or inhibition by Ro 31-8220 (Barrie et al., 1991; Coffey et al., 1993). By monitoring the phosphorylation state of endogenous MARCKS during these experiments it was possible to confirm that PKC activity was being controlled in the expected manner by these agents (Coffey et al., 1993). Thus phosphorylation of a PKC substrate is not necessary for vesicle exocytosis *per se*. In contrast, the use of 4AP or $\alpha$-dendrotoxin to evoke spontaneous action potential firing in the synaptosome preparation resulted in an exocytosis which was totally controlled by PKC activity: phorbol esters could cause a 500% stimulation of release while Ro 31-8220 could virtually abolish the ability of 4AP to evoke release (Coffey et al., 1993).

Since the distinction between 4AP-evoked glutamate release and that evoked by high KCl is the tetrodotoxin-sensitive firing of spontaneous 'action potentials' (Tibbs et al., 1989), the channels involved in this process provides an obvious locus at which to search for an action of PKC. The use of membrane potential-dependent dyes to monitor the time- and population average depolarization of the synaptosomal preparation reveals that phorbol esters enhance the 4AP-evoked depolarization, indicative of an effect on an ion channel implicated in action potential firing (Barrie et al., 1991; Herrero et al., 1992a; Coffey et al., 1993; Coffey et al., 1994). The effect of the phorbol ester is mimicked and occluded by $Ba^{2+}$ or by clofilium (Nicholls and Coffey, 1994); thus the most likely locus is a 4AP-insensitive, α-dendrotoxin-insensitive, $Ba^{2+}$-sensitive, clofilium-sensitive $K^+$ channel.

The PKC isoform mediating the facilitated glutamate release is potentiated by arachidonic acid (AA) at concentrations of 1–2 μM (Herrero et al., 1992a,b). In the presence of AA the threshold concentration of phorbol ester required to observe facilitation is greatly decreased (Herrero et al., 1992a). The dominant physiological agonist activating presynaptic PKC appears to be glutamate itself, since in the presence of AA a facilitation of 4AP-evoked glutamate release is observed (Herrero et al., 1992b) on addition of the mGluR agonist (1S, 3R) 1-aminocyclopentane-1,3-dicarboxylic acid (ACPD). ACPD activates a metabotropic glutamate receptor (mGluR) which is coupled (in the presence of AA) to enhanced phosphorylation of the established PKC substrates MARCKS and GAP-43 (Coffey et al., 1994).

The two established pathways for the formation of the diacylglycerol (DAG) required for the activation of PKC are its direct generation by phospholipase C (PLC) mediated hydrolysis of phosphatidylinositol bisphosphate and the alternative activation of phospholipase D with the initial generation of phosphatidic acid followed by its hydrolysis to DAG (for review see Klein et al., 1995). While a large and transient elevation in DAG has been reported in response to ACPD (Herrero et al., 1994), recent studies in our laboratory (unpublished) indicate a more modest (50–70% elevation) but more prolonged elevation in diacylglycerol (persisting for 15 min even in the absence of AA) in response to 100 μM ACPD. Analysis of the pharmacology of the facilitated glutamate release reveals some anomalous features: firstly ACPD-mediated enhancement is inhibited by (RS)-3,5-dihydroxyphenylglycine (DHPG) which is reported to be a selective group I mGluR agonist (Schoepp et al., 1994). An antagonist activity of DHPG has previously been reported against the ACPD-mediated activation of phospholipase D (PLD) in hippocampal slices (Pellegrini-Giampietro et al., 1996) and preliminary experiments have indicated the presence of a PLD activity in the synaptosomal preparation with this anomalous pharmacology.

Activation of presynaptic PKC by 5 nM PDBu results in a rapid desensitization of the facilitatory mGluR pathway in synaptosomes, and ACPD itself, even in the absence of added AA, causes homologous desensitization of the receptor (Herrero et al., 1994). This rapid desensitization may help to explain some of the difficulties experienced in observing the facilitatory pathway in brain slices, where the presence of extracellular glutamate may be sufficient to cause desensitization. The time-course of recovery of receptor function is slow, taking some 20 min in synaptosomes following ACPD addition, and this delay is further enhanced in the presence of okadaic acid (Herrero et al., 1994).

### Receptor-mediated inhibition of glutamate exocytosis

In addition to the facilitatory pathway, presynaptic glutamatergic terminals possess a variety of presynaptic inhibitory receptors (also see Bruno et al., this volume). Adenosine, acting on A1 receptors located both pre- and postsynaptically, is a potent inhibitor of neurotransmission (Fredholm and Dunwiddie, 1988). Postsynaptically adenosine hyperpolarizes by activating $K^+$-channels (De Mendonça and Ribeiro, 1994) while the presynaptic receptor acts via a pertussis toxin-sensitive G-protein (Dolphin and Prestwich, 1985) to inhibit the release of glutamate and other neurotransmitters (Burke and Nadler, 1988; Barrie and Nicholls, 1993). The exact presynaptic mechanism is still a matter of debate: adenosine A1 inhibition of neurotransmitter release in the avian ciliary ganglion is consistent with an activation of $K^+$-channels resulting in hyperpolarization (Bennett and Ho, 1992), however adenosine is an effective inhibitor of $Ca^{2+}$ elevation and glutamate release in synaptosomes subjected to clamped KCl depolarization (Barrie and Nicholls, 1993). This is in direct contrast to the conditions discussed above for observing PKC-dependent facilitation and strongly suggests a direct inhibition of a release-coupled calcium channel, consistent with hippocampal CA1 neurons where an A1 agonist reduced presynaptic $Ca^{2+}$ transients (Wu and Saggau, 1994).

An inhibitory presynaptic mGluR sensitive to the agonist L-AP4 can be observed in synaptosomal preparations from young (1–3 week PP) but not adult rats (Vázquez et al., 1995) and contrasts with the facilitatory pathway which is not seen until week 3 but remains in terminals from the adult (Vázquez et al., 1995). This is consistent with the developmentally regulated depression of synaptic transmission by L-AP4 in the hippocampus (Baskys and Malenka, 1991), whereas in the striatum L-AP4 inhibits glutamate release from synaptosomes prepared from adult rats (East et al., 1995). The mechanism of the L-AP4-sensitive pathway is closely parallel to that of the adenosine A1 receptor: inhibition of both $Ca^{2+}$ elevation and glutamate release is observed during clamped KCl depolarization and is insensitive to protein kinase inhibitors (Vázquez et al., 1995). The presynaptic L-AP4 receptor isoform is not clearly established: mGluR 4, 7 and 8 are expressed in the cortex (Kristensen et al., 1993; Pin and Duvoisin, 1995). ACPD-sensitive, L-AP4-insensitive inhibition is seen in the striatum acting via a pharmacologically identifiable mGluR2/3 (Lovinger, 1991). However, as discussed above this receptor appears not to be subject to the same developmental control as the cortical/hippocampal isoform/s.

**PKC and heterologous desensitization**

At the presynaptic terminal inhibitory receptor signal transduction pathways appear to involve a direct coupling of G-protein to the release-coupled $Ca^{2+}$ channel (Barrie and Nicholls, 1993) without generation of a detectable second messenger. While the signal transduction pathways do not involve PKC they are very sensitive to PKC activation in situ and a rapid loss of inhibition is observed when PKC is activated by phorbol esters (Barrie and Nicholls, 1993; Budd and Nicholls, 1995). It is important in this context to distinguish between an in vitro suppression of a receptor-mediated pathway due to non-physiological activation of PKC by phorbol esters and one which is agonist evoked and might possibly have a physiological function.

In contrast to the facilitatory pathway, the presynaptic inhibitory receptors do not undergo homologous desensitization. However, the adenosine and L-AP4 mediated inhibitory pathways are each suppressed when PKC is activated either by phorbol esters (Barrie and Nicholls, 1993) or by activation of the facilitatory pathway (Vázquez et al., 1995; Budd and Nicholls, 1995). This indicates incidentally that the facilitatory and inhibitory receptors are present on the same nerve terminals: indeed since no additivity is seen with the two inhibitory agonists in three week rats (Vázquez et al., 1995) the three receptors must co-exist on a high proportion of cortical terminals. The site at which PKC may act to suppress the inhibitory responses has recently been clarified by expression of chimeric $Ca^{2+}$ channel $\alpha_{1A}$ and $\alpha_{1B}$ subunits with $\alpha_2$ and $\beta$ subunits in HEK cells, where they could be inhibited by an endogenous somatostatin receptor. PKC-dependent phosphorylation of residues within the calcium channel's binding site for the modulatory $G\beta\gamma$ disrupted the interaction of the G-protein subunits with the channel (Zamponi et al., 1997).

AA is not required for the ACPD-mediated suppression of the inhibitory pathways (Budd and Nicholls, 1995) but greatly prolongs the duration of the suppression. The rapid reversal of the suppression in the absence of AA indicates that an active phosphatase is present to reverse the PKC-mediated suppression, in contrast to the slow phosphatase activity discussed above for the homologous desensitization of the facilitatory receptor. AA is known to facilitate insertion of PKC into the membrane (Lester and Bramham, 1993), while synergistic activation of protein kinase C by arachidonic acid and diacylglycerol lead to the generation of a stable membrane-bound, cofactor-independent state of protein kinase C activity (Schachter et al., 1996). In the terminal this might be predicted to result in a constitutively active PKC that remains active after the facilitatory mGluR has desensitized. However, to our surprise we find that ACPD in the absence of AA results in a translocation of several PKC isoforms to the synaptosomal membrane fraction, and that this translocation persists for several minutes (unpublished). This together with the evidence that the anomalous pharmacology of the facilitatory receptor is associated with PLD activation implies that the signal transduction pathways in the glutamatergic terminal responsible for PKC activation may be more complex than was originally proposed.

## Physiological correlates

The model system of the synaptosome discussed above shows an extensive but transient PKC-dependent facilitation of glutamate exocytosis in the simultaneous presence of a glutamate agonist and a low concentration of AA. This, together with the parallel suppression of presynaptic inhibitory receptor pathways and the apparently ubiquitous presence of these pathways in cortical and hippocampal glutamatergic terminals is suggestive that these pathways might play a role in synaptic plasticity. An attractive possibility is that the AA

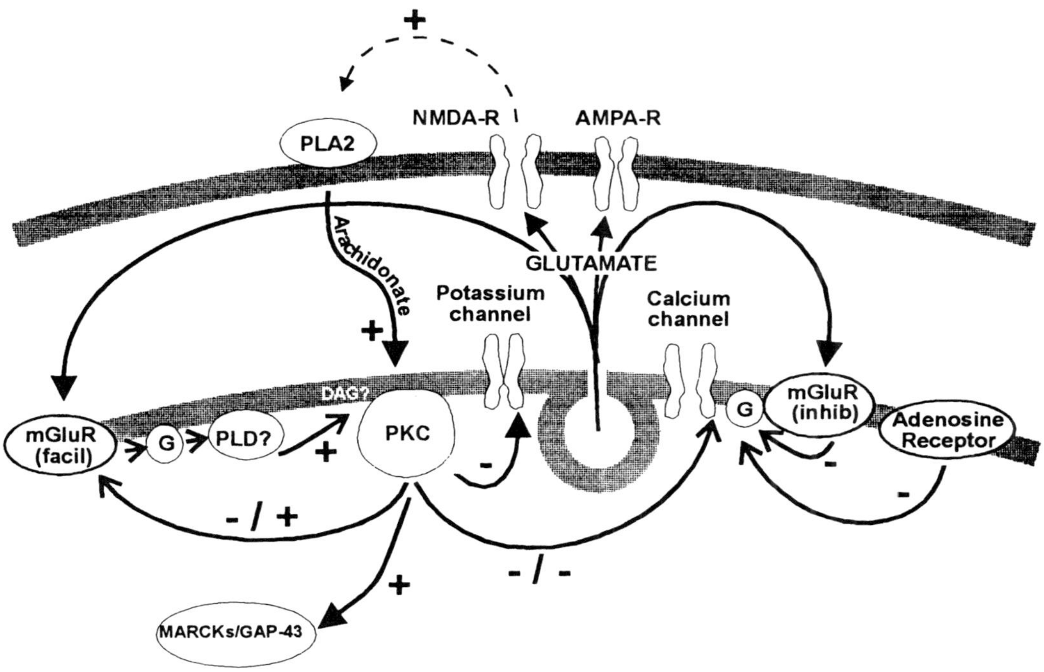


Fig. 1. Schematic representation of the interactions between facilitatory and inhibitory presynaptic receptors controlling glutamate exocytosis. The scheme depicts the pre- and post-synaptic membranes of a hypothetical plastic glutamatergic synapse. A condition of maximal glutamate exocytosis is shown, such as that during the induction of long-term potentiation: post-synaptic NMDA receptor (NMDA-R) activation results in the generation of arachidonic acid (Miller et al., 1992). Overflow of glutamate in the synaptic cleft activates the facilitatory glutamate receptor (mGluR (facil); Herrero et al., 1992b), resulting in the activation of PKC. PKC rapidly desensitizes mGluR (facil), however in the presence of arachidonate as a putative retrograde messenger, PKC continues to be active even after mGluR desensitization. PKC activation results in an enhanced phosphorylation of the presynaptic phosphoproteins GAP-43 and MARCKS, an enhanced presynaptic depolarization (detected in synaptosomal preparations during 4-AP induced excitation (Coffey et al., 1993) and heterologous desensitization or suppression of inhibitory presynaptic receptors, including the adenosine A1 receptor (Barrie and Nicholls, 1993) and the inhibitory mGluR present on terminals from immature rats (Vázquez et al., 1995). This retrograde messenger-dependent facilitation of release and suppression of inhibition could ensure that glutamate exocytosis is maintained at maximal probability to ensure the completion of LTP induction.

originates postsynaptically and that the $PLA_2$ is activated under conditions where the NMDA receptor is active during the establishment of synaptic plasticity. The additional presence of transmitter glutamate in the synaptic cleft would provide the synergy necessary for DAG production and PKC sensitization and could serve to reinforce the release of glutamate occurring during the period required for the establishment of synaptic plasticity. Inhibitory presynaptic receptors become operative when release is enhanced as a consequence of high-frequency stimulation (Scanziani et al., 1997; Takumi et al., this volume); the decoupling of inhibitory presynaptic receptors would thus remove any opposing influence and further enhance the probability of release.

A number of possible sources for AA in vivo can be proposed: one would be phospholipase $A_2$ activated post-synaptically in response to NMDA receptor activation (Dumuis et al., 1990) – thus providing a mechanistic basis for proposals that AA could function as a retrograde messenger during plastic changes at the synapse (Williams et al., 1989). Weak activation of the perforant path paired with application of AA can lead to a slow-onset increase in synaptic efficacy (Williams et al., 1989), while AA and diacylglycerol induce a synergistic facilitation of $Ca^{2+}$-dependent glutamate release from hippocampal mossy fiber nerve endings (Zhang et al., 1996). Furthermore, the synergism between metabotropic glutamate receptor activation and arachidonic acid on glutamate release is occluded by induction of long-term potentiation in the dentate gyrus (McGahon and Lynch, 1996).

While it is unreasonable that such a mechanism would chronically elevate glutamate release, a short-lasting enhancement during the induction of potentiation could act to ensure that glutamate release remained high until a synapse is securely switched to the potentiated form.

## Therapeutic implications

Agonists and antagonists targetting the presynaptic glutamatergic terminal greatly influence the in vitro and in vivo release of the transmitter. The co-existence of facilitatory and inhibitory autoreceptors on the same terminal, their complex interplay and the possible influence of exogenous co-activators of PKC, such as arachidonic acid, imply that the effect of these agents on transmitter release may not always be predictable. One reason underlying the lack of consensus as to whether mGluR agonists are neuroprotective (Bruno et al., this volume) may lie in the complications resulting from these complex signal transduction pathways.

## References


Agoston, D.V., Hargittai, P. and Nagy, A. (1983) Effects of 4-AP in Ca movements and changes of membrane potential in pinched-off nerve terminals from rat cerebral cortex. *J. Neurochem.*, 41: 745–751.

Augustine, G.J., Charlton, M.P. and Smith, S.J. (1985) Calcium entry and transmitter release at voltage-clamped nerve terminals of squid. *J. Physiol. Lond.*, 367: 163–181.

Barrie, A.P. and Nicholls, D.G. (1993) Adenosine A1 receptor inhibition of glutamate exocytosis and protein kinase C-mediated decoupling. *J. Neurochem.*, 60: 1081–1086.

Barrie, A.P., Nicholls, D.G., Sánchez-Prieto, J. and Sihra, T.S. (1991) An ion channel locus for the protein kinase C potentiation of transmitter glutamate release from guinea pig cerebrocortical synaptosomes. *J. Neurochem.*, 57: 1398–1404.

Baskys, A. and Malenka, R.C. (1991) Agonists at metabotropic glutamate receptors presynaptically inhibit EPSCs in neonatal rat hippocampus. *J. Physiol., (Lond)*, 444: 687–701.

Bennett, M.R. and Ho, S. (1992) Adenosine modulation of potassium currents in preganglionic nerve terminals of avian ciliary ganglia. *Neurosci. Lett.*, 137: 41–44.

Budd, D.C. and Nicholls, D.G. (1995) Protein kinase C mediated decoupling of the presynaptic adenosine A1 receptor by a facilitatory metabotropic glutamate receptor. *J. Neurochem.*, 65: 615–621.

Burke, S.P. and Nadler, J.V. (1988) Regulation of glutamate and aspartate release from slices of the hippocampal CA1 area: effects of adenosine and baclofen. *J. Neurochem.*, 51: 1541–1551.

Clements, J.D., Lester, R.A.J., Tong, G., Jahr, C.E. and Westbrook, G.L. (1992) The time course of glutamate in the synaptic cleft. *Science*, 258: 1498–1501.

Coffey, E.T., Herrero, I., Sihra, T.S., Sánchez-Prieto, J. and Nicholls, D.G. (1994) Glutamate exocytosis and MARKS phosphorylation are enhanced by a metabotropic glutamate receptor coupled to a protein kinase C synergistically activated by diacylglycerol and arachidonic acid. *J. Neurochem.*, 63: 1303–1310.

Coffey, E.T., Sihra, T.S. and Nicholls, D.G. (1993) Protein kinase C and the regulation of glutamate exocytosis from cerebrocortical synaptosomes. *J. Biol. Chem.*, 268: 21060–21065.

De Mendonça, A. and Ribeiro, J.A. (1994) Endogenous adenosine modulates long-term potentiation in the hippocampus. *Neurosci.*, 62: 385–390.

Dolezal, V. and Tucek, S. (1983) The effects of 4-aminopyridine and tetrodotoxin on the release of acetylcholine from rat striatal slices. *Naunyn-Schmied. Arch. Pharmacol.*, 323: 90–95.

Dolphin, A.C. and Prestwich, S.A. (1985) Pertussis toxin reverses adenosine inhibition of neuronal glutamate release. *Nature*, 316: 148–150.

Dumuis, A., Pin, J.P., Oomagari, K., Sebben, M. and Bockaert, J. (1990) Arachidonic acid released from striatal neurons by joint stimulation by ionotropic and metabotropic quisqualate receptors. *Nature*, 347: 182–184.

East, S.J., Hill, M.P. and Brotchie, J.M. (1995) Metabotropic glutamate receptor agonists inhibit endogenous glutamate release from rat striatal synaptosomes. *Eur. J. Pharmacol.*, 277: 117–121.

Fredholm, B.B. and Dunwiddie, T.V. (1988) How does adenosine inhibit transmitter release? (TIPS Review). *Trends Pharmacol. Sci.*, 9: 130–135.

Herrero, I., Miras-Portugal, M.T. and Sánchez-Prieto, J. (1992a) Activation of protein kinase C by phorbol esters and arachidonic acid required for the optimal potentiation of glutamate exocytosis. *J. Neurochem.*, 59: 1574–1577.

Herrero, I., Miras-Portugal, M.T. and Sánchez-Prieto, J. (1992b) Positive feedback of glutamate exocytosis by metabotropic presynaptic receptor stimulation. *Nature*, 360: 163–166.

Herrero, I., Miras-Portugal, M.T. and Sánchez-Prieto, J. (1994) Rapid desensitization of the presynaptic metabotropic receptor for glutamate that facilitates glutamate release. *Eur. J. Neurosci.*, 6: 115–120.

Hessler, N.A., Shirke, A.M. and Malinow, R. (1993) The probability of transmitter release at a mammalian central synapse. *Nature*, 366: 569–572.

Huang, H.Y., Hertting, G., Allgaier, C. and Jackisch, R. (1989) 3:4-Diaminopyridine-induced noradrenaline release from CNS tissue as a model for action potential-evoked transmitter release: effects of phorbol ester. *Eur. J. Pharmacol.*, 169: 115–123.

Klein, J., Chalifa, V., Liscovitch, M. and Löffelholz, K. (1995) Role of phospholipase D activation in nervous system physiology and pathophysiology. *J. Neurochem.*, 65: 1445–1455.

Kristensen, P., Suzdak, P.D. and Thomsen, C. (1993) Expression pattern and pharmacology of the rat type IV metabotropic glutamate receptor. *Neurosci. Lett.*, 155: 159–162.

Lester, D.S. and Bramham, C.R. (1993) Persistent, membrane-associated protein kinase C: From model membranes to synaptic long-term potentiation. *Cell Signalling*, 5: 695–708.

Lovinger, D.M. (1991) Trans-1-aminocyclopentane-1,3-dicarboxylic acid (tert-ACPD) decreases synaptic excitation in rat striatal slices through a presynaptic action. *Neurosci. Lett.*, 129: 17–21.

McGahon, B. and Lynch, M.A. (1996) The synergism between metabotropic glutamate receptor activation and arachidonic acid on glutamate release is occluded by induction of long-term potentiation in the dentate gyrus. *Neurosci.*, 72: 847–855.

McMahon, H.T., Foran, P., Dolly, J.O., Verhage, M., Wiegant, V.M. and Nicholls, D.G. (1992) Tetanus toxin and botulinum toxins type A and B inhibit glutamate, $\gamma$-aminobutyric acid, aspartate, and met-enkephalin release from synaptosomes. Clues to the locus of action. *J. Biol. Chem.*, 267: 21338–21343.

McMahon, H.T. and Nicholls, D.G. (1991) Transmitter glutamate release from isolated nerve terminals: Evidence for biphasic release and triggering by localized $Ca^{2+}$. *J. Neurochem.*, 56: 86–94.

Nicholls, D.G. (1995a) The release of glutamate from synaptic terminals. In T.W. Stone (Ed.), *Glutamate*. (pp. 35–52). New York: CRC Press.

Nicholls, D.G. (1995b) Mechanisms of glutamate release. In H. Wheal & A. Thomson (Eds.), *Excitatory Amino Acids and Synaptic Transmission*. (pp. 1–15). London: Academic Press.

Nicholls, D.G. and Coffey, E.T. (1994) Glutamate exocytosis from isolated nerve terminals. In L. Stjarne, P. Greengard, T. Hokfelt, & D. Ottoson (Eds.), *Molecular and Cellular Mechanisms of Neurotransmitter Release*. (pp. 189–204). New York: Raven Press.

Nicholls, D.G. and Sánchez-Prieto, J. (1997) Neurotransmitter release mechanisms. In A.J. Turner & F.A. Stephenson (Eds.), *Amino Acid Neurotransmission*. London: Portland Press.

Nicholls, D.G. and Sihra, T.S. (1986) Synaptosomes possess an exocytotic pool of glutamate. *Nature*, 321: 772–773.

Pellegrini-Giampietro, D.E., Torregrossa, S.A. and Moroni, F. (1996) Pharmacological characterization of metabotropic glutamate receptors coupled to phospholipase D in the rat hippocampus. *Br. J. Pharmacol.*, 118: 1035–1043.

Pin, J.-P. and Duvoisin, R. (1995) The metabotropic glutamate receptors: structure and functions. *Neuropharmacol.* 34: 1–26.

Sánchez-Prieto, J., Sihra, T.S. and Nicholls, D.G. (1987) Characterization of the exocytotic release of glutamate from guinea-pig cerebral cortical synaptosomes. *J. Neurochem.*, 49: 58–64.

Scanziani, M., Salin, P.A., Vogt, K.E., Malenka, R.C. and Nicoll, R.A. (1997) Use-dependent increases in glutamate concentration activate presynaptic metabotropic glutamate receptors. *Nature*, 385: 630–634.

Schachter, J.B., Lester, D.S. and Alkon, D.L. (1996) Synergistic activation of protein kinase C by arachidonic acid and diacylglycerols in vitro: Generation of a stable membrane-bound, cofactor-independent state of protein kinase C activity. *Biochim. Biophys. Acta*, 1291: 167–176.

Schoepp, D.D., Goldsworthy, J., Johnson, B.G., Salhoff, C.R. and Baker, S.R. (1994) 3,5-dihydroxyphenylglycine is a highly selective agonist for phosphoinositide-linked metabotropic glutamate receptors in the rat hippocampus. *J. Neurochem.*, 63: 769–772.

Segal, M. (1989) Synaptic transmission between cultured rat hippocampal neurons is enhanced by activation of protein kinase-C. *Neurosci. Lett.*, 101: 169

Tapia, R. and Sitges, M. (1982) Effect of 4-aminopyridine on transmitter release in synaptosomes. *Brain Res.*, 250: 291–299.

Tibbs, G.R., Barrie, A.P., Van-Mieghem, F., McMahon, H.T. and Nicholls, D.G. (1989) Repetitive action potentials in isolated nerve terminals in the presence of 4-aminopyridine: Effects on cytosolic free $Ca^{2+}$ and glutamate release. *J. Neurochem.*, 53: 1693–1699.

Vázquez, E., Budd, D., Herrero, I., Nicholls, D.G. and Sánchez-Prieto, J. (1995) Co-existence and interaction between facilitatory and inhibitory metabotropic glutamate receptors and the inhibitory adenosine A1 receptor in cerebrocortical nerve terminals. *Neuropharmacol.*, 34: 919–927.

Vázquez, E., Herrero, I., Miras-Portugal, M.T. and Sánchez-Prieto, J. (1995) Developmental change from inhibition to facilitation in the presynaptic control of glutamate exocytosis by metabotropic glutamate receptors. *Neurosci.*, 68: 117–124.

Verhage, M., McMahon, H.T., Ghijsen, W.E.J.M., Boomsma, F., Wiegant, V. and Nicholls, D.G. (1991) Differential release of amino acids, neuropeptides and catecholamines from nerve terminals. *Neuron*, 6: 517–524.

Wilkinson, R. and Nicholls, D.G. (1989) Compartmentation of glutamate and aspartate within cerebral cortical synaptosomes: evidence for a non-cytoplasmic origin for the Ca releasable pool of glutamate. *Neurochem. Int.*, 15: 191–197.

Williams, J.H., Errington, M.L., Lynch, M.A. and Bliss, T.V.P. (1989) Arachidonic acid induces a long-term activity-dependent enhancement of synaptic transmission in the hippocampus. *Nature*, 341: 739–741.

Wu, L. and Saggau, P. (1994) Adenosine inhibits evoked synaptic transmission primarily by reducing presynaptic calcium influx in area CA1 of hippocampus. *Neuron*, 12: 1139–1148.

Zamponi, G.W., Bourinet, E., Nelson, D., Nargeot, J. and Snutch, T.P. (1997) Crosstalk between G proteins and protein kinase C mediated by the calcium channel $\alpha_1$ subunit. *Nature*, 385: 442–446.

Zhang, L., Ruehr, M.L. and Dorman, R.V. (1996) Arachidonic acid and oleoylacetylglycerol induce a synergistic facilitation of $Ca^{2+}$-dependent glutamate release from hippocampal mossy fiber nerve endings. *J. Neurochem.*, 66: 177–185.

O.P. Ottersen, I.A. Langmoen and L. Gjerstad (Eds.)
*Progress in Brain Research*, Vol 116


CHAPTER 3

# Properties and localization of glutamate transporters

N.C. Danbolt*, F.A. Chaudhry, Y. Dehnes, K.P. Lehre, L.M. Levy, K. Ullensvang and J. Storm-Mathisen

*Department of Anatomy, Institute of Basic Medical Sciences, University of Oslo, P.O. Box 1105 Blindern, N-0317 Oslo, Norway*

## Introduction

The glutamate transporters in the plasma membranes of astrocytes and neurons are essential for the normal functioning of the nervous system (for review see: Kanner and Schuldiner, 1987; Nicholls and Attwell, 1990; Danbolt, 1994; Gegelashvili and Schousboe, 1997; Kanai, 1997; Kanai et al., 1997; Kanner, 1993, 1997; Robinson and Dowd, 1997). They represent the only mechanism capable of quickly removing glutamate from the extracellular fluid. It is important to maintain a low concentration of glutamate extracellularly for two reasons. Firstly, glutamate is the major excitatory neurotransmitter and a high signal to noise ratio requires removal of extracellular glutamate so that the concentration will fluctuate with synaptic release. Secondly, glutamate is highly toxic to neurons expressing glutamate receptors and glutamate receptors are found on most neurons and even on many glial cells. Overstimulation of these receptors may kill the cells, particularly if their energy supply is limited (e.g. hypoglycemia, ischemia and mitchondrial malfunction) or if free radical protective systems are inadequate (e.g. defective superoxide dismutase). Because of the above and because the brain contains large amounts of glutamate (about 10 mmol/kg wet weight), it is logical that the glutamate transporters are present in high numbers throughout the brain. As will be outlined below, the functions of the glutamate transporters are more refined than simple transmitter removal. There is experimental evidence for the idea that the transporters may be actively involved in the regulation of synaptic transmission since they can modify the timecourse of synaptic events (Clements et al., 1992; Barbour et al., 1994; Maki et al., 1994; Mennerick and Zorumski, 1994; Tong and Jahr, 1994; Holmes, 1995; Takahashi et al., 1995, 1996; Clements, 1996; Kleinle et al., 1996; Otis et al., 1996, 1997; Zorumski et al., 1996; Asztely et al., 1997; Barbour and Häusser, 1997; Diamond and Jahr, 1997; Kataoka et al., 1997; Kinney et al., 1997; Trussell, this volume).

Five different sodium-dependent glutamate (excitatory amino acid) transporters have been cloned so far: GLAST (EAAT1) (Storck et al., 1992; Tanaka, 1993), GLT (EAAT2) (Pines et al., 1992), EAAC (EAAT3) (Kanai and Hediger, 1992), EAAT4 (Fairman et al., 1995) and EAAT5 (Arriza et al., 1997). The actual meanings of the acronyms (GLAST, glutamate-aspartate transporter; GLT, glutamate transporter; EAAC, excitatory amino acid carrier; EAAT, excitatory amino acid transporter) are not important, as they do not reflect

*Corresponding author.
Tel.: +47 22851260; fax: +47 22851278; e-mail: ncd@pons.uio.no.

functional differences among the transporters (see below). The first two were originally cloned from rat, while the third and the two last were originally cloned from rabbit and man, respectively. Analogues of these five proteins have been sequenced from a number of species including man (Shashidharan and Plaitakis, 1993; Shashidharan et al., 1994a,b; Arriza et al., 1994; Kanai et al., 1994; Kawakami et al., 1994; Manfras et al., 1994), mouse (Kirschner et al., 1994a,b; Mukainaka et al., 1995; Sutherland et al., 1995; Maenoikichi et al., 1997), rat (Kanai et al., 1995; Bjørås et al., 1996; Velaz-Faircloth et al., 1996) and cow (Inoue et al., 1995) as well as nematode (Kawano et al., 1996, 1997; Radice and Lustigman, 1996) and insect (Donly et al., 1997). Within a single species, the five transporter subtypes share over all about 50% of the amino acids, except for EAAT5 which is 30–40% identical with the others. However, in certain regions the sequences are nearly 100% conserved. Between mammals, the five proteins are some 90% identical to the equivalent proteins of another species. The existence of tissue specific variants of a transporter (GLT) has been demonstrated (Utsunomiya-Tate et al., 1997).

While this family bears no significant homology to the "neurotransmitter transporter family" (comprising carriers for GABA, glycine, and monoamines), it includes bacterial carriers for glutamate (proton-dependent), as well as bacterial (Engelke et al., 1989; Tolner et al., 1992) and mammalian carriers for dicarboxylate (Pajor, 1996) and for neutral amino acids (alanine-serine-cysteine-transporters, ASCT1 and ASCT2) (Arriza et al., 1993; Shafqat et al., 1993; Hofmann et al., 1994; Utsunomiya-Tate et al., 1996).

## Mechanism of transport

The sodium-dependent glutamate transporters use the transmembrane gradients of sodium, potassium and pH as driving forces (Billups et al., this volume). They cotransport sodium and probably $H^+$ ions with glutamate into cells in exchange for potassium and possibly hydroxide/hydrogencarbonate ions (Kanner and Sharon, 1978a,b; Sarantis and Attwell, 1990; Bouvier et al., 1992; Zerangue and Kavanaugh, 1996a,b). The transport of glutamate is electrogenic and stimulated by a negative membrane potential. The exact stoichiometry of the process is still a matter of some uncertainty (see discussion in Billups et al., this volume). Zerangue and Kavanaugh (1996a,b) studied human EAAC expressed in oocytes and concluded that 1 glutamate, 3 $Na^+$ and 1 $H^+$ are exchanged with 1 $K^+$. This stoichiometry can support a concentration difference of $10^6$ between the inside and outside of the cell and would allow removal of glutamate under a wide range of ionic conditions (but see Billups et al., this volume). One of the sodium binding sites is less specific, since lithium may replace sodium (Grünewald and Kanner, 1995). This might be the site at which potassium binds during the resetting step that returns the glutamate binding site to the exterior.

In addition to being co-transporters, these proteins also function as chloride and sodium channels (review Sonders and Amara, 1996). There is a constitutive sodium leak and a chloride conductance triggered by sodium dependent glutamate binding, but thermodynamically independent of the transport process. Consequently, the proteins behave as glutamate gated chloride channels. This interesting property is particularly prominent in EAAT4 (where it can carry $>90\%$ of the charge transfer) and EAAT5, while GLT has the lowest chloride conductance (Wadiche et al., 1995a,b; Fairman et al., 1995; Vandenberg et al., 1995; Arriza et al., 1997). It explains observations, particularly in retinal neurons and glia, associating chloride conductance and excitatory amino acid uptake (Sarantis et al., 1988; Eliasof and Werblin, 1993; Billups et al., 1996; Larsson et al., 1996). The channel is opened on binding of sodium plus amino acid, but the direction of chloride flux is independent of the direction of kation coupled amino acid transport. Nevertheless, the pore conducting chloride may be the same that mediates directional flux of glutamate. The ratio between amino acid transport and chloride channel activation appears to be higher for glutamate than for aspartate (Fairman et al.,

1995), implying possiblilites for differential effects of the two putative neuroeffectors. The site may be a target for physiological modulation, and attempts at pharmacological intervention. It may be significant that the chloride channel property is more developed in the neuronal transporters EAAT4 and EAAC1. The mechanism could (provided a negative chloride equilibrium potential) augment glutamate uptake by counteracting its depolarizing effect and could stabilize cells against depolarization by receptor activation. A transporter with this property, if present presynaptically, could perhaps reduce the likelihood of action potentials invading the terminal to cause further transmitter release.

**Affinities and selectivities**

The activity due to glutamate transporters was initially described in brain synaptosomes and "minislices" (Logan and Snyder, 1971; Balcar and Johnston, 1972) and referred to as sodium-dependent *high affinity* glutamate uptake in order to distinguish it from chloride-dependent, sodium-independent and ill defined *low affinity* uptake. However, the affinities are not particularly high with $K_m$-values for glutamate in the range 10–100 $\mu$M, except for EAAT4 which has a $K_m$ of about 2 $\mu$M. [Lower $K_m$ values reported in the past may have resulted from suboptimal assay conditions (cf. Danbolt, 1994).] The carriers have similar affinities for L-glutamate and several analogues, including L-aspartate and D-aspartate (but not D-glutamate), and are therefore more appropriately designated excitatory amino acid transporters. Nonetheless, glutamate is probably the main substrate in vivo.

GLT is distinguished by sensitivity to blocking of transport by the glutamate analogues kainate and dihydrokainate (Arriza et al., 1994, 1997). The most potent general inhibitors of glutamate transport known are themselves transported. Thus L-*trans*-pyrrolidine-2,4-dicarboxylate (Bridges et al., 1991) can cause excitotoxicity by provoking glutamate release via heteroexchange (Volterra et al., 1996). Potent selective blockers that are themselves not transported are wanted as pharmacological tools. Progress is presently being made in the development of such tools (Lebrun et al., 1997; Vandenberg et al., 1997; Shimamoto et al., 1998).

**Molecular structure and properties**

The initial publications on mammalian glutamate transporters proposed three different structures with 6, 8 or 10 putative transmembrane domains (TM). However, since the primary structures are very similar, the proteins probably have the same numbers of TMs. The first 6 TMs are clearly indicated by hydropathy plots. The models propose a long extracellular stretch between TM3 and TM4 and that the N-terminal and C-terminal are both intracellular. The proposed topology is supported by the intracellular staining by antibodies to the N- and C-terminals (Lehre et al., 1995), and the localization of putative glycosylation (between TM3 and 4) and phosphorylation sites (Casado et al., 1993). "Reporter glycosylation scanning" suggests a topology with 10 TMs, TM1-6 being $\alpha$-helices and TM7-10 $\beta$-sheets (Wahle and Stoffel, 1996). The latter comprise highly conserved stretches. They might allow pore loops that could contribute to the formation of a controlled pore (MacKinnon, 1995). This region contains a seven amino acid stretch, AAI(L/V)FIAQ (residues 407–413 in GLT-1), which is conserved between the mammalian and bacterial glutamate transporters as well as in mammalian neutral amino acid transporters and a bacterial dicarboxylate transporter. A 76 residues long segment (corresponding to residues 364–439 of GLT-1) comprising AAX-FIAQ contains the site responsible for the unique sensitivity of GLT type transporters to kainate and dihydrokainate (Vandenberg et al., 1995). Thus a chimeric molecule E1-2-1 consisting of this stretch from human (h) GLT-1 and the remainder from hGLAST combines properties unique to the GLT type and GLAST type, respectively: glutamate transport as well as a substrate independent but $Cl^-$ dependent kation current (GLAST property) are both blocked by kainate (GLT property) in E1-2-1. Out of the 76 residues deriving from hGLT-1 in the chimera only 18 residues differ from hGLAST.

Site directed mutagenesis of GLT-1 has shown that histidine 326 (located in TM6) is required for transport function (Zhang et al., 1994). This positively charged TM residue is shared by all known members of the family, including the dicarboxylate transporters. As histidine residues have been implicated in proton translocation the authors proposed that the residue could be involved in the pH-changing effect of the transporter. Out of five conserved, negatively charged TM residues, aspartate 398, glutamate 404, and aspartate 470 are critical for glutamate transport (Pines et al., 1995). Negative charge is not enough, since the mutated forms D398E (aspartate 398 mutated to glutamate), E404D and D470E are all unable to transport glutamate. However, E404D can still transport D- and L-aspartate, and bind them at increased affinity, while binding of glutamate is unimpaired. The site thus appears involved in substrate recognition as well as translocation. Subsequently, glutamate 404 has been shown (Kavanaugh et al., 1997) to be essential for the building up of a transmembrane glutamate gradient, this important function being dependent on a single methylene group: the mutant E404D cannot sustain sodium/potassium coupled net transport, but does catalyse sodium dependent exchange of excitatory amino acids, as well as showing sodium and amino acid dependent chloride conductance. Serine 113 of GLT-1 is phosphorylated by protein kinase C, causing an increase of Vmax (Casado et al., 1993).

GLAST residues tyrosine 405 and arginine 479, conserved in glutamate and dicarboxylate transporters, are essential for transport, Y405F and R479T being inactive (Conradt and Stoffel, 1995). On the other hand, arginine residues 122 and 280 of GLAST, situated in the first and second putative cytosolic loops close to TM3 and TM5, respectively, appear to be subtly involved in substrate recognition: the mutants R122I and R280V transport glutamate normally, but show increased affinity for aspartate and D,L-*threo*-3-hydroxy aspartate. The N-glycosylation sites have been identified as asparagine 206 and 216, but the glycosylation is apparently without functional significance (Conradt et al., 1995).

In EAAT5 the last 4 C-terminal amino acids form a motif suggesting that this site interacts with the postsynaptic density protein PSD-95 and thus may be targeted to the postsynaptic membrane at excitatory synapses (Arriza et al., 1997).

Chromosomal localizations of the human genes for hGLT, hGLAST and hEAAC are 11p11.2–p13 (Li and Francke, 1995; Takai et al., 1996), 5p11–p12 (Stoffel et al., 1996) and 9p24 (Smith et al., 1994), respectively, with possible relation to genes involved in neuropsychiatric disorders and dicarboxylic aminoaciduria.

### Oligomeric structure

From the report on the first purification of a glutamate transporter (Danbolt et al., 1990), it has been clear that these transporter proteins have a strong tendency to give rise to oligomer bands on SDS-polyacrylamide gel electrophoresis. Since brain transporters only give rise to monomer bands when electrophoresed immediately after solubilization in SDS, the appearance of oligomer bands in older protein extracts have been interpreted as aggregation. This interpretation has been modified by the recent observations that cross-linking of the glutamate transporter proteins prior to solubilization gives rise to oligomer bands on the gels (Haugeto et al., 1996; Dehnes et al., 1998). This must imply that the transporters exist in the brain as oligomers in which the subunits are non-covalently connected. Radiation inactivation analysis indicated that transport activity was dependent on oligomeric structure (Haugeto et al., 1996). Interestingly, crosslinking of GLT gives rise to predominantly trimers, while crosslinking of EAAT4 only gives rise to dimers. Crosslinked GLAST or EAAC presents an equal mixture of dimers and trimers. Members of the other neurotransmitter transporter family, exemplified by the glycine transporter GLYT1, do not form oligomer bands after crosslinking indicating that these proteins are monomers. Further, the different glutamate transporter subtypes do not form oligomeric complexes with each other suggesting that they are homo-oligomers.

**Localization**

*GLT*: GLT is the only glutamate transporter that has so far been purified to near homogeneity from rat brain in active form (Danbolt et al., 1990). Polyclonal antibodies (Danbolt et al., 1992) that were raised against the purified protein (Danbolt et al., 1990) were used for screening expression libraries (Pines et al., 1992) and to localize the GLT protein in brain tissue (Danbolt et al., 1992). GLT was detected only in astrocytes (fibrous and protoplasmic) (Fig. 1) throughout the central nervous system (except in retina, see below), at the highest concentration in hippocampus and cerebral cortex (Fig. 2). These results were confirmed with monoclonal antibodies (Hees et al., 1992; Levy et al., 1993) and with antibodies to synthetic peptides (Rothstein et al., 1994; Lehre et al., 1995; Schmitt et al., 1996). The GLT concentrations in astrocytic membranes are highest in the parts of the plasma membrane that face neuropil and lower in the parts facing pia mater, capillary endothelium, other astrocytes and neuronal cell bodies (Chaudhry et al., 1995). In cerebellum the glutamatergic parallel fiber boutons synapsing on interneuron dendritic shafts were surrounded by lower densities of glutamate transporters than those synapsing on Purkinje cell spines (Chaudhry et al., 1995). In agreement, electrophysiological observations (Barbour et al., 1994) showed an effect of glutamate uptake inhibitors on the timecourse of the synaptic current in the latter, but not the former type of synapse.

In retina, GLT is expressed in different types of bipolar cells and in some amacrine cells, but apparently not in Müller cells or other retinal glia (Rauen and Kanner, 1994; Euler and Wässle, 1995; Rauen et al., 1996). Outside the CNS, GLT and/or its mRNA has been found in the islands of Langerhans in pancreas (Manfras et al., 1994; Manfras and Boehm, 1995), placenta (Matthews et al., 1998) and in liver (Utsunomiya-Tate et al., 1997).

In view of the highly selective labelling of brain astrocytes with anti-GLT antibodies, it was a surprise to find that certain populations of neurons, particularly the pyramidal cells of hippocampus CA3 (origin of Schaffer collateral synapses in hippocampus CA1) and layer VI of the cerebral neocortex express mRNA encoding GLT (Torp et al., 1994, 1997; Schmitt et al., 1996). GLT containing neurons or neuronal processes have been searched for immunocytochemically, but not found (Chaudhry et al., 1995; Lehre et al., 1995). These studies have used different immunocytochemical techniques and an array of different anti-GLT antibodies. Based on the estimated sensitivities of the methods one may conclude that the neuronal concentration of GLT is less than 5% of that in astrocytes (Chaudhry et al., 1995). Since the astrocytic concentration of GLT is very high, a neuronal concentration 5% of this might still be significant. After all, the GLT protein represents about 1% of all brain membrane protein (Danbolt et al., 1990). [Investigation in material fixed by formaldehyde/pH-shift, which gives higher labeling intensity, did not provide further evidence that GLT is localized in the membrane of Schaffer collateral terminals (F.A. Chaudhry and J. Storm-Mathisen, unpublished).] In contrast, GLT has been demonstrated in hippocampal neurons in microculture (Mennerick et al., 1998), showing that the posttranscriptional block of GLT formation can be lifted in certain conditions. The mismatch between neuronal GLT mRNA and protein is a reminder that increases in mRNA do not necessarily imply increases in the level of the encoded protein.

*GLAST*: Like GLT, GLAST has also exclusively been demonstrated in astrocytes throughout the central nervous system (Chaudhry et al., 1995; Lehre et al., 1995; Schmitt et al., 1997) and, like GLT, it is found at the highest concentrations on the membranes that face neuropil. [An early paper reported the presence of GLAST protein in neurons as well as astrocytes (Rothstein et al., 1994), but this has later been corrected (Ginsberg et al., 1995).] In fact, GLAST and GLT are expressed by the same astrocytes (Fig. 3), but in

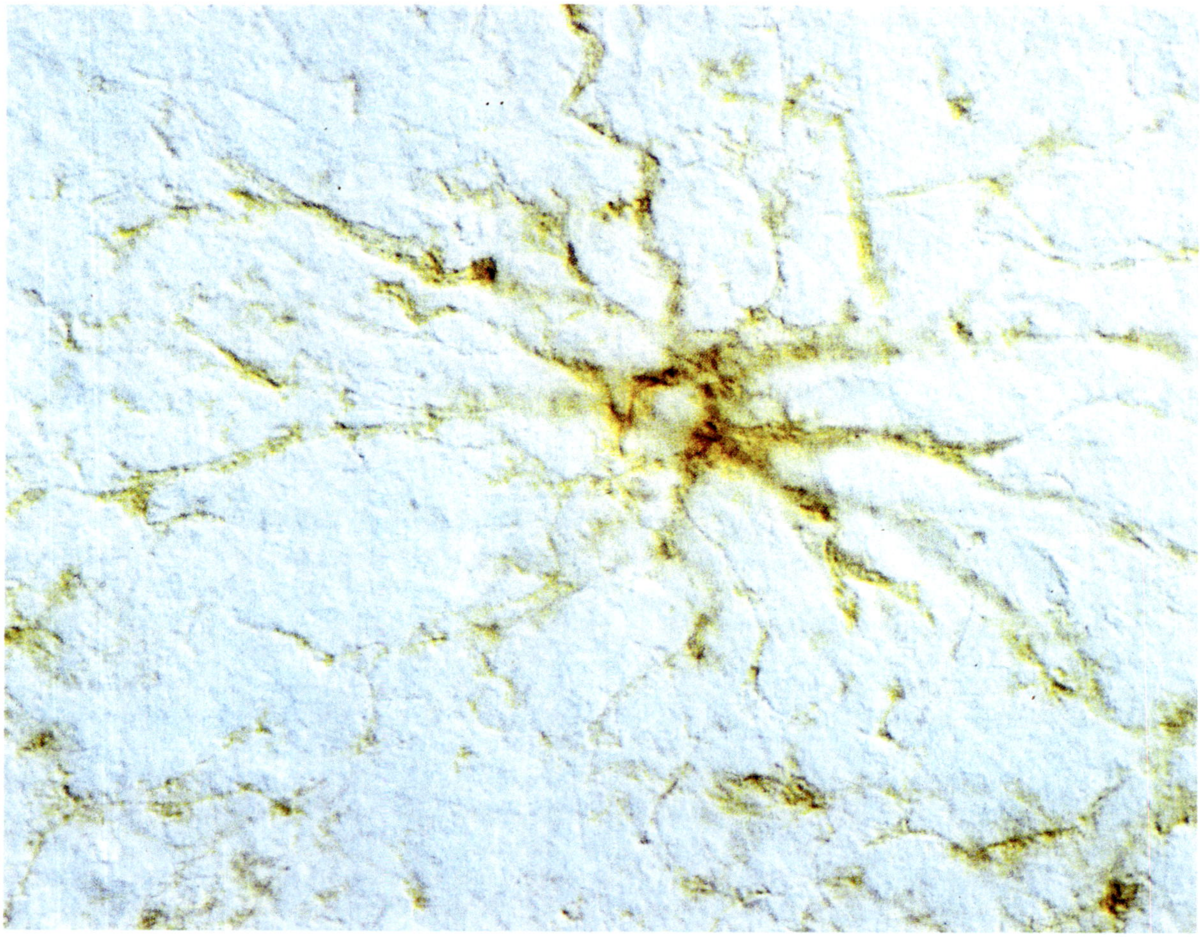

Fig. 1. Astrocyte in the human hippocampus immunoreactive with an antibody to the glutamate transporter GLT-1. Immunoperoxidase stained processes (brown) pervade the grey matter. Oil immersion, differential interference contrast optics.

different proportions in different regions (Chaudhry et al., 1995; Lehre et al., 1995). Because of this colocalization it was speculated that the two proteins could form oligomeric complexes with each other. It has now been shown that the two proteins form homooligomers and do not complex with each other, but exist side by side in the same membranes independently of each other (Haugeto et al., 1996). The highest concentrations of GLAST are found in the Bergmann glia of the cerebellum. It is also relatively higher than GLT in the olfactory bulb, but except for these two regions, GLAST and GLT are distributed similarly among brain regions (Fig. 2). GLAST, but not GLT, is found in the Müller cells and astrocytes in retina (Derouiche and Rauen, 1995; Rauen et al., 1996; Lehre et al., 1997) and in supporting cells in the cochlea (Furness and Lehre, 1997) and vestibular end organ (Takumi et al., 1997). GLAST has been detected in the placenta (Matthews et al., 1998).

*EAAC*: The localization of this protein has, not been studied as thoroughly as the other glutamate transporters because of difficulties with obtaining good antibodies to it. Nevertheless, the protein

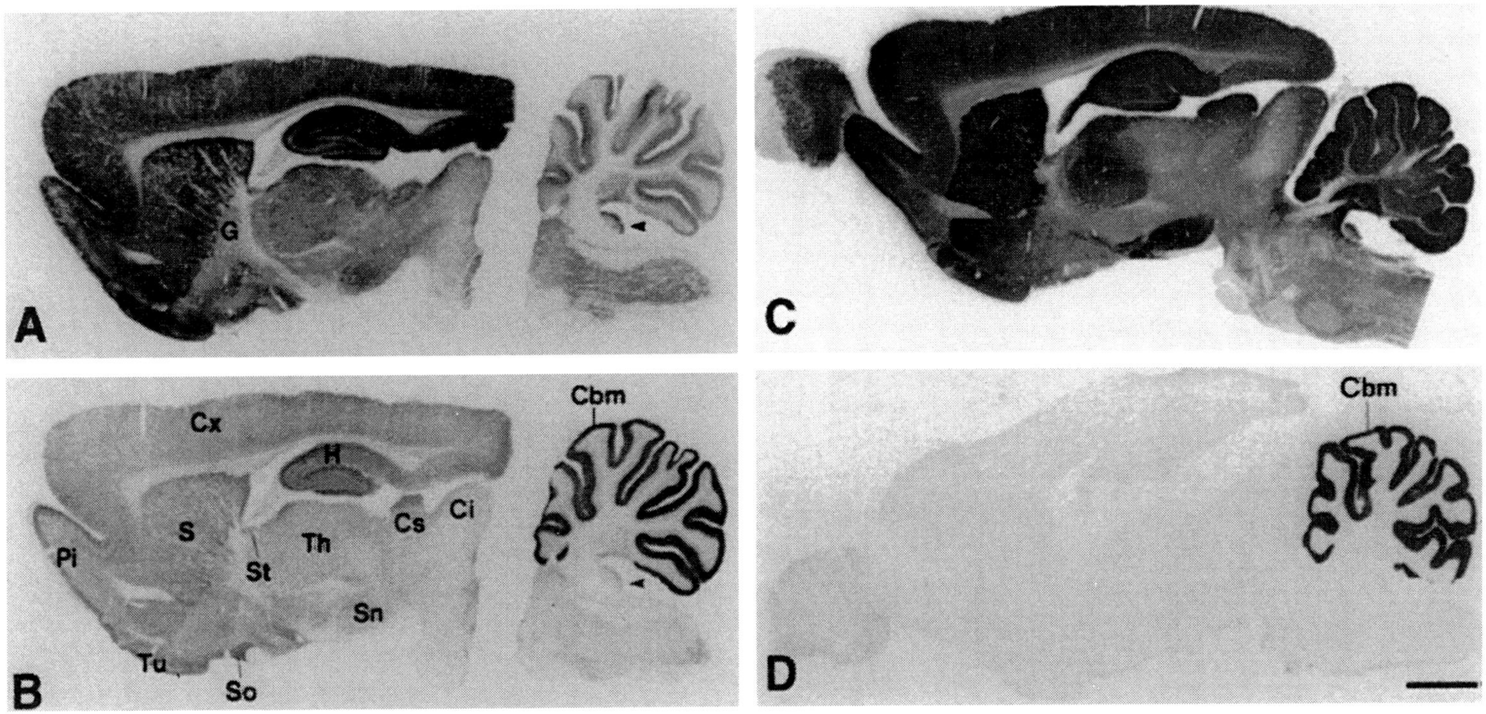


Fig. 2. Regional localizations of four cloned glutamate transporters in rat brain (immunoperoxidase in the presence of Triton X-100, i.e. staining intensities cannot be compared between the different sections/transporters). GLT-1 (A) is highest in hippocampus (H) and other forebrain and brain stem regions rich in glutamatergic nerve terminals, such as regions of the cerebral cortex (Cx, Pi, Tu), striatum (S), and the molecular layer of the dorsal cochlear nucleus (*arrowhead*). GLAST (B) has a similar relative distribution in the forebrain, but is particularly high in the cerebellar molecular layer (Cbm). EAAC (C) is widely distributed, areas strongly stained apart from those in A include targets of GABAergic nerve endings in globus pallidus (G), substantia nigra (Sn), and central cerebellar nuclei. EAAT4 (D) is essentially restricted to the cerebellar molecular layer (Cbm). Other regions indicated: Ci, Cs, inferior and superior colliculus; Pi, piriform cortex; So, supraoptic nucleus; St, stria terminalis; Th, thalamus; Tu, tuberculum olfactorium. Antibodies, raised to synthetic peptides, were affinity purified on the immobilized peptides: amino acid residues 12–26 of GLT-1 (Pines et al., 1992), 522–541 of GLAST (Storck et al., 1992), 510–524 of EAAC1 (Kanai and Hediger, 1992), and 537–564 of EAAT4 (Fairman et al., 1995). The illustrations are modified from Lehre et al. (1995) (A, B), K. Ullensvang and N.C. Danbolt (unpublished) (C), and Dehnes et al. (1998) (D). Scale bar (valid for all), 2 mm.

has been localized immunocytochemically by the peroxidase technique (Rothstein et al., 1994). It has a wide distribution in brain (Fig. 2) and may be found in most if not all glutamatergic neurons, as well as in GABAergic neurons like the cerebellar Purkinje cells and hippocampal interneurons. In GABAergic neurons it is present also in axons and terminals, but in glutamatergic neurons it appears to avoid the nerve terminals. Such a distribution pattern is reported also in hippocampal cultures (Coco et al., 1997). The localization of the protein is in good agreement with the localization of EAAC mRNA which has been studied by several groups (Kanai et al., 1995; Velaz-Faircloth, et al., 1996; Torp et al., 1997). Initially reported to be restricted to neurons (Rothstein et al., 1994), EAAC protein occurs also in some brain astroglial cells (our unpublished observations; Conti et al., 1998). In the retina, EAAC is located in horizontal cells, amacrine cells, displaced amacrine cells and in ganglion cells (Schulz et al., 1996; Rauen et al., 1996). EAAC is found outside the CNS: in the kidneys and in the intestinal mucosa, from where it was cloned

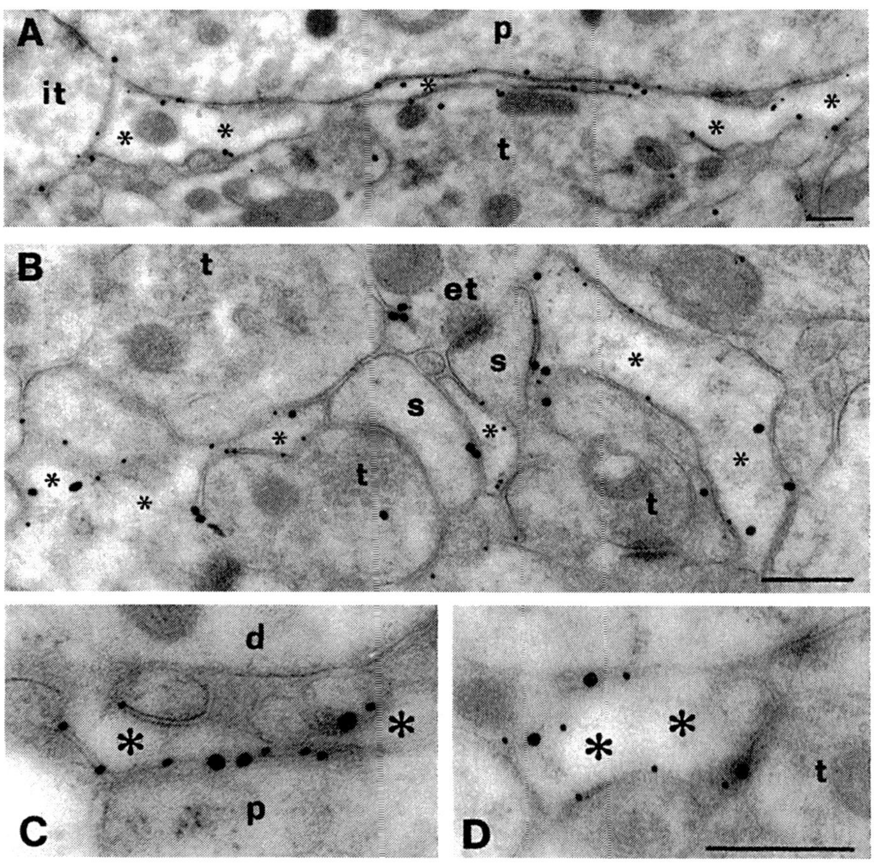


Fig. 3. Colocalization of GLT-1 and GLAST in the membranes of astrocytic processes (*asterisks*) in rat hippocampus. Double labeling immunogold technique, GLT-1 and GLAST are demonstrated by 15 nm and 30 nm particles, respectively. Note that the two proteins are in the same membranes, and that the labeled processes are adjacent to excitatory type as well as inhibitory type synapses. Symbols: d, dendrite; et and it, terminals of "excitatory" and "inhibitory morphology", i.e. making asymmetric contact on a spine and symmetric contact on perikaryon, respectively; t, nerve terminals; s, dendritic spines; p, pyramidal cell perikaryon. From the same material as Haugeto et al. (1996). Bars, 300 nm.

(Kanai and Hediger, 1992), as well as in the intestinal nervous plexuses (Liu et al., 1997).

*EAAT4*: EAAT4 is highly concentrated in the cerebellar molecular layer (Fig. 2), and although it is also present in the forebrain (Dehnes et al., 1998), particularly in thin dendrites in hippocampus and neocortex (Furuta et al., 1997a), the concentrations there are so low that it would be permissible to say that EAAT4 is specific for the Purkinje cells (Yamada et al., 1996; Furuta et al., 1997a; Nagao et al., 1997; Dehnes et al., 1998), in which it is concentrated in the spines (Yamada et al., 1996). Tanaka and coworkers (1997) report that EAAT4 is localized extrasynaptically. Quantitative immunocytochemistry on freeze-substituted material with the postembedding immunogold technique (Dehnes et al., 1998) confirms this, but adds that low, but significant, concentrations are also found in synaptic and perisynaptic membranes. The highest concentrations, however, are found on the parts of the Purkinje cell spines and their parent thin dendrites that are ensheathed by astrocytes. Lower concentrations are found on the parts that are in contact with neurons.

EAAT4 is concentrated in parasagittal zones (Nagao et al., 1997), resembling those expressing zebrin (Brochu et al., 1990; Ahn et al., 1994). Double labelling and study of adjacent sections showed that the highest EAAT4 expression is found in the zebrin + Purkinje cells and that GLT, GLAST and EAAC, in contrast to EAAT4, do not exhibit parasagittal banding in the cerebellar cortex (Dehnes et al., 1998). The functional significance of the banding is therefore unclear. It may be predicted that the Purkinje cells richer in EAAT4 could accept higher frequencies of stimulation via their inputs from parallel and climbing fibers.

*EAAT5*: The mRNA of EAAT5 is selectively expressed in retina (Arriza et al., 1997). In salamander retina, EAAT5 protein in mainly localized in Müller cells, but light microscopy also suggests some neuronal localization (Eliasof et al., 1988).

In Fig. 4 the cellular localizations of GLT-1, GLAST and EAAT4 are schematically summarized as seen in the cerebellum by post embedding immunogold. The precise localization of EAAC1 remains to be determined. None of the known transporters appears to be localized in glutamatergic nerve endings such as the parallel fiber terminals of the cerebellar cortex. Yet, the net uptake of D-aspartate in excitatory nerve endings (Gundersen et al., 1993, 1996; but not in all systems, Gundersen et al., 1995), with properties similar to GLT, GLAST, etc., and the retrograde axonal transport of locally administered D-[$^3$H]aspartate in vivo (Streit, 1980; Baughman and Gilbert, 1980), provide strong evidence for the presence of glutamate transporters at these sites. This suggests that additional excitatory amino acid transporters exist. A candidate is EAAT2B, recently identified in salamander retina (Eliasof et al., 1998). It should be noted that different categories of glutamatergic nerve endings may differ with respect to how avidly they take up excitatory amino acids (Taxt and Storm-Mathisen, 1984; Fischer et al., 1986). This may form a possible target of differential pharmacological intervention.

**Amounts of glutamate transporters in the cerebellum**

When cerebellar proteins are immunoblotted with antibodies to glutamate transporters, the rank order of the labeling signal obtained for the transporter subtypes is: GLAST > GLT > EAAT4 > EAAC (K.P. Lehre, Y. Dehnes and N.C. Danbolt, unpublished). Since the results were obtained with different antibodies to different proteins, direct comparisons are not possible. However, since the same pattern has been observed with several antibodies, it is reasonable to assume that the concentrations of GLAST and EAAC represent the highest and lowest transporter concentrations, respectively, the concentrations of GLT and EAAT4 being somewhere in the middle. The concentration of EAAT4 has now been determined in absolute terms (Dehnes et al., 1998). EAAT4 represents about 0.075% of whole cerebellum tissue protein or about 3500 EAAT4

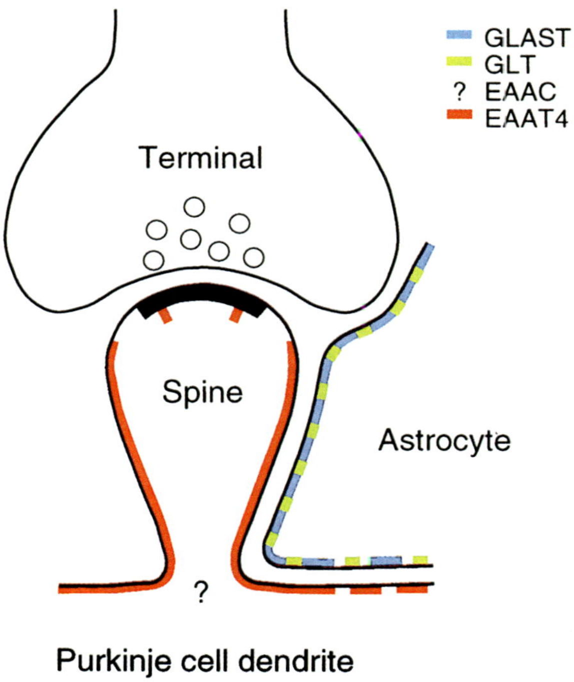


Fig. 4. Schematic representation of the detailed localization of glutamate transporters in the vicinity of a glutamatergic synapse, exemplified by the parallel fiber-to-Purkinje cell dendritic spine synapse in the cerebellar molecular layer. GLT and GLAST are intermingled in the astrocytic membranes, the concentrations being highest near synapses, lower along major dendrites. EAAT4 is concentrated in the glia-covered parts of the membranes of Purkinje cell dendrites, highest at the spines and thinner dendrites. There is a very low concentration of EAAT4 at the postsynaptic density. EAAC1 (not shown) is present in Purkinje cell dendrites and perhaps elsewhere, but the precise localization is not yet known.

molecules per Purkinje cell spine, i.e. about 1400 per $\mu m^2$ spine membrane. This is in agreement with estimates based on electrophysiological techniques indicating about 1000 molecules of postsynaptic glutamate transporter close to a release site (Otis et al., 1997).

### Assay of glutamate uptake

Glutamate uptake activity has traditionally been assayed in synaptosomes, or simply in synaptosome containing homogenates from brain, incubated with radiolabeled substrate in the presence of sodium. This method seems to emphasize uptake into excitatory nerve endings, as evidenced by the loss of uptake after denervation induced loss of such terminals in the area investigated (e.g. Fonnum et al., 1981). As this method is relatively independent of the energy state, it may record preferentially uptake into compartments rich in excitatory amino acids (i.e. excitatory nerve terminals) which can exchange with the radiolabeled amino acid in the medium.

Sodium dependent "binding" of radiolabeled D-aspartate has been used frequently, particularly in human material. (The use of D-aspartate is motivated by its relative resistance to metabolic change and by its being transported at roughly the same rate as glutamate by all known "high affinity" glutamate transporters.) This method was introduced in the belief that binding to the active sites of the transporters could be measured, but in fact the method demonstrates sequestering of the radiolabel inside compartments bounded by membranes containing the transporters (Danbolt and Storm-Mathisen, 1986). The result therefore depends on the internal contents of exchangeable amino acid, as well as on ionic gradients and intactness of the membranes (cf. Danbolt, 1994). The significance of the results obtained is therefore somewhat unclear, but the method may assay mainly transport into membrane vesicles (saccules) of glial origin. Still less clear is the significance of sodium dependent "binding" of D-aspartate recorded in cryostat sections mounted on microscopic slides or coverslips (see discussion in Danbolt, 1994). Using this method (Greenamyre et al., 1990), or a similar method including defatting to avoid resealing of membrane fragments (Anderson et al., 1991), no denervation-induced reductions in "binding" were demonstrated that could reflect the denervation-induced reductions in excitatory amino acid uptake activity and immunoreactivity attributable to the combined loss of nerve terminal uptake sites (Fonnum et al., 1981) and downregulation of glial transporters (Levy et al., 1995). The presence of other proteins binding glutamate could possibly interfere.

Assay of uptake of radiolabeled glutamate or aspartate into liposomes with defined ion gradients removes many of the uncertainties mentioned. With membranes from brain his method measures mainly glial transport, mostly GLT-1 (see above).

Determination of immunoreactivities for the various transporter proteins has other problems, since the epitopes may have different sensitivities to post-mortem breakdown (Beckstrøm et al., 1998).

The use of different methods explains some of the apparent discrepancies between authors who have investigated human pathological material. For instance, glutamate transporter parameters have been reported to be reduced (e.g. Masliah et al., 1996) or not changed (Rothstein et al., 1995) in brains from patients who died in Alzheimer's disease.

**Regulation of glutamate transport activity**

Modulation of glutamate transport activity has been demonstrated in several different model systems. The uptake increases region-selectively in the brains of rats stressed by restraint in narrow cylinders (Gliad et al., 1990). In vivo electrical stimulation (for 10 min) of frontal cortex increases the glutamate uptake in rat striatum by increasing the apparent affinity of the transporter for glutamate (Nieoullon et al., 1983). The animals were killed for uptake measurements 20 min after cessation of electrical stimulation. The increase from basal level can be inhibited by dopaminergic activity (Kerkerian et al., 1987). The same group (Samuel et al., 1996) has also noted an involvement of the glutamatergic metabotropic receptors in the regulation of glutamate uptake and extracellular excitatory amino acid levels in the striatum. Glutamate uptake is increased in neuronal cell cultures surviving chronic hypoxia (Sher and Hu, 1990) and transient retinal ischemia (Otori et al., 1994). However, neonatal hypoxia-ischemia induced loss of neurons and glia in striatum has been related to changes in glutamate transport (Martin et al., 1997). The finding that glucocorticoids inhibit glucose transport and glutamate uptake in hippocampal astrocytes has implications for our understanding of glucocorticoid neurotoxicity (Virgin et al., 1991).

The activity of GLT is stimulated by phosphorylation of serine residue 113 by protein kinase C (Casado et al., 1993). A rapid stimulation of glutamate transport activity, attributed to EAAC1, by phorbol ester has been observed in C6 glioma cells (Dowd and Robinson, 1996). GLAST, however, is inhibited by protein kinase C-mediated phosphorylation at a non-PKC consensus site (Conradt and Stoffel, 1997).

Arachidonic acid has been reported to inhibit several sodium coupled amino acid transporters including the uptake systems for glutamate, glycine and GABA (Rhoads et al., 1983; Chan et al., 1983; Yu et al., 1986; Barbour et al., 1989; Zafra et al., 1990; Lynch et al., 1994; Volterra et al., 1992, 1994a). It is released from both neurons (Dumuis et al., 1988) and glial cells upon activation of glutamate receptors. Free fatty acids, including arachidonic acid, are known to accumulate in the brain under pathological conditions such as ischemia and seizures. Trotti and coworkers (1995) have used partly purified rat transporters reconstituted in liposomes (Danbolt et al., 1990) and shown that the arachidonic acid induced inhibition is due to direct action on the transporter itself rather than an effect through other mechanisms or via the phospholipid membrane. They further showed that the arachidonic acid ethyl ester is inactive, suggesting that the free carboxylic group is required for inhibitory activity. The liposomal uptake in this study is mostly due to GLT. In contrast, transport by hGLT expressed in oocytes is stimulated by arachidonic acid while hGLAST is inhibited (Zerangue et al., 1995). Such differences may underlie differential inhibition of glutamate uptake by arachidonic acid in different brain regions (Manzoni and Mennini, 1997). The activity of the EAAT4 protein is probably subject to regulation by fatty acids (Kataoka et al., 1997). The volatile anesthetics isoflurane (Langmoen et al., 1995; Larsen et al., 1997) and halothane (Miyazaki et al., 1997) increase glutamate uptake. This may contribute to the action of these agents.

Glutamate uptake is inhibited by oxidation effected in various ways (Piani et al., 1993; Palmer et al., 1994; Volterra et al., 1994b; Savolainen et al., 1995; Harris et al., 1995, 1996; Agostinho et al., 1997; Berman and Hastings, 1997; Sorg et al., 1997). The effect is distinct from and additive to the effect of arachidonic acid (Volterra et al., 1994a). The reported (Pogun and Kuhar, 1994) inhibitory effect of NO is probably due to formation of peroxynitrite which potently inhibits GLAST, GLT and EAAC (Trotti et al., 1996). Modification of SH-groups inhibits glutamate uptake (Brookes, 1988; Aschner et al., 1990, 1993; Albrecht et al., 1993; Volterra et al., 1994). SH-groups important for function are found in GLAST, GLT and EAAC and redox interconversion of these groups may be a regulatory mechanism (Trotti et al., 1997). Nagaraja and Brookes (1996) have shown that mercuric chloride uncouples glutamate uptake from the countertransport of hydroxyl equivalents. Keller and coworkers (1997) report that 4-hydroxynonenal, an aldehydic product of membrane lipid peroxidation, impairs glutamate uptake and mitochondrial function in synaptosomes. $\beta$-Amyloid peptide inhibits glutamate uptake in cultured astrocytes (apparently via the formation of free radicals) (Harris et al., 1996), and so does tumor necrosis factor $\alpha$ (Fine et al., 1996), a product of infected macrophages. The observations cited in this paragraph imply possible roles of impaired glutamate uptake in the pathogenesis of neurodegenerative disorders including Alzheimer's disease and AIDS dementia (see also Farber et al., this volume).

## Regulated expression

As one might expect from the highly differentiated localizations and from the findings of regulation of transport activity, the expressions of the transporters are regulated. The first direct demonstration of this was the observation that lesioning of the glutamatergic corticostriatal pathway leads to a down regulation of the expression of the glial glutamate transporters GLT and GLAST in striatum (Ginsberg et al., 1995, 1996; Levy et al., 1995). Transection of the facial nerve leads to an upregulation of mRNA encoding GLAST in the facial nucleus (Yamashita et al., 1996). Interestingly, a circadian variation of mRNA encoding EAAC has been observed in the rat suprachiasmatic nuclei (Cagampang et al., 1996). Differential expression of the glutamate transporter subtypes has also been observed during the development of the rat brain (Sutherland et al., 1996; Furuta et al., 1997b; Ullensvang et al., 1997). At birth, there are significant levels of GLAST in brain, while GLT is barely detectable. During the period of active synaptogenesis a strong increase in the levels of both transporters is seen (Ullensvang et al., 1997).

Earlier observations indicated that neurons release factors inducing glutamate transport activity (Drejer et al., 1983; Voisin et al., 1993) as well as ones inducing GABA transport activity (Nissen et al., 1992) in glial cells. Now it has been shown that the expression of GLAST in astrocyte cultures is increased by activation of glutamate receptors (Gegelashvili et al., 1996; Swanson et al., 1997). Astrocytes do not express GLT when cultured alone, but they do if they are cultured together with neurons (Gegelashvili et al., 1997; Swanson et al., 1997). In fact, it is sufficient to grow the astrocytes in conditioned media from neuronal cultures suggesting that the GLT expression is controlled by novel neuronal soluble factors (Gegelashvili et al., 1997). Thus expressions of GLT and GLAST are differentially regulated. The regulation of the expression of the glutamate transporters may be disturbed in disease. For instance, acute liver failure resuts in lower levels of GLT expression in brain (Knecht et al., 1997; Norenberg et al., 1997).

## Functional consequences of lack of the individual transporters

In the absence of suitable uptake blockers, antisense technology has been used to investigate the consequences of "knock-down" of the synthesis of GLT-1, GLAST and EAAC1 (Rothstein et al.,

1996; Sepukty et al., 1997) and of EAAT4 (Maragakis et al., 1997) in intact rats. By infusion of antisense oligonucleotides into the cerebrospinal fluid for 7–10 days selective reductions by up to 90% could be obtained in the protein levels, suggesting that the proteins have a fairly rapid turnover in the plasma membranes. The results showed, in agreement with immunoprecipitation experiments (Haugeto et al., 1996), that GLT-1 is the quantitatively dominant glutamate transporter. Animals deprived of GLT-1 or GLAST developed neuronal degeneration and motor dysfunction, whereas those deprived of EAAC1 or EAAT4 showed epileptiform fits or ataxic motor disturbance, respectively. The results of antisense knock-down of EAAC1 may be due to impaired GABA synthesis caused by impaired uptake of glutamate in GABAergic neurons in hippocampus and elsewhere (Sepukty et al., 1997). On the other hand, mutant mice in which the gene for EAAC1 was disrupted, showed dicarboxylic aminoaciduria, but apparently no neurological abnormality apart from reduced spontaneous locomotor activity (Peghini et al., 1997). These results are not necessarily inconsistent, since the regulation of GABA synthesis may be compensated in the knock-out mutant mice (in which EAAC1 was absent already during ontogenesis). Mutant mice lacking GLT-1 developed seizures reminiscent of ones induced by NMDA and excitotoxic neuronal loss in hippocampus CA1 (Tanaka et al., 1997). Electrophysiological investigation suggested prolonged elevation of the extracellular glutamate concentration following synaptic release. The mutant animals further exhibited enhanced vulnerability to physically (cold) induced brain injury.

These results suggest that the glial glutamate transporters, notably GLT-1, are the more important for preventing excitotoxic damage, whereas the neuronal ones may be involved in the function of the neurons containing them. The results on GLT-1 are in line with the reports that GLT-1 protein is selectively reduced in patients with amyotrophic lateral sclerosis (Rothstein et al., 1992, 1995). The mRNA levels appear normal, but the neuropathologically affected areas contain abnormal GLT-1 mRNA (Lin et al., 1988).

**Why do we need several different glutamate transporters?**

This question cannot be answered properly until more information is available. An important aspect may be independent control. From the data outlined above it is clear that the presence of different glutamate transporters allows differential regulation of protein expression as well as of transport rate to suit specific demands at a given time point and site. This is illustrated by GLT and GLAST, which have different developmental and localizational patterns. GLT is stimulated by phosphorylation while GLAST is inhibited. The basic functional properties of the known glutamate transporters seem rather similar. The functional consequences of different substrate specificities, exemplified by the sensitivity of GLT to kainate, are not obvious. The higher affinity of EAAT4 should allow this transporter to keep the resting glutamate concentration in the secluded space surrounding Purkinje cells lower than elsewhere in the brain extracellular fluid. Its particularly high glutamate gated chloride conductance adds a property as an inhibitory ionotropic glutamate receptor, even in the presence of an excess of other glutamate transporters (due to its higher glutamate affinity). Whether the kationic leak current of GLAST is functionally important remains to be determined.

**Concluding remarks**

The findings described in the present report give further support to the idea that the tasks of the glutamate transporters are more sophisticated than simple transmitter removal (cf. Danbolt et al., 1994). By binding as well as by transport they can modify the timecourse of the concentration of the excitatory amino acid during synaptic events. On a longer timescale they govern the concentration of excitatory amino acids in the extracellular space. This will affect the levels of activation and desensitization of postsynaptic receptors, as well as of presynaptic and extrasynaptic ones. In agreement, the transporters are regulated at multiple levels: they occur in multiple types, the localization is differentiated, the protein levels are subject to modification, and the activity is modulated through various mechanisms (for review see, Gegelashvili and Schousboe, 1997). The regulatory mechanisms may form future targets for therapeutic intervention.

## Acknowledgements

We thank Ansgar Torvik for providing samples of human brain for immunocytochemistry (Fig. 1). Supported by the Norwegian Research Council, Jahrefondet, Nansenfondet, the Norwegian Academy of Science and Letters, and EU Biomed2 (BMH4-CT95-0571).

## References

Agostinho, P., Duarte, C.B. and Oliveira, C.R. (1997) Impairment of excitatory amino acid transporter activity by oxidative stress conditions in retinal cells: effect of antioxidants. *FASEB. J.*, 11: 154–163.

Ahn, A.H., Dziennis, S., Hawkes, R. and Herrup, K. (1994) The cloning of zebrin II reveals its identity with aldolase C. *Development*, 120: 2081–2090.

Albrecht, J., Talbot, M., Kimelberg, H.K. and Aschner, M. (1993) The role of sulfhydryl groups and calcium in the mercuric chloride-induced inhibition of glutamate uptake in rat primary astrocyte cultures. *Brain Res.*, 607: 249–254.

Anderson, K.J., Bridges, R.J. and Cotman, C.W. (1991) Increased density of excitatory amino acid transport sites in the hippocampal formation following an entorhinal lesion. *Brain Res.*, 562: 285–290.

Arriza, J.L., Eliasof, S., Kavanaugh, M.P. and Amara, S.G. (1997) Excitatory amino acid transporter 5, a retinal glutamate transporter coupled to a chloride conductance. *Proc. Natl. Acad. Sci. USA*, 94: 4155–4160.

Arriza, J.L., Fairman, W.A., Wadiche, J.I., Murdoch, G.H., Kavanaugh, M.P. and Amara, S.G. (1994) Functional comparisons of three glutamate transporter subtypes cloned from human motor cortex. *J. Neurosci.*, 14: 5559–5569.

Arriza, J.L., Kavanaugh, M.P., Fairman, W.A., Wu, Y.N., Murdoch, G.H., North, R.A. and Amara, S.G. (1993) Cloning and expression of a human neutral amino acid transporter with structural similarity to the glutamate transporter gene family. *J. Biol. Chem.* 268: 15329–15332.

Aschner, M., Du, Y.L., Gannon, M. and Kimelberg, H.K. (1993) Methylmercury-induced alterations in excitatory amino acid transport in rat primary astrocyte cultures. *Brain Res.*, 602: 181–186.

Aschner, M., Eberle, N.B., Miller, K. and Kimelberg, H.K. (1990) Interactions of methylmercury with rat primary astrocyte cultures-inhibition of rubidium and glutamate uptake and induction of swelling. *Brain Res.*, 530: 245–250.

Asztely, F., Erdemli, G. and Kullmann, D.M. (1997) Extrasynaptic glutamate spillover in the hippocampus: Dependence on temperature and the role of active glutamate uptake. *Neuron*, 18: 281–293.

Balcar, V. and Johnston, G.A.R. (1972) The structural specificity of the high affinity uptake of L-glutamate and L-aspartate by rat brain slices. *J. Neurochem.*, 19: 2657–2666.

Barbour, B. and Häusser, M. (1997) Intersynaptic diffusion of neurotransmitter. *Trends Neurosci.*, 20: 377–384.

Barbour, B., Keller, B.U., Llano, I. and Marty, A. (1994) Prolonged presence of glutamate during excitatory synaptic transmission to cerebellar Purkinje cells. *Neuron*, 12: 1331–1343.

Barbour, B., Szatkowski, M., Ingledew, N. and Attwell, D. (1989) Arachidonic acid induces a prolonged inhibition of glutamate uptake into glial cells. *Nature*, 342: 918–20.

Baughman, R.W. and Gilbert, C.D. (1980) Aspartate and glutamate as possible neurotransmitters of cells in layer 6 of the visual cortex. *Nature*, 287: 848–850.

Beckstrøm, H., Julsrud, L., Haugeto, Ø., Dewar, D., Graham, D.I., Langmoen, I.A., Lehre, K.P., Storm-Mathisen, J. and Danbolt, N.C. (1998) Large individual differences in the levels of the glutamate transportes GLAST and GLT, but no clear correlation with Alzheimer's disease. *Submitted.*

Berman, S.B. and Hastings, T.G. (1997) Inhibition of glutamate transport in synaptosomes by dopamine oxidation and reactive oxygen species. *J. Neurochem.*, 69: 1185–1195.

Billups, B. and Attwell, D. (1996) Modulation of non-vesicular glutamate release by pH. *Nature*, 379: 171–174.

Billups, B., Rossi, D. and Attwell, D. (1996) Anion conductance behavior of the glutamate uptake carrier in salamander retinal glial cells. *J. Neurosci.*, 16: 6722–6731.

Bjørås, M., Gjesdal, O., Erickson, J.D., Torp, R., Levy, L.M., Ottersen, O.P., Degree, M., Storm-Mathisen, J., Seeberg, E. and Danbolt, N.C. (1996) Cloning and expression of a neuronal rat brain glutamate transporter. *Mol. Brain Res.*, 36: 163–168.

Bouvier, M., Szatkowski, M., Amato, A. and Attwell, D. (1992) The glial cell glutamate uptake carrier countertransports pH-changing anions. *Nature*, 360: 471–474.

Bridges, R.J., Stanley, M.S., Anderson, M.W., Cotman, C.W. and Chamberlin, A.R. (1991) Conformationally defined neurotransmitter analogues. Selective inhibition of glutamate uptake by one pyrrolidine-2,4-dicarboxylate diastereomer. *J. Med. Chem.*, 34: 717–725.

Brochu, G., Maler, L. and Hawkes, R. (1990) Zebrin II: a polypeptide antigen expressed selectively by Purkinje cells reveals compartments in rat and fish cerebellum. *J. Comp. Neurol.*, 291: 538–552.

Brookes, N. (1988) Specificity and reversibility of the inhibition by $HgCl_2$ of glutamate transport in astrocyte cultures. *J. Neurochem.*, 50: 1117–22.

Cagampang, F.R.A., Rattray, M., Powell, J.F., Chong, N.W.S., Campbell, I.C. and Coen, C.W. (1996) Circadian variation of EAAC1 glutamate transporter messenger RNA in the rat suprachiasmatic nuclei. *Mol. Brain Res.*, 35: 190–196.

Casado, M., Bendahan, A., Zafra, F., Danbolt, N.C., Aragon, C., Gimenez and Kanner, B.I. (1993) Phosphorylation and modulation of brain glutamate transporters by protein kinase C. *J. Biol. Chem.*, 268: 27313–27317.

Chan, P.H., Kerlan, R. and Fishman, R.A. (1983) Reductions of $\gamma$-aminobutyric acid and glutamate uptake and ($Na^+$ + $K^+$)-ATPase activity in brain slices and synaptosomes by arachidonic acid. *J. Neurochem.*, 40: 309–316.

Chaudhry, F.A., Lehre, K.P., Campagne, M.V., Ottersen, O.P., Danbolt, N.C. and Storm-Mathisen, J. (1995) Glutamate transporters in glial plasma membranes: Highly differentiated localizations revealed by quantitative ultrastructural immunocytochemistry. *Neuron*, 15: 711–720.

Clements, J.D. (1996) Transmitter timecourse in the synaptic cleft: its role in central synaptic function. *Trends Neurosci.*, 19: 163–171.

Clements, J.D., Lester, R.A.J., Tong, G., Jahr, C.E. and Westbrook, G.L. (1992) The time course of glutamate in the synaptic cleft. *Science*, 258: 1498–1501.

Coco, S., Verderio, C., Trotti, D., Rothstein, J.D., Volterra, A. and Matteoli, M. (1997) Non-synaptic localization of the glutamate transporter EAAC1 in cultured hippocampal neurons. *Eur. J. Neurosci.*, 9: 1902–1910.

Conradt, M. and Stoffel, W. (1995) Functional analysis of the high affinity, $Na^+$-dependent glutamate transporter GLAST-1 by site-directed mutagenesis. *J. Biol. Chem.*, 270: 25207–25212.

Conradt, M. and Stoffel, W. (1997) Inhibition of the high-affinity brain glutamate transporter GLAST-1 via direct phosphorylation. *J. Neurochem.*, 68: 1244–1251.

Conradt, M., Storck, T. and Stoffel, W. (1995) Localization of N-glycosylation sites and functional role of the carbohydrate units of GLAST-1, a cloned rat brain L-glutamate/L-aspartate transporter. *Eur. J. Biochem.*, 229: 682–687.

Conti, F., DeBiasi, S., Minelli, A., Rothstein, J.D. and Melone, M. (1998) EAAC1, a high-affinity glutamate transporter, is localized to astrocytes and gabaergic neurons besides pyramidal cells in the rat cerebral cortext. *Cereb. Cortex*, 8: 108–116.

Danbolt, N.C. (1994) The high affinity uptake system for excitatory amino acids in the brain. *Prog. Neurobiol.*, 44: 377–396.

Danbolt, N.C., Pines, G. and Kanner, B.I. (1990) Purification and reconstitution of the sodium- and potassium-coupled glutamate transport glycoprotein from rat brain. *Biochemistry*, 29: 6734–40.

Danbolt, N.C. and Storm-Mathisen, J. (1986) $Na^+$-dependent "inding" of D-aspartate in brain membranes is largely due to uptake into membrane bounded saccules. *J. Neurochem.*, 47: 819–24.

Danbolt, N.C., Storm-Mathisen, J. and Kanner, B.I. (1992) An [$Na^+$ + $K^+$] coupled L-glutamate transporter purified from rat brain is located in glial cell processes. *Neuroscience*, 51: 295–310.

Danbolt, N.C., Storm-Mathisen, J. and Ottersen, O.P. (1994) Sodium/potassium coupled glutamate transporters, a "new" family of eukaryotic proteins: Do they have "new" physiological roles and could they be new targets for pharmacological intervention? *Prog. Brain Res.*, 100: 53–60.

Dehnes, Y., Chaudhry, F.A., Ullensvang, K., Lehre, K.P., Storm-Mathisen, J. and Danbolt, N.C. (1998) The glutamate transporter EAAT4 in rat cerebellar Purkinje cells: A glutamate-gated chloride channel concentrated near the synapse in parts of the dendritic membrane facing astroglia. *J. Neurosci.*, 18: 3606–3619.

Derouiche, A. and Rauen, T. (1995) Coincidence of l-glutamate/l-aspartate transporter (GLAST) and glutamine synthetase (GS) immunoreactions in retinal glia: Evidence for coupling of GLAST and GS in transmitter clearance. *J. Neurosci. Res.*, 42: 131–143.

Diamond, J.S. and Jahr, C.E. (1997) Transporters buffer synaptically released glutamate on a submillisecond time scale. *J. Neurosci.*, 17: 4672–4687.

Donly, B.C., Richman, A., Hawkins, E., McLean, H. and Caveney, S. (1997) Molecular cloning and functional expression of an insect high-affinity $Na^+$-dependent glutamate transporter. *Eur. J. Biochem.*, 248: 535–542.

Dowd, L.A. and Robinson, M.B. (1996) Rapid stimulation of EAAC1-mediated $Na^+$-dependent L-glutamate transport activity in c6 glioma cells by phorbol ester. *J. Neurochem.*, 67: 508–516.

Drejer, J., Meier, E. and Schousboe, A. (1983) Novel neuron-related regulatory mechanisms for astrocytic glutamate and GABA high affinity uptake. *Neurosci. Lett.*, 37: 301–306.

Dumuis, A., Sebben, M., Haynes, L., Pin, J. and Bockaert, J. (1988) NMDA receptors activate the arachidonic acid cascade system in striatal neurons. *Nature*, 336: 68–70.

Eliasof, S. and Werblin, F. (1993) Characterization of the glutamate transporter in retinal cones of the tiger salamander. *J. Neurosci.*, 13: 402–411.

Eliasof, S., Arriza, J.L., Leighton, B.H., Kavanaugh, M.P. and Amara, S.G. (1998) Excitatory amino acid transporters of the salamander retina: Identification, localization, and function. *J. Neurosci.*, 18: 698–712.

Engelke, T., Jording, D., Kapp, D. and Pühler, A. (1989) Identification and sequence analysis of the Rhizobium meliloti dctA gene encoding the C4-dicarboxylate carrier. *J. Bacteriol.*, 171: 5551–5560.

Euler, T. and Wässle, H. (1995) Immunocytochemical identification of cone bipolar cells in the rat retina. *J. Comp. Neurol.*, 361: 461–478.

Fairman, W.A., Vandenberg, R.J., Arriza, J.L., Kavanaugh, M.P. and Amara, S.G. (1995) An excitatory amino-acid transporter with properties of a ligand-gated chloride channel. *Nature*, 375: 599–603.

Fine, S.M., Angel, R.A., Perry, S.W., Epstein, L.G., Rothstein, J.D., Dewhurst, S. and Gelbard, H.A. (1996) Tumor necosis factor $\alpha$ inhibits glutamate uptake by primary human astrocytes – Implications for pathogenesis of HIV-1 dementia. *J. Biol. Chem.*, 271: 15303–15306.

Fischer, B.O., Ottersen, O.P. and Storm-Mathisen, J. (1986) Implantation of D-[$^3$H] aspartate loaded gel particles permits restricted uptake sites for transmitter-selective axonal transport. *Exp. Brain Res.*, 63: 620–626.

Fonnum, F., Storm-Mathisen, J. and Divac, I. (1981) Biochemical evidence for glutamate as a neurotransmitter in cortico-striatal and cortico-thalamic fibres in rat brain. *Neuroscience*, 6: 863–873.

Furuta, A., Martin, L.J., Lin, C.L., Dykes-Hoberg, M., Rothstein, J.D. (1997a) Cellular and synaptic localization of the neuronal glutamate transporters excitatory amino acid transporter 3 and 4. *Neuroscience*, 81: 1031–1042.

Furuta, A., Rothstein, J.D. and Martin, L.J. (1997b) Glutamate transporter protein subtypes are expressed differentially during rat CNS development. *J. Neurosci.*, 17: 8363–8375.

Furness, D.N. and Lehre, K.P. (1997) Immunocytochemical localization of a high-affinity glutamate-aspartate transporter, GLAST, in the rat and guinea-pig cochlea. *Eur. J. Neurosci.*, 9: 1961–1969.

Gegelashvili, G., Civenni, G., Racagni, G., Danbolt, N.C., Schousboe, I. and Schousboe, A. (1996) Glutamate receptor agonists up-regulate glutamate transporter GLAST in astrocytes. *Neuroreport*, 8: 261–265.

Gegelashvili, G., Danbolt, N.C. and Schousboe, A. (1997) Neuronal soluble factors differentially regulate the expression of the GLT1 and GLAST glutamate transporters in cultured astroglia. *J. Neurochem.*, 69: 2612–2615

Gilad, G.M., Gilad, V.H., Wyatt, R.J. and Tizabi, Y. (1990) Region-selective stress-induced increase of glutamate uptake and release in rat forebrain. *Brain Res.*, 525: 335–338.

Ginsberg, S.D., Martin, L.J. and Rothstein, J.D. (1995) Regional deafferentation down-regulates subtypes of glutamate transporter proteins. *J. Neurochem.*, 65: 2800–2803.

Ginsberg, S.D., Rothstein, J.D., Price, D.L. and Martin, L.J. (1996) Fimbria-fornix transections selectively down-regulate subtypes of glutamate transporter and glutamate receptor proteins in septum and hippocampus. *J. Neurochem.*, 67: 1208–1216.

Greenamyre, J.T., Higgins, D.S. and Young, A.B. (1990) Sodium-dependent D-aspartate 'binding' is not a measure of presynaptic neuronal uptake sites in an autoradiographic assay. *Brain Res.*, 511: 310–318.

Grünewald, M. and Kanner, B. (1995) Conformational changes monitored on the glutamate transporter GLT-1 indicate the existence of two neurotransmitter-bound states. *J. Biol. Chem.*, 270: 17017–17024.

Gundersen, V., Danbolt, N.C., Ottersen, O.P. and Storm-Mathisen, J. (1993) Demonstration of glutamate/aspartate uptake activity in nerve endings by use of antibodies recognizing exogenous D-aspartate. *Neuroscience*, 57: 97–111.

Gundersen, V., Shupliakov, O., Brodin, L., Ottersen, O.P. and Storm-Mathisen, J. (1995) Quantification of excitatory amino acid uptake at intact glutamatergic synapses by immunocytochemistry of exogenous D-aspartate. *J. Neurosci.*, 15: 4417–4428.

Gundersen, V., Ottersen, O.P. and Storm-Mathisen, J. (1996) Selective excitatory amino acid uptake in glutamatergic nerve terminals and in glia in the rat striatum: Quantitative electron microscopic immunocytochemistry of exogenous D-aspartate and endogenous glutamate and GABA. *Eur. J. Neurosci.*, 8: 758–765.

Harris, M.E., Carney, J.M., Cole, P.S., Hensley, K., Howard, B.J., Martin, L., Bummer, P., Wang, Y.N., Pedigo, N.W. and Butterfield, D.A. (1995) $\beta$-amyloid peptide-derived, oxygen-dependent free radicals inhibit glutamate uptake in cultured astrocytes: implications for Alzheimer's disease. *Neuroreport*, 6: 1875–1879.

Harris, M.E., Wang, Y.N., Pedigo, N.W., Hensley, K., Butterfield, D.A. and Carney, J.M. (1996) Amyloid $\beta$, peptide (25–35) inhibits $Na^+$-dependent glutamate uptake in rat hippocampal astrocyte cultures. *J. Neurochem.*, 67: 277–286.

Haugeto, Ø., Ullensvang, K., Levy, L.M., Chaudhry, F.A., Honore, T., Nielsen, M., Lehre, K.P. and Danbolt, N.C. (1996) Brain glutamate transporter proteins form homomultimers. *J. Biol. Chem.*, 271: 27715–27722.

Hees, B., Danbolt, N.C., Kanner, B.I., Haase, W., Heitmann, K. and Koepsell, H. (1992) A monoclonal antibody against a $Na^+$-L-glutamate cotransporter from rat brain. *J. Biol.-Chem.*, 267: 23275–23281.

Hofmann, K., Duker, M., Fink, T., Lichter, P. and Stoffel, W. (1994) Human neutral amino acid transporter ASCT1: structure of the gene (SLC1a4) and localization to chromosome 2p13-p15. *Genomics*, 24: 20–26.

Holmes, W.R. (1995) Modeling the effect of glutamate diffusion and uptake on NMDA and non-NMDA receptor saturation. *Biophys. J.*, 69: 1734–1747.

Inoue, K., Sakaitani, M., Shimada, S. and Tohyama, M. (1995) Cloning and expression of a bovine glutamate transporter. *Mol. Brain Res.*, 28: 343–348.

Kanai, Y. (1997) Family of neutral and acidic amino acid transporters: molecular biology, physiology and medical implications. *Curr. Opin. Cell Biol.*, 9: 565–572.

Kanai, Y. and Hediger, M.A. (1992) Primary structure and functional characterization of a high-affinity glutamate transporter. *Nature*, 360: 467–471.

Kanai, Y., Stelzner, M., Nussberger, S., Khawaja, S., Hebert, S.C., Smith, C.P. and Hediger, M.A. (1994) The neuronal and epithelial human high affinity glutamate transporter – insights into structure and mechanism of transport. *J. Biol. Chem.*, 269: 20599–20606.

Kanai, Y., Bhide, P.G., Difiglia, M. and Hediger, M.A. (1995a) Neuronal high-affinity glutamate transport in the rat central nervous system. *Neuroreport*, 6: 2357–2362.

Kanai, Y., Nussberger, S., Romero, M.F., Boron, W.F., Hebert, S.C. and Hediger, M.A. (1995b) Electrogenic properties of the epithelial and neuronal high affinity glutamate transporter. *J. Biol. Chem.*, 270: 16561–16568.

Kanai, Y., Trotti, D., Nussberger, S. and Hediger, M.A. (1997) The high-affinity glutamate transporter family. Structure, function and physiological relevance. In Reith, M.E.A. (ed.) *Neurotransmitter Transporters: Structure, Function, and Regulation*, 171–213. Humana Press, Totowa, N.J.

Kanner, B.I. (1993) Glutamate transporters from brain – A novel neurotransmitter transporter family. *FEBS Lett.*, 325: 95–99.

Kanner, B.I. (1997) Sodium-coupled GABA and glutamate transporters. In Reith, M.E.A. (ed.) *Neurotransmitter Transporters: Structure, Function, and Regulation*, 151–169. Humana Press, Totowa, N.J.

Kanner, B.I. and Schuldiner, S. (1987) Mechanism of transport and storage of neurotransmitters. *CRC Crit. Rev. Biochem.*, 22: 1–38.

Kanner, B.I. and Sharon, I. (1978a) Solubilization and reconstitution of the L-glutamic acid transporter from rat brain. *FEBS Lett.*, 94: 245–8.

Kanner, B.I. and Sharon, I. (1978b) Active transport of L-glutamate by membrane vesicles isolated from rat brain. *Biochemistry*, 17: 3949–3953.

Kataoka, Y., Morii, H., Watanabe, Y. and Ohmori, H. (1997) A postsynaptic excitatory amino acid transporter with chloride conductance functionally regulated by neuronal activity in cerebellar Purkinje cells. *J. Neurosci.*, 17: 7017–7024.

Kavanaugh, M.P., Bendahan, A., Zerangue, N., Zhang, Y.M. and Kanner, B.I. (1997) Mutation of an amino acid residue influencing potassium coupling in the glutamate transporter GET-1 induces obligate exchange. *J. Biol. Chem.*, 272: 1703–1708.

Kawakami, H., Tanaka, K., Nakayama, T., Inoue, K. and Nakamura, S. (1994) Cloning and expression of a human glutamate transporter. *Biochem. Biophys. Res. Comm.*, 199: 171–176.

Kawano, T., Takuwa, K. and Nakajima, T. (1996) Molecular cloning of a cDNA for the glutamate transporter of the nematode caenorhabditis elegans. *Biochem. Biophys. Res. Comm.*, 228: 415–420.

Kawano, T., Takuwa, K. and Nakajima, T. (1997) Structure and activity of a new form of the glutamate transporter of the nematode caenorhabditis elegans. *Biosci. Biotech. Biochem.*, 61: 927–929.

Keller, J.N., Mark, R.J., Bruce, A.J., Blanc, E., Rothstein, J.D., Uchida, K., Waeg, G. and Mattson, M.P. (1997) 4-Hydroxynonenal, an aldehydic product of membrane lipid peroxidation, impairs glutamate transport and mitochondrial function in synaptosomes. *Neuroscience*, 80: 685–696.

Kerkerian, L., Dusticier, N. and Nieoullon, A. (1987) Modulatory effect of dopamine on high-affinity glutamate uptake in rat striatum. *J. Neurochem.*, 48: 1301–6.

Kinney, G.A., Overstreet, L.S. and Slater, N.T. (1997) Prolonged physiological entrapment of glutamate in the synaptic cleft of cerebellar unipolar brush cells. *J. Neurophysiol.*, 78: 1320–1333.

Kirschner, M.A., Arriza, J.L., Copeland, N.G., Gilbert, D.J., Jenkins, N.A., Magenis, E. and Amara, S.G. (1994a) The mouse and human excitatory amino acid transporter gene (EAAT1) maps to mouse chromosome 15 and a region of syntenic homology on human chromosome 5. *Genomics*, 22: 631–633.

Kirschner, M.A., Copeland, N.G., Gilbert, D.J., Jenkins, N.A. and Amara, S.G. (1994b) Mouse excitatory amino acid transporter EAAT2: Isolation, characterization, and proximity to neuroexcitability loci on mouse chromosome 2. *Genomics*, 24: 218–224.

Kleinle, J., Vogt, K., Luscher, H.R., Muller, L., Senn, W., Wyler, K. and Streit, J. (1996) Transmitter concentration profiles in the synaptic cleft: An analytical model of release and diffusion. *Biophys. J.*, 71: 2413–2426.

Knecht, K., Michalak, A., Rose, C., Rothstein, J.D. and Butterworth, R.F. (1997) Decreased glutamate transporter (GLT-1) expression in frontal cortex of rats with acute liver failure. *Neurosci. Lett.*, 229: 201–203.

Langmoen, I.A., Larsen, M. and Berg-Johnsen, J. (1995) Volatile anaesthetics: Cellular mechanisms of action. *Eur. J. Anaesthesiol.*, 12: 51–58.

Larsen, M., Hegstad, E., Berg-Johnsen, J. and Langmoen, I.A. (1997) Isoflurane increases the uptake of glutamate in synaptosomes from rat cerebral cortex. *Br. J. Anaesth.*, 78: 55–59.

Larsson, H.P., Picaud, S.A., Werblin, F.S. and Lecar, H. (1996) Noise analysis of the glutamate-activated current in photoreceptors. *Biophys. J.*, 70: 733–742.

Lebrun, B., Sakaitani, M., Shimamoto, K., Yasudakamatani, Y. and Nakajima, T. (1997) New β-hydroxyaspartate derivatives are competitive blockers for the bovine glutamate/aspartate transporter. *J. Biol. Chem.*, 272: 20336–20339.

Lehre, K.P., Davanger, S. and Danbolt, N.C. (1997) Localization of the glutamate transporter protein GLAST in rat retina. *Brain Res.*, 744: 129–137.

Lehre, K.P., Levy, L.M., Ottersen, O.P., Storm-Mathisen, J. and Danbolt, N.C. (1995) Differential expression of two glial glutamate transporters in the rat brain: Quantitative and immunocytochemical observations. *J. Neurosci.*, 15: 1835–1853.

Levy, L.M., Lehre, K.P., Rolstad, B. and Danbolt, N.C. (1993) A monoclonal antibody raised against an [$Na^+ + K^+$] coupled L-glutamate transporter purified from rat brain confirms glial cell localization. *FEBS Lett.*, 317: 79–84.

Levy, L.M., Lehre, K.P., Walaas, S.I., Storm-Mathisen, J. and Danbolt, N.C. (1995) Down-regulation of glial glutamate transporters after glutamatergic denervation in the rat brain. *Eur. J. Neurosci.*, 7: 2036–2041.

Li, X. and Francke, U. (1995) Assignment of the gene SLC1a2 coding for the human glutamate transporter EAAT2 to human chromosome 11 bands p13-p12. *Cytogen. Cell Gen.*, 71: 212–213.

Lin, C.L., Bristol, L.A., Jin, L., Dykes-Hoberg, M., Crawford, T., Clawson, L. and Rothstein, J.D. (1998) Aberrant RNA processing in a neurodegenerative disease: The cause for absent EAAT2, a glutamate transporter, in amyotrophic lateral sclerosis. *Neuron*, 20: 589–602.

Liu, M.T., Rothstein, J.D., Gershon, M.D. and Kirchgessner, A.L. (1997) Glutamatergic enteric neurons. *J. Neurosci.*, 17: 4764–4784.

Logan, W.J. and Snyder, S.H. (1971) Unique high affinity uptake systems for glycine, glutamic and aspartic acids in central nervous tissue of the rat. *Nature*, 234: 297–299.

Lynch, M.A., Jaquesberg, W., Lawson, P.R., Voss, K.L. and Bliss, T.V.P. (1994) The effect of arachidonic acid on glutamate uptake in cortical and hippocampal preparations. *Neurosci. Res. Comm.*, 14: 53–61.

Maenohikichi, Y., Tanaka, K., Shibata, T., Watanabe, M., Inoue, Y., Mukainaka, Y. and Wada, K. (1997) Structure and functional expression of the cloned mouse neuronal high-affinity glutamate transporter. *Mol. Brain Res.*, 48: 176–180.

MacKinnon, R. (1995) Pore loops: an emerging theme in ion channel structure. *Neuron*, 14: 889–892.

Maki, R., Robinson, M.B. and Dichter, M.A. (1994) The glutamate uptake inhibitor L-*trans*-pyrrolidine-2,4-dicarboxylate depresses excitatory synaptic transmission via a presynaptic mechanism in cultured hippocampal neurons. *J. Neurosci.*, 14: 6754–6762.

Manfras, B.J. and Boehm, B.O. (1995) Expression of a glutamate transporter cDNA in human pancreatic islets. *Exp. Clin. Endocrinol. Diabet.*, 103: 95–98.

Manfras, B.J., Rudert, W.A., Trucco, M. and Boehm, B.O. (1994) Cloning and characterization of a glutamate transporter cDNA from human brain and pancreas. *Biochem. Biophys. Acta*, 1195: 185–188.

Manzoni, C. and Mennini, T. (1997) Arachidonic acid inhibits H-3-glutamate uptake with different potencies in rodent central nervous system regions expressing different transporter subtypes. *Pharmacol. Res.*, 35: 149–151.

Maragakis, N., Lin, C.G., Jin, L., Dykes-Hoberg, M. and Rothstein, J.D. (1997) Ataxia due to loss of the Purkinje-cell specific glutamate transporter EAAT4. *Soc. Neurosci. Abstr.*, 23: 1484–1484.

Martin, L.J., Brambrink, A.M., Lehmann, C., Porteracailliau, C., Koehler, R., Rothstein, J. and Traystman, R.J. (1997) Hypoxia-ischemia causes abnormalities in glutamate transporters and death of astroglia and neurons in newborn striatum. *Ann. Neurol.*, 42: 335–348.

Masliah, E., Alford, M., DeTeresa, R., Mallory, M. and Hansen, L. (1996) Deficient glutamate transport is associated with neurodegeneration in Alzheimer's disease. *Ann. Neurol.*, 40: 759–766.

Matthews, J.C., Beveridge, M.J., Malandro, M.S., Rothstein, J.D., Campbell Thompson, M., Verlander, J.W., Kilberg, M.S. and Novak, D.A. (1998) Activity and protein localization of multiple glutamate transporters in gestation day 14 vs, day 20 rat placenta. *Am. J. Physiol.*, 43: C603–C614.

Mennerick, S. and Zorumski, C.F. (1994) Glial contributions to excitatory neurotransmission in cultured hippocampal cells. *Nature*, 368: 59–62.

Mennerick, S., Dhond, R.P., Benz, A., Xu, W., Rothstein, J.D., Danbolt, N.C., Isenberg, K.E. and Zorumski, C.F. (1998) Neuronal expression of the glutamate transporter GLT-1 in hippocampal microcultures. *J. Neurosci.*, 18: 4490–4499.

Miyazaki, H., Nakamura, Y., Arai, T. and Kataoka, K. (1997) Increase of glutamate uptake in astrocytes: A possible mechanism of action of volatile anesthetics. *Anesthesiology*, 86: 1359–1366.

Mukainaka, Y., Tanaka, K., Hagiwara, T. and Wada, K. (1995) Molecular cloning of two glutamate transporter subtypes from mouse brain. *Biochem. Biophys. Acta*, 1244: 233–237.

Nagao, S., Kwak, S. and Kanazawa, I. (1997) EAAT4, a glutamate transporter with properties of a chloride channel, is predominantly localized in Purkinje cell dendrites and forms parasagittal compartments in rat cerebellum. *Neuroscience*, 78: 929–933.

Nagaraja, T.N. and Brookes, N. (1996) Mercuric chloride uncouples glutamate uptake from the countertransport of hydroxyl equivalents. *Am. J. Physiol.*, 40: C1487–C1493.

Nicholls, D. and Attwell, D. (1990) The release and uptake of excitatory amino acids. *Trends Pharmacol. Sci.*, 11: 462–468.

Nieoullon, A., Kerkerian, L. and Dusticier, N. (1983) Presynaptic dopaminergic control of high affinity glutamate uptake in the striatum. *Neurosci. Lett.*, 43: 191–196.

Nissen, J., Schousboe, A., Halkier, T. and Schousboe, I. (1992) Purification and characterization of an astrocyte GABA-carrier inducing protein (GABA-CIP) released from cerebellar granule cells in culture. *Glia*, 6: 236–243.

Norenberg, M.D., Huo, Z.F., Neary, J.T. and Roigcantesano, A. (1997) The glial glutamate transporter in hyperammonemia and hepatic encephalopathy: Relation to energy metabolism and glutamatergic neurotransmission. *Glia*, 21: 124–133.

Otis, T.S., Kavanaugh, M.P. and Jahr, C.E. (1997) Postsynaptic glutamate transport at the climbing fiber Purkinje cell synapse. *Science*, 277: 1515–1518.

Otis, T.S., Wu, Y.C. and Trussell, L.O. (1996) Delayed clearance of transmitter and the role of glutamate transporters at synapses with multiple release sites. *J. Neurosci.*, 16: 1634–1644.

Otori, Y., Shimada, S., Tanaka, K., Ishimoto, I., Tano, Y. and Tohyama, M. (1994) Marked increase in glutamate-aspartate transporter (GLAST/glut-1) mRNA following transient retinal ischemia. *Mol. Brain Res.*, 27: 310–314.

Pajor, A.M. (1996) Molecular cloning and functional expression of a sodiumdicarboxylate cotransporter from human kidney. *Am. J. Physiol.*, 39: F642–F648.

Palmer, A.M., Robichaud, P.J. and Reiter, C.T. (1994) The release and uptake of excitatory amino acids in rat brain-effect of aging and oxidative stress. *Neurobiol. Aging*, 15: 103–111.

Peghini, P., Janzen, J. and Stoffel, W. (1997) Glutamate transporter EAAC-1-deficient mice develop dicarboxylic aminoaciduria and behavioral abnormalities but no neurodegeneration. *EMBO J.*, 16: 3822–3832.

Piani, D., Frei, K., Pfister, H.W. and Fontana, A. (1993) Glutamate uptake by astrocytes is inhibited by reactive oxygen intermediates but not by other macrophage-derived molecules including cytokines, leukotrienes or platelet-activating factor. *J. Neuroimmunol.*, 48: 99–104.

Pines, G., Danbolt, N.C., Bjørås, M., Zhang, Y., Bendahan, A., Eide, L., Koepsell, H., Storm-Mathisen, J., Seeberg, E.K. and Kanner, B.I. (1992) Cloning and expression of a rat brain L-glutamate transporter. *Nature*, 360: 464–467.

Pines, G., Zhang, Y.M. and Kanner, B.I. (1995) Glutamate 404 is involved in the substrate discrimination of GLT-1, a ($Na^+ + K^+$)-coupled glutamate transporter from rat brain. *J. Biol. Chem.*, 270: 17093–17097.

Pogun, S. and Kuhar, M.J. (1994) Regulation of neurotransmitter reuptake by nitric oxide. *Ann. N.Y. Acad. Sci.*, 738: 305–315.

Radice, A.D. and Lustigman, S. (1996) Cloning and characterization of cDNAs encoding putative glutamate transporters from caenorhabditis elegans and onchocerca volvulus. *Mol. Biochem. Parasitol.*, 80: 41–53.

Rauen, T. and Kanner, B.I. (1994) Localization of the glutamate transporter GLT-1 in rat and macaque monkey retinae. *Neurosci. Lett.*, 169: 137–140.

Rauen, T., Rothstein, J.D. and Wässle, H. (1996) Differential expression of three glutamate transporter subtypes in the rat retina. *Cell Tissue Res.*, 286: 325–336.

Rhoads, D.E., Ockner, R.K., Peterson, N.A. and Raghupathy, E. (1983) Modulation of membrane transport by free fatty acids: Inhibition of synaptosomal sodium-dependent amino acid uptake. *Biochemistry*, 22: 1965–1970.

Robinson, M.B. and Dowd, L.A. (1997) Heterogeneity and functional properties of subtypes of sodium-dependent glutamate transporters in the mammalian central nervous system. *Adv. Pharmacol.*, 37: 69–115.

Rothstein, J.D., Martin, L.J. and Kuncl, R.W. (1992) Decreased glutamate transport by the brain and spinal cord in amyotrophic lateral sclerosis (see comments). *N. Engl. J. Med.*, 326: 1464–1468.

Rothstein, J.D., Martin, L., Levey, A.I., Dykeshoberg, M., Jin, L., Wu, D., Nash, N. and Kuncl, R.W. (1994) Localization of neuronal and glial glutamate transporters. *Neuron*, 13: 713–725.

Rothstein, J.D., Van Kammen, M., Levey, A.I., Martin, L.J. and Kuncl, R.W. (1995) Selective loss of glial glutamate transporter GLT-1 in amyotrophic lateral sclerosis. *Ann. Neurol.*, 38: 73–84.

Rothstein, J.D., Dykes-Hoberg, M., Pardo, C.A., Bristol, L.A., Jin, L., Kuncl, R.W., Kanai, Y., Hediger, M.A., Wang, Y., Schielke, J.P. and Welty, D.F. (1996) Knockout of glutamate transporters reveals a major role for astroglial transport in excitotoxicity and clearance of glutamate. *Neuron*, 16: 675–686.

Savolainen, K.M., Loikkanen, J. and Naarala, J. (1995) Amplification of glutamate-induced oxidative stress. *Toxicol. Lett.*, 82–3: 399–405.

Samuel, D., Pisano, P., Forni, C., Nieoullon, A. and Kerkerian-Le Goff, L. (1996) Involvement of the glutamatergic metabotropic receptors in the regulation of glutamate uptake and extracellular excitatory amino acid levels in the striatum of chloral hydrate-anesthetized rats. *Brain Res.*, 739: 156–162.

Sarantis, M. and Attwell, D. (1990) Glutamate uptake in mammalian retinal glia is voltage-dependent and potassium-dependent. *Brain Res.*, 516: 322–325.

Sarantis, M., Everett, K. and Attwell, D. (1988) A presynaptic action of glutamate at the cone output synapse. *Nature*, 332: 451–453.

Schmitt, A., Asan, E., Puschel, B., Jons, T. and Kugler, P. (1996) Expression of the glutamate transporter GLT1 in neural cells of the rat central nervous system: Non-radioactive in situ hybridization and comparative immunocytochemistry. *Neuroscience*, 71: 989–1004.

Schmitt, A., Asan, E., Puschel, B. and Kugler, P. (1997) Cellular and regional distribution of the glutamate transporter GLAST in the CNS of rats: Non-radioactive in situ hybridization and comparative immunocytochemistry. *J. Neurosci.*, 17: 1–10.

Schulz, K. and Stell, W.K. (1996) Immunocytochemical localization of the high-affinity glutamate transporter, EAACl, in the retina of representative vertebrate species. *Neurosci. Lett.*, 211: 191–194.

Sepukty, J., Eccles, C.U., Lesser, R.P., Dykes-Hoberg, M. and Rothstein, J.D. (1997) Molecular knockdown of neuronal glutamate transporter EAAT3 produces epilepsy and dysregulation of GABA metabolism. *Soc. Neurosci. Abstr.*, 23: 1484–1484.

Shafqat, S., Tamarappoo, B.K., Kilberg, M.S., Puranam, R.S., Mcnamara, J.O., Guadanoferraz, A. and Fremeau, R.T. (1993) Cloning and expression of a novel $Na^+$-dependent neutral amino acid transporter structurally related to mammalian $Na^+$/glutamate cotransporters. *J. Biol. Chem.*, 268: 15351–15355.

Shashidharan, P. and Plaitakis, A. (1993) Cloning and characterization of a glutamate transporter cDNA from human cerebellum. *Biochem. Biophys. Acta*, 1216: 161–164.

Shashidharan, P., Huntley, G.W., Meyer, T., Morrison, J.H. and Plaitakis, A. (1994a) Neuron-specific human glutamate transporter: molecular cloning, characterization and expression in human brain. *Brain Res.*, 662: 245–250.

Shashidharan, P., Wittenberg, I. and Plaitakis, A. (1994b) Molecular cloning of human brain glutamate/aspartate transporter II. *Biochem. Biophys. Acta*, 1191: 393–396.

Sher, P.K. and Hu, S.X. (1990) Increased glutamate uptake and glutamine synthetase activity in neuronal cell cultures surviving chronic hypoxia. *Glia*, 3: 350–357.

Shimamoto, K., Lebrun, B., Yasudakamatani, Y., Sakaitani, M., Shigeri, Y., Yumoto, N. and Nakajima, T. (1998) DL-*threo-β*-benzyloxyaspartate, a potent blocker of excitatory amino acid transporters. *Mol. Pharmacol.*, 53: 195–201.

Smith, C.P., Weremowicz, S., Kanai, Y., Stelzner, M., Morton, C.C. and Hediger, M.A. (1994) Assignment of the gene coding for the human high-affinity glutamate transporter EAAC1 to 9p24 – potential role in dicarboxylic aminoaciduria and neurodegenerative disorders – brief report. *Genomics*, 20: 335–336.

Sonders, M.S. and Amara, S.G. (1996) Channels in transporters. *Curr. Opin. Neurobiol.*, 6: 294–302.

Sorg, O., Horn, T.F.W., Yu, N.C., Gruol, D.L. and Bloom, F.E. (1997) Inhibition of astrocyte glutamate uptake by reactive oxygen species: role of antioxidant enzymes. *Mol. Med.*, 3: 431–440.

Stoffel, W., Sasse, J., Duker, M., Muller, R., Hofmann, K., Fink, T. and Lichter, P. (1996) Human high affinity, $Na^+$-dependent L-glutamate/L-aspartate transporter GLAST-1 (EAAT-1): gene structure and localization to chromosome 5p11-p12. *FEBS Lett.*, 386: 189–193.

Storck, T., Schulte, S., Hofmann, K. and Stoffel, W. (1992) Structure, expression and functional analysis of a $Na^+$-dependent glutamate/aspartate transporter from rat brain. *Proc. Natl. Acad. Sci. USA*, 89: 10955–10959.

Streit, P. (1980) Selective retrograde labeling indicating the transmitter of neuronal pathways. *J. Comp. Neurol.*, 191: 429–463.

Sutherland, M.L., Delaney, T.A. and Noebels, J.L. (1995) Molecular characterization of a high-affinity mouse glutamate transporter. *Gene*, 162: 271–274.

Sutherland, M.L., Delaney, T.A. and Noebels, J.L. (1996) Glutamate transporter mRNA expression in proliferative zones of the developing and adult murine CNS. *J. Neurosci.*, 16: 2191–2207.

Swanson, R.A., Miller, J.W., Rothstein, J.D., Farrell, K., Stein, B.A. and Longuemare, M.C. (1997) Neuronal regulation of glutamate transporter subtype expression in astrocytes. *J. Neurosci.*, 17: 932–940.

Takahashi, M., Kovalchuk, Y. and Attwell, D. (1995) Pre- and postsynaptic determinants of EPSC waveform at cerebellar climbing fiber and parallel fiber to Purkinje cell synapses. *J. Neurosci.*, 15: 5693–5702.

Takahashi, M., Sarantis, M. and Attwell, D. (1996) Postsynaptic glutamate uptake in rat cerebellar Purkinje cells. *J. Physiol.* (Lond.), 497: 523–530.

Takumi, Y., Matsubara, A., Danbolt, N.C., Laake, J.H., Storm-Mathisen, J., Usami, S., Shinkawa, H. and Ottersen, O.P. (1997) Discrete cellular and subcellular localization of glutamine synthetase and the glutamate transporter GLAST in the rat vestibular end organ. *Neuroscience*, 79: 1137–1144.

Tanaka, K. (1993) Expression cloning of a rat glutamate transporter. *Neurosci. Res.*, 16: 149–153.

Tanaka, J., Ichikawa, R., Watanabe, M., Tanaka, K. and Inoue, Y. (1997a) Extra-junctional localization of glutamate transporter EAAT4 at excitatory Purkinje cell synapses. *Neuroreport*, 8: 2461–2464.

Tanaka, K., Watase, K., Manabe, T., Yamada, K., Watanabe, M., Takahashi, K., Iwama, H., Nishikawa, T., Ichihara, N., Kikuchi, T., Okuyama, S., Kawashima, N., Hori, S., Takimoto, M. and Wada, K. (1997b) Epilepsy and exacerbation of brain injury in mice lacking the glutamate transporter GLT-1. *Science*, 276: 1699–1702.

Taxt, T. and Storm-Mathisen, J. (1984) Uptake of D-aspartate and L-glutamate in excitatory axon terminals in hippocampus: Autoradiographic and biochemical comparison with gamma-aminobutyrate and other amino acids in normal rats and in rats with lesions. *Neuroscience*, 11: 79–100.

Tolner, B., Poolman, B., Wallace, B. and Konings, W.N. (1992) Revised nucleotide sequence of the gltP gene, which encodes the proton-glutamate-aspartate transport protein of Escherichia- coli K-12. *J. Bacteriol.*, 174: 2391–2393.

Tong, G. and Jahr, C.E. (1994) Block of glutamate transporters potentiates postsynaptic excitation. *Neuron*, 13: 1195–1203.

Torp, R., Danbolt, N.C., Babaie, E., Bjørås, M., Seeberg, E., Storm-Mathisen, J. and Ottersen, O.P. (1994) Differential expression of two glial glutamate transporters in the rat brain: An in situ hybridization study. *Eur. J. Neurosci.*, 6: 936–942.

Torp, R., Hoover, F., Danbolt, N.C., Storm-Mathisen, J. and Ottersen, O.P. (1997) Differential distribution of the glutamate transporters GLT1 and rEAAC1 in rat cerebral cortex and thalamus: An in situ hybridization analysis. *Anat. Embryol.*, 195: 317–326.

Trotti, D., Rizzini, B.L., Rossi, D., Haugeto, Ø., Racagni, G., Danbolt, N.C. and Volterra, A. (1997) Neuronal and glial glutamate transporters possess an SH-based redox regulatory mechanism. *Eur. J. Neurosci.*, 9: 1236–1243.

Trotti, D., Rossi, D., Gjesdal, O., Levy, L.M., Racagni, G., Danbolt, N.C. and Volterra, A. (1996) Peroxynitrite inhibits glutamate transporter subtypes. *J. Biol. Chem.*, 271: 5976–5979.

Trotti, D., Volterra, A., Lehre, K.P., Rossi, D., Gjesdal, O., Racagni, G. and Danbolt, N.C. (1995) Arachidonic acid inhibits a purified and reconstituted glutamate transporter directly from the water phase and not via the phospholipid membrane. *J. Biol. Chem.*, 270: 9890–9895.

Ullensvang, K., Lehre, K.P., Storm-Mathisen, J. and Danbolt, N.C. (1997) Differential developmental expression of the two rat brain glutamate transporter proteins GLAST and GLT. *Eur. J. Neurosci.*, 9: 1646–1655.

Utsunomiya-Tate, N., Endou, H. and Kanai, Y. (1996) Cloning and functional characterization of a system ASC-like $Na^+$-dependent neutral amino acid transporter. *J. Biol. Chem.*, 271: 14883–14890.

Utsunomiya-Tate, N., Endou, H. and Kanai, Y. (1997) Tissue specific variants of glutamate transporter GLT-1. *FEBS Lett.*, 416: 312–316.

Vandenberg, R.J., Arriza, J.L., Amara, S.G. and Kavanaugh, M.P. (1995) Constitutive ion fluxes and substrate binding domains of human glutamate transporters. *J. Biol. Chem.*, 270: 17668–17671.

Vandenberg, R.J., Mitrovic, A.D., Chebib, M., Balcar, V.J. and Johnston, G.A.R. (1997) Contrasting modes of action of methylglutamate derivatives on the excitatory amino acid transporters, EAAT1 and EAAT2. *Mol. Pharmacol.*, 51: 809–815.

Velaz-Faircloth, M., McGraw, T.S., Malandro, M.S., Fremeau, R.T., Kilberg, M.S. and Anderson, K.J. (1996) Characterization and distribution of the neuronal glutamate transporter EAAC1 in rat brain. *Am. J. Physiol.*, 39: C67–C75.

Virgin, C.E jr., Ha, T.P.T., Packan, D.R., Tombaugh, G.C., Yang, S.H., Horner, H.C. and Sapolsky, R.M. (1991) Glucocorticoids inhibit glucose transport and glutamate uptake in hippocampal astrocytes: Implications for glucocorticoid neurotoxicity. *J. Neurochem.*, 57: 1422–1428.

Voisin, P., Viratelle, O., Girault, J.M., Morrisonbogorad, M. and Labouesse, J. (1993) Plasticity of astroglial glutamate and $\gamma$-aminobutyric acid uptake in cell cultures derived from postnatal mouse cerebellum. *J. Neurochem.*, 60: 114–127.

Volterra, A., Trotti, D., Cassutti, P., Tromba, C., Salvaggio, A., Melcangi, R.C. and Racagni, G. (1992) High sensitivity of glutamate uptake to extracellular free arachidonic acid levels in rat cortical synaptosomes and astrocytes. *J. Neurochem.*, 59: 600–606.

Volterra, A., Bezzi, P., Rizzini, B.L., Trotti, D., Ullensvang, K., Danbolt, N.C. and Racagni, G. (1996) The competitive transport inhibitor L-*trans*-pyrrolidine-2,4-dicarboxylate triggers excitotoxicity in rat cortical neuron-astrocyte co-cultures via glutamate release rather than uptake inhibition. *Eur. J. Neurosci.*, 8: 2019–2028.

Volterra, A., Trotti, D. and Racagni, G. (1994a) Glutamate uptake is inhibited by arachidonic acid and oxygen radicals via two distinct and additive mechanisms. *Mol. Pharmacol.*, 46: 986–992.

Volterra, A., Trotti, D., Tromba, C., Floridi, S. and Racagni, G. (1994b) Glutamate uptake inhibition by oxygen free radicals in rat cortical astrocytes. *J. Neurosci.*, 14: 2924–32.

Wadiche, J.I., Amara, S.G. and Kavanaugh, M.P. (1995a) Ion fluxes associated with excitatory amino acid transport. *Neuron*, 15: 721–728.

Wadiche, J.I., Arriza, J.L., Amara, S.G. and Kavanaugh, M.P. (1995b) Kinetics of a human glutamate transporter. *Neuron*, 14: 1019–1027.

Wahle, S. and Stoffel, W. (1996) Membrane topology of the high-affinity L-glutamate transporter (GLAST-1) of the central nervous system. *J. Cell Biol.*, 135: 1867–1877.

Yamada, K., Watanabe, M., Shibata, T., Tanaka, K., Wada, K. and Inoue, Y. (1996) EAAT4 is a post-synaptic glutamate transporter at Purkinje cell synapses. *Neuroreport*, 7: 2013–2017.

Yamashita, T., Kohmura, E., Yuguchi, T., Shimada, S., Tanaka, K., Hayakawa, T. and Tohyama, M. (1996) Changes in glutamate/aspartate transporter (GLAST/glut-1) mRNA expression following facial nerve transection. *Mol. Brain Res.*, 38: 294–299.

Yu, A.C.H., Chan, P.H. and Fishman, R.A. (1986) Effects of arachidonic acid on glutamate and $\gamma$-aminobutyric acid uptake in primary cultures of rat cerebral cortical astrocytes and neurons. *J. Neurochem.*, 47: 1181–1189.

Zafra, F., Alcantara, R., Gomeza, J., Aragon, C. and Gimenez, C. (1990) Arachidonic acid inhibits glycine transport in cultured glial cells. *Biochem. J.*, 271: 237–242.

Zerangue, N. and Kavanaugh, M.P. (1996a) Flux coupling in a neuronal glutamate transporter. *Nature*, 383: 634–637.

Zerangue, N. and Kavanaugh, M.P. (1996b) Interaction of L-cysteine with a human excitatory amino acid transporter. *J. Physiol. (Lond.)*, 493: 419–423.

Zerangue, N., Arriza, J.L., Amara, S.G. and Kavanaugh, M.P. (1995) Differential modulation of human glutamate transporter subtypes by arachidonic acid. *J. Biol. Chem.*, 270: 6433–6435.

Zhang, Y.M., Pines, G. and Kanner, B.I. (1994) Histidine 326 is critical for the function of GLT-1, a ($Na^+$, $K^+$) coupled glutamate transporter from rat brain. *J. Biol. Chem.*, 269: 19573–19577.

Zorumski, C.F., Mennerick, S. and Que, J. (1996) Modulation of excitatory synaptic transmission by low concentrations of glutamate in cultured rat hippocampal neurons. *J. Physiol.(Lond)*, 494: 465–477.

O.P. Ottersen, I.A. Langmoen and L. Gjerstad (Eds.)
*Progress in Brain Research*, Vol 116


CHAPTER 4

# Physiological and pathological operation of glutamate transporters

Brian Billups, David Rossi, Takeo Oshima, Orpheus Warr, Michiko Takahashi, Monique Sarantis, Marek Szatkowski and David Attwell*

*Department of Physiology, University College London, Gower St., London, WC1E 6BT, England, UK*

## Introduction

When the extracellular concentration of neurotransmitter glutamate, $[glu]_o$, rises too high in the brain for too long then the death of neurons is triggered (reviewed by Szatkowski & Attwell, 1994). Normally sodium-dependent glutamate uptake carriers maintain $[glu]_o$ below neurotoxic levels. So far, 5 distinct subtypes of such carriers have been cloned (Danbolt et al., this volume). By pumping glutamate out of the extracellular space, they maintain a low $[glu]_o$ around synapses, which allows released glutamate to diffuse rapidly out of the synaptic cleft. The carriers therefore also help to terminate the synaptic action of glutamate and thus enable rapid information processing in the brain. In pathological conditions like brain ischaemia, however, when there is a run-down of the ionic gradients which power glutamate uptake, the carriers may run backwards, releasing glutamate into the extracellular space and thus triggering the death of neurons (Szatkowski & Attwell, 1994; Takahashi et al., 1997). The following review describes recent advances in our understanding of the ion movements powering glutamate transport, of the properties of an anion channel in the transporters, of the localisation of the transporters, and of their role in ischaemia.

*Corresponding author.
Tel.: +44 0171 380 7342; fax: +44 0171 383 7005; e-mail: D.Attwell@ucl.ac.uk.

## The anion channel in glutamate transporters

### *Charge movements produced by glutamate transporters*

Glutamate uptake can be monitored as an inward membrane current (Brew & Attwell, 1987) in part because of the co-transport of ions which power the movement of glutamate. For each glutamate transported, at least 2 $Na^+$ are transported into the cell (see below), while one $K^+$ is transported out (Barbour et al., 1988, 1991; Amato et al., 1994) and either an $OH^-$ is transported out or an $H^+$ enters (Bouvier et al., 1992; Zerangue & Kavanaugh, 1996a). Thus at least one net positive charge enters the cell per glutamate transported. However, Wadiche et al. (1995a,b) found that for cloned glutamate transporters expressed in oocytes, the charge movement was not as predicted by the stoichiometry just described. Extra charge entry occurred which they showed was produced by $Cl^-$ efflux through an anion channel apparently in the structure of the carrier (no such $Cl^-$ efflux was seen in oocytes lacking the glutamate transporters). Thus, activation of glutamate uptake seems to open an anion channel. The direction of $Cl^-$ flow through the channel has little influence on the amount of glutamate transport (Wadiche et al., 1995a; Billups et al., 1996); it appears that progression of the carrier through its transport cycle leads to a channel opening which is physically independent of the permeation pathway for glutamate.

### The cone synaptic terminal transporter and its anion channel

Discovery of this anion channel resolved a puzzling observation made by Sarantis et al. (1988), who characterised an unusual glutamate-evoked current in the synaptic terminal of cone photoreceptors, which was sodium-dependent but was carried largely by chloride ions. A similar response was later found in retinal bipolar cells (Grant & Dowling, 1995). Subsequent work on the cone response suggested it was mediated by an uptake carrier (Eliasof & Werblin, 1993), but controversy centered on the pharmacology of the response: Sarantis et al. (1988) reported that kainate could evoke a response in the cone (suggesting it was not mediated by uptake), while Eliasof & Werblin (1993) did not see this. More recent work by Werblin's group (Picaud et al., 1995) has reported that kainate does indeed sometimes evoke a current, but it turns out that this current is blocked by CNQX (unlike the majority of the glutamate-evoked current) and so is likely to be mediated by a presynaptic non-NMDA receptor (D. Attwell, unpublished). Ignoring the latter current component, the majority of the cone response to glutamate is therefore produced by an uptake carrier which activates a large anion conductance.

### Different glutamate transporters have different anion conductances

For the cone transporter, almost all of the current generated when glutamate is applied is produced by $Cl^-$ flux through the anion channel (Sarantis et al., 1988; Picaud et al., 1995). By contrast for the transporter in retinal glial cells most of the current is generated by the charge movements coupled directly to glutamate movement, since removal of intra- or extra-cellular $Cl^-$ has relatively little effect on the glutamate-evoked current (Barbour et al., 1991; Billups et al., 1996). This variation in the ratio of current generated by $Cl^-$ movement to that generated by co-transport of $Na^+$ and other ions has been characterized for the cloned glutamate transporters: the glial transporters GLT1 (or EAAT2) and GLAST (or EAAT1, the retinal glial transporter), and the neuronal transporter EAAC1 (EAAT3), show relatively small anion channel currents relative to $Na^+$-co-transport currents, while for the neuronal transporters EAAT4 and EAAT5 (the cone transporter), the anion current overwhelmingly dominates the $Na^+$-co-transport current (Wadiche et al., 1995a,b; Fairman et al., 1995; Arriza et al., 1997). A possible basis for these differences is postulated below.

### The anion conductance is activated both during forward and reversed uptake

The anion conductance was originally discovered to be activated during normal uptake of glutamate. However during ischaemia, when the extracellular potassium concentration rises and cells are depolarised, the transport process can be reversed (see below): this can be detected at the single cell level as an outward membrane current in the glutamate releasing cell, or as a glutamate-gated current in a 'sniffer neuron' held near the releasing cell (Szatkowski et al., 1990; Billups & Attwell, 1996). Billups et al. (1996) showed that the anion conductance is also activated during reversed uptake, suggesting that no matter which way the carrier proceeds through its transport cycle, it has to pass through a state which allows opening of the anion channel.

### The anion conductance can be activated, and modulated, independently of glutamate transport

To study further the interaction between the glutamate transport and anion channel functions of glutamate transporters, Billups et al. (1996)

removed extra- and intracellular $K^+$ from retinal glial cells expressing glutamate transporters: this manoeuvre is known to block net glutamate transport because $K^+$-counter-transport is an essential part of the transport cycle (Kanner & Sharon, 1978; Amato et al., 1994; Billups & Attwell, 1996). With glutamate transport thus blocked, the anion channel was still able to open, provided that both glutamate and $Na^+$ were present on both sides of the membrane. Furthermore, while acidifying the extracellular pH greatly reduces glutamate transport (Bouvier et al., 1992; Billups & Attwell, 1996), it had no effect on the anion channel opening seen with glutamate transport blocked by $K^+$-removal.

A further demonstration of the partial independence of glutamate transport and anion channel opening is seen with the effects of zinc: zinc reduces glutamate transport by the retinal glial cell transporter, but potentiates opening of the transporter's anion channel (Spiridon et al., 1997).

**A kinetic scheme for gating of the anion conductance**

Billups et al. (1996) found that the opening of the anion channel seen with glutamate transport blocked had a Michaelis-Menten (1st order) dependence on $[glu]_o$, but a higher order dependence on $[Na^+]_o$ (proportional to $[Na^+]_o^{2.5}$ at low $[Na^+]_o$). Together with the fact that opening can be seen independent of the presence of $K^+$, and independent of external pH, this led to the suggestion of a kinetic scheme for the transporter as shown in Fig. 1. At the outer membrane surface the transporter binds 2 (or more) $Na^+$ and then a glutamate anion, which it translocates to the inner membrane surface. There the translocated species can unbind, before the carrier binds a $K^+$ and $OH^-$ which it transports to the outside of the cell. Anion channel opening is postulated to occur from a state with one glutamate and 2 (or more) $Na^+$ bound. This explains, for experiments with no $K^+$ present, the glutamate- and $[Na^+]$-dependence of the opening, the fact that opening can occur in the absence of $K^+$, the fact that opening is unaffected

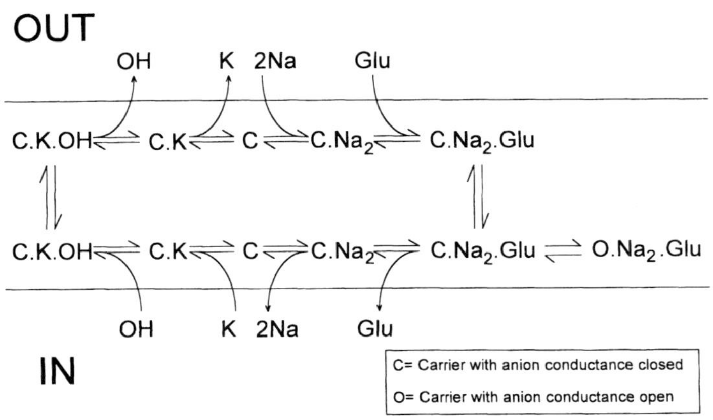


Fig. 1. A possible kinetic scheme for transport of glutamate and gating of the transporter's anion channel. C denotes the carrier in conformations for which the anion channel is closed. In these conformations the carrier can bind extracellular $glu^-$ and $2Na^+$, transport them to the inner face of the membrane, then bind $K^+$ and $OH^-$ at the inner membrane surface and transport them to the outside of the cell. (The order in which $K^+$ and $OH^-$ bind is unknown and, although we show $OH^-$ transport out of the cell, as discussed in the text the carrier might transport $H^+$ into the cell). O denotes a conformation of the transporter in which the anion conductance is open. Here we postulate that the open state can be accessed from the closed conformation which has $Na^+$ and $glu^-$ bound at the inner face of the membrane. Our data would also be consistent with an O state being accessed from the closed conformation with $Na^+$ and $glu^-$ bound at the outer membrane surface (or from both the $C.Na_2.Glu$ conformations). Taken from Billups et al. (1996, Fig. 7). More recent work suggests that 3, rather than 2, $Na^+$ are transported (Zerangue and Kavanaugh, 1996a).

by variations in $[OH^-]$ caused by altered pH, and the fact that $Na^+$ and glutamate need to be present inside the cell as well as outside for opening to occur (since in the absence of internal $Na^+$ and glutamate the transporter would depart from the state with $2Na^+$ and glutamate bound).

The scheme of Fig. 1 suggests a simple possible explanation for the fact that the different cloned transporters, which have quite closely related structures, show widely differing ratios of anion current to transport current. A simple difference in one (or both) of the rate constants for the transporter entering and leaving the state with $2Na^+$ and glutamate bound, in which the anion channel is open, would result in a corresponding change in the amount of anion current generated per glutamate transport cycle.

Consistent with the scheme in Fig. 1, Kavanaugh et al. (1997) have found a mutated form of GLT1 in which the $K^+$-dependent limb of the transport cycle is isolated from the $Na^+$- and glutamate-dependent limb: for this carrier, $K^+$-induced forward and reverse glutamate transport are blocked, yet $Na^+$-dependent exchange of glutamate and anion channel opening still occur. Furthermore, the structurally related transporter of neutral amino acids ASCT1 seems to normally function in this way, being unaffected by $K^+$, being only able to mediate homo- and heteroexchange of substrate amino acids and not net transport, and generating a current by activating an anion channel (Zerangue & Kavanaugh, 1996b).

### Existence of the anion channel weakens evidence for $OH^-$ counter-transport

Following earlier work by Erecinska et al. (1983) and Nelson et al. (1983), Bouvier et al. (1992) showed that the glutamate transporters in retinal glia generate an intracellular acidification and an extracellular alkalinization consistent with the transport of either an $H^+$ into the cell or of an $OH^-$ out of the cell for each glutamate entering the cell (see also Billups et al. (1996) and Zerangue & Kavanaugh (1996a) for evidence that the movement of $H^+/OH^-$ is coupled to the movement of glutamate). Bouvier et al. (1992) opted for an $OH^-$ leaving the cell (as in Fig. 1) after finding that intracellular anions such as $NO_3^-$, $ClO_4^-$ or $SCN^-$ increased the glutamate-evoked current and could be detected leaving the cell with anion-selective electrodes: the idea was that these anions might be competing for transport (at a higher rate) at an $OH^-$-binding site at the inner membrane surface. However, the discovery of the transporters' anion conductance undermines this suggestion, since it turns out that the selectivity of this conductance is such that it allows $NO_3^-$, $ClO_4^-$ and $SCN^-$ to pass much better than $Cl^-$ (Eliasof & Jahr, 1996; Billups et al., 1996) – consequently the increased current and anion efflux seen by Bouvier et al. (1992) were mediated by the anion conductance, and on the basis of these data it is not possible to determine whether glutamate transport is coupled to the pH gradient by co-transport of an $H^+$ or counter-transport of an $OH^-$.

By studying the transport of cysteine in place of glutamate, Zerangue & Kavanaugh (1996a,c) have suggested that for EAAT3 $H^+$ co-transport, rather than $OH^-$ counter-transport, is occurring. This was based on the fact that the transporter is able to generate a current by transporting either anionic cysteine or neutral cysteine (the pK of cysteine's thiol group is 8.5 so the ratio of neutral to anionic cysteine can be increased by making the pH more acid). They suggested that this could be explained by the transport either of anionic cysteine plus a separate $H^+$, or of neutral cysteine (effectively with the $H^+$ attached to the anionic cysteine), so that the number of separate $H^+$ co-transported would be varied (from 0 to 1) according to the cysteine species being transported. (Alternatively, though perhaps more complicatedly, one could postulate that, if $OH^-$ is counter-transported, the $OH^-$ stoichiometry is altered from 1 to 0 when the cysteine species transported is switched from anionic to neutral.) A similar variable proton co-transport stoichiometry has been postulated for a peptide transporter (Temple et al., 1996). To be plausible such a scheme would require that the $H^+$ binds to a site close to where the anionic thiol group binds when anionic cysteine is being transported. An extreme version of this hypothesis is that only protonated amino acids (neutral cysteine and neutral glutamate, i.e. glutamate with one carboxyl group protonated) can be transported: for glutamate (but not cysteine with its more alkaline pK) the proton would then dissociate and acidify the cytoplasm when the transported species unbinds at the inner membrane surface. If this were the case for glutamate then the transporter would actually have an affinity 3 orders of magnitude higher than the normal micromolar value quoted in terms of anionic glutamate concentration, because less than 0.1% of glutamate is in the neutral protonated form at physiological pH.

## Postsynaptic glutamate uptake

### *Localization of glutamate transporters*

Radiotracing experiments have suggested the presence of glutamate transporters in glial cells and in the presynaptic terminals of neurons. These, respectively, might serve to keep a generally low $[glu]_o$ in the CNS, and to recycle released glutamate to neurons for repackaging into vesicles. However, the localization of the cloned transporters is not in perfect accord with this textbook account. Although the transporters GLAST (or EAAT1) and GLT1 (EAAT2) are found in cerebellar and retinal glia and all over the brain in glia respectively (Storck et al., 1992; Pines et al., 1992; Rothstein et al., 1994; Lehre et al., 1995; Danbolt et al., this volume), and the transporter EAAT5 is found in the cone synaptic terminal (Sarantis et al., 1988; Arriza et al., 1997), the neuronal carriers EAAC1 (or EAAT3) and EAAT4 do not always have a presynaptic location (Rothstein et al., 1994; Yamada et al., 1996).

## Postsynaptic glutamate transporters in Purkinje cells

Unexpectedly, EAAC1 and EAAT4 have been reported to be present in cerebellar Purkinje cells postsynaptic to the glutamate-releasing terminals of the parallel and climbing fibres (Rothstein et al., 1994; Yamada et al., 1996; Dehnes et al., 1998), although the presence of EAAC1 has been disputed (Shibata et al., 1996). We demonstrated that the glutamate transporters detected immunocytochemically in Purkinje cells are indeed functional in the cell membrane by showing the presence of a membrane current component with the pharmacology and ion-dependence of a $Na^+$-dependent glutamate transporter which also activates an anion channel (Takahashi et al., 1996). Furthermore, when the rate of glutamate uptake by these transporters was reduced, by including D-aspartate in the pipette used to whole-cell clamp the cell (D-aspartate should bind to the glutamate transport site at the inner face of the membrane, reducing the number of glutamate binding sites at the outside), the decay of the climbing fibre synaptic current was prolonged. This suggests that binding of glutamate to postsynaptic glutamate transporters may help to terminate glutamate's synaptic action at these synapses. Since the cycle time for glutamate transporters has been suggested to be 70 msec (Wadiche et al., 1995b), 10-fold longer than the synaptic current, it may be buffering of $[glu]_o$ by binding to the transporters, rather than transport of glutamate out of the synaptic cleft, that helps to terminate the synaptic current. Calculations presented by Takahashi et al. (1996) indicate that the perhaps 5000 glutamate molecules released in each vesicle could encounter a comparable number of postsynaptic EAAC1 or EAAT4 transporters in the first millisecond after exocytosis has occurred, suggesting that buffering of this sort may well occur.

## Do postsynaptic transporters generate a synaptically-activated current?

Tanaka et al. (1997) and Dehnes et al. (1998) have found that EAAT4 transporters are localized around, but not at, the postsynaptic density at parallel fibre to Purkinje cell synapses. Since the diffusion time from the presynaptic release site to these transporters is less than 1 msec, one might expect them to bind glutamate during the time course of the synaptic current. The question then arises as to whether they could generate a significant component of the postsynaptic current. In the case of EAAT4 this could be particularly important because of the large anion conductance that these transporters have. Even if the 70 msec cycle time mentioned above precludes a large transport current during the synaptic current, it is possible that binding of glutamate might allow the carrier to reach the state where the anion channel opens (see Fig. 1), generating an outward current that opposes the inward current generated by postsynaptic AMPA receptors and thus reduces cell excitability.

In preliminary experiments (Warr, Takahashi & Attwell, unpublished) we have been unable to detect such a current however: applying CNQX essentially abolishes the synaptic current, indicating that it is almost all generated by postsynaptic glutamate-gated channels.

## Stoichiometry of glutamate transport

### *Some data suggest co-transport of 2 $Na^+$*

Until recently the consensus of numerous experiments on different glutamate transporters was that for each glutamate anion transported into the cell, 2 $Na^+$ are transported into the cell, one $K^+$ is transported out, either an $OH^-$ is transported out or an $H^+$ enters, and one net positive charge enters the cell (reviewed by Takahashi et al., 1997). However Zerangue & Kavanaugh (1996a) have suggested that 3 rather than 2 $Na^+$ are carried. The original evidence for transport of 2 $Na^+$ came in several forms. First, radiotracing measurements of the sodium and glutamate fluxes into CNS cell lines gave a ratio of 2 $Na^+$ transported per glutamate (Baetge et al., 1979; Stallcup et al., 1979). Secondly, measurement of the equilibrium accumulation into synaptosomes of D-aspartate, a non-metabolized glutamate analogue that is taken up by glutamate carriers, again showed accumulation proportional to the square of $[Na^+]_o$ (Erecinska et al., 1983: these authors also suggested that 2 net charges entered the cell, but that suggestion ignored counter-transport of $K^+$ and their data are consistent with one charge entering). Finally, Kanai et al. (1995) used radiotracing, pH measurement and voltage-clamping of oocytes expressing EAAC1 to suggest that 2 $Na^+$ and one net charge were transported per glutamate.

### *Could 3 $Na^+$ be co-transported?*

Zerangue and Kavanaugh (1996a) suggested, however, that 3 $Na^+$ and 2 net charges were co-transported by EAAT3 transporters (homologous to the EAAC1 carriers studied by Kanai et al., 1995) expressed in oocytes. Their evidence came from 3 experiments. First, with anion currents suppressed by chloride removal, they found that high doses (up to 20 mM) of the glutamate analogue kainate, said to be non-transported, suppressed glutamate transport and the glutamate-evoked currents generated by EAAT3. Assuming that kainate simply blocks both forward and reversed uptake (rather than abolishing the current by changing the carrier stoichiometry and making the transport electroneutral), the [Na]-dependence of the reversal potential of the kainate-evoked current change indicated co-transport of 3 rather than 2 $Na^+$. Secondly, comparing radiotracing and current measurement experiments gave a ratio of 2 charges transported into the cell with each glutamate anion, consistent with 3 $Na^+$ being transported. Thirdly, oocytes expressing the transporter were shown to be able to pump down the external glutamate concentration to a value apparently below that theoretically possible if only 2 $Na^+$ were transported (although interpretation of this experiment is complicated by the fact that the [glutamate] gradient across the oocyte membrane was not actually measured: instead the effective total concentrations of all the intra- and extra-cellular amino acids which can activate an uptake current were assayed, and some of these substrates may not have reached equilibrium across the membrane).

There is no simple way to account for the difference between the results of Zerangue &

Kavanaugh (1996a) and those of the other workers reviewed above. Although the earlier data on cell lines and synaptosomes might be explained in terms of the transport being mediated by a different glutamate transporter, the conflict between the results of Zerangue & Kavanaugh (1996a) on EAAT3 and the results of Kanai et al. (1995) on the homologous transporter EAAC1 is particularly striking. Below we consider the implications of the possibility of 3 $Na^+$ being transported.

## Reversal of the uptake process in ischaemia

### *Ionic changes in ischaemia promote glutamate release by reversed uptake*

Inhibition of the $Na^+/K^+$ pump by the oxygen and glucose lack occurring in brain ischaemia leads to a rundown of the ion gradients which power glutamate uptake (reviewed by Attwell et al., 1993; Szatkowski & Attwell, 1994). The first ionic change to occur is an acidification produced by the switch to anaerobic respiration – after a few minutes both intra- and extracellular pH shift acid by about one unit. The extracellular $[K^+]$ rises slowly at first, but after a few minutes of ischaemia $[K^+]_o$ rises suddenly to about 60 mM (reviewed by Siesjö, 1990), depolarizing cells to approximately – 20 mV (the anoxic depolarization). At the same time $[K^+]_i$ and $[Na^+]_o$ fall and $[Na^+]_i$ rises, and there is a large release of glutamate to the extracellular space where it triggers the death of neurons.

There has been controversy over the mechanism of this glutamate release, with some finding it to be calcium-independent and others claiming the opposite (see Szatkowski & Attwell, 1994; Takahashi et al., 1997). However, the ionic gradient changes just described reduce the driving force for glutamate uptake, so if there are sufficient glutamate transporters present for $[glu]_o$ to be pumped down to its equilibrium value initially (where influx and efflux are equal on the transporter), then when $[K^+]_o$ rises and the cell is depolarized the transporters must start to run backwards, releasing glutamate until $[glu]_o$ rises to a new equilibrium value (calculated below). We have detected the reversed uptake produced by ion gradient changes simulating ischaemia as an outward current accompanying glutamate release from retinal glial cells and cerebellar Purkinje cells (Szatkowski et al., 1990; Takahashi et al., 1996), and also by sensing the glutamate release using glutamate-gated channels in a 'sniffer neuron' held near to the releasing glial cell (Fig. 2; Billups & Attwell, 1996). During metabolic blockade, glutamate release due to reversed uptake (suppressed by preloading the cells with glutamate transport blockers) has been shown by Madl & Burgesser (1993), Longuemare & Swanson (1995) and Roettger & Lipton (1996).

As a novel approach to determining the mechanism of glutamate release in ischaemia, we have recently whole cell-clamped pyramidal cells in area CA1 of hippocampal slices during metabolic blockade, monitoring glutamate release in the slice as a membrane current suppressible by CNQX and APV which is induced in the cells whose death is being triggered by the glutamate. We find that most of the glutamate release is calcium-independent (Oshima et al., 1997), and that preloading with a glutamate transport blocker suppresses most of the release (Oshima, Rossi & Attwell, unpublished data).

## Is glutamate release by reversed uptake from neurons or glial cells?

Radiotracing experiments, showing that most radioactive glutamate applied to brain tissue ends up in glial cells, suggest that glial cells have more glutamate transporters than neurons (McLennan, 1976). However, in normal conditions at least, glial cells have a lower intracellular glutamate concentration than neurons (Storm-Mathisen et al., 1992) because of conversion to glutamine by glutamine synthetase, so neurons may have more glutamate to release despite having fewer transporters to release it. To complicate matters, in ischaemia the glutamate concentration in glia rises (Storm-Mathisen et al., 1992) because conversion of glutamate to glutamine requires ATP. It is possible that early in ischaemia there is a release of glutamate from neurons by reversed uptake (Madl & Burgesser, 1993) at a time when glia are still taking up glutamate.

**The acid pH shift in ischaemia slows glutamate release by reversed uptake**

The stoichiometry of glutamate transport (see above) is such that during reversed uptake, as glutamate leaves the cell, either an $OH^-$ is transported into the cell or an $H^+$ is transported out. The acid pH shift occurring in ischaemia is expected to deprive the carrier of $OH^-$ to transport in, or hinder the loss of $H^+$ transported out, and thus slow the release of glutamate by reversed uptake. Monitoring glutamate release by reversed uptake either as a membrane current or using a 'sniffer neuron' (Fig. 2), Billups & Attwell (1996) found that the pH shift occurring in ischaemia slows reversed uptake 14-fold. This delay in the release of glutamate could be neuroprotective in the case of transient ischaemic attacks.

**How high will $[glu]_o$ rise in ischaemia?**

The lowest value of $[glu]_o$ that can be maintained by a transporter co-transporting 1 glutamate anion and a number, n, of $Na^+$, as well as counter-transporting 1 $K^+$ and 1 $OH^-$ (or co-transporting 1 $H^+$) is (Attwell et al., 1993; Zerangue & Kavanaugh, 1996a):

$$[glu]_o = [glu]_i\,([Na^+]_i/[Na^+]_o)^n\,([H^+]_i/[H^+]_o) \times ([K^+]_o/[K^+]_i)\,\exp\{(n-1)VF/RT\} \quad (1)$$

where the subscripts o and i denote concentrations outside and inside the cell, V is the membrane potential, R the gas constant, T the temperature

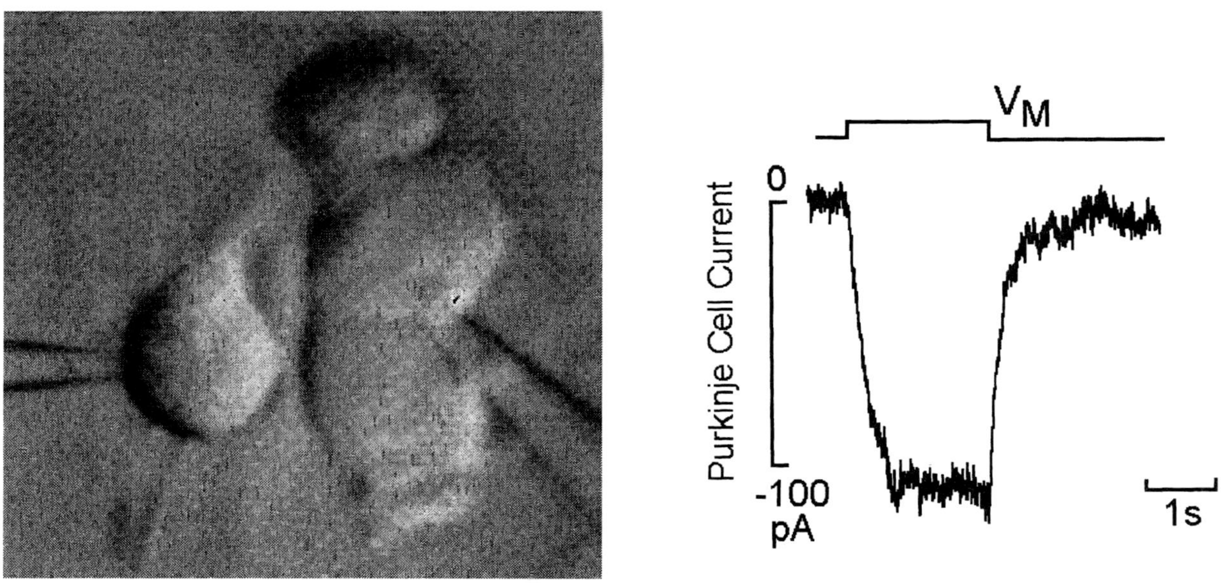


Fig. 2. Sensing glutamate release by reversed uptake using a 'sniffer neuron'. Left: A whole-cell clamped isolated rat cerebellar Purkinje cell (left) being used to sense glutamate release from an isolated salamander retinal glial (Muller) cell (right) whole-cell clamped with an electrode containing glutamate. From the study of Billups & Attwell (1996). Right: The current change produced at -60 mV in a 'sniffer neuron' as in A, when glutamate release by reversed uptake was activated in an apposed Muller cell by mimicking the conditions of ischaemia (depolarizing the Muller cell from $V_M = -60$ to $+20$ mV (top trace) in a solution containing 30 mM $[K^+]_o$). Taken from Billups et al. (1996, Fig. 6).

and F the Faraday constant. Under normal conditions this equation predicts a minimum maintainable $[glu]_o$ of about 2 nM for n=3 and 0.2 $\mu$M for n=2 (Attwell et al., 1993, Zerangue & Kavanaugh, 1996a). To calculate how high this minimum maintainable $[glu]_o$ will rise when $[K^+]_o$ rises in ischaemic conditions we make the following assumptions.

(a) Throughout, $[Na^+]_o + [K^+]_o = [Na^+]_i + [K^+]_i = 144$ mM, (2)

(b) Initially $[K^+]_o = 4$ mM, $[Na^+]_o = 140$ mM, $[Na^+]_i = 25$ mM (Ballanyi et al., 1987; Friedman & Haddad, 1994), and $[K^+]_i = 144$ mM-25 mM = 119 mM. Thus, the increase in $[K^+]_o$ during ischaemia is $\Delta[K^+]_o = [K^+]_o - 4$ mM. (3)

(c) From (2), $[Na^+]_o = 144\text{ mM} - [K^+]_o$ (4) so $[Na^+]_o$ drops by the same amount as $[K^+]_o$ rises (cf. Jiang et al., 1992);

(d) The extracellular volume fraction is ¼ that of the intracellular, so that when $[K^+]_o$ has risen, the new intracellular ion concentrations are:

$$[K^+]_i = 119\text{ mM} - \Delta[K^+]_o/4 \text{ and} \quad (5)$$

$$[Na^+]_i = 144\text{ mM} - [K^+]_i = 25\text{ mM} + \Delta[K^+]_o/4 \quad (6)$$

This implies that when $[K^+]_o$ rises to 60 mM in ischaemia (Siesjo, 1990), $[Na^+]_i$ will rise by 14 mM to 39 mM (cf. Friedman & Haddad, 1994);

(e) the intra- and extracellular pH shift acid at the same rate so that throughout $[H^+]_i/[H^+]_o = 2$ (7)

(f) the membrane potential is the Nernst potential for potassium, so

$$VF/RT = \ln([K^+]_o/[K^+]_i) \quad (8)$$

Using equations (2)–(8), eqn. (1) can be rewritten in terms of the value of $[K^+]_o$ reached as

$$[glu]_o/[glu]_i = (\{25\text{ mM} + ([K^+]_o - 4\text{ mM})/4\}/\{144\text{ mM} - [K^+]_o\})^n \cdot 2 \cdot ([K^+]_o/\{119\text{ mM} - ([K^+]_o - 4\text{ mM})/4\})^n \quad (9)$$

To calculate the minimum value of $[glu]_o$ we need to assume an average value for $[glu]_i$. As noted above, and discussed in Attwell et al. (1993), this is complicated by the likelihood of the value being different in neurons and glia (but becoming less different in ischaemia when glial $[glu]_i$ rises). As a reasonable compromise value we use the value $[glu]_i$ = 3 mM (Attwell et al., 1993; Storm-Mathisen et al., 1992). Using this value, equation (9) is plotted in Figure 3 for either 2 $Na^+$ or 3 $Na^+$ co-transported (n=2 or 3). When $[K^+]_o$ rises in ischaemia to values approaching 60 mM, the lowest that glutamate transporters can hold $[glu]_o$ down to (in equilibrium) is 422 $\mu$M in the case of 2 $Na^+$ being co-transported and 112 $\mu$M in the case of 3 $Na^+$ being co-transported.

For comparison with these values, Choi et al. (1987) found that exposing cultured neurons to a $[glu]_o$ of 100 $\mu$M for 5 minutes led to over half the neurons dying. Thus, the rundown of ionic gradients in ischaemia will eventually lead to $[glu]_o$ rising to neurotoxic levels (although, as described above, the acid pH shift occurring will slow the rate at which this equilibrium is achieved: simple estimates of the rate of rise of $[glu]_o$ are given in Takahashi et al., 1997). Zerangue & Kavanaugh (1996a) suggested that the extra accumulative power provided by co-transporting 3 rather than 2 $Na^+$ would prevent $[glu]_o$ rising in ischaemia. However the calculations for Fig. 3 show that, even for 3 $Na^+$ moving, $[glu]_o$ will still rise to a neurotoxic level (the lower values of $[glu]_o$ predicted in Fig. 4a of Zerangue & Kavanaugh (1996a) result from the use of extracellular ion concentrations which are reasonable for ischaemia, with intracellular ion concentrations, particularly for $Na^+$, which are non-ischaemic). Consistent with these predictions, as described above, glutamate release by reversed uptake during metabolic inhibition has been demonstrated by Madl & Burgesser (1993), Longuemare & Swanson (1995), Roettger & Lipton (1996) and Oshima et al. (1997 and unpublished).

The values predicted in Fig. 3 would scale with whatever value were assumed for $[glu]_i$, but are not dramatically affected by altering the assumed

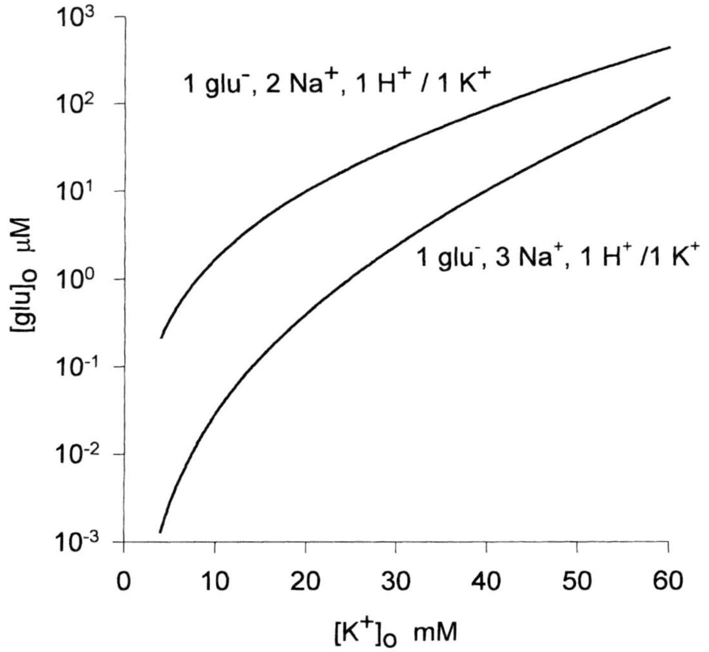


Fig. 3. Prediction of the lowest value at which glutamate transporters can maintain the extracellular glutamate concentration ($[glu]_o$), as a function of extracellular potassium concentration ($[K^+]_o$), for transporters carrying 2 or 3 $Na^+$. In normal conditions (low mM $[K^+]_o$) $[glu]_o$ can be kept below micromolar levels. In ischaemic conditions, when $[K^+]_o$ rises to 50–60 mM, the lowest value of $[glu]_o$ that can be maintained is sufficiently high to trigger neuronal death.

extracellular volume fraction. Fig. 3 was calculated assuming a commonly quoted extracellular volume fraction of 20% (e.g. Nicholson & Phillips, 1981). If this value is reduced to 13%, as measured for hippocampus (McBain et al., 1990), the minimum $[glu]_o$ predicted at $[K^+]_o = 60$ mM is 278 $\mu$M and 60 $\mu$M, respectively, for $n = 2$ or 3 $Na^+$ co-transported, i.e. still in the neurotoxic range.

## $Na^+$-independent glutamate transporters

Although the role of sodium-dependent glutamate transporters in controlling $[glu]_o$ has been well studied, other transporters could contribute to terminating glutamate's synaptic action or keeping $[glu]_o$ below neurotoxic levels. We have carried out preliminary experiments looking at the role of cystine-glutamate transporters in the CNS (Warr et al., 1997). The glutamate binding capacity of these transporters varies with brain region, but can be as high as 75% of that of $Na^+$-dependent glutamate carriers (Anderson et al., 1990) so in principle cystine-glutamate exchange might significantly affect $[glu]_o$. The concentration of cystine inside cells is about 100-fold higher than that outside (Murphy et al., 1989), whereas for glutamate the corresponding factor is > 10,000. Thus, the cystine-glutamate exchanger may normally transport glutamate out of cells and cystine in (Kessler et al., 1987). However, this carrier can also run in the opposite direction (Cho & Bannai, 1990; Kato et al., 1993) and might do so when $[glu]_o$ rises during synaptic transmission.

We found that raising the cystine concentration in the solution superfusing cerebellar or cortical slices resulted in neurons developing an inward current which was blocked by glutamate receptor blockers (Warr et al., 1997). A similar inward current is seen when the glutamate analogue PDC is applied, which both blocks $Na^+$-dependent glutamate uptake and releases glutamate by heteroexchange (Sarantis et al., 1993; Volterra et al., 1996). It is likely that the cystine-evoked current reflects a rise in $[glu]_o$ caused by cystine-glutamate exchange. It remains to be determined whether this exchange can modify $[glu]_o$ enough to modulate synaptic current waveforms or alter tonic activation of glutamate receptors.

## The future

Advances in our molecular knowledge of glutamate transporters are likely to be the focus of the most dramatic advances in this field. For example the generation of mice with the GLT-1 and EAAC1 glutamate transporters knocked out (Peghini et al., 1997; Tanaka et al., 1997) could allow investigation of the question (raised above) of whether the glutamate release in ischaemia is mainly by reversed operation of glial or neuronal transporters. Gaining a better molecular understanding of the factors determining the amount of anion conductance activated by glutamate transporters could offer therapeutic approaches to decreasing CNS excitability in conditions like epilepsy (Billups et al., 1996). Examination of changes in glutamate transporter expression in diseases leading to neuronal death (Rothstein et al., 1992; 1996) may lead to further examples of decreased expression leading to a rise of $[glu]_o$ which triggers neuronal death. Finally, increased understanding of the promoters controlling expression of glutamate transporters (Hagiwara et al., 1996) may allow strategies to be devised to upregulate transporter expression in clinical conditions in which $[glu]_o$ is expected to rise.

## Conclusion

In this review we describe recent advances in understanding the properties and functional role of CNS glutamate transporters, focusing on the properties of an anion channel present in the transporters, on the existence of postsynaptic glutamate uptake, on recent controversy over the ionic stoichiometry of glutamate transport, on the role of glutamate transporters in raising the extracellular glutamate concentration in brain ischaemia, and on the possible importance of glutamate transporters other than the sodium-dependent carriers.

## Acknowledgements

Supported by the Wellcome Trust, M.R.C. and European Community (BMH4-CT95-871).

## References

Amato, A., Barbour, B., Szatkowski, M. and Attwell, D. (1994) Counter-transport of potassium by the glutamate uptake carrier in glial cells isolated from the tiger salamander retina. *J. Physiol.*, 479: 371–380.

Anderson, K.J., Monaghan, D.T., Bridges, R.J., Tavoularis, A.L. and Cotman, C.W. (1990) Autoradiographic characterization of putative excitatory amino acid transport sites. *Neuroscience*, 38: 311–322.

Arriza, J.L., Eliasof, S., Kavanaugh, M.P. and Amara, S. (1997) Excitatory amino acid transporter 5, a retinal glutamate transporter coupled to a chloride conductance. *Proc. Natl. Acad. Sci. USA*, 94: 4155–4160.

Attwell, D., Barbour, B. and Szatkowski, M. (1993) Nonvesicular release of neurotransmitter. *Neuron*, 11: 401–407.

Baetge EE., Bulloch K. and Stallcup WB (1979) A comparison of glutamate uptake in cloned cell lines from the central nervous system. *Brain Res*., 167: 210–214.

Ballanyi, K., Grafe, P. and Ten Bruggencate, G. (1987) Ion activities and potassium uptake mechanisms of glial cells in guinea-pig olfactory cortex slices. *J. Physiol.*, 382: 159–174.

Barbour, B., Brew, H. and Attwell, D (1988) Electrogenic glutamate uptake in glial cells is activated by intracellular potassium. *Nature*, 335: 433–435.

Barbour B., Brew, H. and Attwell, D. (1991) Electrogenic uptake of glutamate and aspartate into glial cells isolated from the salamander retina. *J. Physiol.*, 436: 169–193.

Billups, B. and Attwell, D. (1996) Modulation of non-vesicular glutamate release by pH. *Nature*, 379: 171–174.

Billups, B., Rossi, D. and Attwell, D. (1996) Anion conductance behavior of the glutamate uptake carrier in salamander retinal glial cells. *J. Neurosci.*, 16: 6722–6731.

Bouvier, M., Szatkowski, M., Amato, A. and Attwell, D. (1992) The glial cell glutamate uptake carrier countertransports pH-changing anions. *Nature*, 360: 471–473.

Brew, H. and Attwell, D. (1987) Electrogenic glutamate uptake is a major current carrier in the membrane of axolotl retinal glial cells. *Nature*, 327: 707–709.

Cho, Y. and Bannai, S. (1990) Uptake of glutamate and cysteine in C-6 glioma cells and in cultured astrocytes. *J. Neurochem.*, 55: 2091–2097.

Choi, D.W., Maulucci-Gedde, M. and Kriegstein, A.R. (1987) Glutamate neurotoxicity in cortical cell culture. *J. Neurosci.*, 7: 357–368.

Dehnes, Y., Chaudhry, F.A., Ullensvang, K., Lehre, K.P., Storm-Mathisen, J. and Danbolt, N.C. (1998) The glutamate transporter EAAT4 is localized extrasynaptically on dendritic spines of cerebellar Purkinje cells *J. Neurosci.*, 18: 3606–3619.

Eliasof, S. and Jahr, S. (1996) Retinal glial cell glutamate transporter is coupled to an anionic conductance. *Proc. Natl. Acad. Sci. USA.*, 93: 4153–4158.

Eliasof, S. and Werblin, F. (1993) Characterization of the glutamate transporter in retinal cones of the tiger salamander. *J. Neurosci.*, 113: 402–411.

Erecinska, M., Wantorsky, D. and Wilson, D.F. (1983) Aspartate transport in synaptosomes from rat brain. *J. Biol. Chem.*, 258: 9069–9077.

Fairman, W.A., Vandenberg, R.J., Arriza, J.L. and Kavanaugh, M.P., Amara, S.G. (1995) An excitatory amino-acid transporter with properties of a ligand-gated chloride channel. *Nature*, 375: 599–603.

Friedman, J.E. and Haddad, G.G. (1994) Anoxia induces an increase in intracellular sodium in rat central neurons in vitro. *Brain Res.*, 663: 329–334.

Grant, G.B. and Dowling, J.E. (1995) A glutamate-activated chloride current in cone-driven ON bipolar cells of the white perch retina. *J. Neurosci.*, 15: 3852–3862.

Hagiwara, T., Tanaka, K., Takai, S., Maeno-Hikichi, Y., Mukainaka, Y. and Wada, K. (1996) Genomic organization, promoter analysis, and chromosomal localization of the gene for the mouse glial high-affinity glutamate transporter Slc1a3. *Genomics*, 33: 508–515.

Jiang, C., Agulian, S. and Haddad, G.G. (1992) $Cl^-$ and $Na^+$ homeostasis during anoxia in rat hypoglossal neurons: Intracellular and extracellular in vitro studies. *J. Physiol.*, 448: 697–708.

Kanai, Y., Nussberger, S., Romero, M.F., Boron, W.F., Hebert, S.C. and Hediger, M.A. (1995) Electrogenic properties of the epithelial and neuronal high affinity glutamate transporter. *J. Biol. Chem.*, 270: 16561–16568.

Kanner., BI. and Sharon, I. (1978) Active transport of L-glutamate by membrane vesicles isolated from rat brain. *Biochemistry*, 17: 3949–3953.

Kato, S., Ishita, S., Sugawara, K. and Mawatari, M. (1993) Cystine/glutamate antiporter expression in retinal Muller glial cells: Implications for DL-α-aminoadipate toxicity. *Neuroscience*, 57: 473–482.

Kavanaugh, M.P., Bendahan, A., Zerangue, N., Zhang, Y.M. and Kanner, B.I. (1997) Mutation of an amino acid residue influencing potassium coupling in the glutamate transporter GLT-1 induces obligate exchange. *J. Biol. Chem.*, 272: 1703–1708.

Kessler, M., Petersen, G., Vu, H.M., Baudry, M. and Lynch, G. (1987) L-phenylalanyl-L-glutamate-stimulated, chloride-dependent glutamate binding represents glutamate sequestration mediated by an exchange system. *Journal of Neurochemistry*, 48: 1191–1200.

Lehre, K.P., Levy, L.M., Ottersen, O.P., Storm-Mathisen, J. and Danbolt, N.C. (1995) Differential expression of two glial glutamate transporters in the brain: Quantitative and immunocytochemical observations. *J. Neurosci.*, 15: 1835–1853.

Longuemare, M.C. and Swanson, R.A. (1995) Excitatory amino acid release from astrocytes during energy failure by reversal of sodium-dependent uptake. *J. Neurosci. Res.*, 40: 379–385.

Madl, J.E. and Burgesser, K. (1993) Adenosine triphosphate depletion reverses sodium-dependent, neuronal uptake of glutamate in rat hippocampal slices. *J. Neurosci.*, 13: 4429–4444.

McBain, C.J., Traynelis, S.F. and Dingledine, R. (1990) Regional variation of extracellular space in the hippocampus. *Science*, 249: 674–677.

McLennan, H. (1976) The autoradiographic localization of [$^3$H] glutamate in rat brain. *Brain Res.*, 115: 139–144.

Murphy, T.H., Miyamoto, M., Sastre, A., Schnaar, R.L. and Coyle, J.T. (1989) Glutamate toxicity in a neuronal cell line involves inhibition of cystine transport leading to oxidative stress. *Neuron*, 2: 1547–1558.

Nelson, P.J., Dean, G.E., Aronson, P.S. and Rudnick, G. (1983) Hydrogen ion cotransport by the renal brush border glutamate transporter. *Biochemistry*, 22: 5459–5463.

Nicholson, C. and Phillips, J.M. (1981) Ion diffusion modified by tortuosity and volume fraction in the extracellular microenvironment of the rat cerebellum. *J. Physiol.*, 321: 225–257.

Oshima, T., Rossi, D. and Attwell, D. (1997) Non-vesicular release of glutamate in simulated ischaemia, sensed by whole-cell clamping CA1 pyramidal cells in rat hippocampal slices. *J. Physiol. Proceedings*, In press.

Peghini, P., Janzen, J. and Stoffel, W. (1997) Glutamate transporter EAAC-1 deficient mice develop dicarboxylic aminoaciduria and behavioral abnormalities but no neurodegeneration. *EMBO Journal*, 16: 3822–3832.

Picaud, SA., Larsson, H.P., Grant, G.B., Lecar, H. and Werblin, F.S. (1995) Glutamate-gated chloride channel with glutamate-transporter-like properties in cone photoreceptors of the tiger salamander. *J. Neurophysiol.*, 74: 1760–1771.

Pines, G., Danbolt, N.C., Bjoras, M., Zhang, Y., Bendahan, A., Eide, L., Koepsell, H., Storm-Mathisen, J., Seeberg, E. and Kanner, B.I. (1992) Cloning and expression of a rat brain L-glutamate transporter. *Nature*, 360: 464–467.

Roettger, V. and Lipton, P. (1996) Mechanism of glutamate release from rat hippocampal slices during in vitro ischemia. *Neuroscience*, 75: 677–685.

Rothstein, J.D., Dykes-Hoberg, M., Pardo, C.A., Bristol, L.A., Jin, L., Kuncl, R.W., Kanai, Y., Hediger, M.A., Wang, Y., Schielke, J. and Welty, D.F. (1996 ) Knockout of glutamate transporters reveals a major role for astroglial transport in excitotoxicity and clearance of glutamate. *Neuron*, 16: 675–686.

Rothstein, J.D., Martin, L., Dykes-Hoberg, M., Lin, L., Wu, D., Nash, N. and Kumcl, R.W. (1994) Localization of neuronal and glial glutamate transporters. *Neuron*, 13: 713–725.

Sarantis, M., Everett, K. and Attwell, D. (1988) A presynaptic action of glutamate at the cone output synapse. *Nature*, 332: 451–453.

Shibata, T., Watanabe, M., Tanaka, K., Wada, K. and Inoue, Y. (1996) Dynamic changes in expression of transporter mRNAs in developing brain. *NeuroReport*, 7: 705–709.

Siesjo, B.K. (1990) Calcium, excitotoxins and brain damage. *News Physiol. Sci.*, 5: 120–125.

Spiridon, M., Kamm, D., Billups, B., Mobbs, P. and Attwell, D. (1998) Modulation by zinc of the glutamate transporter in glial cells isolated from the tiger salamander retina. Submitted for publication.

Stallcup, W.B., Bulloch, K. and Baetge, E.E. (1979) Coupled transport of glutamate and sodium in a cerebellar nerve cell line. *J. Neurochem.*, 32: 57–65.

Storck, T., Schulte, S., Hofmann, K. and Stoffel, W. (1992) Structure, expression, and functional analysis of a $Na^+$-dependent glutamate/aspartate transporter from rat brain. *Proc. Natl. Acad. Sci, U.S.A.*, 89: 10955–10959.

Storm-Mathisen, J., Danbolt, N.C., Rothe, F., Torp, R., Zhang, N., Aas, J-E., Kanner, B.I., Langmoen, I. and Ottersen, O.P. (1992) Ultrastructural immunocytochemical observations on the localization, metabolism and transport of glutamate in normal and ischemic brain tissue. *Prog. Brain. Res.*, 94: 225–241.

Szatkowski, M. and Attwell, D. (1994) Triggering and execution of neuronal death in brain ischaemia: Two phases of glutamate release by different mechanisms. *Trends in Neurosci.*, 17: 359–365.

Szatkowski, M., Barbour, B. and Attwell, D. (1990) Non-vesicular release of glutamate from glial cells by reversed electrogenic glutamate uptake. *Nature*, 348: 443–446.

Takahashi, M., Billups, B., Rossi, D., Sarantis, M., Hamann, M. and Attwell, D. (1997) The role of glutamate transporters in glutamate homeostasis in the brain. *J. Exp. Biol.*, 200: 401–409.

Takahashi, M., Sarantis, M. and Attwell, D. (1996) Postsynaptic glutamate uptake in rat cerebellar Purkinje cells. *J. Physiol.*, 497: 523–530.

Tanaka, J., Ichikawa, R., Watanabe, M., Yanaka, K. and Inoue, Y. (1997) Extra-junctional localization of glutamate transporter EAAT4 at excitatory Purkinje cell synapses. *NeuroReport*, In Press.

Tanaka, K., Watase, K., Manabe, T., Yamada, K., Watanabe, M., Takahashi, K., Iwama, H., Nishikawa, T., Ichihara, N., Kikuchi, T., Okuyama, S., Kawashima, N., Hori, S., Takimoto, M. and Wada, K. (1997) Epilepsy and exacerbation of brain injury in mice lacking the glutamate transporter GLT-1. *Science*, 276: 1699–1702.

Temple, C.S., Bronk, J.R., Bailey, P.D. and Boyd, C.A.R. (1995) Substrate-charge dependence of stoichiometry shows membrane potential is the driving force for proton-peptide cotransport in rat renal cortex. *Pflugers Archiv*, 430: 825–829.

Volterra, A., Bezzi, P., Rizzini, B.L., Trotti, D., Ullensvang, K., Danbolt, N.C. and Racagni, G. (1996) The competitive transport inhibitor L-trans-pyrrolidine-2,4-dicarboxylate triggers excitotoxicity in rat cortical neuron-astrocyte co-cultures via glutamate release rather than uptake inhibition. *Eur. J. Neurosci.*, 8: 2019–2028.

Wadiche, J.I., Amara, S.G. and Kavanaugh, M.P. (1995a) Ion fluxes associated with excitatory amino acid transport. *Neuron*, 15: 721–728.

Wadiche, J.I., Arriza, J.L., Amara, S.G. and Kavanaugh, M.P. (1995b) Kinetics of a human glutamate transporter. *Neuron*, 14: 1019–1027.

Warr, O., Takahashi, M., Billups, B. and Attwell, D. (1997) Cystine-glutamate exchange and the regulation of extracellular glutamate concentration, $[glu]_o$, in rat brain slices. *J. Physiol. Proceedings*: In Press.

Yamada, K., Watanabe, M., Shibata, T., Tanaka, K., Wada, K. and Inoue, Y. (1996) EAAT4 is a post-synaptic glutamate transporter at Purkinje cell synapses. *NeuroReport*, 7: 2013–2017.

Zerangue, N. and Kavanaugh, M.P. (1996a) Flux coupling in a neuronal glutamate transporter. *Nature*, 383: 634–637

Zerangue, N. and Kavanaugh, M.P. (1996b) ASCT-1 is a neutral amino acid exchange with chloride channel activity. *J. Biol. Chem.*, 271: 27991–27994.

Zerangue, N. and Kavanaugh, M.P. (1996c) Interaction of L-cysteine with a human excitatory amino acid transporter. *J. Physiol.*, 493: 419–423.

O.P. Ottersen, I.A. Langmoen and L. Gjerstad (Eds.)
*Progress in Brain Research*, Vol 116


CHAPTER 5

# Control of time course of glutamatergic synaptic currents

Laurence Trussell

*Department of Physiology, University of Wisconsin, 1300 University Avenue, Madison WI 53706, USA*

## Introduction

The duration of the synaptic current is a key element in determining the overall function of glutamatergic synapses. For example, one-to-one transmission of signals at these synapses is carried out by large, brief excitatory synaptic currents (EPSC's). Such rapidly rising and decaying EPSC's are generally mediated by AMPA (α-amino-3-hydroxy-5-methyl-4-isoxazolepropionate) receptors (e.g., Dale and Roberts, 1985; Zhang and Trussell, 1994a). By contrast, EPSC's mediated by NMDA (*N*-methyl-D-aspartate) receptors are quite slow and either induce long bursts of spikes or bring the cell close to threshold for long periods of time, thereby enhancing excitability (e.g., Dale and Roberts, 1985; Hu and Bourque, 1992). The duration of the postsynaptic response is also of importance in determining the consequence of repetitive synaptic activity. Temporal summation, a fundamental component in the integration of synaptic signals, is prominent when excitatory postsynaptic potentials (EPSP's) are long enough to overlap. Conversely, precise neural coincidence detection requires the minimization of temporal summation, so that only near-simultaneous, narrow synaptic events are able drive a cell to action potential threshold.

In addition to mediating electrical computation, glutamatergic synaptic events drive long-term changes in the biochemical state of postsynaptic cells. NMDA receptors, and to some extent AMPA receptors, activate channels permeable to the second messenger $Ca^{2+}$ (Burnashev, 1996). The number of $Ca^{2+}$ ions injected into the postsynaptic neuron will increase with the amplitude or duration of the synaptic current and so more effectively initiate events underlying intermediate or long-term synaptic plasticity (Teyler et al., 1994).

Studies in the last eight years have identified many of the major parameters that influence the duration of the synaptic current. These include synaptic geometry, receptor binding affinity and channel gating kinetics, glutamate transporters, and quantal release kinetics (for related reviews, see Jonas and Spruston, 1994; Edmonds et al., 1995; Trussell and Otis, 1996). Additional and important factors shaping the EPSP are the postsynaptic membrane properties, such as the electrotonic structure of the dendrite and its voltage-sensitive conductances. However, this review will focus exclusively on factors influencing the synaptic current itself. Remarkably, the extent to which any one factor is utilized to shape the EPSC varies widely at different synapses. Thus, these factors may seen as variables which sculpt the profile of EPSP's and in this way enable diverse modes of information transfer through different neuronal circuits. Moreover, the extent to which a given type of synapse is unique in its particular combination of receptors, glutamate transporters, and so forth, may dictate how sensitive these synapses are to pharmacological interventions and how broad are the side-effects of those treatments. Indeed, inasmuch as glutamate is the most ubiquitous of CNS transmitters, development of selectively acting drugs, i.e., ones which act only on subtypes of glutamatergic synapse, may require a complete understanding of all the key players which combine to shape the EPSC.

### NMDA receptor EPSC's

The conceptual basis of many recent studies of the physiology and biophysics of glutamatergic transmission may be traced to work on the neuromuscular junction and the acetylcholine receptor. One lesson from these early studies is the importance of receptor/channel kinetics in determining the time course of the EPSC. In particular, since the lifetime of acetylcholine in the neuromuscular synaptic cleft is shorter than the channel burst duration of the acetylcholine receptors, the latter determine the duration of the postsynaptic current (Magleby and Stevens, 1972; Anderson and Stevens, 1973). This case is clearly reproduced at CNS glutamatergic synapses mediated by NMDA receptors. In two classic studies, Lester et al. (1990) and Lester and Jahr (1992) showed that brief ($< 5$ ms) application of glutamate to an outside-out membrane patch containing NMDA receptors produced a current response that closely mimicked the time course of the NMDA receptor-mediated EPSC. Because the glutamate had been removed long before the response ended, the time course of the response must have reflected the intrinsic kinetics of the receptors, rather than the duration of exposure to transmitter.

Later work confirmed that indeed the time course of glutamate at cultured hippocampal synapses is generally far briefer than NMDA receptor decay kinetics (Clements et al., 1992; Clements, 1996). These conclusions by Clements and colleagues derived from experiments in which weak competitive antagonists were used to set up a dynamic competitive interaction with synaptically-released glutamate; because the transmitter is never in equilibrium with the receptors, the outcome of this competition is acutely sensitive to the lifetime of the transmitter. The fleeting nature of the transmitter was also demonstrated by simple models of diffusion in cleft-like structures (Eccles and Jaeger, 1958; Holmes, 1995, Otis et al., 1996a, Clements, 1996). To summarize, analyses of channel kinetics and transmitter diffusion indicate that the decay phase of the patch response/EPSC can be accounted for primarily by several processes completely intrinsic to the receptor: 1) the channel opening and closing rates, 2) the rate of desensitization, and most importantly, 3) the unbinding rate of glutamate (Patneau and Mayer, 1990; Clements and Westbrook, 1991; Gibb and Colquhoun, 1991, 1992; Lester and Jahr, 1992).

Some controversy has arisen regarding certain details of this model, in particular whether the slow NMDA receptor current decay is purely an outcome of gradual glutamate unbinding or is also a reflection of a long delay before a bound channel first opens (Edmonds and Colquhoun, 1992; Jahr, 1992). Dzubay and Jahr (1996) addressed this problem by taking advantage of the rapid channel blocking (and slow unblocking) action of the non-competitive antagonist MK-801. They showed that most channels could be blocked by a brief exposure to MK-801 soon after glutamate binding. Since only open, non-desensitized channels can be blocked, these data support the notion that most channels open within 10 ms, and thus the decay of the current response must reflect the gradual reduction in the number of open channels. It should be noted, though, that the time needed for all the channels to open, again about 10 ms, is still long when compared to other types of synaptic current (see below); thus, the latency to opening plays a key role in determining the characteristically sluggish rise time of the NMDA receptor EPSC's.

As noted, other factors besides channel opening and ligand unbinding rates contribute to shaping the EPSC. Lester and Jahr (1992) showed that brief pulses of glutamate to patches induce a partial desensitization of the receptors, as determined by repeatedly testing the sensitivity of the patch during the falling phase of the inital conditioning response. Assuming receptors in patches behave like synaptic receptors (see Lester and Jahr, 1992; Lester et al., 1993; Clark et al., 1997), one would expect some desensitization to occur as the EPSC decays. However, it is difficult to separate clearly those contributions to the decay which represent unbinding and those that represent desensitization. Glycine is a necessary co-agonist of the NMDA receptor, which is normally at concentrations adequate to permit receptor activation (McBain and Mayer, 1994). In later work, Lester and colleagues examined the effect of different agonists of the glycine binding site of the NMDA receptor on the decay of the EPSC, taking advantage of the observation that one form of NMDA receptor desensitization is dependent on the affinity of the agonist (Lester et al., 1993). Both EPSC's and desensitization were faster when low-affinity L-alanine replaced glycine, despite maintained glutamate binding to the receptor. When glycine binding is reduced by an antagonist HA-966, EPSC's also decayed faster, indicating that glycine-dependent desensitization could shape the EPSC decay. While the glycine site on NMDA receptors is a site of modulation in vitro (Wilcox et al., 1996), it remains unclear the extent to which this will occur in vivo, as physiological levels glycine may nearly saturate its binding site. Other, glycine-resistant forms of desensitization are also part of the complex repertoire of the NMDA receptor (see McBain and Mayer, 1994); their participation in shaping the EPSC decay has yet to be determined.

The decay of NMDA receptor EPSC's has been shown to vary with brain region and with development (reviewed by Hestrin, 1996). For example, at early postnatal ages, EPSCs are significantly slower than at later ages; more prolonged currents may serve to enhance the chances of temporal coincidence of synaptic activity, which plays a major role in refining circuitry (Carmignoto and Vicini, 1992; Hestrin, 1992; Farrant et al., 1994; Ramoa and McCormick, 1994; Crair and Malenka, 1995). As these changes are paralleled by changes in the kinetics of the receptors which underlie the EPSC (Hestrin, 1992; Carmignoto and Vicini, 1992), developmental changes in receptor structure, rather than structural changes at the level of the whole synapse, may be key to dynamic control of EPSC duration. Indeed, reciprocal developmental changes in the expression of the NR2A and NR2B subunits have been well documented (Williams, et al., 1993; Monyer et al., 1994), with NR2B being dominant at earlier ages. When each of these subunits is co-expressed with the NR1 subunit, the NR2B-containing receptors have slower channel kinetics than those with NR2A (Monyer et al., 1994).

Finally, conditions that slow the clearance of transmitter have been observed to prolong slightly NMDA receptor EPSC's, an effect which might be significant in terms of changes in free intracellular calcium. Mennerick and Zorumski (1994) observed that block of glutamate uptake lengthened NMDA receptor EPSC's in cultured hippocampal neurons. Moreover, at the granule cell/CA3 pyramidal cell synapse (mossy fiber synapse) of the hippocampus, repetitive presynaptic stimulation produces a large potentiation of release which is associated with a small prolongation of the NMDA receptor EPSC (Scanziani et al., 1997). Thus, factors related to synaptic geometry and glial apposition, density of release sites, and rate of transporter activity may contribute to shaping the EPSC.

## AMPA receptor EPSC's

### *Kinetics and the single quantum*

Following rapid application and removal of glutamate, the current produced by receptors in excised membrane patches decays with a characteristic time course known as "deactivation". Application of a prolonged glutamate step results in a more gradual current decay termed "desensitization". As seen above, the decay of the NMDA receptor EPSC is dominated by deactivation. Which of these kinetic components figures more prominently in the AMPA receptor EPSC has been an outstanding issue, in part because the decay phase of rapid EPSC's is sensitive to factors affecting the clearance of glutamate, the time course of glutamate release, as well as filtering along dendrites, which confounds the analysis. Many studies indicate that the decay phase of the quantal or miniature EPSC (mEPSC) is largely dependent on the intrinsic deactivation kinetics of the AMPA receptor, simply because mEPSCs, when adequately voltage-clamped, decay with a time constant nearly identical to that seen during deactivation of channels activated by AMPA receptors, recorded in the same cell types (Colquhoun et al., 1992; Hestrin, 1992; Otis et al., 1996a; Silver et al., 1996a), as seen in Fig. 1 and Fig. 2A. While it is possible to interpret slower mEPSC's as decaying according to transmitter

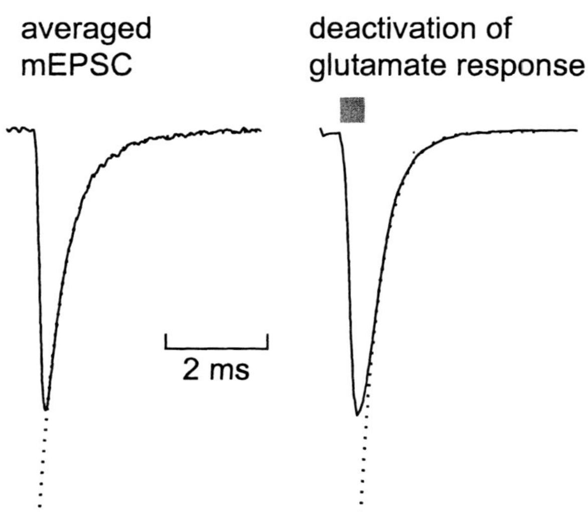


Fig. 1. The time course of the mEPSC's matches the deactivation time course of the AMPA receptor. Shown on the left is an average of 35 mEPSC's recorded from a voltage clamped neuron in the chick nucleus magnocellularis. The dashed line is a biexponential curve fitted to the decay phase having a dominant component of 0.5 ms. On the right is the response of a patch excised from a different nucleus magnocellularis neuron to rapid application of 10 mM glutamate during the time marked by the grey bar. The exponential fit to the decay of this current had a dominant time constant of 0.45 ms. The amplitudes of the waveforms have been normalized. Both recordings were made at a holding potential of −70 mV at room temperature.

diffusion rates or receptor desensitization kinetics (e.g., Trussell and Fischbach, 1989; Livsey et al., 1993), unless cable distortion of mEPSC's can be rigorously excluded, there is little reason to suppose glutamate can significantly *re*bind and *re*activate receptors following release of a single quantum (although a small amount of rebinding might occur; Silver et al., 1996a). Again, since receptor kinetics are a dominant factor in shaping the mEPSC, an implication of the short duration of the mEPSC is that the lifetime of the transmitter in the cleft is extremely short. For example, at somatic synapses of the cochlear nucleus, mEPSC's rise in under 100 μs and and decay in 180 μs at 30°C (Zhang and Trussell, 1994b). Thus, transmitter concentration must fall significantly below $K_d$ for activation of AMPA receptors within 100–200 μs following its release.

The link between channel deactivation kinetics and EPSC time course has an important ramification in the fact that the biophysical properties of AMPA receptors that determine their kinetics are under genetic control (see Borges and Dingledine, this volume), thus leading to a molecular mechanism for the varied kinetics of AMPA receptor EPSC's throughout the CNS. The diverse expression of AMPA receptor genes, and their processing, throughout the CNS may explain part of the heterogeneity in synaptic kinetics seen at different synapses. For example, Fig. 2A shows a comparison of mEPSC decays and AMPA receptor deactivation rates for several cell types in the CNS. There is a clear correlation between these parameters, although the relationship has a slope of less than one, presumably because of dendritic filtering of EPSC's. While other aspects of AMPA receptor kinetics, such as desensitization, may also contribute to the shape of the EPSC (see below), they are regulated in parallel with deactivation, as indicated in Fig. 2B. Thus, the channel gating kinetics of the AMPA receptor are among the most important variables in producing "fast" synapses and "slow" synapses. Although the molecular underpinnings of this kinetic variation is discussed in other chapters in this volume (see chapters by Wenthold and Roche and by Borges and Dingledine), it is worth noting here some of the close matches between the kinetics of native receptors and those of AMPA receptor subunits known to be expressed in the same neurons (Fig. 2B).

*Release time course*

Again, drawing on studies from the neuromuscular junction, we know that vesicles at different release sites do not all fuse with the plasma membrane at the same time. Instead, there is a measurable "release time course" which smears the overall synaptic wave form (Barrett and Stevens, 1972; Isaacson and Walmsley, 1995), as illustrated in Fig. 3A. This asynchrony of release most likely arises from the probabilistic nature of the release process, as well as the non-simultaneous arrival of action potentials at different terminal branches. At AMPA synapses, evoked EPSC's are generally longer-lasting than mEPSC's (Diamond and Jahr, 1995; Isaacson and Walmsley, 1995; Mennerick and Zorumski, 1995; Borst and Sakmann, 1996; Otis et al., 1996a; Otis and Trussell, 1996; Taschenberger and Grantyn, 1995; Geiger et al., 1997), which may be an indication of this "smearing" effect of the release time course. In cases in which the duration of the EPSC is independent of release probability, a release time course can be extracted either by measuring the first latency of release of single quanta or by deconvolution of the evoked and quantal EPSC wave forms (Isaacson and Walmsley, 1995; Diamond and Jahr, 1995; Borst and Sakmann, 1996; Geiger et al., 1997). The release function so obtained decays with a time constant from 100's of microseconds (Isaacson and Walmsley, 1995; Borst and Sakmann, 1996; Geiger et al., 1997) to milliseconds (Diamond and Jahr, 1995), and explains well the difference between the decay time of the mEPSC and EPSC at these synapses. At synapses in which the kinetics of the quantal response is far slower than the release time course, the decay of the single and multiquantal responses might be expected to be more similar, as shown in Fig. 3B. Such is probably the case with NMDA receptor- (see above) and GABA receptor- (e.g., Geiger et al., 1997) mediated currents.

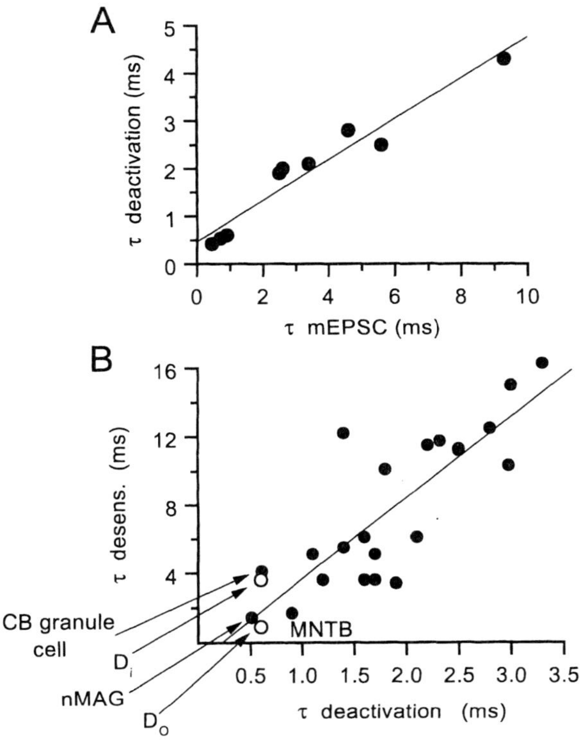


Fig. 2. Relation between different kinetic components of AMPA receptor and synaptic kinetics. **A.** The exponential decay time of mEPSCs varies with the channel deactivation time constant in different cell types. **B.** Correlation between exponential time constants for deactivation and desensitization in different cell types (filled circles). Open circles show decay kinetics for D-flop ($D_o$) and D-flip ($D_i$) subunits expressed in HEK 293 cells. These are similar to those of auditory neurons (nMAG and MNTB) and of cerebellar granule cells, respectively. A similar correlation has been previously described by Otis and Trussell, 1996 and Partin et al., 1996. Data for these plots taken from the following sources: Colquhoun et al., 1992; Hestrin, 1993; Partin et al., 1993; Raman et al., 1994; Geiger et al., 1995; Raman and Trussell, 1995b; Otis et al., 1996a; Partin et al., 1996; Silver et al., 1996a; Götz et al., 1997.

*Delayed clearance of transmitter*

At other synapses, however, the decay of EPSCs may be significantly influenced by delayed clearance of transmitter. This is particularly true at synapses in which multiple, nearby release sites share a contiguous synaptic cleft or where synaptic structures somehow restrict diffusion. Here, reduction of transmitter release by lowering extracellular $Ca^{2+}$ or application of $Cd^{2+}$, baclofen, or adenosine may narrow the EPSC (Trussell et al., 1993; Mermerick and Zorumski, 1995; Takahashi et al., 1995; Otis and Trussell, 1996; Silver et al., 1996b), a result originally seen in esterase-blocked neuromuscular junctions (Hartzell et al., 1975; Magelby and Terrar, 1975). The broadening of the EPSC under high-release conditions at synapses appears to result from the pooling of transmitter released from adjacent active sites (Trussell et al., 1993; Rossi et al., 1995; Otis et al., 1996a; Silver et al., 1996b; Kinney et al., 1997; see review by Barbour and Häusser, 1997). Delayed clearance has been proposed at spine synapses in cerebellar Purkinje cells (Barbour et al., 1994; Takahashi et al., 1995), in which transmitter reaches a "bottle-neck" by diffusing down around the base of a spine before leaving the synaptic region, and at mature hippocampal cultures in which glial/neural associations may limit diffusion (Mennerick and Zorumski, 1995).

However, even among the largest, multi-release site synapses, the extent of slowed clearance varies widely. At the mossy fiber-unipolar brush cell synapse, the EPSC has two components, an initial, fast phase and a second slow phase (Rossi et al., 1995). The second phase, which decays over several seconds, is most likely a consequence of the enormous mossy fiber terminal and its high

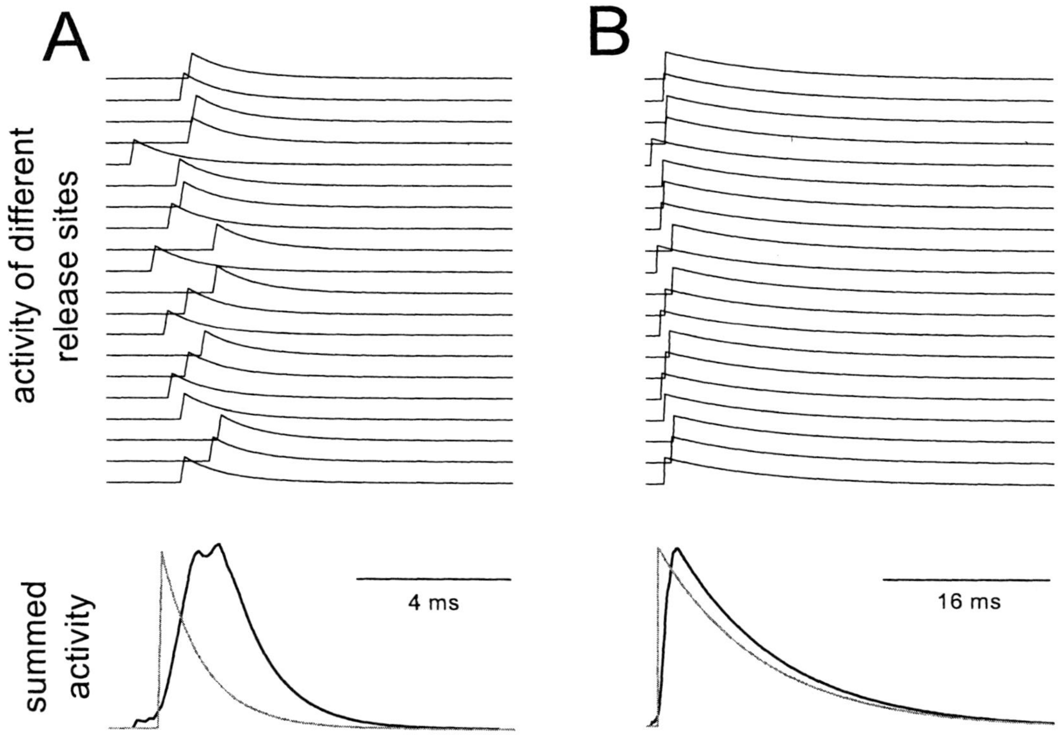


Fig. 3. The effects of the time course of quantal release a different synaptic sites on the time course of the postsynaptic response. Shown are simulations in which single quantal currents are modeled as rising instantaneously but decaying with a time constant of either 1 ms (A) or 12 ms (B). In A and B, the upper traces show individual quanta occurring with variable latencies, determined by random sampling from a Gaussian distribution having a standard deviation of 0.5 ms. The lower traces show a summation of 100 of the individual trials. Superimposed on these in grey is a representative quantal current. Note the more dramatic effects of the release period on the shape of the compound event when the quantal current is brief.

density of release sites (Rossi et al., 1995; Kinney et al., 1997). A biphasic EPSC was also observed in chick cochlear nuclear end-bulb synapses (Otis et al., 1996a). In both of these cases, and particularly when release levels are high and uptake is reduced, the second, low phase of the EPSC does not decay exponentially, but rather shows either a plateau (Otis et al., 1996a) or a torpid, secondary rising phase (Rossi et al., 1995). The explanation for the plateau and the second rising phase hinges on a peculiarity of the equilibrium concentration-response curve for the postsynaptic AMPA receptor, in which glutamate-evoked current decreases with increase in glutamate above about 50 $\mu$M (Raman and Trussell, 1992; Kinney et al., 1997). The consequence of this biphasic relation is a resurgence in postsynaptic current as transmitter levels fall over a range of concentrations between 1000 and 50 $\mu$M (Rossi et al., 1995; Otis et al., 1996a; Kinney et al., 1997).

What physiological function could the secondary phase serve? Probably in end-bulb and mossy fiber/unipolar brush cell synapses they serve a role as supplementary to NMDA receptor EPSCs, permitting a steady depolarization to occur following a single presynaptic action potential, although without the voltage-sensitivity characteristic of NMDA receptors. In the unipolar brush cell, the slow phase induces a flurry of action potentials, and thus amplifies the action of low frequency action potential activity in single mossy fibers (Rossi et al., 1995). We proposed that, at the auditory synapse, the plateau potential produced by the slow EPSC does not by itself induce action potentials, but causes the activation of voltage-sensitive $K^+$ currents which shorten the membrane time constant. The shorter time constant would then speed EPSP's produced during the early phases of closely spaced synaptic responses (Zhang and Trussell, 1994a).

By contrast, at other large auditory synapses the clearance of transmitter is apparently more rapid, and there is no change in EPSC time course upon reduction of release (Isaacson and Walmsley, 1995; Borst and Sakmann, 1996). The most probable explanation for these differences among the large synapses is that, in the latter cases, the density of active release sites is not as high as might be supposed from the size of the synapse. Lower release intensity could result from either a low density of sites, or from a lower release probability at each site.

*Desensitization at AMPA receptor synapses*

Given the broadening of EPSC at synapses with restricted diffusion it is likely that the time course of the EPSC is determined by the diffusion time of glutamate or, if clearance is gradual enough, the desensitization rate of receptors. Which process dominates? The best answer is that both probably contribute. Few EPSC's decay with a time course similar to that of responses to a step in glutamate concentration (Trussell and Fischbach, 1989; Barbour et al., 1994; Otis et al., 1996a). Nevertheless, the onset of desensitization is sufficiently rapid at some synapses that it is hard to argue that *no* desensitization occurs during the decay of the EPSC.

Another approach to the issue is to ask, apart from a role in shaping the EPSC, does *any* synaptically-induced desensitization occur at AMPA receptor synapses? Studies of the responsiveness of AMPA receptors to pairs of extremely brief, exogenous pulses of glutamate indicates that desensitization will always occur to some extent, even if glutamate is present only for brief periods at the receptor (Trussell and Fischbach, 1989; Hestrin, 1992; Colquhoun et al., 1992; Raman and Trussell, 1995a). Thus, some reduction in sensitivity might generally be expected following EPSCs, although this conclusion is tempered by the absence of postsynaptic paired-pulse depression at putatively single release site synapses (Stevens and Wang, 1995). Drugs which slow or block desensitization also slow EPSC's (Isaacson and Nicoll, 1991; Tang et al., 1991; Vycklicky et al., 1991; Thio et al., 1993; Trussell et al., 1993; Yamada and Tang, 1993; Barbour et al., 1994; Takahashi et al., 1995) which might suggest a role for desensitization in shaping the EPSC. However these studies are confounded by complex and diverse actions of these modulators (e.g., Diamond and Jahr, 1995); thus, without more complete analysis of the effect, drug-induced slowing of an EPSC is nearly an uninterpretable result.

However, desensitization blockers also reduce synaptic depression at calyceal synapses (Trussell et al., 1993; Isaacson and Walmsley, 1996), a result which is more easily interpreted in terms of a reduction in desensitization. Desensitization can be more directly assayed by measuring glutamate sensitivity immediately after an EPSC. Several assays of sensitivity have been used to accomplish this, including measuring the response to rapid uncaging of a caged form of glutamate, the size of steady kainate responses evoked during the EPSC, and the amplitude of miniature synaptic currents just after the EPSC (Otis et al., 1996b). Single EPSC's at large end-bulb synapses were observed to induce 35–40% reduction in the sensitivity of AMPA receptors (Otis et al., 1996b). Whether such desensitization also occurs at other types of glutamatergic synapses remains to be tested.

*Glutamate transporters and the shape of EPSC's*

The role of glutamate uptake in the control of EPSC size and shape has yet to be fully understood and its role will likely vary widely at different synapses (see discussion by Billups et al., in this volume). The relevant parameters here are manifold: What is the density, affinity, and kinetics of transporters? Where are they in relation to release sites? What diffusion barriers are present? Block of transporter function broadens EPSC's at some synapses (Barbour et al., 1994; Takahashi et al., 1995; Mennerick and Zorumski, 1994, 1995; Otis et al., 1996a; Kinney et al., 1997), indicating that transporters can play a key role in clearing transmitter and limiting the EPSC duration. However, among the examples cited, some illustrate slowing of the entire EPSC wave form (Barbour et al., 1994) while others show only slowing of the smaller, later phases of the EPSC (Otis et al., 1996a; Kinney et al., 1997), highlighting the difficulty that exists in generalizing about the effects of transporters. Indeed, at some synapses there is virtually no detectable effect of transporter inhibition on the EPSC time course (Isaacson and Nicoll, 1993; Sarantis et al., 1993). Jahr and colleagues have emphasized the action of transporters at controlling the amplitude of EPSCs, which is interpreted to be a consequence of rapid buffering of glutamate by a high, local density of binding sites provided by transporters (Tong and Jahr, 1994; Diamond and Jahr, 1997). Recent quantitative analysis of uptake at climbing fiber-Purkinje cells synapses indicates that over 20% of released glutamate is rapidly taken up by the postsynaptic Purkinje cells (Otis et al., 1997). Kinetics of the transporter-induced current show that uptake is quite rapid, with an onset comparable to that of the postsynaptic AMPA receptor current. Measurements of glutamate uptake activity in adjacent Bergmann glial cells also indicates a rapid onset of transporter activity (Clark and Barbour, 1997). Presumably, this early influence on glutamate concentration could act physiologically to control response size even when clearance is rapid enough that receptor deactivation kinetics, diffusion/desensitization or release time course dominate in determining the EPSC decay time.

## Conclusions

Synaptic transmission mediated by glutamate and glutamate receptors draws upon a wide variety of factors to determine the duration of the synaptic current. Because these factors can be varied independently, glutamate synapses can, with appropriate parallel variation in postsynaptic cable properties, be "specialized" for integration, coincidence detection, or the passive relay of signals. Ironically, the availability of powerful electrophysiological and pharmacological tools has left us in an awkward position in that our knowledge of the control of the synaptic waveform has now far outpaced our understanding of what the synapses actually do. Without a clearer picture of the role of particular synapses in information processing, the details of synaptic specializations will remain phenomenological. We can hope, though, that these synaptic "details" may themselves give insight into the larger function of neural circuits, and that the behavior of circuits, and how signals are transferred through them, will be seen as critically dependent on the character of single synapses. Accordingly, any wholistic evaluation of therapeutic approaches that are targeted to specific molecular components of the glutamatergic synapse will inevitably have to account for drug actions on a synapse-specific basis. It may be possible to turn this requirement around and use it to some advantage: knowing the relative importance of different components of specific synapses may aid in identifying which should be down- or up-regulated to achieve normalized function.

## Acknowledgements

The data in Figure 1 were obtained by Drs. I. Raman and T. Otis. I thank my colleagues S. Brenowitz, J. David, J. Lawrence, T. Lu, and I. Raman for comments on the manuscript. Work in my laboratory is supported by NIH grants NS28901 and DC02004.

## References

Anderson, C.R. and Stevens, C.F. (1973) Voltage clamp analysis of acetylcholine produced end-plate current fluctuations at frog neuromuscular junction. *J. Physiol.*, 235: 655–691.

Barbour, B. and Häusser, M. (1997) Intersynaptic diffusion of transmitter. *Trends Neurosci.*, 20: 377–384.

Barbour, B., Keller, B.U., Llano, I. and Marty, A. (1994) Prolonged presence of glutamate during excitatory synaptic transmission to cerebellar Purkinje cells. *Neuron*, 12: 1331–1343.

Barrett, E.F. and Stevens, C.F. (1972) The kinetics of transmitter release at the frog neuromuscular junction. *J. Physiol.*, 227: 691–708.

Borst, J.G. and Sakmann, B. (1996) Calcium influx and transmitter release in a fast CNS synapse. *Nature*, 383: 431–434.

Burnashev, N. (1996) Calcium permeability of glutamate-gated channels in the central nervous system. *Curr. Op. Neurobiol.*, 6: 311–317.

Carmignoto, G. and Vicini, S. (1992) Activity-dependent decrease in NMDA receptor responses during development of the visual cortex. *Science*, 258: 1007–1011.

Clark, B.A. and Barbour, B. (1997) Currents evoked in Bergmann glial cells by parallel fibre stimulation in rat cerebellar slices. *J. Physiol.*, 502: 335–350.

Clark, B.A., Farrant, M. and Cull-Candy, S.G. (1997) A direct comparison of the single-channel properties of synaptic and extrasynaptic NMDA receptors. *J. Neurosci.*, 17: 107–116.

Clements, J.D. (1996) Transmitter timecourse in the synaptic cleft: its role in central synaptic function. *Trends Neurosci.*, 19: 163–171.

Clements, J.D., Lester, R.A.J., Tong, G., Jahr, C.E. and Westbrook, G.L. (1992) The time course of glutamate in the synaptic cleft. *Science*, 258: 1498–1501.

Clements, J.D. and Westbrook, G.L. (1991) Activation kinetics reveal the number of glutamate and glycine binding sites on the *N*-methyl-D-aspartate receptor. *Neuron*, 7: 605–613.

Colquhoun, D., Jonas, P. and Sakmann B. (1992) Action of brief pulses of glutamate on AMPA/kainate receptors in patches from different neurones of rat hippocampal slices. *J., Physiol.*, 458: 261–287.

Crair, M.C. and Malenka, R.C. (1995) A critical period for long-term potentiation at thalamocortical synapses. *Nature*, 375: 325–328.

Dale, N. and Roberts, A. (1985) Dual-component amino-acid-mediated synaptic potentials: Excitatory drive for swimming in Xenopus embryos. *J. Physiol.*, 363: 35–59.

Diamond, J.S. and Jahr, C.E. (1995) Asynchronous release of synaptic vesicles determines the time course of the AMPA receptor-mediated EPSC. *Neuron*, 15: 1097–1107.

Diamond, J.S. and Jahr, C.E. (1997) Transporters buffer synaptically released glutamate on a submillisecond time scale. *J. Neurosci.*, 17: 4672–4687.

Dzubay, J.A. and Jahr, C.E. (1996) Kinetics of NMDA channel opening. *J. Neurosci.*, 16: 4129–4134.

Eccles, J.C. and Jaeger, J.C. (1958) The relationship between the mode of operation and the dimensions of the junctional regions at synapses and motor end-organs. *Proc. R. Soc. Lond., B.*, 148: 38–56.

Edmonds, B. and Colquhoun, D. (1992) Rapid decay of averaged single-channel NMDA receptor activations recorded at low agonist concentration. *Proc. Royal Soc. London, Series B*, 250: 279–286.

Edmonds, B., Gibb, A.J. and Colquhoun, D. (1995) Mechanisms of activation of glutamate receptors and the time course of excitatory synaptic currents. *Annu. Rev. Physiol.*, 57: 495–519.

Farrant, M., Feldmeyer, D., Takahashi, T. and Cull-Candy, S.G. (1994) NMDA-receptor channel diversity in the developing cerebellum. *Nature*, 368: 335–339.

Geiger, J.R.P., Lubke, J., Roth, A., Frotcher, M. and Jonas, P. (1997) Submillisecond AMPA-receptor-mediated signaling at a principal neuron-interneuron synapse. *Neuron*, 18: 1009–1023.

Geiger, J.R.P., Melcher, T., Koh, D.S., Sakmann, B., Seeburg, P.H., Jonas, P. and Monyer, H. (1995) Relative abundance of subunit mRNAs determines gating and $Ca^{2+}$ permeability of AMPA receptors in principal cells and interneurons in rat CNS. *Neuron*, 15: 193–204.

Gibb, A.J. and Colquhoun, D. (1991) Glutamate activation of a single NMDA receptor-channel produces a cluster of channel openings. *Proc. Royal Soc. London B*, 243: 39–45.

Gibb, A.J. and Colquhoun, D. (1992) Activation of *N*-methyl-D-aspartate receptors by L-glutamate in cells dissociated from adult rat hippocampus. *J. Physiol.*, 456: 143–179.

Götz, T., Kraushaar, U., Geiger, J., Lübke, J., Berger, T. and Jonas, P. (1997) Functional properties of AMPA and NMDA receptors expressed in identified types of basal ganglia neurons. *J. Neurosci.*, 17: 204–215.

Hartzell, H.C., Kuffler, S.W. and Yoshikami, D. (1975) Post-synaptic potentiation: Interaction between quanta of acetylcholine at the skeletal neuromuscular synapse. *J. Physiol.*, 251: 427–463.

Hestrin, S. (1992) Activation and desensitization of glutamate-activated channels mediating fast excitatory synaptic currents in the visual cortex. *Neuron*, 9: 991–999.

Hestrin, S. (1993) Different glutamate receptor channels mediate fast excitatory synaptic currents in inhibitory and excitatory cortical neurons. *Neuron*, 11: 1083–1091.

Hestrin, S. (1996) Physiology of NMDA receptors and synaptic currents. In F. Conti and T.P. Hicks (Eds.), *Excitatory Aminos Acids and the Cerebral Cortex*, MIT Press, Cambridge, pp. 53–62.

Holmes, W.R. (1995) Modeling the effect of glutamate diffusion and uptake on NMDA and non-NMDA receptor saturation. *Biophys. J.*, 69: 1734–1747.

Hu, B. and Bourque, C.W. (1992) NMDA receptor-mediated rhythmic bursting activity in rat supraoptic nucleus neurones in vitro. *J. Physiol.*, 458: 667–687.

Isaacson, J.S. and Nicoll, R.A. (1991) Aniracetam reduces glutamate receptor desensitization and slows the decay of fast excitatory synaptic currents in the hippocampus. *Proc. Nat'l. Acad. Sci. USA*, 88: 10396–10940.

Isaacson, J.S. and Nicoll, R.A. (1993) The uptake inhibitor L-trans-PDC enhances responses to glutamate but fails to alter the kinetics of excitatory synaptic currents in the hippocampus. *J. Neurophysiol.*, 70: 2187–2191.

Isaacson, J.S. and Walmsley, B. (1995) Counting quanta: direct measurements of transmitter release at a central synapse. *Neuron*, 15: 875–84

Isaacson, J.S. and Walmsley, B. (1996) Amplitude and time course of spontaneous and evoked excitatory postsynaptic current in bushy cells of the anteroventral cochlear nucleus. *J. Neurophysiol.*, 76: 1566–1571

Jahr, C.E. (1992) High probability opening of NMDA receptor channels by L-glutamate. *Science*, 255: 470–472.

Jonas, P. and Spruston, N. (1994) Mechanisms shaping glutamate-mediated excitatory postsynaptic currents in the CNS. *Curr. Op. Neurobiol.*, 4: 366–372.

Kinney, G.A., Overstreet, L.S. and Slater, N.T. (1997) Prolonged physiological entrapment of glutamate in the synaptic cleft of cerebellar unipolar brush cells. *J. Neurophysiol.*, 78: 1320–1333.

Lester, R.A. and Jahr, C.E. (1992) NMDA channel behavior depends on agonist affinity. *J. Neurosci.*, 12: 635–643.

Lester, R.A., Tong, G. and Jahr, C.E. (1993) Interactions between the glycine and glutamate binding sites of the NMDA receptor. *J. Neurosci.*, 13: 1088–1096.

Lester, R.A., Clements, J.D., Westbrook, G.L. and Jahr, C.E. (1990) Channel kinetics determine the time course of NMDA receptor-mediated synaptic currents. *Nature*, 346: 565–567.

Livsey, C.T., Costa, E. and Vicini, S. (1993) Glutamate-activated currents in outside-out patches from spiny versus aspiny hilar neurons of rat hippocampal slices. *J. Neurosci.*, 13: 5324–5333.

Magleby, K.L. and Stevens, C.F. (1972) A quantitative description of end-plate currents. *J. Physiol.*, 223: 173–97.

Magleby, K.L. and Stevens, C.F. (1972) The effect of voltage on the time course of end-plate currents. *J. Physiol.*, 223: 151–171.

Magleby, K.L. and Terrar, D.A. (1975) Factors affecting the time course of decay of end-plate currents: A possible cooperative action of acetylcholine on receptors at frog neuromuscular junction. *J. Physiol.*, 244: 467–495.

McBain, C.J. and Mayer, M.L. (1994) *N*-methyl-D-aspartic acid receptor structure and function. *Physiol. Rev.*, 74: 723–760

Mennerick, S. and Zorumski, C.F. (1994) Glial contributions to excitatory neurotransmission in cultured hippocampal cells. *Nature*, 355: 50–55.

Mennerick, S. and Zorumski, C.F. (1995) Presynaptic influence on the time course of fast excitatory synaptic currents in cultured hippocampal cells. *J. Neurosci.*, 15: 3178–3192.

Monyer, H., Burnashev, N., Laurie, D.J., Salmann, B. and Seeburg, P.H. (1994) Developmental and regional expression in the rat brain and functional properties of four NMDA receptors subtypes. *Neuron*, 12: 529–540.

Otis, T.S. and Trussell, L.O. (1996) Inhibition of transmitter release shortens the duration of the excitatory synaptic current at a calyceal synapse. *J. Neurophysiol.*, 76: 3584–3588.

Otis, T.S., Kavanaugh, M.P. and Jahr, C.E. (1997) Postsynaptic glutamate transport at the climbing fiber-Purkinje cell synapse. *Science*, 277: 1515–1518.

Otis, T.S., Wu, Y.C. and Trussell, L.O. (1996a) Delayed clearance of transmitter and the role of glutamate uptake at synapses with multiple release sites. *J. Neurosci.*, 16: 1633–1643.

Otis, T.S., Zhang, S. and Trussell., L.O. (1996b) Direct measurement of AMPA receptor desensitization induced by glutamatergic synaptic transmission. *J. Neurosci.*, 16: 7496–7504.

Partin, K.M., Fleck, M.W. and Mayer, M.L. (1996) AMPA receptor flip/flop mutants affecting deactivation, desensitization, and modulation by cyclothiazide, aniracetam, and thiocyanate. *J. Neurosci.*, 16: 6634–6647.

Patneau, D.K. and Mayer, M.L. (1990) Structure-activity relationships for amino acid transmitter candidates acting at *N*-methyl-D-aspartate and quisqualate receptors. *J. Neurosci.*, 10: 2385–2399.

Raman, I.M. and Trussell, L.O. (1995a) The mechanism of AMPA receptor desensitization after removal of glutamate. *Biophys. J.*, 68: 137–146.

Raman, I.M. and Trussell, L.O. (1995b) Concentration jump analysis of voltage dependent conductances activated by glutamate and kainate in neurons of the avian cochlear nucleus. *Biophys. J.*, 69: 1869–1879.

Raman, I.M., Zhang, S. and Trussell, L.O. (1994) Pathway-specific expression of AMPA receptors and their contribution to neuronal signalling. *J. Neurosci.*, 14: 4998–5010.

Ramoa, A.S. and McCormick, D.A. (1994) Enhanced activation of NMDA receptors responses at the immature retinogeniculate synapse. *J. Neurosci.*, 14: 2098–2105.

Rossi, D.J., Alford, S., Mugnaini, E. and Slater, N.T. (1995) Properties of transmission at a giant glutamatergic synapse in cerebellum: the mossy fiber-unipolar brush cell synapse. *J. Neurophysiol.*, 74: 24–42.

Sarantis, M., Ballerini, L., Miller, B., Silver, R.A., Edwards, M. and Attwell, D. (1993) Glutamate uptake from the synaptic cleft does not shape the decay of the non-NMDA component of the synaptic current. *Neuron*, 11: 541–549.

Scanziani, M., Salin, P.A., Vogt, K.E., Malenka, R.C. and Nicoll, R.A. (1997) Use-dependent increases in glutamate concentration activate presynaptic metabotropic glutamate receptors. *Nature*, 385: 630–634.

Silver, R.A., Colquhoun, D., Cull-Candy, S.G. and Edmonds, B. (1996a) Deactivation and desensitization of non-NMDA receptors in patches and the time course of EPSCs in rat cerebellar granule cells. *J. Physiol.*, 493: 167–173.

Silver R.A., Cull-Candy, S.G. and Takahashi, T. (1996b) Non-NMDA glutamate receptor occupancy and open probability at a rat cerebellar synapse with single and multiple release sites. *J. Physiol.*, 494: 231–250.

Stevens, C.F. and Wang, Y. (1995) Facilitation and depression at single central synapses. *Neuron*, 14: 795–802.

Takahashi, M., Kovalchuk, Y. and Attwell, D. (1995) Pre- and postsynaptic determinants of EPSC waveform at cerebellar climbing fiber and parallel fiber to Purkinje cell synapses. *J. Neurosci.*, 15: 5693–5702.

Tang, C.-M., Shi, Q.-Y., Katchman, A. and Lynch, G. (1991). Modulation of the time course of fast EPSCs and glutamate channel kinetics by aniracetam. *Science*, 254: 288–290.

Taschenberger, H., Engert, F and Grantyn, R. (1995) Synaptic current kinetics in a solely AMPA-receptor-operated glutamatergic synapse formed by rat retinal ganglion neurons. *J. Neurophysiol.*, 74: 1123–1136.

Teyler, T.J, Cavus, I., Coussens, C., DiScenna, P., Grover L., Lee, Y.P. and Little, Z. (1994) Multideterminant role of calcium in hippocampal synaptic plasticity. *Hippocampus.*, 4: 623–634.

Thio, L.L., Clifford, D.B. and Zorumski, C.F. (1993) Concanavalin A enhances excitatory synaptic transmission in cultured rat hippocampal neurons. *Synapse*, 13: 94–107

Tong, G. and Jahr, C.E. (1994) Block of glutamate transporters potentiates postsynaptic excitation. *Neuron*, 13: 1195–1203.

Trussell, L.O. and Otis, T.S. (1996) Physiology of AMPA receptors: Biophysical properties that subserve the integrative roles of synapses. In F. Conti and T.P. Hicks (Eds.) *Excitatory amino acids and the cerebral cortex*, MIT Press, Cambridge, pp. 63–72.

Trussell, L.O., Zhang, S. and Raman, I.M. (1993) Desensitization of AMPA receptors upon multi-quantal neurotransmitter release. *Neuron*, 10: 1185–1196.

Trussell, L.O. and Fishbach, G.D. (1989) Glutamate receptor desensitization and its role in synaptic transmission. *Neuron*, 3: 209–218.

Vyklicky, L. Jr., Patneau, D.K. and Mayer, M.L. (1991) Modulation of excitatory synaptic transmission by drugs that reduce desensitization at AMPA/Kainate receptors. *Neuron*, 7: 971–984.

Wilcox, K.S., Fitzsimonds, R.M., Johnson, B. and Dichter, M.A. (1996) Glycine regulation of synaptic NMDA receptors in hippocampal neurons. *J. Neurophysiol.*, 76: 3415–3424.

Williams, K., Russell, S.L., Shen, Y.M. and Molinoff, P.B. (1993) Developmental switch in the expression of NMDA receptors occurs in vivo and vitro. *Neuron*, 10: 267–278.

Yamada, K.A. and Tang, C.-M. (1993) Benzothiadiazides inhibit rapid glutamate receptor desensitization and enhance glutamatergic synaptic currents. *J. Neurosci.*, 13: 3904–3915.

Zhang, S. and Trussell, L.O. (1994a) A characterization of excitatory postsynaptic potentials in the avian nucleus magnocellularis. *J. Neurophysiol.*, 72: 705–718.

Zhang, S. and Trussell, L.O. (1994b) Voltage clamp analysis of excitatory synaptic transmission in the avian nucleus magnocellularis. *J. Physiol.*, 480: 123–136.

SECTION II

# Glutamate synthesis and translocation to vesicles

O.P. Ottersen, I.A. Langmoen and L. Gjerstad (Eds.)
*Progress in Brain Research*, Vol 116


CHAPTER 6

# Synthesis of glutamate and its regulation

Elling Kvamme

*Neurochemical Laboratory, Institute of Basic Medical Sciences, University of Oslo, PO Box 1115 Blindern, N-0317, Oslo, Norway*

## Introduction

Glutamate (Glu) is a key substance in the brain, being involved in metabolism, pathology and neurotransmission and is an essential source of energy. It is involved in the synthesis of proteins and peptides (Weil-Malherbe 1950) as well as other compounds and in the detoxication of ammonia in the brain (Kvamme, 1983). Furthermore, Glu has an important function as a neurotransmitter and Glu is in addition the precursor for the equally important neurotransmitter GABA. Since there is no net uptake of Glu and glutamine (Gln) from blood (see Peng et al., 1993), these compounds must be synthesized in the brain. The overall Glu concentration in the brain is high, about 10 mM and it is localized intracellularly. The extracellular Glu concentration is extremely low, about 2–3 $\mu$M which is maintained by high affinity Na-dependent uptake systems (see Yudkoff et al., 1993). Estimates of the concentration of Glu in glutamatergic terminals vary from 20–30 mM (Ottersen et al., 1992) to about 45 mM, and that of Gln 4–11 mM, whereas the concentration of Glu and Gln in astrocytes is estimated to be about 7 and 22 mM, respectively (see Fonnum, 1993).

Important regulatory processes are cellular uptake mechanisms for Glu and precursor substances (which will not be dealt with in this presentation), compartmentalization and enzymes involved in the synthesis and degradation of Glu. Glu is compartmentalized in at least two metabolic pools in the brain, a small glial pool and a large neuronal pool. This is based on the so called "Waelsch effect", the finding that the ratio between the specific activities of labelled Glu and Gln in brain is different when incubated with various labelled precursors, e.g. glucose, acetate and bicarbonate (Berl et al., 1961; Van den Berg, 1973; Nicklas, 1983). Thus, the concept of the Glu-Gln cycle was developed which postulates that Glu is taken up by glial cells from the extracellular environment and converted to Gln which is transported to neurons where Glu and GABA are synthesized for possible use in nervous transmission. This concept is later supported by the finding that Gln synthetase, that synthesizes Gln from Glu, is preferentially if not exclusively found to be localized in glial cells (Martinez-Hernandez et al., 1977; Patel et al., 1982). In addition, pyruvate carboxylase which forms oxaloacetate from glucose-derived pyruvate, is detected only in astrocytes (Yu et al., 1983; Shank et al., 1985). When oxaloacetate enters the tricarboxylic acid cycle, net synthesis of ATP occurs. Glucose is the ultimate source of fuel for the energy requiring Gln synthetase reaction in glial cells, but it has been estimated that only about 10% of the glucose oxidatively metabolized by brain is metabolized via pyruvate carboxylase in the astrocytes (Yu et al., 1983).

It is now generally accepted that the Glu-Gln cycle has an important function in the brain.

However, the cycle does not operate in a stoichiometric manner, and Glu and Gln are taken up and metabolized both by cultured astrocytes and neurons (see Hertz and Schousboe, 1988; Peng et al., 1993).

Gln and glucose-derived 2-oxoglutarate (2-OG) are considered to be main precursors of metabolic and neurotransmitter Glu both in vitro and in vivo. These substrates and the corresponding enzymes involved in Glu formation will be discussed below. However, since Gln appears to be the preferred precursor for transmitter Glu (see Fonnum, 1993; Peng et al., 1993; Yudkoff et al., 1993), the key enzyme in this reaction, phosphate activated glutaminase (PAG) (Gln amidohydrolase), also called phosphate dependent glutaminase, will be dealt with in more detail here.

**Glucose-derived 2-oxoglutarate as precursor for glutamate**

Shank and coworkers have in a series of papers provided evidence that 2-OG is an important precursor for metabolic and transmitter Glu. They showed that net amounts of 2-OG are taken up by a Na-dependent high-affinity system ($K_m$ in the range of 1 $\mu$M) and converted to Glu in synaptosomes prepared from rat and guinea pig brain tissue (see Shank and Aprison, 1988) as well as from bovine, rat retina and cerebral cortex (Lehmann et al., 1993). Furthermore, they have shown that Glu, aspartate (Asp), Gln, and malate modulate the uptake. It is inhibited by Glu and Asp, whereas Gln inhibits the uptake at high concentrations, but enhances 2-OG uptake at concentrations in the physiological range (0.1–1.0 mM). However, Hassel et al. (1992) reported that 2-OG is mainly taken up by glial cells in the brain and that the synthesis of Glu from 2-OG is reduced in fluorocitrate-treated brains. This indicates that a major fraction of 2-OG that has been taken up by the astrocytes, is first converted to Glu and then to Gln by the energy-requiring Gln synthetase followed by the formation of Glu. The authors therefore suggest that 2-OG may be an indirect precursor of transmitter Glu in vivo, but this may also apply to metabolic Glu.

The most important enzymes for the synthesis of Glu from 2-OG are glutamate dehydrogenase and 2-OG aminotransferases. A very low flux of glutamate dehydrogenase has been found in synaptosomes in the direction of synthesis of Glu from 2-OG (see Yudkoff et al., 1993), whereas an active flux in astrocytes and in the glia was observed (Cooper et al., 1979). Immunocytochemical (Rothe et al., 1994) and histochemical studies (Kugler, 1993) indicate that glutamate dehydrogenase is enriched or exclusively localized in astrocytes throughout the brain. However, in rat brain in vivo under mild hyperammonia, glutamate dehydrogenase may catalyse a minimum of 19% of the Glu pool that is recycled from neurons to astrocytes through the Glu-Gln cycle (Kanamori and Ross, 1995a). Under this condition glutamate dehydrogenase may be important for replenishment of Glu.

2-OG aminotransferase reactions are quite active in the brain (Benuck et al., 1972) and will be focused on in what follows.

**Amino acid substrates for 2-oxoglutarate aminotransferases as precursors for Glu**

*Aspartate*

As described above, Asp inhibits the synaptosomal uptake of 2-OG and is therefore unlikely to be essential for replenishing the transmitter pool of Glu under normal conditions. Asp aminotransferase (ASAT) which is the main enzyme involved in the metabolism of Asp, apparently functions predominantly in the direction of synthesis of Asp and 2-OG from Glu and oxaloacetate. Thus, Glu is to a great extent metabolized by ASAT in the brain (Erecinska et al., 1988). The activity of the cytoplasmic and mitochondrial isoenzymes of ASAT which are present in this organ, appears to be high (Fonnum, 1968; Magee and Phillips, 1971), and ASAT is found both in cultured neurons and astrocytes (see Hertz and Schousboe 1988). The distribution of ASAT and PAG activities parallels that of Asp in the olfactory bulb and Glu in the cochlear nucleus (Godfrey et al., 1994). ASAT has been proposed to be a marker for glutamatergic structures (Inagaki et al., 1987), but Donoghue et al. (1985) found that the highest enzyme immunoreactivity was localized in GABAergic structures of the cerebral cortex. Moreover, in cerebellum ASAT was detected both in glutamatergic and GABAergic neurons (Würdig and Kugler, 1991). Therefore, ASAT is not specific for glutamatergic structures and is believed to be mostly important for the energy metabolism in brain.

*Alanine*

Synaptosomes have high as well as low-affinity uptake systems for alanine (Ala) (Yudkoff et al., 1993). Ala is formed to a great extent from pyruvate by Ala aminotransferase (ALAT) in brain homogenate and neuronal and astrocytic cultures (see Peng et al., 1993), but Ala was found to be essentially released from astrocytes and not from neurons (Sonnewald et al., 1991). Westergaard et al. (1993) showed that the activity of ALAT in cultured astrocytes is almost four times greater than that of cortical neurons or cerebellar granule cells which support the assumption that Ala is essentially produced in glial cells. A high extracellular potassium concentration increased the $CO_2$ production from [U-$^{14}$C]Ala in the glutamatergic cerebellar granule cells incubated in the presence of 2-oxoglutarate (Peng et al., 1993), and ouabain did not inhibit the increase. The authors therefore suggested that the excess Ala is used for formation of transmitter Glu. However, Yudkoff et al. (1993) found that the conversion of Ala to Glu in the presence of excess 2-OG in synaptosomes is much less than the rate of Ala uptake, and that the rate of Ala formation from Gln is a 7–8 times greater than the Glu formation from Ala. Thus, there seems to be a discrepancy between synaptosomes and cultured neurons which has not been clarified. Other authors claim that ALAT, which is much slower than PAG, is unlikely to play a major role during normal neuronal activity (Erecinska et al., 1994), but that the enzyme might serve as a source of Glu during recovery from ischemia/hypoxia.

*Branched chain amino acids (BCAA)*

The three branched chain amino acids, leucine isoleucine and valine, which are important nitrogen donors particularly in skeletal muscle, cross the blood-brain barrier, and the influx of leucine is greater than that of other amino acids (Smith et al., 1987). Thus, the brain may obtain a constant supply of BCAA from the peripheral tissues. Since Gln may be released when BCAA are extracted from the periphery, BCAA are likely to furnish nitrogen for Gln synthesis (Grill et al., 1992). It has been reported that one third of the Gln and Glu nitrogen in whole cerebellar explants originates from leucine and valine, and that at least 25% of 2-N of Gln in primary cultures of astrocytes is leucine-derived (Yudkoff et al., 1993). The branched chain aminotransferases are active and reversible, and the rate of leucine consumption is much greater than the rate of oxidation of its corresponding keto acid, α-ketoisocaproate (Yudkoff et al., 1993).

In conclusion, BCAA and particularly leucine, may represent important precursors of Glu and Gln in astrocytes and therefore indirectly also of transmitter Glu.

*Ornithine and GABA*

Earlier suggestions that ornithine could be an important precursor of Glu and GABA, are rendered less likely because of the low concentration of ornithine and low activity of ornithine 2-OG-aminotransferase in the brain (Wong et al., 1981; Wroblewski et al., 1985).

It is generally agreed that GABA is not an important precursor of Glu in the brain.

**Glutamine as precursor for glutamate**

As discussed above, glucose and Gln are the main precursors for Glu, of which Gln is generally considered to play a major role (see Peng et al., 1993). This applies also to the synthesis of transmitter Glu in hippocampus (Laake et al., 1995), cerebral cortex (Georgopoulos et al., 1995) and synaptosomes (Bradford et al., 1978). The PAG content of rat brain synaptosomes is very high (Bradford and Ward, 1976; Weiler et al., 1979; Kvamme and Lenda, 1981). These organelles contain about 40% of the total tissue PAG activity (assay: 25 mM phosphate, 0 05 mM K-borate buffer, 16 mm Gln), but only 10% of the tissue protein (Bradford and Ward, 1976). Glu is synthesized from Gln by glutaminases which catalyse the hydrolytic deamidation of Gln: Gln + $H_2O$ = Glu + $NH_3$. The following enzymes are members of the glutaminase family:

Glutamine aminotransferase, previously called glutaminase II, operating in sequence with ω-amidase and presumably Glu dehydrogenase, maleate-activated glutaminase, also called phosphate-independent glutaminase, PAG.

*Glutamine aminotransferase*

Glutamine aminotransferase K (cytosolic) and ω-amidase which metabolize Gln to 2-OG, are widespread throughout the brain (see Cooper, 1988). The 2-OG may be further converted to Glu by Glu dehydrogenase. It has been suggested that transmitter Glu can be formed in this way. However, the highest specific activities of glutamine aminotransferase and ω-amidase are found in the choroid plexus which indicates that metabolism of Gln by these enzymes is an important function in the choroid plexus (Cooper et al., 1993).

*Maleate-activated glutaminase*

Maleate-activated glutaminase has been shown to be identical with γ glutamyl transferase (γ -GT); also called γ-glutamyltranspeptidase (Tate and Meister, 1975; Curthoys and Kuhlenschmidt, 1975). Katunuma (1966) detected this enzyme in kidney and it is located in several brain structures (see Huseby, 1988), but the maleate activation property of the enzyme could not be observed in brain synaptosomes (Kvamme and Olsen, 1981). However, γ-GT may be important for regulation of the pool of endogenous Glu since the development profile of endogenous Glu in cultured cerebral cortex interneurons and cerebellar granule cells correspond closely with that of γ-GT and not with other Glu-metabolizing enzymes. Furthermore, the development profile of endogenous Glu in mouse cerebellum in vivo corresponds with those of γ-GT and PAG during the first 14 days post partum (Kvamme et al., 1985b).

*PAG*

*Enzymatic properties*

PAG is considered to be the most important enzyme of the glutaminase family for the synthesis of metabolic as well as transmitter Glu (see Kvamme et al., 1988a, 1997). The enzyme was detected by Krebs (1935) who distinguished between a "brain type" (brain and kidney PAG) and a "hepatic type". It is a mitochondrial phosphate-activated enzyme (Errera and Greenstein, 1949). Activation both by phosphate and organic anions (e.g. succinate, malate, citrate) has been confirmed by several authors (Kvamme et al., 1965; O'Donovan and Lotspeich, 1966; Svenneby et al., 1970; Svenneby, 1971), but the latter compounds are less potent. This demonstrates that the enzyme is not phosphate-dependent and that the original term phosphate-activated glutaminase therefore is more appropriate than phosphate-dependent glutaminase.

PAG was purified to apparent homogeneity (ca. 10.000 fold) from pig kidney (Kvamme et al., 1970), pig brain (Svenneby et al., 1973, Kvamme et al., 1985a), and later from rat kidney (Curthoys et al., 1976) and rat brain (Haser et al., 1985). PAG has also been partially purified from cow brain (Chiu et al., 1979).

Electron micrographs demonstrated that in Tris buffer a soluble form, of pig kidney and brain PAG, the T-form (mol.wt. 130–150 kD), is found which consists of particles with a diameter of 60 Å and a length of about 90 Å. Following addition of phosphate, the enzyme dimerizes, and by addition of phosphate + borate an insoluble polymerized form is produced, the PB form (mol. wt, 1,500–2.000 kD), that consists of long, double-stranded, rope-like polymers (Olsen et al., 1970, 1973). The number of free SH⁻ groups was greatly reduced by conversion of the T-form to the PB-form (Kvamme et al., 1968). By sodium dodecyl sulfate-polyacrylamide gel electrophoresis two subunits were detected in the pig kidney PAG with molecular weights of 63 and 57 kD respectively, and in the pig brain enzyme one subunit with molecular weight of 63 kD (see Kvamme et al., 1985a). Rat kidney and brain PAG are heterotetrameric enzymes with three 66 kD subunits and one 68 kD subunit (Shapiro et al., 1987; Perera et al., 1991). The rat PAG has been cloned and the amino acid sequence is known (Shapiro et al., 1991).

The monomeric T-form of purified pig kidney and brain PAG has low activity, and the specific activity of the dimeric and polymeric forms of pig kidney and brain PAG is three times that of the monomeric form (see Kvamme et al., 1988a). Dimerization and polymerization of the monomeric T-form is accompanied by a corresponding time-dependent activation in the presence of phosphate, phosphate-borate (Svenneby et al., 1970; Svenneby, 1972; Nimmo and Tipton, 1981) or acetyl CoA (Kvamme and Torgner, 1974). It should be noted that the monomeric T-form of purified pig kidney and brain PAG is less susceptible to activation by low concentrations of phosphate than the polymeric PB-form.

PAG from kidney and brain have rather similar properties. In addition to phosphate and organic anions, nucleoside triphosphates strongly activate PAG (Weil-Malherbe, 1972). PAG is also activated by mitochondrial swelling (Kovacevic et al., 1980; Welbourne et al., 1976; Kvamme et al., 1991a), by increased oxygen pressure (Kritschevskaya et al., 1959) and by thyroxine (Hovhannissian et al., 1970).

A special group of compounds sensitize the enzyme to phosphate activation in low concentrations, whereas it has inhibitory effect when the concentration is raised. Bromthymol blue represents a prototype of this type of activation (Kvamme et al., 1965, 1970). Some physiological compounds such as certain fatty acids and acyl CoA derivatives have similar effects, since they activate purified renal and brain PAG in $\mu$M concentrations and inhibit at higher concentrations (Kvamme and Torgner, 1974, 1975) Furthermore, thyroxin has also been reported to have this effect (Sahagian and Hovhannasian, 1982).

$Ca^{2+}$ activates the Gln hydrolysis in brain synaptosomes (Kvamme and Olsen, 1979; Kvamme et al., 1983), kidney mitochondria, (Kvamme et al., 1991a) and brain tissue (Benjamin, 1981). It should be noted that $Ca^{2+}$ activates only under "normal" assay condition (pH 7.4, Gln 2 mM, sodium phosphate 5–10 mM), and not when "optimal" assay conditions are used (pH 8.6, Gln 20 M, sodium phosphate 150 mM) (Kvamme et al., 1988b). In the presence of the reaction products Glu and $[NH_4]^+$, $Ca^{2+}$ has no effect (Kvamme et al., 1983), and calcium activates only the NEM-insensitive fraction of PAG (see below). Moreover, $Ca^{2+}$ does not act on purified PAG (Kvamme, et al., 1983), indicating that the activation is an indirect effect. $Ca^{2+}$ sensitizes PAG to phosphate, e.g. by liberating an activator of PAG, e.g. a lipid derivate, which may activate PAG by causing a conformational change.

The extracellular concentration of Gln is about 0.5 mM (Gjessing et al., 1972) and the activity of pig brain PAG in vivo under this condition and with 5–8 mM phosphate is calculated by the author to be at most 5–10% of that obtained with maximal activation in the presence of 20 mM Gln and 100 mM phosphate. In vivo measurement of PAG in rat brain has confirmed that only a small fraction of the maximal activity is utilized (Kanamori and Ross, 1995b).

Main inhibitors of purified PAG are protons and the reaction product Glu (Kvamme et al., 1970; Svenneby et al., 1970; Tveit et al., 1970; Svenneby, 1971), whereas structurally bound PAG in addition is inhibited by $[NH_4]^+$ (Bradford and Ward, 1976; Kvamme and Olsen, 1981; Kvamme and Lenda, 1982). pH optimum for PAG is about 8–9, and by lowering the pH of the incubation medium from 7.6 to 7.0, the activity of pig kidney mitochondrial PAG (Kvamme et al., 1988b) and pig brain synaptosomal PAG (Kvamme and Olsen, 1981) is reduced by 50 to 60%. This is of interest, because acute acidosis increases the activity of renal PAG in some species (see Kvamme et al., 1988a). It is remarkable that phosphate is particularly effective in protecting the purified PB-form against inhibition by Glu (Svenneby, 1971; Tveit et al., 1970).

Rat brain synaptosomal PAG is very sensitive to inhibition by low concentrations of Glu and ammonia (Bradford and Ward, 1976; Kvamme and Lenda, 1982) and so is PAG in rat brain cortex in vitro (Benjamin, 1981). Synaptosomal PAG activity is thus reduced by 40 to 50% when the Glu or $[NH_4]^+$ concentration is raised to 0.5 mM (Kvamme and Lenda, 1982). On the other hand, purified PAG is activated by high concentrations (10 to 20 mM) of $[NH_4]^+$ (Kvamme et al., 1970), which indicates that the inhibition by $[NH_4]^+$ is an indirect effect. If not compartmentalized, the high concentration of endogenous Glu (more than 4 mM) in the synaptosomes (Kvamme and Lenda, 1981) would have considerably reduced the PAG activity. However, the lack of inhibition is explained by the finding that Glu is compartmentalized in the synaptosomes and therefore unavailable to the mitochondrially localized PAG. Thus, Glu which constitutes a large fraction of Glu in the glutamatergic nerve endings (Storm-Mathisen et al., 1992), as well as transmitter Glu (Fonnum et al., 1992) is found to be localized to synaptic vesicles.

Moreover, PAG is inhibited by c-AMP and c-GMP (Weil-Malherbe, 1972) as well as by long-chain acyl CoA derivates in mM concentrations (Kvamme and Torgner, 1975), mersalyl (see Kvamme et al., 1988a; Roberg et al., 1995) and 6-diazo-5-oxo-L-norleucine (Goldstein et al., 1967).

N-Ethylmaleimide (NEM) exerts a partial inhibition of PAG in pig kidney cortex mitochondria and rat brain synaptosomes, which has given rise to distinction between a NEM-sensitive and a NEM-insensitive fraction of PAG with different properties (Kvamme and Olsen, 1979, 1981). Thus, phosphate and $[H]^+$ preferentially affect the NEM-sensitive fraction and $Ca^{2+}$ (see above) the NEM-insensitive one.

Purified pig kidney and brain PAG are allosteric enzymes (Kvamme et al., 1965, 1970; Svenneby, 1971). Double reciprocal plots of enzyme activity against the concentration of the activators phosphate (Svenneby et al., 1970; Svenneby, 1971; Tveit et al., 1970) or some acyl CoA derivatives (Kvamme and Torgner, 1974, 1975) are concave upwards and the cooperativity increases when the inhibitor Glu is added. Interestingly, human renal mitochondria (Kvamme et al., 1985c) and human brain $P_2$ fraction (Svenneby et al., 1986) show similar kinetic properties as that of pig kidney and brain. The inhibition of PAG by Glu in pig and human brain synaptosomal fractions is counteracted competitively by phosphate (Svenneby, 1971; Svenneby et al., 1986). The kinetics of Glu inhibition of the Gln activation of PAG is controversial and both competitive and noncompetitive inhibitions have been described. The reason for this might be due to methodological differences (see Kvamme, 1988a). However, it has been demonstrated that PAG in the supernatant of sonicated rat brain $P_2$ fractions shows linear double inverse plots and competitive inhibition with Glu, whereas the membrane fraction shows linear noncompetitive inhibition with Glu. Both glutaminases are phosphate-activated, but the soluble enzyme has lower affinity for Gln ($K_m$ 30 to 40 mM) than the membrane-bound form ($K_m$ about 2 mM) (Kvamme et al., 1988a). It is of interest that when calculated from the data of Tveit et al. (1970) the ratio of the $K_m$ value for the purified T-form of PAG to the PB form is about 18 which is about the same as the ratio between the $K_m$ values for the soluble and membrane bound forms which is 15 to 20. This indicates similarities between the soluble enzyme and the monomeric T-form and the membrane bound enzyme and the polymeric PB-form, respectively.

*Structural localization in the brain*

Immunocytochemical studies do not show a specific regional distribution of PAG in the brain, but it has been reported that the enzyme activity is regionally regulated by phosphate and calcium (Wallace and Dawson, 1993). The activity is also higher in cultured glutamatergic cerebellar granule cells than in cultured GABAergic cortical neurons (Hogstad et al., 1988). Most GABAergic, cholinergic and peptidergic nonpyramidal neurons in neo- and mesocortex do not contain PAG when measured by use of double immunofluorescence method (Kaneko et al., 1992). By immunogold labelling for Glu and Gln in hippocampal slice cultures, it was found that Glu accumulates in glia when the flux through Gln synthetase is blocked, and that Gln-derived Glu through the glutaminase reaction is necessary for the maintenance of normal levels of Glu in the terminals (Laake et al., 1995).

*Comparison of PAG in cultured mouse brain astrocytes with that of cortex neurons, cerebellar granule cells and synaptosomes*

The activity of PAG in bulk isolated glial cells has been reported to be low (Weiler et al., 1979) in contrast to that of PAG in cultured mouse astrocytes which has been found to be relatively high (Kvamme et al., 1982), but significantly lower than the activities of PAG in synaptosomes, cultured mouse cerebral cortex neurons and cerebellar granule cells which show the highest activity (Hogstad et al., 1988). A recent study demonstrates that infusion of quinolinic acid into rat neostriatum, which causes selective, damage to medium-sized spiny neurons, resulted in 65–75% decrease in Type I inositol (1,4,5) trisphosphate receptor and PAG levels and no change in glutamate dehydrogenase level, another mitochondrial enzyme (Haug et al., 1997). Thus PAG is highly enriched in these neurons. However, cultured astrocytes appear to contain both the Gln synthesizing enzyme Gln synthetase and the Gln hydrolysing enzyme PAG. Evidence has been produced that the two enzyme systems are compartmentalized in the astrocytes (McKenna et al., 1996). It can be argued that the high PAG activity of astrocytes is an artifact due to induction of PAG since astrocytes are cultured in a medium which contains a high Gln concentration. However, a tenfoldness reduction in the Gln content of the culture medium (from 2 mM to 0.2 mM) did not affect the activity (Kvamme et al., 1982), but this may not exclude the possibility of induction. Phosphate activates PAG in all these cell types, and the phosphate activation is sensitized by calcium.

It is of interest that PAG in astrocytes cultured from brain is not inhibited by $[NH_4]^+$ (Kvamme et al., 1982). This is in contrast to synaptosomal PAG, PAG in cultured neurons from mouse brain cortex and cultured mouse cerebellar granule cells, whereas the inhibitory effect of Glu in astrocytes is similar to that of synaptosomes. Moreover, PAG in astrocytes requires a much higher phosphate concentration for half maximal activation (Hogstad et al., 1988) than the neuronal and synaptosomal PAG. These findings demonstrate that PAG in cultured astrocytes do not have similar properties as PAG in the other tissues examined.

*Submitochondrial localization of PAG*

The submitochondrial localization of PAG has been controversial. PAG has been claimed to be preferentially localized to the matrix region (Crompton et al., 1973; Kalra and Brosnan, 1974), or to the inner face of the inner mitochondrial membrane (Strzelecki and Schoolwerth, 1984; Shapiro et al., 1985). On the contrary, we have reached the conclusion that both the kidney (Kvamme et al., 1991b) and brain PAG (Roberg et al., 1995) are *functionally* localized to the outer face of the inner mitochondrial membrane. Thus, we were originally struck by the observation that Glu when added to the incubation medium of renal mitochondria, synaptosomes or cultured mouse brain astrocytes, exerts a similar inhibition of PAG, as when Glu in the same concentrations is added to disrupted preparations (Kvamme and Lenda, 1981) in spite of the fact that Glu has a very restricted permeability to the inner mitochondrial membrane. The same holds true for protons which also have a very restricted permeability to the inner membrane, so that PAG appears to be regulated by the extramitochondrial concentration of these compounds. Moreover, PAG is inhibited by the SH-reagent mersalyl which is impermeable to the inner mitochondrial membrane, and mitochondrial Gln transport does not limit the activity of PAG. In addition, the specific activity of extramitochondrial [$^{14}$C]Gln-derived Glu is similar to that of added [$^{14}$C]Gln and much higher than that of pig kidney (Kvamme et al., 1991b) and rat brain mitochondrial Glu (Kvamme et al., 1995).

Srinivasan et al. (1995) reported that a major fraction of the mature PAG is localized to the matrix region and a minor fraction to the inner mitochondrial membrane, following in vitro transcription and incorporation into liver mitochondria of PAG precursor protein. Previously, a soluble and a membrane-bound form of renal PAG with distinct kinetic properties are isolated from disrupted mitochondria (Kovacevic, 1976; O'Donovan, 1985, 1991). Similar results are reported by us (Roberg et al., 1997), and both the kinetics and the pH curves resemble that of the soluble purified enzyme (T-form) and the polymerized form (PB-form), respectively. In spite of that, in intact mitochondria PAG behaves kinetically as one species because double inverse plots of Gln concentration vs enzyme activity is a straight line (Kvamme et al., 1991b), which indicates that the species that is internally located is dormant when the Gln concentration is kept below 10 mM. This view is supported by the finding that disruption of the mitochondria yields a break in the Arrhenius plots (Kovacevic, 1976; Strzelecki and Schoolwerth, 1984), and furthermore that two kinetically distinguishable forms of PAG are found in homogenate of cultured mouse brain cortex neurons, cerebellar granule cells and astrocytes (Hogstad et al., 1988) when the Gln concentration is raised above 10 mM.

Therefore, under normal condition the enzymically active species of PAG in intact mitochondria must either have an external location in the inner membrane or an internal location in the mitochondria, e.g. in the matrix region If the dominant species is internally located, there must exist a rapid channelling mechanism of Gln and Glu, similar to that described for ornithine transcarbamylase (Cohen et al., 1987). Moreover, mersalyl must be able to penetrate the inner mitochondrial membrane which appears to be contrary to the experimental findings. We have recently provided evidence that the membrane-bound form is the enzymically dominant form.

External localization in the inner mitochondrial membrane of functionally active PAG has implications for PAG as producer of transmitter Glu which in that case is readily released to the cytoplasm. On the other hand, Palaiologos et al. (1988) suggest that Glu produced by PAG located on the outer face of the inner membrane is transported to the matrix and transaminated by ASAT to 2-oxoglutarate which is transported to the cytoplasm by the ketocarboxylate carrier and then transaminated by ASAT to transmitter Glu. Essential for this hypothesis is the finding that the formation of transmitter Glu is inhibited by phenylsuccinate, an inhibitor of the ketocarboylate carrier. However, as pointed out by Fonnum (1993), phenylsuccinate could affect release by complex formation with $Ca^{2+}$. In addition, Peng et al. (1993) claim that synthesis of transmitter Glu from 2-OG and alanine by ALAT is likely to occur exclusively in the cytosol and does not involve any transmitochondrial transport. Thus, it seems difficult to understand why Glu in the cytosol synthesized by PAG and ALAT should be treated differently. If so, it must involve some compartmentation of cytosolic Glu.

### Conclusion and therapeutical strategies

Glucose and Gln are both precursors for metabolic and transmitter Glu. These substrates and the corresponding enzymes involved in Glu formation, may all contribute in maintaining the endogenous as well as the transmitter pool of Glu, most likely depending on the brain structure, cell type and also on environmental conditions. Although PAG is considered to be the most important enzyme for production of transmitter Glu, little is known whether this production can be restored by other Glu producing enzymes when the PAG activity is inhibited. Transmitter Glu mediates its effect by a variety of receptors and subtypes of receptors and is likely to be involved in a great number of pathological conditions including epilepsy, stroke and perhaps also schizophrenia (See Kvamme, 1988, Glutamine and glutamate in mammals, Vol II). However, little has been done to investigate how transmitter Glu production can be affected in vivo, and the relation between Glu produced for metabolic use and nervous transmission. Most workers agree that Glu mediated transmission is effected by release of vesicle bound Glu, but more information is needed about whether Glu transmission in vivo can be affected by regulating the production of Glu aimed for vesicle binding, and also if it is possible selectively to regulate the availability of Glu for special receptor types. In this respect it should be kept in mind that Glu is a potent inhibitor of PAG, that this enzyme also is regulated by some natural lipids and that a large fraction of PAG is likely to be dormant under physiological conditions (see above).

## References


Benjamin, A.M. (1981) Control of glutaminase activity in rat brain cortex in vitro: Influence of glutamate, phosphate, ammonium, calcium and hydrogen ions. *Brain Res.*, 208: 363–377.

Bradford, H.F. and Ward, H.K. (1976) On glutaminase activity in mammalian synaptosomes. *Brain Res.*, 110: 115–125.

Bradford, H.F., Ward, K.H. and Thomas, A.J. (1978) Glutamine – A major substrate for nerve endings. *J. Neurochem.*, 30: 1453–1459.

Benuck, M., Stern, F. and Lajtha, A. (1972) Regional and subcellular distribution of aminotransferases in rat brain. *J. Neurochem.*, 510: 19: 949–957.

Berl, S., Lajtha, A. and Waelsch, H. (1961) Amino acid and protein metabolism. VI. Cerebral compartments of glutamic acid metabolism. *J. Neurochem.*, 7: 186–197.

Chiu, J.F. and Boeker, E.A. (1979) Cow brain glutaminase: partial purification and mechanism of action. *Arch. Biochem. Biophys.*, 196: 493–500.

Cohen, N.S., Cheung, C.-W. and Raijman, L. (1987) Channelling of extramitochondrial ornithine to matrix ornithine transcarbamylase. *J. Biol. Chem.*, 262: 203–208.

Cooper, A.J.L., McDonald, J.M., Gelbard, A.S., Gledhill, R.F. and Duffy, T.E. (1979) The metabolic fate of $^{13}$N-labeled ammonia in rat brain. *J. Biol. Chem.*, 253: 4982–4992.

Cooper, A.J.L. (1988) Glutamine aminotransferases and ω-amidases. In E. Kvamme (Ed.), *Glutamine and Glutamate in Mammals*. Vol. 1, CRC Press, Boca Raton, Florida, pp. 33–52.

Cooper, A.J.L., Abraham, D.G., Gelbard, A.S., Lai, J.C. and Petito, C.K. (1993) High activities of glutamine transaminase K (dichlorovinylcysteine β-lyase) and ω-amidase in the choroid plexus of rat brain. *J. Neurochem.*, 61: 1731–1741.

Crompton, M., McGivan, J.D. and Chappell, J.B. (1973) The intramitochondrial location of the glutaminase isoenzymes of pig kidney. *Biochem. J.*, 132: 27–34.

Curthoys, N.P. and Kuhlenschmidt, T. (1975) Phosphate independent glutaminase from rat kidney. Partial purification and identity with γ-glutamyltranspeptidase. *J. Biol. Chem.*, 250: 2099–2105.

Curthoys, N.P., Kuhlenschmidt, T. and Godfrey, S.S. (1976) Regulation of renal ammoniagenesis. Purification and characterization of phosphate-dependent glutaminase from rat kidney. *Arch. Biochem. Biophys.*, 174: 82–89.

Donoghue, J.P., Wenthold, R.J. and Altschuler, R.A. (1985) Localization of glutaminase-like and aspartate aminotransferase-like immunoreactivity in neurons of cerebral cortex. *J. Neurosci.*, 5: 2597–2608.

Erecinska, M., Nelson, D., Nissim, I., Daikhin, Y. and Yudkoff, M. (1994) Cerebral alanine transport and alanine aminotransferase reaction: Alanine as a source of neuronal glutamate. *J. Neurochem.*, 62: 1953–1964.

Erecinska, M., Zaleska, M.M., Nissim, I., Nelson. D., Dagani, F. and Yudkoff, M. (1988) Glucose and synaptosomal glutamate metabolism. Studies with [$^{15}$N]glutamate. *J. Neurochem.*, 51: 892–902.

Errera, M. and Greenstein, J.P. (1949) Phosphate activated glutaminase in kidney and other tissues. *J. Biol. Chem.*, 178: 495–502.

Fonnum, F. (1968) The distribution of glutamate decarboxylase and aspartate transaminase in subcellular fractions of rat and guinea pig brain. *Biochem. J.*, 106: 401–412.

Fonnum, F. (1993) Regulation of the synthesis of the transmitter glutamate pool. *Prog. Biophys. molec. Biol.*, 60: 47–57.

Fonnum, F., Christensen, H. and Fykse, E.M. (1992). The role of synaptic vesicles defining the transmitter pool amino acids. *Molec. Neuropharmacol.*, 2: 3–6.

Georgopoulos, A., Svarna, R. and Palaiologos, G. (1995) Regulatory sites and effectors of D-[3H]aspartate release from rat cerebral cortex. *Neurochem. Res.*, 20: 45–49.

Gjessing., L.R., Gjesdahl, P. and Sjaastad, O. (1972) The free amino acids in human cerebrospinal fluid. *J. Neurochem.*, 19: 1807–1808.

Godfrey, D.A., Ross, C.D., Parli, J.A. and Carlson, L. (1994) Aspartate aminotransferase and glutaminase activities in rat olfactory bulb and cochlear nucleus; comparisons with retina and with concentrations of substrate and product amino acids. *Neurochem. Res.*, 19: 693–703.

Goldstein, L. (1967) Pathways of glutamine deamidation and their control in rat kidney. *Am. J. Physiol.*, 213: 983–989.

Grill, V., Bjørkheim, M., Gutniak, M. and Lindqvist, M. (1992) Brain uptake and release of amino acids in nondiabetic and insulin-dependent diabetic subjects: Important role of glutamine release for nitrogen balance. *Metabolism*, 41: 28–32.

Haser, W.G., Shapiro, R.A. and Curthoys, N.P. (1985) Comparison of the phosphate-dependent glutaminase obtained from rat kidney and brain. *Biochem. J.*, 229: 399–408.

Hassel, B., Paulsen, R.E., Johnsen, A. and Fonnum, F. (1992) Selective inhibition of glial cell metabolism *in vivo* by fluorocitrate. *Brain. Res.*, 576: 120–124.

Haug, L.S., Østvold, A.C., Torgner, I., Roberg, B., Dvorakaova, L. and Walaas, S.I. (1997) Type I inositol (1,4,5)trisphosphate receptor and phosphate-activated glutaminase are highly enriched in neurons as compared to glial cells in rat neostriatum. In A.W. Telken and J. Kerf (Eds.), *Neurochemistry*, Plenum Press, New York, pp. 389–394.

Hertz, L. and Schousboe, A. (1988) Metabolism of glutamate and glutamine in neurons and astrocytes in primary cultures. In E. Kvamme (Ed.), *Glutamine and Glutamate in Mammals*, Vol. 2, CRC Press, Boca Raton, Florida, pp. 39–55.

Hogstad, S., Svenneby, G., Torgner, I.Aa., Kvamme, E., Hertz, L. and Schousboe, A. (1988) Glutaminase in neurons and astrocytes cultured from mouse brain: Kinetic properties and effects of phosphate, glutamate and ammonia. *Neurochem. Res.*, 13: 383–388.

Hovhannissian, V.S., Buniatian, H.C., Ukrdumova, G.S. and Badalian, L.L. (1970) The participation of thyroxine in the interaction of the isoenzymes of brain glutaminase and certain features of its action. *Vopr. Biokhim. Mozga Akad. Nauk. Armenjan SSR*, 6: 5–13.

Huseby, N.-E. (1988) $\gamma$-glutamyltransferase. In E. Kvamme (Ed.), *Glutamine and Glutamate in Mammals*. Vol. 1, CRC Press, Boca Raton, Florida, pp.153–163.

Inagaki, N., Kamisaki, Y., Kiyama, H., Horio, Y., Tonyama, M. and Wada, H. (1987) Immunocytochemical localizations of cytosolic and mitochondrial glutamic oxaloacetic transaminase isozymes in rat primary sensory neurons as markers for the glutamate neuronal system. *Brain Res.*, 402: 197–200.

Kalra, J. and Brosnan, J.T. (1974) The subcellular localization of glutaminase isoenzymes in rat kidney cortex. *J. Biol. Chem.*, 249: 3255–3260.

Kanamori, K. and Ross, B.D. (1995a) Steady-state in vivo glutamate dehydrogenase activity in rat brain measured by 15N NMR. *J. Biol. Chem.*, 270: 24805–24809.

Kanamori, K. and Ross, B.D. (1995b) In vivo activity of glutaminase in the brain of hyperammonaemic rats measured by $^{15}N$ nuclear magnetic resonance. *Biochem. J.*, 305: 329–336.

Kaneko, T., Nakaya, Y. and Mizuno, N. (1992) Paucity of glutaminase-immunoreactive nonpyramidal neurons in the rat cerebral cortex. *J. Comp. Neurol.*, 322: 181–190.

Katunuma, N., Tomino, L. and Nishino, H. (1966). Glutaminase isoenzymes in rat kidney. *Biochim. Biophys. Res. Commun.*, 22: 321–328.

Kovacevic, Z. (1976) Importance of the flux of phosphate across the inner membrane of kidney mitochondria for the activation of glutaminase and the transport of glutamine. *Biochim. Biophys. Acta*, 430: 399–412.

Kovacevic, Z, Bajin, K. and Pavlovic, M. (1980) Volume changes of rat kidney mitochondria, transport of glutamine and its inhibition by mersalyl. *Int. J. Biochem.*, 12: 139–143.

Krebs, H.A. (1935) Metabolism of amino acids. IV. The synthesis of glutamine from glutamic acid and ammonia, and the enzymic hydrolysis of glutamine in animal tissue. *Biochem. J.*, 29: 1951–1969.

Kritschevskaya, A.A., Gerschenovitsch, Z. and Scherbatzker, S. (1959) Ammonia formation from amides by the brain and liver homogenates under increased oxygen pressure. *Biochimija*, 64: 459–464.

Kugler, P. (1993) *In situ* measurement of enzyme activities in the brain. *Histochem. J.*, 25: 329–338.

Kvamme, E. (1983) Ammonia metabolism in the CNS. *Prog. Neurobiol.*, 20: 109–132.

Kvamme, E. and Lenda, K. (1981) Evidence for compartmentalization of glutamate in rat brain synaptosomes using the glutamate sensitivity of phosphate activated glutaminase as a functional test. *Neurosci. Lett.*, 25: 193–198.

Kvamme, E. and Lenda, K. (1982) Regulation of glutaminase by exogenous glutamate, ammonia and 2-oxoglutarate in synaptosomal enriched preparations from rat brain. *Neurochem. Res.*, 7: 667–678.

Kvamme, E. and Olsen, B.E. (1979) Evidence for two species of mammalian phosphate-activated glutaminase having different regulatory properties. *FEBS Lett.*, 107: 33–36.

Kvamme, E. and Olsen, B.E. (1981) Evidence for compartmentation of synaptosomal phosphate-activated glutaminase. *J. Neurochem.*, 36: 1916–1923.

Kvamme, E., Roberg, B., Johansen, L. and Torgner, I.Aa. (1988b) Interrelated effects of calcium and sulfhydryl reagents on renal phosphate-activated glutaminase. *Contr. Nephrol.*, Basel, 63: 156–160.

Kvamme, E., Roberg, B. and Torgner, I.Aa. (1991a) Effects of mitochondrial swelling and calcium on phosphate activated glutaminase in pig renal mitochondria. *Eur. J. Biochem.*, 197: 675–680.

Kvamme, E., Roberg, B. and Torgner. I.Aa. (1997) An overview of mitochondrial glutamine transport and phosphate activated glutaminase in the kidney. *Contr. Nephrol.*, Basel 121: 69–78.

Kvamme, E., Schousboe, A., Hertz, L., Torgner, I.Aa. and Svenneby, G. (1985b) Developmental change of endogenous glutamate and $\gamma$-glutamyl transferase in cultured cerebral cortical interneurons and cerebellar granule cells, and in mouse cerebellar cortex and cerebellum in vivo. *Neurochem. Res.*, 10: 993–1008.

Kvamme, E., Svenneby, G., Hertz, L. and Schousboe, A. (1982) Properties of phosphate activated glutaminase in astrocytes cultured from mouse brain. *Neurochem. Res.*, 7: 761–770.

Kvamme, E., Svenneby, G. and Torgner, I.Aa. (1983) Calcium stimulation of glutamine hydrolysis in synaptosomes from rat brain. *Neurochem. Res.*, 8: 25–38.

Kvamme, E., Svenneby, G. and Torgner, I.Aa. (1988a) Glutaminases. In E. Kvamme (Ed.), *Glutamine and Glutamate in Mammals. Vol. 1*, CRC Press, Boca Raton, Florida pp. 53–67.

Kvamme, E., Svenneby, G. and Tveit, B. (1968) Molecular aspects of kidney glutaminase. In E. Kvamme and A. Pihl (Eds.), *Regulation of enzyme activity and allosteric interactions*, Oslo, Universitetsforlaget, pp. 89–108.

Kvamme, E., Svenneby, G., Tveter, K.J. and Torgner, I.AA. (1985c) Phosphate-activated glutaminase in human kidney. *Contr. Nephrol.*, Basel, 47: 145–149.

Kvamme, E. and Torgner, I.Aa. (1974) The effect of acetyl-coenzyme A on phosphate-activated glutaminase from pig kidney and brain. *Biochem. J.*, 137: 525–530.

Kvamme, E. and Torgner, I.Aa. (1975) Regulatory effects of fatty acyl-coenzyme A derivatives on phosphate-activated pig brain and kidney glutaminase *in vitro*. *Biochem. J.*, 149: 83–91.

Kvamme, E., Torgner, I.Aa. and Roberg, B. (1991b) Evidence indicating that pig renal phosphate-activated glutaminase has a functionally predominant external localization in the inner mitochondrial membrane. *J. Biol. Chem.*, 266: 13185–13192.

Kvamme, E., Torgner, I. Aa. and Svenneby, G. (1985a) Glutaminase from mammalian tissues. In S.P. Colowick and N.O. Kaplan (Eds.), *Methods in Enzymology, Vol. 113*, Academic Press, New York, pp 241–256.

Kvamme, E., Tveit, B. and Svenneby, G. (1965) Glutaminase from pig kidney, an allosteric protein. *Biochem. Biophys. Res. Commun.*, 20: 566–572.

Kvamme, E., Tveit, B. and Svenneby, G. (1970) Glutaminase from pig renal cortex I. Purification and general properties. *J. Biol. Chem.*, 245: 1871–1877.

Laake, J.H., Slyngstad, T.A., Haug, F.-M.S. and Ottersen, O.P. (1995) Glutamine from glial cells is essential for the maintenance of the nerve terminal pool of glutamate: Immunogold evidence from hippocampal slice cultures. *J. Neurochem.*, 65: 871–881.

Lehmann, J.C., Kapov, D. and Shank, R.P. (1993) Kinetics of 2-oxoglutarate uptake by synaptosomes from bovine and rat retina and cerebral cortex and regulation by glutamate and glutamine. *Dev. Neurosci.*, 15: 330–335.

Magee, S.C. and Phillips, A.T. (1971) Molecular properties of the multiple aspartate aminotransferases purified from rat brain. *Biochemistry*, 10: 3397–3405.

Martinez-Hernandez, A., Bell, K.P. and Norenberg, M.D. (1977) Glutamine synthetase: Glial localization in brain. *Science*, 195: 1356–1358.

McKenna, M.C., Tildon, J.T., Stevenson, J.H. and Huang, X.L. (1996) New insights into the compartmentation of glutamate and glutamine in cultured rat brain astrocytes. *Dev. Neurosci.*, 18: 380–390.

Niclas, W.L. (1983) Relative contributions of neurons and glia to metabolism of glutamate and GABA. In L. Hertz, E. Kvamme, E. McGeer and A. Schousboe (Eds.), *Glutamine,Glutamate and GABA in the central nervous system*, Alan R. Liss, New York, pp. 219–227.

Nimmo, G.A. and Tipton, K.F. (1981) Kinetic comparisons between soluble and membrane-bound glutaminase preparations from pig brain. *Eur. J. Biochem.*, 117: 57–64.

O'Donovan, J.D. and Lotspeich, W.D. (1966) Activation of kidney mitochondrial glutaminase by inorganic phosphate and organic acids. *Nature*, 212: 930–932.

O'Donovan, D.J. (1985) Effects of solubilization on the kinetics of rat kidney mitochondrial glutaminase. *Contrib. Nephrol.*, Basel, 47: 165–169.

O'Donovan, D.J. (1991) Changes in rat renal glutaminase activity on isolation from the mitochondrial membrane. *Contrib. Nephrol.*, Basel, 92: 13–17.

Olsen, B.R, Svenneby, G., Kvamme, E., Tveit, B. and Eskeland, T. (1970) Formation and ultrastructure of enzymically active polymers of pig renal glutaminase. *J. Mol. Biol.*, 52: 239–245.

Olsen, B.R., Torgner, I.Aa., Christensen, T.B. and Kvamme, E. (1973) Ultrastructure of pig renal glutaminase. Evidence for conformational changes during polymer formation. *J. Mol. Biol.*, 74: 239–251.

Ottersen, O.P., Zhang, N. and Walberg, T. (1992) Metabolic compartmentation of glutamate and glutamine: Morphological evidence obtained by quantitative immunocytochemistry in rat cerebellum. *Neurosci.*, 46: 519–534.

Palaiologos, G., Hertz, L. and Schousboe, A. (1988) Evidence that aspartate aminotransferase activity and ketodicarboxylate carrier function are essential for biosynthesis of transmitter glutamate. *J. Neurochem.*, 51: 317–320.

Patel, A.J., Hunt, A., Gordon, R.D. and Balazs, R. (1982) The activities in different neural types of certain enzymes associated with the metabolic compartmentation glutamate. *Dev. Brain Res.*, 4: 3–11.

Peng, L., Hertz, L., Huang, R., Sonnewald, U., Petersen, S.B., Westergaard, N., Larsson, O. and Schousboe, A. (1993) Utilization of glutamine and of TCA cycle constituents as precursors for transmitter glutamate and GABA. *Dev. Neurosci.*, 15: 367–376.

Perera, S.Y., Voith, D.M. and Curthoys, N.P. (1991) Biosynthesis and processing of mitochondrial glutaminase in HTC hepatoma cells. *Biochem. J.*, 273: 265–270.

Roberg, B., Torgner, I.Aa. and Kvamme, E. (1995) The orientation of phosphate activated glutaminase in the inner mitochondrial membrane of synaptic and non-synaptic mitochondria. *Neurochem Int.*, 27: 367–376.

Roberg, B., Torgner, I.Aa. and Kvamme, E. (1997) Soluble and membrane bound forms of phosphate activated glutaminase with different kinetic characteristics in pig and rat kidney mitochondria. *Contrib. Nephrol.*, Basel, 121: 11–18.

Rothe, F., Brosz, M. and Storm-Mathisen, J. (1994) Quantitative ultrastructural localization of glutamate dehydrogenase in the rat cerebellar cortex. *Neurosci.*, 62: 1133–1146.

Sahagian, J.J. and Hovhannisian, V.S. (1982) Effect of different modulators on the activity of rat kidney mitochondrial glutaminase and role of thyroid hormones in this process. *Biol. J. Armenii*, 35: 264–271.

Shank, R.P. and Aprison, M.H. (1988) Glutamate as a neurotransmitter. In E. Kvamme, (Ed.), *Glutamine and Glutamate in Mammals*. Vol. 2, CRC Press, Boca Raton, Florida, pp. 3–19.

Shank, R.P., Bennett, G.S., Freytag, S.O. and Campbell, G.L. (1985) Pyruvate carboxylase: An astrocyte-specific enzyme implicated in the replenishment of amino acid neurotransmitter pools. *Brain Res.*, 329: 364–367.

Shapiro, R.A., Haser, W.G. and Curthoys, N.P. (1985) The orientation of phosphate-dependent glutaminase of the inner membrane on rat renal mitochondria. *Arch. Biochem. Biophys.*, 243: 1–7.

Shapiro, R.A., Haser, W.G. and Curthoys, N.P. (1987) Immunoblot analysis of glutaminase peptides in intact and solubilized mitochondria isolated from various rat tissues. *Biochem. J.*, 242: 743–747.

Shapiro, R.A., Farrell, L., Srinivasan, M and Curthoys, N. (1991) Isolation, characterization, and *in vitro* expression of a cDNA that encodes for the kidney isoenzyme of the mitochondrial glutaminase. *J. Biol. Chem.*, 266: 18792–18796.

Smith. Q.R., Momma, S., Aoyaga, M. and Rapoport, S.I. (1987) Kinetics of neutral amino acid transport across the blood-brain barrrier. *J. Neurochem.*, 49: 1651–1658.

Sonnewald, U., Westergaard, N., Krane, J., Unsgard, G., Petersen, S.B. and Schousboe, A. (1991) First direct demonstration of preferential release of citrate from astrocytes using [$^{13}$C]NMR spectroscopy of cultured neurons and astrocytes. *Neurosci. Lett.*, 128: 235–239.

Srinivasan, M., Kalousek, F. and Curthoys, N.P. (1995) *In vitro* characterization of the mitochondrial processing and the potential function of the 68-kDa subunit of renal glutaminase. *J. Biol. Chem.*, 270: 1185–1190.

Storm-Mathisen, J., Danbolt, N.C., Rothe, F., Torp, R., Zhang, N., Aas, J.-E., Kanner, B.I., Langmoen, I. and Ottersen, O.P. (1992) Ultrastructural immunocytochemical observations on the localization, metabolism and transport of glutamate in normal and ischemic tissue. *Prog. Brain. Res.*, 94: 225–241.

Strzelecki, T. and Schoolwerth, A.C. (1984) The significance of the attachment of rat kidney glutaminase to the inner mitochondrial membrane. *Biochim. Biophys. Acta*, 801: 334–341.

Svenneby, G. (1971) Activation of pig brain glutaminase. *J. Neurochem.*, 18: 2201–2208.

Svenneby, G. (1972) Time and temperature dependent activation of pig brain glutaminase. *J. Neurochem.*, 19: 165–174.

Svenneby, G., Roberg, B., Hogstad, S., Torgner, I.AA. and Kvamme, E. (1986). Phosphate-activated glutaminase in the crude mitochondrial fraction (P2 fraction) from human brain cortex. *J. Neurochem.*, 47: 1351–1355.

Svenneby, G., Torgner, I. and Kvamme, E. (1973) Purification of phosphate-dependent pig brain glutaminase. *J. Neurochem.*, 20: 1217–1224.

Svenneby, G., Tveit, B. and Kvamme, E. (1970) Glutaminase from pig renal cortex II. Activation by inorganic and organic acids. *J. Biol. Chem.*, 245: 1878–1882.

Tate, S.S. and Meister, A. (1975) Identity of maleate-stimulated glutaminase activity with $\gamma$-glutamyl transpeptidase in rat kidney. *J. Biol. Chem.*, 250: 4619–4627.

Tveit, B., Svenneby, G. and Kvamme, E. (1970) Kinetic properties of glutaminase from pig renal cortex. *Eur. J. Biochem.*, 14: 337–344.

Van den Berg, C.J. (1973) A model of compartmentation in mouse brain based on glucose and acetate metabolism. In R. Balazs and J.E. Cremer (Eds.), *Metabolic Compartmentation in the Brain*, MacMillian, pp. 137–166.

Wallace, D.R. and Dawson, R.Jr. (1993) Regional differences in glutaminase activation by phosphate and calcium in rat brain: Impairment in aged rats and implications for regional glutaminase isozymes. *Neurochem. Res.*, 18: 1271–1279.

Weil-Malherbe, H. (1950) Significance of glutamic acid for the metabolism of nervous tissue. *Physiol. Rev.*, 30: 549–555.

Weil-Malherbe, H. (1972) Modulators of glutaminase activity. *J. Neurochem.*, 19: 2257– 2267.

Weiler, C.T., Nystrøm, B. and Hamberger, A. (1979) Glutaminase and glutamine synthetase activity in synaptosomes, bulk-isolated glia and neurons. *Brain Res.*, 160: 539–543.

Welbourne, T.C., Francoeur, D., Thornley-Brown, G. and Welbourne, C.J. (1976) Ammonia production and pathways of glutamine utilization in rat kidney slices. *Biophys. Acta*, 444: 644–652.

Westergaard, N., Varming, T., Peng, L., Sonnewald, U., Hertz, L. and Schousboe, A. (1993) Uptake, release and metabolism of alanine in neurons and astrocytes in primary cultures. *J. Neurosci. Res.*, 35: 540–545.

Wong, P.T.-H., McGeer, E.G. and McGeer, P.L. (1981) A sensitive radiometric assay for ornithine aminotransferase: Regional and subcellular distributions in rat brain. *J. Neurochem.*, 36: 501–515.

Wroblewski, J.T., Blaker, W.D. and Meek, J.L. (1985) Ornithine as a precursor of neurotransmitter glutamate: Effect of canaline on ornithine aminotransferase activity and glutamate content in the septum of rat brain. *Brain. Res.*, 329: 161–168.

Wührdig, S. and Kugler, P. (1991) Histochemistry of glutamate metabolizing enzyme in the rat cerebellar cortex. *Neurosci. Lett.*, 130: 165–168.

Yu, A.C.H., Drejer, J., Henry, L. and Schousboe, A. (1983) Pyruvate carboxylase activity in primary cultures of astrocytes and neurons. *J. Neurochem.*, 41: 1484–1487.

Yudkoff, M., Nissim, I., Daikhin, Y., Lin, Z.-P., Nelson, D., Pleasure, D. and Erecinska, M. (1993) Brain glutamate metabolism: Neuronal-astroglial relationships. *Dev. Neurosci.*, 15: 343–350.

O.P. Ottersen, I.A. Langmoen and L. Gjerstad (Eds.)
*Progress in Brain Research*, Vol 116


CHAPTER 7

# Uptake of glutamate into synaptic vesicles

Frode Fonnum[1,2,*], Else Marie Fykse[1] and Svein Roseth[1,3]

[1]*Norwegian Defence Research Establishment, Division for Environmental Toxicology, PO Box 25, N-2007 Kjeller, Norway*
[2]*VISTA (Norwegian Academy of Science and Statoil Company)*
[3]*Present address: ASTRA Norway*

## Introduction

Communication between nerve cells or nerve cells and effector cells occurs either by electrical or chemical transmission (Jessel and Kandel, 1993). In electrically excitable cells (for example horizontal cells in the retina), gap junctions allow electrical charges in one cell to pass directly to the adjoining cell (Brightman and Reese, 1969). In most parts of the vertebrate nervous system however, the communication takes place by chemical neurotransmitter molecules. Neurotransmitters are often classified in two major groups: classical low molecular weight transmitters and peptides. Quantitatively, the most important class of transmitters are the amino acids L-glutamate (later termed glutamate), γ-amino butyric acid (GABA) and glycine which are present in the brain in millimolar concentrations (Fonnum, 1984). The most important excitatory neurotransmitter is glutamate, of which about 10 mM is present in the brain. In glutamatergic nerveterminals the concentration of glutamate is estimated to be about 40 mM (Fonnum, 1991; Bramham et al., 1990). Glutamate has other functions in the brain besides neurotransmission such as general metabolism and protein synthesis. Neurotransmitters are stored in specialized organelles called synaptic vesicles in the brain as well as in the peripheral nervous system. The advantage of storing neurotransmitters in synaptic vesicles is that the vesicles allow the selective storage of high concentrations of neurotransmitters. In the case of glutamate the concentration is estimated to about 100 mM (Fonnum, 1991). This allows a separation of the transmitter pool of glutamate from other pools. However, as discussed below, there is probably a rapid exchange between the different pools of glutamate. In addition this method of storage gives the nerve cells a better way of controlling the release of neurotransmitter.

Several different forms and sizes of synaptic vesicles have been described in fixed brain tissue. These different forms may indicate differences in both protein content and mechanical stability towards the fixation treatment. In general, the excitatory neurotransmitters glutamate and acetylcholine are stored in nerve terminals with small clear round vesicles (Bodian, 1972; Iliakis et al., 1996), the catecholamines in terminals with granule vesicles (Fried et al., 1981) and the inhibitory amino acid transmitters in terminals with pleomorphic vesicles (Iliakis et al., 1996). The gases NO and CO which are also candidates for neurotransmitters, are freely diffusible gases and are not stored in vesicles. Although immunohistochemical methods allow identification of glutamatergic terminals, it has been more difficult for technical

*Corresponding author.
Tel.: +47 63807 803; fax: +47 63807 811; e-mail: frode.fonnum@ffi.no.

reason to demonstrate labelling of glutamate in synaptic vesicles. In the giant reticulo-spinal terminals of lamprey the density of glutamate immuno-staining correlates well with the packing density of the synaptic vesicle (Shupliakov et al., 1992). The diameter of synaptic vesicles were recently estimated to be 35.2 = 3.4 nM in a CA1 hippocampal excitatory synapse that use glutamate as neurotransmitter. The mean inner diameter was subsequently estimated to be 23 nm corresponding to only 500 to 600 molecules per vesicle. The number of vesicles in such a terminal ranged from 41 to 870 (279 ± 176) (Schikorski and Stevens, 1997). In most other reports the size of vesicles ranged between 35–50 nm with 1000–2000 glutamate molecules per vesicle (Edwards, 1995).

The other class of neurotransmitters, peptides, consists of more than 100 neuropeptides which are grouped in several peptide families. They are synthesised in the cell soma and packaged into secretory granules (~100 nm) via the Golgi apparatus and are present in nanomoles to picomoles per gram wet weight. In addition to a peptide, such granules may also contain low molecular weight transmitter. This co-existence was first described by Hökfelt et al. (1980) and today it is believed that most neurons store and secrete more than one messenger, usually one low molecular weight transmitter and a peptide. In the mossy fibre terminals of the hippocampus, glutamate is stored together with dynorphin (Gannon and Terrian, 1991; Salin et al., 1995).

### Vesicular uptake of glutamate

The synthesis of neurotransmitters takes place extravesicularly necessitating an uptake of neurotransmitters into synaptic vesicles. The catecholamines, acetylcholine and transmitter amino acids are all taken up into vesicles by an energy dependent uptake system. The uptake is driven by an electrochemical proton gradient generated by a $Mg^{2+}$-activated vacuolar ATPase (V-ATPase) (Philippu and Matthaei, 1975; Toll and Howard, 1978; Anderson et al., 1982; Naito and Ueda, 1983, 1985; Fykse and Fonnum, 1988; Maycox et al., 1988; Fykse et al., 1989; Kish et al., 1989; Christensen et al., 1990). The kinetic properties, the substrate specificity and the ion dependency clearly distinguish the vesicular uptake of glutamate from the high affinity plasma membrane uptake of glutamate. The high affinity uptake across the plasma membrane is driven by the transmembrane $Na^+$-electrochemical gradient (Fonnum et al., 1980). The absence of $Na^+$-dependency is one of the main criteria to distinguish vesicular uptake from the high affinity plasma membrane uptake (see Billups et al., this volume). In the last years, the mechanisms of the vesicular uptake of glutamate have been described in detail (Fig. 1) (Naito and Ueda, 1985; Maycox et al., 1988; Fykse et al., 1989; Cidon and Shira, 1989; Moriyama et al., 1990; Hartinger and Jahn, 1993; Moriyama and Yamamoto, 1995).

The $Mg^{2+}$-activated V-ATPase, responsible for generating the energy gradient across the vesicle membrane, has been characterised (Nelson, 1992). This ATPase belongs to the vacuolar class of proton pumps. When protons are pumped into the vesicle lumen, a proton gradient (ΔpH), acidic inside, and a membrane potential (ΔΨ), positive inside, are built across the membrane. The relative proportions of ΔpH and ΔΨ vary. In the absence of a permeant anion (e.g. $Cl^-$) the proton charge is not counterbalanced, and thus ΔΨ predominates

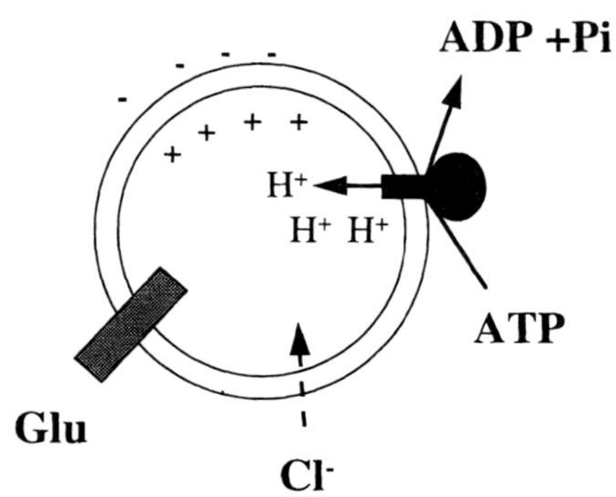


Fig. 1. A model of the uptake mechanisms for glutamate. ATP hydrolysis causes protons to be pumped into the vesicle lumen, a ΔpH, acidic inside, and a ΔΨ, positive inside, are built across the membrane. The driving force for the glutamate uptake is primarily the ΔΨ.

over ΔpH. When a high concentration of $Cl^-$ is present there is a progressive fall of the ΔΨ, and a significant ΔpH is formed across the membrane (Van Dyke, 1988). The V-ATPase is specifically inhibited by the macrolide antibiotics bafilomycin $A_1$ and folimycin. They bind irreversibly, but non-covalently, to the proton channel region of the ATPase (Bowman et al., 1988; Mattsson and Keeling, 1996). The peptide antibiotic destruxin B is also reported to be a specific inhibitor of the V-ATPase, but its inhibitory effect is weaker than that of bafilomycin $A_1$ and folimycin (Muroi et al., 1994). V-ATPase is distinguished from the $E_1$-$E_2$ plasma membrane ATPase and the mitochondrial F1-F0-ATPase in that it is not inhibited by azide, vanadate, oligomycin or ouabain (Nelson, 1991). In common with other ATPases, V-ATPase is inhibited by sulphydryl reagents such as N-ethylmaleimide (Nelson, 1991). Recently it has been shown that V-ATPase is also inhibited by nitric oxide (NO) and that the inhibitory mechanism involves S-nitrosylation of critical sulfhydryl groups in the enzyme which accelerates disulfide bond formation (Wolosker et al., 1996a).

While it is known that all vesicular uptake of neurotransmitters is driven by an electrochemical proton gradient generated by V-ATPase, the precise mechanisms for the uptake of amino acids are still not known. There is disagreement over whether the uptake of glutamate is driven by the ΔΨ only (Maycox et al., 1988; Cidon and Shira, 1989; Moriyama et al., 1990; Hartinger and Jahn, 1993; Moriyama and Yamamoto, 1995), or by both the ΔΨ and ΔpH components of the electrochemical proton gradient (Naito and Ueda, 1985; Shioi and Ueda, 1990; Tabb et al., 1992). The uptake of GABA (Fykse and Fonnum, 1988) and glycine (Kish et al., 1989; Christensen et al., 1990) which have no net charge at neutral pH are supposedly driven by both the ΔpH and ΔΨ components of the electrochemical proton gradient (Hell et al., 1990).

The uptake of glutamate, which is an anion at neutral pH, is stimulated (3- to 6-fold) by low concentrations (1–5 mM) of $Cl^-$, $Br^-$ and to a lesser extent $I^-$. The apparent $K_m$ for $Cl^-$ was 0.2 mM (Roseth et al., 1995). On the other hand, low concentration of $F^-$ (5 mM) inhibits the uptake of glutamate (Naito and Ueda, 1985; Fykse et al., 1989; Fykse and Fonnum, 1991). At high $Cl^-$ concentrations (50–100 mM) the uptake of glutamate is inhibited due to collapse of the ΔΨ as discussed above (Maycox et al., 1988). The uptake of GABA and glycine are not activated by $Cl^-$ (Fykse and Fonnum, 1988; Christensen et al., 1990). However, in some studies a weak stimulation (50 to 100%) has been observed (Kish et al., 1989; Hell et al., 1990; Burger et al., 1991).

The stimulation of the uptake of glutamate by $Cl^-$ is not fully understood, although different mechanisms have been suggested. Chloride ions may activate the transporter molecule directly by influx into the vesicle lumen or through binding to an anion sensitive site. Different anion channel blockers, 4-acetamido-4′-isothiocyano-stilbene-2,2′-disulphonic acid (SITS), 4,4′-diiso-thiocyano-2,2′-stilbene-disulphonic acid (DIDS) and 5-nitro-2-(3-phenylpropylamino)-benzoic acid (N144), are known to inhibit the uptake of glutamate (Fykse and Fonnum, 1991). Subsequent work, however, has shown that SITS and N144 are competitive inhibitors of glutamate and are non competitive towards $Cl^-$ (Roseth el al., 1995). DIDS has also later been used by Hartinger and Jahn (1993) and Moriyama and Yamamoto (1995) to inhibit the vesicular uptake of glutamate. Moriyama and Yamamoto (1995) have shown that pinealocytes (endocrine cells that synthesise and secrete melatonin) possess a large number of synaptic-like vesicles which accumulate glutamate. They find the same anion dependency in this uptake as Fykse et al. (1989) found for vesicular glutamate uptake. $Cl^-$ flux did not occur through either the glutamate transporter or a $Cl^-$ channel of the microvesicles. Therefore, Moriyama and Yamamoto concluded that $Cl^-$ regulates uptake probably via an anion binding site on the transporter. A similar conclusion was also reached by Hartinger and Jahn (1993) who also found that DIDS inhibited glutamate uptake more than the the ΔΨ, ΔpH and the V-ATPase activity. Furthermore, when the proton pump of synaptic vesicles was substituted by the proton pump of bacteriorhodopsin, uptake of glutamate was still dependent on $Cl^-$ (Maycox et al., 1990). Recently, Wolosker et al. (1996b) proposed that $Cl^-$ could affect the uptake of glutamate by two different mechanisms. One is related to changes of the proportions between the $\Delta\Psi$ and the $\Delta$pH components of the electrochemical proton gradient, whereas the other involves a direct interaction of $Cl^-$ with the glutamate transporter. They proposed that the $\Delta\Psi$ component controls the apparent affinity for glutamate whereas the $\Delta$pH component is important for opposing the effect of high chloride concentration. In addition the, $\Delta$pH component would be important for retaining glutamate inside the vesicle.

Recently, Xu et al. (1996) demonstrated that phosphocreatine (PCr) could substitute as an energy source for the uptake of low concentration of glutamate (50 $\mu$M). The PCr-dependent uptake of glutamate is NEM-insensitive, as opposed to $Mg^{2+}$- or $Cl^-$-dependent uptake. It was suggested that PCr donates its high energy phosphate group to a novel kinase or acts as an allosteric effector for a novel enzyme that regulates the glutamate transporter. At physiological concentrations of ATP and PCr in the brain, the latter uptake would be 30% of ATP supported uptake. The PCr-driven uptake was not inhibited by DIDS which also indicates that a different transport mechanism is involved.

The storage of neurotransmitters in vesicles has varying stability. Synaptic vesicles containing acetylcholine and catecholamines can be isolated by hyperosmotic treatment (Fried et al., 1981; Whittaker, 1986). Proton NMR and $^{31}$P-NMR show that acetylcholine and ATP are present in free solution in synaptic vesicles (Stadler and Fuldner, 1980; Fuldner and Stadler, 1982). In contrast, the vapour pressure osmometer shows that osmolarity was reduced in a solution of catecholamines and ATP indicating a complex binding between the two (Kopell and Westhead, 1982). No such investigation has been undertaken for the transmitter amino acids. The storage of amino acids in synaptic vesicles seems to be more labile than the other transmitters, requiring the preservation of an energy gradient across the vesicle membrane (Burger et al., 1989; Carlson and Ueda, 1990; Roseth et al., 1998a). Storage of glutamate in the absence of ATP at 37° leads to a 50% loss within two minutes, the storage of GABA is even more labile (Roseth et al., 1998a). Glutamate stored in synaptic vesicles turns over with a half-time of minutes, even during active uptake under physiological conditions (Carlson and Ueda, 1990; Wang and Floor, 1994). The efflux of glutamate from the vesicles is probably caused both by a transport-dependent and a transport-independent efflux mechanism.

**Specificity**

*Substrate*

The vesicular uptake of glutamate is substrate specific and clearly different from the uptake of GABA/glycine, acetylcholine and catecholamines into synaptic vesicles. The $K_m$ value for the uptake of L-glutamate is 1 mM (Naito and Ueda, 1985). It appears that L- and D-aspartate are not taken up into synaptic vesicles (Naito and Ueda, 1985; Christensen et al., 1991; Moriyama and Yamamoto, 1995). However, certain cyclic amino acids that are metabotropic glutamate receptor agonists, are taken up by the vesicular transport mechanism. Of all the substrates described this far, the one showing the highest affinity is trans-1-aminocyclopentane-1,3 dicarboxylate with a $K_m$ value of 40 $\mu$M. Other substrate analogues such as trans and cis-1-aminocyclobutane-1,3-dicarboxylate and cis-1-aminocyclopentane-1,3-dicarboxylate are also substrates with slighly higher affinity than glutamate (Winter and Ueda, 1993). Substrate analogues such as D-glutamate, L-aminoadipate and L-cysteate are poor inhibitors and are therefore most likely poor substrates (Dunlop et al., 1991; Fykse et al., 1992). The specificity of the proton-driven vesicular uptake is much higher than that of high affinity plasma membrane uptake. Vesicular uptake seems to be the most important property for identification of a glutamatergic terminal.

*Species*

The vesicular proteins including the transport system for glutamate are well preserved throughout the vertebrate class both with respect to chloride dependency, ATP dependency and substrate specificity. In two recent studies of several different species such as fish, avian, amphibian and mammals, the $K_m$ value for glutamate uptake was about 1 mM and the uptake being dependent upon low concentrations of $Cl^-$ (Tabb and Ueda, 1991; Roseth og Fonnum, 1995). Aspartate was neither a substrate nor an inhibitor in any of the species.

**Vesicular inhibitors and their targets**

Synaptic vesicles and their content are potential target for a wide spectrum of compounds. Inhibition of vesicular uptake leads to a less efficient neurotransmission (Van der Klot, 1991). There are several ways of inhibiting the uptake (Fig. 2). A

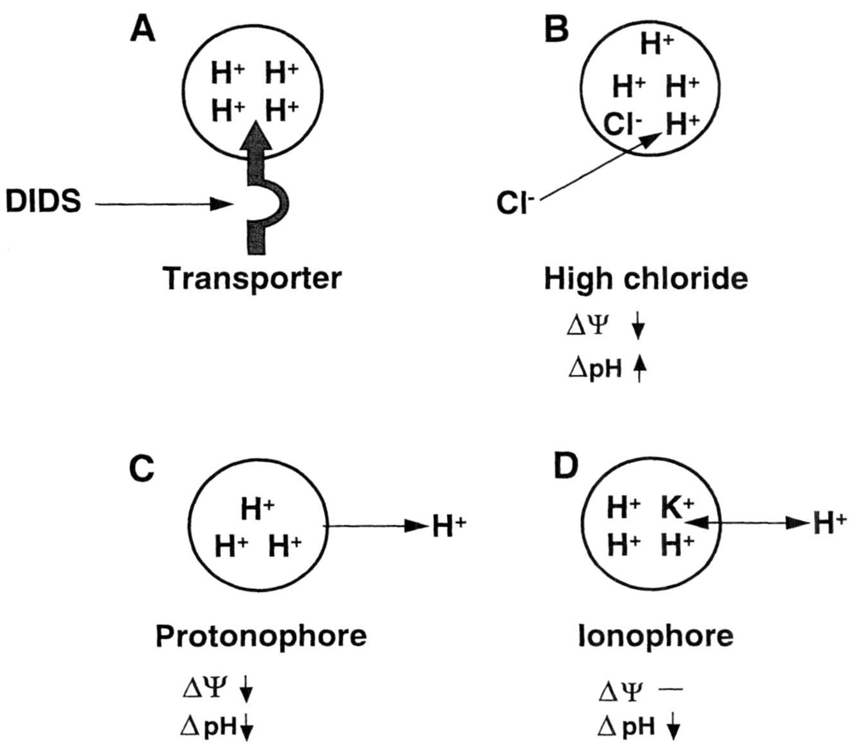


Fig. 2. Schematic outline of the different ways to inhibit vesicular uptake. A) A transmitter specific inhibitor binds to the transporter molecule. Evans Blue and Trypan Blue are examples of such inhibitors. A $Cl^-$ binding site on the transporter has been proposed, and DIDS, a potent inhibitor of glutamate uptake, binds to this site. B) A high concentration of $Cl^-$ inhibits the vesicular uptake of glutamate. When protons are pumped into the vesicle lumen, a $\Delta pH$, acidic inside, and a $\Delta\Psi$, positive inside, are built across the membrane. The relative proportion of $\Delta pH$ and $\Delta\Psi$ varies. A high concentration of $Cl^-$ gives a progressive fall in the $\Delta\Psi$ and a $\Delta pH$ is formed which causes the inhibition of the vesicular uptake of glutamate. C) Protonophores, for example FCCP, act as mobile carriers of $H^+$ ions. Both the transmembrane $\Delta pH$ and $\Delta\Psi$ are abolished and this will non-selectively inhibit uptake of all neurotransmitters. D). Nigericin, an example of an ionophore, causes exchange of $H^+$ ions for $K^+$ ions across the membrane in the presence of extravesicular $K^+$. The transmembrane pH gradient will be abolished without affecting the $\Delta\Psi$. Nigericin is a selective inhibitor for uptake of the different neurotransmitters. The uptake of dopamine and GABA are inhibited to a larger extent that the uptake of glutamate.

transmitter specific inhibition requires an effect on the different transporter molecules of the synaptic vesicle. An ionophore which abolishes either the $\Delta$pH or the $\Delta\Psi$ can be partly specific, since the uptake of catecholamines are mainly dependent on the $\Delta$pH while the uptake of glutamate is mainly driven by the $\Delta\Psi$. Inhibitors of V-ATPase, or an ionophore which destroys the proton electrochemical gradient, will inhibit the uptake of all transmitters to the same extent and will therefore be unspecific.

*Inhibitors of vesicular glutamate transporter*

The best inhibitors of vesicular glutamate uptake are dyes with a biphenyl group and amino and sulphonic acid-substituted naphthyl groups at the end such as Trypan Blue and Evans Blue (Figure 3). Trypan Blue and Evans Blue inhibit the vesicular uptake of glutamate with $K_i$ values at approximately 50 and 90 nM (Roseth et al., 1995, 1998b). At these concentrations, the inhibitors have no significant effect on the ATP dependent

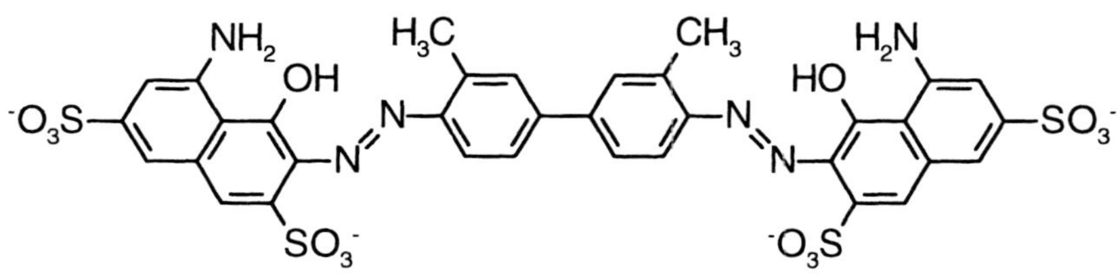

Trypan Blue

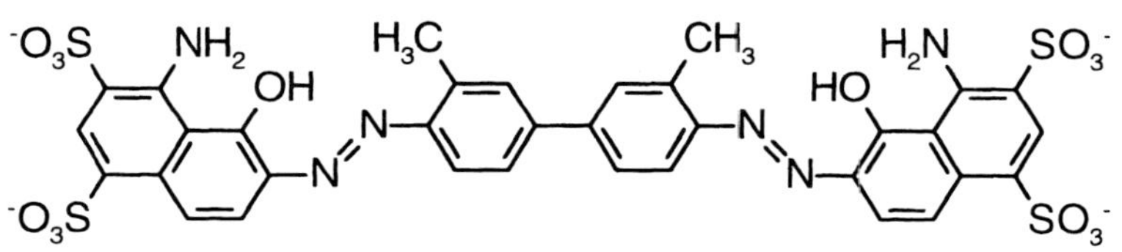

Evans Blue

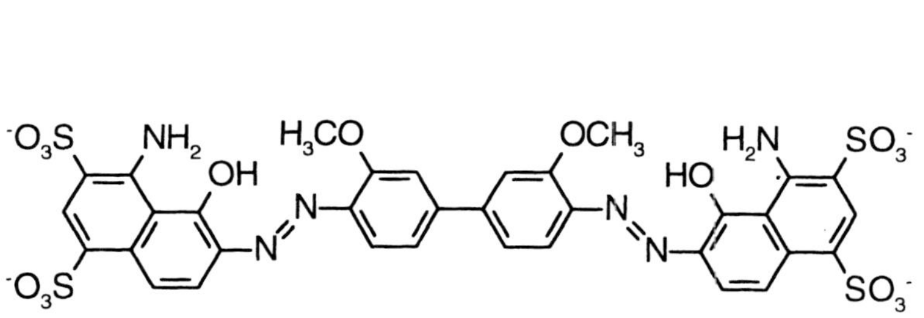

Chicago Skye Blue 6B

Fig. 3. Structures of inhibitors of vesicular glutamate uptake. The structures show the most likely conformation in gas phase.

generation of a transmembrane electrochemical gradient. Kinetic studies show that Trypan Blue and Evans Blue are competitive inhibitors of glutamate uptake. Hence, the affinity of the glutamate transporter for Trypan Blue and Evans Blue is about 10 000–20 000 fold greater than for glutamate itself. Trypan Blue and Evans Blue also inhibit the uptake of GABA, but this is only at high concentrations ($>10$ $\mu$M) (Roseth et al., 1995, 1998b). Trypan Blue and Evans Blue did not affect the plasma membrane uptake of glutamate at millimolar concentrations. In contrast, the AMPA/kainate receptor is inhibited by 50% at 355 nM of Evans Blue (Keller et al., 1993). This could indicate that there are some structural similarities between this receptor and the vesicular glutamate transporter. The two dyes have conjugated double bonds and data modelling studies have shown that these bonds stabilise the molecule in a planar conformation. Insertion of different functional groups in the naphthalene rings does not seem to change this planar conformation. Several other dyes related to Trypan Blue and Evans Blue were also potent inhibitors of glutamate uptake (Roseth et al., 1998b). A related dye, Chicago Skye Blue 6B, which has two methoxy groups instead of two methyl groups, is a far less potent inhibitor than Trypan Blue and Evans Blue (Roseth et al., 1995). A possible explanation may be that the outward orientation of the methyl groups in Trypan Blue and Evans Blue fits more precisely into the glutamate binding site than the methoxy group in Chicago Skye Blue 6B.

Other inhibitors of glutamate uptake that have been described include a series of $Cl^-$ channel blockers such as SITS, N-144, furosemide and bumetamide (Fykse and Fonnum, 1991). They all contain two anionic groups and are reversible and competitive inhibitors of glutamate uptake. They were more potent inhibitors of glutamate uptake than of GABA uptake. Other inhibitors of the vesicular uptake of glutamate have been described, but these have low specificity and potency. These include bromocriptine, sulphur amino acids and glutamate analogues (Carlson et al., 1989; Dunlop et al., 1991; Fykse et al., 1992; Winter and Ueda, 1993). Initially, bromocriptine was reported to inhibit the vesicular uptake of glutamate with no detectable effect on the vesicular membrane potential or pH gradient (Carlson et al., 1989). However, a later study by Moriyama et al. (1990) showed that bromocriptine also affected the $\Delta$pH and the $\Delta\Psi$. Several glutamate analogues such as L- and D-aspartate, D-glutamate, kainate and L-cysteate are poor inhibitors of glutamate uptake (Table 1).

*The effect of ionophores on the vesicular uptake of glutamate*

The synaptic vesicle membrane consists of a lipid bilayer which functions as a barrier in order to concentrate specific molecules inside the vesicle.

TABLE 1

Effects of different inhibitors on the vesicular uptake of glutamate

| Inhibitors | $IC_{50}$ values |
|---|---|
| Trypan Blue | 49 nM |
| Evans Blue | 87 nM |
| Naphthol Blue Black | 327 nM |
| Reactive Blue | 428 nM |
| Ponceau SS | 795 nM |
| DIDS | 1400 nM |
| Chicago Skye Blue | 3000 nM |
| SITS | 3100 nM |
| $N_{144}$ | 15000 nM |
| **Less potent inhibitors** | **% uptake at 10 mM** |
| Kainate | 8 ± 1 |
| L-glutamate | 22 ± 2 |
| L-Homocysteate | 44 ± 5 |
| D-glutamate | 47 ± 8 |
| D-Aspartate | 72 ± 6 |
| L-Cysteate | 89 ± 12 |
| L-Aspartate | 95 ± 8 |

The data are taken from Fykse and Fonnum, 1988; Fykse et al., 1992; Roseth et al., 1995; Roseth et al., 1998b. The concentration of glutamate was 1 mM during these experiments.

The importance of the electrochemical proton gradient across the vesicle membrane as a driving force for the vesicular uptake of neurotransmitters is well established. Hence, vesicular uptake is highly sensitive to lipid-soluble molecules which may alter chemical and electrical gradients across the vesicle membrane. Protonophores act as mobile carriers of $H^+$-ions. They are lipid-soluble weak acids that selectively increase the permeability of the membrane to $H^+$-ions. Several studies (Cidon and Sihra, 1989; Fykse et al., 1989; Naito and Ueda, 1985; Carlson and Ueda, 1990) have shown that protonophores, such as carbonylcyanide-*p*-(trifluoromethoxy)phenylhydrazone (FCCP) and carbonylcyanid-*m*-chlorophenylhydrazone (CCCP), inhibit the vesicular uptake of different neurotransmitters to the same extent.

Valinomycin and nigericin are other examples of mobile ionophores. Valinomycin forms a lipid-soluble complex with $K^+$ which readily passes through the membrane and partially abolishes the $\Delta\Psi$ without affecting the $\Delta$pH. Valinomycin has been shown to inhibit the vesicular uptake of glutamate (Cidon and Sihra, 1989; Moriyama et al., 1990), but the $\Delta$pH driven uptake of catecholamines was not affected (Toll and Howard, 1978). Nigericin exchanges $H^+$-ions for $K^+$-ions across the membrane and will discharge any pH gradient in the presence of $K^+$-ions without changing the membrane potential. Fykse and Fonnum (1996) showed that in the presence of different concentrations of $K^+$-ions nigericin proved to be a more potent inhibitor of the uptake of dopamine and GABA than of glutamate. Dopamine and GABA uptake are more dependent on the $\Delta$pH whereas the glutamate uptake is more dependent on the $\Delta\Psi$. Some ionophores can act as channel formers. Both mobile carriers and channel formers operate by shielding the charge of the transported ion so it can penetrate the hydrophobic interior of the lipid bilayer. A well-known channel former is gramicidin A which selectively allows monovalent cations ($H^+$, $Na^+$, $K^+$) to flow down their electrochemical gradients, thereby abolishing both the $\Delta$pH and the $\Delta\Psi$. Gramicidin has an equal effect on the vesicular uptake of glutamate and GABA (Fykse et al., 1989). The ionophores have been extensively used both to investigate the mechanisms of the driving force of the vesicular uptake of different neurotransmitters and in determining which component of the electrochemical proton gradient is important for the uptake.

*Endogenous inhibitors*

Polyunsaturated fatty acids such as arachidonic acid, eicosapentanoic acid and linolenic acid inhibited the vesicular uptake of glutamate and GABA to the same extent (Roseth et al., 1998a). It was also found that long-chain polyunsaturated fatty acids strongly, and in a dose-dependent manner, inhibit the generation of a proton gradient across the vesicular membrane and also rapidly collapse an existing proton gradient. The monounsaturated fatty acids, such as oleic acid, were weak inhibitors. Furthermore, the saturated fatty acid arachidic acid, had no effect on the electrochemical proton gradient nor did it inhibit vesicular uptake. Therefore, it seems that in vesicles, polyunsaturated fatty acids like arachidonic acid, act like protonophores such as FCCP and CCCP. Protonophoric effects of arachidonic acids have also been reported in other membrane systems. Wang et al. (1995) reported that arachidonic acid affects glutamate-induced intracellular pH changes in cultured hippocampal neurons in a manner similar to FCCP. Furthermore, Heaton and Nicholls (1976) and Husebye and Flatmark (1984) concluded that polyunsaturated fatty acids like arachidonic acid act as uncouplers, dissipating pH gradients in mitochondria and chromaffin granule ghosts, respectively. We believe that the inhibitory effect of the polyunsatuated fatty acid compared to the saturated fatty acid may be explained by the lower melting point and therefore the higher fluidity of the former. The polyunsaturated fatty acids are probably intercalated into membranes.

Fatty acids such as arachidonic acid are released in high concentrations during such conditions as ischemia, hypoxia and hypoglycemia (Agardh et al.,

1980; Gardiner et al., 1981). During long term potentiation fatty acids are also released (Williams et al., 1989). The polyunsaturated fatty acids released during such conditions could influence the vesicular function. An inhibition of the vesicular uptake could increase the cytosolic concentration of neurotransmitters. Polyunsaturated fatty acids are also shown to inhibit the high affinity uptake of neurotransmitters over the plasma membrane (Rhoads et al., 1982; Chan et al., 1983; Yu et al., 1986; Volterra et al., 1992; Lundy and McBean, 1995; Trotti et al., 1995). Blockage of both uptake systems could lead to a high extracellular concentration of glutamate. This could contribute to the toxic effects of glutamate as well as the long term potentiation. Fatty acids are barely soluble in aqueous solutions and accumulate in the lipid phase; furthermore, the concentration needed to inhibit the uptake depends on the amount of vesicular material rather than its nominal solution concentration. Therefore, reports that express the fatty acid concentrations as nominal concentrations may be misleading.

Another endogenous inhibitor of the vesicular uptake is bilirubin. Bilirubin is a well-known neurotoxic agent having multiple effects on neuronal function that is primarily seen in the neonate; however, it may appear later in life as hyperbilirubinemia. During in vitro experiments, several different effects have been observed, including decreased neuronal viability, uncoupling of oxidative phosphorylation, inhibition of protein phosphorylation reactions, decreased DNA synthesis, decreased protein synthesis, increased membrane permeability and inhibition of a large number of enzymes (Constantopoulos and Matsaniotis, 1976; Karp, 1979; Morphis et al., 1982; Hansen et al., 1988a, 1996a). It was recently reported for the first time that bilirubin has an effect on the vesicular uptake of neurotransmitters (Roseth et al., 1998c). Bilirubin inhibits glutamate, GABA, dopamine and noradrenaline uptake to the same extent with an $IC_{50}$ value of approximately 75 $\mu$M, yet it has no effect on the generation of the vesicular pH gradient. Bilirubin is a lipophilic compound and its effect is dependent on the lipid storage. The inhibitory action of bilirubin may be due to an interaction between bilirubin and the transport proteins and/or lipids in the vesicular membrane. The results on vesicular uptake inhibition are in agreement with Hansen et al. (1996b) who conclude from release experiments that bilirubin results in reduced neurotransmitter content of the vesicle. The inhibition of the vesicular neurotransmitter uptake may also explain the reduced synaptic activation shown in transverse hippocampal slices (Hansen et al., 1988b), and may agree with the alterations in auditory evoked potentials in vivo (Nwaesei et al., 1984).

Recently, Özkan et al. (1997) isolated three proteins of molecular weight 135000 that probably are derived from fodrin, and they inhibit both glutamate and GABA uptake into vesicles. The authors suggest that these proteins could be important in the regulation of uptake of amino acid neurotransmitters. In the study, the $IC_{50}$ value was about 26 nM for all three proteins. Considering the high molecular weight, considerable amounts of the protein must be present to have any physiological effect.

*Environmental toxins*

Several persistent organochloride compounds, such as polychlorinated biphenyls (PCB) and the pesticides DDT and toxaphene, are lipophilic and widely distributed in nature. They accumulate in wildlife, particularly in the arctic regions, and bioconcentrate in the food chain (Evans et al., 1991). Several of these compounds have neurotoxic effects, including on human beings (Seegal, 1996). It was recently shown that PCB in concentrations found in the brain of arctic wild life, will inhibit the uptake of neurotransmitters such as glutamate and dopamine into synaptic vesicles. Thus 10 $\mu$M 2,2-dichlorobiphenyl inhibits vesicular uptake more than 50 percent (Andersen et al., unpublished). Similar effects may be expected of several of these environmental toxins. The concentration of PCB found in the brain of dead animals was 10 times higher than the concentration inhibiting vesicular uptake.

**Glutamate and the vesicular hypothesis**

The exocytotic release of neurotransmitters is believed to be quantal, that is the release of packets of neurotransmitters of unitary size of individual synaptic vesicles. This hypothesis is based on two important and independent observations of the peripheral nervous system from the 1950s. First, the smallest single post-synaptic response in the cholinergic frog neuromuscular junction was found to be composed of multiples of the smallest recorded potentials, the quantal response (Fatt and Katz, 1952). Secondly, electronmicroscopic studies showed that nerve endings contained a large number of small round particles now known as synaptic vesicles (Sjøstrand, 1953; De Robertis and Bennett, 1954). These studies led to the hypothesis of Del Castillo and Katz (1956) that synaptic vesicles contain neurotransmitters and that their release is the equivalent of the electrophysiologically observed quanta. This hypothesis was strongly supported by evidence of the isolation of cholinergic synaptic vesicles from the brain (De Robertis et al., 1963; Whittaker et al., 1963, 1964). It is not apparent, however, whether the term quantal is a unit of response or a unit of release. In studying the peripheral nervous system, Van Der Klot (1991) shows that the size of the quantum depends on the transmitter content of the vesicle. Increased vesicular content gives increased quantal response.

It has been more difficult to provide electrophysiological evidence from the CNS of miniature events that vary in steps and are of unit size. Recently, however, such evidence has been obtained from glutamatergic synapses in the hippocampus. Nevertheless, the increase in steps tends to be larger than the average size of the miniature events (Kullman and Nicoll, 1992; Stern et al., 1992; Jonas et al., 1993). Later Paulsen and Hegglund (1994) showed that increase in steps is equal to the average size of miniatures from retinogeniculate synapses in cats. They suggested that the skew distribution of amplitude events observed by others as well as in their own preparation might be due to multiquantal miniatures. Multiquantal miniatures are explained as the coupled release of vesicles from several release sites in the same terminal. Other authors have shown a skewed distribution of the miniature events and explained this distribution as derived from vesicles of different sizes and therefore different transmitter content (Bekkers et al., 1990). In principle this series of papers is consistent with the hypothesis that the transmitter content of a vesicle represents the quantum.

An alternative hypothesis forwarded by Edwards (1995) is that receptors are saturated by the release of a single vesicle. Miniature events of different steps might then be due to the release of one vesicle against one, two or three clusters of receptors. In agreement, several authors have suggested that receptor saturation occurs at glutamatergic synapses (Faber et al., 1992, Kullman, 1993). By using glutamate antagonists Tong and Jahr (1994) were able to show that multivesicular release occurs within a single active zone. Despite the increase in transmitter concentration, NMDA and non-NMDA receptors of the synaptic response do not change in their relative amplitude. The hypothesis also concurs with the vesicular release of transmitter glutamate but the quantum is determined by receptor numbers.

The first morphological evidence for vesicular release of neurotransmitters was provided by Ceccarelli and Hurlbut (1980). Other evidence has evolved from studies of larger particles such as granulocytes, mast cells and chromaffin cells utilising of the patch clamp technique and the sensitive high resolution amperometry (Lindau and Neher, 1988; Wightman et al., 1991). The fusion of a granule or a vesicle with the plasma membrane gives an increase of cell membrane that corresponds to the granule or vesicle membrane area (Lindau and Neher, 1988). The combination of the patch clamp and amperometric techniques demonstrated that upon fusion of mast cells with the plasma membrane, a spike in the release of neurotransmitters was seen, which supports the vesicle hypothesis (Alvarez de Toledo, 1993).

The importance of vesicular storage of glutamate for glutamate release has also been shown by using inhibitors of vesicular uptake. Bafilomycin $A_1$ inhibits the potassium-stimulated calcium-dependent release of glutamate from cerebellar granule cells (Fykse et al., unpublished).

**Perspectives**

The development of specific inhibitors of vesicular neurotransmitter uptake systems may in the future be of clinical importance in treating of several diseases in the CNS. The specific inhibitor of vesicular catecholamine uptake, the *Rauvolfia* alkaloid reserpine, was the first effective remedy in the treatment of psychoses. Its effect is caused by the selective depletion of noradrenaline and serotonin stores in neurons followed by lowered catecholaminergic neurotransmission. In theory, diseases caused by imbalance in the neurotransmitter distribution could be treated by a regulation of the vesicular content of the current transmitter.

The potential therapeutic agents would have to fulfil several important criteria. First, the substance must be very specific. If the drug has a low specificity, it may interfere with other proteins or targets and create undesirable effects. Secondly, the dose of the agent must be monitored carefully to achieve the desired effect without too many unwanted or even toxic effects. Finally, the therapeutic agent has to be lipid soluble to be able to cross the blood-brain barrier. There are several pathological conditions such as epilepsy, stroke and ischemia or neurological diseases such as a amyloid lateral sclerosis, Alzheimers diseases and Huntingtons disease where regulation of glutamate release could be relevant.

It seems likely that the inhibition of vesicular uptake by both arachidonic acid and bilirubin will be of physiological importance. The simultaneous inhibition of both vesicular and plasmamembrane uptake may well contribute to a high extracellular toxic level of glutamate during ischemia and hypoglycemia. Similarly, the neurotoxic effect of bilirubin is well known and a high extracellular level of glutamate may contribute to the toxic effect.

**Conclusion**

Glutamate is taken up and stored in synaptic vesicles by a $Mg^{2+}$-ATP dependent process. The driving force for vesicular glutamate uptake is primarily the membrane potential. Uptake is stimulated by low concentrations of $Cl^-$ or $Br^-$ and inhibited by $F^-$. Uptake is inhibited by several anion channel blockers. The mechanisms for the stimulation of $Cl^-$ are not well understood, but one explanation suggests that it may be due to an effect on the transporter.

Vesicular uptake is highly specific for L-glutamate. D-glutamate, L-, D-aspartate, L-aminoadipate and kainate are poor substrates. Certain cyclic glutamate analogues, that are metabotropic glutamate agonists, are good substrates for vesicular uptake. The high specificity of the vesicular uptake of glutamate plays an important role in the identification of glutamatergic neurons.

Several dyes with biphenyl groups and naphtalene amino- and sulphonic acid substituted groups at the end such as Trypan blue and Evans blue are excellent inhibitors in the nanomolar range. Unfortunately, they do not easily penetrate membranes. Polyunsaturated fatty acids such as arachidonic acid and bilirubin are good but not specific inhibitors of the vesicular uptake of glutamate. This may be significant under adverse conditions such as ischemia or hypoglycemia when large concentrations of such fatty acids are released or else in the case of hyperbilirubinemia. Recently, environmental toxins such as PCB have also been shown to inhibit vesicular uptake and they may play a role in the neurotoxic effect of such compounds.

The synaptic vesicular uptake system for glutamate is highly conserved throughout the vertebrate class both with respect to energy requirement and substrate specificity. This concurs with other studies which claim that the overall architecture of synaptic vesicle proteins appears to be highly conserved during evolution, and homologues of these proteins govern the constitutive secretion in yeast (Ferro-Novick and Jahn, 1994).

## Acknowledgements

The present work was supported by a Research Fellowship awarded to Dr Svein Roseth by the Research Council of Norway and a Research Professorship awarded to Dr Frode Fonnum from VISTA (Norwegian Academy of Science and Statoil Company, Norway).


## References


Agardh, C.D., Westerberg, E. and Siesjö, B.K. (1980) Severe hypoglycemia leads to accumulation of arachidonic acid in brain tissue. *Acta. Physiol. Scand.*, 109: 115–116.

Alvarez de Toledo, G., Fernandez Chacon, R. and Fernandez, J.M. (1993) Release of secretory products during transient vesicle fusion. *Nature*, 363: 554–558.

Anderson, D.C., King, S.C. and Parson, S.M. (1982) Proton gradient linkage to active uptake of [3H]acetylcholine by *Torpedo* electric organ synaptic vesicles. *Biochem.*, 21: 3037–3043.

Bekkers, J.M., Richerson, G.B. and Stevens, C.F. (1990) Origin of variability in quantal size in cultured hippocampal neurons and hippocampal slices. *Proc. Natl. Acad. Sci., USA*, 87: 5359–5362.

Bodian, D. (1972) Synaptic diversity and characterization by electronmicroscopy. In Pappas, G.D. and Purpura, P.D. (eds.) Structure and function of synapses. New York: Raven pp 45–65.

Bowman, E.J., Siebers, A. and Altendorf, K. (1988) Bafilomycins: A class of inhibitors of membrane ATPases from microorganisms, animal cells, and plant cells. *Proc. Natl. Acad. Sci. US*, 85: 7972–7976.

Bramham, C.R., Torp, R., Zhang, N., Storm Mathisen, J. and Ottersen, O.P. (1990) Distribution of glutamate-like immunoreactivity in excitatory hippocampal pathways: A semiquantitative electron microscopic study in rats. *Neurosci.*, 39: 405–17.

Brightman, M.W. and Reese, T.S. (1969) Junctions between intimately apposed cell membranes in the vertebrate brain. *J. Cell. Biol.*, 40: 648–677.

Burger, P.M., Hell, J., Mehl, E., Krasel, C., Lottespleich, F. and Jahn, R. (1989) Synaptic vesicles immunoisolated from rat cerebral cortex contain high levels of glutamate. *Neuron*, 3: 715–720.

Burger, P.M., Hell, J., Mehl, E., Krasel, C., Lottspeich, F., Jahn, R. (1991) GABA and glycine in synaptic vesicles: Storage and transport characteristics. *Neuron*, 7: 287–293.

Carlson, M.D., Kish, P.E. and Ueda, T. (1989) Glutamate uptake into synaptic vesicles: Competitive inhibition by bromocriptine. *J. Neurochem.*, 53: 1889–1894.

Carlson, M.D. and Ueda, T. (1990) Accumulated glutamate levels in the synaptic vesicle are not maintained in the absence of active transport. *Neurosci. Lett.*, 110: 325–330.

Castillo, J. Del and Katz, B. (1956) Biophysical aspects of neuro-muscular transmission. *Prog. Biophys. Chem.*, 6: 121–170.

Ceccarelli, B. and Hurlbut, W. P. (1980) Vesicle hypothesis of the release of quanta of acetylcholine. *Physiol. Rev.*, 60: 396–441.

Chan, P.H., Kerlan, R. and Fishman, R.A. (1983) Reductions of GABA and glutamate uptake and $Na^+/K^+$ ATPase activity in brain slices and synaptosomes by arachidonic acid. *J. Neurochem.*, 40: 309–316.

Christensen, H., Fykse, E.M., and Fonnum, F. (1990) Uptake of glycine into synaptic vesicles isolated from rat spinal cord. *J. Neurochem.*, 54: 1142–1147.

Christensen, H., Fykse, E.M. and Fonnum, F. (1991) Inhibition of $\gamma$ aminobutyric acid and glycine uptake into synaptic vesicles. *Eur. J. Pharmacol.*, 207: 73–79.

Cidon, S. and Sihra, T. (1989) Characterization of a $H^+$-ATPase in rat brain synaptic vesicles. *J. Biol. Chem.*, 264: 8281–8288.

Constantopoulos, A. and Matsaniotis, N. (1976) Bilirubin inhibition of protein kinase: Its prevention by cyclic AMP. *Cytobios.*, 17: 17–20.

De Robertis, E.D.P and Bennett, HS (1954) Submicroscopic vesicular component in the synapse. *Fed. Proc.* 13: 85.

De Robertis, E.D.P., Rodriguez de Lores Arnaiz, G., Salganic-off, L., Pellegrino de Iraldi, A. and Zieher, L.M. (1963) Isolation of synaptic vesicles and structural organization of the acetylcholine system within brain nerve endings. *J. Neurochem.*, 10: 225–235.

Del Castillo, J. and Katz, B. (1956) Biophysical aspects of neuromuscular transmission. *Prog. Biophys. Biophys. Chem.*, 6: 121–170.

Dunlop, J., Fear, A. and Griffiths, R. (1991) Glutamate uptake into synaptic vesicles-inhibition by sulphur amino acids. *Neuroreport*, 2: 377–379.

Edwards, F.A. (1995) Anatomy and electrophysiology of fast central synapses lead to a structural model for long term potentiation. *Physiol. Rev.*, 75: 759–787.

Evans, M.S., Noguchi, G.E. and Rice, C.P. (1991) The biomagnification of polychlorinated biphenyls, toxaphene and DDT compounds in a Lake Michigan offshore food web. *Arch. Environ. Contam. Toxicol.*, 20: 87– 93.

Faber, D.S., Young, W.S., Legendre, P. and Korn, H. (1992) Intrinsic quantal variability due to stochastic properties of receptor-transmitter interactions. *Science*, 258: 1494–1498.

Fatt, P. and Katz, B. (1952) Spontaneous subthreshold activity at motor endings. *J. Physiol. (London)*, 117: 109–128.

Ferro-Novick, S. and Jahn, R. (1994) Vesicle fusion from yeast to man. *Nature*, 370: 191–193.

Fonnum, F., Lund Karlsen, R., Malte-Sørensen, D., Sterri, S.H. and Walaas, I. (1980) High affinity transport systems and their role in transmitter action. In: Cotman, C.W., Poste,

G. and Nicolson, G.L. (eds) The cell surface and the neuronal function. Elsevier/North-Holland Biomedical Press, 445–504.

Fonnum, F. (1984) Glutamate: A neurotransmitter in mammalian brain. *J. Neurochem.*, 42: 1–11.

Fonnum, F. (1991) Neurochemical studies of glutamate-mediated neurotransmission. *Fidia Research Foundation Symposium series* 5: 15–25.

Fried, G., Thureson-Klein, A.K. and Lagercrantz, H. (1981) Noradrenaline content correlated to matrix density in small noradrenergic vesicles from the rat seminal ducts. *Neuroscience*, 6: 787–800.

Fuldner, H.H. and Stadler, H. (1982) $P^{31}$-NMR analysis of synaptic vesicles: Status of ATP and internal pH. *Eur. J. Biochem.*, 121: 519–524.

Fykse, E.M. and Fonnum, F. (1988) Uptake of $\gamma$-aminobutyric acid by a synaptic vesicle fraction isolated from rat brain. *J. Neurochem.*, 50: 1237–1242.

Fykse, E.M., Christensen, H., and Fonnum, F. (1989) Comparison of the properties of $\gamma$-aminobutyric acid and L-glutamate uptake into synaptic vesicles isolated from rat brain. *J. Neurochem.*, 52: 946–951.

Fykse, E.M. and Fonnum, F. (1991) Transport of $\gamma$-aminobutyrate and L-glutamate into synaptic vesicles: Effect of different inhibitors on the vesicular uptake of neurotransmitters and on the $Mg^{2+}$ATPase. *Biochem. J.*, 276: 363–367.

Fykse, E.M., Iversen, E.G., and Fonnum, F. (1992) Inhibition of L-glutamate uptake into synaptic vesicles. *Neurosci. Lett.*, 135: 125–128.

Fykse, E.M. and Fonnum, F. (1996) Amino acid neurotransmission: Dynamics of vesicular uptake. *Neurochem. Res.*, 9: 1053–1060.

Gannon, R.L. and Terrian, D.M. (1991) Presynaptic modulation of glutamate and dynorphin release by excitatory amino acids in the guinea pig hippocampus. *Neurosci.*, 41: 401–410.

Gardiner, M., Nilsson, B., Rehncrona, S. and Siesjo, B.K. (1981) Free fatty acids in the brain in moderate and severe hypoxia. *J. Neurochem.*, 36: 1500–1505.

Hansen, T.W.R., Bratlid, D. and Walaas, S.I. (1988a) Bilirubin decreases phosphorylation of synapsin I, a synaptic vesicle-associated neuronal phosphoprotein, in intact synaptosomes from rat cerebral cortex. *Pediatr. Res.*, 23: 219–223.

Hansen, T.W.R., Paulsen, O., Gjerstad, L. and Bratlid, D. (1988b) Short-term exposure to bilirubin reduces synaptic activation in rat transverse hippocampal slices. *Pediatr. Res.*, 23: 453–456.

Hansen, T.W.R., Mathiesen, S.B.W. and Walaas, S.I. (1996a) Bilirubin has widespread inhibitory effects on protein phosphorylation. *Pediatr. Res.*, 39: 1072–1077.

Hansen, T.W.R., Mathiesen, S.B.W., Sefland, I. and Walaas, S.I. (1996b) Bilirubin inhibits $Ca^{2+}$-dependent release of noradrenaline from permeabilized nerve terminals. *Pedatr. Res.*, 39: 60A.

Hartinger, J. and Jahn, R. (1993) An anion binding site that regulates the glutamate transporter of synaptic vesicles. *J. Biol. Chem.*, 268: 23122–23127.

Heaton, G.M. and Nicholls, D.G. (1976) Hamster brain – adipose tissue mitochondria. The role of fatty acids and the proton conductance of the inner membrane. *J. Biochem.*, 67: 511–517.

Hell, J.W., Maycox, P.R. and Jahn, R. (1990) Energy dependence and functional reconstitution of the $\gamma$-aminobutyric acid carrier from synaptic vesicles. *J. Biol. Chem.*, 265: 2111–2117.

Hökfelt, T., Lundberg, J.M., Schultzberg, M., Johansson, O., Ljungdahl, A. and Rehfeld, J. (1980) Coexistence of peptides and putative transmitters in neurons. *Adv. Biochem. Psychopharmacol.*, 22: 1–23.

Husebye, E.S. and Flatmark, T. (1984) The content of long-chain free fatty acids and their effect on energy transduction in chromaffin granule ghosts. *J. Biol. Chem.*, 259: 15272–15276.

Iliakis, B., Anderson, N.L., Irish, P.S., Henry, M.A. and Westrum, L.E. (1996) Electron microscopy of immunoreactivity patterns for glutamate and $\gamma$-aminobutyric acid in synaptic glomeruli of feline spinal trigeminal nucleus (Subnucleus Caudalis). *J. Comp. Neurol.*, 11: 465–477.

Jessel, T.M. and Kandel, E.R. (1993) Synaptic transmission: A bidirectional and self-modifiable form of cell-cell communication, *Cell*, 72: 1–30.

Jones, P., Major, G. and Sakmann, B. (1993) Quantal components of unitary EPSC at the mossy fibre synapse on CA3 pyramidal cells of the rat hippocampus. *J. Physiol.*, 472: 610–613.

Karp, W.B. (1979) Biochemical alterations in neonatal hyperbilirubinemia and bilirubin encephalopathy: A review. *Pediatrics*, 64: 361–368.

Keller, B.U., Blaschke, M., Rivosecchi, R., Hollmann, A.M., Heinemann, S.F. and Konnerth, A. (1993) Identification of a subunit-specific antagonist of $\alpha$-amino-3-hydroxy-5-methyl isoxazolepropionate/kainate receptor channels. *Proc. Natl. Acad. Sci. USA.*, 90: 605–609.

Kish, P.E., Fischer-Bovenkerk, C. and Ueda, T. (1989) Active transport of $\gamma$-aminobutyric acid and glycine into synaptic vesicles. *Proc. Natl. Acad. Sci. USA.*, 86: 3877–3881.

Kopell, W.N. and Westhead, E.W. (1982) Osmotic pressure of solutions of ATP and catecholamines to storage in chromaffin granules. *J. Biol. Chem.*, 287: 5707–5710.

Kullmann, D.M. and Nicoll, R.A. (1992) Long-term potentiation is associated with increases in quantal content and quantal amplitude. *Nature*, 357: 1111–1120.

Kullmann, D.M. (1993) Quantal variability of excitatory transmission in the hippocampus: Implications for the opening probability of fast glutamate-gated channels. *Proc. R. Soc. Lond. B. Biol.Sci.*, 253: 107–116.

Lindau, M. and Neher, E. (1988) Patch-clamp techniques for time-resolved capacitance measurements in single cells. *Pflugers. Arch.*, 411: 137–146.

Lundy, D.F. and McBean, G.J. (1995) Pre-incubation of synaptosomes with arachidonic acid potentiates inhibition of [$^3$H]D-aspartate transport. *Eur. J. Pharmacol.*, 291: 273–279.

Mattsson, J.P. and Keeling, D.J. (1996) [3H]Bafilomycin as a probe for the transmembrane proton channel of the osteoclast vacuolar $H^+$-ATPase. *Biochim. Biophys. Acta.*, 1280: 98–106.

Maycox, P.R., Deckwerth, T., Hell, J.W. and Jahn, R. (1988) Glutamate uptake by brain synaptic vesicles. *J. Biol. Chem.*, 263: 15423–15428.

Maycox, P.R., Deckwerth, T., Hell, J. and Jahn, R. (1990) Bacteriorhodopsin drives the glutamate transporter of synaptic vesicles. *EMBO J.*, 9: 1465–1469.

Moriyama, Y., Maeda, M. and Futai, M. (1990) Energy coupling of L-glutamate transport and vacuolar $H^+$-ATPase in brain synaptic vesicles. *J. Biochem.*, 108: 689–693.

Moriyama, Y. and Yamamoto, A. (1995) Vesicular L-glutamate transporter in microvesicles from bovine pineal glands. *J. Biol. Chem.*, 270: 22314–22320.

Morphis L., Constantopoulos A. and Matsaniotis N. (1982) Bilirubin-induced modulation of cerebral protein phosphorylation in neonate rabbits in vivo. *Science*, 218: 156–158.

Muroi, M., Shiragami, N. and Takatsuki, A. (1994) Destruxin B, a specific and readily reversible inhibitor of vacuolar-type H(+)-ATPase. *Biochem. Biophys. Res. Commun.*, 205: 1358–1365.

Naito, S. and Ueda, T. (1983) Adenosine triphosphate-dependent uptake of glutamate into protein I-associated synaptic vesicles. *J. Biol. Chem.*, 258: 696–699.

Naito, S. and Ueda, T. (1985) Characterization of glutamate uptake into synaptic vesicles. *J. Neurochem.*, 44: 99–109.

Nelson, N. (1991) Structure and pharmacology of the proton ATPases. *Trends Pharmacol. Sci.*, 12: 71–75.

Nelson, N. (1992) Structural conservation and functional diversity of V-ATPases. *J. Bioenerg. Biomembr.*, 24: 407–414.

Nwaesei, C.G., Van Aerde, J., Boyden, M. and Perlman, M. (1984) Changes in auditory brainstem responses in hyperbilirubinemic infants before and after exchange transfusion. *Pediatrics*, 74: 800–803.

Özkan, E.D., Lee, F.S. and Ueda, T. (1997) A protein factor that inhibits ATP-dependent glutamate and γ-aminobutyric acid accumulation into synaptic vesicles: Purification and initial chracterization. *Proc. Natl. Acad. Sci.*, 94: 4137–4142.

Paulsen, O. and Heggelund, P. (1994) The quantal size at retinogeniculate synapse determined from spontaneous and evoked EPSCs in guinea pig thalamic slices. *J. Physiol.*, 480: 505–511.

Philippu, A. and Matthaei, H. (1975) Uptake of serotonin, γ-aminobutyric acid and histamine into synaptic vesicles of the pig caudate nucleus. *Naunyn Schiederbergs Arch. Pharmacol.*, 287: 191–204.

Rhoads, D.E., Kaplan, N.A., Peterson, M.A. and Raghupathy, E. (1982) Effects of fatty acids on synaptosomal amino acid uptake systems. *J. Neurochem.*, 38: 1255–1260.

Roseth, S. and Fonnum, F. (1995) A study of the uptake of glutamate, γ-aminobutyric acid (GABA), glycine and β-alanine in synaptic brain vesicles from fish and avians. *Neurosci. Lett.*, 183: 62–66.

Roseth, S., Fykse, E.M. and Fonnum, F. (1995) Uptake of L-glutamate into rat brain synaptic vesicles: Effect of inhibitors that binds specifically to the glutamate transporter. *J. Neurochem.*, 65: 96–103.

Roseth, S., Fykse, E.M. and Fonnum, F. (1998a) The effect of arachidonic acid and free fatty acids on vesicular uptake of glutamate and GABA. *Eur. J. Pharmacol.*, 34: 281–288.

Roseth, S., Fykse, E.M. and Fonnum, F. (1998b) Uptake of L-glutamate into synaptic vesicles: Competitive inhibition by dyes with biphenyl and amino and sulphonic acid substituted naphthyl groups. *Biochem. Pharmacol.* (in press).

Roseth, S., Hansen T.W.R., Fonnum, F. and Walaas, S.I. (1998c) Bilirubin inhibits transport of neurotransmitters in synaptic vesicles. *Pedatr. Res.*, (in press).

Salin, P.A., Weisskopf, M.G. and Nicoll, R.A. (1995) A comparison of the role of dynorphin in the hippocampal mossy fiber pathway in guinea pig and rat. *J. Neurosci.*, 15: 6939–6945.

Schikorski, T. and Stevens, C. (1997) Quantitative ultrastructural analysis of hippocampal excitatory synapses. *J. Neurosci.*, 17: 5858–5867.

Seegal, R.F. (1996) Epidemiological and laboratory evidence of PCB-induced neurotoxicity. *Crit. Rev. Toxicol.*, 26: 709–737.

Shioi, J. and Ueda, T. (1990) Artificially imposed electrical potentials drive L-glutamate uptake into synaptic vesicles of bovine cerebral cortex. *Biochem. J.*, 267: 63–68.

Shupliakov, O., Brodin, L., Cullheim, S., Ottersen, O.P. and Storm Mathisen, J. (1992) Immunogold quantification of glutamate in two types of excitatory synapse with different firing patterns. *J. Neurosci.*, 12: 3789–3803.

Sjøstrand, F.S. (1953) The ultrastructure of the retinal rod synapses of the guinea pig eye. *J. Appl. Phys.*, 24: 1422.

Stadler, H. and Fuldner, H.H. (1980) Proton NMR detection of acetylcholine in vesicles. *Nature*, 286: 293–294.

Stern, P., Edwards, F.A. and Sakmann, B. (1992) Fast and slow components of unitary EPSCs on stellate cells elicited by focal stimulation in slices of rat visual cortex. *J. Physiol.*, 449: 247–278.

Tabb, N.S. and Ueda, T. (1991) Phylogenetic studies on the synaptic vesicle glutamate transport system. *J. Neurosci.*, 11: 1822–1828.

Tabb, J.S., Kish, P.E., Van Dyke, R. and Ueda, T. (1992) Glutamate transport into synaptic vesicles: Roles of membrane potential, pH gradient, and intravesicular pH. *J. Biol. Chem.*, 267: 15412–15418.

Toll, L. and Howard, B.D. (1978) Role of $Mg^{2+}$-ATPase and a pH gradient in the storage of catecholamines in synaptic vesicles. *Biochemistry*, 17: 2517–2523.

Tong, G. and Jahr, C.E. (1994) Multivesicular release from excitatory synapses of cultured hippocampal neurons. *Neuron*, 12: 51–9.

Trotti, D., Volterra, A., Lehre, K.P., Rossi, D., Gjesdal, O., Racagni, G. and Danbolt, N.C. (1995) Arachidonic acid inhibits a purified and reconstituted glutamate transporter directly from the water phase via the phospholipid membrane. *J. Biol. Chem.*, 270: 9890–9895.

Van Dyke, R.W. (1988) Proton pump-generated electrochemical gradients in rat liver multivesicular bodies. Quantitation and effects of chloride. *J. Biol. Chem.*, 263: 2603–2611.

Van der Klot, W. (1991) The regulation of quantal size. *Prog. Neurobiol.*, 36: 93–130.

Volterra, A., Trotti, D., Cassutti, P., Tromba, C., Salvaggio, A., Melcangi, C. and Racagni, G. (1992a) High sensitivity of glutamate uptake to extracellular free arachidonic acid levels in rat cortical synaptosomes and astrocytes. *J. Neurochem.*, 59: 600–606.

Wang, Y. and Floor, E. (1994) Dynamic storage of glutamate in rat brain synaptic vesicles. *Neurosci. Lett.*, 180: 175–178.

Wang, G.J, Richardson S.R. and Thayer S.A. (1995) Intracellular acidification is not a prerequisite for glutamate-triggered death of cultured hippocampal neurons. *Neurosci. Lett.*, 186: 139–144.

Whittaker, V.P., Michaelson, J.A. and. Kirkland, R.J.A (1963) The separation of synaptic vesicles from disrupted nerve-ending particles. *Biochem. Pharmacol.*, 12: 300–302.

Whittaker, V.P., Michaelson, J.A. and. Kirkland, R.J.A (1964) The separation of synaptic vesicles from nerve-ending particles (synaptosomes). *Biochem. J.*, 90: 293–303.

Whittaker, V.P. (1986) The storage and release of acetylcholine. *Trends Pharmacol. Sci.*, 7: 312–315.

Wightman, R.M., Jankowski, J.A., Kennedy, R.T., Kawagoe, K.T., Schroeder, T.J., Leszczyszyn, D.J., Near, J.A., Diliberto, E.J. Jr. and Viveros, O.H. (1991) Temporally resolved catecholamine spikes correspond to single vesicle release from individual chromaffin cells. *Proc. Natl. Acad. Sci. USA.*, 88: 10754–10758.

Williams, J.H., Errington, M.L., Lynch, M.A. and Bliss, T.V.P. (1989) Arachidonic acid induced a long activity-dependent enhancement of synaptic transmission in the rat hippocampus. *Nature*, 341: 739–742.

Winter, H.C. and Ueda, T. (1993) Glutamate uptake system in the presynaptic vesicle: Glutamic acid analogs as inhibitors and alternate substrates. *Neurochem. Res.*, 18: 79–85.

Wolosker, H., Reis, M., Assreuy, J. and De Meis, L. (1996a) Inhibition of glutamate uptake and proton pumping in synaptic vesicles by S-nitrosylation. *J. Neurochem.*, 66: 1943–1948.

Wolosker, H., De Souza, D.O. and De Meis, L. (1996b) Regulation of glutamate transport into synaptic vesicles by chloride and proton gradient. *J. Biol. Chem.*, 271: 11726–11731.

Xu, C.J., Klunk, W.E., Kanfers, J.N., Xiong, Q., Miller, G. and Pettegrew, J.W. (1996) Phosphocreatine-dependent glutamate uptake by synaptic vesicles. *J. Biol. Chem.*, 271: 13435–13440.

Yu, A.C.H., Chan, P.H. and Fishman, R.A. (1986) Effects of arachidonic acid on glutamate and $\gamma$-aminobutyric acid uptake in primary cultures of rat cerebral cortical astrocytes and neurons. *J. Neurochem.*, 47: 1181–1189.

SECTION III

# Glutamate receptors

O.P. Ottersen, I.A. Langmoen and L. Gjerstad (Eds.)
*Progress in Brain Research*, Vol 116


CHAPTER 8

# Synaptic arrangement of glutamate receptors

Yutaka Takumi[1,2], Linda Bergersen[1], Alf S. Landsend[1], Eric Rinvik[1] and Ole P. Ottersen[1,*]

[1]*Department of Anatomy, Institute of Basic Medical Sciences, University of Oslo, POB 1105 Blindern, N-0317 Oslo, Norway*
[2]*Department of Otorhinolaryngology, Hirosaki University School of Medicine, 5 Zaifu-cho, Hirosaki 036, Japan*

## Introduction

Consistent with its role as a fast transmitter, glutamate has to be rapidly cleared from the extracellular space following an exocytotic event. This occurs by lateral diffusion, and by binding to membrane molecules including neuronal and glial glutamate transporters (Billups et al., this volume; Danbolt et al., this volume). As the synaptic glutamate transient is brief and spatially restricted the postsynaptic response will depend not only on the identity and intrinsic properties of the postsynaptic glutamate receptors but also on their precise arrangement. Factors of importance are their distance to the release site, their spatial relation to transporters, and the geometry of the synapse in which they are expressed.

The precise localization of receptors is also a matter of concern in the development of therapies aimed at the glutamate synapse. It is of obvious relevance whether a given type of receptor is expressed pre- or postsynaptically or in perisynaptic glial membranes. Its exact distribution may determine its accessibility to drugs and its coupling to intracellular signalling cascades.

There is now compelling evidence that different types of glutamate receptor are expressed at distinct sites within an individual synapse (see Luján et al., 1997, for a recent account of this issue). The aim of the present chapter is to review this evidence and to discuss some of the functional implications. We will draw mainly on studies based on pre- or postembedding immunogold procedures, as these are the only procedures in current use that offer a spatial resolution compatible with reliable localization of membrane proteins at the nanometer level (Ottersen and Landsend, 1997). An obvious advantage of these techniques is that the distance between the epitope and the immunocytochemical signal (i.e., gold particle) is well defined, depending solely on the size of the interposed immunoreagents (Merighi et al., 1992).

A segregated distribution of glutamate receptors at the synaptic level is compatible with our current understanding of the mechanisms responsible for receptor anchoring (see Hsueh and Sheng, this volume). The three major types of glutamate receptors (NMDA, AMPA, and metabotropic receptors) appear to be anchored to different proteins (PSD-95, GRIP, and Homer, respectively; also see Wyszynski et al., 1998) although in each case the interaction occurs through PDZ or PDZ-like domains (Kornau et al., 1995; Dong et al., 1997; Brakeman et al., 1997). Different anchoring mechanisms may allow for a differential reorgani-

*Corresponding author.
Tel.: +4722851270; fax: +4722851299; e-mail: o.p.ottersen@basalmed.uio.no.

zation of glutamate receptors, e.g., under conditions of synaptic plasticity. Indeed, regulation of the precise subsynaptic localization of glutamate receptors may prove to be an important mechanism for the modulation of glutamatergic transmission.

## Methodological considerations

Before embarking on a description of the individual receptors it would seem reasonable to outline the major differences between the pre- and postembedding immunogold techniques. Use of the former entails immunoincubation of several μm thick sections which are subsequently embedded and resectioned for electron microscopy (Priestley et al., 1992). To facilitate penetration of the immunoreagents most protocols recommend the application of very small gold particles (1 nm or less). Even when such particles are used in combination with detergents or freeze thawing the labelling extends only a short distance into the tissue. Thus antigen accessibility becomes a critical factor, which may explain why certain GABA and glutamate receptor epitopes have proved difficult to visualize with this approach (Baude et al., 1995; Nusser et al., 1995).

Bias due to differences in antigen accessibility and differential antibody penetration should be less of a problem when the antibodies are applied directly to ultrathin sections of resin-embedded tissue, as with the postembedding immunogold technique. With postembedding incubation the resin serves to restrict the labelling to those epitopes that are exposed at the surface of the section (Merighi, 1992). In this case the labelling is not affected by differential tissue penetration, and all receptor pools should in principle stand an equal chance of being detected. The properties of the postembedding immunogold technique favour quantitative analyses of glutamate receptors, a potential that is beginning to be exploited (Nusser et al., 1994; Popratiloff et al., 1996; Matsubara et al., 1996; Bernard et al., 1997; Landsend et al., 1997; Luján et al., 1997).

Still, restriction of labelling to the section surface represents a major drawback in terms of sensitivity, as only a minute fraction of the glutamate receptors in the ultrathin section is exposed at the cut surface. It is understandable, therefore, that postembedding immunogold data are still scarce, particularly in regard to some of the metabotropic glutamate receptors (but see Luján et al., 1997). Recent modifications of the postembedding technique (Matsubara et al., 1996) have led to an increased sensitivity and invite a wider application of this approach.

This notwithstanding, enzyme based immunocytochemical procedures (such as the peroxidase-antiperoxidase procedure) will undoubtedly remain the most sensitive approach for the detection of glutamate receptors in situ. The considerable amplification provided by the enzymatic reaction, combined with the good tissue penetration offered by standard pre-embedding protocols, allows ready visualization even of intracellular and extrasynaptic pools of glutamate receptors that are difficult to demonstrate in immunogold preparations. However, high sensitivity is achieved at the expense of spatial resolution. Thus the peroxidase reaction product is prone to diffusion and may be deposited at a distance from the epitope (Priestley et al., 1992). The maximum diffusion distance depends on the incubation parameters and local penetration barriers and cannot be precisely determined. While enzyme-based procedures are the first choice when high sensitivity is the prime concern, they are less useful for the precise subsynaptic localization of receptors (see Ottersen and Landsend, 1997, for a discussion of methodological aspects).

A possible pitfall in receptor localization studies that has not received appropriate attention so far is linked to the common practice of using C-terminal peptides for generating receptor specific antibodies. This strategy is understandable given the rather low homology and high hydrophilicity of this part of the molecule, but is also a cause of concern since it is the C-terminus that interacts with the intracellular anchoring proteins (Hsueh and Sheng, this volume). Whether this interaction is robust enough to survive the preparative steps and then to mask the epitope is not known but should be considered a source of bias in quantitative studies.

### Receptors expressed at the postsynaptic specialization

A distinct feature of glutamatergic synapses (Fig. 1) is a prominent electron dense zone located beneath the postsynaptic membrane. This zone is referred to as the postsynaptic density (Peters et al., 1991; Kennedy, 1997). The extent of the synapse is often taken to coincide with the postsynaptic density although it is now established (see below) that glutamate receptors also occur in a narrow zone around it ("the perisynaptic annulus") and even more laterally (in "extrasynaptic" membranes). The postsynaptic specialization is directly apposed to the presynaptic active site with no intervening glial lamellae. The peak concentration of exocytotically released glutamate at the

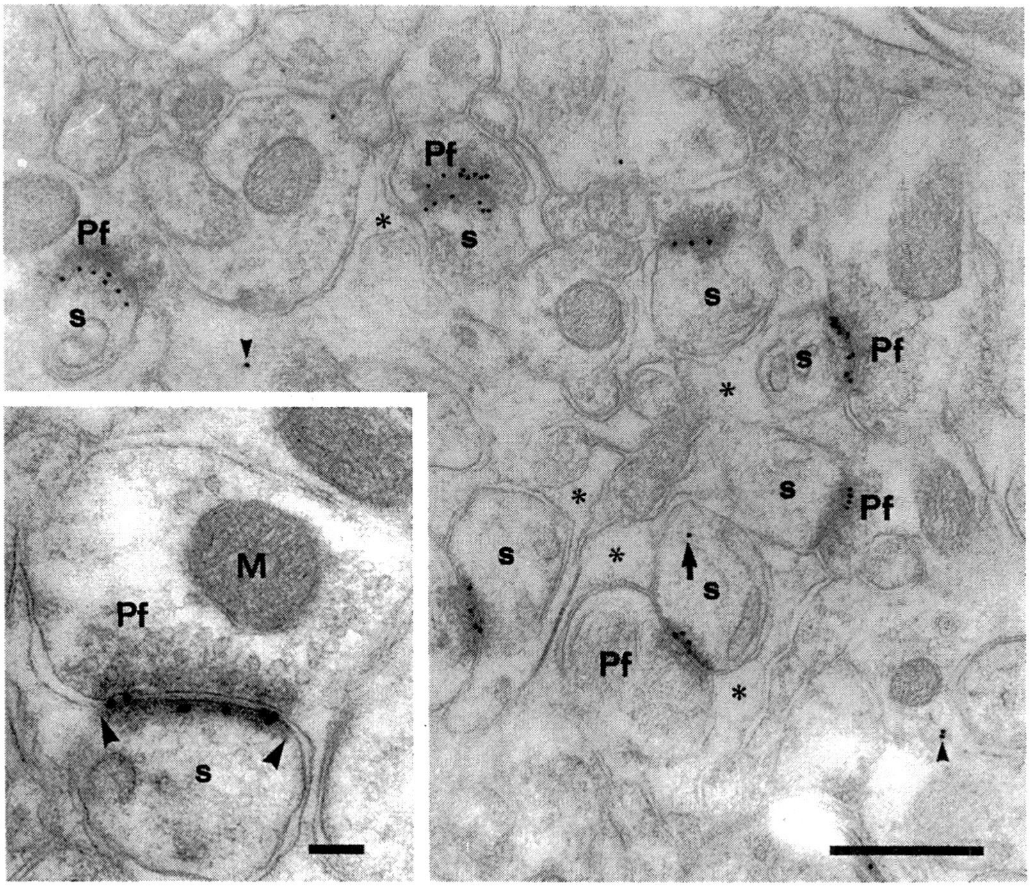


Fig. 1. Postembedding immunogold labelling of $\delta 2$ glutamate receptors in the cerebellar molecular layer (rat). The gold particles (15 nm) reveal expression of $\delta 2$ receptors at the synapses between parallel fiber terminals (Pf) and Purkinje cell spines (s). Note that the extent of receptor expression coincides with the extent of the postsynaptic specialization (indicated by arrowheads in inset). The synapse at top center is obliquely cut, the two rows of particles representing receptors exposed at opposite surfaces of the section. Only very few gold particles (arrows and arrowheads in main micrograph) occur outside the postsynaptic densities. Asterisks represent glial profiles. M, mitochondrion. Bars: 0.5 $\mu$m; inset, 0.1 $\mu$m. Antiserum kindly provided by Dr. R.J. Wenthold. Reproduced with permission from Landsend et al. (1997).

postsynaptic specialization should thus be largely unattenuated by lateral diffusion or by binding to glial glutamate transporters. These factors may, however, affect the transmitter time course in the synaptic cleft (Clements, 1996).

*Types of receptor*

Immunogold analyses have so far revealed three classes of glutamate receptor that are concentrated at the postsynaptic specialization (Figs. 1–4): NMDA, AMPA, and δ2 receptors (review: Ottersen and Landsend, 1997). The localization of NMDA and AMPA receptors at this site is consistent with their role as the prime mediators of fast glutamatergic transmission in the CNS. It is also in line with physiological data (Bekkers and Stevens, 1989) that NMDA and AMPA receptors are coexpressed at the synaptic level (Kharazia et al., 1996; but see Rao and Craig, 1997) although the proportion of double labelled vs. single labelled synapses has not been determined for any fiber system so far. Resolving this issue would seem particularly important in the hippocampus, where synapses found to be "silent" at resting membrane potentials have been claimed to contain NMDA receptors but no functional AMPA receptors (Isaac et al., 1995; Liao et al., 1995; Malenka and Nicoll, 1997; MacDonald et al., this volume). One hypothesis that has been put forward to explain aspects of long term potentiation is that such silent synapses may be brought to express AMPA currents, by recruitment of new AMPA receptors to the postsynaptic specialization or by conversion of nonfunctional AMPA receptors into functional ones (Kullmann, 1994; Liao et al., 1995). It should be noted, however, that the data supporting the existence of postsynaptically silent synapses have been subjected to alternative explanations (Malenka and Nicoll, 1997), one of which involves the possible effect of transmitter "spillover" to adjacent synapses (Kullmann et al., 1996; Barbour and Häusser, 1997).

The NMDA receptor distribution may similarly be regulated by activity. Rao and Craig (1997) reported that chronic blockade of NMDA receptors of cultured hippocampal neurons led to a pronounced increase in the number of NMDA receptor clusters and an increased ratio of synaptic to extrasynaptic receptors. The mechanism underlying this redistribution is not known. The distribution of the NMDA receptor anchoring protein, PSD-95, was not affected by the receptor blockade.

The third type of receptor that is localized preferentially at the postsynaptic specialization is the δ2 receptor (Landsend et al., 1997; Figs. 1, 2). Early studies in expression systems indicated that this receptor was nonfunctional (Araki et al., 1993; Lomeli et al., 1993) and it was therefore difficult to understand why it should be localized at this privileged site. However, recent knockout studies have provided evidence that the δ2 receptor (which is strongly expressed in the cerebellum) is critically involved in long term depression and motor coordination (Kashiwabuchi et al., 1995). It was also found that the mutation responsible for the neurological defect in the atactic Lurcher mice resided in the δ2 receptor gene (Zuo et al., 1997). The mutated δ2 receptors sustained a large, constitutive inward current, which was proposed to underlie the Purkinje cell degeneration that is a characteristic feature of the Lurcher mice. However, the mode of action of the wild-type receptor has not been elucidated.

High resolution immunogold data are still wanting in the case of kainate receptors, whose contribution to synaptic transmission has now been demonstrated by use of the novel 2,3-benzodiazepine antagonists (Vignes and Collingridge, 1997; Castillo et al., 1997). These antagonists are selective for AMPA receptors and thus allow specific activation of kainate receptors by exogenous ligands. Immunoelectron microscopic investigations of the kainate receptor subunits KA2 and GluR6/7 (Petralia et al., 1994) or GluR5/6/7 (Huntley et al., 1993) showed deposits of peroxidase reaction product in postsynaptic densities, and immunofluorescence studies localized the KA2 subunit on dendritic spines of cultured hippocampal neurons (Roche and Huganir,

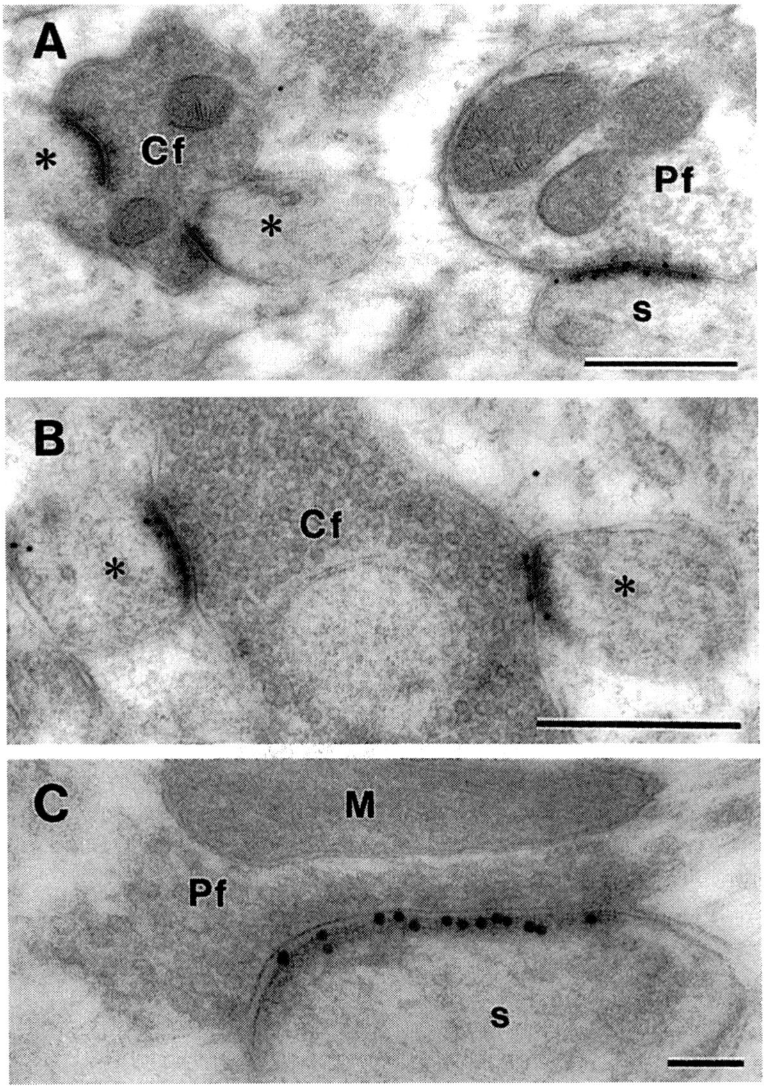


Fig. 2. A: Postembedding immunogold labelling shows that δ2 glutamate receptors occur postsynaptic to parallel fibers (Pf; also see Fig. 1), but not postsynaptic to climbing fibers (Cf). This is not due to a general masking of receptors at the latter synapse since labelling can be obtained with an antibody to the AMPA receptor GluR2/3 (B). C: Labelling for δ2 receptors as in A. High magnification micrograph of parallel fiber synapse reveals even density of gold particles along the extent of the postsynaptic density (cf. Fig. 3). M, mitochondrion; s, Purkinje cell spine; asterisk; thorn postsynaptic to climbing fiber. Bars: A and B, 0.5 μm; C, 0.1 μm.

1995). However, it is clear that there are significant extrasynaptic pools of kainate receptors as well as of kainate binding proteins (Somogyi et al., 1990; Van Bockstaele and Colago, 1996).

*Do the different glutamate receptors populate distinct parts of the postsynaptic specialization?*

In cases of coexpression at the level of the postsynaptic specialization the question arises whether the different receptors are segregated to separate microdomains. Spines postsynaptic to cerebellar parallel fibers exhibit AMPA as well as δ2 receptors, and tangential sections through the receptor matrix of these spines showed that the two types of receptor were intermingled (Landsend et al., 1997). Double labelling with different gold particle sizes was used to distinguish between the two receptor proteins. Whether AMPA and NMDA receptors occupy distinct parts of the postsynaptic specialization remains to be determined. Of direct relevance to this issue is the spatial relationship between the respective anchoring proteins, GRIP and PSD-95. It was recently shown (Hsueh et al., 1997; Hsueh and Sheng, this volume) that PSD-95 may form multimers by disulfide crosslinking of their N-termini. PSD-95 may also form heteromultimers with chapsyn-110 (Kim et al., 1996). If such homo- or heteromultimers sterically exclude other anchoring proteins at the synapse one would expect the NMDA receptors to cluster at separate subsynaptic domains rather than being mixed with AMPA receptors or other PSD95-independent proteins.

*Heterogeneous distribution of receptors along the postsynaptic specialization*

Even if evidence for a distinct subcompartmentation of the postsynaptic specialization is still lacking, the question may still be asked whether the receptors at this site are homogeneously or heterogeneously distributed. One synapse that lends itself to an analysis of this issue is the large synapse between the inner hair cells and afferent dendrites in the organ of Corti. This synapse is assumed to use glutamate as transmitter (review: Ottersen et al., 1998) and is characterized by a presynaptic dense body that is thought to represent the preferential site of transmitter release. A quantitative immunogold analysis indicated that the concentration of AMPA receptors increased on moving away from the presynaptic body towards the lateral margin of the postsynaptic specialization (Matsubara et al., 1996; Fig. 3). This mediolateral gradient was observed with antibodies to GluR2/3 as well as with antibodies recognizing GluR4. No immunogold signal could be detected for GluR1. One interpretation of these observations is that the AMPA receptor density is

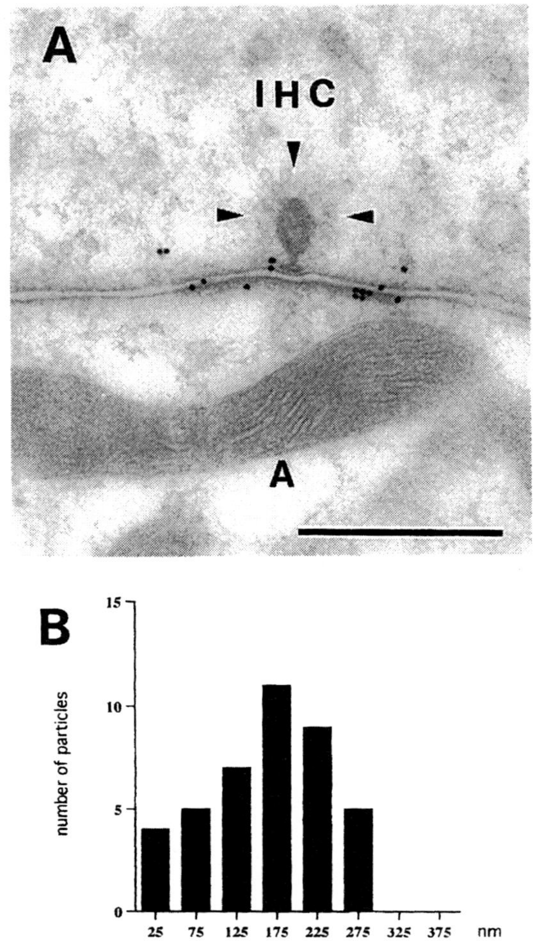


Fig. 3. Postembedding immunogold labelling of afferent hair cell synapse in the inner ear of the rat. A: Gold particles signalling AMPA receptor GluR4 occur at higher concentrations laterally in the synapse than centrally, as shown quantitatively in B (zero is defined as the point opposite the center of the presynaptic body, which is thought to represent the preferential site of exocytotic release). Arrowheads in A indicate synaptic vesicles associated with the presynaptic dense body. IHC, inner hair cell; A, afferent dendrite. The histogram in B shows the distribution of GluR4 receptors along the postsynaptic specialization. Values along the x-axis indicate the centers of bins (bin width 50 $\mu$m). Values along the ordinate represent the number of particles in each bin. Bar: 0.5 $\mu$m. Reproduced with permission from Matsubara et al. (1996).

tuned to the glutamate concentration in the cleft in such a way that the lateral attenuation of glutamate is compensated for by an increased number of receptors. During synaptic transmission, this arrangement would favour an even density of open receptor channels throughout the receptor matrix. A more trivial explanation of the central trough in receptor density is that this part of the synapse is occupied by another protein (e.g., a different type of glutamate receptor) but no candidate protein has been identified so far.

In the central nervous system, evidence for heterogeneous AMPA receptor distribution in the postsynaptic specialization was obtained for striatal synapses (Bernard et al., 1997), but the mediolateral gradient was less pronounced than in the case discussed above. The synapses between parallel fibers and Purkinje cell spines were reported to exhibit an even mediolateral density of AMPA receptors (Nusser et al., 1994) as well as of $\delta 2$ receptors (Fig. 2C). Any correlation between synaptic glutamate concentration and receptor density would be less evident in central synapses than in the afferent hair cell synapse since the site of exocytosis is less well defined and the synapses are smaller.

## Receptors expressed outside the postsynaptic specialization

### *Metabotropic receptors: Differential distribution at the synapse*

The properties and physiological roles of the metabotropic glutamate receptors are reviewed by Bruno et al., elsewhere in this volume. Briefly, this class of receptor consists of three groups, which are coupled to polyphosphoinositide hydrolysis (group I) or negatively coupled to adenylyl cyclase (groups II and III).

All of the metabotropic glutamate receptors that have been studied so far seem to be excluded from the postsynaptic specialization (Fig. 4). This conclusion is based on evidence from immunogold studies. Admittedly, several immunoelectron microscopic investigations utilizing the peroxidase-antiperoxidase technique have demonstrated deposits of reaction product at the postsynaptic density (Martin et al., 1992; van den Pol, 1994; Romano et al., 1995; Petralia et al., 1996b), but the possibility cannot be excluded that these deposits reflect diffusion of reaction product from extrasynaptic sites (see Luján et al., 1997, for a discussion of this point).

One could argue that the absence of labelling at the postsynaptic specialization following pre-embedding immunogold incubation might be due to problems of receptor accessibility (see above). However, recent studies employing postembedding immunogold techniques have confirmed the exclusive peri- or extrasynaptic localization of metabotropic receptors and shown that the two immunogold procedures may produce matching data even at the quantitative level (Luján et al., 1997).

Taken together, the pre- and postembedding immunogold data suggest that the different groups of metabotropic receptors are expressed in distinct membrane domains. Group I receptors (mGluR 1$\alpha$ and 5) are concentrated in an annulus surrounding the postsynaptic specialization (Baude et al., 1993; Nusser et al., 1994; Luján et al., 1996, 1997; Vidnyánszky et al., 1996; Négyessy et al., 1997). Detailed analyses of Purkinje cell spines revealed that 50% of the immunogold particles signalling mGluR1$\alpha$ were found within 60 nm of the edge of the postsynaptic density (the remaining particles occurring at more distant positions) while analyses of hippocampal pyramidal cells showed a corresponding but less pronounced compartmentation of mGluR5 (25% in the 60 nm wide perisynaptic annulus; Luján et al., 1997). It remains to be determined whether the observed difference in mGluR1$\alpha$ and mGluR5 distribution depends on factors related to cell type or whether it reflects distinct targeting or anchoring mechanisms for the two receptor isoforms. A resolution of this issue must await the identification of a neuronal population that expresses both receptors at concentrations sufficient for quantitative analysis. The finding that both mGluR1$\alpha$ and mGluR5 bind to Homer (Brakeman et al., 1997) is not easily

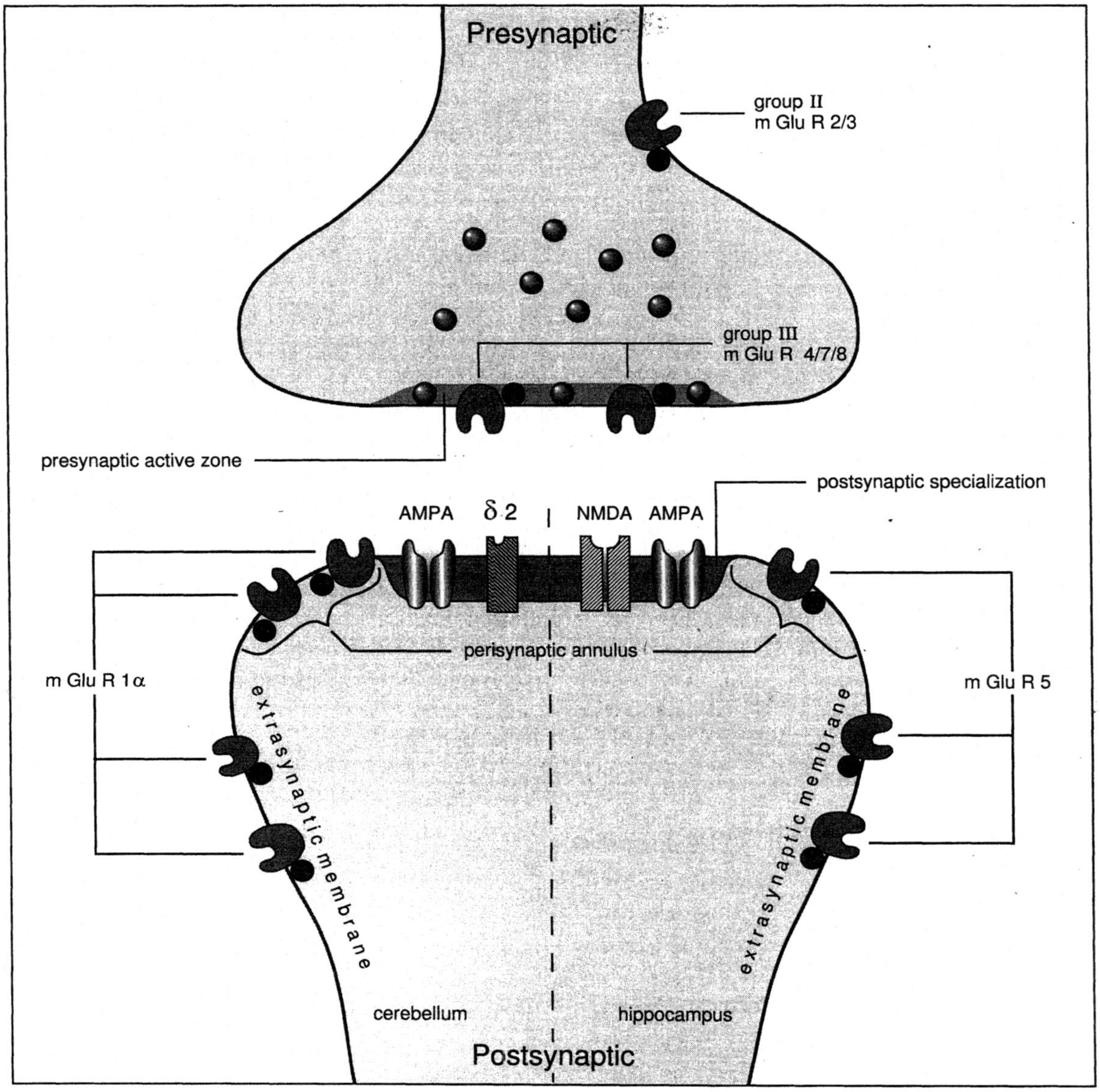


Fig. 4. Diagram illustrating central principles for the organization of glutamate receptors at the synapse. The drawing is based mainly on data obtained by pre- and postembedding immunogold procedures since these are the most reliable procedures for analyses at high level of anatomical resolution (see text). As different glutamate synapses may differ considerably in regard to the types and subsynaptic localization of the receptors that they contain, the drawing will not be representative of all glutamate synapses. Most of the studies on which the diagram was based were performed on the hippocampus or cerebellum (parallel fiber – Purkinje cell synapses). The data from the hippocampus are reflected in the presynaptic element and in the right half of the spine while those from the cerebellum are reflected in the left half. Ionotropic receptors (NMDA and AMPA) and δ2 receptors are concentrated in the postsynaptic specialization (coinciding with the postsynaptic density) while metabotropic receptors are found peri- or extrasynaptically or in the presynaptic element. Group III presynaptic mGluRs typically occur in the presynaptic active zone (shown in grey) while the group II presynaptic mGluRs are expressed in the preterminal axolemma. Extrasynaptic ionotropic receptors are not included in the drawing. The differential postsynaptic distribution of the group I metabotropic receptors (mGluR1α and mGluR5) is according to Luján et al. (1997). The filled circles associated with the mGluRs indicate G proteins. Kainate receptors are not included.

reconciled with the presence of distinct molecular mechanisms for receptor anchoring. Knowledge of the processes that govern mGluR1α and mGluR5 expression is important in view of their differential kinetics of intracellular $Ca^{2+}$ mobilization (Bruno et al., this volume).

The general picture that has emerged from immunoelectron microscopic studies is that the group II receptors (mGluR2 and 3) primarily occur in extrasynaptic membrane domains. The cerebellar Golgi cells, which express particularly high concentrations of group II receptors (Petralia et al., 1996b), displayed mGluR2/3 immunoreactivity in somatodendritic as well as in axonal membranes (Ohishi et al., 1994; Neki et al., 1996). In agreement, recent quantitative analyses showed that immunogold particles signalling mGluR2 were randomly distributed along the Golgi cell plasmalemma with no apparent correlation to synaptic sites (Luján et al., 1997). mGluR2 immunoreactivity has also been demonstrated in the hippocampal mossy fiber system where it was concentrated in the preterminal axolemma (Yokoi et al., 1996; Petralia et al., 1996b; Shigemoto et al., 1997). As discussed below the distribution of group II receptors indicates that their activation may depend on diffusion of glutamate from distant synapses (Scanziani et al., 1997) or from non-synaptic glutamate pools.

The group III receptors have been localized to presynaptic elements. In contrast to mGluR2, which was demonstrated in the preterminal axolemma (see above), those of the group III receptors that are expressed in the brain (mGluR4, 7, and 8) are typically concentrated in the presynaptic grid (Shigemoto et al., 1996, 1997; Li et al., 1997). Obviously, this is the position expected of autoreceptors, and is in accord with their assumed role in counteracting excessive glutamate release (Conn et al., 1994). The synaptic localization of the group III receptors should ensure that they are exposed to the same concentration of exocytotically released glutamate as the postsynaptic ionotropic receptors, and with the same delay.

By combining immunocytochemistry with specific lesions it has been shown that different fiber systems in the hippocampus are equipped with different sets of presynaptic metabotropic receptors (Shigemoto et al., 1997). While the perforant path contains type 7a and 8, the mossy fibers express type 7a, and 7b, and the Schaffer collaterals type 7a. The perforant path and mossy fibers also contain mGluR2, a member of group II receptors (see above). A further heterogeneity was noted within the perforant path where mGluR2 and mGluR8 were enriched in its medial and lateral components, respectively – a finding that is in line with pharmacological data (Macek et al., 1996). The inner zone of the dentate molecular layer was found to express mGluR4a (Shigemoto et al., 1997). The latter receptor was assumed to be preferentially associated with the axons of the hilar mossy cells since this would agree with the known projections and the high mGluR4a mRNA content of these cells (Ohishi et al., 1995). The functional implications of the differentiated distribution of presynaptic group III receptors need to be explored.

It should be noted that the present account of the synaptic localization of metabotropic glutamate receptors is biased in the sense that it is focused on immunogold data primarily obtained in the hippocampus and cerebellum. While the general principles outlined above may be valid for many fiber systems there are numerous examples of divergent patterns of expression. A complete coverage of this issue would be beyond the scope of the present review. Suffice it to mention, as one example, that the mGluR7 receptor, known to occur presynaptically in the hippocampus, is expressed pre- as well as postsynaptically in the retina (Brandstätter et al., 1996).

### *Extrasynaptic distribution of metabotropic receptors: Functional implications*

Baude et al. (1993) were the first to discuss the functional implications of the peri- and extrasynaptic localization of metabotropic glutamate receptors. The picture that has emerged from quantitative immunogold studies is that postsynaptic metabotropic receptors are expressed, in a graded manner, along a wide span of membrane lateral to the synaptic cleft. Glutamate released from the parallel fibers will have to diffuse about 900 nm to reach the distant pool of mGluR1α receptors at this type of synapse (Luján et al., 1997), and even longer diffusion distances would be required for synaptically released glutamate to reach the metabotropic receptors on the Golgi cells. Modelling studies (Barbour and Häusser, 1997) clearly show that diffusion over such distances will entail a pronounced attenuation of transmitter, which can be compensated for by increasing the amount of released glutamate. In other words, the higher the frequency of presynaptic stimulation, the more distally the activation of metabotropic receptors will extend. While this prediction is difficult to test experimentally, it is easily reconciled with the large number of investigations that have emphasized the activity dependency of metabotropic receptor activation (Charpak and Gähwiler 1991; McCormick and von Krosigk, 1992; Bashir et al., 1993; Miles and Poncer, 1993; Batchelor and Garthwaite, 1997). The arrangement of the postsynaptic metabotropic receptors would make them uniquely suited as sensors of presynaptic firing frequency and the gain of the their response could be changed by regulating their precise distribution along the spine membrane. It is of interest in this regard that serial sections of hippocampal pyramidal cells revealed that about 50% of the mGluR5 immunopositive spines were equipped with a perisynaptic as well as an extrasynaptic pool of receptors while the remaining 50% contained only an extrasynaptic pool (Luján et al., 1997). The expression of the metabotropic receptor binding protein Homer is modulated by activity (Brakeman et al., 1997), and it has been suggested (Luján et al., 1997) that the heterogeneous distribution of mGluR5 could reflect different physiological states of the parent neurons.

The extent to which glutamate diffuses to extrasynaptic sites is not only a function of the amount of transmitter release but is also dependent on the anatomy and molecular organization of the synapse. More specifically, the lateral attenuation of glutamate is influenced by the geometry of the synapse and extracellular space, and the number of glutamate-binding membrane molecules (transporters and receptors) in the synapse and its vicinity (see Barbour and Häusser, 1997, for a mathematical analysis of transmitter diffusion). It is obvious that the significance of these factors will vary considerably among different types of synaptic contact. In some, such as those between parallel fibers and dendritic stems of cerebellar interneurons, the synaptic cleft appears to communicate rather freely with the surrounding extracellular space, while in others, such as those between parallel fibers and Purkinje cell spines, the outlet of the synapse is restricted by glial lamellae (Palay and Chan-Palay, 1974; Chaudhry et al., 1995). In the latter type of synapse the glial lamellae extend almost as near as to the lateral edge of the synaptic specialization and serve to bring the glial glutamate transporters (most notably GLAST) in close apposition to the synaptic cleft (Chaudhry et al., 1995; Ottersen et al., 1997). Such an arrangement would tend to impede the outflow of glutamate from the cleft and reduce the concentration of glutamate at extrasynaptic receptors. Hence one would assume that the activation of metabotropic receptors would be sensitive to any synaptic remodelling that might occur in response to physiological or pathophysiological processes. That glial processes serve to limit glutamate distribution to distant receptors is in line with the observations of Asztely et al. (1997) indicating that glial glutamate transporters restrict cross-talk between neighboring excitatory synapses.

The above discussion is relevant not only for the postsynaptic metabotropic receptors but also for those presynaptic receptors that are expressed at a distance from the synapse. It has been shown that the preterminal mGluR2 receptors on hippocampal mossy fibers could be activated by high but not low frequency stimulation (Scanziani et al., 1997). In this fiber system, as in other fiber systems that express preterminal metabotropic receptors, there may be a significant component of heterosynaptic receptor activation.

*Extrasynaptic ionotropic receptors*

While the ionotropic receptors are preferentially localized at the postsynaptic specialization they are not restricted to this site. In the case of AMPA receptors, studies based on immunogold (Baude et al., 1995; Bernard et al., 1997) or pre-embedding, peroxidase-antiperoxidase procedures (e.g., Petralia et al., 1996a) have pointed to the presence of AMPA receptors in extrasynaptic membranes. It is likely, however, that the latter procedure has served to overestimate the extrasynaptic pools of receptors at the expense of the synaptic pools. This is because the diffusibility of the peroxidase reaction product may tend to blur the contrast between nonsynaptic and synaptic membranes, in particular if saturating levels of labelling are reached at synaptic sites.

According to published immunogold images (Baude et al., 1995; Matsubara et al., 1996; Popratiloff et al., 1996; Bernard et al., 1997; Morrison et al., 1998; Wang et al., 1998) the concentration of AMPA receptors may be up to orders of magnitude higher at the postsynaptic specialization than in extrasynaptic membrane domains. Detailed quantitative analyses of synaptic vs. extrasynaptic AMPA receptors are still pending. An interesting possibility is that there may be an exchange of receptors between these two membrane domains and that AMPA receptors may be recruited to the center of the synapse in the induction phase of long term potentiaton (Xie et al., 1997). In the case of δ2 receptors it was found that only 3% of the immunogold particles in the cerebellar cortex were associated with nonsynaptic membranes, indicating a receptor density in the order of 1/1000 of that at the postsynaptic specialization (Landsend et al., 1997). In another quantitative immunogold study, of the shell region of nucleus accumbens, about half of the particles signalling NMDAR1 receptor subunits were found to be associated with extrasynaptic neuronal membranes (Gracy and Pickel, 1997).

The latter study also raised the issue of presynaptic ionotropic receptors. The pre-embedding protocol used by Gracy and Pickel (1997) revealed some silver enhanced gold particles in tyrosine hydroxylase-containing terminals (identified by the peroxidase-antiperoxidase procedure), indicating that these terminals were equipped with a small presynaptic pool of NMDA receptors. This would be consistent with the assumed coupling between NMDA receptor activation and presynaptic dopamine release (Krebs et al., 1991). The occurrence of presynaptic NMDA receptors has also been proposed for several other fiber systems (Liu et al., 1997) but definitive immunocytochemical evidence has been scarce (however, see Aoki et al., 1994). There is little evidence for presynaptic AMPA receptors in the brain while physiological and pharmacological data have pointed to a presynaptic localization of kainate receptors in some fiber systems (Chittajalu et al., 1996; Rodríguez-Moreno et al., 1997). A presynaptic action of kainate at GABAergic synapses in the hippocampus was consistent with an increased failure rate of transmitter release and a decreased frequency of miniature inhibitory postsynaptic potentials (Rodríguez-Moreno et al., 1997).

## Intracellular sorting of glutamate receptors

Many of the above observations attest to the highly precise targeting of glutamate receptors to specific pre- and postsynaptic membrane domains. As pointed out in the Introduction, the fact that different classes of glutamate receptors seem to be anchored by different proteins (Hsueh and Sheng, this volume) help explain how the neurons can maintain such a segregated receptor distribution. The question that will be addressed in the present paragraph is whether an individual neuron is capable of directing different types of glutamate receptor to different populations of glutamate synapses. This would add a new level of complexity to glutamatergic neurotransmission and would call for caution whenever attempts are made to extrapolate from the cellular level to the level of the individual synapse.

The Purkinje cell is a suitable model for investigations of this issue. The Purkinje cells receive two distinct glutamatergic inputs, the parallel and climbing fibres, that can be distinguished on morphological grounds (Palay and Chan-Palay, 1974). Both types of input terminate on spines or thorns and are associated with AMPA receptors, as judged by immunogold labelling with antibodies recognizing GluR2/3 (Nusser et al., 1994; Landsend et al., 1997; also see Fig. 2). However, only the parallel fiber synapses reveal a detectable immunogold signal after incubation with antibodies to the $\delta 2$ receptor (Landsend et al., 1997; Figs. 1, 2). (The antibodies used also recognize the $\delta 1$ receptor which is weakly expressed in the cerebellum according to in situ hybridization data.) From this study it appears that neurons may target a given receptor to one subset of postsynaptic spines. In a subsequent investigation (Rubio and Wenthold, 1997; Wenthold and Roche, this volume) it was found that the fusiform cells of the dorsal cochlear nucleus express GluR4 and mGluR1$\alpha$ postsynaptic to auditory nerve terminals, but not postsynaptic to another category of glutamatergic input (the parallel fibers). The sorting mechanisms underlying the postsynaptic receptor segregation in Purkinje and fusiform cells are not known. As a first step it would seem important to clarify whether the sorting occurs at the level of the anchoring proteins, the receptor proteins, or their respective mRNAs.

Glutamate receptors have been found to be subject to an analogous segregation at the presynaptic level. A striking example was reported in the hippocampus by Shigemoto et al. (1996) who found that different terminals of the same pyramidal cell axon exhibited different densities of mGluR7, depending on the nature of the postsynaptic target. Similarly, mossy fibers expressed mGluR7a and 7b almost exclusively at their contacts with mGluR1$\alpha$ positive neurons, and not at their contacts with hippocampal pyramidal cells (Shigemoto et al., 1997). Other group III metabotropic receptors appeared to be subject to a similar target cell-specific segregation (Shigemoto et al., 1997). Segregation of metabotropic receptors have been found to occur even between two different active sites of a single synapse (Brandstätter et al., 1996).

The apparent ability of the postsynaptic element to specify a unique glutamate receptor profile in the presynaptic element, and vice versa, is likely to depend on a molecular interaction across the synaptic cleft. This may be of a type similar to that described by Irie et al. (1997). In the latter study PSD-95 was found to bind to neuroligin, a neuronal cell adhesion molecule, through the third PDZ domain (NMDAR2 receptors and $K^+$ channels bind to the first and second PDZ domains of PSD-95). Neuroligin is known to interact with $\beta$-neurexins which in turn may be anchored to specific PDZ domain proteins in the adjacent cell (Irie et al., 1997). That molecular complexes of this kind may serve to maintain specific combinations of pre- and postsynaptic proteins receives support from the recent observations of Thomas et al. (1997) on type I glutamatergic synapses of the Drosophila neuromuscular junction. At this junction, the cell adhesion molecule Fasciclin II (FASII) binds to the PDZ1-2 domains of Discs-Large (DLG; a protein belonging to the PSD-95 family). Both FASII and DLG occur pre- as well as postsynaptically and may thus constitute a scaffold for the attachment of ion channels and receptors at both sides of the synaptic cleft.

## The term "synapse" – is a redefinition required?

The data dealt with in the present review invite some thoughts about the very definition of the term "synapse", which recently celebrated its 100th anniversary (Sherrington, 1897). Ever since its introduction the use of this term has followed a tortuous course, influenced by technological and conceptual breakthroughs in the fields of physiology and morphology. With the advent of the electron microscope it was quite natural that the synapse came to be defined by the extent of the pre- and postsynaptic specializations. The use of the terms "perisynaptic" and "extrasynaptic" (as in Luján et al., 1997) to designate membrane domains peripheral to the postsynaptic density is in line with this tradition. But Sherrington, in his later writings, saw the synapse as a physiological concept and emphasized that synapsis means a "process of contact" rather than an "instrument of contact" (reviewed by Shepherd and Erulkar, 1997). On this background it would seem logical to include the receptor-containing peri- and extrasynaptic membrane domains in the synapse, even if they do not show any increased electron density that qualifies to the term "postsynaptic specialization". The problem then arises as to how far away from the release site a receptor can be localized and still be considered part of the synapse. In fact, at least under some circumstances, transmitter released from one active site may extend not only to the most distant "extrasynaptic" receptors but also to receptors in neighboring synapses (review: Barbour and Häusser, 1997). As a further complication the lateral diffusion of glutamate may be influenced by the efficacy of glutamate transporters (Asztely et al., 1997; also see above). It is difficult to envisage an anatomical definition of the synapse that adequately reflects such a dynamic area of functional contact. Thus, in view of the above complications it seems reasonable to maintain a definition based on the extent of the pre- and postsynaptic specializations (see Schikorski and Stevens, 1997). This means that the synapse should be taken to include those receptors that are enriched at the postsynaptic specialization and to exclude those receptors that are situated peripheral to this. One is left with the concept that there are two parallel routes of interneuronal signal transfer – one synaptic (mediated by ionotropic receptors) and one parasynaptic (mediated by metabotropic receptors) – although these routes certainly converge downstream of the receptors (see Bruno et al., this volume).

## Conclusion

Following the initial cloning studies it soon became clear that the brain is equipped with a family of proteins and splice variants that combine to form a large number of functional glutamate receptors with different properties and cellular localization (Seeburg, 1993; Hollmann and Heinemann, 1994; Nakanishi and Masu, 1994). The recent studies discussed here have introduced additional levels of complexity, by revealing a highly selective targeting of glutamate receptors to different subsets of synapses and to distinct membrane domains at individual synaptic contacts. An important challenge for future research is to convert our static picture of receptor distribution into a more realistic and dynamic one. Thus it is likely that the synaptic arrangement of receptors will prove to be subject to regulation – the activity dependent expression of one of the glutamate receptor anchoring proteins (Brakeman et al., 1997) and the effect of activity on the clustering of NMDA receptors (Rao and Craig, 1997) are but two clues in this direction. Redistributions of receptors may be superimposed on a structural reorganization of the entire synapse, as proposed to occur during LTP (Geinisman et al., 1993; Edwards, 1995; Bolshakov et al., 1997). More than ever before can one appreciate the truth in Bernard Katz' statement (Katz, 1966): "The more one finds out about the properties of different synapses, the less grows one's inclination to make general statements about their mode of action". If we append "..and molecular organization" to this statement it would aptly describe the remarkable heterogeneity of glutamate synapses in the central nervous system.

## Acknowledgements

This work was supported by the Norwegian Research Council, the EU Biomed Program (BMH-CT96-0851), and the Sasakawa Foundation. We are grateful to Professor T.W. Blackstad for helpful discussions and critical reading of the manuscript.

## References

Aoki, C., Venkatesan, C., Go, C.G., Mong, J.A. and Dawson, T.M. (1994) Cellular and subcellular localization of NMDA-R1 subunit immunoreactivity in the visual cortex of adult and neonatal rats. *J. Neurosci.*, 14: 5202–5222.

Araki, K., Meguro, H., Kushiya, E., Takayama, C., Inoue, Y. and Mishina, M. (1993) Selective expression of the glutamate receptor channel δ2 subunit in cerebellar Purkinje cells. *Biochem. Biophys. Res. Commun.*, 197: 1267–1276.

Asztely, F., Erdemli, G. and Kullmann, D.M. (1997) Extrasynaptic glutamate spillover in the hippocampus: Dependence on temperature and the role of active glutamate uptake. *Neuron*, 18: 281–293.

Barbour, B. and Häusser, M. (1997) Intersynaptic diffusion of neurotransmitter. *Trends Neurosci.*, 20: 377–384.

Bashir, Z.I., Bortolotto, Z.A., Davies, C.H., Berretta, N., Irving, A.J., Seal, A.J., Henley, J.M., Jane, D.E., Watkins, J.C. and Collingridge, G.L. (1993) Induction of LTP in the hippocampus needs synaptic activation of glutamate metabotropic receptors. *Nature*, 363: 347–350.

Batchelor, A.M. and Garthwaite, J. (1997) Frequency detection and temporally dispersed synaptic signal association through a metabotropic glutamate receptor pathway. *Nature*, 385: 74–77.

Baude, A., Nusser, Z., Roberts, J.D.B., Mulvihill, E., McIlhinney, R.A.J. and Somogyi, P. (1993) The metabotropic glutamate receptor (mGluR1α) is concentrated at perisynaptic membrane of neuronal subpopulations as detected by immunogold reaction. *Neuron*, 11: 771–787.

Baude, A., Nusser, Z., Molnár, E., McIlhinney, R.A.J. and Somogyi, P. (1995) High-resolution immunogold localization of AMPA type glutamate receptor subunits at synaptic and non-synaptic sites in rat hippocampus. *Neuroscience*, 69: 1031–1055.

Bekkers, J.M. and Stevens, C.F. (1989) NMDA and non-NMDA receptors are co-localized at individual excitatory synapses in cultured rat hippocampus. *Nature*, 341: 230–233.

Bernard, V., Somogyi, P. and Bolam, J.P. (1997) Cellular, subcellular, and subsynaptic distribution of AMPA-type glutamate receptor subunits in the neostriatum of the rat. *J. Neurosci.*, 17: 819–833.

Bolshakov, V.Y., Golan, H., Kandel, E.R. and Siegelbaum, S.A. (1997) Recruitment of new sites of synaptic transmission during the cAMP-dependent late phase of LTP at CA3-CA1 synapses in the hippocampus. *Neuron*, 19: 635–651.

Brakeman, P.R., Lanahan, A.A., O'Brien, R., Roche, K., Barnes, C.A., Huganir, R.L. and Worley, P.F. (1997) Homer: A protein that selectively binds metabotropic glutamate receptors. *Nature*, 386: 284–288.

Brandstätter, J.H., Koulen, P., Kuhn, R., van der Putten, H. and Wässle, H. (1996) Compartmental localization of a metabotropic glutamate receptor (mGluR7): Two different active sites at a retinal synapse. *J. Neurosci.*, 16: 4749–4756.

Castillo, P.E., Malenka, R.C. and Nicoll, R.A. (1997) Kainate receptors mediate a slow postsynaptic current in hippocampal CA3 neurons. *Nature*, 388: 182–186.

Charpak, S. and Gahwiler, B.H. (1991) Glutamate mediates a slow synaptic response in hippocampal slice cultures. *Proc. R. Soc. Lond. B. Biol. Sci.*, 243: 221–226.

Chaudhry, F.A., Lehre, K.P., van Lookeren Campagne, M., Ottersen, O.P., Danbolt, N.C. and Storm-Mathisen, J. (1995) Glutamate transporters in glial plasma membranes: Highly differentiated localizations revealed by quantitative ultrastructural immunocytochemistry. *Neuron*, 15: 711–720.

Chittajallu, R., Vignes, M., Dev, K.K., Barnes, J.M., Collingridge, G.L. and Henley, J.M. (1996) Regulation of glutamate release by presynaptic kainate receptors in the hippocampus. *Nature*, 379: 78–81.

Clements, J.D. (1996) Transmitter timecourse in the synaptic cleft: its role in central synaptic function. *Trends Neurosci.*, 19: 163–171.

Conn, P.J., Winder, D.G. and Gereau IV, R.W. (1994) Regulation of neuronal circuits and animal behavior by metabotropic glutamate receptors. In: P.J. Conn and J. Patel, (Eds.), *The metabotropic glutamate receptors.*, Humana Press Inc., Totowa, NJ, pp. 195–229.

Dong, H., O'Brien, R.J., Fung, E.T., Lanahan, A.A., Worley, P.F. and Huganir, R.L. (1997) GRIP: A synaptic PDZ domain-containing protein that interacts with AMPA receptors. *Nature*, 386: 279–284.

Edwards, F.A. (1995) LTP – A structural model to explain the inconsistencies. *Trends Neurosci.*, 18: 250–255.

Geinisman, Y., de Toledo-Morrell, L., Morrell, F., Heller, R.E., Rossi, M. and Parshall, R.F. (1993) Structural synaptic correlate of long-term potentiation: Formation of axospinous synapses with multiple, completely partitioned transmission zones. *Hippocampus*, 3: 435–445.

Gracy, K.N. and Pickel, V.M. (1997) Ultrastructural localization and comparative distribution of nitric oxide synthase and *N*-methyl-D-aspartate receptors in the shell of the rat nucleus accumbens. *Brain Res.*, 747: 259–272.

Hollmann, M. and Heinemann, S. (1994) Cloned glutamate receptors. *Annu. Rev. Neurosci.*, 17: 31–108.

Hsueh, Y.P., Kim, E. and Sheng, M. (1997) Disulfide-linked head-to-head multimerization in the mechanism of ion channel clustering by PSD-95. *Neuron*, 18: 803–814.

Huntley, G.W., Rogers, S.W., Moran, T., Janssen, W., Archin, N., Vickers, J.C., Cauley, K., Heinemann, S.F. and Morrison, J.H. (1993) Selective distribution of kainate receptor subunit immunoreactivity in monkey neocortex revealed by a monoclonal antibody that recognizes glutamate receptor subunits GluR5/6/7. *J. Neurosci.*, 13: 2965–2981.

Irie, M., Hata, Y., Takeuchi, M., Ichtchenko, K., Toyoda, A., Hirao, K., Takai, Y., Rosahl, T.W. and Südhof, T.C. (1997) Binding of neuroligins to PSD-95. *Science*, 277: 1511–1515.

Isaac, J.T., Nicoll, R.A. and Malenka, R.C. (1995) Evidence for silent synapses: Implications for the expression of LTP. *Neuron*, 15: 427–434.

Kashiwabuchi, N., Ikeda, K., Araki, K., Hirano, T., Shibuki, K., Takayama, C., Inoue, Y., Kutsuwada, T., Yagi, T., Kang, Y., Aizawa, S. and Mishina, M. (1995) Impairment of motor coordination, Purkinje cell synapse formation, and cerebellar long-term depression in GluR δ2 mutant mice. *Cell*, 81: 245–252.

Katz, B. (1966) Nerve, muscle and synapse. McGraw-Hill, New York.

Kennedy, M.B. (1997) The postsynaptic density at glutamatergic synapses. *Trends Neurosci.*, 20: 264–268.

Kharazia, V.N., Phend, K.D., Rustioni, A. and Weinberg, R.J. (1996) EM colocalization of AMPA and NMDA receptor subunits at synapses in rat cerebral cortex. *Neurosci. Lett.*, 210: 37–40.

Kim, E., Cho, K.O., Rothschild, A. and Sheng, M. (1996) Heteromultimerization and NMDA receptor-clustering activity of Chapsyn-110, a member of the PSD-95 family of proteins. *Neuron*, 17: 103–113.

Kornau, H.C., Schenker, L.T., Kennedy, M.B. and Seeburg, P.H. (1995) Domain interaction between NMDA receptor subunits and the postsynaptic density protein PSD-95. *Science*, 269: 1737–1740.

Krebs, M.O., Trovero, F., Desban, M., Gauchy, C., Glowinski, J. and Kemel, M.L. (1991) Distinct presynaptic regulation of dopamine release through NMDA receptors in striosome- and matrix-enriched areas of the rat striatum. *J. Neurosci.*, 11: 1256–1262.

Kullmann, D.M. (1994) Amplitude fluctuations of dual-component EPSCs in hippocampal pyramidal cells: Implications for long-term potentiation. *Neuron*, 12: 1111–1120.

Kullmann, D.M., Erdemli, G. and Asztely, F. (1996) LTP of AMPA and NMDA receptor-mediated signals: Evidence for presynaptic expression and extrasynaptic glutamate spillover. *Neuron*, 17: 461–474.

Landsend, A.S., Amiry-Moghaddam, M., Matsubara, A., Bergersen, L., Usami, S., Wenthold, R.J. and Ottersen, O.P. (1997) Differential localization of δ glutamate receptors in the rat cerebellum: Coexpression with AMPA receptors in parallel fiber-spine synapses and absence from climbing fiber-spine synapses. *J. Neurosci.*, 17: 834–842.

Li, H., Ohishi, H., Kinoshita, A., Shigemoto, R., Nomura, S. and Mizuno, N. (1997) Localization of a metabotropic glutamate receptor, mGluR7, in axon terminals of presumed nociceptive, primary afferent fibers in the superficial layers of the spinal dorsal horn: An electron microscope study in the rat. *Neurosci. Lett.*, 223: 153–156.

Liao, D., Hessler, N.A. and Malinow, R. (1995) Activation of postsynaptically silent synapses during pairing-induced LTP in CA1 region of hippocampal slice. *Nature*, 375: 400–404.

Liu, H., Mantyh, P.W. and Basbaum, A.I. (1997) NMDA-receptor regulation of substance P release from primary afferent nociceptors. *Nature*, 386: 721–724.

Lomeli, H., Sprengel, R., Laurie, D.J., Kohr, G., Herb, A., Seeburg, P.H. and Wisden, W. (1993) The rat δ-1 and δ-2 subunits extend the excitatory amino acid receptor family. *FEBS Lett.*, 315: 318–322.

Luján, R., Nusser, Z., Roberts, J.D.B., Shigemoto, R. and Somogyi, P. (1996) Perisynaptic location of metabotropic glutamate receptors mGluR1 and m GluR5 on dendrites and dendritic spines in the rat hippocampus. *Eur. J. Neurosci.*, 8: 1488–1500.

Luján, R., Roberts, J.D.B., Shigemoto, R., Ohishi, H. and Somogyi, P. (1997) Differential plasma membrane distribution of metabotropic glutamate receptors mGluR1α, mGluR2 and mGluR5, relative to neurotransmitter release sites. *J. Chem. Neuroanat.*, 13: 219–241.

Macek, T.A., Winder, D.G., Gereau, R.W.t., Ladd, C.O. and Conn, P.J. (1996) Differential involvement of group II and group III mGluRs as autoreceptors at lateral and medial perforant path synapses. *J. Neurophysiol.*, 76: 3798–3806.

Malenka, R.C. and Nicoll, R.A. (1997) Silent synapses speak up. *Neuron*, 19: 473–476.

Martin, L.J., Blackstone, C.D., Huganir, R.L. and Price, D.L. (1992) Cellular localization of a metabotropic glutamate receptor in rat brain. *Neuron*, 9: 259–270.

Matsubara, A., Laake, J.H., Davanger, S., Usami, S. and Ottersen, O.P. (1996) Organization of AMPA receptor subunits at a glutamate synapse: A quantitative immunogold analysis of hair cell synapses in the rat organ of Corti. *J. Neurosci.*, 16: 4457–4467.

McCormick, D.A. and von Krosigk, M. (1992) Corticothalamic activation modulates thalamic firing through glutamate "metabotropic" receptors. *Proc. Natl. Acad. Sci. USA*, 89: 2774–2778.

Merighi, A. (1992) Post-embedding electron microscopic immunocytochemistry. In: J.M. Polak and J.Y. Priestley, (Eds.), *Electron microscopic immunocytochemistry-principles and practice.*, Oxford University Press, New York, pp. 51–87.

Miles, R. and Poncer, J.C. (1993) Metabotropic glutamate receptors mediate a post-tetanic excitation of guinea-pig hippocampal inhibitory neurones. *J. Physiol. (Lond)*, 463: 461–473.

Morrison, B.M., Janssen, W.G., Gordon, J.W. and Morrison, J.H. (1998) Light and electron microscopic distribution of the AMPA receptor subunit, GluR2, in the spinal cord of control and G86R mutant superoxide dismutase transgenic mice. *J. Comp. Neurol.*, 395: 523–534.

Nakanishi, S. and Masu, M. (1994) Molecular diversity and functions of glutamate receptors. *Annu. Rev. Biophys. Biomol. Struct.*, 23: 319–348.

Neki, A., Ohishi, H., Kaneko, T., Shigemoto, R., Nakanishi, S. and Mizuno, N. (1996) Metabotropic glutamate receptors mGluR2 and mGluR5 are expressed in two non-overlapping populations of Golgi cells in the rat cerebellum. *Neuroscience*, 75: 815–826.

Négyessy, L., Vidnyánszky, Z., Kuhn, R., Knöpfel, T., Görcs, T.J. and Hámori, J. (1997) Light and electron microscopic demonstration of mGluR5 metabotropic glutamate receptor immunoreactive neuronal elements in the rat cerebellar cortex. *J. Comp. Neurol.*, 385: 641–650.

Nusser, Z., Mulvihill, E., Streit, P. and Somogyi, P. (1994) Subsynaptic segregation of metabotropic and ionotropic glutamate receptors as revealed by immunogold localization. *Neuroscience*, 61: 421–427.

Nusser, Z., Roberts, J.D., Baude, A., Richards, J.G., Sieghart, W. and Somogyi, P. (1995) Immunocytochemical localization of the α 1 and β 2/3 subunits of the GABAA receptor in relation to specific GABAergic synapses in the dentate gyrus. *Eur. J. Neurosci.*, 7: 630–646.

Ohishi, H., Ogawa-Meguro, R., Shigemoto, R., Kaneko, T., Nakanishi, S. and Mizuno, N. (1994) Immunohistochemical localization of metabotropic glutamate receptors, mGluR2 and mGluR3, in rat cerebellar cortex. *Neuron*, 13: 55–66.

Ohishi, H., Akazawa, C., Shigemoto, R., Nakanishi, S. and Mizuno, N. (1995) Distributions of the mRNAs for L-2-amino-4-phosphonobutyrate-sensitive metabotropic glutamate receptors, mGluR4 and mGluR7, in the rat brain. *J. Comp. Neurol.*, 360: 555–570.

Ottersen, O.P., Chaudhry, F.A., Danbolt, N.C., Laake, J.H., Nagelhus, E.A., Storm-Mathisen, J. and Torp, R. (1997) Molecular organization of cerebellar glutamate synapses. *Prog. Brain Res.*, 114: 113–123.

Ottersen, O.P. and Landsend, A.S. (1997) Organization of glutamate receptors at the synapse. *Eur. J. Neurosci.*, 9: 2219–2224.

Ottersen, O.P., Takumi, Y., Matsubara, A., Landsend, A.F., Laake, J.H. and Usami, S. (1998) Molecular organization of a type of peripheral glutamate synapse: The afferent synapses of hair cells in the inner ear. *Prog. Neurobiol.*, 54: 127–148

Palay, S.L. and Chan-Palay, V. (1974) *Cerebellar cortex: cytology and organization*, Springer-Verlag, New York.

Peters, A., Palay, S.L. and Webster, H.D. (1991) *The fine structure of the nervous system: neurons and their supporting cells, 3rd edn.* Oxford University Press, New York.

Petralia, R.S., Wang, Y.X. and Wenthold, R.J. (1994) Histological and ultrastructural localization of the kainate receptor subunits, KA2 and GluR6/7, in the rat nervous system using selective antipeptide antibodies. *J. Comp. Neurol.*, 349: 85–110.

Petralia, R.S., Wang, Y.X., Zhao, H.M. and Wenthold, R.J. (1996a) Ionotropic and metabotropic glutamate receptors show unique postsynaptic, presynaptic, and glial localizations in the dorsal cochlear nucleus. *J. Comp. Neurol.*, 372: 356–383.

Petralia, R.S., Wang, Y.X., Niedzielski, A.S. and Wenthold, R.J. (1996b) The metabotropic glutamate receptors, mGluR2 and mGluR3, show unique postsynaptic, presynaptic and glial localizations. *Neuroscience*, 71: 949–976.

Popratiloff, A., Weinberg, R.J. and Rustioni, A. (1996) AMPA receptor subunits underlying terminals of fine-caliber primary afferent fibers. *J. Neurosci.*, 16: 3363–3372.

Priestley, J.V., Alvarez, F.J. and Averill, S. (1992) Pre-embedding electron microscopic immunocytochemistry. In: J.M. Polak and J.Y. Priestley, (Eds.) *Electron microscopic immunocytochemistry-principles and practice*, Oxford University Press, New York, pp. 89–121.

Rao, A. and Craig, A.M. (1997) Activity regulates the synaptic localization of the NMDA receptor in hippocampal neurons. *Neuron*, 19: 801–812.

Roche, K.W. and Huganir, R.L. (1995) Synaptic expression of the high-affinity kainate receptor subunit KA2 in hippocampal cultures. *Neuroscience*, 69: 383–393.

Rodríguez-Moreno, A., Herreras, O. and Lerma, J. (1997) Kainate receptors presynaptically downregulate GABAergic inhibition in the rat hippocampus. *Neuron*, 19: 893–901.

Romano, C., Sesma, M.A., McDonald, C.T., O'Malley, K., Van den Pol, A.N. and Olney, J.W. (1995) Distribution of metabotropic glutamate receptor mGluR5 immunoreactivity in rat brain. *J. Comp. Neurol.*, 355: 455–469.

Rubio, M.E. and Wenthold, R.J. (1997) Glutamate receptors are selectively targeted to postsynaptic sites in neurons. *Neuron*, 18: 939–950.

Scanziani, M., Salin, P.A., Vogt, K.E., Malenka, R.C. and Nicoll, R.A. (1997) Use-dependent increases in glutamate concentration activate presynaptic metabotropic glutamate receptors. *Nature*, 385: 630–634.

Schikorski, T. and Stevens, C.F. (1997) Quantitative ultrastructural analysis of hippocampal excitatory synapses. *J. Neurosci.*, 17: 5858–5867.

Seeburg, P.H. (1993) The TINS/TiPS Lecture. The molecular biology of mammalian glutamate receptor channels. *Trends Neurosci.*, 16: 359–365.

Shepherd, G.M. and Erulkar, S.D. (1997) Centenary of the synapse: From Sherrington to the molecular biology of the synapse and beyond. *Trends Neurosci.*, 20: 385–392.

Sherrington, C.S. (1897) In: M. Foster, (Eds.) *Textbook of physiology*, pp. 60.

Shigemoto, R., Kulik, A., Roberts, J.D.B., Ohishi, H., Nusser, Z., Kaneko, T. and Somogyi, P. (1996) Target-cell-specific concentration of a metabotropic glutamate receptor in the presynaptic active zone. *Nature*, 381: 523–525.

Shigemoto, R., Kinoshita, A., Wada, E., Nomura, S., Ohishi, H., Takada, M., Flor, P.J., Neki, A., Abe, T., Nakanishi, S. and Mizuno, N. (1997) Differential presynaptic localization of metabotropic glutamate receptor subtypes in the rat hippocampus. *J. Neurosci.*, 17: 7503–7522.

Somogyi, P., Eshhar, N., Teichberg, V.I. and Roberts, J.D. (1990) Subcellular localization of a putative kainate receptor in Bergmann glial cells using a monoclonal antibody in the chick and fish cerebellar cortex. *Neuroscience*, 35: 9–30.

Thomas, U., Kim, E., Kuhlendahl, S., Koh, Y.H., Gundelfinger, E.D., Sheng, M., Garner, C.C. and Budnik, V. (1997) Synaptic clustering of the cell adhesion molecule fasciclin II by discs-large and its role in the regulation of presynaptic structure. *Neuron*, 19: 787–799.

Van Bockstaele, E.J. and Colago, E.E. (1996) Ultrastructural localization of the kainate selective glutamate receptor in noradrenergic perikarya and dendrites of the nucleus locus coeruleus in the rat brain. *Brain Res.*, 732: 223–231.

Van den Pol, A.N. (1994) Metabotropic glutamate receptor mGluR1 distribution and ultrastructural localization in hypothalamus. *J. Comp. Neurol.*, 349: 615–632.

Vidnyánszky, Z., Görcs, T.J., Négyessy, L., Borostyánkio, Z., Knöpfel, T. and Hámori, J. (1996) Immunocytochemical visualization of the mGluR1α metabotropic glutamate receptor at synapses of corticothalamic terminals originating from area 17 of the rat. *Eur. J. Neurosci.*, 8: 1061–1071.

Vignes, M. and Collingridge, G.L. (1997) The synaptic activation of kainate receptors. *Nature*, 388: 179–182.

Wang, Y.X., Wenthold, R.J., Ottersen, O.P. and Petralia, R.S. (1998) Endbulb synapses in the anteroventral cochlear nucleus express a specific subset of AMPA-type glutamate receptor subunits. *J. Neurosci.*, 18: 1148–1160

Wyszynski, M., Kharazia, V., Shanghvi, R., Rao, A., Beggs, A.H., Craig, A.M., Weinberg, R. and Sheng, M. (1998) Differential regional expression and ultrastructural localization of α-actinin-2, a putative NMDA receptor-anchoring protein, in rat brain. *J. Neurosci.*, 18: 1383–1392.

Xie, X., Liaw, J.S., Baudry, M. and Berger, T.W. (1997) Novel expression mechanism for synaptic potentiation: Alignment of presynaptic release site and postsynaptic receptor. *Proc. Natl. Acad. Sci. USA.*, 94: 6983–6988.

Yokoi, M., Kobayashi, K., Manabe, T., Takahashi, T., Sakaguchi, I., Katsuura, G., Shigemoto, R., Ohishi, H., Nomura, S., Nakamura, K., Nakao, K., Katsuki, M. and Nakanishi, S. (1996) Impairment of hippocampal mossy fiber LTD in mice lacking mGluR2. *Science*, 273: 645–647.

Zuo, J., De Jager, P.L., Takahashi, K.A., Jiang, W., Linden, D.J. and Heintz, N. (1997) Neurodegeneration in Lurcher mice caused by mutation in δ2 glutamate receptor gene. *Nature*, 388: 769–773.

O.P. Ottersen, I.A. Langmoen and L. Gjerstad (Eds.)
*Progress in Brain Research*, Vol 116


CHAPTER 9

# Anchoring of glutamate receptors at the synapse

Yi-Ping Hsueh and Morgan Sheng*

*Howard Hughes Medical Institute and Department of Neurobiology, Massachusetts General Hospital and Harvard Medical School, Boston, MA 02114, USA*

## Introduction

Ion channels and receptors are not randomly distributed over the surface of neurons, but are typically concentrated at specific subcellular sites. Glutamate receptors are no exception to this rule. The ionotropic glutamate receptors (NMDA receptors and AMPA receptors) are predominantly localized at postsynaptic sites, where they are suitably positioned to bind to the neurotransmitter glutamate released from the presynaptic terminal. A subset of metabotropic glutamate receptors is also concentrated postsynaptically, though many members of this class of glutamate receptors are present at presynaptic locations (see Takumi et al., this volume). Moreover, it almost goes without saying that glutamate receptors are found specifically in glutamatergic synapses (as opposed to for instance GABAergic synapses). Understanding the molecular mechanisms of clustering of glutamate receptors in glutamatergic synapses is important because the correct localization of these receptors is critical for efficient excitatory transmission. This review will concentrate on recent progress in the identification of proteins that are involved in clustering and anchoring of glutámate receptors at postsynaptic sites in glutamatergic synapses.

## NMDA receptor interactions with the PSD-95 family of proteins

A breakthrough in this field came with the discovery that NMDA receptors bind directly to a family of proteins of which the prototype is the abundant postsynaptic density protein PSD-95/SAP90 (Cho et al., 1992; Kistner et al., 1993). Including PSD-95, the family presently numbers four, the other members being SAP97/hdlg (Lue et al., 1994; Müller et al., 1995); chapsyn-110/PSD-93 (Brenman et al., 1996; Kim et al., 1996); and SAP102 (Müller et al., 1996). This family of peripheral membrane proteins contains three N-terminal PDZ domains, a central SH3 domain, and a C-terminal guanylate kinase-like domain (Sheng, 1996). It has become clear in the past two years that PDZ domains are modular protein–protein interaction domains, specialized for the binding of specific peptide sequences at the very C-terminus of interacting proteins. This conclusion of wide cell biological significance was reached largely based on research on PSD-95 interactions with the C-terminal tails of ion channels and receptors. Specifically, it was found that the C-terminal intracellular tail of the NR2 subunits of the NMDA receptor binds directly to the PDZ domains of the PSD-95 family of proteins (Kornau

*Corresponding author.
Howard Hughes Medical Institute, Mass. General Hospital (Wel 423) 50 Blossom Street, Boston, MA 02114
Tel.: +1 617 724 2800; fax: +1 617 724 2805; e-mail: sheng@helix.mgh.harvard.edu.

et al., 1995; Niethammer et al., 1996). Moreover, concurrent work uncovered an interaction between the PDZ domains of PSD-95 family proteins and the C-terminal tail of Shaker-type $K^+$ channels (Kim et al., 1995). Since Shaker $K^+$ channels (-ETDV) share an almost identical C-terminal peptide sequence with NR2 subunits of the NMDA receptor (-ESDV), these concomitant discoveries consolidated the idea that PDZ domains recognize specific C-termini of ion channel proteins. The importance of the last four amino acids of the intracellular tail in specific binding to PDZ domains of PSD-95 was confirmed by mutagenesis (Kim et al., 1995) and by X-ray crystallography of a PDZ domain bound to its cognate peptide ligand (Doyle et al., 1996). Based on in vitro assays and yeast two-hybrid analysis, it appears that the N-terminal two PDZ domains (PDZ1 and PDZ2) of PSD-95 can both bind to the C-terminal tails of NR2 subunits, though PDZ2 has higher affinity (Müller et al., 1996). PDZ3 does not show detectable binding to either NR2 subunits or Shaker $K^+$ channels (Kim et al., 1995; Müller et al., 1996; Niethammer et al., 1996).

### Clustering/anchoring function of PSD-95

Although the above studies showed that PSD-95 and its relatives have specific and high affinity for NR2 proteins, they did not demonstrate a functional significance of this interaction. A prerequisite for an anchoring function of PSD-95 in vivo is that PSD-95 should colocalize with NMDA receptors at glutamatergic synapses. This has now been confirmed by double labeling immunocytochemistry in cultured hippocampal neurons (Kornau et al., 1995; Müller et al., 1996). In addition, evidence that PSD-95 family proteins and NMDA receptors exist in brain extracts as a complex has been obtained by biochemical cofractionation and by co-immunoprecipitation (Kim et al., 1996; Lau et al., 1996; Müller et al., 1996; Wyszynski et al., 1997).

A remarkable feature of PSD-95 family proteins is that they can cluster their ion channel binding partners (including NMDA receptors) when coexpressed with them in heterologous cells (Hsueh et al., 1997; Kim et al., 1996; Kim et al., 1995). This coclustering is dependent on the interaction between PSD-95 and the C-terminal PDZ-binding motif of the ion channels. PSD-95 or its relative chapsyn-110 can cluster NR2 subunits in COS-7 cells in the absence of NR1 (Kim et al., 1996). They can also cluster NR1 when NR2 subunits are coexpressed, but they cannot cluster NR1 directly since the common splice variants of NR1 do not interact with PSD-95 (Kim et al., 1996). The mechanism of membrane protein clustering by PSD-95 depends on a head-to-head multimerization of PSD-95 that is mediated by disulfide bonds between the conserved N-terminal regions of PSD-95 family proteins (Hsueh et al., 1997). The stoichiometry of PSD-95 multimerization is not fully established but preliminary evidence is consistent with a tetrameric configuration (Hsueh et al., 1997).

The exquisite punctate colocalization of PSD-95 and NMDA receptors at postsynaptic sites is suggestive of an anchoring function for PSD-95, but confirmation of this would ideally require genetic experiments. That is to say, loss of function of PSD-95 (e.g. through targeted gene disruption) might be expected to cause loss of clustering and localization of NMDA receptors in glutamatergic synapses. This genetic knock-out experiment remains to be done in mice. Interpretation of such an experiment may be confounded by the probable redundancy that exists among the four members of the PSD-95 family. These proteins are highly similar to each other at the primary sequence level, and this is reflected in the fact that their binding specificities for interacting proteins are indistinguishable, at least in vitro. In addition, three members of this family, PSD-95, chapsyn-110/PSD-93 and SAP102, all appear to be predominantly postsynaptic in glutamatergic synapses (Hunt et al., 1996; Müller et al., 1996) (and M. Sheng, unpublished observations) and hence are likely to be subserving similar functions in neurons. The notable exception in the PSD-95 family is SAP97, which is not localized postsynaptically but is instead concentrated in axons and presynaptic terminals (Müller et al., 1996). The specific in vivo function of SAP97 remains to be determined. One possibility is that SAP97 interacts preferentially with Shaker family $K^+$ channels, which tend to have an axonal and presynaptic subcellular distribution in neurons.

Although the key genetic experiments are lacking in mammals, an important genetic finding in Drosophila strongly supports the idea that PSD-95 family proteins are involved in synaptic clustering of their ion channel binding partners. In Drosophila, a single gene called discs large (*dlg*) appears to represent the homolog of the mammalian PSD-95 family of proteins (Woods and Bryant, 1991). The Dlg protein has an identical domain organization as PSD-95 and its PDZ domains have indistinguishable binding specificities for NMDA receptor NR2 and Shaker $K^+$ channel C-termini. In heterologous cells, Dlg can cluster Shaker $K^+$ channels just like its mammalian counterparts, and in vivo Dlg is specifically localized to the glutamatergic synapses of the larval neuromuscular junction where it colocalizes with Shaker $K^+$ channels (Tejedor et al., 1997). The availability of fly mutants of dlg allowed one to test the idea that Dlg is essential for the synaptic clustering of its partner ion channels. And indeed, loss of function mutations in Dlg abolished the normal clustering of Shaker $K^+$ channels at the neuromuscular junction (Tejedor et al., 1997). These results clearly implicate Dlg as being essential for localization of its ion channel binding partner in vivo. By extrapolation to mammalian systems, it is reasonable to expect that the PSD-95 family of proteins will also have similar functions in localizing postsynaptic clusters of NMDA receptors.

How the PSD-95 family interacts with the cytoskeleton to anchor NMDA receptors at the postsynaptic site is unclear. One member of the PSD-95 family, SAP97/hDlg, binds to the actin-binding protein 4.1 (Lue et al., 1996; Lue et al., 1994; Marfatia et al., 1996; Marfatia et al., 1994). In neurons, however, SAP97 appears to be the only member of the PSD-95 family that is presynaptic and axonal in distribution, so it is unlikely to be interacting significantly with NMDA receptors in vivo. It remains to be determined whether PSD-95 and its postsynaptic relatives (chapsyn-110 and SAP102) also interact with protein 4.1. If they do, this would represent one mechanism by which NMDA receptors can be indirectly linked to the actin cytoskeleton.

All four members of the PSD-95 family bind via their guanylate kinase-like domain to a postsynaptic density protein called guanylate kinase-associated protein or GKAP (Kim et al., 1997; Naisbitt et al., 1997). The primary structure of GKAP shows no homology to proteins of known function, and its role in the PSD is unknown. It does appear to have multivalent ability to bind to PSD-95, however; thus GKAP could function to raft together multimers of PSD-95 beneath the postsynaptic membrane, thereby clustering NMDA receptors at high concentration. A family of GKAP-like proteins has been recently cloned and termed SAPAPs (for Synapse Associated Protein Associated Protein) (Takeuchi et al., 1997).

In addition to GKAP, PSD-95 family proteins also interact with other intracellular proteins, including neuronal nitric oxide synthase (nNOS). Interestingly, nNOS itself possesses a PDZ domain at its N-terminal end, and it binds to PSD-95 through a PDZ-PDZ domain interaction, rather than a C-terminus-PDZ interaction (Brenman et al., 1996). Through this association, nNOS can be brought in close proximity to the NMDA receptor. Bridging of NMDA receptors and nNOS by PSD-95 could explain how the activity of nNOS (a calcium-regulated enzyme) is functionally coupled to the activation of NMDA receptors (glutamate-gated channels with high calcium permeability). The importance of this observation is its implication that the PSD-95 family of proteins act not only to cluster and localize ion channels and receptors, but also to nucleate a signalling complex physically linked to these membrane proteins. The physiological importance of a signalling complex organized by a multivalent PDZ domain protein (InaD) has recently been demonstrated in Drosophila photoreceptors (Tsunoda et al., 1997).

## NMDA receptor anchoring by α-actinin

NMDA receptors are heteromeric channels composed of NR1 and NR2 subunits (Hollmann and Heinemann, 1994; Monyer et al., 1994; Nakanishi, 1992; Wisden and Seeburg, 1993). As discussed above, NR2 subunits bind to the PSD-95 family of proteins via their C-terminus. Although some splice variants of NR1 have been reported to bind to PSD-95 (Kornau et al., 1995), the major splice forms do not. Some attention has focused on the NR1 subunit because it can by itself form clusters when overexpressed in heterologous cells (Ehlers et al., 1995), suggesting that it may directly interact with the cytoskeleton. The "self-clustering" activity is dependent on the presence of the C1 exon in the C-terminal tail of NR1, which contains multiple phosphorylation sites for protein kinase C (Tingley et al., 1994). Interestingly, activation of PKC results in the dispersal of the NR1 clusters (Ehlers et al., 1995).

To identify potential proteins that might link NR1 to the cytoskeleton, the yeast two-hybrid system has been used to isolate proteins that bind to the intracellular C-terminal tail of NR1. One protein identified in this way was α-actinin, a ubiquitous actin binding protein of the dystrophin/spectrin family (Wyszynski et al., 1997). α-actinin binds via its central rod-like domain to the membrane proximal region of NR1's C-terminal tail. This region, now termed C0, is common to all the splice variants of NR1. Thus it is presumed that α-actinin can bind to all NR1 splice forms. α-actinin is concentrated in the PSD of glutamatergic synapses and can be co-immunoprecipitated with NMDA receptors from brain extracts. Immunocytochemically, it is specifically localized to dendritic spines (Wyszynski et al., 1997). Thus the evidence for an in vivo interaction between NR1 and α-actinin is quite strong. Given that α-actinin is an actin-binding protein, such an interaction would connect NMDA receptors to the actin cytoskeleton that is abundant in dendritic spines.

An intriguing feature of the cytoskeletal anchoring of NMDA receptors by α-actinin is its potential regulation by calcium-calmodulin. Calcium-calmodulin binds directly to two sites in the NR1 C-terminal tail, and competes with binding of α-actinin (Ehlers et al., 1996; Wyszynski et al., 1997). If this competition occurs in vivo (and calmodulin is abundant in synapses), then the possibility arises that NMDA receptor attachment will be regulated by synaptic activity and calcium influx, perhaps through activated NMDA receptors themselves. In this way, NMDA receptor clustering and localization may be affected by synaptic activity.

## Synaptic anchoring of AMPA receptors

AMPA receptors have the same overall membrane topology as NMDA receptors, and they are similarly localized in glutamatergic synapses, but they do not bind to the same proteins that interact with NMDA receptors. For instance, AMPA receptor subunits end in a C-terminal peptide sequence that is incompatible with binding to the PDZ domains of PSD-95. Very satisfyingly, however, recent studies (again using yeast two-hybrid screens) revealed that AMPA receptor subunits bind to their own specific PDZ-containing protein (Dong et al., 1997; Sheng, 1997). This AMPA receptor binding protein (named GRIP for Glutamate Receptor Interacting Protein) contains seven PDZ domains and no other recognizable protein domain. The fourth and fifth of these PDZ domains (PDZ4 and PDZ5) are involved in binding to the C-terminus of GluR2 and GluR3, which terminate in the sequence –ESVKI. This is a good illustration of the differential binding specificities of distinct PDZ domains.

GRIP has been immunocytochemically colocalized with AMPA receptors at synaptic sites in cultured neurons (Dong et al., 1997), suggesting that GRIP may play a role in the clustering and anchoring of AMPA receptors at postsynaptic sites. This is supported by the finding that overexpression of the GluR2 C-terminal tail causes loss of AMPA receptor clustering in transfected neurons (Dong et al., 1997). A genetic test of GRIP's anchoring function is still awaited, though the situation is likely to be complicated by the

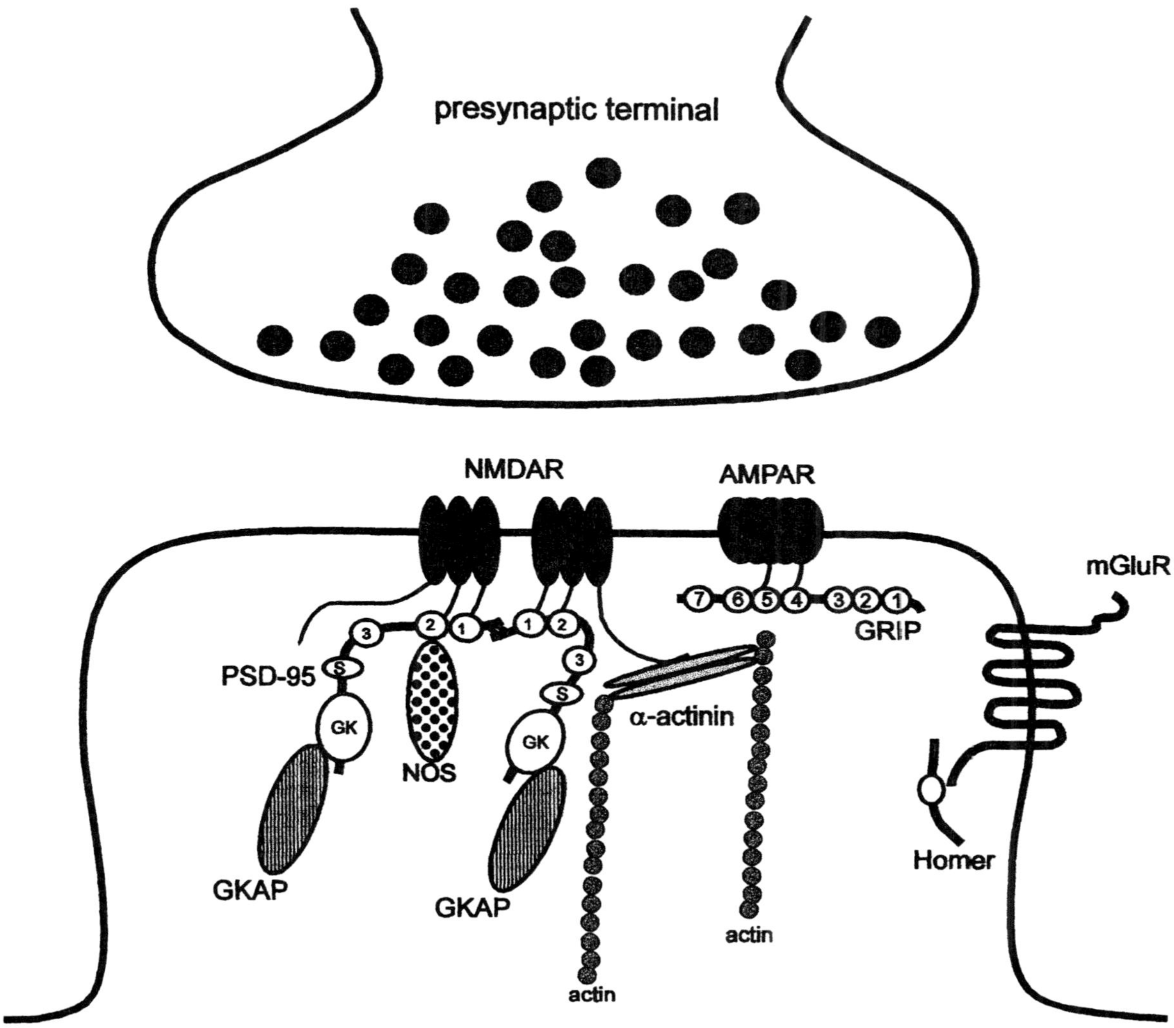


Fig. 1. Postsynaptic metabotropic glutamate receptors (mGluR), NMDA receptors (NMDAR), and AMPA receptors (AMPAR) are shown binding to specific PDZ domains of Homer, PSD-95, and GRIP, respectively. The remaining PDZ domains of PSD-95 and GRIP presumably bind to other synaptic proteins. Subsynaptic segregation of receptors is illustrated: mGluRs are located at the periphery of the synapse, whereas NMDA AMPA receptors are located more centrally. One subunit of NMDAR (NR1) is shown interacting with the central rod domain of α-actinin, which in turn binds actin. PDZ domains are shown as white circle (numbers indicate their relative position from the N-terminus). S, SH3 domain; GK, guanylate kinase domain.

presence of multiple GRIP-like genes expressed in mammalian brain (Wyszynski, M. and Sheng, M., unpublished results).

The presence of seven PDZ domains in GRIP raises the obvious question: what proteins does GRIP bind to in addition to the AMPA receptors? A reasonable prediction is that GRIP functions as a "super-adaptor" protein that couples AMPA receptors with cytoskeletal and signalling elements, in analogous fashion to PSD-95 and NMDA receptors. Identification of these other GRIP-binding proteins will undoubtedly shed further light on the molecular organization of glutamatergic synapses.

## Metabotropic glutamate receptors

The NMDA receptors and AMPA receptors discussed above are major classes of ionotropic glutamate receptor. Glutamate also acts on G-protein coupled seven-transmembrane receptors or metabotropic glutamate receptors (mGluRs). mGluRs are themselves divided into several classes based on pharmacology and downstream signalling mechanisms see Bruno et al., this volume. Partly because of this heterogeneity, the subcellular localization of mGluRs is less comprehensively understood. Nevertheless, it is clear that unlike the ionotropic glutamate receptors, which are concentrated predominantly in postsynaptic sites, mGluRs are found both presynaptically and postsynaptically. The mechanisms for presynaptic localization of mGluRs is unknown, but recent studies provide insight into the molecular basis of postsynaptic anchoring of mGluR1α and mGluR5. These two mGluRs (the group I class of mGluRs) are found exclusively at postsynaptic sites in hippocampal neurons (Lujan et al., 1996; Martin et al., 1992) also see Takumi et al., this volume) and are coupled to phospholipase C and phosphoinositide turnover.

Metabotropic glutamate receptors mGluR1α and mGluR5 uniquely possess long cytoplasmic tails that are 67% identical over the last 55 amino acids and whose C-terminal sequences are highly related (mGluR1α,-QSSSTL; mGluR5, -QSSSSL). By yeast two-hybrid screening it was determined that these C-terminal peptide sequences are involved in binding to Homer, a 28 kD protein that was initially isolated as an immediate early gene inducible by synaptic activity (Brakeman et al., 1997). Homer does not bind to the C-terminal tails of mGluR2 or mGluR4. Intriguingly, the primary sequence of Homer contains a peptide sequence -GLGF- that is a signature sequence found in many PDZ domains (Brakeman et al., 1997). Whether Homer really contains a PDZ domain, however, is open to question. The sequence surrounding the -GLGF- motif in Homer bears minimal resemblance (less than 10% amino acid identity) to known PDZ domains. Nevertheless, the fact that Homer binds to a C-terminal peptide in a sequence-specific manner is reminiscent of PDZ domain mediated interactions.

Homer has been shown to be localized in glutamatergic synapses, and it can be co-immunoprecipitated from brain extracts with mGluR5 (Brakeman et al., 1997). These findings support the idea that Homer and mGluR5 are physiological partners in vivo. A reasonable speculation is that Homer is involved in the localization and anchoring of a specific subset of mGluRs in glutamatergic synapses. In the context of this model, a most remarkable feature is that Homer is dramatically inducible by physiological neuronal activity. The possibility is raised that the synaptic levels or localization of mGluRs might be regulated by neuronal activity, through the induction of expression of Homer protein (Brakeman et al., 1997). Whether this is the case remains to be determined. It should also be emphasized that Homer may have functions different from receptor anchoring. For instance, it may be involved in signal transduction by the mGluRs via binding to their C-terminus. Clearly, a great deal more research is needed to confirm that interacting proteins such as Homer (or GRIP and PSD-95) are essential for the clustering and anchoring of their receptor partner in glutamatergic synapses.

## Diversity of glutamate receptor anchoring mechanisms

The above discussion reveals common themes in glutamate receptor interactions with intracellular proteins. Notably, a variety of glutamate receptor subunits use their intracellular C-terminal tails to bind to specific proteins that are associated with the cytoskeleton or that function as multidomain scaffold proteins. PDZ domains, modular domains specialized for specific binding of C-terminal peptides, play a prominent role in mediating these interactions. A natural question that arises is why NMDA receptors, AMPA receptors, and mGluR1α/5 metabotropic receptors have gone through so much trouble to interact with different putative anchoring proteins, when they are all ultimately localized in the same synapses. Three advantages are afforded by differential interactions with specific anchoring/organizing molecules: 1) the different classes of glutamate receptor can be differentially localized within the synapse; 2) the levels of the various classes of receptor in the synapse can be differentially regulated by altering the activity or abundance of a specific anchoring protein; and 3) each class of receptor can be physically and functionally linked to distinct signal transduction pathways in the postsynaptic cell.

Hints of the complexity of glutamate receptor organization in a synapse have already surfaced. For instance, immuno-EM studies have revealed that group I mGluRs are concentrated in a ring around the edge of the synapse (perisynaptic distribution), whereas ionotropic glutamate receptors are concentrated more centrally with respect to the active zone (Baude et al., 1993; Kharazia et al., 1996; Lujan et al., 1996; Nusser et al., 1994; Takumi et al., this volume). This subsynaptic segregation would predict that mGluRs would interact with different anchoring proteins than do NMDA and AMPA receptors. Whether NMDA receptors and AMPA receptors are segregated within glutamatergic synapses remains to be established. However, the specific interactions of NMDA receptors and AMPA receptors with PSD-95 and GRIP, respectively, provides a potential mechanism for their differential distribution in the postsynaptic specialization.

### Glutamate receptor anchoring and therapeutic implications

Although it remains to be formally proven, it is a reasonable assumption that the proper function of glutamatergic synapses depends on the correct localization of glutamate receptors, not only with respect to the presynaptic transmitter release site, but also with respect to postsynaptic downstream signalling pathways. The correct targeting and anchoring of glutamate receptors will certainly involve their interactions with specific intracellular proteins, some of which are discussed above and some of which undoubtedly remain to be discovered. Genetic mutations that affect PDZ-containing proteins in invertebrate model organisms result in mislocalization of channels and receptors and cause developmental and behavioral defects (Simske et al., 1996; Tejedor et al., 1997; Tsunoda et al., 1997). It is within the realms of possibility that defective localization of glutamate receptors might also give rise to neurological or psychiatric disease in humans.

From a therapeutic viewpoint, one plausible avenue for inhibiting glutamate receptor functions is to disrupt their normal anchoring and localization, rather than to block the receptor per se with classical antagonists. In this case, the drugs would be directed against specific protein–protein interactions that are involved in the anchoring of glutamate receptors or in their coupling to downstream signalling pathways.

Among the protein–protein interactions involving glutamate receptors, the C-terminus-PDZ domain interaction looks most promising for pharmacological intervention. It involves a well-defined domain of 80–100 amino acids that has been crystallized (the PDZ domain) binding to a C-terminal few amino acids (perhaps no more than four residues) in a highly specific manner (Doyle et al., 1996; Songyang et al., 1997; Stricker et al., 1997). Indeed, different PDZ domains can bind either to quite distinct C-terminal sequences, or discriminate between related sequences in a highly selective fashion (Songyang, 1997; M. Niethammer and M. Sheng, unpublished observations). The specificity of PDZ domain binding makes it possible to select for drugs that selectively target a single (or single class) of PDZ interactions. The relative simplicity of the C-terminal peptide-PDZ interface may make it possible to find effective small molecule drugs that can specifically interfere with the interaction. Screening for such molecules could be conveniently done using the yeast two-hybrid system as a reporter assay. Clearly, the utility of drugs directed against interactions of glutamate receptor with other proteins remains an academic concept at the present time. However, the recent exciting developments described in this chapter represent not only an advance in the cell biology of glutamate receptors, they also represent an opportunity to develop a new class of therapeutics directed against glutamatergic synaptic transmission.

**Acknowledgements**

We thank Elaine Aidonidis for help with the manuscript. Morgan Sheng is an Assistant Investigator of the Howard Hughes Medical Institute. Some of the studies from the authors' laboratory described in this review were supported by NIH Grant NS35050 (to M. S.)

**References**

Baude, A., Nusser, Z., Roberts, J.D.B., Mulvihill, E., McIlhinney, R.A.J. and Somogyi, P. (1993) The metabotropic glutamate receptor (mGluRlα) is concentrated at perisynaptic membrane of neuronal subpopulations as detected by immunogold reaction. *Neuron*, 11: 771.

Brakeman, P.R., Lanahan, A.A., O'Brien, R., Roche, K., Barnes, C.A., Huganir, R.L. and Worley, P.F. (1997) Homer, a PDZ-domain protein selectively binds metabotropic glutamate receptors. *Nature*, 386: 284–288.

Brenman, J.E., Chao, D.S., Gee, S.H., McGee, A.W., Craven, S.E., Santillano, D.R., Wu, Z., Huang, F., Xia, H., Peters, M.F., Froehner, S.C. and Bredt, D.S. (1996) Interaction of nitric oxide synthase with the postsynaptic density protein PSD-95 and α1-syntrophin mediated by PDZ domains. *Cell*, 84: 757–767.

Brenman, J.E., Christopherson, K.S., Craven, S.E., McGee, A.W. and Bredt, D.S. (1996) Cloning and characterization of postsynaptic density 93, a nitric oxide synthase interacting protein. *The Journal of Neurosicence*, 16: 7407–7415.

Cho, K.-O., Hunt, C.A. and Kennedy, M. (1992) The rat brain postsynaptic density fraction contains a homolog of the Drosophila discs-large tumor suppressor protein. *Neuron*, 9: 929–942.

Dong, H., O'Brien, R.J., Fung, E.T., Lanahan, A.A., Worley, P.F. and Huganir, R.L. (1997) GRIP: A synaptic PDZ domain-containing protein that interacts with AMPA receptors. *Nature*, 386: 279–284.

Doyle, D.A., Lee, A., Lewis, J., Kim, E., Sheng, M. and MacKinnon, R. (1996) Crystal structures of a complexed and peptide-free membrane protein-binding domain: Molecular basis of peptide recognition by PDZ. *Cell*, 85: 1067–1076.

Ehlers, M.D., Tingley, W.G. and Huganir, R.L. (1995) Regulated subcellular distribution of the NR1 subunit of the NMDA receptor. *Science*, 269: 1734–1737.

Ehlers, M.D., Zhang, S., Bernhardt, J.P. and Huganir, R.L. (1996) Inactivation of NMDA receptors by direct interaction of calmodulin with the NR1 subunit. *Cell*, 84: 745–755.

Hollmann, M. and Heinemann, S. (1994) Cloned glutamate receptors. *Annu. Rev. Neurosci.*, 17: 31–108.

Hsueh, Y.-P., Kim, E. and Sheng, M. (1997) Disulfide-linked head-to-head multimerization in the mechanism of ion channel clustering by PSD-95. *Neuron*, 18: 803–814.

Hunt, C.A., Schenker, L.J. and Kennedy, M.B. (1996) PSD-95 is associated with the postsynaptic density and not with the presynaptic membrane at forebrain synapses. *Journal of Neuroscience*, 16: 1380–1388.

Kharazia, V.N., Phend, K.D., Rustioni, A. and Weinberg, R.J. (1996) EM colocalization of AMPA and NMDA receptor subunits at synapses in rat cerebral cortex. *Neuroscience Letters*, 210: 37–40.

Kim, E., Cho, K.-O., Rothschild, A. and Sheng, M. (1996) Heteromultimerization and NMDA receptor-clustering activity of chapsyn-110, a member of the PSD-95 family of proteins. *Neuron*, 17: 103–113.

Kim, E., Naisbitt, S., Hsueh, Y.-P., Rao, A., Rothschild, A., Craig, A.M. and Sheng, M. (1997) GKAP, a novel synaptic protein that interacts with the guanylate kinase-like domain of the PSD-95/SAP90 family of channel clustering molecules. *Journal of Cell Biology*, 136: 669–678.

Kim, E., Niethammer, M., Rothschild, A., Jan, Y.N. and Sheng, M. (1995) Clustering of shaker-type $K^+$ channels by interaction with a family of membrane-associated guanylate kinases. *Nature*, 378: 85–88.

Kistner, U., Wenzel, B.M., Veh, R.W., Cases-Langhoff, C., Garner, A.M., Appeltaner, U., Voss, B., Gundelfinger, E.D. and Garner, C.C. (1993) SAP90, a rat presynaptic protein related to the product of the Drosophila tumor suppressor gend dlg-A. *J. of Biol. Chem.*, 268: 4580–4583.

Kornau, H.-C., Schenker, L.T., Kennedy, M.B. and Seeburg, P.H. (1995) Domain interaction between NMDA receptor subunits and the postsynaptic density protein PSD-95. *Science*, 269: 1737–1740.

Lau, L.-F., Mammen, A., Ehlers, M.E., Kindler, S., Chung, W.J., Garner, C.C. and Huganir, R.L. (1996) Interaction of the *N*-methyl-D-aspartate receptor complex with a novel synapse-associated protein, SAP102. *The Journal of Biological Chemistry*, 271: 21622–21628.

Lue, R.A., Brandin, E., Chan, E.P. and Branton, D. (1996) Two independent domains of hDlg are sufficient for subcellular targeting: The PDZ1-2 conformational unit and an alternatively spliced domain. *The Journal of Cell Biology*, 135: 1125–1137.

Lue, R.A., Marfatia, S.M., Branton, D. and Chishti, A.H. (1994) Cloning and characterization of hdlg: The human homologue of the *Drosophila*, discs large tumor suppressor binds to protein 4.1. *Proceedings of the National Academy of Science*, 91: 9818–9822.

Lujan, R., Nusser, Z., Roberts, J.D.B., Shigemoto, R. and Somogyi, P. (1996) Perisynaptic location of metabotropic glutamate receptors mGluR1 and mGluR5 on dendrites and dendritic spines in the rat hippocampus. *Eur. J. Neurosci.*, 8: 1488–1500.

Marfatia, S.M., Cabral, J.H.M., Lin, L., Hough, C., Bryant, P.J., Stolz, L. and Chishti, A.H. (1996) Modular organization of the PDZ domains in the human discs-large protein suggests a mechanism for coupling PDZ domain-binding proteins to ATP and the membrane cytoskeleton. *The Journal of Cell Biology*, 135: 753–766.

Marfatia, S.M., Lue, R.A., Branton, D. and Chishti, A.H. (1994) In vitro binding studies suggest a membrane-associated complex between erythroid p55, protein 4.1, and glycophorin C. *Journal of Biological Chemistry*, 269: 8631–8634.

Martin, L.J., Blackstone, C.D., Huganir, R.L. and Price, D.L. (1992) Cellular localization of a metabotropic glutamate receptor in rat brain. *Neuron*, 9: 259–270.

Monyer, H., Burnashev, N., Laurie, D.J., Sakmann, B. and Seeburg, P.H. (1994) Developmental and regional expression in the rat brain and functional properties of four NMDA receptors. *Neuron*, 12: 529–540.

Müller, B.M., Kistner, U., Kindler, S., Chung, W.J., Kuhlendahl, S., Lau, L.-F., Veh, R.W., Huganir, R.L., Gundelfinger, E.D. and Garner, C.C. (1996) SAP102, a novel postsynaptic protein that interacts with the cytoplasmic tail of the NMDA receptor subunit NR2B. *Neuron*, 17: 255–265.

Müller, B.M., Kistner, U., Veh, R.W., Cases-Langhoff, C., Becker, B., Gundelfinger, E.D. and Garner, C.C. (1995) Molecular characterization and spatial distribution of SAP97, a novel presynaptic protein homologous to SAP90 and the *Drosophila*, discs-large tumor suppressor protein. *J. Neurosci.*, 15: 2354–2366.

Naisbitt, S., Kim, E., Weinberg, R.J., Rao, A., Yang, F.-C., Craig, A.M. and Sheng, M. (1997) Characterization of guanylate kinase-associated protein, a postsynaptic density protein at excitatory synapses that interacts directly with postsynaptic density-95/synapse-associated protein 90. *The Journal of Neuroscience*, 7: 5687–5696.

Nakanishi, S. (1992) Molecular diversity of glutamate receptors and implications for brain function. *Science*, 58: 597–603.

Niethammer, M., Kim, E. and Sheng, M. (1996) Interaction between the C terminus of NMDA receptor subunits and multiple members of the PSD-95 family of membrane-associated guanylate kinases. *J. Neurosci.*, 16: 2157–2163.

Nusser, Z., Mulvihill, E., Streit, P. and Somogyi, P. (1994) Subsynaptic segregation of metabotropic and ionotropic glutamate receptors as revealed by immunogold localization. *Neuroscience*, 61: 421–427.

Sheng, M. (1996) PDZs and receptor/channel clustering: rounding up the latest suspects. *Neuron*, 17: 575–578.

Sheng, M. (1997) Glutamate receptors put in their place. *Nature*, 386: 221–223.

Simske, J.S., Kaech, S.M., Harp, S.A. and Kim, S.K. (1996) LET-23 receptor localization by the cell junction protein LIN-7 during C. elegans vulval induction. *Cell*, 85: 195–204.

Songyang, Z., Fanning, A.S., Fu, C., Xu, J., Marfatia, S.M., Chishti, A.H., Crompton, A., Chan, A.C., Anderson, J.M. and Cantley, L.C. (1997) Recognition of unique carboxyl-terminal motifs by distinct PDZ domains. *Science*, 275: 73–77.

Stricker, N.L., Christopherson, K.S., Yi, B.A., Schatz, P.J., Raab, R.W., Dawes, G., Bassett, D.E.J., Bredt, D.S. and Li, M. (1997) PDZ domain of neuronal nitric oxide synthase recognizes novel C-terminal peptide sequences. *Nature-Biotechnology*, 15: 336–342.

Takeuchi, M., Hata, Y., Hirao, K., Toyoda, A., Irie, M. and Takai, Y. (1997) SAPAPs: A family of PSD-95/SAP90-associated proteins localized at postsynaptic density. *Journal of Biological Chemistry*, 272: 11943–11951.

Tejedor, F.J., Bokhari, A., Rogero, O., Gorczyca, M., Zhang, J., Kim, E., Sheng, M. and Budnik, V. (1997) Essential role for dlg in synaptic clustering of shaker $K^+$ channels *in vivo.*, *The Journal of Neuroscience*, 17: 152–159.

Tingley, J.M., Blake, D.J., Zuellig, R.A. and Davies, K.E. (1994) Increasing complexity of the dystrophin-associated protein complex. *Proceedings of the National Academy of Science*, 91: 8307–8313.

Tsunoda, S., Sierralta, J., Sun, Y., Bodner, R., Suzuki, E., Becker, A., Socolich, M. and Zuker, C.S. (1997) A multivalent PDZ-domain protein assembles signalling complexes in a G-protein-coupled cascade. *Nature*, 388: 243–249.

Wisden, W. and Seeburg, P.H. (1993) Mammalian ionotropic glutamate receptors. *Current Opinion in Neurobiology*, 3: 291–298.

Woods, D.F. and Bryant, P.J. (1991) The discs-large tumor suppressor gene of drosophila encodes a guanylate kinase homolog localized at septate junctions. *Cell*, 66: 451–464.

Wyszynski, M., Lin, J., Rao, A., Nigh, E., Beggs, A.H., Craig, A.M. and Sheng, M. (1997) Competitive binding of $\alpha$-actinin and calmodulin to the NMDA receptor. *Nature*, 385: 439–442.

O.P. Ottersen, I.A. Langmoen and L. Gjerstad (Eds.)
*Progress in Brain Research*, Vol 116


CHAPTER 10

# The organization and regulation of non-NMDA receptors in neurons

Robert J. Wenthold* and Katherine W. Roche

*Laboratory of Neurochemistry, National Institute on Deafness and Other Communication Disorders, NIH, Bethesda, MD, USA*

## Introduction

Glutamate receptors are the predominant mediators of excitatory neurotransmission in the central nervous system, and are expressed in nearly every neuron. While an extraordinary amount of information has been obtained on these receptors since the first members were cloned in 1989, much of this research has characterized recombinant receptors expressed in non-neuronal cells. In contrast, relatively little is known about the regulation of the expression and function of glutamate receptors in neurons. The number, type, and functional state of receptors on the postsynaptic membrane will determine if the postsynaptic neuron responds and the nature of the response to neurotransmitter release from the presynaptic terminal. Multiple complex and finely-regulated steps, including synthesis of mRNA and protein, assembly of subunits into complexes, posttranslational modification of the receptor, targeting of the receptor complex to the proper synaptic sites, and association of the receptor with interacting proteins, precede the actual insertion of receptors into the synaptic membrane. A complete understanding of each of these steps is required to understand the mechanisms regulating glutamate receptor expression in neurons. Because of the critical role they play in neurotransmission in the central nervous system, all aspects of glutamate receptors are currently being aggressively investigated, and our knowledge of glutamate receptor regulation in neurons is expanding rapidly.

In this review, we will focus on the molecular properties of native, non-NMDA receptors and discuss several of the key steps leading to formation of functional receptors in neurons which include subunit assembly, intracellular targeting, synaptic expression, and regulation of activity by phosphorylation. An understanding of the biochemical properties of non-NMDA receptors, as determined largely from recombinant receptors, is necessary to appreciate the biochemistry of glutamate receptors in neurons, and this information is reviewed briefly. Two related topics, the synaptic localization of receptors and the interaction of receptors with synaptic anchoring proteins, are covered elsewhere in this volume (Takumi et al., Hsueh and Sheng) and will not be addressed in detail here. We will cover all three non-NMDA receptor subtypes, AMPA, kainate and delta receptors, but will concentrate on the AMPA receptors which have been the most

*Corresponding author.
Tel.: 1 301-496-6530; fax: 1 301-480-2324; e-mail: wenthold@nidcd.nih.gov.

thoroughly characterized thus far, and, in many cases, are representative of all non-NMDA ionotropic glutamate receptors.

## Biochemical properties of non-NMDA receptors

### *AMPA, kainate and delta receptor subunits*

Three subtypes of non-NMDA receptors, AMPA, kainate, and delta receptors, have been identified in mammalian brain (Table 1). Mammalian AMPA receptors are comprised of four subunits, GluR1-4, and in expression systems, these subunits can form functional homomeric or heteromeric receptor complexes (Borges and Dingledine, this volume). Each subunit exists in alternatively spliced forms, flip and flop, which differ in a 38 amino acid segment in the region preceding the fourth transmembrane region, TM4 (Sommer et al., 1990). Although both forms are widely expressed in brain, their patterns of expression and developmental profiles differ. Most neurons express both flip and flop variants and presumably native receptor complexes contain mixtures of flip and flop. The only neuron reported to express a single variant is the principal neuron of the medial nucleus of the trapezoid body (MNTB) which expresses only flop variants of GluR1, 2 and 4 (Geiger et al., 1995). Glutamate is 4–5 times more effective in activating flip variants than flop (Sommer et al., 1990), while the flop variants have considerably faster desensitization constants (Mosbacher et al., 1994). Therefore, the splice variants confer unique functional properties on the receptor. GluR2 is responsible for controlling the calcium permeability of AMPA receptor ion channels; channels without GluR2 are calcium permeable and those with GluR2 are calcium impermeable (Hume et al., 1991; Verdoorn et al., 1991). A single amino acid in the TM2 region controls calcium permeability: an arginine in GluR2 and a glutamine in the other subunits. Arginine, however, is not encoded in GluR2 DNA, but is generated through RNA editing (Sommer et al., 1991; Burnashev et al., 1992; Lomeli et al., 1994). Additional editing sites have been identified on other AMPA and kainate receptor subunits (Table 1); these affect the functional properties of the receptors, but not to the extent that editing of the TM2 region has on GluR2.

Five subunits of kainate receptors are known, GluR5, 6, 7, and KA 1 & 2 (Table 1). GluR5 and 6 are functional when expressed in vitro, and the remaining subunits form functional receptors when coexpressed with GluR5 or 6. In neurons it is believed that kainate receptors exist as heteromeric complexes made up of at least two different subunits. Although they are widely distributed (Huntley et al., 1993; Petralia et al., 1994), kainate receptors are less abundant than AMPA receptors with 50-fold fewer ligand binding sites for kainate in whole brain (Hampson et al., 1987; Hunter et al., 1990). Functional kainate receptors have been identified in a limited number of cases including dorsal root ganglia and a population of hippocampal neurons (Huettner, 1990; Lerma et al., 1993; Vignes and Collingridge, 1997; Castillo et al., 1997), but their widespread expression, as seen with in situ hybridization histochemistry and immunocytochemistry, would suggest that functional kainate receptors are present on virtually every neuron. The role of kainate receptors remains unclear, since they are generally present with AMPA receptors which are considerably more abundant and appear to be predominant during normal transmission. Kainate receptors on CA3 pyramidal neurons, postsynaptic of mossy fiber terminals, were shown to be use dependent and may provide a mechanism for augmenting pyramidal cell excitability during high levels of activity (Castillo et al., 1997).

Delta receptor subunits, $\delta1$ and $\delta2$, do not form functional homomeric channels when expressed in vitro (Yamazaki et al., 1992; Araki et al., 1993; Lomeli et al., 1993). Expression of the two subunits together or with subunits of other glutamate receptors also does not produce functional receptors. Immunoprecipitation of delta receptors from detergent solubilized cerebellum does not co-immunoprecipitate other glutamate receptor subunits and does not immunoprecipitate $^3$H-AMPA, $^3$H-kainate or $^3$H-glutamate ligand binding activ-

TABLE 1

Non-NMDA receptor subunits and their properties

| Subunit | Alternative Nomenclature | MW (daltons) | Splice Variants | Editing Sites | Homomeric Function | Binding ligand/Kd (nM) | Reference |
|---|---|---|---|---|---|---|---|
| AMPA Receptors | | | | | | | |
| GluR1 | GluR-A, $\alpha$1 | 99,796 | flip, flop | | yes | $^{3}$H-AMPA, 33 (flop) | 4, 9, 12, 14, 17, 19, 24, 27 |
| GluR2 | GluR-B, $\alpha$2 | 96,400 | flip, flop | Q/R, R/G | yes | $^{3}$H-AMPA, 4.7, 28 (flop) 2.3, 109 (flip) | 4, 9, 13, 14, 18, 19, 24, 27, 28 |
| GluR3 | GluR-C | 98,000 | flip, flop | R/G | yes | $^{3}$H-AMPA, 52 (flop) | 4, 9, 14, 18, 19, 24, 27 |
| GluR4 | GluR-D | 101,034 | flip, flop, -4c | R/G | yes | $^{3}$H-AMPA, 14 (flop) | 4, 7, 9, 14, 18, 19, 24, 27 |
| Kainate Receptors | | | | | | | |
| GluR5 | | 93,891 | -1, -2a, -2b, -2c, -2d | Q/R | yes | $^{3}$H-kainate, 67 (GluR5-2a) 73 (GluR5-2b) | 2, 8, 11, 14, 29 |
| GluR6 | $\beta$2 | 93,891 | -2 | Q/R, I/V, Y/C | yes | $^{3}$H-kainate, 12.9, 36 | 5, 6, 8, 14, 20, 25, 29, 30 |
| GluR7 | | 100,000 | | | no | $^{3}$H-kainate, 63, 77 | 3, 14, 21 |
| KA1 | EAA1 | 105,000 | | | no | $^{3}$H-kainate, 4.7 (KA1) 2.3 (EAA1) | 14, 16, 31 |
| KA2 | $\gamma$2, EAA2 | 109,000 | | | no | $^{3}$H-kainate, 15.2 (KA2) 2.6 (EAA2) | 10, 14, 15, 25, 26 |
| Delta Receptors | | | | | | | |
| $\delta$1 | | 110,440 | | | no | none | 1, 14, 23, 32 |
| $\delta$2 | | 113,229 | | | no | none | 1, 14, 21, 23 |

1. Araki et al., 1993
2. Bettler et al., 1990
3. Bettler et al., 1992
4. Boulter et al., 1990
5. Egebjerg et al., 1991
6. Egebjerg and Heinemann 1993
7. Gallo et al., 1992
8. Gregor et al., 1993
9. Hennegriff et al., 1997
10. Herb et al., 1992
11. Herb et al., 1996
12. Hollmann et al., 1989
13. Hollmann et al., 1991
14. Hollmann and Heinemann 1994
15. Kamboj et al., 1992
16. Kamboj et al., 1994
17. Kawamoto et al., 1994
18. Kawamoto et al., 1995
19. Keinanen et al., 1990
20. Köhler et al., 1993
21. Lomeli et al., 1992
22. Lomeli et al., 1993
23. Mayat et al., 1995
24. Monyer et al., 1991
25. Morita et al., 1992
26. Sakimura et al., 1992
27. Sommer et al., 1990
28. Sommer et al., 1991
29. Sommer et al., 1992
30. Tygesen et al., 1994
31. Werner et al., 1991
32. Yamazaki et al., 1992

ity (Mayat et al., 1995). Initially, these results raised questions as to whether or not delta is a functional receptor in brain. However, a recent study (Zuo et al., 1997) localized the mutation of the Lurcher mouse to the $\delta 2$ gene; this mutation results in a single amino acid change in the TM3 region that creates a constitutively active receptor that leads to degeneration of Purkinje neurons. These results demonstrate that $\delta 2$ can form an ion channel and that the delta receptor can be activated under certain conditions. One possibility is that glutamate is not an agonist for delta receptors and the correct ligand has not yet been identified.

Although they have similar structural properties, subunits of the AMPA, kainate, and delta receptor subtypes do not appear to combine across subtypes to form hybrid receptor complexes. Immunoprecipitation studies with detergent solubilized brain, neuronal cultures, CD4 cells, and transfected cells show no appreciable interaction between different subtypes, and physiological studies have found no evidence of such hybrid receptors (Wenthold et al., 1994; Puchalski et al., 1994; Roche and Huganir, 1995). Based on studies of the nicotinic acetylcholine receptor (nAChR), assembly of glutamate receptor subunits is likely to be controlled by the interactions of domains on the subunits themselves (Yu and Hall, 1994).

The topology of glutamate receptors, essential for understanding many of their properties such as agonist binding and interactions with other proteins, has been extensively studied. This topic is addressed in detail elsewhere in this volume (Borges and Dingledine) and will only be summarized here. The first reports of the cloning of glutamate receptors assigned a four transmembrane topology to the molecules, although there was disagreement on the position of one of the transmembrane regions (Gregor et al., 1989; Hollmann et al., 1989; Wada et al., 1989). The structure now generally accepted places the N-terminus and the large loop between TM3 and TM4 extracellular and the C-terminus intracellular (Wo and Oswald, 1994; Hollmann et al., 1994; Bennett and Dingledine, 1995). The original nomenclature with four transmembrane domains has been retained in most publications, although it is now assumed that there are only three true transmembrane domains.

### *The structure of the ionotropic glutamate receptor*

Based on the structure of other ligand-gated ion channels, it was initially assumed that functional glutamate receptors are pentamers. Subsequent experiments addressed the size of the AMPA receptor complex but the exact number of subunits in a functional receptor complex has not been completely resolved. Gel filtration and sucrose density centrifugation of detergent solubilized membranes from rat brain showed an $M_r$ of 425,000 to 610,000 for AMPA receptors (Hunter et al., 1990; Blackstone et al., 1992) and an $M_r$ of 650,000 for kainate receptors (Hampson et al., 1987). A more detailed analysis of the size of the AMPA receptor complex using chemical cross-linking (Wenthold et al., 1992) showed three high molecular weight bands, identified with anti-receptor antibodies, migrating with $M_r$ of 330,000, 470,000, and 590,000. With a monomer $M_r$ of 108,000, the largest component would most closely correspond to a pentamer. But the limited accuracy of molecular weight measurements of proteins of this size, as well as the nonlinear structure and aberrant migration in SDS of the cross-linked proteins, makes this conclusion somewhat tentative. Complexes of 4 or 6 subunits should also be considered. The lower molecular weight bands probably arise from incomplete cross-linking, but the presence of a population of receptors with fewer subunits cannot be ruled out. A functional approach, using an assay based on the differential sensitivity of mutant subunits to channel blockers, also suggested a pentameric structure for AMPA receptors (Ferrier-Montiel and Montal, 1996).

## Receptors in neurons

### *Biochemical properties of native non-NMDA receptors*

The molecular properties of receptors expressed in the central nervous system are usually studied either functionally, such as with ligand binding, or by using antibodies that selectively interact with the receptor subunits, since the low abundance of these proteins does not permit more direct chemical analysis. AMPA and kainate receptors were extensively characterized with ligand binding analyses several years before the first subunits were cloned (Honore et al., 1982; Simon et al., 1976), and these studies showed a clear distinction between AMPA and kainate receptors, which physiological studies were unable to do because of the lack of selective agonists and antagonists for the two receptor types. After their cDNAs were cloned, the binding properties of native AMPA receptors were compared to those of recombinant receptors expressed in transfected cells, and interesting differences emerged. Binding of $^3$H-AMPA to brain membranes or detergent-solubilized preparations is characterized by curvilinear Scatchard plots which indicate the presence of at least two populations of binding sites with dissociation constants ranging from 10 nM to greater than 500 nM (Hunter et al., 1990). Of the recombinant receptors, only GluR2 shows more than one binding site by Scatchard analysis, and the dissociation constants of all populations are less than 100 nM (Hennegriff et al., 1997). The authors of this study suggest that native receptors undergo a modification in neurons that does not occur for recombinant receptors, and propose that this modification takes place when the receptor is inserted into the postsynaptic membrane. If this proves to be the case, the yet unidentified modification could be a key step in the regulation of receptor expression at the postsynaptic membrane.

It has generally been assumed that ionotropic receptors in brain assemble into functional receptor complexes as they do in transfected cells or oocytes. This seems to be the case since neuronal receptors are physiologically similar to recombinant receptors, and immunoprecipitation studies on brain show that both AMPA and kainate receptors are heteromeric complexes (Wenthold et al., 1992; 1994). But the mechanisms regulating assembly in neurons are likely to be far more complicated than those in transfected cells or oocytes since neurons are both architecturally and functionally more complex. The subunit composition of a receptor complex dramatically influences the properties of the receptor, and a variety of postsynaptic responses can be obtained with different combinations of glutamate receptors and receptor subunits. Glutamate receptor expression in neurons is complex, with multiple subtypes, subunits, and splice variants expressed together in most neurons. Since most neurons receive many excitatory inputs, it seems reasonable to hypothesize that neurons have the capability of targeting functionally distinct receptors or combinations of receptors to different populations of synapses on the neuron.

Neurons that express more than one subunit of an ionotropic receptor theoretically could produce multiple physiologically distinct receptor complexes by varying the relative number of subunits in a complex. Physiological studies were the first to attempt to determine if this is actually the case in neurons by determining the distribution of calcium permeable (lacking GluR2) and calcium impermeable (containing GluR2) AMPA receptors on a neuron. Lerma et al. (1994) reported a greater density of calcium permeable AMPA receptors on dendrites of hippocampal pyramidal neurons compared with cell bodies, suggesting a differential distribution of receptor complexes that contain GluR2. However, a later study (Spruston et al., 1995) on the same neurons failed to confirm these findings.

To address the issue of the biochemical nature of receptors in neurons more directly, a relatively homogeneous population of neurons, the CA1/CA2 pyramidal neurons of the hippocampus, was investigated biochemically using antibodies selective for the AMPA receptor subunits (Wenthold et al., 1996). These neurons, which express GluR1,

2 and 3, can be dissected in relatively pure form and in sufficient abundance for biochemical analysis. AMPA receptors can be efficiently solubilized with Triton X-100 to yield intact receptor complexes which bind $^{3}$H-AMPA (Hunter et al., 1990). By immunoprecipitating the solubilized receptor complexes with antibodies selective for the individual subunits, the nature of the receptor complex can be determined. These studies demonstrate that CA1/CA2 neurons contain two predominant AMPA receptor complexes, those with GluR1 and GluR2 and those with GluR3 and GluR2. About 8% of the complexes contained homomeric GluR1 and only a few complexes were found that contain all three subunits, GluR1, 2, and 3. These studies suggest that although these neurons express all three subunits, the complexes formed do not simply reflect a random assortment of the subunits expressed, but that a mechanism exists which regulates the composition of the complexes.

The subcellular distribution of receptors can also reveal the nature of receptor complexes in neurons. The fusiform neuron of the dorsal cochlear nucleus is an ideal model to investigate receptor targeting. It expresses multiple ionotropic subunits and metabotropic receptors and has two major glutamatergic inputs, a parallel fiber input to its apical dendrites and the eighth nerve input to its basal dendrites. A study designed to determine if the same or different receptors are present postsynaptic of the two inputs showed two key differences (Rubio and Wenthold, 1997). While most receptor subunits and subtypes were found at both synaptic populations, the metabotropic receptor, mGluR1α, and the AMPA receptor subunit, GluR4, were both present at the eighth nerve synapses but were absent from the parallel fiber synapses, clearly demonstrating a differential targeting of the receptors (also see Takumi et al., this volume). The presence of GluR4 at one synaptic population, but not at the other, not only shows selective targeting, but shows there must be at least two different AMPA receptor complexes, those containing GluR4 and those lacking GluR4.

The functional properties of the synaptic populations on fusiform cells are consistent with the immunocytochemical results showing homomeric GluR4 at the basal synapses. AMPA receptors at major relay synapses in the lower auditory system display very fast channel kinetics in both chick and rat. Studies on the chick cochlear nucleus have shown that synapses between the auditory nerve and principal neurons have exceptionally fast desensitization and deactivation rates (Raman et al., 1994). Single cell PCR of neurons of the rat medial nucleus of the trapezoid body (MNTB) showed that these neurons have high levels of GluR4 and exclusively flop splice variants of all AMPA receptor subunits (Geiger et al., 1995). Analysis of receptors in transfected cells showed that GluR4 flop has the fastest channel kinetics and that combinations, particularly with GluR2 flip, produced a slower channel. Since the fusiform neuron receives auditory nerve input to its basal dendrites, this synapse might be expected to have AMPA receptors with fast channel kinetics, and the expression of homomeric GluR4 receptors would be one mechanism to obtain such receptors.

Based on similar studies on nicotinic receptors and $GABA_A$ receptors (Laurie et al., 1992; Conroy et al., 1992: Vernallis et al., 1993; Conroy and Berg, 1995), the assembly of multiple receptor complexes in neurons is not unexpected, but questions arise concerning the mechanisms by which this could be achieved. Two mechanisms could account for the differential targeting of AMPA receptor subunits in fusiform cells. In the first scenario, receptor complexes are formed which contain GluR2, 3 &4 or GluR2 & 3. Such a mechanism would require control of the assembly process to ensure that both types of complexes are produced and in the correct proportions. Alternatively, complexes containing GluR2 & 3 and those containing only GluR4 could be made, and the selective targeting affects only GluR4. This latter scenario could be achieved either by regulating assembly in the cell body such that GluR4 does not form complexes with any other subunit, or by targeting the mRNA to an area of the cell devoid of the other AMPA mRNAs, such as the dendrite. A dendritic localization has been convincingly demonstrated for some mRNAs, although little evidence exists to support glutamate receptor mRNA in dendrites. One study reported that the NMDAR1 (NR1) mRNA is localized in dendrites and that the amount increases after neurons are denervated (Gazzaley et al., 1997). The clearest example of receptor mRNA in dendrites was recently shown for the glycine receptor (Racca et al., 1997). In these studies it was shown that the $\alpha$ subunit mRNA was found in cell bodies and dendrites while the $\beta$ subunit was predominantly found in the cell body. If GluR4 mRNA is present in dendrites, and other AMPA receptor mRNA is not, it would support the idea that homomeric GluR4 receptors are present in basal dendrites. The alternative, that GluR4 is synthesized in the dendrite and is in a complex with the other AMPA receptor subunits, would appear unlikely since the other subunits would need to be synthesized in the cell body, transported in a fully or partially unassembled state to the dendrite and then assemble with GluR4, or mRNA for all subunits would need to be present in the dendrite.

*Targeting of receptors in neurons*

Since most protein synthesis occurs in the cell body, an efficient mechanism is needed to assure that the proteins are properly targeted to their appropriate location in the neuron. The selective transport to somatodendritic or axonal compartments has been clearly established for a number of proteins including, for example, some neurotransmitter receptors which are predominantly postsynaptic and neurotransmitter synthesizing enzymes which are predominantly axonal (Craig and Banker, 1994; Kelley and Grote, 1993). For glutamate receptors, most ionotropic receptors are postsynaptic and confined to the somatodendritic compartment, whereas some metabotropic receptors are presynaptic (Petralia, 1997; Takumi et al., this volume). The simplest situation would be for some receptors to be exclusively postsynaptic and others to be exclusively presynaptic. With such a scenario, targeting to the axon or dendrite could be achieved by a signal sequence contained in the receptor itself. However, this does not appear to be the case for glutamate receptors. There are scattered reports of ionotropic receptors in presynaptic compartments (Petralia, 1997), but the data are not as rigorous as those for the postsynaptic localization where multiple antibodies to different epitopes on the receptor molecule have been used. The most convincing evidence for both a pre- and postsynaptic localization of a receptor comes from studies on the metabotropic receptor, mGluR2; using two different antibodies, along with knockout mice to establish the antibody specificity, it was demonstrated that mGluR2 is both pre- and postsynaptic (Neki et al., 1996; Luján et al., 1997). The simultaneous targeting of mGluR2 to axons and dendrites occurs in the same population of neurons, since both axons and dendrites of cerebellar Golgi cells contain the receptor. Such a distribution could be achieved if mGluR2 contains sorting signals for both the somatodendritic and axonal compartments, or if it contains neither signal and by default it is targeted to both compartments.

Most ionotropic glutamate receptors are found within the somatodendritic compartment, but their distributions vary with receptor type and are by no means uniform (Petralia, 1997). Receptors are highly concentrated at postsynaptic sites (Takumi et al., this volume), but nonsynaptic receptors are present as is an intracellular pool which is likely to be associated with organelles involved in the addition and removal of receptors from the plasma membrane. Within the same neuron, glutamate receptors can be differentially distributed among glutamatergic synapses as discussed above. How might glutamate receptors (and other dendritic proteins) be selectively targeted to sites within the somatodendritic compartment? Local protein synthesis in the dendrite would be one mechanism to selectively target proteins to a postsynaptic site, as discussed above for GluR4 in fusiform cell basal dendrites. While dendritic synthesis of glutamate receptors may occur under some conditions, it does not appear to be the predominant mechanism for the synthesis of glutamate receptors, whose mRNA is mainly in the cell body (Eshhar et al., 1993). If we assume that most glutamate receptor synthesis occurs in the cell body, two possible mechanisms could be considered to obtain a selective combination of receptors at various populations of postsynaptic sites (Fig. 1). In the first case, the mature or nearly mature receptor is selectively targeted to a particular synaptic population. This would require some type of targeting signal either on the receptor itself or on a protein associated with the receptor. The receptor could be associated with the dendritic endoplasmic reticulum, which has been reported to form a continuous membrane system within the dendrite (Martone et al., 1993; Terasaki et al., 1994), or small organelles which bud from the endoplasmic reticulum either in the dendrite or cell body. Since these organelles are selectively targeted to a particular population of synaptic sites, only those proteins found at that synapse, along with various others that may play a role in transport, would be associated with the organelle. In the second case, protein synthesis and transport would occur like that in the first, but targeting would be random. All dendritic membrane proteins would be mixed together in intracellular organelles which would associate and dissociate continuously with the plasma membrane along the length of the dendrite. Receptor specificity would be achieved by the selective retention of receptors at a particular postsynaptic site through an interaction with an anchoring protein. The discovery of postsyn-

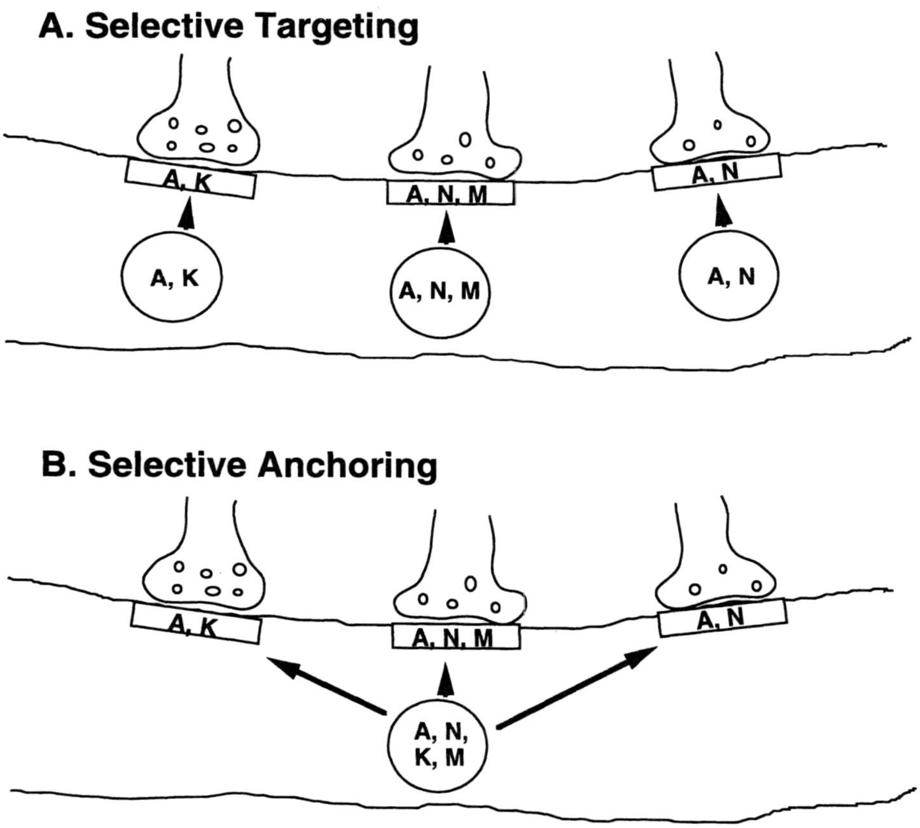


Fig. 1. Two models for delivering glutamate receptors to postsynaptic membranes. In A, a specific combination of receptors is prepackaged in intracellular organelles for delivery to their respective postsynaptic membranes. In B, packaging and targeting is random, but selective receptor expression is achieved by an interaction of the receptor with a specific anchoring molecule. Abbreviations: A, AMPA receptor; K, kainate receptor; N, NMDA receptor; M, metabotropic glutamate receptor.

aptic density proteins which bind to the C-termini of several glutamate receptors tends to support this mechanism (Sheng, 1996; Dong et al., 1997).

While it remains to be determined if either of these two mechanisms is active in sorting glutamate receptors in neurons, it is noteworthy that somewhat related mechanisms occur in polarized epithelial cells. For example, in MDCK cells, adrenergic receptors are sorted to the basolateral membrane by different mechanisms (Wozniak and Limbird, 1996). The $\alpha_{2A}$ receptor is delivered directly to the basolateral membrane and the intracellular pool of this subtype is minimal. The $\alpha_{2C}$ receptor is also delivered directly to the basolateral membrane but a substantial intracellular pool is present. The $\alpha_{2B}$ receptor is delivered randomly to both apical and basolateral membranes, but is retained preferentially at the basolateral membrane.

*Regulation of non-NMDA glutamate receptors by phosphorylation*

Posttranslational modifications, such as phosphorylation, play a central role in the regulation of glutamate receptors. Phosphorylation of ligand-gated ion channels alters the channel function of many neurotransmitter receptors including the nAChR, $GABA_A$ receptor, and more recently both NMDA and non-NMDA glutamate receptors (reviewed in Swope et al., 1992). Unlike other regulatory mechanisms controlling the expression and activity of receptors, phosphorylation can directly modify receptors in the postsynaptic membrane in both a rapid and reversible manner. The phosphorylation of AMPA and kainate receptors has been characterized using a variety of biochemical and electrophysiological techniques in primary neuronal cultures and, following the cloning of these glutamate receptor subunits, in heterologous cells transiently expressing glutamate receptor subunits (see Borges and Dingledine, this volume). By comparing results obtained from these various methods, we get a clearer picture of how phosphorylation modulates glutamate receptor function.

*Phosphorylation of AMPA/kainate receptors by cAMP-dependent protein kinase (PKA)*

Over the past ten years substantial evidence has accumulated indicating that PKA modulates non-NMDA receptor function. One of the earliest studies suggesting an important regulatory role for PKA demonstrated that AMPA/kainate receptor currents were potentiated upon dopamine (DA) receptor activation of white perch retinal horizontal cells. This potentiation was due to increases in cAMP levels leading to PKA activation (Knapp and Dowling, 1987). Subsequent studies from Dowling's laboratory showed that direct injection of the catalytic subunit of PKA into the horizontal cells had the same effect (Liman et al., 1989). Several years later, two groups independently reported that PKA phosphorylation elicited increases in the current amplitude of kainate responses recorded from cultured hippocampal neurons (Greengard et al., 1991; Wang et al., 1991). More recently, Rosenmund et al. (1994) also characterized this effect in hippocampal cultures and found that not only is PKA activation critical to the potentiation of AMPA/kainate responses, but that other proteins called AKAPs (A-kinase-anchoring proteins) are also required. These proteins bind with high affinity to the regulatory subunit of PKA and are thought to target or dock PKA to the appropriate subcellular location for substrate interaction. AKAP 79 is localized to the postsynaptic density (PSD) (Carr et al., 1992), and when its interaction with PKA is disrupted in hippocampal neurons, the PKA-mediated potentiation of kainate-evoked currents is eliminated (Rosenmund et al., 1994).

The cloning of the AMPA and kainate receptor subunits made it possible to characterize direct phosphorylation of non-NMDA receptor subunits and to determine if the previously-described effects of PKA phosphorylation on channel properties were direct or indirect. The first experiments focused on the kainate receptor subunit GluR6 which, unlike the AMPA receptor subunits, contains a strong PKA consensus phosphorylation site, Arg-Arg-X-Ser (where X is any amino acid) in the loop between TM 3 and 4, which was previously thought to be intracellular, as discussed elsewhere in this volume (Borges and Dingledine, this volume). Two laboratories independently demonstrated that application of PKA increased kainate-evoked whole cell currents from cells expressing GluR6 homomeric channels (Raymond et al., 1993; Wang et al., 1993; Fig. 2). Importantly, these studies both demonstrated that mutation of serine 684 to alanine reduced or completely eliminated the PKA modulation. In addition, Raymond et al. (1993) presented biochemical evidence of a major phosphopeptide after PKA phosphorylation of GluR6, and this phosphopeptide was no longer observed when serine 684 was mutated to alanine. These were the first direct reports of PKA phosphorylation of a glutamate receptor subunit modulating function.

Similar experiments were conducted on AMPA receptor subunits (also see Borges and Dingledine, this volume). Keller et al. (1992) observed that Xenopus oocytes, coexpressing GluR1 and GluR3 AMPA receptor subunits, displayed potentiated agonist-induced currents and $Ca^{2+}$ influx upon bath application of membrane permeable cAMP. This finding demonstrated that, like native AMPA receptors, AMPA receptors expressed in oocytes were potentiated upon PKA activation. A detailed biochemical characterization of the AMPA receptor subunit GluR1 revealed that this subunit was basally phosphorylated on a single serine. Furthermore, with forskolin treatment or cotransfection with the catalytic subunit of PKA, an additional serine was phosphorylated in both cortical cultures and in HEK-293 cells expressing

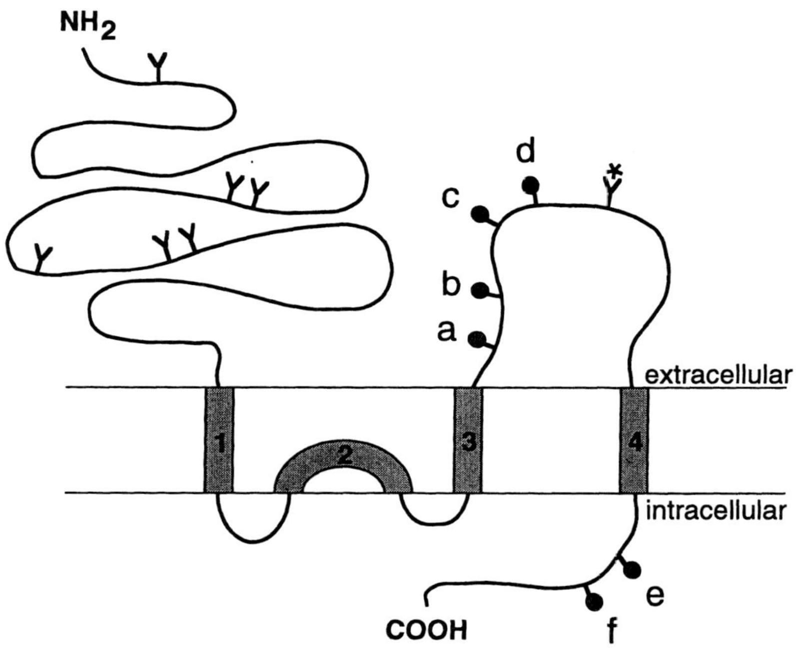


Fig. 2. Illustration of the current transmembrane topology model for ionotropic glutamate receptors includes the endogenous glycosylation and phosphorylation sites which have been demonstated on AMPA or kainate receptor subunits. Demonstrated phosphorylation sites are indicated (⚲) and are labeled (a–f): a) serine 627 on GluR1 (Yakel et al., 1995); b) serine 666 on GluR6 (Wang et al., 1993); c) serine 684 on GluR6 (Raymond et al., 1993; Wang et al., 1993); d) serine 696 on GluR2 (Nakazawa et al., 1995); e) serine 831 on GluR1 (Roche et al., 1996; Mammen et al., 1997); and f) serine 845 on GluR1 (Roche et al., 1996; Mammen et al., 1997). Sites demonstrated to be glycosylated on the N-terminus of GluR1 (Y ; Hollmann et al., 1994) and the TM3-TM4 loop of GluR6 (Y*; Roche et al., 1994; Taverna et al., 1994) are indicated.

GluR1 (Blackstone et al., 1994). The PKA phosphorylation site on GluR1 was identified as serine 845 (Roche et al., 1996). Importantly, phosphorylation of homomeric GluR1 channels expressed in HEK-293 cells by PKA resulted in a 40% potentiation of peak current which was eliminated by the mutation of serine 845 to alanine (Roche et al., 1996). Thus, PKA phosphorylation of GluR1 may, at least in part, account for the potentiated agonist-induced currents observed in primary neuronal cultures.

*Phosphorylation of AMPA/kainate receptors by protein kinase C (PKC) and calcium/calmodulin-dependent protein kinase II (CAM-KII)*

There is ample evidence that AMPA receptors are phosphorylated by PKC and CAM-KII, and that these kinases modulate AMPA receptor currents. Initially, McGlade-McCulloh et al. (1993) demonstrated that the AMPA receptor subunit GluR1 is phosphorylated by both PKC and CAM-KII using in vitro phosphorylation techniques. In addition, agonist treatment of hippocampal neurons induced autophosphorylation of CAM-KII and phosphorylation of AMPA receptors, and the AMPA receptor phosphorylation was blocked by APV and other NMDA channel blockers (Tan et al., 1994). Activation of PKC with phorbol ester also increased AMPA receptor phosphorylation in these cultures. The functional effect of phosphorylation was investigated by introducing activated CAM-KII into the hippocampal neurons, which resulted in a three-fold increase in AMPA receptor currents (McGlade-McCulloh et al., 1993).

Using recombinant receptors expressed in oocytes, Yakel et al. (1995) demonstrated that CAM-KII phosphorylation potentiated currents through GluR1 homomeric channels and that the mutation of serine 627 to alanine of GluR1 removed this potentiation. Serine 627 is located within the TM3-TM4 loop which, according to the present transmembrane topology model, is extracellular. Similarly, Nakazawa et al. (1995) characterized the phosphorylation of a conserved serine also within the TM3-TM4 loop of AMPA receptors by using phosphorylation-state specific antibodies to probe cerebellar slice preparations.

Another group demonstrated that activation of PKC leads to phosphorylation of the AMPA receptor subunit GluR1 in cortical cultures (Blackstone et al., 1994). By first pre-labeling cells expressing GluR1 with $^{32}$P-orthophosphate and then using phosphopeptide map analysis of phosphorylated GluR1, it was demonstrated that synaptic activity resulted in phosphorylation of the same site on GluR1 that was phosphorylated with PKC activation using phorbol treatment. Thus, physiological stimuli, at least in culture, were able to stimulate phosphorylation of a residue not phosphorylated under basal conditions. Later this residue was identified as serine 831 which, in contrast to the previously identified phosphorylation sites, is present within the C-terminal domain of GluR1 (Roche et al., 1996). Phosphorylation of serine 831 has been characterized further using phosphorylation-state specific antibodies to serine 831, and it was demonstrated that this residue was phosphorylated by CAM-KII, as well as PKC, both in transfected cells and in hippocampal slices (Mammen et al., 1997). Both Roche et al. (1996) and Mammen et al. (1997) found no evidence for phosphorylation of serine 627 of GluR1, and the finding that serine 831 is an excellent substrate for CAM-KII (Mammen et al., 1997) suggests that modification of this residue may underlie the functional modulation previously characterized in primary cultures.

*Phosphorylation: The center of the transmembrane topology debate*

As discussed above, the ionotropic glutamate receptor subunits initially were proposed to contain four transmembrane domains, with an extracellular N- and C-terminus, analogous to the $GABA_A$ and nAChR receptors. However, several studies (also discussed by Borges and Dingledine, this volume) make a compelling case that the originally-proposed topology is incorrect and that the glutamate receptors have a unique topology including only three true transmembrane domains and an intracellular C-terminus. The major unresolved data which support the original model show phosphorylation of residues that are now believed to be extracellular (Fig. 2). These data include: 1) evidence that GluR6 was phosphorylated on serine 684 which is contained within the TM3-TM4 loop (Raymond et al., 1993; Wang et al., 1993); 2) a report that the GluR1 AMPA receptor subunit was directly phosphorylated on serine 627 in a region contained within the TM3-TM4 loop (Yakel et al., 1995); and 3) a report that demonstrated increases in AMPA receptor phosphorylation upon agonist exposure in cerebellar slices using a phosphorylation-state specific antibody to a peptide contained within the TM3-TM4 loop of the GluR2 AMPA receptor subunit (Nakazawa et al., 1995). However, as discussed above, there are several studies that show phosphorylation of residues in the C-terminus of both NMDA (Tingley et al., 1993; Omkumar et al., 1996) and non-NMDA receptors (Roche et al. 1996; Mammen et al., 1997), consistent with the new, three transmembrane model of the glutamate receptor (Fig. 2).

After reviewing the glutamate receptor phosphorylation literature, it appears to be impossible to reconcile the data showing phosphorylation of the TM3-TM4 loop with the newly revised topology model. Possible explanations range from technical flaws in the experiments to unusual properties of the glutamate receptor molecule which would allow the TM3-TM4 loop to reside either inside or outside of the cell, depending on its state of activation (see Hollmann, 1997 and Soderling, 1997, for more complete discussions). Resolution of these inconsistencies awaits further studies.

*Glutamate receptor phosphorylation and synaptic plasticity*

Glutamate receptor phosphorylation is of potential importance in models of synaptic plasticity (reviewed in Roche et al., 1994b). It is now well accepted that phosphorylation plays a critical role in both long-term potentiation (LTP; Bliss and Collingridge, 1993) and long-term depression (LTD; Linden, 1994). Many laboratories have reported that kinase activity is essential for the induction and maintenance of LTP and that phosphatases play an analogous role in LTD (Malenka et al., 1989; Malinow et al., 1989; Wang and Feng, 1992; O'Dell et al., 1991; Mulkey et al., 1993). In addition, mice lacking the $\alpha$ subunit of CAM-KII, the $\gamma$ isoform of PKC, or the tyrosine kinase fyn have altered LTP (Silva et al., 1992; Abeliovich et al., 1993; Grant et al., 1992). These data, coupled with the finding that AMPA receptor sensitivity is increased during LTP (Davies et al., 1989; Liao et al., 1992; Manabe et al., 1992), make direct phosphorylation of AMPA receptors a likely candidate for one of the mechanisms underlying LTP. The kinase best suited to play a role in synaptic plasticity is CAM-KII. It is extremely abundant in the PSD and, once activated, CAM-KII remains active in the absence of $Ca^{2+}$ due to autophosphorylation. There have been several reports of CAM-KII's role in LTP (reviewed in Lisman et al., 1997), and recently Barria et al. (1997) reported increased phosphorylation of both CAM-KII and AMPA receptors during LTP. In their experiments, hippocampal slices were pre-labeled with $^{32}P$-orthophosphate, and LTP was induced using theta-burst stimulation. Analysis of the immunoprecipitated CAM-KII and AMPA receptors revealed that both autophosphorylation of CAM-KII and phosphorylation of AMPA receptors were increased after induction of LTP and persisted at least an hour after induction. These results provide the first evidence that AMPA receptor phosphorylation is coincident with the expression of synaptic plasticity. Based on these findings and the evidence that phosphorylation of glutamate receptors can potentiate agonist-induced currents, it is logical to conclude that the increase in synaptic strength observed during the maintenance phase of LTP involves phosphorylation of AMPA receptors.

Phosphorylation of AMPA receptors could lead directly to the potentiation of agonist-induced currents during LTP in vivo, as has been described earlier in a variety of recombinant expression systems. This would provide an explanation for the increased synaptic responsiveness during LTP.

However, it also is possible that phosphorylation of glutamate receptor subunits contributes to LTP in ways other than direct functional modulation. For example, it has been demonstrated that phosphorylation of the NMDA receptor subunit NR1 alters the subcellular distribution of the NR1A and NR1D splice variants (Ehlers et al., 1995) suggesting that the subcellular localization of glutamate receptors could be influenced by phosphorylation. One postulated mechanism for the postsynaptic changes observed during LTP is that there is an increase in the number of functional AMPA receptors in the postsynaptic membrane. This would require either an activation of existing dormant receptors or the addition of new receptors into the membrane. The most likely source of additional receptors is the intracellular pool that has been described for all glutamate receptors, including AMPA receptors (Petralia, 1997). As discussed below, little is known about the addition and removal of synaptic receptors, but phosphorylation of the receptor itself or of interacting proteins remains a leading potential mechanism for regulating this process.

**Synaptic turnover of glutamate receptors**

The insertion and removal of receptors from the postsynaptic membrane is a critical, although poorly understood, process. It has been proposed that LTP is generated by the conversion of synapses that previously did not contain AMPA receptors or contained electrophysiologically silent AMPA receptors to synapses that contain functional AMPA receptors through the activation of NMDA receptors (Liao et al., 1995; Isaac et al., 1995). A similar up-regulation of AMPA receptors may occur during synapse development (Wu et al., 1996; Isaac et al., 1997). The intracellular pool of receptor represents a likely source for receptor molecules to be inserted into the postsynaptic membrane. It is possible that receptor insertion and removal is a dynamic, rapidly occurring process, with synaptic receptors recycling on the scale of seconds or minutes. Such a mechanism would allow for a rapid up- or down-regulation of synaptic receptors by simply altering insertion or removal rates. However, there is no evidence to support this scenario and, in fact, the data support a rather static situation for surface glutamate receptors (Huh and Wenthold, 1996; Mammen et al., 1997).

The majority of the research done on membrane trafficking in neurons has focused on recycling of synaptic vesicles in presynaptic terminals (reviewed in De Camilli and Takei, 1996). The major constituent of brain-derived clathrin-coated vesicles are synaptic vesicle proteins (Maycox et al., 1992), and internalization of synaptic vesicles from the presynaptic membrane is generally believed to be a clathrin-dependent process. In addition to the heavy and light chain of clathrin, the adaptor complex AP2 is also required for clathrin-mediated endocytosis from the plasma membrane. Adaptor complexes bind directly to membrane proteins destined to be packaged in clathrin-coated vesicles (Ohno et al., 1995), and they promote clathrin assembly around the budding vesicle. The GTPase dynamin is also essential to endocytosis at the nerve terminal. Furthermore, dynamin has also been demonstrated to interact directly with one of the AP2 subunits (Wang et al., 1995), and mutations of dynamin prevent synaptic vesicle internalization (Herskovits et al., 1993). Finally, it has been demonstrated that the synaptic vesicle protein synaptotagmin I binds to the AP2 complex (Li et al., 1995) and therefore may play a role in coat assembly on synaptic vesicles.

In contrast, very little is known about the trafficking of membrane proteins localized to the PSD in neurons. Neurotransmitter receptor recycling has been best characterized for the G-protein coupled receptors (GPCR) of which the $\beta_2$-adrenergic receptor is the prototype. It has been demonstrated that many GPCRs are internalized upon prolonged exposure to agonist including the $\beta_2$-adrenergic receptor, bombesin receptor, and thrombin receptor, just to name a few. Many of the GPCRs contain the tyrosine internalization motif Asn-Pro-X-X-Tyr (X being any amino acid) at the border between the seventh transmembrane domain and the cytosolic C-terminus. This motif is similar to the Asn-Pro-X-Tyr signal present in the LDL receptor which mediates internalization via clathrin-coated vesicles. This tyrosine has been demonstrated to be essential for some GPRC agonist-induced internalization (Barak et al., 1994), but not all (Thomas et al., 1995). In addition, although most of the GPCRs display clathrin-mediated endocytosis, some are internalized in a clathrin-independent manner (Zhang et al., 1996).

Phosphorylation is critically involved in the desensitization and recycling of GPCRs. Upon agonist exposure, the $\beta_2$-adrenergic receptor is phosphorylated, leading to desensitization which is characterized by uncoupling from the G-proteins. $\beta$-arrestin binds to the phosphorylated $\beta_2$-adrenergic receptor and acts as an adaptor protein to mediate internalization via clathrin-coated vesicles (Goodman et al., 1996). Once the receptor is sequestered in an endosomal compartment, the low pH of the endosomes allows a conformational change resulting in dephosphorylation and recycling to the surface, or resensitization (Krueger et al., 1997). Thus phosphorylation plays a central role in the surface expression of the $\beta_2$-adrenergic receptor.

It is not clear how much of the information revealed about GPCR recycling can be applied to the non-NMDA glutamate receptors. These receptors do not appear to have any of the well-characterized internalization signals within their C-termini and there is no evidence that these proteins associate with known adaptor proteins or clathrin. However, although the signals and proteins involved in glutamate receptor targeting may be different, it is quite possible that posttransla-

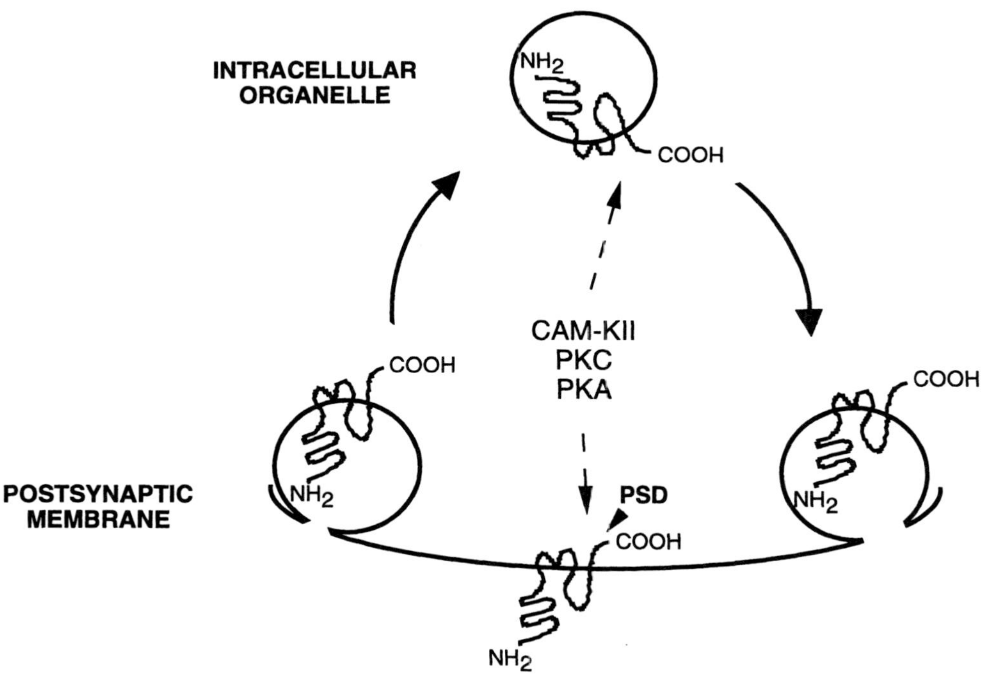


Fig. 3. A model for synaptic turnover of ionotropic glutamate receptors. An intracellular organelle delivers receptor to the postsynaptic membrane, the surface receptor is docked to the postsynaptic density and then the synaptic receptors are internalized within vesicles. Direct phosphorylation of the glutamate receptors by CAM-KII, PKC, and PKA may influence the insertion and removal of receptor from the postsynaptic membrane as well as modulating receptor activity while expressed on the postsynaptic surface. Abbreviations: CAM-KII, calcium/calmodulin-dependent protein kinase II; PKC, protein kinase C; PKA, cAMP-dependent protein kinase; PSD, postsynaptic density.

tional modifications such as phosphorylation may mediate this process (Fig. 3) as has been demonstrated for some GPCRs.

Another factor which may be relevant to the mechanisms of synaptic receptor turnover is that glutamate receptors are believed to be anchored through their C-termini to the PSD through a number of PDZ domain-containing proteins (Sheng, 1996). If this linkage is dynamic with the interactions being continually formed and broken, receptor complexes may then diffuse throughout the postsynaptic membrane and be available for recycling. In contrast, if the interaction is static, the receptor would be locked into the postsynaptic density.

## Conclusions

One of the most interesting recent discoveries in the glutamate receptor field is that the neurodegeneration in the Lurcher mouse is caused by a mutation in the $\delta_2$ glutamate receptor gene (Zuo et al., 1997). This study demonstrates the functional importance of the delta receptor family as well as TM3 in regulating the channel, but more importantly demonstrates that a genetic abnormality in the glutamate receptor system can lead to neurodegeneration. This has long been inferred from the many studies demonstrating the excitotoxic properties of glutamate, but never before demonstrated. This finding raises obvious questions about whether or not neurodegenerative disorders in humans are caused by mutations in the glutamate receptor system. As we have presented in this review, the expression and regulation of glutamate receptors in neurons is a complex process involving multiple steps. An error in the receptor itself or in one of the multiple proteins with which it interacts in any one of these steps could conceivably lead to an inappropriate expression of a glutamate receptor resulting in neurodegeneration, aberrant synapse formation, defective learning or memory, or any one of many critical neuronal processes which are known to involve glutamate receptors. The findings of Zuo et al. (1997) provide a strong incentive for the continued study of the multiple steps involved in the expression of neuronal glutamate receptors.

## References


Abeliovich, A., Chen, C., Goda, Y., Silva, A.J., Stevens, C.F. and Tonegawa, S. (1993) Modified hippocampal long-term potentiation in PKC $\gamma$ mutant mice. *Cell*, 75: 1253–1262.

Araki, K., Meguro, H., Kushiya, E., Takayama, C., Inoue, Y. and Mishina, M. (1993) Selective expression of the glutamate receptor channel $\delta 2$ subunit in cerebellar Purkinje cells. *Biochem. Biophys. Res. Commun.*, 197: 1267–1276.

Barak, L.S., Tiberi, M., Freedman, N.J., Kwatra, M.M., Lefkowitz, R.J. and Caron, M.G. (1994) A highly conserved tyrosine residue in G protein-coupled receptors is required for agonist-mediated $\beta_2$-adrenergic receptor sequestration. *J. Biol. Chem.*, 269: 2790–2795.

Barria, A., Muller, D., Derkach, V., Griffith, S.C. and Soderling, T.R. (1997) Regulatory phosphorylation of AMPA-type glutamate receptors by CaM-KII during long-term potentiation. *Science*, 276: 2042–2045.

Bennett, J.A. and Dingledine, R. (1995) Topology profile for a glutamate receptor: Three transmembrane domains and a channel-lining reentrant membrane loop. *Neuron*, 14: 373–384.

Bettler, B., Boulter, J., Hermans-Borgmeyer, I., O'Shea-Greenfield, A., Deneris, E.S., Moll, C., Borgmeyer, U., Hollmann, M. and Heinemann, S. (1990) Cloning of a novel glutamate receptor subunit, GluR5: Expression in the nervous system during development. *Neuron*, 5: 583–595.

Bettler, B., Egebjerg, J., Sharma, G., Pecht, G., Hermans-Borgmeyer, I., Moll, C., Stevens, C.F. and Heinemann, S. (1992) Cloning of a putative glutamate receptor: A low affinity kainate-binding subunit. *Neuron*, 8: 257–265.

Blackstone, C.D., Moss, S.J., Martin, L.J., Levey, A.I., Price, D.L. and Huganir, R.L. (1992) Biochemical characterization and localization of a non-NMDA glutamate receptor in rat brain. *J. Neurochem.*, 58: 1118–1126.

Blackstone, C., Murphy, T.H., Moss, S.J., Baraban, J.M. and Huganir, R.L. (1994) Cyclic AMP and synaptic activity-dependent phosphorylation of AMPA-preferring glutamate receptors. *J. Neurosci.*, 14: 7585–7593.

Bliss, T.V.P. and Collingridge, G.L. (1993) A synaptic model of memory: Long-term potentiation in the hippocampus. *Nature*, 361: 31–39.

Boulter, J., Hollmann, M., O'Shea-Greenfield, A., Hartley, M., Deneris, E., Maron, C. and Heinemann, S. (1990) Molecular cloning and functional expression of glutamate receptor genes. *Science*, 249: 1033–1037.

Burnashev, N., Monyer, H., Seeburg, P.H. and Sakmann, B. (1992) Divalent ion permeability of AMPA receptor channels is dominated by the edited form of a single subunit. *Neuron*, 8: 189–198.

Carr, D.W., Stofko-Hahn, R.E., Fraser, I.D., Cone, R.D. and Scott, J.D. (1992) Localization of the cAMP-dependent protein kinase to the postsynaptic densities by A-kinase anchoring proteins. Characterization of AKAP 79. *J. Biol. Chem.*, 267: 16816–16823.

Castillo, P.B., Malenka, R.C. and Nicoll, R.A. (1997) Kainate receptors mediate a slow postsynaptic current in hippocampal CA3 neurons. *Nature*, 388: 182–186.

Conroy, W.G. and Berg, D.K. (1995) Neurons can maintain multiple classes of nicotinic acetylcholine receptors distinguished by different subunit compositions. *J. Biol. Chem.*, 270: 4424–4431.

Conroy, W.G., Vernallis, A.B. and Berg, D.K. (1992) The $\alpha5$ gene product assembles with multiple acetylcholine receptor subunits to form distinctive subtypes in brain. *Neuron*, 9: 679–691.

Craig, A.M. and Banker, G. (1994) Neuronal polarity. *Annu. Rev. Neurosci.*, 17: 267–310.

De Camilli, P. and Takei K. (1996) Molecular mechanisms in synaptic vesicle endocytosis and recycling. *Neuron*, 16: 481–486.

Dong, H., O'Brien, R.J., Fung, E.T., Lanahan, A.A., Worley, P.F. and Huganir, R.L. (1997) GRIP: A PDZ domain-containing protein that interacts with AMPA receptors. *Nature*, 386: 279–284.

Egebjerg, J., Bettler, B., Hermans-Borgmeyer, I. and Heinemann, S. (1991) Cloning of a cDNA for a glutamate receptor subunit activated by kainate but not AMPA. *Nature*, 351: 745–748.

Egebjerg, J. and Heinemann, S. (1993) $Ca^{2+}$ permeability of an unedited and edited version of the kainate selective glutamate receptor GluR6. *Proc. Natl. Acad. Sci. USA*, 90: 755–759.

Ehlers, M.D., Tingley, W.G. and Huganir, R.L. (1995) Regulated subcellular distribution of the NR1 subunit of the NMDA receptor. *Science*, 269: 1734–1737.

Eshhar, N., Petralia, R.S., Winters, C.A., Niedzielski, A.S. and Wenthold, R.J. (1993) The segregation and expression of glutamate receptor subunits in cultured hippocampal neurons. *Neuroscience*, 57: 943–964.

Ferrer-Montiel, A.V. and Montal, M. (1996) Pentameric subunit stoichiometry of a neuronal glutamate receptor. *Proc. Natl. Acad. Sci. USA*, 93: 2741–2744.

Gallo, V., Upson, J.M., Hayes, W.P., Vyklicky, L.J., Winters, C.A. and Buonanno, A. (1992) Molecular cloning and development analysis of a new glutamate receptor subunit isoform in cerebellum. *J. Neurosci.*, 12: 1010–1023.

Gazzaley, A.H., Benson, D.L., Huntley, G.W. and Morrison, J.H. (1997) Differential subcellular regulation of the NMDAR1 protein and mRNA in dendrites of dentate gyrus granule cells after perforant path transection. *J. Neurosci.*, 17: 2006–2017.

Geiger, J.R.P., Melcher, T., Koh, D.-S,, Sakmann, B., Seeburg, P.H., Jonas, P. and Monyer, H. (1995) Relative abundance of subunit mRNAs determines gating and $Ca^{2+}$ permeability of AMPA receptors in principal neurons and interneurons in rat CNS. *Neuron*, 15: 193–204.

Grant, S.G.N., O'Dell, T.J., Karl, K.A., Stein, P.L., Soriano, P. and Kandel, E.R. (1992) Impaired long-term potentiation, spatial learning, and hippocampal development in fyn mutant mice. *Science*, 258: 1903–1910.

Goodman, O.B. Jr., Krupnick, J.G., Santini, F., Gurevich, V.V., Penn, R.B., Gagnon, A.W., Keen, J.H. and Benovic, J.L. (1996) $\beta$-arrestin acts as a clathrin adaptor in endocytosis of the $\beta_2$-adrenergic receptor. *Nature*, 383: 447–450.

Greengard, P., Jen, J., Nairn, A.C. and Stevens, C.F. (1991) Enhancement of the glutamate response by cAMP dependent protein kinase in hippocampal neurons. *Science*, 253: 1135–1138.

Gregor, P., Mano, I., Maoz, I., McKeown, M. and Teichberg, V.I. (1989) Molecular structure of the chick cerebellar kainate-binding subunits of a putative glutamate receptor. *Nature*, 342: 689–692.

Gregor, P., O'Hara, B.F., Yang, X. and Uhl, G.R. (1993) Expression and novel isoforms of glutamate receptor genes GluR5 and GluR6. *NeuroReport*, 4: 1343–1346.

Hampson, D.R., Huie, D. and Wenthold, R.J. (1987) Solubilization of kainic acid binding sites from rat brain. *J. Neurochem.*, 49: 1209–1215.

Hennegriff, M., Aria, A., Kessler, M., Vanderklish, P., Mutneja, M.S., Rogers, G., Neve, R.L. and Lynch, G. (1997) Stable expression of recombinant AMPA receptor subunits: Binding affinities and effects of allosteric modulators. *J. Neurochem.*, 68: 2424–2434.

Herb, A., Burnashev, N., Werner, P., Sakmann, B., Wisden, W. and Seeburg, P.H. (1992) The KA-2 subunit of excitatory amino acid receptors shows widespread expression in brain and forms ion channels with distantly related subunits. *Neuron*, 8: 775–785.

Herb, A., Higuchi, M., Sprengel, R. and Seeburg, P. (1996) Q/R site editing in kainate receptor GluR5 and GluR6 pre-mRNAs requires distant intronic sequences. *Proc. Natl. Acad. Sci. USA.*, 93: 1875–1880.

Herskovits, J.S., Burgess, C.C., Obar, R.A. and Vallee, R.B. (1993) Effects of mutant rat dynamin on endocytosis. *J. Cell Biol.*, 122: 565–578.

Hollmann, M. (1997) The topology of glutamate receptors: sorting through the domains. In D.T. Monaghan and R.J. Wenthold (eds): *The ionotropic glutamate receptors.*, Humana Press, Totowa, NJ pp 39–79.

Hollmann, M., Hartley, M. and Heinemann, S. (1991) $Ca^{2+}$ channels of KA-AMPA-gated glutamate receptor channels depends on subunit composition. *Science*, 252: 851–853.

Hollmann, M. and Heinemann, S. (1994) Cloned glutamate receptors. *Ann. Rev. Neurosci.*, 17: 31–108.

Hollmann, M., Maron, C. and Heinemann, S. (1994) N-glycosylation site tagging suggests a three transmembrane domain topology for the glutamate receptor GluR1. *Neuron*, 13: 1331–1343.

Hollmann, M., O'Shea-Greenfield, A., Rogers, S.W. and Heinemann, S. (1989) Cloning by functional expression of a member of the glutamate receptor family. *Nature*, 342: 643–648.

Honoré, T., Lauridsen, J. and Krogsgaard-Larsen, P (1982) The binding of [$^3$H]AMPA, a structural analogue of glutamic acid, to rat brain membranes. *J. Neurochem.*, 38: 173–178.

Huettner, J.E. (1990) Glutamate receptor channels in rat DRG neurons: Activation by kainate and quisqualate and blockade of desensitization by Con A. *Neuron*, 5: 255–266.

Huh, K.H. and Wenthold, R.J. (1996) NMDA receptors in cultured cerebellar granule cells: Subunit expression and metabolic turnover. *Neurosci.*, Abst. 22: 593.

Hume, R.I., Dingledine, R. and Heinemann, S.F. (1991) Identification of a site in glutamate receptor subunits that controls calcium permeability. *Science*, 253: 1028–1031.

Hunter, C., Wheaton, K.D. and Wenthold, R.J. (1990) Solubilization and partial purification of an AMPA binding protein from rat brain. *J. Neurochem.*, 54: 118–125.

Huntley, G.W., Rogers, S.W., Moran, T., Janssen, W., Archin, N., Vickers, J.C., Cauley, K., Heinemann, S.F. and Morrison, J.H. (1993) Selective distribution of kainate receptor subunit immunoreactivity in monkey neocortex revealed by a monoclonal antibody that recognizes glutamate receptor subunits GluR5/6/7. *J. Neurosci.*, 13: 2965–2981.

Isaac, J.T.R., Crair, M.C., Nicoll, R.A. and Malenka, R.C. (1997) Silent synapses during development of thalamocortical inputs. *Neuron*, 18: 269–280.

Isaac, J.T.R., Nicoll, R.A. and Malenka, R.C. (1995) Evidence for silent synapses: Implications for the expression of LTP. *Neuron*, 15: 427–434.

Kamboj, R.K., Schoepp, D.D., Nutt, S., Shekter, L., Korczak, B., True, R.A., Rampersad, V., Zimmerman, D.M. and Wosnick, M.A. (1994) Molecular cloning, expression, and pharmacological characterization of humEAA1, a human kainate receptor subunit. *J. Neurochem.*, 62: 1–9.

Kamboj, R.K., Schoepp, D.D., Nutt, S., Shekter, L., Korczak, B., True, R.A., Zimmerman, D.M. and Wosnick, M.A. (1992) Molecular structure and pharmacological characterization of humEAA2, a novel human kainate receptor subunit. *Mol. Pharmacol.*, 42: 10–15.

Kawamoto, S., Hattori, S., Oiji, I., Hamajima, K., Mishina, M. and Okuda, K. (1994) Ligand-binding properties and N-glycosylation of α1 subunit of the α-amino-3-hydroxy-5-methyl-4-isoxazole-propionate (AMPA)-selective glutamate receptor channel expressed in a baculovirus system. *Eur. J. Biochem.*, 223: 665–673.

Kawamoto, S., Hattori, S., Sakimura, K., Mishina, M. and Okuda, K. (1995) N-linked glycosylation of the α-amino-3-hydroxy-5-methylisoxazole-4-propionate (AMPA)-selective glutamate receptor channel α2 subunit is essential for the acquisition of ligand-binding activity. *J. Neurochem.*, 64: 1258–1266.

Keinänen, K., Wisden, W., Sommer, B., Werner, P., Herb, A., Verdoorn, T.A., Sakmann, B. and Seeburg, P.H. (1990) A family of AMPA-selective glutamate receptors. *Science*, 249: 556–560.

Keller, B.U., Hollmann, M., Heinemann, S. and Konnerth, A. (1992) Calcium influx through subunits GluR1/GluR3 of kainate/AMPA receptor channels is regulated by cAMP dependent protein kinase. *EMBO J.*, 11: 891–896.

Kelley, R.B. and Grote, E. (1993) Protein targeting in the neuron. *Annu. Rev. Neurosci.*, 16: 95–127.

Knapp, A.G. and Dowling, J.E. (1987) Dopamine enhances excitatory amino acid-gated conductances in cultured retinal horizontal cells. *Nature*, 325: 437–439.

Köhler, M., Burnashev, N., Sakmann, B. and Seeburg, P.H. (1993) Determinants of $Ca^{2+}$ permeability in both TM1 and TM2 of high affinity kainate receptor channels: Diversity by RNA editing. *Neuron*, 10: 491–500.

Krueger, K.M., Daaka, Y., Pitcher, J.A. and Lefkowitz, R.J. (1997) The role of sequestration in G protein-coupled receptor resensitization. *J. Biol. Chem.*, 272: 5–8.

Laurie, D.J., Seeburg, P.H. and Wisden, W. (1992) The distribution of 13 $GABA_A$ receptor subunit mRNAs in the rat brain. II. Olfactory bulb and cerebellum. *J. Neurosci.*, 12: 1063–1076.

Lerma, J., Morales, M., Ibarz, J.M. and Somohano, F. (1994) Rectification properties and $Ca^{2+}$ permeability of glutamate receptor channels in hippocampal cells. *Eur. J. Neurosci.*, 6: 1080–1088.

Lerma, J. Paternain, A.V., Naranjo, J.R. and Mellström, B. (1993) Functional kainate-selective glutamate receptors in cultured hippocampal neurons. *Proc. Natl. Acad. Sci. USA*, 90: 11688–11692.

Li, C., Ullrich, B., Zhang, J.Z., Anderson, R.G., Brose, N. and Südhof, T.C. (1995) $Ca^{2+}$-dependent and -independent activities of neural and non-neural synaptotagmins. *Nature*, 375: 594–599.

Liao, D., Hessler, N.A. and Malinow, R. (1995) Activation of postsynaptically silent synapses during pairing-induced LTP in CA1 region of hippocampal slice. *Nature*, 375, 400–404.

Liao, D., Jones, A. and Malinow, R. (1992) Direct measurement of quantal changes underlying long-term potentiation of CA1 hippocampus. *Neuron*, 9: 1089–1097.

Liman, E.R., Knapp, A.G. and Dowling, J.E. (1989) Enhancement of kainate-gated currents in retinal horizontal cells by cyclic AMP-dependent protein kinase. *Brain Res.*, 481: 399–402.

Linden, D.J. (1994) Long-term synaptic depression in the mammalian brain. *Neuron*, 12: 457–472.

Lisman, J., Malenka, R.C., Nicoll, R.A. and Malinow, R. (1997) Learning mechanisms: The case for CaM-KII. *Science*, 276: 2001–2002.

Lomeli, H., Mosbacher, J , Melcher, T., Höger, T., Geiger, J.R.P., Kuner, T., Monyer, H., Higuchi, M., Bach, A. and Seeburg, P.H. (1994) Control of kinetic properties of AMPA receptor channels by nuclear RNA editing. *Science*, 266: 1709–1713.

Lomeli, H., Sprengel, R., Laurie, D.J., Köhr, G., Herb, A., Seeburg, P.H. and Wisden, W. (1993) The rat δ-1 and δ-2 subunits extend the excitatory amino acid receptor family. *FEBS Lett.*, 315: 318–322.

Lomeli, H., Wisden, W., Kohler, M., Keinänen, K., Sommer, B. and Seeburg, P.H. (1992) High-affinity kainate and domoate receptors in rat brain. *FEBS Lett.*, 307: 139–143.

Luján, R., Roberts. J.D., Shigemoto, R., Ohishi, H. and Somogyi, P. (1997) Differential plasma membrane distribution of metabotropic glutamate receptors mGluR1α, mGluR2 and mGluR5, relative to neurotransmitter release sites. *J. Chem. Neuroanatomy*, 13: 219–241.

Malenka, R.C., Kauer, J.A., Perkel, D.J., Mauk, M.D., Kelley, P.T., Nicoll, R.A. and Waxham, M.N. (1989) An essential role for postsynaptic calmodulin and protein kinase activity in long-term potentiation. *Nature*, 340: 554–557.

Malinow, R., Schulman, H. and Tsien, R.W. (1989) Inhibition of postsynaptic PKC or CAMKII blocks induction but not expression of LTP. *Science*, 245: 862–866.

Mammen, A.L., Huganir, R.L. and O'Brien, R.J. (1997) Redistribution of cell surface glutamate receptors during synapse formation. *J. Neurosci.*, 17: 7351–7358.

Mammen, A.L., Kameyama, K., Roche, K.W. and Huganir, R.L. (1997) Phosphorylation of the AMPA receptor GluR1 subunit by calcium/calmodulin-dependent kinase II. *J. Biol. Chem.*, In Press.

Manabe, T., Renner, P. and Nicoll, R.A. (1992) Postsynaptic contribution to long-term potentiation revealed by the analysis of miniature synaptic currents. *Nature*, 355: 50–55.

Martone, M.E., Zhang, Y., Simpliciano, V.M., Carragher, B.O. and Ellisman, M.H. (1993) Three-dimensional visualization of the smooth endoplasmic reticulum in Purkinje cell dendrites. *J. Neurosci.*, 13: 4636–4646.

Mayat, E., Petralia, R.S., Wang, Y-X. and Wenthold, R.J. (1995) Immunoprecipitation, immunoblotting and immunocytochemistry studies suggest that glutamate receptor δ subunits form novel postsynaptic receptor complexes. *J. Neurosci.*, 15: 2533–2546.

Maycox, P.R., Link, E., Reetz, A., Morris, S.A. and Jahn, R. (1992) Clathrin-coated vesicles in nervous tissue are involved primarily in synaptic vesicle recycling. *J. Cell Biol.*, 118: 1379–1388.

McGlade-McCulloh, E., Yamamoto, H., Tan, S.-E., Brickey, D.A. and Soderling, T.R. (1993) Phosphorylation and regulation of glutamate receptors by calcium/calmodulin-dependent protein kinase II. *Nature*, 362: 640–642.

Monyer, H., Seeburg, P.H. and Wisden, W. (1991) Glutamate operated channels: Developmentally early and mature forms arise by alternative splicing. *Neuron*, 6: 799–810.

Morita, T., Sakamura, K., Kushiya, E., Yamazaki, M., Meguro, H., Araki, K., Abe, T., Mori, K.J. and Mishina, M. (1992) Cloning and functional expression of a cDNA encoding the mouse β2 subunit of the kainate-selective glutamate receptor channel. *Mol. Brain Res.*, 14: 143–146.

Mosbacher, J., Schoepfer, R., Monyer, H., Burnashev, N. and Seeburg, P.H., Ruppersberg, J.P. (1994) A molecular determinant for submillisecond desensitization in glutamate receptors. *Science*, 266: 1059–1062.

Mulkey, M.M., Herron, C.E. and Malenka, R.C. (1993) An essential role for protein phosphatases in hippocampal long-term depression. *Science*, 261: 1051–1055.

Nakazawa, K., Mikawa, S., Hashikawa, T. and Ito, M. (1995) Transient and persistent phosphorylation of AMPA-type glutamate receptor subunits in cerebellar Purkinje cells. *Neuron*, 15: 697–709.

Neki, A., Ohishi, H., Kaneko, T., Shigemoto, R., Nakanishi, S. and Mizuno, N. (1996) Pre- and postsynaptic localization of a metabotropic glutamate receptor, mGluR2, in the rat brain: An immunocytochemical study with a monoclonal antibody. *Neurosci. Lett.*, 202: 197–200.

O'Dell, T.J., Kandel, E.R. and Grant, S.G.N. (1991) Long-term potentiation in the hippocampus is blocked by tyrosine kinase inhibitors. *Nature*, 353: 558–560.

Ohno, H., Stewart, J., Fournier, M.-C., Bosshart, H., Rhee, I., Miyatake, S., Saito, T., Gallusser, A., Kirchhausen, T. and Bonifacino, J.S. (1995) Interaction of tyrosine-based sorting signals with clathrin-associated proteins. *Science*, 269: 1872–1875.

Omkumar, R.V., Kiely, M.J., Rosenstein, A.J., Min, K.T. and Kennedy, M.B. (1996) Identification of a phosphorylation site for calcium/calmodulin dependent protein kinase II in the NR2B subunit of the *N*-methyl-D-aspartate receptor. *J. Biol. Chem.*, 271: 31670–31680.

Petralia, R.S. (1997) Immunocytochemical localization of ionotropic glutamate receptors (GluRs) in neuronal circuits. In D.T. Monaghan and R.J. Wenthold (eds): *The ionotropic glutamate receptors.*, Humana Press, Totowa, NJ pp 219–263.

Petralia, R.S., Wang, Y-X. and Wenthold, R.J. (1994) Histological and ultrastructural localization of the kainate receptor subunits, KA2 and GluR6/7, in the rat nervous system using selective antipeptide antibodies. *J. Comp. Neurol.*, 349: 85–110.

Petralia, R.S. and Wenthold, R.J. (1992) Light and electron immunocytochemical localization of AMPA-selective glutamate receptors in the rat brain. *J. Comp. Neurol.*, 318: 329–354.

Puchalski, R.B., Louis, J-C., Brose, N., Traynelis, S.F., Egebjerg, J., Kukekov, V., Wenthold, R.J., Rogers, S.W., Lin, F., Moran, T., Morrison, J.H. and Heinemann, S.F. (1994) Selective RNA editing and subunit assembly of native glutamate receptors. *Neuron*, 13: 131–147.

Racca, C., Gardiol, A. and Triller, A. (1997) Dendritic and postsynaptic localizations of glycine receptor a subunit mRNAs. *J. Neurosci.*, 17: 1691–1700.

Raman, I.M., Zhang, S. and Trussell, L.O. (1994) Pathway-specific variants of AMPA receptors and their contribution to neuronal signalling. *J. Neurosci.*, 14: 4998–5010.

Raymond, L.A., Blackstone, C.D. and Huganir, R.L. (1993) Phosphorylation and modulation of reçombinant GluR6 glutamate receptors by cAMP-dependent protein kinase. *Nature*, 361: 637–641.

Roche, K.W. and Huganir, R.L. (1995) Synaptic expression of the high-affinity kainate receptor subunit KA2 in hippocampal cultures. *Neuroscience*, 69: 383–393.

Roche, K.W., O'Brien, R.J., Mammen, A. L., Bernhardt, J. and Huganir, R.L. (1996) Characterization of multiple phosphorylation sites on the AMPA receptor GluR1 subunit. *Neuron*, 16: 1179-1188.

Roche, K.W., Raymond, L.A., Blackstone, C. and Huganir, R.L. (1994) Transmembrane topology of the glutamate receptor subunit GluR6. *J. Biol. Chem.*, 269: 11679–11682.

Roche, K.W., Tingley, W.G. and Huganir, R.L. (1994b) Glutamate receptor phosphorylation and synaptic plasticity. *Curr. Opin. Neurol.*, 4: 383–388.

Rosenmund, C., Carr, D.W., Bergeson, S.E., Nilaver, G., Scott, J.D. and Westbrook, G.L. (1994) Anchoring of protein kinase A is required for modulation of AMPA/kainate receptors on hippocampal neurons. *Nature*, 368: 853–856.

Rubio, M.E. and Wenthold, R.J. (1997) Glutamate receptors are selectively targeted to postsynaptic sites in neurons. *Neuron*, 18: 939–950.

Sakimura, K., Morita, T., Kushiya, E. and Mishina, M. (1992) Primary structure and expression of the $\gamma 2$ subunit of the glutamate receptor channel selective for kainate. *Neuron*, 8: 267–274.

Sheng, M. (1996) PDZs and receptor/channel clustering: Rounding up the latest suspects. *Neuron*, 17: 575–578.

Silva, A., Stevens, C.F., Tonegawa, S. and Wang, Y. (1992) Deficient hippocampal long-term potentiation in $\alpha$-calcium-calmodulin kinase II mutant mice. *Science*, 257: 201–206.

Simon, J.R., Contrera, J.F. and Kuhar, M.J. (1976) Binding of [$^3$H]kainic acid, an analogue of L-glutamate, to brain membranes. *J. Neurochem.*, 26: 141–147.

Soderling, T.R. (1997) Phosphorylation of non-NMDA glutamate receptor ion channels: Implications for synaptic plasticity and their membrane topology. In D.T. Monaghan and R.J. Wenthold (eds): *The ionotropic glutamate receptors.* Humana Press, Totowa, NJ, pp 121–134.

Sommer, B., Burnashev, N., Verdoorn, T.A., Keinänen, K., Sakmann, B. and Seeburg, P.H. (1992) A glutamate receptor channel with high affinity for domoate and kainate. *EMBO J.*, 11: 1651–1656.

Sommer, B., Keinänen, K., Verdoorn, T.A., Wisden, W., Burnashev, N., Herb, A., Köhler, M., Takagi, T., Sakmann, B. and Seeburg, P.H. (1990) Flip and flop: A cell-specific functional switch in glutamate-operated channels of the CNS. *Science*, 249: 1580–1585.

Sommer, B., Köhler, M., Sprengel, R. and Seeburg, P.H. (1991) RNA editing in brain controls a determinant of ion flow in glutamate gated channels. *Cell*, 67: 11–19.

Spruston, N., Jonas, P. and Sakmann, B. (1995) Dendritic glutamate receptor channels in rat hippocampal CA3 and CA1 pyramidal neurons. ci., 482: 325–352.

Swope S.L., Moss S.J., Blackstone C.D. and Huganir R.L. (1991) Phosphorylation of ligand-gated ion channels: A possible mode of synaptic plasticity. *FASEB J.*, 6: 2514–2523.

Tan, S.E., Wenthold, R.J. and Soderling, T.R. (1994) Phosphorylation of AMPA-type glutamate receptors by calcium/calmodulin-dependent protein kinase II and protein kinase C in cultured hippocampal neurons. *J. Neurosci.*, 14: 1123–1129.

Taverna, F.A., Wang, L-Y., Macdonald, J.F. and Hampson, D.R. (1994) A transmembrane model for an ionotropic glutamate receptor predicted on the basis of the location of asparagine-linked oligosaccharides. *J. Biol. Chem.*, 269: 14159–14164.

Terasaki, M., Slater, N.T., Fein, A., Schmidek, A. and Reese, T.S. (1994) Continuous network of endoplasmic reticulum in cerebellar Purkinje neurons. *Proc. Natl. Acad. Sci. USA*, 91: 7510–7514.

Thomas, W.G., Baker, K.M., Motel, T.J. and Thekkumkara, T.J. (1995) Angiotensin II receptor endocytosis involves two distinct regions of the cytoplasmic tail. A role for residues on the hydrophobic face of a putative amphipathic helix. *J. Biol. Chem.*, 270: 22153–22159.

Tingley, W.G., Roche, K.W., Thompson, A.K. and Huganir, R.L. (1993) Regulation of NMDA receptor phosphorylation by alternative splicing of the C-terminal domain. *Nature*, 364: 70–73.

Tygesen, C.K., Rasmussen, J.S., Jones, S.V.P., Hansen, A., Hansen, K. and Andersen, P.H. (1994) Stable expression of a functional GluR6 homomeric glutamate receptor channel in mammalian cells. *Proc. Natl. Acad. Sci. USA*, 91: 13018–13022.

Verdoorn, T.A., Burnashev, N., Monyer, H., Seeburg, P.H. and Sakmann, B. (1991) Structural determinants of ion flow through recombinant glutamate receptor channels. *Science*, 252: 1715–1718.

Vernallis, A.B., Conroy, W.G. and Berg, D.K. (1993) Neurons assemble acetylcholine receptor with as many as three kinds of subunits while maintaining subunit segregation among receptor subtypes. *Neuron*, 10: 451–464.

Vignes, M. and Collingridge, G.L. (1997) The synaptic activation of kainate receptors. *Nature*, 388: 179–182.

Wada, K., Dechesne, C.J., Shimasaki, S., King, R.G., Kusano, K., Buonanno, A., Hampson, D.R., Banner, C., Wenthold, R.J. and Nakatani, Y. (1989) Sequence and expression of a frog complementary DNA encoding a kainate-binding protein. *Nature*, 342: 684–689.

Wang, J.-H. and Feng, D.-P. (1992) Postsynaptic protein kinase C essential to induction and maintenance of long-term potentiation in the hippocampal CA1 region. *Proc. Natl. Acad. Sci., USA*, 89: 2576–2580.

Wang, L.H., Südhof, T.C. and Anderson, R.G. (1995) The appendage domain of α-adaptin is a high affinity binding site for dynamin. *J. Biol. Chem.*, 270: 10079–10083.

Wang, L.-Y., Salter, M.W. and MacDonald, J.F. (1991) Regulation of kainate receptors by cAMP-dependent protein kinase and phosphatases. *Science*, 253: 1132–1135.

Wang, L.Y., Taverna, F.A., Huang, X.P., MacDonald, J.F. and Hampson, D.R. (1993) Phosphorylation and modulation of a kainate receptor (GluR6) by cAMP-dependent protein kinase. *Science*, 259: 1173–1175.

Wenthold, R.J., Petralia, R.S., Blahos, J. and Niedzielski, A.S. (1996) Evidence for multiple AMPA receptor complexes in hippocampal CA1/CA2 neurons. *J. Neurosci.*, 16: 1982–1989.

Wenthold, R.J., Trumpy, V.A., Zhu, W-S. and Petralia, R.S. (1994) Biochemical and assembly properties of GluR6 and KA2, two members of the kainate receptor family, determined with subunit-specific antibodies. *J. Biol. Chem.*, 269: 1332–1339.

Wenthold, R.J., Yokotani, N., Doi, K. and Wada, K. (1992) Immunochemical characterization of the non-NMDA glutamate receptor using subunit-specific antibodies: Evidence for a hetero-oligomeric structure in rat brain. *J. Biol. Chem.*, 267: 501–507.

Werner, P., Voigt, M., Keinänen, K., Wisden, W. and Seeburg, P.H. (1991) Cloning of a putative high-affinity kainate receptor expressed predominantly in hippocampal CA3 cells. *Nature*, 351: 742–744.

Wo, Z.G. and Oswald, R.E. (1994) Transmembrane topology of two distinct kainate receptor subunits revealed by N-glycosylation. *Proc. Natl. Acad. Sci. USA*, 91: 7154–7158.

Wozniak, M. and Limbird, L.E. (1996) The three α2-adrenergic receptor subtypes achieve basolateral localization in Madin-Darby canine kidney II cells via different targeting mechanisms. *J. Biol. Chem.*, 271: 5017–5024.

Wu, G.-Y., Malinow, R. and Cline, H.T. (1996) Maturation of a central glutamatergic synapse. *Science*, 274: 972–976.

Yakel, J.L., Vissavajjhala, P., Derkach, V.A., Brickey, D.A. and Soderling, T.R. (1995). Identification of a $Ca^{2+}$/calmodulin-dependent protein kinase II regulatory phosphorylation site in non-*N*-methyl-D-aspartate glutamate receptors. *Proc. Natl. Acad. Sci. USA*, 92: 1376–1380.

Yamazaki, M., Araki, K., Shibata, A. and Mishina, M. (1992) Molecular cloning of a cDNA encoding a novel member of the mouse glutamate receptor channel family. *Biochem. Biophys. Res. Comm.*, 183: 886–892.

Yu, X-M. and Hall, Z.W. (1994) A sequence in the main cytoplasmic loop of the α subunit is required for assembly of mouse muscle nicotinic acetylcholine receptor. *Neuron*, 13: 247–255.

Zhang, J., Ferguson, S.S.G., Barak, L.S., Menard, L. and Caron, M.G.(1996) Dynamin and β-arrestin reveal distinct mechanisms for G protein-coupled receptor internalization. *J. Biol. Chem.*, 271: 18302–18305.

Zuo, J., DeJager, P.L., Takahashi, K.A., Jiang, W., Linden, D.J. and Heinz, N. (1997) Neurodegeneration in Lurcher mice caused by a mutation in δ2 glutamate receptor gene. *Nature*, 388: 767–773.

O.P. Ottersen, I.A. Langmoen and L. Gjerstad (Eds.)
*Progress in Brain Research*, Vol 116


CHAPTER 11

# AMPA receptors: Molecular and functional diversity

Karin Borges* and Raymond Dingledine

*Department of Pharmacology, Emory University School of Medicine, 1510 Clifton Road, Atlanta, GA 30322, USA*

The AMPA (α-amino-3-hydroxy-5-methyl-4-isoxazolepropionic acid) subtypes of glutamate receptor are the workhorse excitatory neurotransmitter receptors of the brain. AMPA receptors are postsynaptic ion channels that, when gated open by synaptically released glutamate, mediate the influx of $Na^+$ (and in some cases $Ca^{2+}$) into the neuron, and the efflux of $K^+$. All or nearly all CNS neurons express AMPA receptors, whereas these receptors are sparse outside the CNS. Thus these essential postsynaptic components of excitatory synaptic transmission are largely brain-specific in their distribution.

Whereas the influx of monovalent cations through AMPA receptor channels causes a transient postsynaptic depolarization, the influx of $Ca^{2+}$ may trigger in addition cytoplasmic signaling cascades that lead to synaptic plasticity and other longer-lasting changes. Under some conditions, AMPA receptor activation can produce pathophysiological or toxic effects. For example, the ingestion of mussels that had concentrated domoic acid caused, in several well-publicized cases, severe convulsions, memory impairment and death (Olney, 1994). Neurolathyrism is a crippling neurodegenerative condition caused by ingestion of a plant which naturally contains an excitotoxic glutamate analog, β-*N*-methylamino-L-alanine (BMAA), that activates AMPA receptors. AMPA receptor antagonists are effective neuroprotectant drugs in animal models of global ischemia (Sheardown et al., 1990). Excessive AMPA receptor activation may also play a role in the excitotoxic damage associated with some chronic degenerative disorders such as Huntington's Disease, Parkinson's Disease and amyelotrophic lateral sclerosis (Greene and Greenamyre, 1996; Lancelot and Bear, this volume). Rasmussen's encephalitis, which is accompanied by severe seizures and neurodegeneration, is associated with the specific production of antibodies to one of the AMPA receptor subunits (Rogers et al., 1994). The growing appreciation of pharmacological diversity among AMPA receptors (see below) suggests that new therapeutic approaches for certain neurologic disorders based on selective AMPA receptor antagonism may be a realistic goal.

As described below, different combinations of recombinant subunits produce a plethora of receptors with different functional properties. In this review we will summarize current knowledge of the pharmacology and molecular biology of AMPA receptors, with special attention to understanding the link between molecular and functional diversity and the resulting insights for drug discovery.

## Four mammalian genes in one family

AMPA receptors belong to the larger family of ionotropic glutamate receptors, which also include

*Corresponding author.
Tel.: +1 404 727 5612; fax: 1 404 727 0365; e-mail: kborges@-bimcore.emory.edu.

NMDA and kainate receptors. These three receptor subtypes were originally distinguished on pharmacological grounds by selective agonists. They are now known to be multimeric assemblies of subunits in five distinct molecular families defined by sequence homologies (Fig. 1 also see Wenthold and Roche, this volume). As far as we know, the four AMPA receptor subunits GluR1-GluR4 do not form functional interactions with subunits from either NMDA or kainate receptor classes, but can only assemble with one another. All AMPA receptor subunits are structurally similar. The membrane inserted regions M1–M4, the loop preceding M1 and the region between M3 and M4 containing the agonist binding sites as well as the flip/flop regions controlling desensitization are the most conserved regions throughout the family (Fig. 2). The GluR2 and GluR4 subunits appear to determine much of the functional diversity among AMPA receptors. The presence of edited GluR2 suppresses $Ca^{2+}$ permeability, blocks inward rectification, and reduces single channel conductance, all due to an arginine at the Q/R editing site in M2 (see below, Burnashev et al., 1992; Hume et al., 1991; Verdoorn et al., 1991) GluR4 confers a particularly fast rate of desensitization in the submillisecond range (Mosbacher et al., 1994). In addition to transcriptional regulation, functional diversity among AMPA receptors is generated by posttranscriptional modifications, such as alternative splicing, RNA editing and phosphorylation.

### Activation and block of AMPA receptors

The pharmacology of AMPA receptors is very similar to that of kainate receptors but distinct from NMDA receptors (for excellent review see Fletcher and Lodge, 1996). Thus, drugs that discriminate between AMPA and kainate receptors are especially prized. Typically, AMPA receptors show a high affinity to [$^3$H]AMPA ($K_D$ about 5–40 nM depending on the solubilization state and ionic conditions), and the electrophysiological response to AMPA but not kainate is rapidly desensitizing. Desensitization is reversibly inhibited by cyclothiziade. Kainate receptors, on the other hand, are a heterogenous group of high and low affinity [$^3$H] kainate binding sites ($K_D$ = 5 nM and 50–100 nM) that rapidly desensitize after kainate application; this desensitization is irreversibly relieved by the lectin concanavalin A.

**Ionotropic Glutamate Receptors**

| Functional Classes | NMDA | AMPA | Kainate |
|---|---|---|---|
| Gene families | NR1 | GluR1 | GluR5 |
| | -- | GluR2 | GluR6 |
| | NR2A | GluR3 | GluR7 |
| | NR2B | GluR4 | -- |
| | NR2C | | KA1 |
| | NR2D | | KA2 |

Fig. 1. The 16 mammalian genes encoding functional glutamate receptor subunits. The amino acid sequences of the core membrane domains of individual subunits show 20–40% homology between functional classes, and 75–80% homology within a gene family. Five gene families in three functional classes are illustrated.

*Agonists*

AMPA and kainate receptors are activated by the same agonists and therefore may have a very similar agonist binding site. The rank order of potency for AMPA receptors is quisqualate > AMPA∼domoate > glutamate > kainate, whereas kainate receptors prefer kainate and domoate over quisqualate, glutamate and last AMPA (Fletcher and Lodge, 1996; Hollmann and Heinemann, 1994). However, the rank order of agonist potency is often difficult to determine because of fast receptor desensitization in response to some agonists. If a rapid application system is used, the different desensitization kinetics can help to distinguish between AMPA and kainate receptors. AMPA receptors desensitize >90% within milliseconds after application of glutamate, quisqualate and AMPA, whereas kainate receptors only weakly desensitize after AMPA application. On the other hand kainate and domoate evoke rapidly desensitizing currents at kainate but not AMPA receptors. Thus kainate and domoate are efficacious, non-desensitizing agonists at AMPA receptors despite their low affinity. Other specific desensitizing agonists for AMPA receptors include drugs of the willardiine group. The rank order of potency is (S)-5-fluoro-willardiine > AMPA > (S)-5-iodo-willardiine > kainate for mouse hippocampal cells rich in AMPA receptors. For dorsal root ganglia, which express the kainate receptor subunits GluR5 and KA2, the rank order changes to 5-iodo-willardiine > kainate > 5-fluoro-willardiine > AMPA (Wong et al., 1994). Currently, there are no agonists that distinguish among different AMPA receptor subunit combinations.

*Allosteric modulators*

Interestingly, certain diuretic drugs specifically decrease desensitization, which can circumvent the need for rapid application of agonists. Benzothiazides, such as cyclothiazide and with less potency diazoxide, attenuate desensitization of AMPA receptors responding to glutamate and AMPA (Partin et al., 1993; Randle et al., 1993). Cyclothiazide reduces the extent of desensitization and increases the decay time constant. In addition cyclothiazide and diazoxide increase the apparent potency for kainate and domoate (Partin et al., 1995; Randle et al., 1993). The nootropic drug aniracetam also reduces desensitization at AMPA receptors, but aniracetam is much less potent than cyclothiazide and its action is very agonist dependent (Fletcher and Lodge, 1996). 4-[2-(Phenylsulfonylamino)ethylthio]-2,6-difluoro-phenoxyacetamide (PEPA) is a member of a new structural class of allosteric potentiators that, like aniracetam, preferentially slows the rate of onset of desensitization and potentiates steady-state glutamate-evoked currents in flop splice variants of AMPA receptors (Sekiguchi et al., 1997).

Thiocyanate is a chaotropic modulator acting at AMPA receptors in millimolar concentrations. In binding studies it enhances [$^3$H]AMPA but not [$^3$H]kainate binding by increasing the proportion of high affinity [$^3$H]AMPA binding sites. In electrophysiological studies thiocyanate increases the desensitization of AMPA-evoked responses at recombinant GluR1flip but slows desensitization at GluR1flop receptors (Partin et al., 1996).

In contrast to the selective effect of cyclothiazide on AMPA receptors, the plant lectin concanavalin A (conA) blocks desensitization of kainate- and glutamate-induced responses at kainate receptors, such as in dorsal root ganglia or recombinant GluR6 (Partin et al., 1993). Thus the combination of cyclothiazide with AMPA or conA with kainate can help to distinguish between and efficiently activate the two receptor subtypes.

The best noncompetitive AMPA-receptor antagonists are found among the 2,3-benzodiazepines, which in contrast to the conventional 1,4-benzodiazepines lack sedative-hypnotic activity and do not bind to $GABA_A$ receptors. GYKI 53655, a derivative of GYKI 52466, is a powerful AMPA receptor-selective blocker in this class. GYKI 53655 blocks glutamate responses in cortical cells with an $IC_{50}$ of 1 $\mu$M, but is ineffective against kainate receptors in dorsal root ganglia or recombinant GluR6 receptors ($IC_{50}$ > 200 $\mu$M)

(Wilding and Huettner, 1995). Desensitizing kainate responses could be unmasked in cultured hippocampal cells using this blocker (Paternain et al., 1995).

A number of compounds with related chemical structures, including two other GYKI 52466 derivatives and substituted 1,2 dihydrophtalazines, display variable degrees of selectivity between AMPA and kainate receptors. Some are also potent anticonvulsants when administered systemically (Chimirri et al., 1997; Pelletier et al., 1996). Structurally similar to the 2,3 benzodiazepines, the substituted 1,2 dihydrophtalazines are potent selective noncompetitive AMPA receptor inhibitors. Attention has been focused on the *n*-propyl substituted derivative, which blocks AMPA receptors in rat cortical neurons with an $IC_{50}$ of 2.8 $\mu$M, only weakly blocks recombinant GluR6 receptors (10% block at 100 $\mu$M), and appears to be a potent systemic anticonvulsant (Pelletier et al., 1996).

*Competitive antagonists*

The development of specific antagonists is crucial to discriminate between AMPA and kainate receptor-mediated synaptic responses. Moreover, many glutamate receptor antagonists have neuroprotective and anticonvulsant effects that could potentially lead to new clinical therapies. Again the antagonist pharmacology of kainate and AMPA receptors overlaps, with both receptor subtypes sharing some competitive antagonists of the quinoxaline group. The most discriminating quinoxaline is 2,3-dihydroxy-6-nitro-7-sulfamoyl-benz(F)quinoxaline (NBQX), which is about 30-fold more selective for cortical AMPA receptors ($IC_{50}$ = 0.15 $\mu$M) than cortical kainate receptors ($IC_{50}$ = 4.8 $\mu$M) in binding assays (Sheardown et al., 1990). 6-Cyano-7-nitro-(CNQX) and 6,7-dinitro-quinoxalindiones (DNQX) are only about five times more potent on AMPA receptors ($IC_{50}$ =0.3 and 0.5 $\mu$M respectively) than on kainate receptors ($IC_{50}$ = 1.5 and 2.0 $\mu$M, respectively); these compounds also block the glycine site of NMDA receptors (Kleckner and Dingledine, 1989). YM90K, an imidazole-substituted quinoxalinedione with 84 nM affinity for brain AMPA receptors, is a particularly effective neuroprotectant in ischemic models at an intravenous dose that does not cause precipitation in the kidney like other quinoxaline-derived drugs (Yatsugi et al., 1996).

Newer compounds with somewhat better selectivity have been developed recently, such as the competitive decahydroisoquinoline antagonist LY 293558 (Ornstein et al., 1993). Apart from its antagonism at GluR5, LY293558 seems to be the most selective AMPA receptor antagonist available, with about 200 fold difference between the $IC_{50}$ at recombinant GluR6 receptors and native AMPA receptors (Fletcher and Lodge, 1996). Other 6-substituted decahydroisoquinoline-3-carboxylic acids, a series of AMPA analogs (ATOA and ATPO), and the sulfonamide-derived NS 257, are also promising competitive AMPA receptor antagonists (Madsen et al., 1996; Ornstein et al., 1996 a,b; Wätjen et al., 1994), but the selectivity of these drugs needs to be further evaluated on a wider range of heteromeric recombinant receptors.

*Channel blockers*

Many different voltage-dependent channel blockers are found among synthetic spermine derivatives and polyamine-containing toxins, e.g., N-(4-hydroxyphenyl-propanoyl)-spermine, (NHPP-spermine), argiotoxin, Joro spider toxin, philanthotoxin-433, *Nephila* spider toxin and ageltoxin-489 (Fletcher and Lodge, 1996). These positively charged compounds obstruct the pore of calcium channels, nicotinic acetylcholine receptors, NMDA receptors, and, depending on the subunit composition and the editing state in M2, kainate and AMPA receptors. Polyamines enter the open glutamate receptor channel from the external side and become trapped in the pore, perhaps by interaction with carbonyl oxygens on the side chains of the asparagine in M2 of the NMDA receptors or the glutamine at the Q/R site in AMPA or kainate receptors (see below). Thus after external polyamine application, AMPA receptors that lack GluR2 are selectively blocked.

The $IC_{50}$ values for blockade of GluR1 receptors at −80 mV are 40 nM for Joro spider toxin, 3.4 μM for agriotoxin and 2.8 μM for philanthotoxin. GluR2-containing receptors are insensitive to Joro spider toxin and have 100 fold lower affinity for Argiotoxin and philanthotoxin (Blaschke et al., 1993; Brackley et al., 1993). A relatively inexpensive GluR2-discriminating channel blocker is the monoacylated spermine analog, NHPP-spermine, which reversibly blocks GluR3 receptors in oocytes with an $IC_{50}$ of 80 nM at −70 mV. 1 μM NHPP-spermine had no effect on GluR2-containing receptors in both oocytes and hippocampal neurons (Laezza and Dingledine, 1996; Washburn and Dingledine, 1995).

**Regulation by post-transcriptional modifications**

AMPA receptor subunits are subject to posttranscriptional alterations, i.e., alternative splicing, RNA editing and phosphorylation. Depending on the subunit these changes at the RNA and protein level are subject to regulation and determine some of the most important functional properties of the AMPA receptors.

*Alternative splicing: Flip and flop exons influence desensitization*

All four AMPA receptor subunits exist in two alternatively spliced variants termed flip and flop, which largely control the desensitization kinetics in response to glutamate and AMPA (Sommer et al., 1990). The flop and flip versions (exon 14 and 15 in the mouse GluR2 gene) differ by 9 to 11 amino acids within the 38 amino acid exon cassette preceding M4. The consensus sequences in mammals are S-$(X)_{21}$-KDSG for flip and N-$(X)_{21}$-GGGD for flop. The attenuation of desensitization by cyclothiazide is stronger on the less desensitizing flip variant than the flop versions. The structural basis of this difference in GluR1 involves serine 750 in flip or asparagine 750 in the flop variant (Partin et al., 1995).

The flip versions are already present before birth and their slower desensitization is speculated to be important for synaptogenesis. The faster desensitizing flop forms begin to increase in abundance during rat development from p8 until p14, thereafter they are expressed at about the same level as the flip forms (Monyer et al., 1991). However, the expression patterns of flip and flop vary regionally. For example GluR4flip is only expressed in the cerebellum, whereas GluR4flop is additionally found in other structures of the brain, e.g., the hippocampal CA1 region. Another case of differential expression are adult CA1 pyramidal cells, which coexpress the flip and flop versions of GluR1–3 and GluR4flop. In contrast CA3 pyramidal cells only express the flip variants of GluR1–3; they do not seem to undergo the developmental switch to the flop versions. The regulation of alternative splicing of the flip-flop exons is clearly subunit dependent in the same neuron. For example, cerebellar granule cells express only the flip form of one subunit (GluR2) and the flop form of another subunit (GluR4). Changes in the flip to flop ratio are activation dependent and might underlie synaptic plasticity. For example, kainate-induced seizures increased the GluR2flip to flop ratio in the hippocampal CA1 region and the dentate gyrus of rats (Pollard et al., 1993).

*Alternative splicing: Short and long C-tails influence targeting*

Two alternative C-termini, short and long, have been found among AMPA receptors (Fig. 2). The longer C-tail is generated by addition of an exon within the short C-tail of GluR2 and GluR4 (Gallo et al., 1992; Köhler et al., 1994). Most GluR4 and a minority of GluR2 transcripts contain this additional exon, which resembles the long GluR1 (67 AA) C-terminus. GluR3 and most of the GluR2 subunits occur in the 36 AA short version. In addition some GluR4 transcripts, termed GluR4c, have a short C-terminus and are mainly expressed in the cerebellum (Gallo et al., 1992, Fig. 2).

The short C-tails seem to be involved in receptor clustering by interaction with a putative subcellular targeting protein GRIP (Dong et al., 1997).

```
----MPYIFA FFCTGFLGAV VG ANFPNNIQ IGGLFPNQQS QEHAAFRFAL SQLTEPP--- ----KLLPQI DIVNISDTFE MTYRFCSQFS 61 GluR1
-MQKIMHISV LLSPVLWGLI F. V-SSNSIQ .....PRGAD ..YS...VGM VQFSTSE--- ---FR.TPHI .NLEVANS.A V.NA....F. 61 GluR2
MGQSVLRAVF FLVLGLLGHS H. G-FPNTIS .....MRNTV ..HS...FAV QLYNTNQNTT EKPFH.NYHV .HLDSSNS.S V.NA....F. 67 GluR3
-MRIICRQIV LLFSGFWGLA M. A-FPSSVQ .....IRNTD ..YT...LAI FLHNTSPNAS EAPFN.VPHV .NIETANS.A V.NA....Y. 67 GluR4
.......... .......... .G.........IGGLF......QE..AFR.............--- ----.L.....D........F....T..FCSQ.S consensus
                            LIVBP homology

KGVYAIFGFY ERRTVNMLTS FCGALHVCFI TPSFPVDTSN QFVLQLRPEL QEALISIIDH YKWQTFVYIY DADRGLSVLQ RVLDTAAEKN 151
R..Y....F. DKKSVNTI.. ..GT..VSFI .....TDGTH P..I.M..D. KG..L.LIEY .Q.DK.AYL. .SD..L.T.. AVLDS.AEKK 151
R..Y....F. DQMSMNTL.. ..GA..TSFV .....TDADV Q..I.M..A. KG..L.LLSY .K.EK.VYL. .TE..F.V.. AIMEA.VQNN 157
R..F....L. DKRSVHTL.. ..RR..ISLI .....TEGES Q..L.L..S. RG..L.LLDH .E.NC.VFL. .TD..Y.I.. AIMEK.GQNG 157
..GV.AIFG.Y..........TS.FC..LH.....TPSFP.......FV.Q.RP.L...AL.S.....Y.W..F...Y.D..RG.S.LQ.......A.....

WQVTAVNI-- --LTTTEEGY RMLFQDLEKK KERLVVVDCE SERLNAILGQ IVKLEKNGIG YHYILANLGF MDIDLNKFKE SGRNVTGFQL 237
.Q.T.INVGN INNDKKDETY .SLFQDLELK K.RRVIL... RDKVND.VD. VITIG.HVK. ...II..... T.GD.LKIQF G.A.VS...I 241
.Q.T.RSVGN I---KDVQEF .RIIEEMDRR Q.KRYLI... VERINT.LE. VVILG.HSR. ...ML..... T.IL.ERVMH G.A.IT...I 244
.H.S.ICVEN FN----DVSY .QLLEELDRR Q.KKFVI... IERLQN.LE. IVSVG.HVK. ...II..... K.IS.ERFIH G.A.VT...L 243
W.V.A.....-- --............R.............E......DCE........I..Q........K...G.YHY..ANLGF..D..L..........G.N...GFQ.

VNYTDTIPAR IMQQWRTSDS RDHTRVDWKR PKYTSALTYD GVKVMAEAFQ SLRRQRIDIS RRGNAGDCLA NPAVPWGQGI DIQRALQQVR 327
.DYDDSLVSK FIER.STLEE KEYPGAHTAT I.......Y. AVQ.MT.A.R N..K.RIEI. ...N...... ...V..G..V EIE.A.KQ.Q 331
.NNENPMVQQ FIQR.VRLDE REFPEAKNAP L.......H. AIL.IA.A.R Y..R.RVDV. ...S...... ...V..S..I DIE.A.KM.Q 334
.DFNTPMVTK LMDR.KKLDQ REYPGSETPP -.......Y. GVL.MA.T.R S..R.KIDI. ...N...... ...A..G..I DME.T.KQ.R 332
V................W.......................KYTSALT.D.....V...E.F....LR.Q.....S.RRG.AGDCLA.NPA.PW.QG.......R.L...V.

                                                                       (398)
FEGLTGNVQF NEKGRRTNYT LHVIEMKHDG IRKIGYWNED DKFVPAATDA QAGGDNSSVQ NRTYIVTTIL EDPYVMLKKN ANQFEGNDRY 417
VE.LS..IK. DQN.K.I... INIM.L.TN. P..I...SEV DKMVVTLTEL PSGN.TSGLE .K.VV....L .S....M... HEML...ER. 421
VQ.MT..IQ. DTY.R.T... IDVY.M.VS. S..A...NEY ERF-VPFSDQ QISN.SSSSE .R.IV....L .S....Y... HEQL...ER. 423
IQ.LT..VQ. DHY.R.V... MDVF.L.ST. P..V...NDM DKLVLIQDMP TLGN.TAAIE .R.VV....M .S....Y... HEMF...DK. 422
...G...GN...F.....G.R.NYT.......E.K...G...RK.GYW... .......... ....D..... N.T...VTTI...E.PYVM.KKN.......EGN...Y
                                                                             QBP homology
                         (445)                                           (481)
EGYCVELAAE IAKHVGYSYR LEIVSDGKYG ARDPDTKAWN GMVGELVYGR ADVAVAPLTI TLVREEVIDF SKPFMSLGIS IMIKKPQKSK 507
.....D..A. ....CGFK.K LT..G..... ...AD..I.. .........K .DI.I..... .......... .......... .......... 511
.....D..Y. ....VRIK.K LS..G..... ...PE..I.. .........R .DI.V..... .......... .......... .......... 513
.....D..S. ....IGIK.K IA..P..... ...AD..I.. .........K .EI.I..... .......... .......... .......... 512
EGYCV.LA.E.IAKH.....Y......IV.DGKYG.ARD...TK.WN.GMVGELVYG...A...A.APLTI.TLVREEVIDF.SKPFMSLGIS.IMIKKPQKSK

                                                                                         *
PGVFSFLDPL AYEIWMCIVF AYIGVSVVLF LVSRFSPYEW HSEE-FEEGR D-QTTSDQSN EFGIFNSLWF SLGAFMQQGC DISPRSLSGR 595
.......... .......... .......... .......... .T.E-F.DGR ETQS-SEST. .......... ......R... .......... 599
.......... .......... .......... .......... .L.DNN.EPR DPQSPPDPP. .......... ......Q... .......... 603
.......... .......... .......... .......... .T.EP-.DGK EGPS-DQPP. .......... ......Q... .......... 600
PGVFSFLDPL AYEIWMCIVF AYIGVSVVLF LVSRFSPYEW H.E...... .........N EFGIFNSLWF SLGAFM.QGC DISPRSLSGR
                     M1                                                      M2
                                 + (627)                $         $              @      $
IVGGVWWFFT LIIISSYTAN LAAFLTVERM VSPIESAEDL AKQTEIAYGT LEAGSTKEFF RRSKIAVYEK MWTYMKSAEP SVFTRTTEEG 685
.......... .......... .......... .......... S......... .DS....... .......FD. ..T..R.... ...VR..AE. 689
.......... .......... .......... .......... A......... .DS....... .......YE. ..S..K.... ...TK..AD. 693
.......... .......... .......... .......... A......... .DS....... .......YE. ..T..R.... ...TR..AE. 690
IVGGVWWFFT LIIISSYTAN LAAFLTVERM VSPIESAEDL..KQTEIAYGT.L...GSTKEFF.RRSKIAV...K.MW.YM.SAEP.SVF...TT...G
          M3                                               QBP homology

(689) @                                                       *         (750)
MIRVRKSKGK YAYLLESTMN EYIEQRKPCD TMKVGGNLDS KGYGIATPKG SALRNPVNLA VLKLNEQGLL DKLKNKWWYD KGECGTGGGD 775
VA........ Y.Y....... .......... .......... ....I..... .S.GNA.... ....N.Q.L. .......... .....SGGGD 779
VA........ F.F....... .......... .......... ....V..... .A.GNA.... ....N.Q.L. .......... .....SGGGD 783
VA........ F.F....... .......... .......... ....V..... .S.RTP.... ....S.A.V. .......... .....PKDSG 780
...RVRKSKGK...A.LLESTMN EYIEQRKPCD TMKVGGNLDS KGYG.ATPKG S.L...VNLA VLKL.E.G.L DKLKNKWWYD KGECG.....
                                                                 flip (GluR4)/flop (GluR1-3)

                                                     SHORT C-TAILS:
SKDKTSALSL SNVAGVFYIL IGGLGLAMLV ALIEFCYKSR SESKRMK----------------- ---------- ---------- ---
..E. ........... ..... V.......L. .......... A.A....---VAK NPQNINPSSS QNSQNFATYK EGYNVYGIES VKI 862 GluR2
..D. ........... ..... V.......M. .......... A.S....---LT. NT.NFKPAPA T.TQ.Y...R ...........V.. 866 GluR3
..D. ........... ..... V.......L. .......... A.A....---VA. SA.TFNPTSS Q.TH.L...R .......... I.. 863 GluR4c
SK.KTSALSL SNVAGVFYIL .GGLGLAM.V ALIEFCYKSR .E.KRMK---..K ..Q....... .N..N.ATY. EGYNVYGIES .KI
                         M4                                                          GRIP bdg site

LONG C-TAILS:
          $               # (845)
GFCLIPQQSI NEAIRTSTLP RNSGAGASGG GGSGENGRVV SQDFPKSMQS IPSMSHSSGM PLGATGL  889 GluR1
-------MTL SA.T.NK--- ----.RL.IT .ST......M TPEF..AVHA VPYVSPGMGM NVSVTDLS 880 GluR2long

-------LTF SE.I.NK--- ----.RL.IT .SV......L TPDC..AVHT GTAIRQSSGL AVIASDLP 881 GluR4

.......... ..A.R..... ....A..S.. G..GENGRV. ....PK.... .......... ........  consensus
```

GRIP was cloned by using the short GluR2 C-tail as bait in the yeast two-hybrid system. Only the short C-termini of GluR2 and 3, but not the long GluR1 C-tail, interact with the protein. GRIP is a clustered dendritic protein that contains seven PDZ domains, a protein motif found in many proteins important for subcellular receptor targeting. The same region of the GluR2 C-tail that is involved in GRIP binding, namely the last seven amino acids, may also play a role in receptor clustering, because microinjection of a peptide consisting of the complete short GluR2 tail, but not the long GluR1 C-tail, could inhibit AMPA receptor clustering in cultured hippocampal neurons (Dong et al., 1997). Whether the short and long C-termini have additional functions awaits further study.

*Editing of AMPA receptors*

The glutamate receptor mRNAs were the first neurotransmitter receptor RNAs found to be edited. Editing controls the most critical properties of AMPA receptor permeation. Discrepancies between the cDNA and genomic DNA sequences led to the discovery of post-transcriptional modifications of single bases in the pre-mRNA. There are two edited sites in the coding region of the glutamate receptor pre-mRNA, the Q/R site in GluR2 and the R/G site in GluR2-4. At all these sites the genomically encoded adenosine appears to be deaminated to inosine, which basepairs like a guanosine (Higuchi et al., 1993; Rueter et al., 1995). Thus at the amino acid level an arginine (codon CIG) is predicted instead of a glutamine (CAG ) at the Q/R site in GluR2. At the R/G site glycine (IGA) substitutes for the arginine (AGA) after editing in GluR2, 3 and 4 (reviewed by Seeburg, 1996).

Fig. 2. Sequence alignment of the four rat glutamate receptor subunits and the location of functionally important residues. The subunits GluR1–3 are depicted in the flop form, GluR4 is in the flip form and the alternative C-termini of the GluR2 long and GluR4c splice variants are shown. Numbering is from the mature protein, startings after the signal peptides (in italics) have been removed. The membrane-inserted regions, M1–M4, are double-underscored. Dotted lines mark homologous regions to bacterial amino acid binding proteins, LIVBP and QBP. Also underlined are potential glycosylation sites (N), the flip/flop module and the GRIP-binding site. Bold residues were shown to influence pharmacological properties, such as ligand binding, ion permeation and desensitization (see text). Edited sites are marked by an asterisk (*) and putative phosphorylation sites for CaMKII (+), PKC ($), PKA (#) as well as the one reported by Nakazawa et al. (1995) (@) are indicated.

*Q/R site*

The Q/R editing site, which is located in the M2 pore loop, influences single channel conductance and divalent ion permeability of the AMPA receptor channels (see below). In contrast to GluR2, the mRNAs encoding the other AMPA receptor subunits, GluR1, 3 and 4, are not edited at the Q/R site. So far, in all animal species analysed to date (rat, mouse, fish, pigeon, chicken) and in all brain regions analysed, >99% of mature GluR2 transcripts contain an arginine codon at the Q/R site throughout postnatal development. Only in fish is the arginine encoded genomically (Kung et al., 1996). In humans GluR2 editing may be regulated during development, aging or in specific brain areas. Evidence for some unedited mRNAs were found in fetal brain (4%), in the corpus striatum (11%) and the substantia nigra (18%) of 57 to 63 year old males and females by DNA sequencing and Bbv1 restriction site analysis of PCR products from cDNA libraries (Nutt and Kamboj, 1994). In rats, kindling and ischemia did not reduce the extent of GluR2 Q/R editing (Kamphuis et al., 1995; Kamphuis and Lopes da Silva, 1995).

*R/G site*

The R/G site is found at the end of the exon that precedes the alternatively spliced flop and flip exons (exon 13 in GluR2). R-to-G codon editing at this site is found in all mammalian and fish AMPA receptor subunits except GluR1. The extent of editing at the R/G site is regulated during development and depends on the subunit and the flip/

flop splice variant. For example in E14-P0 rats GluR3flop is already edited 70–75%, whereas the other subunits GluR2, GluR3flip and GluR4 are only edited by 5–25%. However, in adult animals most of these subunits and splice variants are edited (Lomeli et al, 1994). Again, editing of the R/G site has functional consequences. The edited subunits, except for GluR4, impart less desensitization and recover about two times faster from desensitization compared to the non-edited receptors. Thus, it seems that during development the appearance of the faster desensitizing flop versions is offset by the increased occurrence of the less desensitizing R/G edited receptor variants.

*Mechanism of editing*

Editing of AMPA receptor subunits occurs in primary RNA transcripts before splicing and depends on complementary sequences in the 3′ adjacent intron. These intronic sequences basepair with the exonic sequences to form double stranded intron-exon RNA duplexes (Fig. 3). Specific adenosines within the duplexes are substrates for the recently cloned dsRNA adenosine deaminases, DRADA (dsRNA-dependent adenosine deaminase) and RED1 (dsRNA-specific editase), which deaminate selected adenosines to inosines specifically in dsRNA. These editing enzymes, which have slightly different dsRNA substrate preferences, do not recognize ssRNA, ssDNA or dsDNA as substrates.

DRADA (or ds-RAD) is a ubiquitously expressed editing enzyme that prefers mismatched adenosines as substrates (Kim et al., 1994; O'Connell et al., 1995, for review Bass et al., 1997). RED-1 is homologous to DRADA, but can edit adenosines in perfectly matched duplexes and is highly expressed in brain (Melcher et al., 1997, O'Connell et al., 1997). Recently, four different splice variants of human RED1 (also named DRADA2a-d) were cloned that differ somewhat in substrate specificity and editing activity (Gerber et al., 1997; Lai et al., 1997). The function of these splice variants is still enigmatic.

Editing of the GluR2 Q/R site depends on two sequences within intron 11, a 10 nucleotide long editing complementary sequence (ECS) and a preceding imperfect inverted repeat (Higuchi et al., 1993). These complementary sequences are not found in the homologous intron of unedited GluR1, -3 and 4. Mutational analyses indicated that editing takes place in a stretch of perfectly matched 10 bases at the stem of a proposed RNA duplex structure formed by about 300 nucleotides containing the inverted repeat with about 85% base pairing (Fig. 3). The edited adenosine is basepaired, rendering it a poor substrate for DRADA. Instead, rat and human RED1 (specifically the human splice variants DRADA2a and b) can efficiently edit the GluR2 Q/R site in vitro (Lai et al., 1997; Melcher et al., 1996; O'Connell et al., 1997). Interestingly not only exonic but also intronic adenosines, also called hot spots (the intronic $A_{+60}$ and the triplett $A_{+262}AA$), are edited within the inverted repeat of GluR2 intron 11 (Higuchi et al., 1993). It seems that the secondary structure of the RNA renders these adenosines good substrates for a dsRNA adenosine deaminase, e.g. DRADA (Melcher et al., 1996).

Editing at the R/G site in GluR2-4 is also dependent on an intronic inverted repeat sequence followed by the ECS sequence (intron 13 in GluR2), which again are not found in the unedited GluR1 (Lomeli et al., 1994). The RNA is thought to form an intron-exon hairpin structure with a 31 basepair duplex region of about 80% perfect matches, the ECS site being located at the end of the hairpin stem (Fig. 3). The edited adenosines are mismatched making them a good substrate for DRADA (Maas et al., 1996). However, RED1 can also edit this site in vitro (Melcher et al., 1996; O'Connell et al., 1997).

Editing is not confined to ionotropic glutamate receptors. Recently the 5-$HT_{2C}$ receptor was also found to be edited at 4 sites in the second intracellular loop, which mediates binding to G-proteins and subsequent phospholipase C activation. DRADA and RED1 were sufficient to edit these sites in vitro. A mixture of non-edited, partially and fully edited receptors is found in different brain regions, but only the fully edited receptors activate phospholipase C more poorly (Burns et al., 1997).

*Phosphorylation*

AMPA receptor activation is potentiated by protein kinases A and C, and $Ca^{2+}$/calmodulin kinase II (CAMKII), but the mechanism and the location of the phosphorylation sites are controversial. Immunoprecipitation of recombinant and native AMPA receptors together with phosphopeptide analysis showed that AMPA receptor subunits are phosphorylated by CaMKII and PKC after appropriate stimulation, such as glutamate/glycine treatment of hippocampal neuronal cultures or induction of long-term potentiation (LTP) (McClade-McCulloh et al., 1993; Tan et al., 1994). In several preparations CaMKII activation leads to enhanced AMPA receptor currents, a mechanism that might contribute to LTP (reviewed by Soderling, 1996). Indeed, LTP induction in hippocampal CA1 pyramidal cells resulted in AMPA receptor phosphorylation by CaMKII that correlated temporally with an increase in AMPA receptor responsiveness (Barria et al., 1997a). The phosphorylation consensus sites for CaMKII (Ser627 in GluR1) and PKC (serines 650, 658 and 676) are located between M3 and M4 (Fig. 2). According to

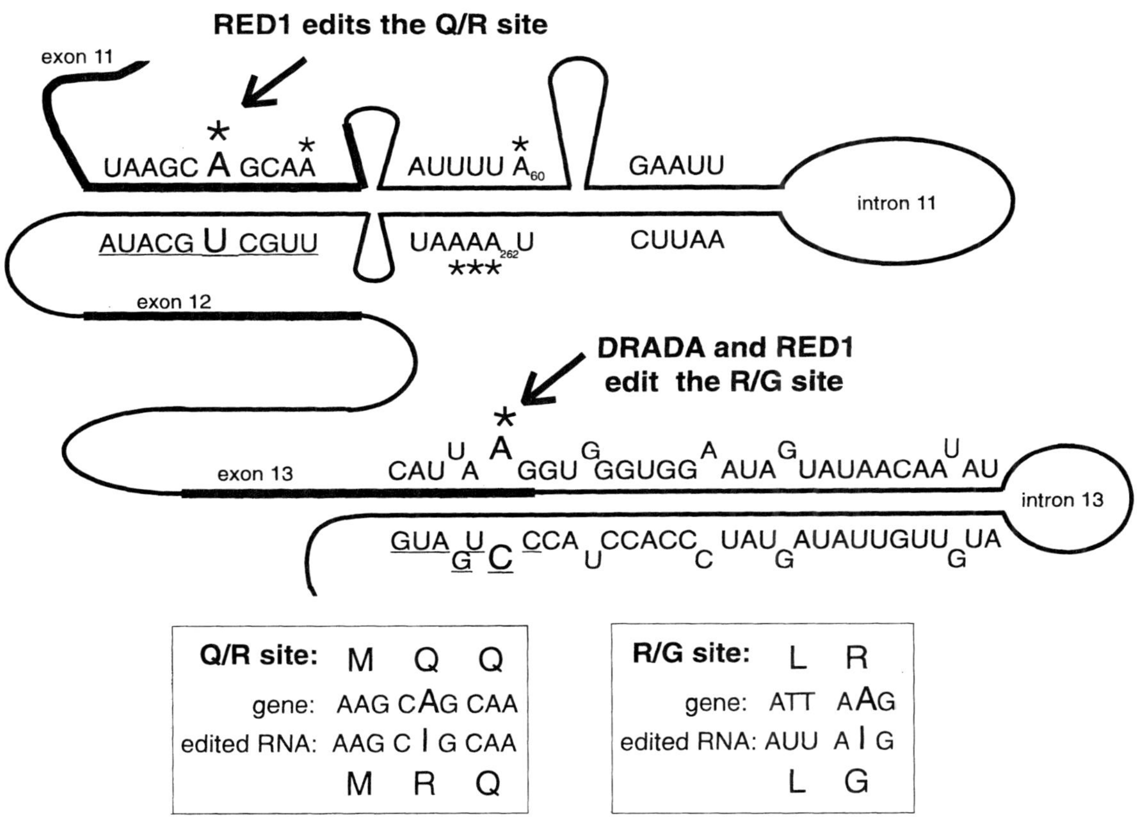


Fig. 3. Proposed pre-mRNA secondary structure for editing of the GluR2 mRNA. The proposed base-pairing between exon 11 containing the Q/R site and intron 11 as well as the exon 13 with the R/G site and intron 13 of rat GluR2 is shown. The exons are indicated by thickened lines, the editing complementary sequences are underlined and the edited adenosines are marked by stars. Note the intron/exon boundary after the R/G site.

the present topology model (see below), these sites would be extracellular. However, Ser 627 is crucial for the CaMKII-induced increase of the current, since the S627A mutation abolished the effect (Yakel et al., 1995) but it is not phosphorylated. Instead, Ser831 within the C-terminus of GluR1 seems to be the PKC and CAMKII phosphorylation site as determined by mutagenesis and western blots with anti-phosphoserine831 antibodies (Barria et al., 1997b; Mammen et al., 1997; Roche et al., 1996; Wenthold and Roche, this volume). In addition, an antiphosphopeptide antibody against a peptide containing the phosphoserine 696 of GluR2 showed that phosphorylation at this site between the M3 and the flip/flop region occurs in Purkinje cells (Nakazawa et al., 1995). It is difficult to reconcile these data with the currently accepted topology, unless there is translocation of either the phosphorylation sites or the kinase.

Forskolin, cAMP analogs and PKA potentiated AMPA receptor-mediated currents in various preparations including excised patches (Greengard et al., 1991; Roche et al., 1996). The C-terminal serine 845 seems to be phosphorylated, since in transfected HEK 293 cells only wildtype receptors but not mutant GluR1(S845A) receptors were potentiated by forskolin and gave rise to the appropriate phosphopeptide in a tryptic digest (Roche et al., 1996). In addition, forskolin treatment of hippocampal slices increased the Western Blot signal obtained with antibodies directed against phosphoserine845 (Mammen et al., 1997). In contrast Tan et al. (1994) did not observe forskolin-stimulated phosphorylation of glutamate receptors in cultured hippocampal neurons. These discrepancies might be resolved by the different preparations studied.

**Topology of AMPA receptors**

When AMPA receptor cDNAs were first isolated it was expected that their subunits would have a similar transmembrane topology to nicotinic acetylcholine receptor subunits, i.e., four transmembrane-spanning domains with both N- and the C-termini located extracellularly. However, subsequent studies showed that the ionotropic glutamate receptors represent a structurally distinct ligand-gated ion channel superfamily, perhaps similar to certain potassium channels. Glutamate receptor subunits contain only three transmembrane-spanning domains, M1, M3 and M4, plus a reentrant pore loop segment (M2) that contributes to the open ion channel (Fig. 4). This arrangement places the N-terminus and the flip–flop domains on the extracellular side of the membrane and the C-terminus on the inside. Several lines of evidence led to this topology model. Analysis of native and introduced N-glycosylation sites throughout GluR1 and GluR3 indicated that both the N-terminus preceding M1, and the region between M3 and M4 that contains the agonist binding half-site and the flip/flop module, are extracellular (Bennett and Dingledine, 1995; Hollmann et al., 1994; Wo and Oswald, 1994). The M2 region has the lowest homology to a transmembrane domain and indeed, deletion of the M2 region did not interfere with the glycosylation pattern at native or introduced glycosylation sites, suggesting that M2 does not span the membrane (Hollmann et al., 1994; Wo and Oswald, 1994). In another approach, the proteolytic sensitivity of a prolactin epitope fused to GluR3 in different sites was used to infer topology. After translation of GluR3-prolactin chimeras in the presence of microsomes, protection of intramicrosomal and membrane-inserted domains from proteinase K digestion served as a marker for location (Bennett and Dingledine, 1995). Prolactin fused to either end of the M2 sequence was digested, thus demonstrating that M2 is a reentrant pore-loop domain with both ends cytoplasmic. The M2 loop is thus similar to the P-segment in potassium channels, with which it shares sequence homology (Wo and Oswald, 1995).

The very N-terminal region is homologous to the bacterial leucine-isoleucine-valine-binding protein (LIVBP, Fig. 2). When the N-terminal domain was swapped between GluR2 and GluR6, quinoxaline antagonist but not agonist pharmacology was altered (Tygesen et al., 1995). Exchanging domains between GluR2 and GluR6 or GluR3 and GluR6, followed by pharmacological analysis of the chimeras revealed a bipartite agonist binding site for glutamate receptors that has been visualized as a clam shell consisting of part of the N-terminal region and the M3-M4 loop preceding the flip/flop site (Stern-Bach et al., 1994; Tygesen et al., 1995; Figs. 2, 4). These domains show homology to bacterial amino acid binding proteins, e.g., glutamine binding proteins (QBP, O'Hara et al., 1993; Sutcliffe et al., 1996). Simultaneous exchange of both domains is required to convert the agonist affinities of the recipient to that of the donor subunit, although the N-terminal half of the clam shell appears more important for the agonist pharmacology than the C-terminal region.

Mutation of glutamate 398 in the N-terminal domain to lysine in mouse GluR1 or replacement of lysine 445 by glutamate, differentially changed the affinity for different agonists.The affinities for glutamate and AMPA but not for kainate were decreased by the K445E mutation (Li et al., 1995; Uchino et al., 1992). Thus, these amino acids might be involved in selective interaction with the agonists, perhaps by an electrostatic interaction because mutation to neutral amino acids had smaller effects. Interestingly in heteromeric GluR1/GluR2 receptors the K445E mutation increased the EC50 for AMPA more if the mutation was introduced into GluR1 than GluR2, whereas the EC50 for glutamate was affected equally by both mutations. Thus, the two subunits contribute differently to agonist binding (Li et al., 1995). The mutation of amino acids at position 481, 486 and 502 severely decreased channel activity, implicating these sites in proper folding or agonist binding. Indeed, R481 is completely conserved among all AMPA receptor subunits and kainate-binding proteins, and is homologous to R77 in the bacterial lysine-arginine-ornithine-binding protein that is involved in substrate binding. Based on these results, Sutcliffe et al. (1996) modeled a glutamate binding site in rat GluR1 as consisting of $Glu^{398}$, $Lys^{445}$, $Arg^{481}$ and $Arg^{689}$.

**Structural determinants of ion permeation**

Two amino acid residues determine the major permeation properties of AMPA receptors: a ring of aspartates that are found in all AMPA subunits and are predicted by transmembrane topology to be located near the cytoplasmic mouth of the channel, plus the Q/R site located four residues

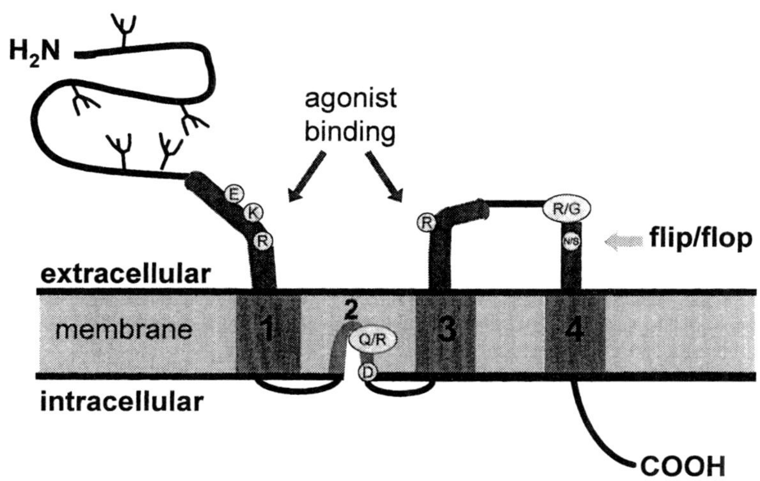


Fig. 4. AMPA receptor topology. The three transmembrane domains and the pore-loop domain are shown. N-glycosylation sites of GluR3 are denoted by trees. The agonist binding site is highlighted with the main amino amino acids denoted in the one amino acid letter code. The edited Q/R and R/G sites are shown in ovals. The location of the flip/flop module with the amino acids Asn and Ser 750 influencing the sensitivity to cyclothiazide is indicated.

upstream (Figs. 4 and 5). An arginine or glutamine residing in the Q/R site of all known kainate receptor subunits also strongly affects the degree of $Ca^{2+}$ permeability, and asparagines in the homologous site account for the high $Ca^{2+}$ permeability of NMDA receptors. Thus it is likely that the critical portion of the pore that forms the cation "selectivity filter" is similar among all glutamate receptors.

Three features of AMPA receptor phenotype are strongly influenced by the number of arginines in the Q/R site, which in turn is determined by the number of GluR2 subunits in a receptor. First, the presence of GluR2 in an AMPA receptor sharply reduces its $Ca^{2+}$ permeability (Geiger et al., 1995; Hollmann et al., 1991). Second, internal block of the AMPA receptor channel by cytosolic polyamines, which causes inward rectification, is opposed by the arginine residue in the Q/R site of GluR2 (e.g., Bowie and Mayer, 1995). Recent evidence indicates that, in contrast to the steep dependence of calcium permeability on GluR2, the affinity of the channel blocking site for internal polyamines is a *graded* function of relative GluR2 abundance (Washburn et al., 1997). This indicates that the number of GluR2 subunits in AMPA receptors is variable, which could contribute to extensive physiological diversity at glutamatergic synapses. Third, GluR2-containing AMPA receptors, by virtue of the positively-charged arginine residue poking into the channel, have a smaller single channel cation conductance than GluR2-lacking receptors (Swanson et al., 1997). Thus EPSPs mediated by GluR2-deficient receptors will likely be larger than those mediated by GluR2-rich receptors. Changes in EPSP amplitude at a

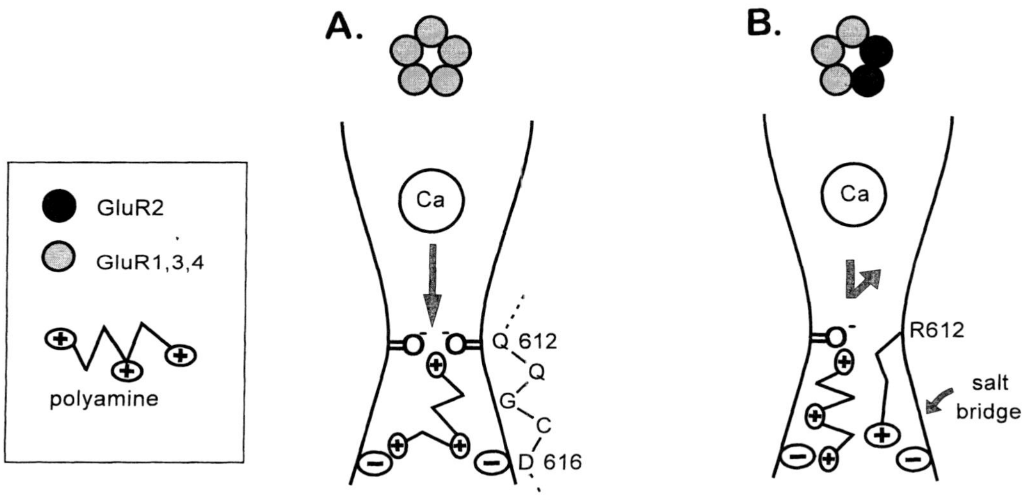


Fig. 5. Proposed model for the suppression by GluR2 of both calcium permeability and inward rectification. In the model two amino acid residues determine the major permeation properties: a ring of aspartates that are found in all AMPA subunits and are predicted by transmembrane topology to be located near the cytoplasmic mouth of the channel, and the Q/R site located four residues upstream. The ring of carbonyl oxygens donated by glutamines in the Q/R site of GluR2-lacking AMPA receptors (A) is proposed to contribute to or form a binding site for permeating divalent cations. Internal polyamines (e.g., spermine) also interact with this ring of polar residues and, by electrostatic interactions of the positively charged amine groups, with the carboxyl groups of the ionized aspartates at D616. Positively charged arginines in the Q/R site donated by GluR2 subunits (B) both disrupt the ring of carbonyl groups, and also neutralize negative charges at the internal channel mouth by formation of a salt bridge between the guanidinium groups and the carboxyl groups of the aspartates. The former effect is proposed to destroy the calcium binding site, and the latter to reduce the number of anionic binding sites for internal polyamines and thus decrease their affinity for the channel. As more arginines are incorporated into the Q/R site, the number of negative charges at the channel mouth, and hence polyamine affinity for the internal blocking site, is progressively reduced.

synapse have major physiological consequences, and for all these reasons even small to moderate changes in GluR2 abundance are expected to be functionally significant.

How do the aspartates and the Q/R site work together to determine permeation properties? The ring of carbonyl oxygens on the glutamines residing in the Q/R site of GluR2-lacking AMPA receptors is thought to contribute to or form a binding pocket within the channel for permeating calcium ions (Fig. 5A). Cytoplasmic polyamines probably block the channel by binding to this ring of polar residues as well as to the carboxyl groups of the ionized aspartates at the cytoplasmic mouth of the channel. A single subunit bearing a positively charged arginine at the Q/R site appears sufficient to prevent $Ca^{2+}$ permeation (Geiger et al., 1995), presumably by disrupting the ring of carbonyl groups. Evidence suggests that arginines in the Q/R site also neutralize negative charges at the internal channel mouth by formation of salt bridges between the guanidinium groups and the carboxyl groups of aspartates (Fig. 5B), thereby reducing the affinity of cytoplasmic polyamines for the channel and thus reducing the degree of inward rectification (Washburn et al., 1997).

**Regulation of AMPA receptor expression**

With four AMPA receptor genes, multiple splice variants and RNA editing, it comes as no surprise that excitatory synaptic transmission is profoundly influenced by regulation of AMPA receptor subunit gene expression. The abundance of GluR2 mRNA relative to other AMPA receptor subunit mRNAs is highly variable in many types of GABAergic interneurons, which endows these interneurons with a wide spectrum of AMPA receptor phenotypes (Bochet et al., 1994; Geiger et al., 1995; Washburn et al., 1997). GluR2 mRNA and protein levels decline in certain vulnerable neuron populations following intense seizure activity or transient ischemia (Gorter et al., 1997; Prince et al., 1995; Pollard et al., 1993), apparently before histological signs of cell damage. The ratio of AMPA receptor subunit mRNA levels also changes during development (Pelligrini-Giampietro et al., 1992). The expected dynamic remodeling of AMPA receptor properties during these conditions is likely to influence synaptic physiology and consequently brain functions. For example, Gorter et al. (1997) report that the down-regulation of GluR2 levels in gerbils subjected to transient global ischemia is accompanied by an increase in $Ca^{2+}$ influx through AMPA receptors expressed by the affected hippocampal neurons. Similarly, Fitzgerald et al. (1996) showed that chronic exposure of rats to morphine causes a selective increase in GluR1 mRNA levels in the ventral tegmental area (VTA). They subsequently showed that injection of a GluR1-expressing virus into the VTA sensitized the animals to the acute effects of morphine (Carlezon et al., 1997), pointing to a causal relation between increased GluR1 expression in the VTA and sensitization to morphine's behavioral effects.

All of these findings indicate that the mechanisms responsible for moderate changes in AMPA receptor expression, both developmentally and after traumatic insult in the adult, are important neuronal regulatory controls governing synaptic phenotype. To date all pharmacologic approaches for manipulating glutamatergic synaptic transmission have involved the use of receptor antagonists or modulators of desensitization. Given that different subunit combinations confer specific functional features to AMPA receptors, an alternative strategy would involve judicious manipulation of subunit expression with the objective of more selectively influencing the properties of synaptic AMPA receptors. Indeed, given sufficient knowledge of the functional consequences of changing subunit stoichiometry, this approach perhaps offers a more systematic means to alter receptor function than the search for new antagonists. A promising new research direction then is the identification and characterization of the key mechanisms by which AMPA receptor gene expression is regulated in neurons. These studies are now underway in our laboratory (Myers et al., 1996; Borges et al., 1997). Understanding the transcription factor pharmacology of AMPA receptors would open the door to new therapeutic strategies for several neurologic disorders based on adjusting the relative abundance of the four AMPA receptor subunits.

## Gene targeting: GluR2 function in the brain

To study GluR2 function in vivo, mice were genetically engineered to either express unedited GluR2 subunits or to completely lack GluR2. Brusa et al. (1995) constructed heterozygous mice in the C57BL/6 background with one editing incompetent GluR2 allele, by replacing the Q/R editing complementary sequence with a loxP sequence in 129 embryonic stem cells. The mice showed an overall 30% reduction in the GluR2 mRNA level as judged from northern blots, and 25% of the transcripts were not edited. In hippocampal CA1 cells the $Ca^{2+}$ permeability of AMPA receptors was increased by 7.3 fold. Up to postnatal day 12 the mice were healthy, but then started to develop spontaneous and recurrent seizures and died by day 30 with severe degeneration in the CA3 region. This suggests that reduction of GluR2 and/or inefficient editing can cause epileptic seizures. However, calcium permeable AMPA receptors might not be responsible alone, since Jia et al. (1996) did not observe any epileptiform seizures in GluR2-null mice constructed by deleting the M1 and M2 region in 129 embryonic stem cells crossed into CD1 mice. At three weeks the $GluR2^{-/-}$ mice were smaller, showed reduced weight, and increased mortality, probably due to poorer suckling in competition with littermates. Surviving $GluR2^{-/-}$ mice appeared healthy as adults but exhibited reduced exploration and disrupted motor coordination. Whether this behavior arises from other alleles of the 129 mice, which are normally passive in these behavioral tests (Gerlai, 1996), or is caused by the lack of GluR2, is unclear at this point. As with the heterozygous $GluR2^{+/deltaECS}$ mice, morphological abnormalities were not observed in the $GluR2^{-/-}$ mice. In hippocampal CA1 pyramidal neurons isolated from $GluR2^{-/-}$ mice, inward rectification of AMPA receptor currents was enhanced, the calcium permeability increased by about 9-fold, and AMPA receptor-mediated LTP in the CA1 region was pronounced even though the AMPA receptor component of synaptic currents was reduced by about 50%. The enhanced LTP was observed in $GluR2^{-/-}$ mice in two different genetic backgrounds suggesting that it is due to the lack of GluR2. Activation of natively occurring GluR2-lacking AMPA receptors in cultured dorsal horn neurons also produces synaptic strengthening (Gu et al., 1996). It is currently unclear why $GluR2^{-/-}$ mice did not develop epileptiform seizures, but the answer may lie either in the different genetic backgrounds used by Brusa et al. (1995) and Jia et al. (1996) (C57BL/6 vs CD1), or in the observation that the amplitude of AMPA receptor currents was unusually low in the Jia mouse. Regardless of the outcome of this issue, both GluR2-impaired mice clearly show that the presence of edited GluR2 affects crucial brain functions.

## Conclusion

The understanding of AMPA receptor functions has benefited greatly from, and contributed to, the expansion of molecular neuroscience research in the past decade. Indeed, the appreciation that AMPA receptors are involved in numerous physiological and pathophysiological phenomena continues to provide important research avenues. With the cloning and mutagenesis of AMPA receptors, the application of the patch voltage clamp, and the refinement of neuroanatomical techniques, new opportunities have arisen to address the basic functions of these nearly ubiquitous neurotransmitter receptors. The application of molecular biological approaches has been particularly instructive. For example, the transmembrane topology of subunits is known, the major structural domains responsible for ion permeation and ligand recognition have been identified, and we know that different combinations of subunits assemble into receptors with different functional properties. The molecular diversity of these receptors raises the possibility that drugs targeted to specific subunit combinations might be useful therapeutically.

Current and future research will seek to determine which subunit combinations are permissible in situ, the genetic regulation of AMPA receptor functions, and the precise roles of glutamatergic pathways and receptor subtypes in neurological disorders, development, and learning. Realization of these goals will require a coordinated effort from multiple disciplines.

## References


Barria, A., Muller, D., Derkach, V., Griffith, L.C. and Soderling, T.R. (1997a) Regulatory phosphorylation of AMPA-type glutamate receptors by CaM-KII during long-term potentiation. *Science*, 276: 2042–2044.

Barria, A., Derkach, V. and Soderling, T. (1997b) Identification of the $Ca^{2+}$/calmodulin-dependent protein kinase II regulatory phosphorylation site in the α-amino-3-hydroxyl-5-methyl-4-isoxazole-propionate-type glutamate receptor. *J. Biol. Chem.*, 272: 32727–32730.

Bass, B.L. (1997) RNA editing and hypermutation by adenosine deamination. *Trends Biochem. Sci.*, 22: 157–162.

Bennett, J.A. and Dingledine, R. (1995) Topology profile for a glutamate receptor: Three transmembrane domains and a channel-lining reentrant membrane loop. *Neuron*, 14: 373–384.

Blaschke, M., Keller, B.U., Rivosecchi, R., Hollmann, M., Heinemann, S. and Konnerth, A. (1993) A single amino acid determines the subunit-specific spider toxin block of α-amino-3-hydroxy-5-methylisoxazole-4-propionate/kainate receptor channels. *Proc. Natl. Acad. Sci. USA*., 90: 6528–6532.

Bochet, P., Audinat, E., Lambolez, B., Crépel, F., Rossier, J., Iino, M., Tsuzuki, K. and Ozawa, S. (1994) Subunit composition at the single-cell level explains functional properties of a glutamate-gated channel. *Neuron*, 12: 383–388.

Borges, K., Payne, S. and Dingledine, R. (1997) Cloning and promoter activity of the 5′ flanking region of the rat GluR1 gene. *Soc. Neurosci. Abstr*., 23: 369.2.

Bowie, D. and Mayer, M.L. (1995) Inward rectification of both AMPA and kainate subtype glutamate receptors generated by polyamine-mediated ion channel block. *Neuron*, 15: 453–462.

Brackley, P.T.H., Bell, D.R., Choi, S.-K., Nakanishi, K. and Usherwood, P.N.R. (1993) Selective antagonism of native and cloned kainate and NMDA receptors by polyamine-containing toxins. *J. Pharmacol. Exp. Ther*., 266: 1573–1580.

Brusa, R., Zimmerman, F., Koh, D.S., Feldmeyer, D., Gass, P., Seeburg, P.H. and Sprengel, R. (1995) Early-onset epilepsy and postnatal lethality associated with an editing-deficient *GluR-B* allele in mice. *Science*, 270: 1677–1680.

Burnashev, N., Monyer, H., Seeburg, P.H. and Sakmann, B. (1992) Divalent ion permeability of AMPA receptor channels is dominated by the edited form of a singled subunit. *Neuron*, 8: 189–198.

Burns, C.M., Chu, H., Rueter, S.M., Hutchinson, L.K., Canton, H., Bush, E.S. and Emeson, R.B. (1997) Regulation of serotonin-2C receptor G-protein coupling by RNA editing. *Nature*, 387: 303–308.

Carlezon, W.A., Boundy, V.A., Haile, C.N., Lane, S.B., Kalb, R.G., Neve, R.L. and Nestler, E.J. (1997) Sensitization to morphine induced by viral-mediated gene transfer. *Science*, 277: 812–814.

Chimirri, A., De Sarro, G., De Sarro, A., Gitto, R., Grasso, S., Quartarone, S., Zappalà, Giusti, P., Libri, V., Constanti, A. and Chapman, A.G. (1997) 1-Aryl-3,5-dihydro-4*H*-2,3-benzodiazepin-4-ones: Novel AMPA Receptor Antagonists. *J. Med. Chem*., 40: 1258–1269.

Dong, H., O'Brien, R.J., Fung, E.T., Lanahan, A.A., Worley, P.F. and Huganir, R.L. (1997) GRIP: A synaptic PDZ domain-containing protein that interacts with AMPA receptors. *Nature*, 386: 279–284.

Fitzgerald, L.W., Ortiz, J., Hamedani, A.G. and Nestler, E.J. (1996) Drugs of abuse and stress increase the expression of GluR1 and NMDAR1 glutamate receptor subunits in the rat ventral tegmental area: Common adaptations among across-desensitizing agents. *J. Neurosci*., 16: 274–282.

Fletcher, E.J. and Lodge, D. (1996) New developments in the molecular pharmacology of α-amino-3-hydroxy-5-methyl-4-isoazole propionate and kainate receptors. *Pharmacol. Ther*., 70: 65–89.

Gallo, V., Upson, M.L., Hayes, W.P., Vyklicky, L. Jr., Winters, C.A. and Buonanno, A. (1992) Molecular cloning and developmental analysis of a new glutamate receptor subunit isoform in cerebellum. *J. Neurosci*., 12: 1010–1023.

Geiger, J.R.P., Melcher, T., Koh, D.-S., Sakmann, B., Seeburg, P.H., Jonas, P. and Monyer, H. (1995). Relative abundance of subunit mRNAs determines gating and $Ca^{2+}$ permeability of AMPA receptors in principal neurons and interneurons in rat CNS. *Neuron*, 15: 193–204.

Gerber, A., O'Connell, M.A. and Keller, W. (1997) Two forms of human double-stranded RNA-specific editase 1 (hRED1) generated by the insertion of an Alu cassette. *RNA*, 3: 453–463.

Gerlai, R. (1996) Gene-targeting studies of mammalian behavior: Is it the mutation or the background genotype? *Trends Neurosci*., 19: 177–181.

Gorter, J.A., Petrozzino, J.J., Aronica, E.M., Rosenbaum, D.M., Opitz, T., Bennett, M.V.L., Connor, J.A. and Zukin, R.S. (1997) Global ischemia induces downregulation of GluR2 mRNA and increases AMPA receptor-mediated $Ca^{2+}$ influx in hippocampal CA1 neurons of gerbils. *J. Neurosci*., 17: 6179–6188.

Green, J.G. and Greenamyre, J.T. (1996) Bioenergetics and glutamate excitotoxicity. *Prog. Neurobiol.*, 48: 613–634.
Greengard, P., Jen, J., Nairn, A.G. and Stevens, C.F. (1991) Enhancement of the glutamate response by cAMP-dependent protein kinase in hippocampal neuons. *Science*, 253: 1135–1138.
Gu, J.G., Albuquerque, C., Lee, C.J. and MacDermott, A.B. (1996) Synaptic strengthening through activation of $Ca^{2+}$-permeable AMPA receptors. *Nature*, 381: 793–796.
Higuchi M., Single F.N., Köhler, M., Sommer, B., Sprengel, R. and Seeburg, P.H. (1993) RNA editing of AMPA receptor subunit GluR-B: A base-paired intron-exon structure determines position and efficiency. *Cell*, 75: 1361–1370.
Hollmann, M., Hartley, M. and Heinemann, S. (1991) $Ca^{2+}$ permeability of KA-AMPA-gated glutamate receptor channels depends on subunit composition. *Science*, 252: 851–853.
Hollmann, M. and Heinemann, S. (1994) Cloned glutamate receptors. *Annu. Rev. Neurosci.*, 17: 31–108.
Hollmann, M., Maron, C. and Heinemann, S. (1994) N-Glycosylation site tagging suggests a three transmembrane domain topology for the glutamate receptor GluR1. *Neuron*, 13: 1331–1343.
Hume, R.I., Dingledine, R. and Heinemann, S.F. (1991) Identification of a site in glutamate receptor subunits that controls calcium permeability. *Science*, 253: 1028–1031.
Jia, Z., Agopyan, N., Miu, P., Xiong, Z., Henderson, J., Gerlai, R., Taverna, F.A., Velumian, A., Macdonald, J., Carlen, P., Abramow-Newerly, W. and Roder, J. (1996) Enhanced LTP in mice deficient in the AMPA receptor GluR2. *Neuron*, 17: 945–956.
Kamphuis, W. and Lopes da Silva, F.H. (1995) Editing status at the Q/R site of glutamate receptor-A, -B, -5 and -6 subunit mRNA in the hippocampal kindling model of epilepsy. *Mol. Brain Res.*, 29: 35–42.
Kamphuis, W., de Leeuw, F.E. and Lopes da Silva, F.H. (1995) Ischaemia does not alter the editing status at the Q/R site of glutamate receptor-A, -B, -5 and -6 subunit mRNA. *Neuroreport*, 6: 1133–1136.
Kim, U., Wang, Y., Sanford, T., Zeng, Y. and Nishikura, K. (1994) Molecular cloning of cDNA for double-stranded RNA adenosine deaminase, a candidate enzyme for nuclear RNA editing. *Proc. Acad. Natl. Acad. Sci. USA.*, 91: 11457–11461.
Kleckner, N.W. and R. Dingledine (1989) Selectivity of quinoxalines and kynurenines as antagonists of the glycine site on NMDA receptors in Xenopus oocytes. *Mol. Pharmacol.*, 36: 430–436.
Köhler, M., Burnashev, N., Sakmann, B. and Seeburg, P.H. (1993) Determinants of $Ca^{2+}$ permeability in both TM1 an TM2 of high affinity kainate receptor channels: Diversity by RNA editing. *Neuron*, 10: 491–500.
Köhler, M., Kornau, H.-C. and Seeburg, P.H. (1994) The organization of the gene for the functionally dominant α-amino-3-hydroxy-5-methyl-isoxazole-4-propionic acid receptor subunit GluR-B. *J. Biol. Chem.*, 269: 17367–17370.
Kung, S.-S., Wu, Y.-M. and Chow, W.-Y. (1996) Characterization of two fish glutamate receptor cDNA molecules: Absence of RNA editing at the Q/R site. *Mol. Brain Res.*, 35: 119–130.
Laezza, F. and R. Dingledine, R. (1996) Tetanic stimulation of excitatory synaptic inputs to CA3 stratum radiatum interneurons. *Soc. Neurosci. Abstr.*, 22: 143.15.
Lai, F., Chen, C.-X., Carter, K.C. and Nishkura, K. (1997) Editing of glutamate receptors B subunit ion channel RNAs by four alternatively spliced DRADA2 double-stranded RNA adenosine deaminases. *Mol. Cell. Biol.*, 17: 2413–2424.
Li, F., Owens, N. and Verdoorn, T.A. (1995) Functional effects of mutations in putative agonist binding region of recombinant α-amino-3-hydroxy-5-methyl-4- isoxazolepropionic acid receptors. *Mol. Pharmacol.*, 47: 148–154.
Lomeli, H., Mosbacher, J., Melcher, T., Höger, Geiger, J.R.P., Kuner, T., Monyer, H., Higuchi, M., Bach, A. and Seeburg, P.H. (1994) Control of kinetic properties of AMPA receptor channels by nuclear RNA editing. *Science*, 266: 1709–1713.
Maas, S., Melcher, T., Herb, A., Seeburg, P.H., Keller, W., Krause, S., Higuchi, M. and O'Connell, M.A. (1996) Structural requirements for RNA editing in glutamate receptor pre-mRNAs by recombinant double-stranded RNA adenosine deaminase. *J. Biol. Chem.*, 271: 12221–12226.
Madsen, U., Andersen, B.B., Brehm, L., Christensen, I.T., Ebert, B., Kristoffersen, T.S., Lang, Y. and Krogsgaard-Larsen, P.K. (1996) Synthesis and pharmacology of highly selective carboxy and phosphono isoxazole amino acid AMPA receptor antagonists. *J. Med. Chem.*, 39: 1682–1691.
Mammen, A.L., Kameyama, K., Roche, K.W. and Huganir, R.L. (1997) Phosphorylation of the α-amino-3-hydroxyl-5-methylisoxazole-4-propionic acid receptor GluRl subunit by calcium/calmodulin-dependent kinase II. *J. Biol. Chem.*, 272: 32528–32533.
McClade-McCulloh, E., Yamamoto, H., Tan, S.-E., Brickey, D.A. and Soderling, T.R. (1993) Phosphorylation and regulation of glutamate receptors by calcium/calmodulin-dependent protein kinase II. *Nature*, 362: 640–642.
Melcher, T., Maas, S., Herb, A., Sprengel, R., Seeburg, P.H. and Higuchi, M. (1996) A mammalian RNA editing enzyme. *Nature*, 379: 460–463.
Monyer, H., Seeburg, P.H. and Wisden, W. (1991) Glutamate-operated channels: Developmentally early and mature forms arise by alternative slicing. *Neuron*, 6: 799–810.
Mosbacher, J., Schoepfer, R., Monyer, H., Burnashev, N., Seeburg, P.H. and Ruppersberg, J.P. (1994) A molecular determinant for submillisecond desensitization in glutamate receptors. *Science*, 266: 1059–1061.
Myers, S.J., Peters, J., Huang, Y., Comer, M.B., Barthel, F. and Dingledine, R. (1998) Transcriptional regulation of the

GluR2 gene: Neural-specific expression, multiple promoters, and regulatory elements. *J. Neurosci.*, in press
Nakazawa, K., Mikawa, S., Hashikawa, T. and Ito, M. (1995) Transient and persistent phosphorylation of AMPA-type glutamate receptor subunits in cerebellar Purkinje cells. *Neuron*, 15: 697–709.
Nutt, S.L. and Kamboj, R.K. (1994) Differential RNA editing efficiency of AMPA receptor subunit GluR-2 in human brain. *Neuroreport*, 5: 1679–1683.
O'Connell, M.A., Krause, S., Higuchi, M., Hsuan, J.J., Totty, N.F., Jenny, A. and Keller, W. (1995) Cloning of cDNAs encoding mammalian double-stranded RNA-specific adenosine deaminase. *Mol. Cell. Biol.*, 15: 1389–1397.
O'Connell, M.A., Gerber, A. and Keller, W. (1997) Purification of human double-stranded RNA-specific editase 1 (hRED1) involved in editing of brain glutamate receptor B pre-mRNA. *J. Biol. Chem Mol. Biol.*, 272: 473–478.
O'Hara, P.J., Sheppard, P.O., Thøgersen, H., Venezia, D., Haldeman, B.A., McGrane, V., Houamed, K.M., Thomsen, C., Gilbert, T.L. and Mulvihill, E.R. (1993) The ligand-binding domain in metabotropic glutamate receptors is related to bacterial periplasmic binding proteins. *Neuron*, 11: 41–52.
Olney, J.W. (1994) Excitotoxins in foods. *Neurotoxicology*, 15: 535–544.
Ornstein, P.L., Arnold, M.B., Augenstein, N.K., Lodge, D., Leander, J.D. and Schoepp, D.D. (1993) (3*SR*, 4a*RS*,6*R*-*S* ,8a*RS*)-6-[2-(1*H*-tetrazol-5-yl)ethyl]decahydroisoquinoline-3-carboxylic acid: A structurally novel, systemically active, competitive AMPA receptor antagonist. *J. Med. Chem.*, 36: 2046–2048.
Ornstein, P.L., Arnold, M.B., Allen, N.K., Bleisch, T., Borromeo, P.S., Lugar, C.W., Leander, J.D., Lodge, D. and Schoepp, D.D. (1996a) Structure-activity studies of 6-(tetrazolylalkyl)-substituted decahydroisoquinoline-3-carboxylic acid AMPA receptor antagonists. 1. Effects of stereochemistry, chain length, and chain substitution. *J. Med. Chem.*, 39: 2219–2231.
Ornstein, P.L., Arnold, M.B., Allen, N.K., Bleisch, T., Borromeo, P.S., Lugar, C.W., Leander, J.D., Lodge, D. and Schoepp, D.D. (1996b) Structure-activity studies of 6-substituted decahydroisoquinoline-3-carboxylic acid AMPA receptor antagonists. 2. Effects of distal acid bioisosteric substitution, absolute stereochemical preferences, and in vivo activity. *J. Med. Chem.*, 39: 2232–2244.
Partin, K.M., Patneau, D.K., Winters, C.A., Mayer, M.L. and Buonanno, A. (1993) Selective modulation of desensitization at AMPA versus kainate receptors by cyclothiazide and concanavalin A. *Neuron*, 11: 1069–1082.
Partin, K.M., Bowie, D. and Mayer, M.L. (1995) Structural determinants of allosteric regulation in alternatively spliced AMPA receptors. *Neuron*, 14: 833–843.
Partin, K.M., Fleck, M.W. and Mayer, M.L. (1996) AMPA receptor flip/flop mutants affecting deactivation, desensitization, and modulation by cyclothiazide, aniracetam, and thiocyanate. *Journal of Neuroscience*, 16(21): 6634–6647.
Paternain, A.V., Morales, M. and Lerma, J. (1995) Selective antagonism of AMPA receptors unmasks kainate receptor-mediated responses in hippocampal neurons. *Neuron*, 14: 185–189.
Pelletier, J.C., Hesson, D.P., Jones, K.A. and Costa, A.-M. (1996) Substituted 1,2-dihydrophtalazines: potent, selective, and noncompetitive inhibitors of the AMPA receptor. *J. Med. Chem.*, 39: 343–346.
Pelligrini-Giampietro, D.E.P., Bennett, M.V.L. and Zukin, R.S. (1992) Are $Ca^{2+}$-permeable kainate/AMPA receptors more abundant in immature brain? *Neurosci. Lett.*, 144: 65–69.
Pollard, H., Héron, A., Moreau, J., Ben-Ari, Y. and Khrestchatisky, M. (1993) Alterations of the GluR-B AMPA receptor subunit flip/flop expression in kainate-induced epilepsy and ischemia. *Neuroscience*, 57: 545–554.
Prince, H.K., Conn, P.J., Blackstone, C.D., Huganir, R.L. and Levey, A.I. (1995) Down-regulation of AMPA receptor subunit GluR2 in amygdaloid kindling. *J. Neurochem.*, 64(1): 462–465.
Randle, J.C.R., Biton, C. and Lepagnol, J.M. (1993) Allosteric potentiation by diazoxide of AMPA receptor currents and synaptic potentials. *Eur. J. Pharmacol.*, 247: 257–265.
Roche, K.W., O'Brien, R.J., Mammen, A.L., Bernhardt, J. and Huganir, R.L. (1996) Characterization of multiple phosphorylation sites on the AMPA receptor GluR1 subunit. *Neuron*, 16: 1179–1188.
Rogers, S.W., Andrews, P.I., Gahring, L.C., Whisenand, T., Cauley, K., Crain, B., Hughes, T.E., Heinemann, S.F. and McNamara, J.O. (1994) Autoantibodies to glutamate receptor GluR3 in Rasmussen's encephalitis. *Science*, 265: 648–651.
Rueter, S.M., Burns, C.M., Coode, S.A., Mookherjee, P. and Emeson R.B. (1995) Glutamate receptor RNA editing in vitro by enzymatic conversion of adenosine to inosine. *Science*, 267: 1491–1494.
Seeburg, P.H. (1996) The role of RNA editing in controlling glutamate receptor channel properties. *J. Neurochem.*, 66: 1–5.
Sekiguchi, M., Fleck, M.W., Mayer, M.L., Takeo, J., Chiba, Y., Yamashita, S. and Wada, K. (1997) A novel allosteric potentiator of AMPA receptors: 4-2(phenylsulfonylamino)ethylthio]-2,6-difluoro-phenoxyacetamide. *J. Neurosci.*, 17: 5760–5771.
Sheardown, M.J., Nielsen, E.Ø., Hansen, A.J., Jacobsen, P. and Honoré, T. (1990) 2.3-dihydroxy-6-nitro-7-sulfamoyl-benzo(F)quinoxaline: A neuroprotectant for cerebral ischemia. *Science*, 247: 571–574.
Soderling, T.R. (1996) Modulation of glutamate receptors by calcium/calmodulin-dependent protein kinase II. *Neurochem. Int.*, 28: 359–361.
Sommer, B., Keinänen, K., Verdoorn, T.A., Wisden, W., Burnashev, N., Herb, A., Köhler, M., Takagi, T., Sakmann,

B. and Seeburg, P.H. (1990) Flip and flop: A cell-specific functional switch in glutamate-operated channels of the CNS. *Science*, 249: 1580–1585.

Sommer, B., Köhler, M., Sprengel, R. and Seeburg, P.H. (1991) RNA editing in brain controls a determinant of ion flow in glutamate-gated channels. *Cell*, 67: 11–19.

Stern-Bach, Y., Bettler, B., Hartley, M., Sheppard, P.O., O'Hara, P.J. and Heinemann, S.F. (1994) Agonist selectivity of glutamate receptors is specified by two domains structurally related to bacterial amino acid-binding proteins. *Neuron*, 13: 1345–1357.

Sutcliffe, M.J., Wo, Z.G. and Oswald, R.E. (1996) Three-dimensional models of non-NMDA glutamate receptors. *Biophysic. J.*, 70: 1575–1589.

Swanson, G.T., Kamboj, S.K. and Cull-Candy, S.G. (1997) Single-channel properties of recombinant AMPA receptors depend on RNA editing, splice variation, and subunit composition. *J. Neurosci.*, 17: 58–69.

Tan, S-T., Wenthold, R.J. and Soderling, T.R. (1994) Phosphorylation of AMPA-type glutamate receptors by calcium/calmodulin-dependent protein kinase II and protein kinase C in cultured hippocampal neurons. *J. Neurosci.*, 14(3): 1123–1129.

Tygesen, C.K., Jørgensen, M. and Andersen, P.H. (1995) The importance of two specific domains in ligand binding to the AMPA/kainate glutamate receptors GluR2 and GluR6. *FEBS Lett.*, 363: 184–188.

Uchino, S., Sakimura, K., Nagahari, K. and Mishina, M. (1992) Mutations in a putative agonist binding region of the AMPA-selective glutamate receptor channel. *FEBS Lett.*, 308(3): 252–257.

Verdoorn, T.A., Burnashev, N., Monyer, H., Seeburg P.H. and Sakmann, B. (1991) Structural determinants of ion flow through recombinant glutamate receptor channels. *Science*, 252: 1715–1718.

Wätjen, F., Bigge, C.F., Jensen, L.H., Boxer, P.A., Lescosky, L.J., Nielsen, E.Ø., Malone, T.C., Campbell, G.W., Coughenour, L.J., Rock, D.M., Drejer, J. and Marcoux, F. (1994) NS 257 (1,2,3,6,7,8-hexahydro-3-(hydroxyimino)-*N*,*N*,7-trimethyl-2-oxobenzobenzo[2,1-b:3,4-c′]dipyrrole-5-sulfonamide) is a potent, systemically active AMPA receptor antagonist. *Biomed. Chem. Lett.*, 4: 371–376.

Washburn, M.S. and Dingledine, R. (1996) Block of α-amino-3-hydroxy-5-methyl-4-isoxazolepropionic acid (AMPA) receptors by polyamines and polyamine toxins. *J. Pharmacol. Exp. Ther.*, 278: 669–678.

Washburn, M.S., Numberger, M., Zhang, S. and Dingledine, R. Differential dependence on GluR2 expression of three characteristic features of AMPA receptors. *J. Neurosci.*, 17: 9393–9406.

Wilding, T.J. and Huettner, J.E. (1995) Differential antagonism of α-Amino-3-hydroxy-5-methyl-4-isoxazolepropionic acid-preferring and kainate-preferring receptors by 2,3-benzodiazepines. *Mol. Pharmacol.*, 47: 582–587.

Wo, Z.G. and Oswald, R.E. (1994) Transmembrane topology of two kainate receptor subunits revealed by N-glycosylation. *Proc. Natl. Acad. Sci. USA*. 91: 7154–7158.

Wo, Z.G. and Oswald, R.E. (1995) A topological analysis of goldfish kainate receptors predicts three transmembrane segments. *J. Biol. Chem.*, 270: 2000–2009.

Wong, L.A., Mayer, M.L., Jane, D.E. and Watkins, J.C. (1994) Willardiines differentiate agonist binding sites for kainate- vs. AMPA-preferring glutamate receptors in DRG and hippocampal neurons. *J. Neurosci.*, 14: 3881–3897.

Yakel, J.L., Vissavajjhala, P., Derkach, V.A., Brickey, D.A. and Soderling, T. (1995) Identification of a $Ca^{2+}$/calmodulin-dependent protein kinase II regulatory phosphorylation site in non-*N*-methyl-D-aspartate glutamate receptors. *Proc. Natl. Acad. Sci. USA*., 92: 1376–1380.

Yatsugi, S., Takahashi, M., Kawasaki-Yatsugi, S., Koshiya, K., Sakamoto, S., Uematsu, D. and Sasamata, M.S. (1996) Neuroprotective effect of YM90K, a novel AMPA/kainate receptor antagonist, in focal cerebral ischemia in cats. *J. Cereb. Blood Flow Metabol.*, 16: 959–966.

O.P. Ottersen, I.A. Langmoen and L. Gjerstad (Eds.)
*Progress in Brain Research*, Vol 116


CHAPTER 12

# Molecular determinants of NMDA receptor pharmacological diversity

Daniel T. Monaghan*, Vincent J. Andaloro and Donald A. Skifter

*Department of Pharmacology, University of Nebraska Medical Center, 600 S. 42nd Street, Omaha, NE 68198-6260, USA*

Over the past decade there has been an intensive effort to develop therapeutic agents that act at the *N*-methyl-D-aspartate (NMDA) receptor. In preclinical studies, the blockade of NMDA receptors has been shown to significantly reduce seizure activity and neuronal loss following focal ischemia, head trauma, and spinal cord injury (Lipton and Rosenberg, 1994; Meldrum and Garthwaite, 1990). Furthermore, the potent excitotoxic effect of NMDA receptor overactivation has made this receptor a prime suspect in various neurodegenerative diseases such as Alzheimer's, ALS, Parkinson's, and AIDS dementia (Lancelot and Beal, this volume). However, many agents being developed for stroke and epilepsy have been found to have unacceptable side effects (Rogawski, 1993). This is consistent with the widespread distribution of NMDA receptors and their involvement in a variety of neuronal processes such as long term potentiation (LTP), long term depression (LTD), experience-dependent formation of synaptic connections in development (Artola and Singer, 1994), neuronal differentiation/migration (Komuro and Rakic, 1993), pain modulation (Meller and Gebhart, 1993), locomotion (Headley and Grillner, 1990; Hochman et al., 1994), baroreceptor and respiratory reflexes, peristalsis in the colon (Cosentino et al., 1995), and other functions in various neuronal systems.

Two general approaches have been taken to develop agents with a higher therapeutic index – using compounds that act at different regulatory domains on the NMDA receptor or identifying compounds that act at subtypes of NMDA receptors. NMDA receptors display a rich diversity of sites at which pharmacological agents can modify activity. In addition to the glutamate and glycine agonist binding sites, there are sites for channel blockers, polyamines, redox reagents, ifenprodil, protons, steroids, $Zn^{2+}$, $Mg^{2+}$, and histamine. It is hoped that inhibition of NMDA receptor activity via one of these domains may be associated with fewer adverse effects. This may be feasible since the nature of the blockade is different at these various sites and thus the specific set of NMDA receptors blocked in vivo need not be the same for the different types of blockers. For example blockade at the glutamate binding site would be expected to be reversed at those receptors exposed to a steady, high concentration of extracellular glutamate. Channel blockers, on the other hand, would be expected to become more effective due to their use-dependency. In contrast to both of these sites, glycine site antagonism would be relatively unaltered by extracellular glutamate levels. Further selectivity is possible with channel blockers wherein low affinity channel

* Corresponding author.
Tel.: +1 402 559 7196; fax: +1 402 559 7495; e-mail: dtmonagh@mail.unmc.edu.

blockers more rapidly block, and reverse from block, than high affinity channel blockers (for review see Rogawski, 1993). The low affinity channel blockers have thus been proposed to have fewer effects upon normal synaptic activation of NMDA receptors, an observation that is consistent with clinical data. For example the low affinity channel blockers ketamine and dextromethorphan are better tolerated than the high affinity blockers phencyclidine (PCP) and MK-801.

The other approach to develop therapeutically useful NMDA receptor antagonists is to find agents that work selectively on discrete subtypes of NMDA receptors. At the moment, however, it is not known how many different types of NMDA receptors are found in vivo. Two families of NMDA receptor subunits have been cloned, together these consist of twelve different subunits. Since NMDA receptors are a hetero-oligomeric structure presumably containing 4 or 5 subunits each, there is a vast number of potential subunit combinations. In spite of this large number, there appear to be only a small number of pharmacologically-distinct NMDA receptor subtypes. By targeting drugs at these subtypes, it should be possible to generate NMDA receptor antagonists with varied therapeutic and adverse effects.

G-Protein coupled receptors are single peptides and thus display a straightforward 1:1:1 relationship between transmitter recognition binding site, receptor subtype, and pharmacological subtype. Ligand-gated ion channels are more complicated because they consist of multiple subunits, multiple transmitter binding sites, and, generally, unknown subunit stoichiometry. Thus, the theoretical limit of pharmacological diversity among NMDA receptors is not clear. The objective of the remainder of this review is to describe the molecular basis of the known pharmacologically-distinct NMDA receptor subtypes. That is, what is the relationship between NMDA receptor subtypes and specific NMDA receptor subunits or subunit composition, and what specific domains on receptor subunits are responsible for pharmacological heterogeneity?

## NMDA receptor subtypes in brain

In the 20 years following the identification of NMDA receptors, a variety of physiological and radioligand binding studies have shown that NMDA receptors display differing properties in differing brain regions. Most information regarding native NMDA receptor subtypes has come from radioligand binding studies since this approach readily provides quantitative information on both the distribution and pharmacology of receptor subtypes. Using various radioligands, it has been possible to demonstrate that NMDA receptors are heterogenous at each of the recognition sites for glutamate, glycine, channel blockers, $Zn^{2+}$, $Mg^{2+}$, and ifenprodil-like compounds. Of these different modulator binding domains, only glutamate and channel blocker sites have been found to display more than two pharmacologically-distinguishable subtypes. Importantly, the distributions of these subpopulations of NMDA receptor ligand binding sites correspond well to the distribution and pharmacology of individual NMDA receptor subunits.

There are at least four subtypes of NMDA receptor glutamate recognition sites. Radioligands for the glutamate recognition site (L-[$^3$H]glutamate and [$^3$H]CPP) can distinguish four subpopulations of binding sites. There are two major forebrain subpopulations; one subpopulation is preferentially labelled by the agonist L-[$^3$H]glutamate and the other by the antagonist [$^3$H]-CPP (Monaghan et al., 1988; Buller et al., 1994). In addition, there are two pharmacologically-distinct sites labelled by L-[$^3$H]glutamate, one found in the cerebellum and the other in the diencephalon (Monaghan and Beaton, 1991; 1992; Beaton et al., 1992). It is important to keep in mind that evidence for subtypes of NMDA receptor binding sites indicates that the specific recognition sites are heterogeneous; however their relationship to receptor complexes of known composition remains to be determined.

*Agonist-preferring and antagonist-preferring NMDA receptor binding sites*

Pharmacologically, the two major forebrain subpopulations appear very similar with a small, yet significant, difference in their affinities for agonists and antagonists. Hence we have referred to the major L-[$^3$H]glutamate binding site which has high affinity for agonists as the "agonist-preferring" binding site and the high affinity [$^3$H]CPP binding site as the "antagonist-preferring" binding site. These two subpopulations may correspond to the low and high affinity sites observed for [$^3$H]antagonist binding (Murphy et al., 1988; Porter et al., 1992; and van Amsterdam et al., 1992). While these two populations have otherwise similar pharmacological profiles, it may be possible to develop antagonists that distinguish between these subtypes (other than the few fold higher affinity generally observed for the "antagonist-preferring" subtype).

NMDA receptors in forebrain regions having agonist-preferring and antagonist-preferring NMDA receptor glutamate binding sites do not differ in their channel blocker pharmacological profiles (at least for the compounds tested to date; Beaton et al., 1992; Porter and Greenamyre, 1995). While the channel blocker site itself appears to be pharmacologically similar at these two populations, radiolabelled channel blockers can be used to distinguish the glutamate binding sites at these two populations. Using [$^3$H]MK-801 under non-equilibrium conditions which results in the selective labelling of activated NMDA receptors, low concentrations of L-glutamate promotes [$^3$H]MK-801 binding in agonist-preferring regions (Monaghan, 1991) while glutamate site antagonists selectively inhibit [$^3$H]MK-801 binding in "antagonist-preferring" regions (Sakurai et al., 1993).

*Cerebellar NMDA receptors*

The distinct cerebellar L-[$^3$H]glutamate binding site represents approximately half of the NMDA-specific L-[$^3$H]glutamate binding in the cerebellum (Monaghan and Beaton, 1991). The other cerebellar NMDA-specific L-[$^3$H]glutamate binding site has a pharmacological profile similar to that of the predominant forebrain sites. The site that is specific for the cerebellum displays a relatively low affinity for certain compounds such as quinolinate, homoquinolinate, and 4-(*p*-phenyl)benzoyl piperazine 2,3 dicarboxylic acid (Monaghan and Beaton, 1991; 1992; Andaloro et al., 1996). This site is likely to correspond to the physiologically-defined NMDA receptor subtype of the cerebellum that displays a low affinity for quinolinate (Perkins and Stone, 1983). Other studies have demonstrated that cerebellar NMDA receptors display a polyamine sensitivity that differs from forebrain NMDA receptors (Yoneda and Ogita, 1991; Reynolds and Palmer, 1991).

Cerebellar NMDA receptors also display a distinct channel blocker pharmacological profile (Ebert et al., 1991; Beaton et al., 1992; Porter and Greenamyre, 1995). For example dextromethorphan (Ebert et al., 1991; Beaton et al., 1992) and select remacemide derivatives (Porter and Greenamyre, 1995) display a higher relative affinity for cerebellar [$^3$H]-MK801 binding sites than for forebrain [$^3$H]-MK801 binding sites. In contrast, [$^3$H]MK-801 and [$^3$H]TCP generally display a lower affinity for cerebellar NMDA receptors (Vignon et al., 1986; Maragos et al., 1988; Quarum et al., 1990; Ebert et al., 1991; Reynolds and Palmer, 1991; Yoneda and Ogita, 1991; Beaton et al., 1992; Porter and Greenamyre, 1995; Bresink et al., 1995; although the relative potency in the cerebellum varies widely between studies). These observations are consistent with other indications that cerebellar NMDA receptors have differing channel properties – e.g., lower channel conductance (Farrant et al., 1994) and reduced $Mg^{2+}$-sensitivity (Yoneda and Ogita, 1991; Reynolds and Palmer, 1991).

*Midline thalamic NMDA receptors*

In specific nuclei of the midline thalamus (e.g., the paraventricular and intermediodorsal thalamic nuclei), and other diencephalic structures, NMDA receptors display a distinct pharmacological profile at both the glutamate recognition site and the channel blocker binding site (Beaton et al., 1992). For many compounds, this site is pharmacologically similar to the one found in the cerebellum, however at least two classes of compounds can distinguish between midline thalamic and cerebellar NMDA receptors. For example, the classic NMDA antagonists D-AP5 and D-CPP both have low affinities in the midline thalamus and cerebellum, but the bicyclic compound LY233536 and the biphenyl compound 4-(*p*-phenyl)benzoyl piperazine 2,3 dicarboxylic acid display higher affinities for the midline thalamic receptor type than the cerebellar subtype (Andaloro et al., 1995). Currently, there are no data regarding the physiological properties or effects of $Mg^{2+}$ and polyamines at native receptors of this subtype in the midline thalamus. This subtype may correspond to the NR2D-containing receptor recently characterized in neonatal cerebellar Purkinje cells (Momiyama et al., 1996).

*Antagonist-preferring L-[$^3$H]glutamate binding sites*

In addition to the four sites described above, there is a significant population of forebrain L-[$^3$H]glutamate binding sites that has a high affinity for antagonists (Monaghan et al., 1988). This site may, or may not, be the same as the "antagonist-preferring" site defined by high affinity [$^3$H]CPP binding. While both binding sites have some similarities in their distributions (e.g., deep cerebral cortical layers and ventral lateral thalamus) there are regions, such as the posterior nucleus of the thalamus, that have one site and relatively little of the other.

*Neonatal subtype of NMDA receptors*

Multiple lines of evidence indicate that the properties of NMDA receptors in neonates differ from those of NMDA receptors in adult tissue. Early in development, NMDA receptor-mediated synaptic responses display a longer duration (Carmignoto and Vicini, 1992; Hestrin, 1992), a reduced $Mg^{2+}$-sensitivity and a reduced voltage-sensitivity (Ben Ari, 1988; Nabekura et al., 1994; Morrisett et al., 1990; Kato and Yoshimura, 1993). Neonatal NMDA receptors also differ pharmacologically in having a lower apparent affinity for MK-801 (Kato, 1993) and quinolinate (Foster et al., 1983), a differing radioligand specificity (Kalb et al., 1992), and a high affinity for the antagonist ifenprodil (Williams et al., 1993). Cull-Candy and colleagues (Momiyama et al., 1996) have demonstrated unique physiological properties of NMDA receptors expressed in neonatal Purkinje cells. The lower affinity of $Mg^{2+}$ in neonates is likely to be due to the early expression of NR2D subunits (Watanabe et al., 1992) which are less sensitive to $Mg^{2+}$ (Monyer et al., 1994), whereas the enhanced sensitivity of neonates to ifenprodil is more likely due to the relatively early expression of NR2B subunits in development (Watanabe et al., 1992) which, in recombinant systems, show a greater ifenprodil sensitivity than NR2A subunits (Williams et al., 1993).

**NMDA receptor subunits**

Expression cloning by Dr. Nakanishi and colleagues, and homology screening by Dr. Mishina and coworkers, identified cDNAs encoding NMDA receptor subunits from rat and mouse, respectively (Moriyoshi et al., 1991; Yamazaki et al., 1992). The corresponding protein, NR1 (also termed NMDAR1 for rat or $\zeta 1$ for mouse), is over 900 amino acids in length and displays 22–26% identity with AMPA and kainate receptor subunits. The NR1 gene consists of 22 exons (Hollmann et al., 1993); exons 5, 21 and 22 can be alternatively spliced resulting in eight distinct NR1 isoforms (Sugihara et al., 1992; Hollmann et al., 1993). For clarity, below we use the nomenclature recommended by Durand and colleagues (1993); the presence or absence of a splice site is designated as 1 or 0 (respectively, $x$ if unknown) and they are listed in N-terminal to C-terminal order. Hence NR1-001 is the NR1 form having only the third alternatively spliced cassette.

In studies using the *Xenopus* oocyte expression system, injection of NR1 subunit cRNA alone was found to generate functional NMDA receptors that respond to glutamate, glycine, channel blockers, and $Mg^{2+}$ (Moriyoshi et al., 1991). This is generally interpreted as an indication that the NR1 subunit has both glutamate and glycine recognition sites. However, expression of NR1 subunits alone in mammalian cell lines does not result in functional NMDA receptors (e.g., Lynch et al., 1994).

Fully functional NMDA receptors appear to require an additional type of NMDA receptor subunit. Subsequent to the cloning of the NR1 subunit, a second NMDA receptor subunit family, the NR2 subunits, was identified (Monyer et al., 1992, 1994; Meguro et al., 1992; Kutsuwada et al., 1992; Ikeda et al., 1992; Ishii et al., 1993). The four members of the NR2 subunit family (NR2A-NR2D for rat and ε1–ε4 for mouse) are the products of four separate genes. As will be discussed below, the properties of recombinant NMDA receptors expressed in oocytes vary with the specific NR2 subunit present in the heteromeric complex and show significant parallels to the properties of NMDA receptor subtypes observed in brain (e.g., Williams et al., 1993; Buller et al., 1994).

An additional potential member of the glutamate subunit family has been identified (χ-1, Ciabarra et al., 1995; NMDAR-L, Sucher et al., 1995). This subunit may be an NMDA receptor subunit; among glutamate receptors, it has highest identity with NR1 and NR2 subunits (27%) and when co-expressed with NR1/NR2B in oocytes, it reduces the magnitude of current responses.

**Correspondence between receptor subtypes and receptor subunits**

NMDA receptor subunits critical to subtypes of native NMDA receptor radioligand binding sites can be identified, potentially, by anatomical and pharmacological correlations. In the former, one compares the distribution of individual subunits to the distribution of distinct NMDA receptor subtypes; in the latter their respective pharmacological profiles are compared. This approach assumes that a single subunit is primarily responsible for the pharmacological properties of an NMDA receptor pharmacological binding site. If single subunits are not responsible for the pharmacologically-distinct sites, then one would expect a mismatch between the anatomical and pharmacological correlations. As discussed below, generally there is a good anatomical and pharmacological correlation between individual subunits and subtypes of NMDA receptor radioligand binding sites.

*Subunits responsible for NMDA receptor subtypes: Anatomical constraints*

The anatomical distribution of the individual NMDA receptor radioligand binding sites (Monaghan et al., 1988; Beaton et al., 1992; Buller et al., 1994) correspond well to the distribution of mRNA for specific NMDA receptor subunits (Watanabe et al., 1993; Buller et al., 1994; Laurie and Seeburg, 1994b). Using subunit-specific antibodies, it is apparent that the distinct radioligand binding sites also correlate to the distribution of specific subunit protein (Wenzel et al., 1995; 1996).

TABLE 1

| Radioligand Binding Site | Subunit colocalization. |
|---|---|
| High affinity [$^3$H]CPP binding sites | NR2A |
| Low affinity [$^3$H]-CPP binding sites | NR2B |
| L-[$^3$H]Glutamate high agonist affinity (agonist-preferring) | NR2B + $NR1_{0xx}$ |
| L-[$^3$H]Glutamate high affinity for antagonists | NR2B + $NR1_{1xx}$ |
| L-[$^3$H]Glutamate, low affinity for D-CPPene (cerebellar subtype) | NR2C |
| L-[$^3$H]Glutamate, low affinity for D-CPPene and high affinity for biphenylpiperazines (midline thalamic subtype) | NR2D |

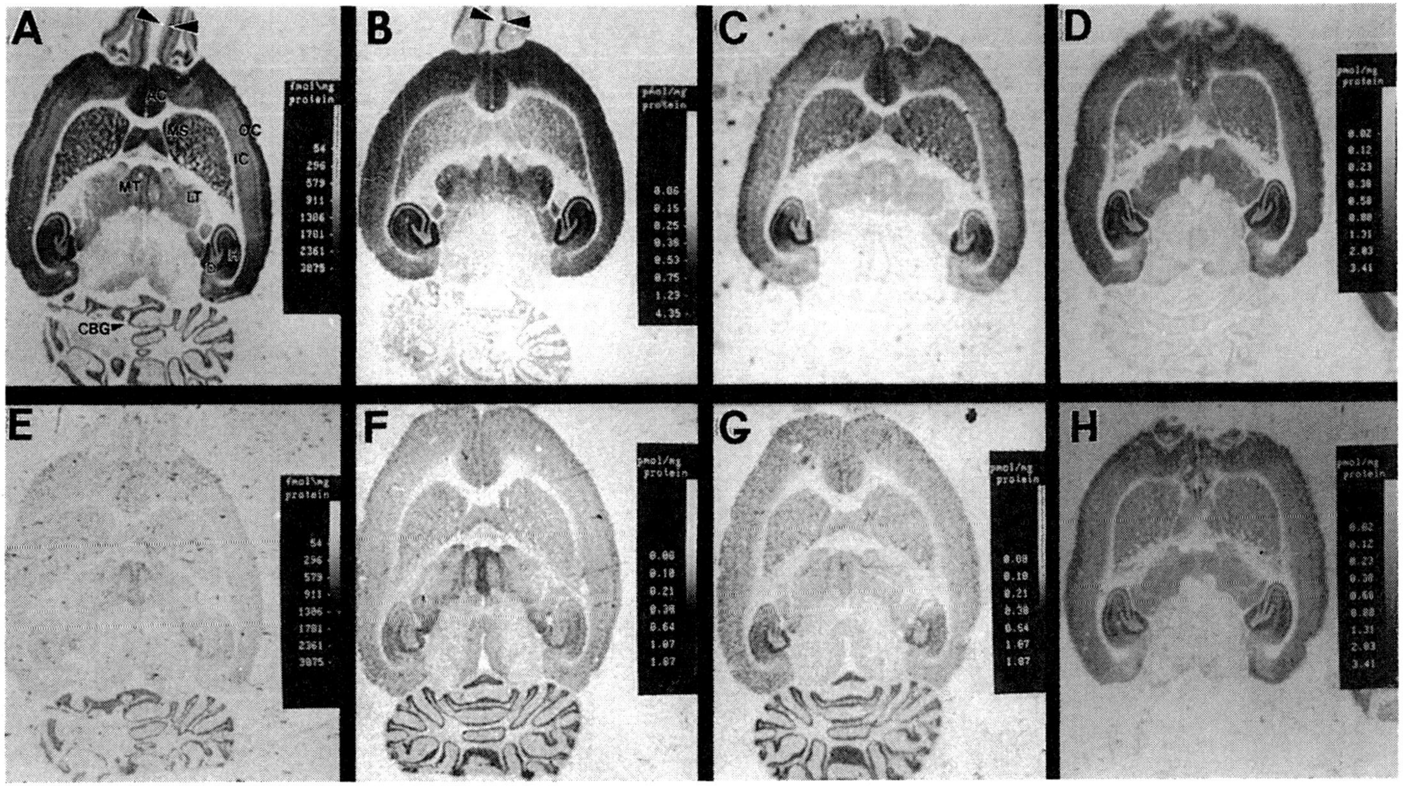


Fig. 1. Anatomical localization of NMDA receptor subpopulations in horizontal sections of rat brain. Autoradiographs of NMDA-sensitive L-[$^3$H]glutamate binding sites (A) and [$^3$H]CPP binding sites (B) display differing distributions for "agonist-preferring" and "antagonist-preferring" subtypes, respectively. Note that [$^3$H]CPP binding distribution is similar to NR2A mRNA distribution (Fig. 3a). NMDA-specific [$^3$H]homoquinolinate binding (C) has a distribution similar to NR2B subunits (Fig. 3b). Cerebellar NMDA receptors are pharmacologically distinct, [$^3$H]MK-801 binding (D, H) is preferentially displaced by 10 $\mu$M dextromethorphan (H) at NMDA receptors in the cerebellum, in contrast 10 $\mu$M D-CPPene (E) selectively spares cerebellar L-[$^3$H]glutamate binding sites. CPP at 100 $\mu$M (F) does not readily displace L-[$^3$H]glutamate binding sites in the midline thalamus and cerebellum; a distribution similar to NR2C and NR2D mRNA (Fig. 3). In contrast, 50 $\mu$M LY233536 (G) displays a higher affinity in displacing L-[$^3$H]glutamate binding sites in the midline thalamus than in the cerebellum. Abbreviations: AC, anterior cingulate cortex; CBG, cerebellar granule cell layer; D, dentate gyrus; H, hippocampus; IC inner parietal cortex; LT, lateral thalamus; MT, midline thalamic nuclei; MS, medial striatum; OC, outer parietal cortex; arrowheads indicate the glomerular layer of the olfactory bulb. Data from Beaton et al., 1992; Buller et al., 1994; and Skifter et al., 1997.

As indicated in Table 1, high affinity [$^3$H]CPP binding sites display a distribution parallel to NR2A mRNA. Quantitatively, the correlation between [$^3$H]CPP binding and NR2A mRNA distributions among 32 brain regions was $r = 0.89$ (Buller et al., 1994). Furthermore, only those regions that were positive for [$^3$H]CPP binding sites were positive for NR2A mRNA, and visa versa. NR1-1xx subunits display a partial anatomical correspondence to high affinity [$^3$H]CPP binding sites (antagonist-preferring receptors; Buller et al., 1994). However, NR1-1xx mRNA and high affinity [$^3$H]CPP binding sites clearly do not colocalize in several brain regions (Buller et al., 1994). There are brain regions with high NR1-1xx mRNA expression but a low density of [$^3$H]CPP binding sites (e.g., posterior nucleus of thalamus). Thus, although homomeric NR1-1xx receptors are associated with higher relative antagonist affinity than NR1-0xx receptors (Hollmann et al., 1993), they cannot account for the distribution of high affinity antagonist binding sites (Buller et al., 1994). The distributions of the various NR1 alternative splice cassettes have been determined (e.g., Laurie and Seeburg, 1994a; Standaert et al., 1993; see Fig. 2); the other alternative splice sites do not display anatomical correlations to any of the known subtypes of NMDA receptor radioligand binding sites.

In contrast to [$^3$H]CPP, L-[$^3$H]glutamate binding sites correlate well to the total distribution of NR2B, NR2C and NR2D mRNA and protein (Watanabe et al., 1993; Buller et al., 1994; Laurie and Seeburg, 1994b; Wenzel et al., 1996). Of the pharmacological subtypes of L-[$^3$H]glutamate binding sites, the cerebellar subtype correlates with NR2C subunit distribution. This correlation is apparent with the essentially exclusive localization of NR2C to the cerebellum and the lateral thalamus (approximately 5% of the L-[$^3$H]glutamate binding in the lateral thalamus displays cerebellum-like pharmacology; Monaghan and Beaton, unpublished observations). The midline thalamic pharmacological subtype of NMDA receptor displays a distinctive distribution (Beaton et al., 1992) that is identical to the distribution of NR2D mRNA (Buller et al., 1994; Monyer et al., 1994) and NR2D protein (as determined by NR2D-specific antibodies; Wenzel et al., 1995). The remaining L-[$^3$H]glutamate binding (after excluding the cerebellar and midline thalamic subtypes) displays a distribution that is parallel to NR2B subunits. However, this binding pattern does not appear to be fully uniform; L-[$^3$H]glutamate binding in regions that contain both NR2B and $NR1_{0xx}$ subunits displays higher affinity for agonists and lower affinity for antagonists than does binding in regions that express NR2B and $NR1_{1xx}$ subunits. The X-1 subunit has a distribution in the adult rat brain that most closely matches the distribution of the NR2D subunit. These findings are summarized in Table 1.

*Pharmacological correspondence between subtypes of NMDA receptor radioligand binding sites and NMDA receptor subunits*

*NR1 contribution to subtype pharmacology*

Among the three alternatively spliced cassettes, only the N-terminal cassette has been reported to modify pharmacological properties of NMDA receptors. This observation is consistent with the currently accepted topology model for glutamate ionotropic receptors; of the three cassettes, only the N-terminal cassette is extracellular. Recombinant NMDA receptors containing the N-terminal cassette (exon 5), display differing sensitivities to polyamines, pH, $Zn^{2+}$, redox agents, and histamine (Durand et al., 1992; Hollmann et al., 1993; Traynelis et al., 1995; Williams et al., 1994). Presumed homomeric receptors composed of NR1 subunit missing exon 5 are more potently inhibited by protons (a decrease in pH), and potentiated by polyamines and $Zn^{2+}$. It appears that positive charge in the N-terminal insert can act like the positively charged polyamines in reversing proton inhibition (Traynelis et al., 1995). Currently, there is very little data regarding how NR1 splice variants alter the sensitivities of native NMDA receptors to pH, $Zn^{2+}$, polyamines, and redox agents.

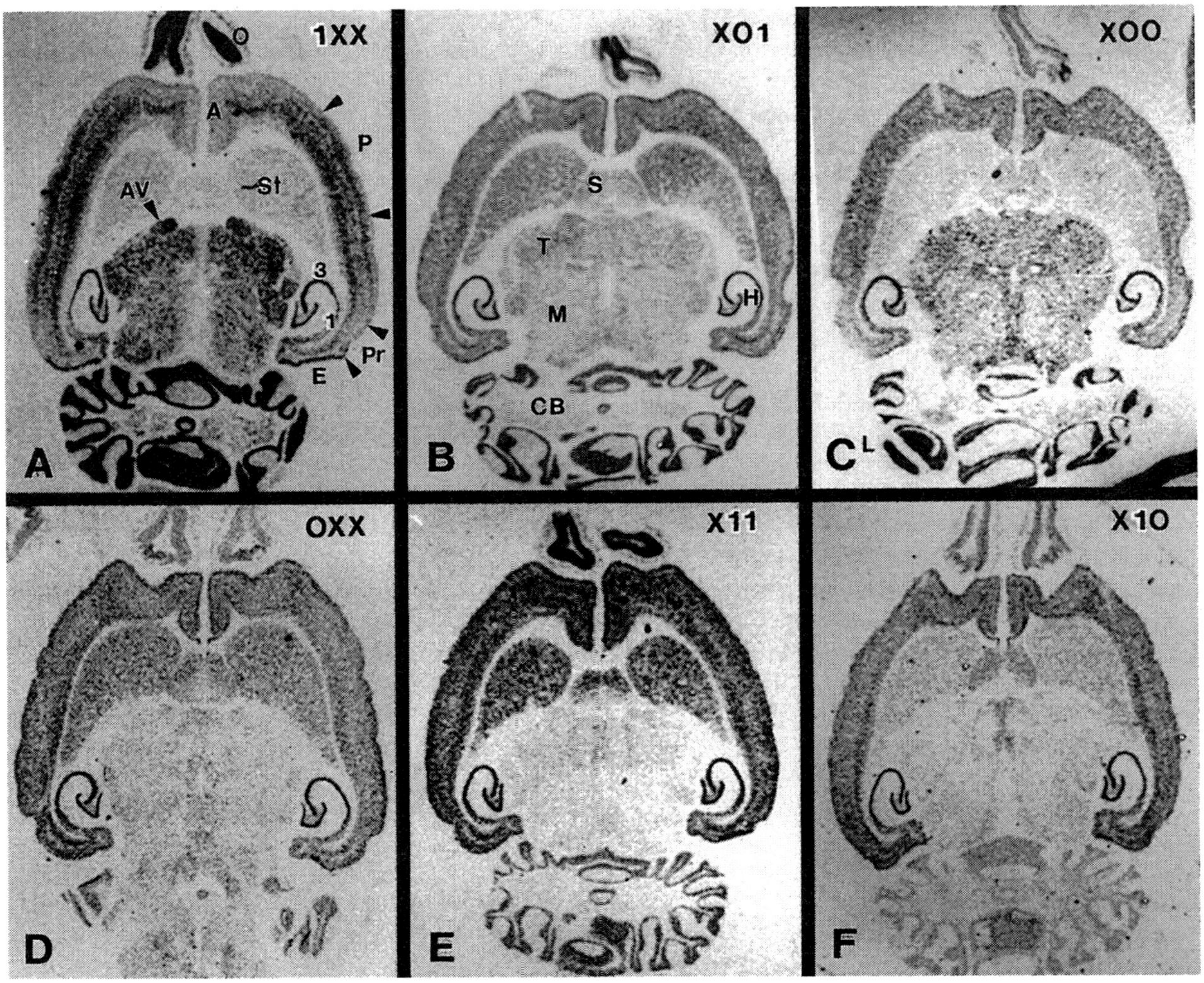


Fig. 2. Distribution of NR1 subunit mRNA splice variants in horizontal sections of rat brain determined by in situ hybridization of $[^{35}S]$ antisense oligonucleotides. Probe specificity is indicated in the figure. Abbreviations: 1, CA1 hippocampus; 3, CA3 hippocampus; A, anterior cingulate cortex; AV, anteroventral thalamus; E, entorhinal cortex; CB, cerebellum; H, hippocampus; M, midbrain; O, olfactory bulb; P, parietal cortex; S, septum, St, striatum; and T, thalamus. (Data from Monaghan, Larson, and Schneider, unpublished observations).

The N-terminal cassette (exon 5) also influences the glutamate recognition site, increasing antagonist affinity while decreasing agonist affinity (Nakanishi et al., 1992; Sugihara et al., 1992; Hollman et al., 1993). This finding is consistent with the anatomical localization of high affinity L-$[^3H]$glutamate binding sites in the medial striatum (which expresses predominately NR2B and $NR1_{0xx}$ subunits) that have a high affinity for agonists and binding sites in the posterior nucleus of the lateral thalamus (predominately NR2B and $NR1_{1xx}$-containing) that have a high affinity for antagonists. The N-terminal insert has little (Nakanishi et al., 1992) or no (Durand et al., 1992) affect on glycine affinity and modestly increases glycine antagonist and channel blocker potency (Hollmann et al.,

1993) in recombinant expression systems. The increase in the apparent potency of channel blockers at $NR1_{1xx}$ containing receptors may be due to increased rate of channel block rather than to an increase in channel blocker affinity (Rodríguez-Paz et al., 1995). In a recent study, channel blockers (of the nine tested) could not distinguish between $NR1_{000}$/NR2A and $NR1_{111}$/NR2A receptors expressed in oocytes (Monaghan and Larson, 1997). In contrast, several significant differences in channel blocker potencies were found between NR2 subunits.

The four NR2 subunits appear to account for most of the pharmacological diversity at the glutamate recognition site (Buller et al., 1994; Lynch et al., 1994, 1995; Laurie and Seeburg, 1994b; Grimwood et al., 1996a, 1996b; Buller and Monaghan, 1997) and the channel blocker site (Yamakura et al., 1993; Monaghan and Larson, 1997). In addition, NR2 subunits confer markedly different affinities for glycine (Ikeda et al., 1992; Buller et al., 1995).

*NR2 contribution to subtype pharmacology: Glutamate recognition site*

*NR2A and NR2B*

Results from several laboratories have now shown that NR2A subunits confer a relatively high affinity for glutamate site antagonists. In physiological studies of the four NR2 subunits coexpressed with NR1 subunits, the rat NR1/NR2A receptor displays the highest affinity for the antagonists D-2-amino-5-phosphonopentanoate (D-AP5) and D-3-(2-carboxypiperazin-4-yl)-1-propenyl-1-phosphonic acid (D-CPPene) (Ikeda et al., 1992; Kutsuwada et al., 1992, Buller et al., 1994; Buller and Monaghan, 1997). These antagonists display decreasing affinity in the order of NR2A, NR2B, NR2C, and NR2D. In contrast, these studies show that NR2B-containing NMDA receptors consistently display a higher affinity for glutamate site agonists than NR1/NR2A receptors. As found for rat NMDA receptors, human NR1/NR2A receptors display higher affinity for glutamate site antagonists than NR1/NR2B receptors (Priestley et al., 1995). This result is consistent with studies of $^{45}Ca^{2+}$ uptake into cultured mammalian cells stably transfected with human NR1/NR2A receptors (Grimwood et al., 1996a).

Radioligand binding studies also indicate that glutamate site antagonists show a preference for recombinant rat and human NR2A-containing receptors transfected into mammalian cells. Indeed, most studies of transfected rat and human NMDA receptors have only been able to label NR2A-containing NMDA receptors with the use of radiolabelled antagonists (e.g. [$^3$H]CGP 39653, Grimwood et al., 1996b; Lynch et al., 1994; Laurie and Seeburg, 1994b). Recently, Kendrick and colleagues (1996) have shown that the addition of $Mg^{2+}$ ions, which increases antagonist affinity (Reynolds, 1994), enables NR2B-containing NMDA receptors to be labelled by [$^3$H]CGP 39653. Using L-[$^3$H]glutamate (Laurie and Seeburg, 1994b) and [$^3$H]CGP 39653 (Kendrick et al., 1996) the NR1a/NR2A heteromer shows the highest affinity of the NR2 heteromers for the antagonists, AP5, CGS19755, CGP39853 and D-CPPene, consistent with physiological observations described above. Photoaffinity labelling studies largely support the conclusion that radiolabelled antagonists selectively label NR2A-containing NMDA receptors; labelling recombinant receptors with the photoaffinity ligand $^{125}$I-CGP 55802A was specific for NR1/NR2A, and not NR1/NR2B and NR1/NR2C receptors (Marti et al., 1993).

In contrast, when expressed in *Xenopus* oocytes or transfected into mammalian cells, both rat and human NR1/NR2B receptors generally display a higher affinity for glutamate site agonists than NR1/NR2A receptors (Ikeda et al., 1992; Kutsuwada et al., 1992; Buller et al., 1994; Laurie and Seeburg, 1994b; Priestley et al., 1995; Grimwood et al., 1996a, 1996b; Buller and Monaghan, 1997). As mentioned above, the glutamate site agonist homoquinolinate preferrentially displaces NMDA-sensitive L-[$^3$H]glutamate binding in regions having NR2B subunits. As expected, NMDA receptors labelled by [$^3$H]homoquinolinate display a distribution identical to that of NR2B subunits (Skifter et al., 1997).

The above studies, together with the colocalization studies, indicate that the high affinity [$^3$H]CPP and [$^3$H]CGP 39653 binding sites represent NR2A-containing NMDA receptors and that high affinity, NMDA-sensitive L-[$^3$H]glutamate binding sites largely represent NR2B-containing NMDA receptors. While radioligands can differentially recognize NR2A- and NR2B-containing native NMDA receptors, the profiles of glutamate site antagonists at these receptors are remarkably similar; generally only a parallel shift in antagonist affinity is seen (consistent with the descriptive terms "antagonist-preferring" or "agonist-preferring"). However, studies of recombinant NR1/NR2A and NR1/NR2B receptors expressed in *Xenopus* oocytes suggests that some structurally-diverse glutamate site antagonists may select for NR1/NR2B receptors (Buller and Monaghan, 1997). If this is confirmed, then it may be possible to develop antagonists for the glutamate recognition site of "agonist-preferring" receptors. Along these lines, is the interesting finding that quinolinate may have lower efficacy (be a partial agonist) at NR1/NR2A receptors than at NR1/NR2B receptors whereas NMDA may have lower efficacy at NR1/NR2B receptors (Priestley et al., 1995).

*NR2C*

Recombinant NR1/NR2C receptors have pharmacological properties that are distinct from NR1/NR2A and NR1/NR2B receptors. Expressed in *Xenopus* oocytes, NR1/NR2C receptors display a markedly lower affinity for homoquinolinate and D-CPPene than NR2A- and NR2B-containing receptors (Buller et al., 1994). This pattern of selectivity matches that of cerebellar NMDA receptors which also displays a relatively low affinity for homoquinolinate and D-CPPene (Monaghan and Beaton, 1992). Since NR2C subunits are almost exclusively expressed in the cerebellum, these subunits are likely to account for the cerebellar subpopulation of L-[$^3$H]glutamate binding sites that have a low affinity for homoquinolinate and D-CPPene. As discussed below, the channel blocker pharmacological properties of cerebellar NMDA receptors also correlate to NR1/NR2C receptors.

*NR2D*

NR2D subunits display a distribution that is identical to a distinct subtype of NMDA receptor glutamate binding site that is enriched in the midline thalamic nuclei (Beaton et al., 1992; Buller et al., 1994; see also discussion above). Consistent with this correlation, compounds that display a relatively low affinity for midline thalamic NMDA-sensitive L-[$^3$H]glutamate binding sites, such as homoquinolinate, D-AP5 and D-CPPene (Beaton et al., 1992; Monaghan and Beaton, 1992) also display a low affinity for NR1/NR2D receptors expressed in *Xenopus* oocytes (Buller and Monaghan, 1997). Conversely, compounds with a relatively high affinity for midline thalamic NMDA receptors (e.g. biphenyl containing antagonists) also display a relatively high affinity for recombinant NR1/NR2D receptors.

Taken together, there is a strong correlation between the pharmacological properties of subtypes of native NMDA receptor glutamate recognition sites and individual NR2 subunits.

*NR2 contribution to subtype pharmacology: Channel blocker binding site*

Channel blockers can distinguish at least 3 and perhaps 4 subtypes of native NMDA receptors (Beaton et al., 1992) and can distinguish between recombinant NR1/NR2A, NR1/NR2B, and NR1/NR2C receptors (Monaghan and Larson, 1997). Studies of native NMDA receptors have not yet shown that channel blockers can distinguish between NMDA receptors presumably containing NR2A and NR2B subunits (Beaton et al., 1992). However, at least one channel blocker compound was found that appeared to differentiate between NR1/NR2A and NR1/NR2B receptors expressed in *Xenopus* oocytes (Monaghan and Larson, 1997).

The NMDA receptor channel blockers MK-801 and PCP have been reported to have lower affinities at NR2C-containing NMDA receptors and cerebellar NMDA receptors. NR1/NR2A receptors expressed in 293 cells have an approximately 50 fold higher affinity for [$^3$H]MK-801 than NR1/NR2C heteromers (Chazot et al., 1994; Laurie and Seeburg, 1994b). Studies using *Xenopus* oocytes indicate that PCP was of slightly higher potency at NR2A and NR2B containing receptors than NR2C and NR2D-containing NMDA receptors (Yamakura et al., 1993). These findings are consistent with reports of low affinity [$^3$H]MK-801 binding at cerebellar NMDA receptors (Ebert et al., 1991; Reynolds and Palmer, 1991; Yoneda and Ogita, 1991; Beaton et al., 1992). Curiously, however, the degree of selectivity of [$^3$H]MK-801 for cerebellar NMDA receptors, compared to forebrain, differs greatly between studies (from a two fold lower affinity in the cerebellum, Yoneda and Ogita, 1991; Beaton et al., 1992, to a 25 fold difference, Ebert et al., 1991). This between-study variation may relate to the recent finding that MK-801 displays an essentially identical affinity at NR1/NR2A and NR1/NR2C receptors expressed in *Xenopus* oocytes, but much slower association and dissociation rates for MK-801 at NR1/NR2C receptors (Monaghan and Larson, 1997). Thus, any factors that might reduce the association rate or prevent equilibration of MK-801 binding for cerebellar or NR1/NR2C NMDA receptors, might result in a lower apparent affinity.

NR2C-containing NMDA receptors do apparently have a unique channel blocker pharmacology, and this pharmacology may correspond to unique pharmacological properties of the cerebellar NMDA receptor. The channel blocker dextromethorphan displays a higher relative affinity for NR1/NR2C receptors than NR1/NR2B and NR1/NR2A receptors (Monaghan and Larson, 1997). This correlates to the higher relative affinity displayed by cerebellar NMDA receptors for dextromethorphan (Ebert et al., 1991; Beaton et al., 1992). Of other channel blockers tested, select remacemide analogues also displayed a higher affinity for NR1/NR2C receptors (Monaghan and Larson, 1997).

*NR2 contribution to subtype pharmacology: Glycine recognition site*

Relatively few studies have examined the ablility of glycine agonists and antagonists to distinguish NMDA receptor subtypes. Glycine has a higher affinity for recombinant NR1/NR2C receptors than for NR1/NR2A or NR1/NR2B receptors (Laurie and Seeburg, 1994b; Kutsuwada et al., 1992; Buller et al., 1995). In contrast, O'Shea and colleagues (1991) reported that [$^3$H]glycine binding to NMDA receptors isolated from the cerebellum displayed a lower affinity than cortical NMDA receptors. While some studies have suggested distinct pharmacological properties of cerebellar NMDA receptors (Sekiguchi et al., 1990; Yoneda and Ogita, 1991; O'Shea et al., 1991), these properties have not yet been correlated to those of recombinant NR1/NR2C receptors.

Priestley and colleagues (1995) found that the glycine antagonist 7-chlorokynurenate displays a modest selectivity for NR1/NR2B receptors over NR1/NR2A receptors, while the antagonist L-701,324 does not distinguish between NMDA receptors. Along these lines, Honer et al. (1997) report that the glycine antagonist CGP 61594 has a significantly higher affinity for recombinant NR1/NR2B receptors than for recombinant NR1/NR2A, NR1/NR2C, or NR1/NR2D receptors. This selectivity is also maintained at native NMDA receptors since the autoradiographic pattern of [$^3$H]CGP-61594 is identical to that of NR2B subunits. This finding suggests the possibility that NR2 subunits may potentially confer distinct glycine-site pharmacological properties on different subpopulations of NMDA receptors and thus raises the possibility of developing therapeutic agents that target individual NR2-containing receptors.

*NR2 contribution to subtype pharmacology: Other recognition sites*

Of pharmacological agents currently available, those of greatest subtype selectivity are the ifenprodil-like compounds that show a high degree of selectivity for NR2B-containing NMDA receptors. In *Xenopus* oocyte expression studies, ifenprodil inhibits NR1/NR2B receptors with a much higher affinity than at NR1/NR2A receptors (Williams et al., 1993). In transfected 293 cells, ifenprodil displayed approximately a 100 fold selectivity for NR1/NR2B over NR1/NR2A receptors (Lynch et al., 1995). This pattern of subtype selectivity is consistent with the findings that ifenprodil becomes less effective at inhibiting [3H]MK-801 binding after neonatal development (Williams et al., 1993) and that NR2A subunits appear later than NR2B subunits during development (Watanabe et al., 1992). That ifenprodil and related compounds, are selective for NR2B subunits is consistent with the autoradiographic distribution of [3H]ifenprodil binding sites (Dana et al., 1991) which is parallel to NR2B subunit distribution (Watanabe et al., 1993; Monyer et al., 1992).

The polyamine binding site that is closely related to the ifenprodil binding site also shows distinctive properties at NR2B and striatal NMDA receptors. Glycine-independent polyamine potentiation of heteromeric recombinant NMDA receptors is specifically displayed only when NR2B subunits are coexpressed with NR1-0xx subunits (Williams et al., 1994; Williams, 1995; Zhang et al., 1994). These results are consistent with regional differences in the effects of polyamines on [$^{3}$H]-MK-801 binding in brain sections (Subramaniam and McGonigle, 1991); spermine causes greater potentiation of [$^{3}$H]-MK-801 binding in striatum (which predominately contains NR1-0xx and NR2B) than in cortex (which is mixed in both NR1 and NR2 composition). Curiously, differential polyamine stimulation was not seen in membrane homogenates prepared from the corresponding regions (Subramaniam and McGonigle, 1993).

Zinc ions are potent and selective inhibitors of NR1/NR2A receptors in the low nanomolar range (Williams, 1996; Chen et al., 1997; Paoletti et al., 1997). At micromolar concentrations $Zn^{2+}$ will also inhibit NR1/NR2B and NR1/NR2C receptors. This effect of $Zn^{2+}$ may account for observations of voltage-independent potentiation of NR1/NR2A receptors by redox reagents that can chelate endogenous zinc (e.g., Köhr et al., 1994). To date, selective $Zn^{2+}$-inhibition has not been correlated with specific native NMDA receptor subpopulations, but subunit specific $Zn^{2+}$ inhibition may account for variations among neurons in the potentiation of NMDA receptor response by metal ion chelators such as dithiothreitol (DTT).

Together these studies indicate that NR2 subunits can account for most of the subtype-specific pharmacological properties of native NMDA receptors. Thus, it is clear that the pharmacological properties of the glutamate recognition site are determined almost exclusively by NR2 subunit expression. Individual NR2 subunits can account for both the pharmacology and anatomical distributions of the observed subtypes of NMDA receptor glutamate recognition sites. NR2 subunits also appear to account for many of the observations of varied NMDA receptor channel pharmacology and variations in ifenprodil and polyamine sensitivity.

*Structural domains underlying pharmacological recognition sites*

NMDA receptors differ in their pharmacological properties as a function of their subunit composition, as discussed above. The precise relationship between distinct subtypes of binding sites for pharmacological agents and NMDA receptors of defined subunit compositions depends upon two general factors: 1) The relationship between individual subunits and pharmacological binding sites and 2) the precise subunit composition of the different native NMDA receptor subtypes. To evaluate the former question, it is necessary to determine the protein domain at which pharmacological agents bind on an NMDA receptor subunit.

Recent studies have shown that the NR1 subunit contains a glycine binding domain (Kuryatov et al., 1994; see Fig. 4). Using site directed mutagenesis of the NR1 subunit, these authors found that mutations in regions homologous to the amino acid binding site of bacterial proteins resulted in a selective shift in glycine binding affinity when coexpressed with wild type NR2 subunits. These studies indicate that aromatic residues immediately preceding the 1st transmembrane domain (TM1) and immediately following

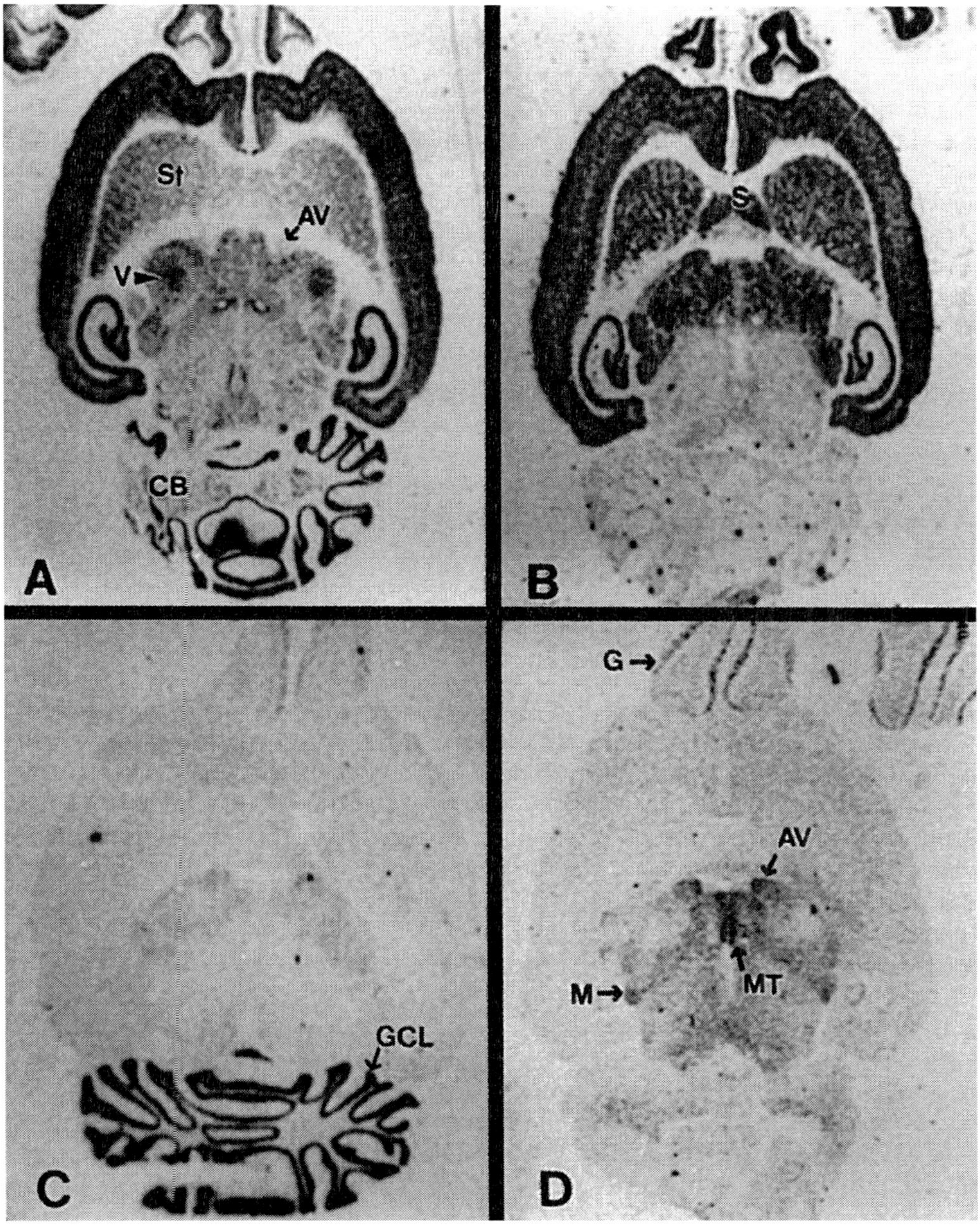


Fig. 3. In situ hybridization of NR2 subunit mRNA. NR2A (A), NR2B (B), NR2C (C), and NR2D (D) mRNA were selectively labeled in horizontal sections of rat brain by [$^{35}$S] antisense oligonucleotides. Abbreviations: AV, anteroventral thalamus; CB, cerebellum; G, glomerular layer of olfactory bulb; GCL, granule cell layer of cerebellum; M, medial geniculate; MT, midline thalamic nuclei (paraventricular, intermediodorsal; mediodorsal lateral); S, Septum; St, striatum; and V, ventroposterior thalamus (Data from Buller et al., 1994).

the 3rd transmembrane domain (TM3) in NR1, are critical to glycine binding. Interestingly, the critical residues include the FXY sequence motif that is found in the a1 subunit of inhibitory glycine receptor (Schmieden et al., 1993). The finding that the NR1 subunit has a glycine binding site is consistent with radioligand binding studies indicating the glycine agents bind to transfected mammalian cells expressing NR1 alone (Lynch et al., 1994; Grimwood et al., 1995).

The homologous site on the NR2 subunit contains a glutamate binding site. Residues on the NR2 subunit homologous to those critical for glycine binding on the NR1 subunit, were found to greatly affect glutamate binding affinity (Laube et al., 1997; Anson et al., 1997; Lummis et al., 1997). This result is thus consistent with the finding of Kendrick et al. (1996) that L-[$^3$H]glutamate binds to membranes of HEK 293 cells transfected with NR2A cRNA alone.

These findings thus support the concept that the subtypes of glutamate recognition sites identified in native brain tissue simply correspond to individual NR2 subunits. Thus, "antagonist-preferring" [$^3$H]antagonist binding sites may be a direct measure of NR2A subunits themselves and the cerebellar-specific subtype of NMDA-sensitive L[$^3$H]glutamate binding site may correspond to NR2C subunits. If this simple hypothesis is correct, then this suggests that native NMDA receptors could potentially display heterogenous pharmacological properties within a single receptor complex. For example, in the adult cerebellum where both NR2A and NR2C receptors are expressed, if these two subunits are coexpressed in the same receptor complex, then they may display differing pharmacological properties at the two subunits even though they are in the same receptor complex. Consistent with this possibility, NMDA-sensitive L[$^3$H]glutamate binding in the cerebellum clearly shows two distinct binding site subpopulations (one unique to the cerebellum and one much like that found in the forebrain; Monaghan and Beaton, 1991).

It is easy to imagine how varied NR2 subunits confer distinct glutamate site pharmacological properties since the glutamate binding site appears to be on the NR2 subunit. However, it is less clear how NR2 subunits contribute to distinct glycine site pharmacological properties. NR2 subunits signficantly alter glycine affinity. We have recently found that an NR2A subunit in which the extracellular loop region between the third and fourth transmembrane segments is replaced by the same segment from the NR2C subunit, displays the glycine affinity of the wild type NR1/NR2C receptor (Skifter, Buller, and Monaghan, unpublished observations). This suggests that while glycine binding may be exclusive to the NR1 subunit, there are domains on the NR2 subunit that alter the properties of the glycine binding site.

The protein interface for channel blocker binding is less well characterized. While residues critical to both $Mg^{2+}$ and channel blocker binding have been identified in the TM2 region of NMDA receptor subunits (Sakurada et al., 1993; Mori et al., 1992; Burnashev et al., 1992; Ferrer-Montiel et al., 1995) these residues do not account for the apparent differences in $Mg^{2+}$ or channel blocker affinity seen for NMDA receptors composed of differing NR2 composition. Kuner and Schoepfer (1996) showed that multiple structural elements are involved in determining NR2 contribution to $Mg^{2+}$ affinity. By replacing sequences in the NR2C subunit (which is less sensitive to $Mg^{2+}$ block than NR2B-containing receptors) with the corresponding NR2B sequences, they were able to reconstruct a NR2C subunit with NR2B-like $Mg^{2+}$ affinity. These findings suggested that $Mg^{2+}$ affinity is determined in part by residues in TM1, TM3, and in, or near, TM2.

Using a similar approach, replacement of the TM2 in NR2A subunits with NR2C's TM2 domain results in a subunit with $Mg^{2+}$ and dextromethorphan affinities that are more similar to that found for wild type NR2A subunits than for NR2C subunits (Skifter, Buller, and Monaghan, unpublished observations). Together, these findings lead to the somewhat surprising conclusion that the TM2 region of NR2 subunits contribute little to the subtype-specific pharmacological diversity of $Mg^{2+}$ and dextromethorphan.

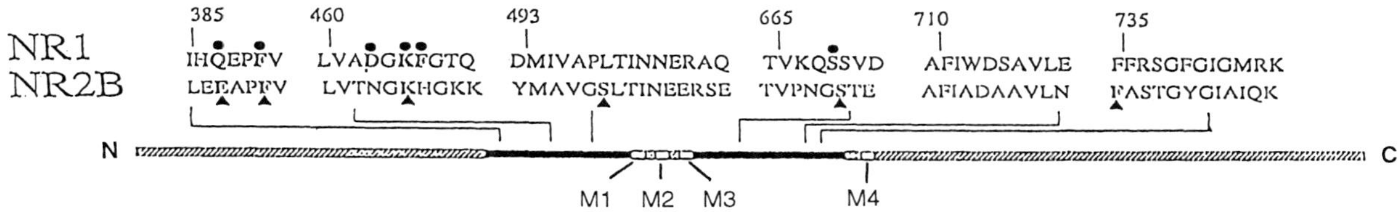


Fig. 4. Amino acid residues in NR1 and NR2 that are important for glycine and glutamate binding. Mutation of individual residues which reduced glycine affinity by more than 20 fold (●) are found on the NR1 subunit while mutated residues altering glutamate affinity (▲) are found at homologous sites on the NR2 subunit. Data from Kuryatov et al., 1994 and Laube et al., 1997.

*Subunit composition of NMDA receptors*

Currently the precise subunit composition is not known for any NMDA receptor complex. There is however, reasonable evidence that various heteromeric assemblies of NMDA receptors can exist. Coimmunoprecipitation studies indicate that NR1/NR2A/NR2B receptors exist in the rat brain (Sheng et al., 1994; Chazot and Stephenson, 1997; Blahos and Wenthold, 1996; Luo et al., 1997), but the estimate of this heterotrimer's prevalence varies widely between studies. Coassembly of NR2A and NR2B subunits is also supported by the finding that Chinese hamster ovary cells cotransfected with NR1/NR2A/NR2B subunits sometimes express channels with properties unlike either NR2A or NR2B (e.g., NR2A-like ifenprodil-site properties but with NR2B-like redox site pharmacological properties, Brimecombe et al., 1997).

There is also evidence that NR2A and NR2C subunits may coassemble. NR2C-specific antibodies coimmunoprecipitate NR2A subunits when they are coexpressed with NR2C subunits in 293 cells (Chazot et al., 1994). Furthermore, glycine activation of *Xenopus* oocytes coexpressing NR1/NR2A/NR2C subunits displays a Hill coefficient indicative of a single class of receptors (Wafford et al., 1993). The glutamate site agonist homoquinolinate was found to have a lower affinity at NR1/NR2B/NR2D receptors expressed in oocytes, than for either NR1/NR2B or NR1/NR2D receptors (Buller and Monaghan, 1997); suggesting the presence of NR1/NR2B/NR2D receptors.

Evidence for heteromeric receptors involving two different NR2 subunits has important implications for NMDA receptor subtype pharmacology. Studies described above indicate that it may be possible to have two pharmacologically-distinct glutamate (or possibly glycine) recognition sites within the same NMDA receptor complex. Thus this complex would display a mixed pharmacology. Since functional activity of an NMDA receptor appears to require the occupation of two glutamate molecules (Patneau et al., 1990) then one might expect receptor antagonism when only one subunit is occupied by an antagonist. Thus, the subunit that displays the highest affinity for an antagonist would determine the anatonist's potency at the heteromeric receptor complex. If this situation exists in the brain, it may limit the degree of pharmacological specificity that one can obtain at NMDA receptors of differing subunit composition.

**Conclusion**

There is now clear evidence that there are multiple populations of NMDA receptors in mammalian brain that can be distinguished by pharmacological techniques. Presently, there are very few compounds available that can discriminate these differing subpopulations. However, now that it is clear that the pharmacological variations that exist are due to molecularly-distinct forms of NMDA receptors, there is great potential for development of highly selective NMDA receptor antagonists (and agonists). Furthermore, since these agents have different selectivities for regions of the CNS, they would be expected to display markedly distinct therapeutic and adverse reaction profiles.

## References

Andaloro, V.J., Tse, H.W., Jane, D.E., Watkins, J.C. and Monaghan, D.T. (1995) Pharmacology of NMDA receptor subtypes. *Soc. Neuroscience, (Abstr.)* 22: 1529.

Anson, L.C., Chen, P.E. and Schoepfer, R. (1997) The NR2A subunit of recombinant NR1/NR2A NMDA receptors contains putative glutamate binding residues. *Soc. Neuroscience, (Abstr.)* 23: 548.2.

Artola, A. and Singer, W. (1994) NMDA receptors and developmental plasticity in visual cortex, in: *The NMDA Receptor*, (Collingridge, G.L. and Watkins, J.C. eds.) Oxford University Press, Oxford, UK, pp. 313–339.

Ben-Ari, Y. (1988) Changes in voltage dependence of NMDA currents during development. *Neurosci. Lett.*, 94: 88–92.

Beaton, J.A., Stemsrud, K. and Monaghan, D.T. (1992) Identification of a novel *N*-methyl-D-aspartate receptor population in the rat medial thalamus. *J. Neurochem.*, 59: 754–757.

Blahos II, J. and Wenthold, R.J. (1996) Relationship between *N*-methyl-D-aspartate receptor NR1 splice variants and NR2 subunits. *J. Biol. Chem.*, 271: 15669–15674.

Bresink, I., Danysz, W., Parsons, C.G. and Mutschler, E. (1995) Different binding affinities of NMDA receptor channel blockers in various brain regions – Indication of NMDA receptor heterogeneity. *Neuropharmacology*, 34: 533–540.

Brimecombe, J.C., Boeckman, F.A. and Aizenman, E. (1997) Functional consequenes of NR2 subunit composition in single recombinant *N*-methyl-D-aspartate receptors. *Proc. Natl. Acad. Sci. USA*, 94: 11019–11024.

Buller, A.L., Larson, H.C., Morrisett, R.A. and Monaghan, D.T. (1995) Glycine modulates ethanol inhibition of heteromeric *N*-methyl-D-aspartate receptors expressed in *Xenopus*, oocytes. *Mol. Pharmacol.*, 48: 717–723.

Buller, A.L., Larson, H.C., Schneider, B.E., Beaton, J.A., Morrisett, R.A. and Monaghan, D.T. (1994) The molecular basis of NMDA receptor sybtypes: Native receptor diversity is predicted by subunit composition. *J. Neurosci.*, 14: 5471–5484.

Buller, A.L. and Monaghan, D.T. (1997) Pharmacological heterogeneity of NMDA receptors: Characterization of NR1a/NR2D heteromers expressed in *Xenopus*, oocytes. *Eur. J. Pharmacol.*, 320: 87–94.

Burnashev, N., Shoepfer, R., Monyer, H., Ruppersberg, J. P., Gunther, W., Seeburg, P.H. and Sakmann, B. (1992) Control by asparagine residues of calcium permeability and magnesium blockade in the NMDA receptor. *Science*, 257: 1415–1419.

Carmignoto, G. and Vicini, S. (1992) Activity-dependent decrease in NMDA receptor responses during development of the visual cortex. *Science*, 258: 1007–1011.

Chazot, P.L. and Stephenson, F.A (1997) Molecular dissection of native mammalian forebrain NMDA receptors containing the NR1 C2 exon: Direct demonstration of NMDA receptors comprising NR1, NR2A, and NR2B subunits within the same complex. *J. Neurochem.*, 69: 2138–2144.

Chazot, P.L., Coleman, S.K., Cik, M. and Stephenson, F.A (1994) Molecular characterization of *N*-methyl-D-aspartate receptors expressed in mammalian cells yields evidence for the coexistence of three subunit types within a discrete receptor molecule. *J. Biol. Chem.*, 269: 24403–24409.

Chen, N., Moshaver, A. and Raymond, L.A. (1997) Differential sensitivity of recombinant *N*-methyl-D-aspartate receptor subtypes to zinc inhibition. *Mol. Pharmacol.*, 51: 1015–1023.

Ciabarra, A.M., Sullivan, J.M., Gahn, L.G., Pecht, G., Heinemann, S. and Sevarino, K.A. (1995) Cloning and characterization of $\chi$-1: A developmentally regulated member of a novel class of the ionotropic glutamate receptor family. *J. Neurosci.*, 15: 6498–6508.

Cosentino, M., De Ponti, F., Marino, F., Giaroni, C., Leoni, O., Lecchini, S. and Frigo, G. (1995) *N*-Methyl-D-aspartate receptors modulate neurotransmitter release and peristalsis in the guinea pig isolated colon. *Neurosci. Lett.*, 183: 139–142.

Dana, C., Benevides, J., Schoemaker, H. and B. Scatton, B. (1991) Pharmacological characterization and autoradiographic distribution of polyamine-sensitive [3H]ifenprodil binding sites in the rat brain. *Neurosci. Lett.*, 125: 45–48.

Durand, G.M., Bennett, M.V.L. and Zukin, R.S. (1993) Splice variants of the *N*-methyl-D-aspartate receptor NR1 identify domains involved in the regulation by polyamines and protein kinase C. *Proc. Natl. Acad. Sci. USA*, 90: 6731–6735.

Durand, G.M., Gregor, P., Zheng, X., Bennett, M.V.L., Uhl, G.R. and Zukin, R.S. (1992) Cloning of an apparent splice variant of the rat *N*-methyl-D-aspartate receptor NMDAR1 with altered sensitivity to polyamines and activators of protein kinase C. *Proc. Natl. Acad. Sci. USA*, 89: 9359–9363.

Ebert, B., Wong, E.H.F. and Krogsgaard-Larsen, P. (1991) Identification of a novel NMDA receptor in rat cerebellum. *Eur. J. Pharmacol.*, 208: 49–52.

Farrant, M., Feldmeyer, D., Takahashi, T. and Cull-Candy, S.G. (1994) NMDA-receptor channel diversity in the developing cerebellum. *Nature*, 368: 335–339.

Ferrer-Montiel, A.V., Sun, W. and Montal, M. (1995) Molecular design of the *N*-methyl-D-aspartate receptor binding site for phencyclidine and dizolcipine. *Proc. Natl. Acad. Sci. USA*, 92: 8021–8025.

Foster, A.C., Collins, J.F. and Schwarcz, R. (1983) On the excitotoxic properties of quinolinic acid, 2,3-piperidine dicarboxylic acids and structurally related compounds. *Neuropharmacol.*, 22: 1331–1342.

Grimwood, S., Gilbert, E., Ragan C.I. and Hutson, P.H. (1996a) Modulation of 45$Ca^{2+}$ influx into cells stably expressing recombinant human NMDA receptors by ligands acting at distinct recognition sites. *J. Neurochem.*, 66: 2589–2595.

Grimwood, S., Le Bourdelles, B., Atack, J.R., Barton, C., Cockett, W., Cook, S.M., Gilbert, E., Hutson, P.H., McKernan, R.M., Myers, J.M., Ragan, C.I., Wingrove, P.B. and Whiting, P.J. (1996b) Generation and characterization of stable cell lines expressing recombinant human *N*-methyl-D-aspartate receptor subtypes. *J. Neurochem.*, 66: 2239–2247.

Grimwood, S., Le Bourdelles, B. and Whiting, P.J. (1995) Recombinant human NMDA homomeric NR1 receptors expressed in mammalian cells form a high-affinity glycine antagonist binding site. *J. Neurochem.*, 64: 525–530.

Headley, P.M. and Grillner, S. (1990) Excitatory amino acids and synaptic transmission: The evidence for a physiological function. *Trends Pharmacolog. Sci.*, 11: 205–211.

Hestrin, S. (1992) Developmental regulation of NMDA receptor-mediated synaptic currents at a central synapse. *Nature*, 357: 686–689.

Hochman, S., Jordan, L.M. and Schmidt, B.J. (1994) TTX-Resistant NMDA receptor-mediated voltage oscillations in mammalian lumbar motoneurons. *J. Neurophysiol.*, 72: 2559–256.

Hollmann, M., Boulter, J., Maron, C., Beasley, L., Sullivan, J., Pecht, G. and Heinemann, S. (1993) Zinc potentiates agonist-induced currents at certain splice variants of the NMDA receptor. *Neuron*, 10: 943–954.

Honer, M., Benke, D., Laube, B., Kuhse, J., Heckendorn, R., Allgeier, H., Angst, C., Monyer, H., Seeburg, P., Betz, H. and Mohler, H. (1997) *Soc. Neuroscience*, (Abstr.) 23: 367.6.

Ikeda, K., Nagasawa, M., Mori, H., Araki, K., Sakimura, K., Watanabe, M., Inoue, Y. and Mishina, M. (1992) Cloning and expression of the ε4 subunit of the NMDA receptor channel. *FEBS Letts.*, 313: 34–38.

Ishii, T., Moriyoshi, K., Sugihara, H., Sakurada, K., Katodani, H., Yokoi, M., Akazawa, C., Shigemoto R., Mizuno, N., Masu, M. and Nakanishi, S. (1993) Molecular characterization of the family of the *N*-methyl-D-aspartate receptor subunits. *J. Biol. Chem.*, 268: 2836–2843.

Kalb, R.G., Lidow, M.S., Halsted, M.J. and Hockfield, S. (1992) *N*-methyl-D-aspartate receptors are transiently expressed in the developing spinal cord ventral horn. *Proc. Natl. Acad. Sci. USA*, 89: 8502–8506.

Kato, N. (1993) Synaptic *N*-methyl-D-aspartate receptors in the neonatal rat visual cortex are less sensitive to MK-801 than in adult. *Brain Res.*, 608: 166–168.

Kato, N. and Yoshimura, H. (1993) Reduced $Mg^{2+}$ block of *N*-methyl-D-aspartate receptor mediated synaptic potentials in developing visual cortex. *Proc. Natl. Acad. Sci. USA*, 90: 7114–7118.

Kawamoto, S., Uhino, S., Hattari, S., Hamajima, K., Mishina, M., Nakajima-Iijima, S. and Okuda, K. (1995) Expression and characterization of the ζ 1 subunit of the *N*-methyl-D-aspartate (NMDA) receptor channel in a baculovirus system. *Mol. Brain Res.*, 30: 137–148.

Kendrick, S.J., Lynch, D.R. and Pritchett, D.B. (1996) Characterization of glutamate binding sites in receptors assembled from transfected NMDA receptor subunits. *J. Neurochem.*, 67: 608–616.

Komuro, H. and Rakic, P. (1993) Modulation of neuronal migration by NMDA receptors. *Science*, 260: 95–97.

Köhr, G., Eckardt, S., Luddens, H., Monyer, H. and Seeburg, P.H. (1994) NMDA receptor channels: Subunit specific potentiation by reducing agents. *Neuron*, 12: 1031–1040.

Kuner, T. and Schoepfer, R. (1996) Multiple structural elements determine subunit specificity of $Mg^{2+}$ block in NMDA receptor channels. *J. Neurosci.*, 16: 3549–3558.

Kuryatov, A., Laube, B., Betz, H. and Kuhse, J. (1994) Mutational analysis of the glycine-binding site of the NMDA receptor: structural similarity with bacterial amino acid-binding proteins. *Neuron*, 12: 1291–1300.

Kutsuwada, T., Kashiwabuchi, N., Mori, H., Sakimura, K., Kushiya, E., Araki, K., Meguro, H., Masaki, H., Kumanishi, T., Arakawa, M. and Mishina, M. (1992) Molecular diversity of the NMDA receptor channel. *Nature*, 358: 36–41.

Laube, B., Hirokazu, H., Sturgess, M., Betz, H. and Kuhse, J. (1997) Molecular determinants of agonist discrimination by NMDA receptor subunits: Analysis of the glutamate binding site on the NR2B subunit. *Neuron*, 18: 493–503.

Laurie, D.J. and Seeburg, P.H. (1994a) Regional and developmental heterogeneity in splicing of the rat brain NMDAR1 mRNA. *J. Neurosci.*, 14: 3180–3194.

Laurie, D.J. and Seeburg, P.H. (1994b) Ligand affinities at recombinant *N*-methyl-D-aspartate receptors depend on subunit composition. *Eur. J. Pharmacol.*, 268: 335–345.

Lipton, S.A. and Rosenberg, P.A. (1994) Excitatory amino acids as a final common pathway for neurologic disorders. *New England J. Med.*, 330: 613–622.

Lummis, S.C.R., Fletcher, E.J.H. and Green, T. (1997) Evedence that glutamate gates NMDA receptors through interactions with the NR2 subunit. *Soc. Neuroscience*, (Abstr.) 23: 372.2.

Luo, J., Wang, Y., Yasuda, R.P., Dunah, A.W. and Wolfe, B.B. (1997) The majority of *N*-methyl-D-aspartate receptor complexes in adult rat cerebral cortex contain at least three different subunits (NR1/NR2A/NR2B). *Mol. Pharmacol.*, 51: 79–86.

Lynch, D.R., Anegawa, N.J., Verdoorn, T. and Pritchett, D.B. (1994) *N*-Methyl-D-aspartate receptors: Different subunit requirements for binding of glutamate antagonists, glycine antagonists, and channel-blocking agents. *Mol. Pharmacol.*, 45: 540–545.

Lynch, D.R., Lawrence, J.J., Lenz, S., Anegawa, N.J., Dichter, M. and Pritchett, D.B. (1995) Pharmacological characterization of heterodimeric NMDA receptors composed of NR 1a and 2B subunits: Differences with receptors formed from NR 1a and 2A. *J. Neurochem.*, 64: 1462–1468.

Maragos, W.F., Penney, J.B. and Young, A.B. (1988) Anatomic correlation of NMDA and [$^3$H] TCP-labelled receptors in rat brain. *J. Neurosci.*, 8: 493–501.

Marti, T., Benke, D., Mertens, S., Heckendorn, R., Pozza, M., Allgeier, H., Angst, C., Laurie, D., Seeburg, P. and Mohler, H. (1993) Molecular distinction of three *N*-methyl-D-aspartate-receptor subtypes in situ and developmental receptor maturation demonstrated with the photoaffinity ligand 125I-labelled CGP 55802A. *Proc. Natl. Acad. Sci. USA*, 90: 8434–8438.

McIlhinney, R.A.J., Molinar, E., Atack, J.R. and Whiting, P.J. (1996) Cell surface expression of the human *N*-methyl-D-aspartate receptor subunit 1a requires the co-expression of the NR2A subunit in transfected cells. *Neuroscience*, 70: 989–997.

Meguro H., Mori H., Araki K., Kushiya E., Kutsuwada T., Yamazaki M., Kumanishi T., Arakawa M., Sakimura T., Mishina M. (1992) Functional characterization of a heteromeric NMDA receptor channel expressed from cloned cDNAs. *Nature*, 357: 70–74.

Meldrum, B. and Garthwaite, J. (1990) Excitatory amino acid neurotoxicity and neurodegenerative disease. *Trends Pharmacol. Sci.*, 11, 379–387.

Meller, S.T. and Gebhart, G.F. (1993) Nitric oxide and nociceptive processing in the spinal cord. *Pain*, 52: 254–262.

Momiyama, A., Feldmeyer, D. and Cull-Candy, S.G. (1996) Identification of a native low-conductance NMDA channel with reduced sensitivity to $Mg^{2+}$ in rat central neurons. *J. Physiol.*, 494: 479–492.

Monaghan, D.T., Olverman, H.J., Nguyen, L., Watkins, J.C. and Cotman, C.W. (1988) Two classes of *N*-methyl-D-aspartate recognition sites: Differential distribution and differential regulation by glycine. *Proc. Natl. Acad. Sci. USA*, 85: 9836–9840.

Monaghan, D.T. (1991) Differential stimulation of [$^3$H]MK801 binding to subpopulations of NMDA receptors. *Neurosci. Letts.*, 122: 21–24.

Monaghan, D.T. and Beaton, J.A. (1991) Quinolinate differentiates between forebrain and cerebellar NMDA receptors. *Eur. J. Pharmacol.*, 194: 123–125.

Monaghan, D.T. and Beaton, J.A. (1992) Pharmacologically-distinct NMDA receptor populations of the cerebellum, medial thalamic nuclei and forebrain. *Mol. Neuropharmacol.*, 2: 71–75.

Monaghan, D.T. and Larson, H. (1997) NR1 and NR2 subunit contributions to *N*-methyl-D-aspartate receptor channel blocker pharmacology. *J. Pharmacol. Exp. Ther.*, 280: 614–620.

Monyer, H., Sprengel, R., Schoepfer, R., Herb, A., Higuchi, M., Lomeli, H., Burnashev, N., Sakmann, B. and Seeburg, P.H. (1992) Heteromeric NMDA receptors: Molecular and functional distinction of subtypes. *Nature*, 256: 1217–1221.

Monyer, H., Burnashev, N., Laurie, D.L., Sakmann, B. and Seeburg, P.H. (1994) Developmental and regional expression in the rat brain and functional properties of four NMDA receptors. *Neuron*, 12: 529–540.

Mori, H., Masaki, H., Yamakura, T. and Mishina, M. (1992) Identification by mutagenesis of a $Mg^{2+}$-block site of the NMDA reeptor channel. *Nature*, 358: 673–675.

Moriyoshi, K., Masu, M., Ishii, T., Shigemoto, R., Mizuno, N. and Nakanishi, S. (1991) Molecular cloning and characterization of the rat NMDA receptor. *Nature*, 354: 31–37.

Morrisett, R.A., Mott, D.D., Lewis, D.V., Wilson, W.A. and Swartzwelder, H.S. (1990) Reduced sensitivity of the *N*-methyl-D-aspartate component of synaptic transmission to magnesium in hippocampal slices from immature rats. *Dev. Brain Res.*, 56: 257–262.

Murphy, D.E., Hutchinson, A.J., Hurt, S.D., Williams, M. and Sills, M.S. (1988) Characterization of the binding of [$^3$H]-CGS19755: A novel *N*-methyl-D-aspartate antagonist with nanomolar affinity in rat brain. *Br. J. Pharmacol.*, 95: 932–938.

Nabekura, J., Kawamoto, I. and Akaike, N. (1994) Developmental change in voltage dependency of NMDA receptor-mediated response in nucleus tractus solitary neurons. *Brain Res.*, 648: 152–156.

Nakanishi, N., Axel, R. and Shneider, N.A. (1992) Alternative splicing generates functionally distinct *N*-methyl-D-aspartate receptors. *Proc. Natl. Acad. Sci.*, 89: 8552–8556.

O'Shea, R.D., Manallack, D.T., Conway, E.L., Mercer, L.D. and Beart, P.M. (1991) Evidence for heterogenous glycine domains but conserved multiple states of the excitatory amino acid recognition site of the NMDA receptor: Regional binding studies with [$^3$H]-glycine and [$^3$H]-glutamate. *Exp. Brain Res.*, 86: 652–662.

Paoletti, P., Ascher, P. and Neyton, J. (1997) High-affinity zinc inhibition of NMDA NR1-NR2A receptors. *J. Neurosci.*, 17: 5711–5725.

Patneau, D.K. and Mayer, M.L. (1990) Structure-activity relationships for amino acid transmitter candidates acting at *N*-methyl-D-aspartate and quisqualate receptors. *J. Neurosi.*, 10: 2385–2399.

Perkins, M.N. and Stone, T.W. (1983a) Pharmacology and regional variations of quinolinic acid-evoked excitations in the rat central nervous system. *J. Pharmacol. Exp. Ther.*, 226: 551–557.

Porter, R.H.P., Cowburn, R.F., Alasuzoff, I., Briggs, R.S.J. and Roberts, P.J. (1992) Heterogeneity of NMDA receptors labelled with [$^3$H]3-(($\pm$)-2-carboxypiperazin-4-yl)propyl-1-phosphonic acid ([$^3$-H]CPP): Receptor status in Alzheimer's disease brains. *Eur. J. Pharmacol.*, 225: 195–201.

Porter, R.H.P. and Greenamyre J.T. (1995) Regional variations in the pharmacology of NMDA receptor channel blockers:

implication for therapeutic potential. *J. Neurohem.*, 64: 614–623.

Priestley, T., Laughton, P., Myers, J., LeBourdellés, B., Kerby, J. and Whiting, P.J. (1995) Pharmacological properties of recombinant human *N*-methyl-D-aspartate receptors comprising NR1a/NR2A and NR1a/NR2B subunit assemblies expressed in permanently transfected mouse fibroblast cells. *Mol. Pharmacol.*, 48: 841–848.

Quarum, M.L., Parker, J.D., Keana, J.F.W. and Wever, E. (1990) (+)-[$^3$H]MK-801 binding sites in postmortem human brain. *J. Neurochem.*, 54: 1163–1168.

Reynolds, I.J. (1994) [$^3$H]CGP 39653 binding to the agonist site of the *N*-methyl-D-aspartate receptor is modulated by $Mg^{2+}$ and polyamines independently of the arcaine-sensitive polyamine site. *J. Neuochem.*, 62: 54–62.

Reynolds, I.J. and Palmer, A.M. (1991) Regional variations in [3H]MK-801 binding to rat brain *N*-methyl-D-aspartate receptors. *J. Neurochem.*, 56: 1731–1740.

Rodríguez-Paz, J., Anatharam, V. and Treistman, S.N. (1995) Block of the *N*-methyl-D-aspartate receptor by phencyclidine-like drugs is influenced by alternative splicing. *Neurosci. Lett.*, 190: 147–150.

Rogawski, M.A. (1993) Therapeutic potential of excitatory amino acid antagonists: Channel blockers and 2,3-benzodiazepines. *Trends Pharmacol. Sci.*, 14: 325–331.

Sakurada, K., Masu, M. and Nakanishi, S. (1993) Alteration of $Ca^{2+}$ permeability and sensitivity to $Mg^{2+}$ and channel blockers by a single amino acid substitution in the *N*-methyl-D-aspartate receptor. *J. Biol. Chem.*, 268: 410–415.

Sakurai, S.Y., Penney, J.B. and Young, A.B. (1993) Regionally distinct *N*-methyl-D-aspartate receptors distinguished by quantitative autoradiography of [$^3$-H]MK-801 binding in rat brain. *J. Neurochem.*, 60: 1344–1353.

Schmieden, V., Kuhse, J. and Betz, H. (1993) Mutation of glycine receptor subunit creates β-alanine receptor responsive to GABA. *Science*, 262: 256–258.

Sekiguchi, M., Okamoto, K. and Sakai, Y. (1990) Glycine-insensitive NMDA receptor expressed in Xenopus oocytes by guinea pig cerebellar mRNA. *J. Neurosci.*, 10: 2148–2155.

Sheng, M., Cummings, J., Roldan, L.A., Jan, Y.N. and Jan, L.Y. (1994) Changing subunit composition of heteromeric NMDA receptors during development of rat cortex. *Nature*, 368: 144–147.

Skifter, D.A., Andaloro, V.J., Brown, J.C., Tse, H.W., Jane, D.E., Watkins, J.C. and Monaghan, D.T. (1997) [$^3$H]Homoquinolinate binds to a subpopulation of NMDA receptors and also to a novel binding site. *Soc. Neuroscience*, (*Abstr.*) 23: 918.

Standaert, D.G., Testa, C.M., Penney, J.B. and Young, A.B. (1993) Alternatively spliced isoforms of the NMDAR1 glutamate receptor subunit: Differential expression in the basal ganglia of the rat. *Neurosci. Lett.*, 152: 161–164.

Subramaniam, S. and McGonigle, P. (1991) Quantitative autoradiographic characterization of (±)-5-methyl-10,11-dihydro-5H-dibenzo[a,d]cyclohepten-5,10-imine ([$^3$-H]MK-801) in rat brain: regional effects of polyamines. *J. Pharmacol. Exp. Ther.*, 256: 811–819.

Subramaniam, S. and McGonigle, P. (1993) Regional heterogeneity of polyamine effects on the *N*-methyl-D-aspartate receptor in rat brain. *J. Neurochem.*, 60: 2276–2284.

Sucher, N.J., Akbarian, S., Chi, C.L., Leclerc, C.L., Awobuluyi, M., Deitcher, D.L., Wu, M.K., Yuan, J.P., Jones, E.G. and Lipton, S.A. (1995) Developmental and regional expression pattern of a novel NMDA receptor-like subunit (NMDAR-L) in the rodent brain. *J. Neurosci.*, 15: 6509–6520.

Sugihara, H., Moriyoshi, K., Ishii, T., Masu, M. and Nakanishi, S. (1992) Structure and properties of seven isoforms of the NMDA receptor generated by alternative splicing. *Biochem. Biophys. Res. Comm.*, 185: 826–832.

Traynelis, S.F., Harley, M. and Heinemann, S.F. (1995) Control of proton sensitivity of the NMDA receptor by RNA splicing and polyamines. *Science*, 268: 873–876.

van Amsterdam, F.T.M., Giberti, A., Mugnaini, M. and Ratti, E. (1992) 3-((±)-2-Carboxypiperazin-4-yl)propyl-1-phosphonic acid recognizes two *N*-methyl-D-aspartate binding sites in rat cerebral cortex membranes. *J. Neurochem.*, 59: 1850–1855.

Vignon, J., Privat, A., Chaudieu, I., Thierry, A., Kamenka, J.M. and Chicheportiche, R. (1986) [$^3$H]Thienyl-phencyclidine ([$^3$H]TCP) binds to two different sites in rat brain: Localization by autoradiographic and biochemical techniques. *Brain Res.*, 378: 133–141.

Wafford, K.A., Bain, C.J., Le Bourdelles, B., Whiting, P.J. and Kemp, J.A. (1993) Preferential co-assembly of recombinant NMDA receptor composed of three different subunits. *Neuroreport*, 4: 1347–1349.

Watanabe, M., Inoue, Y., Sakimura, K. and Mishina, M. (1992) Developmental changes in distribution of NMDA receptor channel subunit mRNAs. *NeuroReport*, 3: 1138–1140.

Watanabe, M., Inoue, Y., Sakimura, K. and Mishina, M. (1993) Distinct distributions of five *N*-methyl-D-aspartate receptor channel subunit mRNAs in the forebrain. *J. Comp. Neurol.*, 338: 377–390.

Wenzel, A., Scheurer, L., Kunzi, R., Fritshy, J.M., Mohler, H. and Benke, D. (1995) Distribution of NMDA receptor subunit proteins NR2A, 2B, 2C and 2D in rat brain. *NeuroReport*, 7: 45–48.

Wenzel, A., Villa, M., Mohler, H. and Benke, D. (1996) Developmental and regional expression of NMDA receptor subtypes containing the NR2D subunit in rat brain. *J. Neurochem.*, 66: 1240–1248.

Williams, K. (1995) Pharmacological properties of recombinant *N*-methyl-D-aspartate (NMDA) receptors containing the ε-4 (NR2D) subunit. *Neurosci. Lett.*, 184: 181–184.

Williams, K. (1996) Separating dual effects of zinc at recombinant *N*-methyl-D-aspartate receptors. *Neurosci. Lett.*, 215: 9–12.

Williams, K., Russell, S.L., Shen, Y.M. and Molinoff, P.B. (1993) Developmental switch in the expression of NMDA receptors occurs in vivo and in vitro. *Neuron*, 10: 267–278.

Williams, K., Zappia, A.M., Pritchett, D.B., Shen, Y.M. and Molinoff, P. (1994) Sensitivity of the *N*-methyl-D-aspartate receptor to polyamines is controlled by NR2 subunits. *Mol. Pharmacol.*, 45: 803–809.

Yamakura, T., Mori, H., Masaki, H., Shimoji, K. and Mishina, M. (1993) Different sensitivities of NMDA receptor subtypes to non-competitive antagonists. *Neuroreport*, 4: 687–690.

Yamazaki, M., Mori, H., Araki, K., Kazuhiro, J.M. and Mishina, M. (1992) Cloning, expression and modulation of a mouse NMDA receptor subunit. *FEBS*, 300: 39–45.

Yoneda, Y. and Ogita, K. (1991) Heterogeneity of the *N*-methyl D-aspartate receptor ionophore complex in rat brain, as revealed by ligand binding techniques. *J. Pharmacol. Exp. Ther.*, 259: 86–96.

Zhang, L., Zheng, X., Paupard, M.C., Wang, A.P., Santchi, L., Friedman, L.K., Zukin, R.S. and Bennett, M.V.L. (1994) Spermine potentiation of recombinant *N*-methyl-D-aspartate receptors is affected by subunit composition. *Proc. Natl. Acad. Sci., USA*, 91: 10883–10887.

O.P. Ottersen, I.A. Langmoen and L. Gjerstad (Eds.)
*Progress in Brain Research*, Vol 116


CHAPTER 13

# Modulation of NMDA receptors

J.F. MacDonald*, X-G. Xiong, W-Y. Lu, R. Raouf and B.A. Orser

*Departments of Physiology and Pharmacology, University of Toronto, 1 King's College Circle, Toronto, Ontario, Canada M5S 1A8*
*Department of Anesthesia, Sunnybrook Health Sciences Center Faculty of Medicine, Toronto, Ontario, Canada M4N 3M5*

Excitatory synapses likely form one of the most important loci for the integration of information in the central nervous system. The degree of synaptic efficacy at one of these synapses depends strongly upon the history of activity at this synapse and in some cases the activity at nearby synapses. For example, relatively high frequency afferent stimulation can evoke a long lasting increase in synaptic efficacy (long-term potentiation, LTP) whilst low frequency stimulation may induce a decrease in transmission (long-term depression, LTD) or cause depotentiation of LTP. In the mammalian central nervous system, L-glutamate is by far the most predominant excitatory transmitter. Therefore, studies of synaptic plasticity require a comprehensive understanding of how L-glutamate is synthesized, stored and released as well as the kinetics of the interactions of L-glutamate with its postsynaptic receptors and uptake transporters (Diamond and Jahr, 1997). The mechanisms responsible for the changes in synaptic efficacy which accompany LTP or LTD likely involve one or more significant alterations in the presynaptic release of L-glutamate or in the properties of postsynaptic receptors.

Postsynaptic ionotropic glutamate receptors fall into three categories: (a) *N*-methyl-D-aspartate (NMDA) receptors (b) a-amino-3-hydroxy-5-methyl-4-isoxazole propionate (AMPA) receptors and (c) kainate receptors (Clements et al., 1992b; Gasic and Hollmann, 1992b; Gasic and Hollmann, 1992a; Seeburg, 1993). Most excitatory postsynaptic currents (epsc's) are generated through activation of both AMPA and NMDA receptors (McBain and Mayer, 1994). The majority of excitatory synapses demonstrate a dominant AMPA receptor-mediated component followed by a much smaller and slower NMDA receptor mediated phase. However, two groups have also recently identified a high affinity kainate receptor-mediated component evoked during strong stimulation of the Mossy Fibre input to CA3 neurons of the hippocampus (Castillo et al., 1997; Vignes and Collingridge, 1997).

The relatively rapid gating kinetics of AMPA channels compared to NMDA (Clements et al., 1992a; Lester et al., 1990), along with the $Mg^{2+}$ block of NMDA channels (Ascher and Nowak, 1988; Nowak et al., 1984; Mayer et al., 1984) means that the largest component of unitary epsc's is mediated by AMPA receptors. NMDA receptors play a much more important role when a more substantial depolarization, due to the summation of postsynaptic potentials, causes the relief of the $Mg^{2+}$ block of these channels (Bliss and Collingridge, 1993). Activation of NMDA receptors can at some cortical and hippocampal synapses induce either LTP or LTD of excitatory transmission (Collingridge and Singer, 1990; Artola and Singer,

* Corresponding author.
Fax: +1 416 978 4940; e-mail: j.macdonald@utoronto.ca.

1993; Collingridge and Bliss, 1995; Bliss and Collingridge, 1993; Oliet et al., 1997). The frequency of the stimulation appears to determine whether or not LTP or LTD is induced and it is believed that the entry of $Ca^{2+}$ through NMDA receptors provides a common signal that induces both of these changes (Collingridge and Bliss, 1995; Bliss and Collingridge, 1993). Therefore, at many synapses NMDA receptors play a key role in switching the level of synaptic efficacy in response to changes in synaptic experience and it is clearly important to understand how NMDA receptors are themselves regulated and/or modulated.

Although the predominant view has been that individual synapses have a complement of both AMPA and NMDA receptors recent evidence suggests that some unitary excitatory postsynaptic currents (uepsc's) apparently arise entirely through the activation of NMDA receptors (Isaac et al., 1995; Liao et al., 1995). Such synaptic activity was previously undetected because at membrane potentials near rest these unitary epsc's were entirely blocked by endogenous $Mg^{2+}$. They were only detected when the cells were depolarized sufficiently to overcome the voltage-dependent blockade by $Mg^{2+}$. The AMPA component of these *Silent Synapses* may be recruited following LTP by a mechanism that involves induction and/or the enhancement of AMPA receptor function (Barria et al., 1997). The influx of $Ca^{2+}$ through NMDA receptors may provide the stimulus required to activate a CaM/CamKII-dependent phosphorylation of AMPA receptors. A persistent phosphorylation of AMPA receptors then either potentiates pre-existing AMPA receptor function or facilitates insertion of new functional receptors into the postsynaptic membrane. If this scenario proves to be true then it emphasizes the potential importance of NMDA receptor regulation in the control of synaptic efficacy.

The objective of the present review is to describe how phosphorylation reactions might regulate the behaviour of the NMDA channels in central neurons. The ubiquity of protein phosphorylation and dephosphorylation necessitates a recognition that whilst NMDA receptors may be directly phosphorylated, there are likely a variety of other proteins that bind to and interact with the NMDA receptor which are themselves also regulated by phosphorylation. Thus, it is important to both understand the functional consequences of direct phosphorylation of NMDA receptor subunits and to identify how phosphorylation of sites on the NMDA receptor influence its binding to support proteins. These support or anchoring proteins are most likely components of the postsynaptic density (PSD) and cytoskeleton.

The effects of serine-threonine protein kinases such as protein kinase C (PKC), $Ca^{2+}$/calmodulin-dependent kinase II (CamKII), cAMP-dependent kinase (PKA) or tyrosine kinases such as Src-tyrosine kinase, on NMDA channels are considered in this review. These enzymes catalyze the transfer of high energy phosphate groups to the serines or threonines residues of substrate proteins or in the case of tyrosine kinases, the transfer to tyrosyl residues. Phosphorylation of a channel protein can alter its physical conformation leading to a variety of potential changes in channel behaviour. However, phosphorylation of anchoring and cytoskeletal proteins can also act indirectly to alter the activity of ion channels (Cantiello, 1995; Prat et al., 1993c; Prat et al., 1993a; Rosenmund et al., 1994). Whether or not a protein is phosphorylated at any given time also depends upon the activity of protein phosphatases including serine-threonine and tyrosine phosphatases. These enzymes catalyze the removal of high energy phosphate groups from the appropriate amino acid residues of substrate proteins. Serine-threonine phosphatases include both $Ca^{2+}$-independent phosphatases such as protein phosphatase 1 and 2A (PP1/2A) as well as the $Ca^{2+}$-dependent phosphatase, protein phosphatase 2B (PP2B, also known as calcineurin). In contrast, tyrosine phosphatases are a family of phosphatases which dephosphorylate, with great specificity, tyrosyl residues.

Each NMDA channel consists of four or five subunits (Sheng et al., 1994; Blahos and Wenthold, 1996) and each subunit is believed to pass through the membrane three times (i.e. three transmembrane domains (Hollmann et al., 1994; Wo and

Oswald, 1995)). The extensive N-terminus (amino terminus) as well as the extracellular loop play a role in the binding of agonists and co-agonists whilst the C-terminal (carboxy-terminal) tail which is directed to the cell's interior plays an important role in interacting with PSD proteins (see below). NMDA receptors likely consist of at least one subunit from the NR1 group as well as one or more subunits from the NR2 subgroup (NR2A to D)(McBain and Mayer, 1994).

*Properties of NR1 subunits*

Three of the twenty-two exons of the NR1 subunit (exons 5, 21, 22) are subject to alternative splicing of their mRNA potentially giving eight forms of the NR1 subunit (NR1A-G or $NR1_{000\text{-}111}$) (Durand et al., 1992; Hollmann et al., 1993; Sugihara et al., 1992). Exon 5 encodes a 21 amino acid cassette (N1) located within the N-terminus whilst exons 21 and 22 encode two cassettes which are inserted into the C-terminus (C1 and C2, respectively). The presence or absence of these cassettes influences some of the functional properties of homomeric channels and the mRNA for a number of these alternatively spliced forms are differentially distributed within the central nervous system (Zukin and Bennett, 1995). Nevertheless, the functional significance of this diversity has not yet been clarified.

*Properties of NR2 subunits*

Native NMDA receptors likely contain at least one NR1 subunit and perhaps no more than a single type of NR2 subunit (Blahos and Wenthold, 1996). The most ubiquitous NR2 subunits are the NR2A and B whilst the NR2C predominates in the cerebellum (Monyer et al., 1992; Kutsuwada et al., 1992). Channels containing the NR2D subunit are found less frequently in the central nervous system and recombinant channels containing this subunit demonstrate relatively slow kinetics of activation and de-activation (McBain and Mayer, 1994). Native NMDA channels are non-selective cation channels (permeable to $Na^+$, $K^+$ and $Ca^{2+}$) and about 11% of the total current flowing through the open channels is carried by $Ca^{2+}$ (Burnashev et al., 1995). This influx of $Ca^{2+}$ plays an important signal transduction role in the regulation of the NMDA receptor itself as well as in the induction of synaptic plasticity. The NMDA receptor is unique in that it requires glycine (or serine) as a co-agonist and the NMDA channel likely cannot gate to an open state in the absence of glycine (McBain and Mayer, 1994).

Both subsynaptic NMDA and AMPA receptors are physically bound to the postsynaptic density (PSD)(Dong et al., 1997; Kornau et al., 1995; Kim et al., 1996; Niethammer et al., 1996; Sheng, 1996; Sheng and Kim, 1996; Hsueh and Shens, this volume). Several proteins of the PSD serve to anchor or cluster glutamate receptors in the membrane at the subsynaptic region (Kornau et al., 1995; Lau et al., 1996; Muller et al., 1996; Kim et al., 1996; Niethammer et al., 1996; Sheng, 1997; Sheng, 1996). In addition, the PSD serves to co-localize many of the enzymes that directly or indirectly respond to the initial $Ca^{2+}$ signal associated with NMDA receptor activation, including PKC (Suzuki et al., 1993), CAMkII (Kennedy, 1993), adenylyl cyclase (Mons and Cooper, 1995), PKA (Carr et al., 1992), PP1 (Dosemeci and Reese, 1993) and PP2B (Coghlan et al., 1995). Thus, NMDA and AMPA receptors are themselves ideally situated to be regulated by these enzymes systems (Raymond et al., 1993). Furthermore enzymes, such as PKC, CamKII and tyrosine kinases are known to play important roles in both LTP and LTD.

*Sites of phosphorylation of NMDA receptors*

Most of the potential sites of phosphorylation by PKC are located in the large C-terminus of the NR1 subunit and it is this regions which is subject to extensive alternative splicing. Phosphorylation of the C-terminus provided the first strong indication that the C-terminus formed an intracellular domain of the receptor (Tingley et al., 1993) The C1 (amino acids 864-890) and C2 domains contain four serines (ser889,890,896,897) each of which can be phosphorylated by exposure to a phorbol ester and activation of an endogenous PKC (Tingley et al., 1997). Mutation of each of the serines to alanines dramatically reduces phosphorylation of the subunit suggesting that most of the PKC-dependent phosphorylation occurs at the C-terminus. The C1 domain is also responsible for physically clustering or aggregating NR1 subunits when they are expressed in QT6 cells (Ehlers et al., 1995). Using tryptic phosphopeptide maps and site specific phosphoantibodies, Tingley et al. (Tingley et al., 1997) demonstrated that both PKC and PKA (stimulated by the application of forskolin) phosphorylate the C-terminus of the NR1 subunit at specific sites. For example, PKC-induced phosphorylation is specific for threonine 879 and serines 890 and 896 whereas PKA is specific for serine 897. Interestingly, although serine 896 was well phosphorylated by PKC alone, the degree of phosphorylation of this site was inhibited by the PKA-induced phosphorylation of serine 897 and vice versa suggesting a potential negative functional interaction between PKA and PKC phosphorylation. This specific phosphorylation of C-terminal sites was also found in homogenates from hippocampal slices suggesting that the NR1 subunit is subject to both in vitro and in vivo phosphorylation by PKC and PKA. Whilst this suggests that phosphorylation of the NR1 subunit is of functional significance, several of the NR2 subunits are also phosphorylated by PKC and PKA (Leonard and Hell, 1997). Both NR2A and NR2B are phosphorylated in vitro by PKA and PKC although two dimensional tryptic phosphopeptide maps suggests that each kinase potentially phosphorylates some non-overlapping sites. NR1, NR2A and NR2B receptors, located in wide spread regions of the central nervous system including the neocortex and hippocampus, are phosphorylated in vivo by these kinases whereas little phosphorylation of NR1 and NR2C was found in the cerebellum where the NR2C subunit appears to substitute for the NR2A,B subunits (Leonard and Hell, 1977). The NR2B subunit is a high affinity substrate for CamKII-induced phosphorylation ($K_m$ = 50 nM) at ser1303 (Kennedy, 1997; Omkumar et al., 1996) and the kinase binds the subunit itself (Kennedy, 1997; Kennedy, 1993). Additional evidence suggests that both NR2A and NR2B subunits are phosphorylated in the hippocampus by CamKII (Lickteig et al., 1995).

Using tyrosine specific phosphopeptide antibodies, Tingley et al. (Tingley et al., 1997) found that NR2A and NR2B but not NR1 subunits are phosphorylated in vivo by tyrosine kinases (see also (Hall and Soderling, 1997)). Estimated endogenous levels of phosphorylation of NR2A and NR2B were relatively low (2.1 and 3.6% respectively) suggesting the potential for considerably greater phosphorylation during stimulation of the kinase (see also (Hall and Soderling, 1997)). Endogenous tyrosine kinase activity found in the homogenates was in fact capable of increasing the phosphorylation of NR2A (but not NR1 or NR2B) by 6 to 8 fold (Tingley et al., 1993; Lau and Huganir, 1995). A PSD protein designated gp180 which can be isolated with the PSD has also been characterized and identified as the NR2B subunit (Kennedy, 1997). The NR2B subunit is the most highly phosphorylated component protein of the PSD (Moon et al., 1994).

In cultured hippocampal neurons, only about half of the cell's complement of NR1 subunits is apparently expressed at the surface whilst almost 90% of the NR2B subunit is expressed on the surface (Hall and Soderling, 1997). This suggests that membrane trafficking of the NR1 subunit is much more dynamic than for the NR2 subunits. NR2 subunits are highly phosphorylated on serines (but not threonines or tyrosines) but there is little basal phosphorylation of NR1 subunits. Applications of glutamate or phorbol esters enhanced phosphorylation of NR1 subunits three to five fold whilst phosphorylation of NR2 subunits was enhanced by less than 2 fold (Hall and Soderling, 1997).

In summary, the native NMDA receptor is directly phosphorylated by a variety of protein kinases. NR1 and NR2 subunits of the receptor are potentially phosphorylated by the same, as well as by different (i.e. serine/threonine versus tyrosine) kinases. Each subunit may be phosphorylated at multiple sites by a given protein kinase and in at least one example, the phosphorylation of one site may modify the phosphorylation of an adjacent site. The degree of basal phosphorylation of the native receptor protein may differ substantially from site to site and is presumably determined by balance between the activity of kinases and phosphatases localized near or bound to the receptor. Thus, phosphorylation of the NMDA receptor may also differ from neuron to neuron. Despite the complexity of this mechanism of regulation of the NMDA receptor the following primary question arises: What are the consequences of phosphorylation to the structure and function of the NMDA receptor?

*Functional studies of the potential phosphorylation of NMDA receptors*

MacDonald et al. (MacDonald et al., 1989) originally suggested that phosphorylation reactions had to be maintained within cultured hippocampal neurons in order to sustain the full activity of NMDA channels. This hypothesis was based upon the observations that during whole cell recordings, responses to brief applications of agonist decline rapidly (time constant of 155 seconds) and irreversibly to about one half of their original amplitude ("run down") (MacDonald et al., 1989). This irreversible "run down" was not prevented the by inclusion of high concentrations of the high affinity $Ca^{2+}$ buffer, BAPTA, within the recording pipette. This suggested that a lack of ATP in the recording pipette depleted endogenous ATP supplies in the cells sufficiently to compromise ongoing phosphorylation of either the NMDA receptor subunits themselves or phosphorylation of related support proteins which were responsible for modulating responses to NMDA. Evidence which supported this hypothesis included the observation that a source of hydrolysable ATP was required in the patch pipette to prevent the "run down" of NMDA-evoked currents (MacDonald et al., 1989). The identity of the endogenous kinases which were responsible for regulating NMDA channels, and the mechanisms by which they modify NMDA channel activity, was not determined.

However, several subsequent studies, using more prolonged applications of agonist concluded, that the "run down" of NMDA-induced currents resulted from an intracellular accumulation of $Ca^{2+}$ which entered through these channels (Rosenmund and Westbrook, 1993b; Vyklicky, Jr., 1993). The gradual elevation in the concentration of intracellular $Ca^{2+}$ was postulated to lead to a depolymerization of the cytoskeleton which in turn reduced the amplitude of these currents by some unspecified mechanism. In contrast to previous reports, this "run down" spontaneously reversed provided the applications of NMDA were suspended during the interim. Moreover, the presence of intracellular BAPTA prevented this "run down" (Rosenmund and Westbrook, 1993b). Both forms of "run down" were reduced by the presence of ATP in the patch pipette, however it was argued that high concentrations of ATP might also serve to directly stabilize the cytoskeleton and the $Ca^{2+}$-dependent form of "run down" did not appear to be dependent upon the hydrolysis of ATP. In addition, the application of a phosphatase inhibitor failed to retard this form of "run down" (Rosenmund and Westbrook, 1993b). Another possible complication was that $Ca^{2+}$ directly depressed the activity of the NMDA channel ($Ca^{2+}$-dependent inactivation, see below). It should be noted that the prolonged applications of agonist would favor the loading of the cell with $Ca^{2+}$. However, Medina et al. (Medina et al., 1996) reported that the irreversible "run down" of NMDA-evoked currents was independent of transient elevations of intracellular $Ca^{2+}$ ($Ca^{2+}$ inactivation, see below) but it was reduced by several phosphatase inhibitors (FK506 and calyculin A, see below).

In a recent study Wang et al. (Wang et al., 1993) examined brief agonist responses whilst simultaneously recording intracellular concentrations of $Ca^{2+}$ in the same cultured spinal cord (dorsal horn) neuron. These authors confirmed that at least one form of "run down" of NMDA-evoked currents was depended upon the hydrolysis of ATP and was entirely unrelated to a progressive elevation of intracellular $Ca^{2+}$. This form of "run down" was observed in the absence of extracellular $Ca^{2+}$ or when high concentrations of BAPTA were included in the patch pipette. The "run down" of NMDA-evoked currents could be prevented by including ATP-$\gamma$-S, an ATP analogue which causes thiophosphorylation of proteins, in the patch pipette, but not by a non-hydrolyzable analogue. Furthermore, applications of the PTK inhibitors genistein or Peptide A, or the inclusion in the patch pipette of a constitutively active form of tyrosine phosphatase mimicked the "run down" in spite of the presence of an excess amount of ATP (Wang et al., 1996). This evidence strongly supports the conclusion that "run down" of NMDA-evoked currents in cultured dorsal horn neurons results from a progressive dephosphorylation of NMDA receptors by tyrosine phosphatases and that it was unrelated to an elevation of intracellular $Ca^{2+}$.

In summary, there are multiple forms of "run down" of NMDA-evoked currents. Some are dependent upon phosphorylation reactions and quite likely phosphorylation of the receptor itself, whilst others are undoubtedly related to the influx of $Ca^{2+}$ which enters via the channels themselves and/or involve interactions of the NMDA channels with cytoskeletal proteins. Altogether this evidence suggests that the function of NMDA channels is influenced both by phosphorylation and by the integrity of the cytoskeletal proteins.

### *Functional regulation of NMDA receptors by PKC*

The influx of $Ca^{2+}$ through NMDA channels, or the activation of a variety of other receptors (i.e. metabotropic glutamate receptors) that stimulate phospholipase C (PLC), can activate PKC. Most, but not all major neuronal isoforms of PKC are activated by DAG and $Ca^{2+}$ (e.g., PKC$\gamma$) (Nakamura et al., 1993; Tanaka and Nishizuka, 1994). PKC is primarily cytosolic in location but upon stimulation it translocates to the cellular and nuclear membranes where it is bound by PKC anchoring proteins or RACKs (receptor for activated C-kinase). Partial proteolysis (e.g., by calpain) may be required to expose the catalytic domain and activate this kinase. Translocated, membrane bound PKC might then be able to phosphorylate glutamate receptors or other proteins located within the PSD (Suzuki et al., 1993) (see also below). This stimulus may be transitory because PKC is thought to be down-regulated, via the activity of $Ca^{2}$ -dependent proteases, although some isozymes may be persistently activated (Tanaka and Nishizuka, 1994).

One subgroup of metabotropic receptors (mGluR1,5) stimulate PLC, and activate PKC while triggering the release of intracellular $Ca^{2+}$ via inositol triphosphate (Schoepp and Conn, 1993). In this respect, the non-specific agonist, 1S, 3R-1-amino cyclopentane-1,3 dicarboxylic acid (ACPD), as well as more selective agonists have been reported to enhance NMDA responses in the Xenopus oocyte expression system (Kelso et al., 1992) and in hippocampal slice neurons (Aniksztejn et al., 1991; Aniksztejn et al., 1992; Ben-Ari et al., 1992) likely via activation of PKC. Activation of the $\mu$ opioid receptor also enhances NMDA-evoked currents in isolated trigeminal neurons via PKC (Chen and Huang, 1991). In contrast, some mGluRs are coupled negatively to adenylate cyclase activity (Pin and Duvoisin, 1995; Joly et al., 1995) and they might also influence NMDA receptors through the modulation of PKA (see below). Another possible scenario involves the activation of the non-receptor protein tyrosine kinase, PYK2. This proline-rich kinase is found in high concentrations in CNS neurons and is selectively activated by either $Ca^{2+}$ or PKC. In turn phosphorylated PYK2 can directly phosphorylate and modulate the function of at least one neuronal $K^{+}$ channel (Lev et al., 1995). This, suggests that PKC could act via phosphorylation of PYK2 to phosphorylate NMDA channels.

Considerable evidence supports the hypothesis that stimulation of PKC results in an enhancement of the activity of NMDA receptors. For example, phorbol esters which activate PKC, dramatically enhance the NMDA-evoked currents expressed in *Xenopus* oocytes (Kelso et al., 1992; Urushihara et al., 1992; Yamazaki et al., 1992; Meguro et al., 1992; Yamakura et al., 1993; Mishina et al., 1993; Durand et al., 1993) and have been reported to enhance NMDA-evoked responses in hippocampal slice (Aniksztejn et al., 1992) and spinal cord neurons. The enhancement in oocytes ranges from several to almost forty fold depending upon the combination of receptor subunits expressed. However, some combinations of heteromeric subunits (i.e. NR1/NR2C) are not enhanced by phorbol esters whilst the presence of the N1 cassette in the NR1 subunit together, with an absence of C1 and C2 greatly enhances the potentiation of homomeric channels (Zukin and Bennett, 1995). Furthermore, mutating the relevant serines in the C-terminus to alanines does not eliminate the phorbol ester-induced enhancement suggesting that phosphorylation of these residues plays no role (Yamakura et al., 1993). This is particularly surprising since the N-terminus is clearly extracellular whilst the major sites of PKC-induced phosphorylation are found in the C1 and C2 cassettes. This evidence suggests that either the phosphorylation by PKC is of residues at alternative locations within the receptor protein or that the effects of phorbol esters are mediated by an associated regulatory protein which is itself phosphorylated by PKC. Removal of potential sites of phosphorylation in the intracellular loop between transmembrane domains TM3 and TM4 in NR1C and NR2A also failed to eliminated the phorbol ester-induced potentiation of NMDA-evoked currents (Sigel et al., 1994).

A recent study by Zheng et al. (Zheng, 1997) has reported that the phorbol ester-induced potentiation depends in part upon the entry of extracellular $Ca^{2+}$ through NMDA channels. For example, a component of the potentiation was blocked by BAPTA or by the substitution of extracellular $Ca^{2+}$ with barium. Furthermore, recombinant receptors lacking a high $Ca^{2+}$ permeability failed to demonstrate this phenomenon. Interestingly, receptors lacking the C1 and C2 cassettes demonstrated a greater degree of phorbol-induced potentiation but this effect was $Ca^{2+}$-independent. These observations may be consistent with a recent finding that phosphorylation of C-terminus by PKC inhibits the binding of $Ca^{2+}$/calmodulin (Hisatsune et al., 1997) (see below). On the other hand, the effects of phorbol esters maybe indirect. For example, evidence suggests that phorbol esters may stimulate the phosphorylation of cytoskeletal proteins that indirectly effect the activity of NMDA channels expressed in oocytes (Wagner and Leonard, 1994). In this regard, phosphorylation of sites within the C-terminus of NR1A,D subunits can disrupt their aggregation at the membrane surface (Ehlers et al., 1995).

The enhancement of NMDA-evoked currents recorded from trigeminal neurons following the intracellular application of PKC has been attributed to a reduction in the affinity of the blocking site within the NMDA channel for $Mg^{2+}$ (Chen and Huang, 1992). However, PKC enhanced the currents by about 50% as opposed to the large potentiations observed for recombinant receptors in oocytes. Also, more recent evidence (Wagner and Leonard, 1996) has demonstrated that a reduction in the $Mg^{2+}$ blockade cannot account for most of the phorbol-ester induced enhancement in oocytes. Furthermore, the intracellular perfusion of cultured hippocampal neurons with the catalytic fragment of PKC (PKCM) potentiated NMDA-evoked currents without changing the affinity of the channels for $Mg^{2+}$ (Raouf et al. 1997). However, the apparent affinity of NMDA receptors in cultured neurons for $Mg^{2+}$ is substantially less than that found in isolated CA1 pyramidal neurons (Raouf et al., 1997). It seems unlikely that this difference can be accounted for simply on the basis of differential expression of $Mg^{2+}$-sensitive subunits particularly as cultured hippocampal neurons and adult hippocampal neurons primarily express NR2A and 2B subunits (Tingley et al., 1997). In this regard, Zhang et al., (Zhang et al., 1996) have recently demonstrated that physical stretching of cultured hippocampal neurons can chronically reduce their sensitivity to $Mg^{2+}$, an effect which can be to some extent prevented by treating the cells with the PKC inhibitor calphostin C. This suggests that activation of PKC during injury may lead to enhanced NMDA currents as a consequence of a reduced blockade of the channels by extracellular $Mg^{2+}$. It also raises the possibility that cultured hippocampal neurons have an abnormally high level of basal phosphorylation of NMDA receptors.

Unexpectedly, phorbol esters have been reported to depress rather than enhance NMDA responses in cultured and hippocampal slice neurones (Bartlett et al., 1989; Markram and Segal, 1992) and in cultured cerebellar granule neurons (Courtney and Nicholls, 1992). The concentrations of phorbol ester used in the slice experiments was much higher than those required to stimulate PKC in vitro and its effect might be attributable to a direct block of the channels (Hockberger et al., 1989). However, a similar depression was observed in cultured neurons using low and relevant concentrations (Bartlett et al., 1989). The depression was not mimicked by an inactive phorbol ester and was blocked by H7 and OAG suggesting that this effect is mediated via stimulation of the kinase. Paradoxically, the intracellular perfusion of PKCM enhanced NMDA currents in the same cultured hippocampal neurones (Wang et al., 1994a). One possible explanation for this paradox is, that activation of PKC-induced phosphorylation can indirectly down regulate NMDA channel activity.

Activation of the tyrosine kinase receptor for platelet-derived growth factor (PDGF) also causes a long-lasting decrease in NMDA-evoked currents recorded in cultured and acutely-isolated hippocampal neurons (Valenzuela et al., 1996). This inhibition can be replicated by co-expressing the PDGF$\beta$ receptor and NMDA receptors in Xenopus oocytes. This inhibition depends upon activation of PLC and requires an initial rise in the concentration of intracellular $Ca^{2+}$. However, the inhibition once established is no longer $Ca^{2+}$-dependent. Surprisingly, the effects of PDGF were entirely blocked by pretreatment with the phosphatase inhibitor calyculin A which effects the $Ca^{2+}$-independent phosphatases PP1 and PP2A whilst inhibitors of the $Ca^{2+}$-dependent phosphatase calcineurin were ineffective (Valenzuela et al., 1996). These results suggest that activation of a $Ca^{2+}$-independent phosphatase results in a depression of NMDA channel activity (see also phosphatases below).

The induction of kindling in rats also chronically modifies the behaviour of NMDA channels recorded from adult dentate gyrus granule cells of the hippocampus (Köhr et al., 1993). Notably there is an increased sensitivity of the NMDA-evoked currents to the application of intracellular ATP and there was a reduction in the sensitivity of the channels to the block by $Mg^{2+}$. Whilst single channel conductance and desensitization of the receptor were not affected there was a substantial increase in the mean open time, burst and cluster duration of channels recorded using the cell attached patch configuration (Köhr et al., 1993).

## The integrity of the cytoskeleton and the activity of NMDA channels

The polymerization state of actin filaments can regulate the activity of various ion channels (Bencherif and Lukas, 1993; Suzuki et al., 1993; Prat et al., 1993a; Prat et al., 1993b; Johnson and Byerly, 1993; Furukawa and Mattson, 1995; Levitan et al., 1995; Reinhart and Levitan, 1995) and potentially the NMDA channel as well. The structural integrity of the cytoskeleton is in turn regulated by $Ca^{2+}$, nucleotides (Janmey, 1994) and by the phosphorylation state of various cytoskeletal proteins (Furman et al., 1993). In epithelial cells, it has been reported that $Na^{+}$ channel activity in inside-out patches is strongly dependent on the presence of PKA-phosphorylated short actin filaments (Prat et al., 1993a; Prat et al., 1993b; Cantiello et al., 1991). Neither fully complexed long filaments, nor monomeric G-actin, were capable of supporting channel activity. It has also been reported that phalloidin, an agent which stabilizes actin filaments, prevented the "run down" of NMDA currents in cultured neurons, whilst cytochalasin D which disrupts filaments, enhanced the "run down" of these currents (Rosenmund and Westbrook, 1993a; Rosenmund and Westbrook, 1993b). The phorbol ester-induced enhancement of NMDA currents recorded in oocytes is also inhibited by injections of phalloidin (Wagner and Leonard, 1994).

The degree of "run down" of NMDA-evoked currents in cultured cortical neurons also influences the sensitivity of NMDA-induced currents to membrane stretch (Paoletti and Ascher, 1994). Initially NMDA-induced currents are insensitive to changes in the osmotic strength of the extracellular solution. However, following "run down" they become highly sensitive to such changes (Paoletti and Ascher, 1994). In outside-out patches where the physical structures anchoring the receptors to the cytoskeleton are substantially disrupted the activity and presumed open probability of NMDA channels become highly stretch sensitive (Paoletti and Ascher, 1994). This evidence suggests that NMDA receptors are held rigidly in place in the membrane by virtue of their binding to cytoskeletal components and thus their activity is protected from modulation by physical deformations in the membrane (stretch).

*Description of NMDA receptor desensitization*

AMPA and kainate channels characteristically undergo a rapid conformational conversion to a *desensitized* or null conductance state despite the continued presence of agonist. The physiological function of desensitization of these channels remains unclear. In the case of the NMDA receptor, desensitization is more complex and likely represents interactions between several different mechanisms. Three major forms of desensitization of NMDA-evoked currents have been observed: (1) a glycine-dependent desensitization (2) a glycine-independent desensitization (3) a $Ca^{2+}$-dependent desensitization or more correctly a $Ca^{2+}$-dependent inactivation. The first two forms of desensitization are related to effects of the co-agonist glycine on NMDA responses. Glycine enhances NMDA-evoked currents and increases the rate of recovery from desensitization (McBain and Mayer, 1994). For example, in the presence of saturating concentrations of glycine, and provided intracellular $Ca^{2+}$ is strongly buffered, NMDA-evoked currents demonstrate little or no decrease in amplitude in the continued presence of agonist (they do not desensitize). Thus, glycine appears to be able to prevent desensitization. However, when NMDA channels are studied in outside-out or nucleated patches, or under other experimental conditions that disrupt the cytosol such as long term whole cell recordings, NMDA-activated currents rapidly desensitize with time constants that can approach those observed with AMPA receptors (McBain and Mayer, 1994). Glycine is still required as a co-agonist but desensitization can no longer be prevented. This suggests that capacity of glycine to facilitate the recovery from desensitization is lost. An underlying glycine-insensitive desensitization is then revealed.

The third form of desensitization is analogous to the inactivation of voltage-dependent $Ca^{2+}$ channels by intracellular $Ca^{2+}$. Whole cell NMDA-evoked currents recorded from cultured neurons or from HEX 293 cells expressing recombinant NMDA receptors (in saturating concentrations of glycine) demonstrate a $Ca^{2+}$-dependent decrease in amplitude during the continuous application of agonist (Medina et al., 1996; Medina et al., 1995; Krupp et al., 1996; Legendre et al., 1993; Rosenmund et al., 1995; Rosenmund and Westbrook, 1993a; Westbrook et al., 1997; Krupp et al., 1996). However, if extracellular $Ca^{2+}$ concentrations are lowered or if intracellular $Ca^{2+}$ is strongly buffered then this form of desensitization is reduced or eliminated. Such evidence suggests that the entry of $Ca^{2+}$ through NMDA channels causes inactivation of the channels. In contrast, the direct application of $Ca^{2+}$ to the cytosolic face of inside-out patches has variable effects on NMDA channel activity although most laboratories have noted that NMDA channel activity cannot be maintained unless $Ca^{2+}$ concentrations are buffered to low values. This inhibition by $Ca^{2+}$ appears to depend upon the duration of the patch recording (Medina et al., 1996).

Recently Ehlers et al. (Ehlers et al., 1996) demonstrated that calmodulin binds directly to two sites on the NR1 subunit; one, a high affinity site in the C1 cassette and the other a low affinity site in a region just proximal to this cassette (designated C0) that is common to all of the NR1 subunits. Furthermore, in the presence of $Ca^{2+}$ the binding of calmodulin reduces NMDA single channel activity in inside-out patches taken from HEK cells providing a possible mechanism to account for the inactivation of NMDA channels following the influx of this divalent cation. More recently, the phorbol-ester-induced phosphorylation of the serines in C1 has been reported to reduce the affinity of the NR1 subunit for $Ca^{2+}$/calmodulin (Hisatsune et al., 1997). Mutation of the serines to alanines blocked the phorbol ester-induced phosphorylation of the NR1 subunit. Furthermore, activation of co-expressed mGluR1$\alpha$ enhanced phosphorylation of these serines. These results suggest that PKC may potentiate NMDA-evoked currents simply by reducing $Ca^{2+}$/calmodulin-dependent inactivation of the channels (Ehlers et al., 1996; Hisatsune et al., 1997).

A number of recent studies has shown that recombinant NR1 and NR2 subunits bind to members of the synapse associate protein (SAPs) family of PSD proteins. For example, PSD-95 (SAP90) and SAP120, possesses a PDZ domain (Kornau et al., 1995) which directly binds a sequence of amino acids (tSXV) found at the very end of the C-terminus of the NR2 subunits (primarily NR2B). This type of binding may provide a general mechanism for anchoring this channel to the PSD. Furthermore, Wyszynski et al. (Wyszynski et al., 1997; Hsueh and Sheng, this volume) have demonstrated that $\alpha$-Actinin-2, a member of the spectrin/dystrophin family of actin-binding proteins, also binds to the C-terminal tails of both NR1 and NR2B subunits. This protein is co-localized with NMDA receptors and PSD-95 at central synapses and potentially provides a substrate to physically link the NMDA receptor to the actin cytoskeleton. Of great interest was the observation that the binding of this protein is directly antagonized by the $Ca^{2+}$/calmodulin complex suggesting that the entry of $Ca^{2+}$ through NMDA channels could transiently modulate the physical anchoring of these channels to the cytoskeleton. Thus, there is the possibility of complex interactions between the competitive binding of $Ca^{2+}$/calmodulin and $\alpha$-Actinin-2, the phosphorylation of the NMDA C-termini which may ultimately determine the level of activity of NMDA channels.

Another PSD protein AKAP79, which is one of a family of PKA anchoring proteins, binds to and co-ordinates the interactions between PKA, calcineurin and PKC (Scott, 1997). These anchoring proteins are required for the PKA induced phosphorylation and regulation of L-type $Ca^{2+}$ (Gao et al., 1997) and AMPA channels (Rosenmund et al., 1994). The activity of both PKC and calcineurin is inhibited by this binding to AKAP79 and AKAP79 also binds, with high affinity and in a $Ca^{2+}$-dependent manner, to calmodulin. Furthermore, both calmodulin and AKAP79 appear to be co-localized in cultured hippocampal neurons. In turn, the binding of calmodulin to AKAP79 reversed the inhibition of PKC (PKC$\beta$II) and approximately doubled the PKC activity present in the PSD (Faux and Scott, 1997). This AKAP signaling complex might also provide a structure for regulating the phosphorylation and dephosphorylation of NMDA channels as well as the inactivation of these channels by $Ca^{2+}$ and calmodulin.

**Phosphatases**

NMDA receptors appear to be regulated by both serine/threonine and tyrosine phosphatases. In cultured hippocampal neurons, when NMDA currents are recorded in the cell attached or in the perforated patch configuration, low concentrations of the serine threonine phosphatase inhibitors okadaic acid and calyculin A enhance NMDA channel activity (Wang et al., 1994b). At nanomolar concentrations these inhibitors are selective for PP1 and PP2A over PP2B (calcineurin). This specificity is lost with higher concentrations of these inhibitors. Furthermore, applications of constitutively active PP1 and PP2A reduced the apparent probability of channel opening in inside-out patches. This inhibition by the phosphatases was reversible and could be blocked during co-application of the phosphatase inhibitors (Wang et al., 1994b). Such evidence is consistent with the presence within the patches of phosphorylating kinase activity. The continued presence of intracellular proteins within inside-out patches is not so surprising with the recognition of the tight binding of a number of PSD proteins directly to NMDA receptor subunits. It should also be emphasized that dephosphorylation of one or a number of these proteins may be responsible for changing the activity of the channels. The enhancement by calyculin A was also associated with a small increase in the apparent affinity of the NMDA receptor for glycine perhaps a consequence of an alteration in glycine-dependent desensitization (Wang et al., 1994b). This evidence strongly suggests that NMDA channel activity can be modulated by protein phosphatases such as PP1 and PP2A in cultured hippocampal neurons. However, the phosphatase inhibitors were without effect on NMDA-evoked currents recorded in cultured dorsal horn neurons (Wang et al., 1996). This cellular selectivity of phosphatase inhibitors was confirmed by Salter (M.W. Salter, personal communication) who showed that calyculin A enhanced NMDA-evoked current in cultured hippocampal but not in cultured dorsal horn neurons.

Mody et al. (Lieberman and Mody, 1994) demonstrated that concentrations of okadaic acid (10 $\mu$M), which non-selectively inhibit calcineurin, and the PP2B selective phosphatase inhibitor FK506 enhanced NMDA single channel activity in cell attached patches recorded from adult dentate gyrus granule neurons. Okadaic acid prolonged channel openings by increasing the duration of bursts, clusters and superclusters but these effects were not observed in the absence of extracellular $Ca^{2+}$ nor when extracellular $Ca^{2+}$ was substituted with barium demonstrating that an influx of $Ca^{2+}$ was required for the enhancement of channel activity. FK506 also prolonged openings whilst no effect on the apparent probability of channel opening was detected. In contrast, direct applications of calcineurin to the cytosolic face of inside-out patches shortened the duration of channel openings. These results provided strong evidence that entry of $Ca^{2+}$ via the channels was likely responsible for activating endogenous calcineurin which in turn might have dephosphorylated NMDA channels and thus reduced their activity. The major effect was an enhancement of burst duration which may better translate into a prolongation in the duration of the NMDA component of the epsc than a change in the open probability of the channels (Lieberman and Mody, 1994).

The influx of $Ca^{2+}$ via NMDA channels, and the subsequent activation of calcineurin, has also been proposed as a mechanism to account for the glycine-independent desensitization of NMDA receptor. In outside-out patches glycine-sensitive desensitization is rapidly lost following formation of the patch (Sather et al., 1992; Ascher et al., 1992). Tong and Jahr (1994) reported that this loss of glycine-sensitive desensitization is retarded if the recording pipette contains ATP-$\gamma$-S and can be blocked if it contains high concentrations of BAPTA suggesting that a $Ca^{2+}$-dependent dephosphorylation of the receptor was responsible (Tong and Jahr, 1994). However, the slow $Ca^{2+}$ buffer EGTA was ineffective in blocking this loss of desensitization. This evidence suggests that a transient influx of $Ca^{2+}$ was responsible for triggering the activation of process which takes place in close physical proximity to the channels themselves. Consistent with this hypothesis, the calcineurin inhibitors FK506 and cyclosporin as well as high concentrations of microcystin (which inhibit calcineurin) or a specific peptide inhibitor of calcineurin also reduced loss of glycine-sensitive desensitization. In contrast, low concentrations of calyculin A or the tryrosine phosphatase inhibitor orthovanadate were without effect on the initial loss of glycine-sensitive desensitization suggesting that PP1 and PP2A and tyrosine phosphatases were not involved (Tong and Jahr, 1994). In contrast, these authors did report that there was a more delayed component which was blocked by low concentrations of the selective PP1/PP2A phosphatase inhibitor microcystin suggesting some role for this group of phosphatases (Tong and Jahr, 1994).

The loss of glycine-dependent desensitization was not prevented by phalloidin which stabilizes the cytoskeleton and it was not related to a sustained elevation in the concentration of intracellular $Ca^{2+}$ which would be required for depolymerization of the cytoskeleton (Tong and Jahr, 1994). Further, the loss of glycine-sensitive desensitization was not prevented by pretreatment of the cells with phorbol esters and thus a loss of phosphorylation by PKC did not appear to be responsible. It has been suggested that a transient elevation of $Ca^{2+}$ following activation of NMDA channels leads to a dephosphorylation of the receptors (Tong and Jahr, 1994). The same group then showed that the NMDA receptor-mediated component of epscs in cultured neurons demonstrates a calcineurin-sensitive desensitization following repetitive activation suggesting that glycine-sensitive desensitization may play a physiological role to limit summation of NMDA synaptic activity (Tong et al., 1995). This synaptic desensitization was reduced by PKA activators and by $\beta$-adrenergic receptor activation suggesting that PKA maybe the kinase required to maintain glycine-sensitive desensitization in cultured hippocampal neurons (Tong et al., 1995; Raman et al., 1996).

*Tyrosine phosphorylation and dephosphorylation*

Perhaps the best established role for kinase regulation of NMDA receptors comes from studies of tyrosine kinase activity and NMDA-evoked currents in cultured dorsal horn neurons (Wang et al., 1993; Wang and Salter, 1994; Wang et al., 1996). In a recent paper, Salter et al. (Yu et al., 1997) demonstrated that Src kinase co-immunoprecipitates with the NR1 subunit suggesting an intimate physical relationship between the receptor and the kinase. These authors had previously shown that inhibitors of tyrosine kinase depressed NMDA-evoked currents in cultured dorsal horn neurons and that intracellular perfusion of recombinant Src enhanced these currents (Wang and Salter, 1994). In addition a truncated form of PTP (T-cell PTP) applied intracellularly in the presence of ATP reduced whole cell currents and mimicked the rundown of these currents whilst the non-selective PTP inhibitor orthovanadate enhanced these currents. Furthermore, this phosphatase reduced channel activity in inside-out patches and reduced the probability of the channels opening (Wang et al., 1996). This evidence established that NMDA receptors are in spinal cord neurons regulated by protein tyrosine kinases and phosphatases and perhaps by PKC, PKA and CamKII (Gerber et al., 1989; Kolaj et al., 1994; Cerne et al., 1993).

In an extensive single channel study from inside-out patches these initial findings have been greatly extended (Yu et al., 1997). A phosphorylated peptide analogue EPQ(pY)EE1P1A which activates tyrosine kinases, but not its unphosphorylated form (does not activate kinases) was shown to increase by 260% the probability of opening of NMDA channels and the mean open time by 152% while having no effect on single channel conductance. The durations of bursts, clusters and superclusters were all increased. In contrast, an antibody which inhibits the Src family of kinases, anti-cst1, decreased the probability of channel opening and decreased the total open time of the channel (reduced burst and cluster durations) also without effecting channel conductance. These results suggested that an endogenous kinase might be present and active in the inside-out patches. A second antibody (anti-src1) which can also specifically inhibit Src kinases was also shown to inhibit the opening and the duration of bursts and clusters demonstrating that the endogenous kinase was likely Src itself (Yu et al., 1997). These effects were blocked when the antibody was incubated with the corresponding immunogen peptide Src (40–58) which corresponds to a non-catalytic region of the kinase. Furthermore, exogenous recombinant $pp60^{c\text{-}src}$ mimicked the effects of EPQ(pY)EE1-P1A. These results demonstrated that it was likely that an endogenous Src kinase was specifically present in the patches and was capable of regulating NMDA channel activity. In addition, these authors (Yu et al., 1997) were able to demonstrate that the NMDA receptor-mediated component of miniature epscs in these neurons were enhanced by EPQ(pY)EE1P1A and inhibited by anti-Src1 and Src (40–58) establishing a potential role for Src in regulating synaptic transmission via phosphorylation of NMDA receptors. The Src (40–58) peptide was then applied to inside-out patches to see if it could compete for the endogenous kinase present in the patch (Yu et al., 1997). This peptide (but not a scrambled version of the peptide) reduced the probability of channel opening and as well the mean open time (reduced bursts and clusters) and it also prevented the effects of EPQ(pY)EE1P1A suggesting that this non-catalytic region of the kinase might require the binding of the receptor protein or some alternative protein in order to facilitate phosphorylation of the receptor protein (Yu et al., 1997). In this regard it has also been shown that NR2A and NR2B subunits bind directly to the SH2 domains of PLC$\gamma$ (Gurd and Bissoon, 1997) emphasizing the potential importance of protein-protein interactions of the NMDA receptor with signal transduction enzymes.

The potential role of tyrosine kinase phosphorylation and dephosphorylation of NMDA receptors is supported by demonstrations that activation of the insulin receptor cause a tyrosine kinase-mediated enhancement of NMDA-evoked currents in hippocampal neurons and of recombinant NMDA receptor responses in Xenopus oocytes (Liu et al., 1995; Chen and Leonard, 1996). The enhancement by tyrosine phosphorylation of NMDA-evoked currents also appears to be specific to recombinant receptors containing the NR2A subunit (Kohr and Seeburg, 1996). Transient ischemic insults to the central nervous system are also associated with an enhanced tyrosine phosphorylation of PSD proteins including the NR2A and NR2B subunit (Takagi et al., 1997).

Considerable interest has centered upon NMDA receptors as potential therapeutic targets for pharmacological intervention in the treatment of epilepsy, stroke, Alzheimer's Disease, Parkinson's Disease and perhaps some psychiatric disturbances. Could drugs be designed for the purpose of regulating the degree of phosphorylation of these receptors and thus provide therapeutic benefit? Drugs which act to modify kinase or phosphatase activity would be unlikely candidates for this role simply because they lack cellular specificity. Another potential approach would be to exploit the specificity of the protein-protein interactions which underlie the binding of NMDA receptor subunits to key anchoring proteins, kinases or phosphatases within the PSD. For example, an agent might compete with a NMDA subunit for binding to an anchoring protein. In turn, this anchoring protein might function to bring the receptor into close proximity with a specific kinase or phosphatase responsible for regulating the phosphorylation state of the NMDA receptor. Using this approach it might be possible to specifically perturb a small number of discrete subcellular events at the PSD (i.e. the activity of the channels, the number of channels expressed etc.).

## References

Ankisztejn, L., Bregestovski, P. and Ben-Ari, Y. (1991) Selective activation of quisqualate metabotropic receptor potentiates NMDA but not AMPA responses. *Eur. J. Pharmacol.*, 205: 327–328.

Aniksztejn, L., Otani, S. and Ben-Ari, Y. (1992) Quisqualate metabotropic receptors modulate NMDA currents and facilitate induction of long term potentiation through protein kinase C. *Eur. J. Neurosci.*, 4: 500–505.

Artola, A. and Singer, W. (1993) Long-term depression of excitatory synaptic transmission and its relationship to long-term potentiation. *Trends Neurosci.*, 16: 480–487.

Ascher, P., Dieudonne, S., MacDonald, J. and Sather W. (1992) Kinetics of activation, deactivation and desensitization of the "NMDA" glutamate receptor. *C.R. Acad. Sci. III.*, 314: 75–77.

Ascher, P. and Nowak, L. (1988) The role of divalent cations in the *N*-methyl-D-aspartate responses of mouse central neurones in culture. *J. Physiol. (Lond)*, 399: 247–266.

Barria, A., Muller, D., Derkach, V., Griffith, L.C. and Soderling, T.R. (1997) Regulatory phosphorylation of AMPA-type glutamate receptors by CaM-KII during long-term potentiation. *Science*, 276: 2042–2045.

Bartlett, M.C., Salter, M.W. and MacDonald, J.F. (1989) Modulation of NMDA currents by intracellular phosphorylation. *Soc. Neurosci. Abstr.*, 218.13: 107.

Ben-Ari, Y., Aniksztejn, L. and Bregestovski, P. (1992) Protein kinase C modulation of NMDA currents: An important link for LTP induction. *Trends Neurosci.*, 15: 333–339.

Bencherif, M. and Lukas, R.J. (1993) Cytochalasin modulation of nicotinic cholinergic receptor expression and muscarinic receptor function in human TE671/RD cells: A possible functional role of the cytoskeleton. *J. Neurochem.*, 61: 852–864.

Blahos, J. and Wenthold, R.J. (1996) Relationship between *N*-methyl-D-aspartate receptor NR1 splice variants and NR2 subunits. *J. Biol. Chem.*, 271: 15669–15674.

Blahos, J. and Wenthold, R.J. (1996) Relationship between *N*-methyl-D-aspartate receptor NR1 splice variants and NR2 subunits. *J. Biol. Chem.*, 271: 15669–15674.

Bliss, T.V. and Collingridge, G.L. (1993) A synaptic model of memory: Long-term potentiation in the hippocampus. *Nature*, 361: 31–39.

Burnashev, N., Zhou, Z., Neher, E. and Sakmann, B. (1995) Fractional calcium currents through recombinant GluR channels of the NMDA, AMPA and kainate receptor subtypes. *J. Physiol. (Lond.)*, 485: 403–418.

Cantiello, H.F. (1995) Role of the actin cytoskeleton on epithelial $Na^+$ channel regulation. *Kidney Int.*, 48: 970–984.

Cantiello, H.F., Stow, J.L., Prat, A.G. and Ausiello, A. (1991) Actin filaments regulate epithelial $Na^+$ channel activity. *Am. J. Physiol.*, 261: C882–C886.

Carr, D.W., Stofko-Hahn, R.E., Fraser, I.D.C., Cone, R.D. and Scott, J.D. (1992) Localization of the cAMP-dependent protein kinase to the postsynaptic densities by A-kinase anchoring proteins. *J. Biol. Chem.*, 267, 16816–16823.

Castillo, P.E., Malenka, R.C. and Nicoll, R.A. (1997) Kainate receptors mediate a slow postsynaptic current in hippocampal CA3 neuron. *Nature*, 388: 182–186.

Cerne, R., Rusin, K.I. and Randic, M. (1993) Enhancement of the *N*-methyl-D-aspartate response in spinal dorsal horn neurons by cAMP-dependent protein kinase. *Neurosci. Lett.*, 161: 124–128.

Chen, C. and Leonard, J.P. (1996) Protein tyrosine kinase-mediated potentiation of currents from cloned NMDA receptors. *J. Neurochem.*, 67: 194–200.

Chen, L. and Huang, L-YM. (1991) Sustained potentiation of NMDA receptor-mediated glutamate responses through activation of protein kinase C by a $\mu$ opioid. *Neuron*, 7: 319–326.

Chen, L. and Huang, L.-Y.M. (1992) Protein kinase C reduces $Mg^{2+}$ block of NMDA-receptor channels as a mechanism of modulation. *Nature*, 356: 521–523.

Clements, J.D., Lester, R.A., Tong, G., Jahr, C.E. and Westbrook, G.L. (1992a) The time course of glutamate in the synaptic cleft. *Science*, 258: 1498–1501.

Clements, J.D., Lester, R.A.J., Tong, G., Jahr, C.E. and Westbrook, G.L. (1992b) The time course of glutamate in the synaptic cleft. *Science*, 258: 1498–1501.

Coghlan, V.M., Perrino, B.A., Howard, M., Langeberg, L.K., Hicks, J.B., Gallatin, W.M. and Scott, J.D. (1995) Association of protein kinase A and protein phosphatase 2B with a common anchoring protein. *Science*, 267: 108–111.

Collingridge, G.L. and Bliss, T.V. (1995) Memories of NMDA receptors and LTP. *Trends Neurosci.*, 18: 54–56.

Collingridge, G.L. and Singer, W. (1990) Excitatory amino acid receptors and synaptic plasticity. *Trends Pharmacol. Sci.*, 11: 290–296.

Courtney, M.J. and Nicholls, D.G. (1992) Interactions between phospholipase C-coupled and *N*-methyl-D-aspartate receptors in cultured cerebellar granule cells: Protein kinase C mediated inhibition of *N*-methyl-D-aspartate responses. *J. Neurochem.*, 59: 983–992.

Diamond, J.S. and Jahr, C.E. (1997) Transporters buffer synaptically released glutamate on a submillisecond time scale. *J. Neurosci.*, 17: 4672–4687.

Dong, H., O'Brien, R.J., Fung, E.T., Lanahan, A.A., Worley, P.F. and Huganir, RL. (1997) GRIP: A synaptic PDZ domain-containing protein that interacts with AMPA receptors. *Nature*, 386: 279–284.

Dosemeci, A. and Reese, T.S. (1993) Inhibition of endogenous phosphatase in a postsynaptic density fraction allows extensive phosphorylation of the major postsynaptic density protein. *J. Neurochem.*, 61: 550–555.

Durand, G.M., Bennett, M.V. and Zukin, R.S. (1993) Splice variants of the *N*-methyl-D-aspartate receptor NR1 identify domains involved in regulation by polyamines and protein kinase C. [published erratum appears in *Proc Natl Acad Sci USA*, 1993 Oct 15; 90 (20): 9739]. *Proc. Natl. Acad. Sci. U.S.A.*, 90: 6731–6735.

Durand, G.M., Gregor, P., Zheng, X., Bennett, M.V.L., Uhl, G.R. and Zukin, R.S. (1992) Cloning of an apparent splice variant of the rat *N*-methyl-D-aspartate receptor NMDAR1 with altered sensitivity to polyamines and activators of protein kinase C. *Proc. Natl. Acad. Sci. USA*, 89: 9359–9363.

Ehlers, M.D., Tingley, W.G. and Huganir, R.L. (1995) Regulated subcellular distribution of the NR1 subunit of the NMDA receptor. *Science*, 269: 1734–1737.

Ehlers, M.D., Zhang, S., Bernhadt, J.P. and Huganir, R.L. (1996) Inactivation of NMDA receptors by direct interaction of calmodulin with the NR1 subunit. *Cell*, 84: 745–755.

Faux, M.C. and Scott, J.D. (1997) Regulation of the AKAP79-protein kinase C interaction by $Ca^{2+}$/Calmodulin. *J. Biol. Chem.*, 272: 17038–17044.

Furman, M.I., Gardner, T.M. and Goldschmidt-Clermont, P.J. (1993). Mechanisms of cytoskeletal reorganization during platelet activation. *Thrombosis and Haemostasis*, 70, 229–232.

Furukawa, K. and Mattson, M.P. (1995) Cytochalasins protect hippocampal neurons against amyloid $\beta$-peptide toxicity: Evidence that actin depolymerization suppresses $Ca^{2+}$ influx. *J. Neurochem.*, 65: 1061–1068.

Gao, T., Yatani, A., Dell'Acqua, M.L., Sako, H., Green, S.A., Dascal, N., Scott, J.D. and Hosey, M.M. (1997) cAMP-dependent regulation of cardiac L-type $Ca^{2+}$ channels requires membrane targeting of PKA and phosphorylation of channel subunits. *Neuron*, 19: 185–196.

Gasic, G.P. and Hollmann, M. (1992a) Molecular neurobiology of glutamate receptors. *Annu. Rev. Physiol.*, 54: 507–536.

Gasic, G.P. and Hollmann, M. (1992b) Molecular neurobiology of glutamate receptors. *Annu. Rev. Physiol.*, 54: 507–536.

Gerber, G., Kangrga, I., Ryu, P.D., Larew, J.S. and Randic, M. (1989) Multiple effects of phorbol esters in the rat spinal dorsal horn. *J. Neurosci.*, 9: 3606–3617.

Gerber, G., Kangrga, I., Ryu, P.D., Larew, J.S.A. and Randic, M. (1989) Multiple effects of phorbol esters in the rat spinal cord. *J. Neurosci.*, 9: 3606–3614.

Gurd, J.W. and Bissoon, N. (1997) The *N*-methyl-D-aspartate receptor subunits NR2A and NR2B bind to the SH2 domains of phospholipase C-γ. *J. Neurochem.*, 69: 623–630.

Hall, R.A. and Soderling, T.R. (1997) Differential surface expression and phosphorylation of the *N*-methyl-D-aspartate receptor subunits NR1 and NR2 in cultured hippocampal neurons. *J. Biol. Chem.*, 272: 4135–4140.

Hisatsune, C., Umemori, H., Inoue, T., Michikawa, T., Kohda, K., Mikoshiba, K. and Yamamoto, T. (1997) Phosphorylation-dependent regulation of *N*-methyl-D-aspartate receptors by calmodulin. *J. Biol. Chem.*, 272: 20805–20810.

Hockberger, P., Toselli, M., Swandulla, D. and Lux, H.D. (1989) A diacylglycerol analogue redues neuronal calcium currents independently of protein kinase C activation. *Nature*, 338, 340–342.

Hollmann, M., Boulter, J., Maron, C., Beasley, L., Sullivan, J., Pecht, G. and Heinemann, S. (1993) Zinc potentiates agonist-induced currents at certain splice variants of the NMDA receptor. *Neuron*, 10: 943–954.

Hollmann, M., Maron, C. and Heinemann, S. (1994) N-glycosylation site tagging suggests a three transmembrane domain topology for the glutamate receptor GluR1. *Neuron*, 13: 1331–1343.

Isaac, J.T.R., Nicoll, R.A. and Malenka, R.C. (1995) Evidence for silent synapses: Implications for the expression of LTP. *Neuron*, 15: 427–434.

Janmey, P.A. (1994) Phosphoinositides and calcium as regulators of cellular actin assembly and disassembly. *Annu. Rev. Physiol.*, 56: 169–191.

Johnson, B.D. and Byerly, L. (1993) A cytoskeletal mechanism for $Ca^{2+}$ channel metabolic dependence and inactivation by intracellular $Ca^{2+}$. *Neuron*, 10: 797–804.

Joly, C., Gomeza, J., Brabet, I., Curry, K., Bockaert, J. and Pin, J.-P. (1995) Molecular, functional and pharmacological characterization of the metabotropic glutamate receptor type 5 splice variants: Comparison with mGluR1. *J. Neurosci.*, 15: 3970–3981.

Kelso, S.R., Nelson, T.E. and Leonard, J.P. (1992) Protein kinase C-mediated enhancement of NMDA currents by metabotropic glutamate receptors in Xenopus oocytes. *J. Physiol. (Lond)*, 449: 705–18: 705–718.

Kennedy, M.B. (1993) The postsynaptic density. *Curr. Opin. Neurobiol.*, 3: 732–737.

Kennedy, M.B. (1997) The postsynaptic density at glutamatergic synapses. *Trends Neurosci.*, 20: 264–268.

Kim, E., Cho, K.O., Rothschild, A. and Sheng, M. (1996) Heteromultimerization and NMDA receptor-clustering activity of Chapsyn-110, a member of the PSD-95 family of proteins. *Neuron*, 17: 103–113.

Kohr, G. and Seeburg, P.H. (1996) Subtype-specific regulation of recombinant NMDA receptor-channels by protein tyrosine kinases of the src family. *J. Physiol. (Lond)*, 492: 445–452.

Kolaj, M., Cerne, R., Cheng, G., Brickey, D.A. and Randic, M. (1994) α Subunit of calcium/calmodulin-dependent protein kinase enhances excitatory amino acid and synaptic responses of rat spinal dorsal horn. *Neurons. J. Neurophysiol.*, 72: 2525–2531.

Kornau, H.-C., Schenker, L.T., Kennedy, M.B. and Seeburg, P.H. (1995) Domain interaction between NMDA receptor subunits and the postsynaptic density protein PSD-95. *Science*, 269, 1737–1739.

Köhr, G., De Koninck, Y. and Mody, I. (1993) Properties of NMDA receptor channels in neurons acutely isolated from epileptic (kindled) rats. *J. Neurosci*, 13: 3612–3627.

Krupp, J.J., Vissel, B., Heinemann, S.F. and Westbrook, G.L. (1996) Calcium-dependent inactivation of recombinant *N*-methyl-D-aspartate receptors is NR2 subunit specific. *Mol. Pharmacol.*, 50: 1680–1688.

Kutsuwada, T., Kashiwabuchi, N., Mori, H., Sakimura, K., Kushiya, E., Araki, K., Meguro, H., Masaki, H., Kumanishi, T., Arakawa, M. and Mishina, M. (1992) Molecular diversity of the NMDA receptor channel. *Nature*, 358: 36–41.

Lau, L.F. and Huganir, R.L. (1995) Differential tyrosine phosphorylation of *N*-methyl-D-aspartate receptor subunits. *J. Biol. Chem.*, 270: 20036–20041.

Lau, L.F., Mammen, A., Ehlers, M.D., Kindler, S., Chung, W.J., Garner, C.C. and Huganir, R.L. (1996) Interaction of the *N*-methyl-D-aspartate receptor complex with a novel synapse-associated protein, SAP102. *J. Biol. Chem.*, 271: 21622–21628.

Legendre, P., Rosenmund, C. and Westbrook, G.L. (1993) Inactivation of NMDA channels in cultured hippocampal neurons by intracellular calcium. *J. Neurosci*, 13: 674–684.

Leonard, A.S. and Hell, J.W. (1997) Cyclic AMP-dependent protein kinase and protein kinase C phosphorylate *N*-methyl-D-aspartate receptors at different sites. *J. Biol. Chem.*, 272: 12107–12115.

Lester, R.A., Clements, J.D., Westbrook, G.L. and Jahr, C.E. (1990) Channel kinetics determine the time course of NMDA receptor-mediated synaptic currents. *Nature*, 346: 565–567.

Lev, S., Moreno, H., Martinez, R., Canoll, P., Peles, E., Musacchio, J.M., Plowman, G.D., Rudy, B. and Schlessinger, J. (1995) Protein tyrosine kinase PYK2 involved in $Ca^{(2+)}$-induced regulation of ion channel and MAP kinase functions. *Nature*, 376: 737–745.

Levitan, I., Almonte, C., Mollard, P. and Garber, S.S. (1995) Modulation of a volume-regulated chloride current by F-actin. *J. Membr. Biol.*, 147: 283–294.

Levitan I.B (1994) Modulation of ion channels by protein phosphorylation and dephosphorylation. *Annu. Rev. Physiol.*, 56: 193–212.

Liao, D., Hessler, N.A. and Malinow, R. (1995) Activation of postsynaptically silent synapses during pairing-induced LTP in CA1 region of hippocampal slice. *Nature*, 375: 400–404.

Lickteig, R.L., Eggemen, K.T. and Browning, M.D. (1995) The NR2A subunit of the NMDA receptor is phosphorylated in situ in the Rat hippocampus. *Soc. Neurosci. Abst.*, 21.83.

Lieberman, D.N. and Mody, I. (1994) Regulation of NMDA channel function by endogenous $Ca^{2+}$-dependent phosphatase. *Nature*, 369: 235–239.

Liu, L., Brown, J.C., III, Webster, W.W., Morrisett, R.A. and Monaghan, DT. (1995) Insulin potentiates *N*-methyl-D-aspartate receptor activity in *Xenopus*, oocytes and rat hippocampus. *Neurosci. Lett.*, 192: 5–8.

MacDonald, J.F., Mody, I. and Salter, M.W. (1989) Regulation of *N*-methyl-D-aspartate receptors revealed by intracellular dialysis of murine neurones in culture. *J. Physiol. (Lond)*, 414: 17–34.

Markram, H. and Segal, M. (1992) Activation of protein kinase C suppresses responses to NMDA in rat CA1 hippocampal neurones. *J. Physiol. (Lond)*, 457: 491–501.

Mayer, M.L., Westbrook, G.L. and Guthrie, PB. (1984) Voltage-dependent block by $Mg^{2+}$ of NMDA responses in spinal cord neurones. *Nature*, 309: 261–263.

McBain, C.J. and Mayer, M.L. (1994) *N*-methyl-D-aspartic acid receptor structure and function. *Physiol. Rev.*, 74: 723–723.

Medina, I., Filippova, N., Bakhramov, A. and Bregestovski, P. (1996) Calcium-induced inactivation of NMDA receptor-channels evolves independently of run-down in cultured rat brain neurones. *J. Physiol. (Lond)*, 495: 411–427.

Medina, I., Filippova, N., Charton, G., Rougeole, S., Ben-Ari, Y., Khrestchatisky, M. and Bregestovski, P. (1995) Calcium-dependent inactivation of heteromeric NMDA receptor-channels expressed in human embryonic kidney cells. *J. Physiol. (Lond.)*, 482: 567–573.

Meguro, H., Mori, H., Araki, K., Kushiya, E., Kutsuwada, T., Yamazaki, M., Kumanishi, T., Arakawa, M., Sakimura, K. and Mishina, M. (1992) Functional characterization of a heteromeric NMDA receptor channel expressed from cloned cDNAs. *Nature*, 357: 70–74.

Mishina, M., Mori, H., Araki, K., Kushiya, E., Meguro, H., Kutsuwada, T., Kashiwabuchi, N., Ikeda, K., Nagasawa, M., Yamazaki, M., Masaki, H., Yamakura, T., Morita, T. and Sakimura, K. (1993) Molecular and functional diversity of the NMDA receptor channel. *Ann. NY Acad. Sci.*, 707: 136–152.

Mons, N. and Cooper, D.M. (1995) Adenylate cyclases: Critical foci in neuronal signaling. *Trends Neurosci.*, 18: 536–542.

Monyer, H., Sprengel, R., Schoepfer, R., Herb, A., Higuchi, M., Lomeli, H., Burnashev, N., Sakmann, B. and Seeburg, PH. (1992) Heteromeric NMDA receptors: Molecular and functional distinction of subtypes. *Science*, 256: 1217–1221.

Moon, I.S., Apperson, M.L. and Kennedy, M.B. (1994) The major tyrosine-phosphorylated protein in the postsynaptic density fraction is *N*-methyl-D-aspartate receptor subunit 2B. *Proc. Natl. Acad. Sci. USA*, 91: 3954–3958.

Muller, B.M., Kistner, U., Kindler, S., Chung, W.J., Kuhlendahl, S., Fenster, S.D., Lau, L.F., Veh, R.W., Huganir, R.L., Gundelfinger, E.D. and Garner, C.C. (1996) SAP102, A novel postsynaptic protein that interacts with NMDA receptor complexes in vivo. *Neuron*, 17: 255–265.

Nakamura, S., Asaoka, Y., Yoshida, K., Sasaki, Y. and Nishizuka, Y. (1993) Protein kinase C for cell signaling: A possible link between phospholipases. *Adv. Second Messenger Phosphoprotein Res.*, 28: 171–178.

Niethammer, M., Kim, E. and Sheng, M. (1996) Interaction between the C terminus of NMDA receptor subunits and multiple members of the PSD-95 family of membrane-associated guanylate kinases. *J. Neurosci*, 16: 2157–2163.

Nowak, L., Bregestovski, P., Ascher, P., Herbet, A. and Prochiantz, A. (1984) Magnesium gates glutamate-activated channels in mouse central neurones. *Nature*, 307: 462–465.

Oliet, S.H., Malenka, R.C. and Nicoll, R.A. (1997) Two distinct forms of long-term depression coexist in CA1 hippocampal pyramidal cells. *Neuron*, 18: 969–982.

Omkumar, R.V., Kiely, M.J., Rosenstein, A.J., Min, K.T. and Kennedy, M.B. (1996) Identification of a phosphorylation site for calcium/calmodulindependent protein kinase II in the NR2B subunit of the *N*-methyl-D-aspartate receptor. *J. Biol. Chem.*, 271: 31670–31678.

Paoletti, P. and Ascher, P. (1994) Mechanosensitivity of NMDA receptors in cultured mouse central neurons. *Neuron*, 13: 645–655.

Pin, J.-P. and Duvoisin, R. (1995) The metabotropic glutamate receptors: Structure and functions. *Neuropharmacology*, 34: 1–26.

Prat, A.G., Ausiello, D.A. and Cantiello, H.F. (1993a) Vasopressin and protein kinase A activate G protein-sensitive epithelial $Na^+$ channels. *AM. J. Physiol.*, 265: C218–23.

Prat, A.G., Bertorello, A.M., Ausiello, D.A. and Cantiello, H.F. (1993b) Activation of epithelial $Na^+$ channels by protein kinase A requires actin filaments. *Am. J. Physiol. Cell Physiol.*, 265: C224–C233.

Prat, A.G., Bertorello, A.M., Ausiello, D.A. and Cantiello, H.F. (1993c) Activation of epithelial $Na^+$ channels by protein kinase A requires actin filaments. *Am. J. Physiol.*, 265: C224–33.

Raman, I.M., Tong, G. and Jahr, C.E. (1996) β-adrenergic regulation of synaptic NMDA receptors by cAMP-dependent protein kinase. *Neuron*, 16: 415–421.

Raouf, R., Xiong, Z.G., Lu, W.Y., Lickteig, R., Dudek, E.M., Browning, M.D., Orser, B.A. and MacDonald, J.F. (1997) Modulation of NMDA-mediated currents by serine/threonine protein kinases. *Soc. Neurosci. Abst.*, 376.12: 946.

Raymond, L.A., Blackstone, C.D. and Huganir, R.L. (1993) Phosphorylation of amino acid neurotransmitter receptors in synaptic plasticity. *Trends Neurosci.*, 16: 147–153.

Reinhart, P.H. and Levitan, I.B. (1995) Kinase and phosphatase activities intimately associated with a reconstituted calcium-dependent potassium channel. *J. Neurosci*, 15: 4572–4579.

Rosenmund, C., Carr, D.W., Bergeson, S.E., Nilaver, G., Scott, J.D. and Westbrook, G.L. (1994) Anchoring of protein kinase A is required for modulation of AMPA/kainate receptors on hippocampal neurons. *Nature*, 368: 853–856.

Rosenmund, C., Feltz, A. and Westbrook, G.L. (1995) Calcium-dependent inactivation of synaptic NMDA receptors in hippocampal neurons. *J. Neurophysiol.*, 73: 427–430.

Rosenmund, C. and Westbrook, G.L. (1993a) Calcium-induced actin depolymerization reduces NMDA channel activity. *Neuron*, 10: 805–814.

Rosenmund, C. and Westbrook, G.L. (1993b) Rundown of *N*-methyl-D-aspartate channels during whole-cell recording in rat hippocampal neurons: Role of $Ca^{2+}$ and ATP. *J. Physiol. (Lond.)*, 470: 705–729.

Sather, W., Dieudonne, S., MacDonald, J.F. and Ascher, P. (1992) Activation and desensitization of *N*-methyl-D-aspartate receptors in nucleated outside-out patches from mouse neurones. *J. Physiol. (Lond)*, 450: 643–672: 643–672.

Schoepp, D.D. and Conn, P.J. (1993) Metabotropic glutamate receptors in brain function and pathology. *Trends Pharmacol. Sci.*, 14: 13–20.

Scott, J.D. (1997) Dissection of protein kinase and phosphatase targeting interactions. *Soc. Gen. Physiol. Ser.*, 52: 227–39: 227–239.

Seeburg, P.H. (1993) The TINS/TiPS Lecture. The molecular biology of mammalian glutamate receptor channels. *Trends Neurosci.*, 16: 359–365.

Sheng, M. (1996) PDZs and receptor/channel clustering: Rounding up the latest suspects. *Neuron*, 17: 575–578.

Sheng, M. (1997) Excitatory synapses. Glutamate receptors put in their place. *Nature*, 386: 221–221.

Sheng, M., Cummings, J., Roldan, LA., Jan, Y.N. and Jan, L.Y. (1994) Changing subunit composition of heteromeric NMDA receptors during development of rat cortex. *Nature*, 368: 144–147.

Sheng, M. and Kim, E. (1996) Ion channel associated proteins. *Curr. Opin. Neurobiol.*, 6: 602–608.

Sigel, E., Baur, R. and Malherbe, P. (1994) Protein kinase C transiently activated heteromeric *N*-methyl-D-aspartate receptor channels independent of the phosphorylatable C-terminal splice domain and of consensus phosphorylation sites. *J. Biol. Chem.*, 269: 8204–8208.

Sugihara, H., Moriyoshi, K., Ishii, T., Masu, M. and Nakanishi, S. (1992) Structures and properties of seven isoforms of the NMDA receptor generated by alternative splicing. *Biochem. Biophys. Res. Commun.*, 185: 826–832.

Suzuki, M., Miyazaki, K., Ikeda, M., Kawaguchi, Y. and Sakai, O. (1993) F-actin network may regulate a $Cl^-$ channel in renal proximal tubule cells. *J. Membr. Biol.*, 134: 31–39.

Suzuki, T., Okumura-Noji, K., Tanaka, R., Ogura, A., Nakamura, K., Kudo, Y. and Tada, T. (1993) Characterization of protein kinase C activities in postsynaptic density fractions prepared from cerebral cortex, hippocampus, and cerebellum. *Brain Res.*, 619: 69–75.

Takagi, N., Shinno, K., Teves, L., Bissoon, N., Wallace, M.C. and Gurd, J.W. (1997) Transient ischemia differentially increases tyrosine phosphorylation of NMDA receptor subunits 2A and 2B. *J. Neurochem.*, 69: 1060–1065.

Tanaka, C. and Nishizuka, Y. (1994) The protein kinase C family for neuronal signaling. *Annu. Rev. Neurosci.*, 17: 551–67: 551–567.

Tingley, W.G., Ehlers, M.D., Kameyama, K., Doherty, C., Ptak, J.B., Riley, C.T. and Huganir, R.L. (1997) Characterization of protein kinase A and protein kinase C phosphorylation of the *N*-methyl-D-aspartate receptor NR1 subunit using phosphorylation site-specific antibodies. *J. Biol. Chem.*, 272: 5157–5166.

Tingley, W.G., Roche, K.W., Thompson, A.K. and Huganir, R.L. (1993) Regulation of NMDA receptor phosphorylation by alternative splicing of the C-terminal domain. *Nature*, 364: 70–73.

Tong, G. and Jahr, C.E. (1994) Regulation of glycine-insensitive desensitization of the NMDA receptor in outside-out patches. *J. Neurophysiol.*, 72: 754–761.

Tong, G., Shepherd, D. and Jahr, C.E. (1995) Synaptic desensitization of NMDA receptors by calcineurin. *Science*, 267: 1510–1512.

Urushihara, H., Tohda, M. and Nomura, Y. (1992) Selectivie potentiation of *N*-methyl-D-aspartate-induced current by protein kinase C in *Xenopus*, oocytes injected with rat brain RNA. *J. Biol. Chem.*, 267: 11697–11700.

Valenzuela, C.F., Xiong, Z., MacDonald, J.F., Weiner, J.L., Frazier, C.J., Dunwiddie, T.V., Kazlauskas, A., Whiting, P.J. and Harris, R.A. (1996) Platelet-derived growth factor induces a long-term inhibition of *N*-methyl-D-aspartate receptor function. *J. Biol. Chem.*, 271: 16151–16159.

Vignes, M. and Collingridge, G.L. (1997) The synaptic activation of kainate receptors. *Nature*, 388: 179–182.

Vyklicky, L., Jr. (1993) Calcium-mediated modulation of *N*-methyl-D-aspartate (NMDA) responses in cultured rat hippocampal neurones. *J. Physiol. (Lond.)*, 470: 575–600.

Wagner, D.A. and Leonard, J.P. (1994) Mechanisms for protein kinase-C potentiation of cloned NMDA receptors. *Soc. Neurosci. Abst.*, 20: 236.

Wagner, D.A. and Leonard, J.P. (1996) Effect of protein kinase-C activation on the $Mg^{2+}$-sensitivity of cloned NMDA receptors. *Neuropharmacology*, 35: 29–36.

Wang, L.Y., Dudek, E.M., Browning, M.D. and MacDonald, J.F. (1994a) Modulation of AMPA/kainate receptors in cultured murine hippocampal neurones by protein kinase C. *J. Physiol. (Lond)*, 475: 431–437.

Wang, L.Y., Orser, B.A., Brautigan, D.L. and MacDonald, J.F. (1994b) Regulation of NMDA receptors in cultured hippocampal neurons by protein phosphatases 1 and 2A. *Nature*, 369: 230–232.

Wang, Y.T., Pak, Y.S. and Salter, M.W, (1993) Rundown of NMDA-receptor mediated currents is resistant to lowering intracellular [$Ca^{2+}$] and is prevented by ATP in rat spinal dorsal horn neurons. *Neurosci. Lett.*, 157: 183–186.

Wang, Y.T. and Salter, M.W. (1994) Regulation of NMDA receptors by tyrosine kinases and phosphatases. *Nature*, 369: 233–235.

Wang, Y.T., Yu, X.M. and Salter, M.W. (1996) $Ca^{(2+)}$-independent reduction of *N*-methyl-D-aspartate channel activity by protein tyrosine phosphatase. *Proc. Natl. Acad. Sci. USA*, 93: 1721–1725.

Westbrook, G.L., Krupp, J.J. and Vissel, B. (1997) Cytoskeletal interactions with glutamate receptors at central synapses. *Soc. Gen. Physiol. Ser.*, 52: 163–75: 163–175.

Wo, Z.G. and Oswald, R.E. (1995) Unraveling the modular design of glutamate-gated ion channels. *Trends Neurosci.*, 18: 161–168.

Wyszynski, M., Lin, J., Rao, A., Nigh, E., Beggs, A.H., Craig, A.M. and Sheng, M. (1997) Competitive binding of α-actinin and calmodulin to the NMDA receptor. *Nature*, 385: 439–442.

Yamakura, T., Mori, H., Shimoji, K. and Mishina, M. (1993) Phosphorylation of the carboxyl-terminal domain of the ζ1 subunit is not responsible for potentiation by TPA of the NMDA receptor channel. *Biochem. Biophys. Res. Commun.*, 196: 1537–1544.

Yamazaki, M., Mori, H., Araki, K., Mori, K.J. and Mishina, M. (1992) Cloning, expression and modulation of a mouse NMDA receptor subunit. *FEBS Lett.*, 300: 39–45.

Yu, X.M., Askalan, R., Keil, G.J. and Salter, M.W. (1997) NMDA channel regulation by channel-associated protein tyrosine kinase Src. *Science*, 275: 674–678.

Zhang, L., Rzigalinski, B.A., Ellis, E.F. and Satin, L.S. (1996) Reduction of voltage-dependent $Mg^{2+}$ blockade of NMDA current in mechanically injured neurons. *Science*, 274: 1921–1923.

Zheng, X., Zhang, L., Wang, A.P., Bennett, M.V.L., Zhukin, R.S. (1997) $Ca^{2+}$ influx amplifies PKC potentiation of recombinant NMDA receptors. *Journal of Neuroscience*. In press.

Zukin, R.S. and Bennett, M.V.L. (1995) Alternatively spliced isoforms of the NMDARI receptor subunit. *Trends Neurosci.*, 18: 306–313.

O.P. Ottersen, I.A. Langmoen and L. Gjerstad (Eds.)
*Progress in Brain Research*, Vol 116


CHAPTER 14

# Metabotropic glutamate receptors and neurodegeneration

Valeria Bruno[1], Giuseppe Battaglia[1], Agata Copani[2], Giacomo Casabona[1], Marianna Storto[1], Valeria Di Giorgi Gerevini[2], Richard Ngomba[1], and Ferdinando Nicoletti[1,2,*]

[1]*I.N.M. Neuromed, Pozzilli; and* [2]*Institute of Pharmacology, School of Pharmacy, University of Catania, Catania, Italy*

## Introduction

Excitatory amino acid receptors contribute to the pathophysiology of neuronal degeneration in a variety of acute and chronic disorders (Choi, 1992; Lancelot and Beal, this volume). Depending on the extent of the insult, excitotoxic degeneration is a combination of necrosis and apoptosis and is triggered by an increase in cytosolic free $Ca^{2+}$ (Ankarcrona, this volume). While very high levels of intracellular free $Ca^{2+}$ induce predominantly a necrotic phenotype, a less sustained increase in intracellular $Ca^{2+}$ may eventually result into apoptotic death. The underlying mechanism is unclear, but mitochondria may have a critical role in this process (Susin et al., 1997). The entry of $Ca^{2+}$ into the mitochondria may generate a death signal by producing a destabilization of the inner membrane potential, with an ensuing release of a caspase activator through a large cyclosporine-sensitive mitochondrial pore (also see Reynolds, this volume). Caspase activation will eventually lead to apoptotic death if energetic reservoirs are not depleted by the concomitant necrotic process (Nicholson and Thornberry, 1997). Whether necrosis and apoptosis are independent processes or the apoptotic program is permissive with respect to the necrotic event, is unclear at present. How does the neuroprotective therapy fit into this scenario? An obvious goal has been for many years to limit the initial event of excitotoxic degeneration, i.e. the increase in intracellular $Ca^{2+}$. This has been achieved experimentally by antagonizing NMDA receptors, the activation of which leads to a massive influx of extracellular $Ca^{2+}$ through a large conductance ion channel. However, results with competitive or non-competitive NMDA receptor antagonists are so far disappointing, due to the low therapeutic index, the narrow therapeutic window and, in some cases, to the intrinsic toxicity of these compounds. To overcome these limitations, the potential use of subunit-specific NMDA receptor antagonists has recently been proposed (Priestley et al., 1994). These drugs, however, will hamper excitatory synaptic transmission, although their effect would be restricted to targeted brain regions. A down regulation of the GluR2 AMPA receptor subunit has been shown after induction of transient global ischemia in rats and gerbils, lending credence to the idea that AMPA receptors becomes $Ca^{2+}$-permeable during the maturation of excitotoxic damage (Pellegrini-Giampietro et al., 1992). However, the efficacy of AMPA receptor antagonists

*Corresponding author.
Institute of Pharmacology, School of Pharmacy, University of Catania, Viale A. Doria, 6, 95125 Catania, Italy
Tel.: +39 95 337167; fax: +39 95 337167.

has been proven only in few models of excitotoxic degeneration, and the impact of these drugs on excitatory synaptic transmission will be even more dramatic than that of NMDA receptor antagonists.

A great variety of drugs has been introduced with the intent of limiting some of the intracellular events that occur downstream of the increase in intracellular $Ca^{2+}$, including free-radical scavengers, protein kinase C inhibitors, calpain inhibitors, nitric oxide synthase (NOS) inhibitors, and, lately, caspase inhibitors (Favaron et al., 1990; Dawson et al., 1991; Dawson et al., 1995; Dawson and Dawson, 1996; Nicholson and Thornberry, 1997; Nakao and Brundin, this volume). None of these drugs, however, has proven to be effective in vivo, perhaps because excitotoxic death is a combination of many intracellular events. Last, and not least, a variety of neurotrophic factors belonging to different families have been shown to reduce excitotoxic death in vitro (Mattson et al., 1995), but they do not easily penetrate the brain, even under conditions in which the integrity of the blood-brain barrier is impaired, such as after ischemia followed by reperfusion. The ideal neuroprotective drug will therefore incorporate most of the characteristics described so far, providing that it is devoid of major side effects and does not interfere significantly with excitatory synaptic transmission. We now discuss how drugs that interact with individual metabotropic glutamate receptor subtypes meet (or not) these criteria (see also Nicoletti et al., 1996).

**Introductory paragraph on metabotropic glutamate receptors**

Metabotropic glutamate (mGlu) receptors belong to a class of G-protein linked receptors, which also includes the $Ca^{2+}$-sensing receptor and the GABAB receptor (Pin and Duvoisin, 1995; Kaupmann et al., 1997). Eight distinct mGlu receptor subtypes (named mGlu1 to -8) have been characterized and classified into three subgroups. Group-I includes mGlu1 (splice variants: a, b, c, d, e, g) and -5 (splice variants: a and b), which are coupled to polyphosphoinositide (PI) hydrolysis in heterologous expression system (Nakanishi, 1994; Pin and Duvoisin, 1995). These two subtypes, however, differ in the kinetics of intracellular $Ca^{2+}$ mobilization, because activation of mGlu5 receptors in recombinant cells results into an oscillatory increase in cytosolic $Ca^{2+}$, whereas activation of mGlu1a receptors induces a single-peaked increase in intracellular $Ca^{2+}$ (Kawabata et al., 1996). The mGlu1 and -5 receptors differ in their anatomical pattern of expression (for example the mGlu1$\alpha$ receptor is particularly abundant in cerebellar Purkinje cells, whereas the mGlu5 receptor is expressed at high levels in the hippocampus, neostriatum and cerebral cortex), but they share the same subcellular localization in the periphery of the postsynaptic densities (Abe et al., 1992; Aramori and Nakanishi, 1992; Nusser et al., 1994; Takumi et al., this volume). Both mGlu1 and -5 receptors serve, therefore, to amplify postsynaptic responses to concentrations of glutamate that are sufficiently high to spread to the periphery of the synapse (Conn et al., 1994). Recent studies suggest that group-I mGlu receptors may also be localized on presynaptic nerve terminals, where thay can either amplify the evoked release of glutamate or inhibit the release of GABA (Gereau and Conn, 1995; Nicholls, this volume). Thus, mGlu1 and -5 receptors are stimulatory in nature, and reinforce the excitatory synaptic transmission. The mGlu5 receptor is also present in cultured astrocytes, where its expression varies as a function of the growth conditions (Miller et al., 1995).

(R,S)-3,5-Dihydroxyphenylglycine (DHPG) is the only known selective agonist of group-I mGlu receptors. A series of compounds are now available that antagonize with a high degree of selectivity mGlu1 receptors. Among these, 1-aminoindan-1,5-ldicarboxylic acid (AIDA), cyclopropan[b] chromen-1a-carboxylate ethylesther (CPPCOEt), and 4-carboxyphenyglycine (4CPG) are active in the low micromolar range (Pellicciari et al., 1995; Roberts, 1995; Annoura et al., 1996). No drugs are available that selectively antagonize mGlu5 receptors. High concentrations of the mixed mGlu receptor antagonist, $\alpha$-methyl-4-carboxyphenylglycine (MCPG) inhibit responses mediated by mGlu5 receptors in recombinant cells (Conquet et al., 1994; Thomsen et al., 1996; but see also Brabet et al., 1995).

Group-II mGlu receptors include the subtypes mGlu2 and -3, which are negatively coupled to adenylyl cyclase through a pertussis-toxin sensitive GTP-binding protein. The mGlu2 receptor is quite diffused throughout the brain, and is preferentially localized in the preterminal region of the axon, far from the active zone of neurotransmitter release (Tanabe et al., 1993; Shigemoto et al., 1995; Shigemoto et al., 1996). Activation of mGlu2 receptors inhibit the release of glutamate, but only in response to high concentrations of glutamate that spread back to the most remote regions of the axon. The mGlu3 receptor is present in neurons, but is also expressed by astrocytes throughout the brain (Tanabe et al., 1993). The function of glial mGlu3 receptors is unknown, although recent evidence indicates that activation of this receptor may lead to the release of neurotrophic factor of glial origin (Bruno et al., 1997). mGlu2 and -3 receptors are potently activated by a series of compounds, one of which, the compound LY 354740, can penetrate the brain and is therefore systemically active. Other agonists include (2R,4R)-aminopyrrolidine-2,4-dicarboxylate (APDC), (2S,1′R,2′R,3′R)-2-(2,3-dicarboxycyclopropyl)glycine (DCG-IV), (2S,3S,4S)-α-(carboxycyclopropyl)glycine (L-CCG-I) and 4-carboxy-3-hydroxyphenylglycine (4C3HPG). While APDC is highly selective, DCG-IV is also able to activate NMDA receptors at high concentrations, L-CCG-I is also a weak group-I mGlu receptor agonist, whereas 4C3HPG also behaves as a mGlu1 receptor antagonist (Pin and Duvoisin, 1995; Toms and Roberts, 1997). The endogenous dipeptide, N-acetylaspartylglutamate (NAAG) is so far the only selective agonist of mGlu3 receptors, with no activity on mGlu2 receptors (Wroblewska et al., 1997). Group-II mGlu receptor antagonists with a sufficient degree of selectivity are now commercially available. These include the α-methyl derivative of L-CCG-I, or MCCG, the (RS)-α-methyl-4-tetrazolylphenylglycine (MTPG), (2S)-α-ethylglutamic acid, and 2-(2′-carboxy-3′-phenylcyclopropyl)glycine (PCCG-IV) (Roberts, 1995; Thomsen et al., 1996; Toms and Roberts, 1997).

Group-III includes the subtypes mGlu4 (splice variants: a and b), -6, -7 (splice variants: a and b) and -8. The mGlu6 receptor is exclusively localized in the "on" bipolar cells of the retina, where it is coupled to cGMP phosphodiesterase (Nakanishi, 1994). mGlu4, -7 and -8 receptors are heterogeneously distributed in the brain, but they all appear localized on nerve terminals, in close proximity to the active zone of neurotransmitter release (Shigemoto et al., 1995; Takumi et al., this volume). It has been hypothesized that these particular subtypes behave as glutamate autoreceptors and act as a primary fail-safe mechanism to avoid an excessive glutamate release (Conn et al., 1994). All group-III mGlu receptor subtypes are activated with a high degree of selectivity by L-2-amino-4-phosphonobutanoate (AP4) and by its analogue, L-serine-O-phosphate (L-SOP), an endogeneous compound that accumulates in the brain of Alzheimer's patients. Among the available antagonists, the most selective is perhaps the α-methyl derivative of L-SOP, (RS)-α-methylserine-O-phosphate (MSOP). (RS)-α-methyl-4-phosphonophenylglycine (MPPG), α-methyl-4-carboxy-3-phenylalanine (MCPA), (S)-2-amino-2-methyl-4-phosphonobutanoic acid (MAP4), and (RS)-α-cyclopropyl-4-phosphonophenylglycine (CPPG) can also antagonize group-III mGlu receptors, but they are less selective (Toms and Roberts, 1997).

## Group-I mGlu receptors and neurodegeneration: An unsolved controversy

A role for mGlu receptors in neurodegeneration was originally proposed by D. Schoepp and his associates, who showed that intracerebral infusion of the mixed mGlu receptor agonist, 1S,3R-ACPD, induces degeneration of striatal or hippocampal neurons. In adult animals, the toxic action of ACPD was attenuated by the NMDA receptor antagonist, MK-801, which was instead ineffective in newborn animals (McDonald and Schoepp,

1992; Sacaan and Schoepp, 1992; McDonald et al., 1993). A possible explanation is that local infusion of ACPD induces neuronal death by activating group-I mGlu receptors coupled to PI hydrolysis. Pharmacological activation of these receptors would directly result into neuronal death in the early postnatal life, due to the high efficiency of the transduction pathway (Nicoletti et al., 1986). In the adult life, activation of group-I mGlu receptors may not be sufficient to produce neuronal damage but may enable the potential toxicity of endogenously activated NMDA receptors. An interaction between mGlu1a and NMDA receptors has been demonstrated in cotransfected oocytes, where mGlu receptor agonists potentiate NMDA currents. This effect is mediated by the activation of protein kinase C, and depends on the subunit composition of NMDA receptors. Homomeric NR1 or dimeric NR1-2A/2B receptors are potentiated by mGlu1a receptor agonists, whereas dimeric NR1-2C/2D receptors are not (Shen et al., 1995). A synergicstic activation of group-I mGlu and NMDA receptors may explain the toxic effect of $\beta$-methylamino-L-alanine (BMAA), a nonproteic aminoacid present in the seed of *Cyca Circinalis*, which has been implicated in the pathogenesis of the amyotrophic lateral sclerosis/parkinsonism-dementia complex of Guam and Rota islands in the Western Pacific (Spencer et al., 1987). In the presence of bicarbonate ions, BMAA forms a carbamate adduct, which enables its interaction with both NMDA and group-I mGlu receptors (Copani et al., 1990; Weiss and Choi, 1988).

More recently, the enhancement of NMDA toxicity by group-I mGlu receptor agonists has been shown by two independent groups in mixed cultures of mouse cortical cells, using the group-I agonists, quisqualate, DHPG and 3-hydroxyphenylglycine (Bruno et al., 1995a; Buisson and Choi, 1995). Quisqualate and DHPG were equally active when applied during or immediately after the NMDA pulse, and their action was attenuated by protein kinase C inhibitors, and mimicked by the protein kinase C activator, $\beta$-phorbol-12,13-myristate acetate (PMA) (Bruno et al., 1995a). Interestingly, 3-hydroxyphenylglycine potentiates neuronal toxicity induced by mild conditions of oxygen-glucose deprivation in culture, suggesting a potential role for group-I mGlu receptors in ischemic brain damage (Buisson and Choi, 1995). The identity of the mGlu receptor subtype that mediates the potentiation of NMDA toxicity has not yet been established. Intrastriatal infusion of "end-capped" phosphorothioate mGlu5 antisense oligonucleotides protects striatal neurons against malonic acid lesions, suggesting that mGlu5 receptors play a permissive role in this particular form of toxicity (Cha et al., 1996). On the other hand, the selective mGlu1 receptor antagonists AIDA and UPF 596 are neuroprotective in gerbils subjected to transient global ischemia (Moroni et al., 1997). It cannot be excluded, however, that the high doses of these drugs required for neuroprotection antagonize mGlu5 receptors in vivo. High affinity mGlu5 receptor antagonists are needed to examine this question in more detail.

One of the reasons for the limited success of NMDA receptor antagonists as neuroprotective agents in brain ischemia is their narrow therapeutic window. In rats subjected to the 4-vessel occlusion model of global ischemia, the efficacy of excitatory amino acids in stimulating PI hydrolysis is substantially enhanced in vulnerable brain regions from 24 hours to 7 days after post-ischaemic recirculation (Seren et al., 1989), lending credence to the idea that a hyperactivity of group-I mGlu receptors contributes to the maturation of ischemic neuronal damage. If so, one expects that potent and systemically active group-I mGlu receptor antagonists show a favorable therapeutic window as antiischemic agents. It remains to be established, however, which (if any) group-I subtype is up-regulated after ischemia. In situ hybridization studies did not show any increase in mGlu1 or -5 mRNA in the hippocampus of ischemic animals. It will be interesting to examine the expression and distribution of group-I mGlu receptor proteins in animals subjected to global and focal ischemia.

In apparent contrast with a neurotoxic role for group-I mGlu receptors, quisqualate, DHPG and the mGlu5 receptor agonist, trans-azetidine-2,4-dicarboxylic acid, protect cultured cerebellar granule cells against NMDA toxicity (Pizzi et al., 1993; Montoliu et al., 1997). Neuroprotection is enhanced by the nootropic drug, aniracetam, which also potentiates the stimulation of PI hydrolysis by mGlu receptor agonists (Pizzi et al., 1993). Thus, it appears that in cultured cerebellar granule cells activation of group-I mGlu receptors is neuroprotective. In this particular model, the synergistic effect between group-I mGlu receptors and NMDA receptors may be lacking, due to the presence of the NR2C subunit in the heteromeric composition of NMDA receptors. The absence of the neurotoxic pathway may disclose a neuroprotective component, which becomes prevalent in cultured cerebellar granule cells. Interestingly, however, quisqualate and DHPG can also protect hippocampal slices against NMDA toxicity (Pizzi et al., 1996). Whereas quisqualate, used at 100 $\mu$M, can also activate mGlu3 receptors, DHPG is a highly selective agonist of group-I mGlu receptors. These results are not therefore consistent with a neurotoxic activity of group-I mGlu receptors. One possible explanation is that activation of group-I mGlu receptors leads either to neurotoxicity or neuroprotection depending on the paradigm of excitotoxicity and, in particular, on the structural and functional properties of the NMDA receptors present in the same synapsis.

**Neuroprotective activity of group-II and -III mGlu receptors**

Activation of group-II mGlu receptors is neuroprotective in a variety of models, including neuronal cultures, brain slices and in vivo models of excitotoxicity. In mixed cultures of mouse cerebral cortex, DCG-IV, 4C3HPG, L-CCG-I and 1S,3R-ACPD protect neurons against the toxic effect of a brief pulse with NMDA (Bruno et al., 1994; 1995b; Buisson et al., 1995; 1996; Ambrosini et al., 1995). Although these compounds are not subtype-selective, their effect is substantially reduced by the group-II mGlu receptor antagonists, PCCG-IV or MCCG-I (Bruno et al., 1995b; Thomsen et al., 1996), suggesting that activation of mGlu2 or -3 receptors protects cortical neurons in culture against NMDA toxicity. Accordingly, the highly selective mGlu2/3 receptor agonist, $2_R$-$4_R$-APDC exhibits neuroprotective acivity in mouse cortical cultures (Bruno et al., unpublished). Neuroprotection by group-II mGlu receptor agonists has also been observed in primary cultures of cerebellar granule cells, which predominantly express mGlu3 receptors, as well as in primary cultures of mesencephalic neurons (Pizzi et al., 1993; Ambrosini et al., 1995). Whether or not group-II mGlu receptor agonists attenuate kainate-induced toxicity is matter of controversy. Whereas Bruno et al. (1995b) found that DCG-IV or L-CCG-I are highly protective against kainate-induced neurodegeneration in mouse cultured cortical cells, Buisson et al., using the same model, showed no overall activity of DCG-IV, whereas Turetsky et al. (1995) found that protection by DCG-IV is restricted to the small percentage of neurons that respond to kainate with an enhanced influx of $Co^{2+}$. On the other hand, i.c.v. infusion of DCG-IV protects vulnerable neurons against toxicity induced by systemic or local injection of kainate (Kwak et al., 1994).

As discussed above, a safe neuroprotective drug with a favourable therapeutic window will be particularly advantageous in the experimental therapy of ischemic brain damage. Group-II mGlu receptor agonists might meet these criteria, because (i) they are neuroprotective against degeneration induced by oxygen-glucose deprivation in culture; (ii) they are not expected to impair excitatory synaptic transmission to the same extent as NMDA or AMPA receptor antagonists; and (iii) at least DCG-IV and 4C3HPG protects cultured neurons even if applied 60 min after a toxic pulse with NMDA, a time at which MK-801 is no longer protective (Bruno et al., 1995b). The availability of potent and systemically active mGlu2/3 receptor agonists, such as LY354740, allows now to examine whether activation of group-II mGlu receptors is neuroprotective in in vivo models of focal or global ischemia.

The mechanism underlying group-II mGlu receptor-mediated neuroprotection has been examined in cultured cortical cells exposed to a brief pulse with NMDA, where toxicity is induced by the influx of extracellular $Ca^{2+}$ through the NMDA channel during the pulse, but is amplified by the endogenous glutamate, which is secondarily released shortly after the termination of the pulse (Monyer et al., 1992). Hence, it has been suggested that activation of group-II mGlu receptors protects neurons by reducing the release of endogenous glutamate. However, this mechanism is not sufficient to explain the delayed rescue effect produced by DCG-IV applied 60 min after the NMDA pulse, when large amounts of endogenous glutamate have already accumulated in the extracellular medium (Bruno et al., 1995b). Searching for an additional mechanism, it has been observed that the protein synthesis inhibitor cycloheximide (CHX) applied after the NMDA pulse prevents the neuroprotective activity of DCG-IV in cultured cortical cells (Bruno et al., 1997a). Thus, one can speculate that activation of group-II mGlu receptors triggers a specific program of neuroprotection, which requires new protein synthesis. Once activated, this program can provide a long-lasting neuroprotection, because a 10-min exposure to DCG-IV, L-CCG-I or 4C3HPG is sufficient to rescue cortical neurons from the toxic action of NMDA applied several hours later. We have demonstrated that this particular form of neuroprotection takes place in glial cells, which express mGlu3 receptors. The conditioned medium collected from pure cultures of astrocytes transiently exposed to DCG-IV, L-CCG-I, 4C3HPG, or to α-NAAG (a selective mGlu3 receptor agonist) is highly neuroprotective when transferred to mixed cultures challenged with NMDA. Protection is mediated by a CHX- and heat-sensitive factor, the identity of which is unknown at present (Bruno et al., 1997a; see Fig. 1). One member of the transforming-growth factor β (TGF-β) family may be a candidate, because neutralizing antibodies against TGF-β attenuate the protective effect of the glial conditioned medium. An alternative explanation is that astrocytes synthesize and release a factor that requires the permissive action of TGF-β to be neuroprotective (Bruno et al., unpublished observation). This speculation is based on the assumption that astrocytes are able to secrete proteic factors in response to mGlu3 receptor activation. Evidence in this line is provided by the observation that DCG-IV inceases the secretion of both nerve-growth factor and protein S-100 from rat cultures astrocytes (Caciagli, unpublished observation).

If this mechanism is proven to be correct, glial mGlu3 receptors will become a novel target for neuroprotective drugs. Activation of glial mGlu3 receptors is expected to produce a minimal impact on excitatory synaptic transmission and may provide a broad spectrum mechanism or protection in different paradigm of toxicity. It is perhaps through this mechanism that DCG-IV attenuates neuronal apoptosis induced by the protein kinase inhibitor staurosporine (Buisson et al., 1994), β-amyloid peptide (Copani et al., 1995), or oxygen-glucose deprivation in the presence of a cocktail of ionotropic glutamate receptor antagonists (Buisson and Choi, 1995), i.e. under conditions in which the endogenous glutamate has no role in the excitotoxic damage. This is particularly relevant because β-amyloid peptide is the major constituent of the senile plaques in the brain of Alzheimer's disease, a chronic degenerative disease in which apoptosis is considered to contribute to neuronal death (Loo et al., 1993).

Activation of group-III mGlu receptors by L-AP4 or L-SOP protects both hippocampal and cortical neurons against excitotoxic death. Neuroprotection is prevented by the group-III mGlu receptor antagonists, MPPG or α-methyl-3-carboxyphenylalanine (Bruno et al., 1996; Maiese et al., 1995). Although group-III mGlu receptors share the same transduction pathway as group-II receptors in recombinant cells (i.e. inhibition of adenylyl cyclase and voltage-operated $Ca^{2+}$ channels), the mechanism underlying neuroprotection differs from these groups of subtypes. As opposed to group-II mGlu receptor agonists, L-AP4 does not protect cultured neurons when applied before the NMDA pulse, and the conditioned medium collected from astrocytes treated with L-AP4 is not neuroprotective (Bruno et al., 1997a). Thus, while neuroprotection by group-II mGlu receptor agonists is mediated by a combination of glial and neuronal events, protection by group-III mGlu receptors may be exclusively dependent on the inhibition of glutamate release. mGlu4, -7 and -8 receptors are presynaptically located in the vicinity of the active zone of transmitter release, and their activation reduces glutamate release through the inhibition of voltage-operated $Ca^{2+}$ channels (Herrero et al., 1996). However, it is noteworthy that the function of group-III mGlu receptors is not as established as other mGlu receptor subtypes. For example, mGlu4 and mGlu2 receptors differ in their coupling to GTP-binding proteins, and only the mGlu4 receptor appears to be coupled to the $\alpha$ subunit of the Gz protein in recombinant cells (Pin, personal communication). How this may contribute to the neuroprotective activity of native mGlu4 receptors may be the subject of future investigation.

**Conclusions and implications for new therapeutical strategies**

The experimental therapy of acute or chronic neurodegenerative disorders can be addressed to the ethiology of the specifc disorder, to the pathophysiology of the degenerative process, or to the reparative phonomena with the aim of preventing the cause, slowing or atteuating the progression, or reinforcing the function of com-

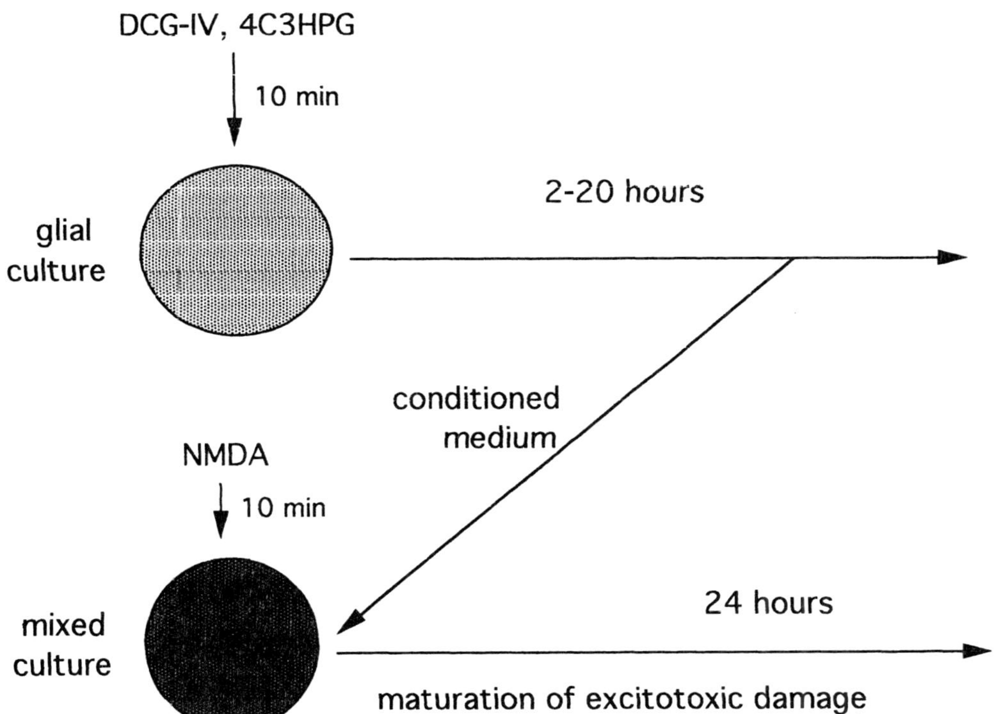


Fig. 1. Experimental strategy designed to uncover the role of astrocytes in the neuroprotective activity of group-II mGlu receptor agonists. Pure cultures of mice cortical glial cells were transiently exposed to DCG-IV or 4C3HPG. The conditioned serum-free medium, collected 2–20 later, was transferred to mixed cortical cultures challenged with NMDA. The astrocyte medium was highly neuroprotective when applied immediately after the NMDA pulse (Bruno et al., 1997a).

pensatory neuronal circuitries. mGlu receptors may be considered as a potential target for drugs aimed at reducing the progression of neuronal degeneration (see Fig. 2). Agonists of group-II receptors may well subserve this function without the undesirable side effects associated with a chronic use of ionotropic glutamate receptor antagonists. We await the results of the clinical trials with the Lilly compound, LY 354740, to know whether group-II mGlu receptor agonists are well tolerated by patients.

The consequences of a chronic use of group-III mGlu receptor agonists in living animals or humans are unknown, because the current agonists available, i.e. L-AP4 and L-SOP, may not cross the blood-brain barrier and any effect is confounded by their ability to interact with the $Ca^{2+}/Cl^{-}$-dependent glutamate transporter. One can predict, however, that a sustained activation of mGlu4, -7 or -8 receptors leads to a significant reduction in the release of glutamate, due to the localization of these particular subtypes in the vicinity of the active core of transmitter release (see above).

The potential use of drugs that interact with group-I mGlu receptors is more controversial because activation of mGlu1 and -5 receptors may result into neurotoxicity or neuroprotection in different models and different circumstances. In addition, in vivo studies are limited by the absence of potent group-I mGlu receptor agonists (DHPG is active in the micromolar range), and by the lack of a selective mGlu5 receptor antagonists. Knockout mice lacking the mGlu1a or -5 receptors have selective deficits in activity-dependent forms of synaptic plasticity, including long-term potentiation (LTP) and long-term depression (LTD) (Aiba et al., 1994a,b; Conquet et al., 1994; Lu et al., 1997). One can therefore predict that a chronic use with group-I antagonists leads to an impairment of

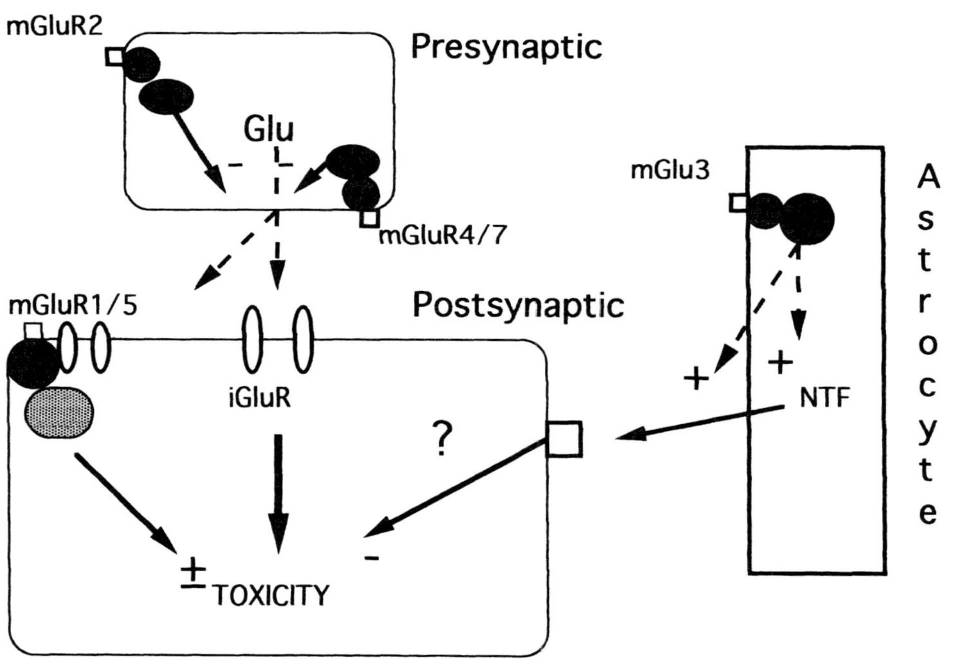


Fig. 2. Modulation of neuronal toxicity by individual mGlu receptor subtypes. Note that activation of glial mGlu3 receptors stimulates the synthesis and/or release of a putative neurotrophic factor (NTF) that supports neuronal survival. GluR = ionotropic glutamate receptor.

synaptic plasticity, thus limiting both the learning process and the formation of new synaptic contacts that compensate for the functional damage. Whether group-I agonists increase the efficacy of new synaptic contacts in the chronically damaged brain has never been addressed; however, group-I agonists deteriorate rather than improving learning in experimental animals, possibly because these drugs increase the background level of synaptic activity, thus making the formation of a "signal-to-noise" more difficult (Riedel, 1996).

In conclusion, we prefer to consider group-II mGlu receptors as potential drug targets for neuroprotective drugs (see Table 1), although further studies are needed to consolidate this preference.

TABLE 1

mGlu receptors: Potential role in neurodegeneration

| Subtype | transduction mechanism | agonists | antagonists | effect on toxicity | lacking |
|---|---|---|---|---|---|
| mGlu1/5 | PI hydrolysis<br>↓K channels | DHPG | CPCCOEt<br>4CPG | ± | mGlu5 antagonists;<br>more potent agonists |
| mGlu2/3 | cAMP<br>↓Ca channels | APDC<br>LY354740 | EGlu<br>PCCG-IV | protection | – |
| mGlu4/7/8 | cAMP<br>↓Ca channels others ? | L-AP4<br>L-SOP | MSOP<br>CPPG | protection? | more potent and selective agonists |

## List of abbreviations

| | |
|---|---|
| NMDA | *N*-methyl-D-aspartate |
| AMPA | α-amino-3-hydroxy-5-methyl-4-is oxazole propionic acid |
| NOS | nitric oxide synthase |
| mGlu | metabotropic glutamate receptor |
| PI | polyphosphoinositide |
| DHPG | (R,S)-3,5-Dihydroxyphenylglycine |
| AIDA | 1-aminoindan-1,5-dicarboxylic acid |
| CPPOEt | cyclopropan[b]chromen-1a-carboxylate ethylesther |
| 4CPG | 4-carboxyphenylglycine |
| MCPG | α-methyl-4-carboxyphenylglycine |
| APDC | (2R,4R)-aminopyrrolidine-2,4-dicarboxylate |
| DCG-IV | (2S,1′R,2′R,3′R)-2-(2,3-dicarboxycyclopropyl)glycine |
| L-CCG-I | (2S,3S,4S)-a-(carboxycyclopropyl)glycine |
| 4C3HPG | 4-carboxy-3-hydroxyphenylglycine |
| NAAG | N-acetylaspartylglutamate |
| MCCG | 2S,1′S,2′S-2-methyl-2-(2′carboxycyclopropyl)glycine |
| MTPG | (RS)-a-methyl-4-tetrazolylphenylglycine |
| PCCG-IV | 2-(2′-carboxy-3′-phenylcyclopropyl)glycine |
| L-SOP | L-serine-O-phosphate |
| MSOP | (RS)-α-methylserine-O-phosphate. |
| MPPG | (RS)-α-methyl-4-phosphonophenylglycine |
| MCPA | α-methyl-4-carboxy-3-phenylalanine |
| MAP4 | (S)-2-amino-2-methyl-4-phosphonobutanoic acid |

| | |
|---|---|
| CPPG | (RS)-α-cyclopropyl-4-phosphonophenylglycine |
| BMAA | β-methylamino-L-alanine |
| MK-801 | dizocilpine |
| $1S,3_R$-ACPD | 1-aminocyclopentane-$1_S,3_R$-dicarboxylic acid |
| CHX | cycloheximide |
| TGF-β, | Transforming growth factor β |
| L-AP4 | L-2-amino-4-phosponobutanoate |
| LTP | long term potentiation |
| LTD | long term depression |
| iGluR | ionotropic glutamate receptor |
| NTF | neurotrophic factor |

## References


Abe, T., Shigemoto, R. and Nakanishi, S. (1992) Molecular characterization of a novel metabotropic glutamate receptor mGluR5 coupled to inositol phosphate/$Ca^{2+}$ signal transduction. *J. Biol. Chem.*, 267: 13361–13367.

Aiba, A., Chen, C., Herrup, K., Rosenmund, C., Stevens, C.F. and Tonegawa, S. (1994a) Reduced hippocampal long-term potentiation and context-specific deficit in associative learning in mGluR1 mutant mice. *Cell*, 79: 365–375.

Aiba, A., Kano, M., Chen, C., Stanton, M.E., Fox, G.D., Herrup, K., Zwingman, T.A. and Tonegawa, S. (1994b) Deficient cerebellar long-term depression and impaired motor learning in mGluR1 mutant mice. *Cell*, 79: 377–388.

Ambrosini, A.L., Bresciani, S., Fracchia, N., Brunello, N. and Racagni, G. (1995) Metabotropic glutamate receptors negatively coupled to adenylate cyclase inhibit *N*-methyl-D-aspartate receptor activity and prevent neurotoxicity in mesencephalic neurons in vitro. *Mol. Pharmacol.*, 47: 1057–1064.

Annoura, H., Fukunaga, A., Uesugi, M., Tatsuoka, T. and Horikawa, Y. (1996) A novel class of antagonists for metabotropic glutamate receptors, 7-(hydroxymino)cyclopropa[b]chromen-1α-carboxylates. *Bioorg. Med. Chem. Lett.*, 6: 763–766.

Aramori, I. and Nakanishi, S. (1992) Signal transduction and pharmacological characteristics of a metabotropic glutamate receptor, mGluR1, in transfected CHO cells. *Neuron*, 5: 757–765.

Brabet, I., Mary, S., Bockaert, J. and Pin, J.P. (1995) Phenyglycine derivatives discriminate between mGluR1- and mGluR5-mediated responses. *Neuropharmacol.*, 34: 895–904.

Bruno, V., Battaglia, G., Copani, A., Giffard, R.G., Raciti, G., Raffaele, R., Shinozaki, H. and Nicoletti, F. (1994) Protective effect of the metabotropic glutamate receptor agonist, DCG-IV, against excitotoxic neuronal death. *Eur. J. Pharmacol.*, 256: 109–112.

Bruno, V., Copani, A., Knopfel, T., Kuhn, R., Casabona, G., Dell'Albani, P., Condorelli, D.F. and Nicoletti, F. (1995a) Activation of metabotropic glutamate receptors coupled to inositol phospholipid hydrolysis amplifies NMDA-induced neuronal degeneration in cultured cortical cells. *Neuropharmacol.*, 34: 1089–1098.

Bruno, V., Battaglia, G., Copani, A., Giffard, R.G., Raciti, G., Raffaele, R., Shinozaki, H. and Nicoletti, F. (1995b) Activation of class II or III metabotropic glutamate receptors protects cultured cortical neurons aganist excitotoxic degeneration. *Eur. J. Neurosci.*, 7: 1906–1913.

Bruno, V., Copani, A., Bonanno, L., Knoepfel, T., Kuhn, R., Roberts, P.J. and Nicoletti, F. (1996) Activation of group III metabotropic glutamate receptors is neuroprotective in cortical cultures. *Eur. J. Pharmacol.*, 310: 61–66.

Bruno, V., Sureda, F.X., Storto, M., Casabona, G., Caruso, A., Knopfel, T., Kuhn, R. and Nicoletti, F. (1997a) The neuroprotective activity of group-II metaotropic glutamate receptors requires new protein synthesis and involves a glial-neuronal signalling. *J. Neurosci.*, 17: 1891–1897.

Bruno, V., Wroblewska, B., Wroblewski, J.T., Fiore, L. and Nicoletti, F. (1997b) Neuroprotective activity of N-acetylaspartylglutamate in cultured cortical cells. *Neurosci.*, in press.

Buisson, A., Yu, S. and Choi, D.W. (1994) Role of metabotropic glutamate receptors in excitotoxic and apoptotic cell death. *Soc. Neurosci. Abstr.*, 20: 198.5.

Buisson, A. and Choi, D.W. (1995) The inhibitory mGluR agonist, S-4-carboxy-3-hydroxyphenylglycine selectively attenuates NMDA neurotoxicity and oxygen-glucose deprivation-induced neuronal death. *Neuropharmacology*, 34: 1081–1087.

Buisson, A., Yu, S. and Choi, D.W. (1996) DCG-IV selectively attenuates rapidly triggered NMDA-induced neurotoxocity in cortical neurons. *Eur. J. Neurosci.*, 8: 138–143.

Cha, J.J., Talati, A., Kerner, J.A., Lichtenbaum, R.A. and Young, A.B. (1996) Antisense oligonucleotides directed against mGluR5 glutamate receptor protect against malonic acid lesions in rat striatum. *Neuropharmacol.*, 35, A7 (28).

Choi, D.W. (1992) Excitotoxic cell death. *J. Neurobiol.*, 23: 1261–1276.

Conn, P.J., Winder, D.G. and Gereau, IV R.W. (1994) Regulation of neuronal ciruits and animal behavior by metabotropic glutamate receptors. In: *The Metabotropic Glutamate Receptors*, (P.J. Conn and J. Patel, eds.) Humana Press Inc. (Totowa, NJ), pp. 195–229.

Conquet, F., Bashir, Z.I., Davies, C.H., Daniel, H., Ferraguti, F., Bordi, F., Franz-Bacon, K., Reggiani, A., Matarese, V., Conde, F., Collingridge, G.L. and Crepel, F. (1994) Motor deficit and impairment of synaptic plasticity in mice lacking mGluR1. *Nature*, 372: 237–243.

Copani, A., Canonico, P.L. and Nicoletti, F. (1990) *β-N*-methylamino-L-alanine (L-BMAA) is a potent agonist of "metabolotropic" glutamate receptors. *Eur. J. Pharmacol.*, 181: 327–328.

Copani, A., Bruno, V., Battaglia, G., Leanza, G., Pellitteri, R., Russo, A., Stanzani, A. and Nicoletti, F. (1995) Activation of metabotropic glutamate receptors protects neurons against apoptosis induced by *β*-amyloid peptide. *Mol. Pharmacol.*, 47: 890–897.

Dawson, T.M., Hung, K., Dawson, V.L., Steiner, J.P. and Snyder, S.H. (1995) Neuroprotective effects of gangliosides may involve inhibition of nitric oxide synthase. *Ann. Neurol.*, 37: 115–118.

Dawson, V.L. and Dawson, T.M. (1996) Free radicals and neuronal cells death. *Cell Death Differ.*, 3: 71–78.

Dawson, V.L., Dawson, T.M., London, E.D., Bredt, D.S. and Snyder, S.H. (1991) Nitric oxide mediates glutamate neurotoxicity in primary cortical cultures. *Proc. Natl. Acad. Sci.*, USA 88: 6368–6371.

Favaron, M., Manev, H. and Siman, R. (1990) Down regulation of protein kinase C protects cerebellar granule neurons in primary culture from glutamate-induced neuronal death. *Proc. Natl. Acad. Sci. USA*, 87: 1983–1987.

Gereau, R.W. and Conn, P.J. (1995) Multiple presynaptic metabotropic glutamate receptors modulate excitatory and inhibitory synaptic transmission in hippocampal area CA1. *J. Neurosci.*, 15: 6879–6889.

Herrero, I., Vazquez, E., Miras-Portugal, M.T. and Sanchez-Prietro, J. (1996) A decrease in $[Ca^{2+}]_c$ but not in cAMP mediates the L-AP4 inhibition of glutamate release: PKC-mediated suppression of this inhibitory pathway. *Eur. J. Neurosci.*, 8: 700–709.

Kaupmann. K., Huggel. K., Heid. J., Flor, P.J., Bishoff, S., Mickel, S.J., McMaster, G., Angst, C., Bittiger, M., Froestl, W. and Bettler, B. (1997) Expression cloning of $GABA_B$ receptors uncovers similarity to metabotropic glutamate receptors. *Nature*, 386: 239–246.

Kawabata, S., Tsutsumi, R., Kohara, A., Yamaguchi, T., Nakanishi, S. and Okada, M. (1996) Control of calcium oscillations by posphorylation of metabotropic glutamate receptors. *Nature*, 383: 89–92.

Kingston, A.E., Burnett, J.P., Mayne, N.G. and Lodge, D. (1995) Pharmacological analysis of 4-carboxyphenylglycine derivatives: Comparison of effects on mGluR1a and mGluR5 subtypes. *Neuropharmacol.*, 34: 887–894.

Koroshetz, W.J. and Moskowitz, M.A. (1996) Emerging treatments for stroke in humans. *Trends Pharmacol. Sci.*, 17: 227–234.

Kwak, S., Ishida, M., Miyamoto, M. and Shinozaki, H. (1994) DCG-IV, a potent agonist for metabotropic receptors alleviates excitotoxin-induced neuronal damage in vivo. Abstract of the *Ninth Rinshoken International Conference on Excitatory Amino Acid: Approaches to Clinical Use*, November 28–30, p. 16.

Loo, D.T., Copani, A., Pike, C.J., Whittemore, E.R., Walencewiczs, A.J. and Cotman, C.W. (1993) Apoptosis is induced by *β*-amyloid in cultured central nervous system neurons. *Proc. Natl. Acad. Sci. USA*, 90: 7951–7955.

Lu, Y.M., Jia, Z., Janus, C., Henderson, J.T., Gerlai, R., Wojtowicz, J.M. and Roder, J.C. (1997) Mice lacking metabotropic glutamate receptor 5 show impaired learning and reduced CA1 long-term potentiation (LTP) but normal CA3 LTP. *J. Neurosci.*, 17: 5196–5205.

Maiese, K., Greenberg, R., Boccone, L. and Swiriduk, M. (1995) Activation of metabotropic glutamate receptors is neuroprotective during nitric oxide toxicity in primary hippocampal neurons of rats. *Neurosci. Lett.*, 194: 173–176.

Mattson, M.P., Lowell, M.A., Furukowa, K. and Markesberym, W.R. (1995) Neurotrophic factors attenuate glutamate-induced accumulation of peroxides, elevation of intracellular $Ca^{2+}$ concentration, and neurotoxicity and increase antioxidant enzyme activities in hippocampal neurons. *J. Neurochem.*, 65: 1740–1745.

McDonald, J.W. and Schoepp, D.D. (1992) The metabotropic excitatory amino acid receptor agonist, 1S,3R-ACPD selectively potentiates *N*-methyl-D-aspartate-induced brain injury. *Eur. J. Pharmacol.*, 215: 353–354.

McDonald, J.W., Fix, A.S., Tizzano, J.P. and Schoepp, D.D. (1993) Seizures and brain injury in neonatal rats induced by 1S,3R-ACPD, a metabotropic glutamate receptor agonist. *J. Neurosci.* 13: 4445–4455.

Miller, S., Romano, C. and Cotman, C.W. (1995) Growth factor up-regulation of a phosphoinositide-coupled metabotropic glutamate receptor in cortical astrocytes. *J. Neurosci.*, 15: 6103–6109.

Montoliu, C., Llansola, M., Cucarella, C., Grisolia, S. and Felipo, V. (1997) Activation of the metabotropic glutamate receptor mGluR5 prevents glutamate toxicity in primary cultures of cerebellar neurons. *J. Pharmacol. Exp. Ther.*, 281: 643–647.

Monyer, H., Giffard, R.G., Hartley, D.M., Dugan, L.L., Goldberg, M.P. and Choi, D.W. (1992) Oxygen and glucose deprivation-induced neuronal injury in cortical cell cultures is reduced by tetanus toxin. *Neuron*, 8: 967.

Moroni, F., Cozzi, A., Pellegrini-Giampietro, D., Leonardi, P., Attucci, S., Peruginelli, F., Mannaioni, G. and Pellicciari R. (1997) Group 1 and group 2 metabotròpic glutamate receptor agents: Pharmacological studies in vitro and in vivo. *The Journal of the European College of Neuropsychopharmacology*, 7 (suppl.2): S95.

Nakanishi, S. (1992) Molecular diversity of glutamate receptors and implications for brain function. *Science*, 258: 597–603.

Nakanishi, S. (1994) Metabotropic glutamate receptor: Synaptic transmission modulation and plasticity. *Neuron*, 13: 1031–1037.

Nicholson, D.W. and Thornberry, N.A. (1997) Caspases: Killer proteases. *Trends Biochem. Sci.*, 22: 299–306.

Nicoletti, F., Iadarola, M.J., Wroblewski, J.T. and Costa, E. (1986) Excitatory amino acid recognition sitescoupled with inositol phospholipid metabolism: Developmental changes and interaction with α1-adrenoceptors. *Proc. Natl. Acad. Sci. USA*, 83: 1931–1935.

Nusser, Z., Mulvihill, E., Treit, P. and Somogyi, P. (1994) Subsynaptic segregation of metabotropic and ionotropic glutamate receptors as revealed by immunogold localization. *Neuroscience*, 61: 421–427.

Orlando, L.R., Standaert, D.G., Cha, J.-H., Penney, J.B. and Young, A.B. (1995) Metabotropic receptors in excitotoxicity: (S)-4C3HPG protects against striatal quinolinic acid lesions. *Soc. Neurosci. Abstr.*, 21: 529.19.

Pellegrini-Giampietro, D.E., Zukin, R.S., Bennett, M.V.L., Cho, S.L. and Pulsinelli, W.A. (1992) Switch in glutamate receptor subunit gene expression in CA1 subfield of hippocampus following global ischemia in rats. *Proc. Natl. Acad. Sci. USA*, 89: 10499–10503.

Pellicciari, R., Luneia, R., Costantino, G., Marinozzi, M., Natalini, B., Jakobsen, P., Kanstrup, A., Lombardi, G., Moroni, F. and Thomsen, C. (1995) 1-Aminoindan-1,5-dicarboxylic acid, a novel antagonist at phospholipase C-linked metabotropic glutamate receptors. *J. Med. Chem.*, 38: 3717–3719.

Petralia, R.S., Wang, W.-X., Niedzielski, A.S. and Wenthold, R.J. (1996) The metabotropic glutamate receptors, mGluR2 and mGluR3, show unique postsynaptic, presynaptic and glial localizations. *Neuroscience*, 71: 949–976.

Pin, J.-P. and Duvoisin, R. (1995) The metabotropic glutamate receptors: structure and functions. *Neuropharmacol.*, 34: 1–26.

Pizzi, M., Fallecara, C., Arrighi, V., Memo, M. and Spano, P.F. (1993) Attenuation of excitatory amino acid toxicity by metabotropic glutalmate receptor agonists and aniracetam in primary cultures of cerebellar granule cells. *J. Neurochem.*, 61: 683–689.

Pizzi, M., Consolandi, O., Memo, M. and Spano, P.F. (1996) Activation of multiple metabotropic glutamate receptor subtypes prevents NMDA-induced excitotoxicity in rat hippocampal slices. *Eur. J. Neurosci.*, 8: 1516–1521.

Priestley, T., Ochu, E. and Kemp, J.A. (1994) Subtypes of NMDA receptor in neurons cultured from rat brain. *NeuroReport*, 5: 1763–1765.

Riedel, G. (1996) Function of metabotropic glutamate receptors in learning and memory. *Trends Neurosci.*, 19: 219–224.

Roberts, P.J. (1995) Pharmacological tools for the investigation of metabotropic glutamate receptors (mGluRs): Phenylglycine derivatives and other selective antagonists–an update. *Neuropharmacol.*, 34: 813–820.

Sacaan, A.I. and Schoepp, D.D. (1992) Activation of hippocampal metabotropic excitatory amino acid receptors leads to seizures and neuronal damage. *Neurosci. Lett.*, 139: 79–82.

Schoepp, D.D. and Conn, P.J. (1993) Metabotropic glutamate receptors in brain function and pathology. *Trends Pharmacol. Sci.*, 14:13–20.

Seren, M.S., Aldinio, C., Zanoni, R., Leon, A. and Nicoletti, F. (1989) Stimulation of inositol phospholipid hydrolysis by excitatory amino acids is enhanced in brain slices from vulnerable regions after transient global ischaemia. *J. Neurochem.*, 53: 1700–1705.

Shen, H., Gorter, J.A., Aronica, E., Zheng, X., Zhang, L., Bennett, M.V.L. and Zukin, R.S. (1995) Potentiation of NMDA responses by metabotropic receptor activation depends on NMDA receptor subunit composition. *Soc. Neurosci. Abstr.*, 21: 39.3.

Shigemoto, R., Wada, E., Ohishi, H., Takada, M., Mizuno, M., Roberts, J.D.B. and Somogyi, P. (1995) Differential presynaptic localization of metabotropic glutamate receptor subtypes, mGluR2/3 and mGluR7 in the rat hippocampus. *Soc. Neurosci. Abstr.*, 21: 3380.20

Shigemoto, R., Kulik, A., Roberts, J.D.B., Ohishi, H., Nusser, Z., Kaneko, T. and Somogyi, P. (1996) Target-cell-specific concentration of a metabotropic glutamate receptor in the presynaptic active zone. *Nature*, 381: 523–525.

Spencer, P.S., Nunn, P.B., Hugon, J., Ludolph, A.C., Ross, S.M., Roy, D.N. and Robertson, R.C. (1987) Guam amyotrophic lateral sclerosis: Dementia linked to a plant excitant neurotoxin. *Science*, 237: 517–522.

Susin, S.A., Zamzami, N., Castedo, M., Daugas, E., Wang, H.-G., Geley, S., Fassy, F., Reed, J. and Kroemer, G. (1997) The central executioner of apoptosis: Multiple connections between protease activation and mitochondria in Fas/APO-1/CD95-and ceramide-induced apoptosis. *J. Exp. Med.*, 1: 25–37.

Tanabe, Y., Masu, M., Ishii, I., Shigemoto, R., Mizuno, N. and Nakanisgi, S. (1992) A family of metabotropic receptord. *Neuron*, 8: 169–179.

Tanabe, Y., Nomura, A., Masu, M., Shigemoto, R., Mizuno, N. and Nakanishi, S. (1993) Signal transduction, pharmacological properties, and expression patterns of two rat metabotropic glutamate receptors, mGluR3 and mGluR4. *J. Neurosci.*, 13: 1372–1378.

Thomsen, C., Bruno, V., Nicoletti, F., Marinozzi, M. and Pellicciari, R. (1996) (2S,1′S,2′S,3′R)-2-(2-carboxy-3′-phenyl-cyclopropyl)glycine, a potent and selective antagonist of type-2 metabotropic glutamate receptors. *Mol. Pharmacol.*, 50: 6–9.

Toms, N.J. and Roberts, P.J. (1997) Metabotropic glutamate receptors. 1997 Receptor & ion channel nomenclature supplement. *Trends Pharmacol. Sci.*

Turetsky, D.M., Buisson, A. and Choi, D.M. (1995) The metabotropic glutamate receptor agonist, 4C3HPG, reduces kainate-induced death of cortical neurons expressing calcium-permeable AMPA/kainate receptors. *Soc. Neurosci.*, Abstr. 21: 590.20.

Weiss, J.H. and Choi, D.W. (1988) β-N-methylamino-L-alanine neurotoxicity: Requirement for bicarbonate as a cofactor. *Science*, 241: 973–975.

Wroblewska, B., Wroblewski, J.T., Pschenichkin, S., Surin A., Sullivan, S.E. and Neale, J.H. (1997) *N*-acetylaspartylglutamate selectively activates mGluR3 receptors in transfected cells. *J. Neurochem.*, 69: 174–181.

SECTION IV

# Glutamate toxicity and cell death in the central nervous system

O.P. Ottersen, I.A. Langmoen and L. Gjerstad (Eds.)
*Progress in Brain Research*, Vol 116


CHAPTER 15

# Intracellular calcium and magnesium: Critical determinants of excitotoxicity?

Ian J. Reynolds

*Department of Pharmacology, University of Pittsburgh, E1354 Biomedical Science Tower, Pittsburgh, PA 15261-0001, USA*

## Introduction

It is widely recognized that glutamate receptor agonists make up an efficacious family of neurotoxins. Glutamate itself is less effective under normal circumstances because of the careful control of extracellular concentrations of this neurotransmitter that is documented elsewhere in this volume by Billups et al. However, under circumstances of bioenergetic compromise the dark side of glutamate signaling is revealed, and glutamate transforms from an essential excitatory neurotransmitter to a lethal over-excitant that can kill neurons following exposures as short as a few minutes. In vitro, glutamate alone is sufficient to kill neurons. While it is unlikely that elevated extracellular glutamate is the sole injurious factor found in pathophysiological circumstances, injury induced by glutamate receptor activation represents a convenient model for the investigation of the mechanisms by which neurons may die in acute disease states like cerebral ischemia and head trauma. Accordingly, many investigators have focused on glutamate mediated signaling to gain insight into the mechanisms of neuronal injury that occur in ischemic injury (Crépel et al., this volume).

The pioneering work of Olney, followed by that of Choi and Rothman (Rothman and Olney, 1987a; Choi, 1988) provided essential insights into the critical features of acute glutamate-induced neuronal injury. These studies identified several forms of toxicity that could be associated with the activation of different subtypes of ionic glutamate receptors. The most rapidly induced form of injury occurred as the result of *N*-methyl-D-aspartate (NMDA) receptor activation. NMDA mediated injury could be induced with agonist exposure lasting as little as five minutes, and was expressed, in the form of loss of viability, a number of hours later (Choi et al., 1987; Rothman et al., 1987b). This form of injury could be completely prevented by omission of $Ca^{2+}$ from the media during agonist exposure (Choi, 1987; Rothman et al., 1987b). Activation of non-NMDA receptors also proved to be toxic to neurons, but kainate-induced injury, for example, usually requires a much longer agonist exposure to induce a lethal (Koh et al., 1990). To provide the link to ischemic injury in vivo, parallel experiments can be performed using oxygen and glucose deprivation to induce injury. This process kills neurons by a mechanism that is considerably slower than glutamate exposure, due to the time required to consume available metabolic substrates after ischemia induction. However, the process still seems to be mediated by NMDA receptor activation, and is thus conceptually similar to rapid glutamate-induced injury (Rothman, 1984; Weiss et al., 1986; Marcoux et al., 1988; Goldberg and Choi, 1993).

The recognition of the $Ca^{2+}$-dependence of NMDA receptor-mediated injury was clearly a key step in the understanding of the mechanisms underlying this form of glutamate-induced neurotoxicity (Choi, 1987). Moreover, there was a broad appreciation of the potential cytotoxic effects of an excessive $Ca^{2+}$ load, and even of a range of $Ca^{2+}$-dependent processes that could be responsible for killing neurons. However, the last ten years have yielded surprisingly few unequivocal answers to the basic questions that these studies raised about the neurotoxicity of intracellular $Ca^{2+}$. The first goal of this review is to address the questions that surround the properties of $Ca^{2+}$ as a neurotoxin in rapidly-induced neuronal injury: how does it get into cells, how much $Ca^{2+}$ is toxic, where does it go inside cells, and what does it do when it gets there? The second goal of the review is to ask whether $Ca^{2+}$ is the only important divalent cation that contributes to injury, or whether there could be an important role for other ions, with a particular emphasis on $Mg^{2+}$ and $Zn^{2+}$.

## Calcium as a neurotoxin

### *What are the important sources of calcium for neurotoxicity?*

There can be no question that the most acute form of glutamate-induced injury is mediated by NMDA receptor activation and requires extracellular $Ca^{2+}$ for expression. NMDA receptors are permeable to both $Ca^{2+}$ and $Na^{+}$ (MacDermott et al., 1986; Kudo and Ogura, 1986; Murphy et al., 1987; Mayer et al., 1987; Pinelis et al., 1994; Kiedrowski et al., 1994b; Stout et al., 1996), so the simplest pathway for delivering $Ca^{2+}$ in order to kill neurons is by influx through the NMDA receptor. NMDA receptors are well suited to deliver large amounts of $Ca^{2+}$ because in addition to their $Ca^{2+}$ permeability they show only modest desensitization or inactivation during prolonged agonist exposure (Clark et al., 1990). This is in contrast to the more marked desensitization typically associated with glutamate-stimulated non-NMDA receptors and the essentially complete inactivation of voltage sensitive $Ca^{2+}$ channels that occurs within seconds of activation. However, it is interesting to note that neurons that express $Ca^{2+}$-permeable forms of non-NMDA receptors (Müller et al., 1992; Burnashev et al., 1992) can be killed by kainate following a relatively short agonist exposure that is qualitatively similar to NMDA receptor mediated injury (Brorson et al., 1994; Weiss et al., 1994; Lu et al., 1996). This suggests that the expression of non-desensitizing $Ca^{2+}$-permeable ion channels is the key feature of acutely mediated neuronal injury.

A rough estimation of the $[Ca^{2+}]_i$ after five minutes of stimulation of NMDA receptors in typical forebrain neurons with modest sized NMDA receptor currents suggests that, if unbuffered, there would be tens of millimolar free $Ca^{2+}$ in the cytoplasm (Brocard et al., 1993). It is not unreasonable to suppose, then, that NMDA receptor mediated $Ca^{2+}$ entry is sufficient to deliver a lethal $Ca^{2+}$ load. However, there is evidence that other mechanisms can contribute to the NMDA-triggered cytoplasmic $Ca^{2+}$ load, and given that circumstances in vivo are unlikely to be able to deliver glutamate to NMDA receptors as efficiently as it is possible to achieve in cell culture, it is certainly reasonable to suggest that other mechanisms that amplify NMDA receptor-mediated $[Ca^{2+}]_i$ changes could contribute to cell injury.

One possible amplification mechanism is $Ca^{2+}$-induced $Ca^{2+}$ release (CICR). This process releases $Ca^{2+}$ from the endoplasmic reticulum in neurons. $Ca^{2+}$ release is triggered by a combination of a small elevation in $[Ca^{2+}]_i$ and the presence of cyclic ADP ribose, which combine to activate a ligand gated $Ca^{2+}$ channel, the ryanodine receptor (Hua et al., 1994; Lee et al., 1994). It is clear that this is an important amplification mechanism in neurons under circumstances where $Ca^{2+}$ transients are small or of a brief duration, so that within certain experimental paradigms blocking ryanodine receptors can almost completely eliminate $[Ca^{2+}]_i$ changes (Friel and Tsien, 1992; Llano et al., 1994; Jacobs and Meyer, 1997). It is much less clear whether CICR can deliver sufficient

$Ca^{2+}$ to kill neurons because of the finite magnitude of this intracellular $Ca^{2+}$ pool. Nevertheless, several studies have shown that the CICR inhibitor, dantrolene, can both decrease NMDA receptor-mediated increases in $[Ca^{2+}]_i$ and protect neurons from excitotoxic injury (Mody et al., 1991; Frandsen and Schousboe, 1992; Lei et al., 1992; Simpson et al., 1993). Fluorescent dye measurements with dantrolene need to interpreted cautiously because the drug can interfere with fluorescence recordings. However, the neuroprotection afforded by dantrolene is unequivocal, and if the protection occurs as a consequence of inhibition of CICR this clearly suggests an important role for this intracellular $Ca^{2+}$ pool. Confirmation of this putative mechanism of action of dantrolene by demonstration of neuroprotection with other agents that deplete the reticular $Ca^{2+}$ stores (e.g., thapsigargin), release the stores (e.g., caffeine) or modulate CICR (e.g., ryanodine) is, at present, lacking.

A distinct mechanism for amplification of NMDA-induced increases in $[Ca^{2+}]_i$ may be mediated by the plasma membrane $Na^+/Ca^{2+}$ exchange ($NCE_p$). Although the $NCE_p$ is normally considered to be an important extrusion mechanism for $Ca^{2+}$, like many $Na^+$-driven transporters $NCE_p$ can operate in reverse if the appropriate gradients exist (Eisner and Lederer, 1985; Philipson, 1985; Lederer et al., 1996; Blaustein et al., 1996). The $NCE_p$ transports $3Na^+$ for each $Ca^{2+}$, and normally uses the $Na^+$ gradient and resting membrane potential to extrude $Ca^{2+}$. When glutamate receptors are activated $[Na^+]_i$ rises and neurons are depolarized, which could promote $NCE_p$-mediated $Ca^{2+}$ entry. Thus, it has been speculated that $NCE_p$ may contribute to injurious $Ca^{2+}$ loading of neurons (Hartley and Choi, 1989; Mattson et al., 1989; Stys et al., 1991; Andreeva et al., 1991; Kiedrowski et al., 1994a). Direct evidence to test this hypothesis has been difficult to obtain because of a paucity of drugs that selectively inhibit $NCE_p$. However, a recent report documented the effects of KB-R7943, which appears to preferentially block the reverse mode (i.e., the $Ca^{2+}$ uptake mode) of $NCE_p$ over the normal $Ca^{2+}$ extrusion mode (Watano et al., 1996; Iwamoto et al., 1996). We recently tested KB-R7943 in cultured neurons and were able to confirm that it selectively blocks $NCE_p$-mediated $Ca^{2+}$ uptake compared to $Ca^{2+}$ extrusion (Hoyt et al., 1997). Interestingly, this drug was an efficacious inhibitor of $[Ca^{2+}]_i$ changes induced by brief agonist exposure, but became progressively less effective as the exposure to glutamate increased in duration. KB-R7943 also lacked neuroprotective activity, which is perhaps predictable given its inability to block $[Ca^{2+}]_i$ increases resulting from prolonged glutamate exposure (Hoyt et al., 1997). These studies certainly support the proposal that $NCE_p$ can mediate increases in $[Ca^{2+}]_i$ but suggest that the contribution to intracellular $Ca^{2+}$ loading made by this pathway does not make a significant contribution to injury.

*How much intracellular calcium causes damage?*

The acute injury process that is initiated by glutamate is quite specific to NMDA receptors, and requires extracellular $Ca^{2+}$. $Ca^{2+}$ is a ubiquitous signaling mechanism, however, and there are a multitude of stimuli that promote $Ca^{2+}$ entry through voltage gated calcium channels (VGCC) or release $Ca^{2+}$ from intracellular stores in neurons that elevate $[Ca^{2+}]_i$ but do not acutely induce injury. One obvious basis for this specificity could be the magnitude of the $[Ca^{2+}]_i$ change that results from NMDA receptor activation, such that NMDA would putatively stimulate larger $[Ca^{2+}]_i$ changes than other, non-toxic agonists. This hypothesis was readily testable with the advent of $Ca^{2+}$-sensitive fluorescent dyes that allow reasonably quantitative measurements of $[Ca^{2+}]_i$ if used appropriately (Grynkiewicz et al., 1985). Early studies on glutamate-mediated $[Ca^{2+}]_i$ changes using these dyes revealed that both NMDA and non-NMDA receptors increase $[Ca^{2+}]_i$ (Murphy et al., 1987). The majority of the former signal is attributable to the receptor itself, while under most circumstances the latter receptors depolarize cells and activate VGCC (Murphy and Miller, 1989; Rajdev and Reynolds, 1994). VGCC can be activated independently of glutamate receptors and this also results in increases in $[Ca^{2+}]_i$ (Murphy et al., 1987). Increases in $[Ca^{2+}]_i$ can also occur as the result of receptor-mediated generation of inositol trisphosphate, and this pathway has been associated with a number of neurotransmitter receptors including glutamate (Sugiyama et al., 1987; Murphy and Miller, 1988; Llano et al., 1991).

A number of laboratories have subsequently compared the properties of $[Ca^{2+}]_i$ changes associated with each of these stimuli in an attempt to reveal a signal specific to NMDA receptor activation. There are some differences in the properties of the $[Ca^{2+}]_i$ changes triggered by some of the above mechanisms that allow them to be clearly distinguished from neurotoxic signals. For example, $[Ca^{2+}]_i$ changes associated with inositol trisphosphate-mediated intracellular $Ca^{2+}$ release are only slightly attenuated by extracellular $Ca^{2+}$ removal. As injury is blocked by extracellular $Ca^{2+}$ removal, it is unlikely that this source of $Ca^{2+}$ is the cause of injury. However, when the comparison is restricted to ion-channel mediated $Ca^{2+}$ entry pathways there are surprisingly few differences that are apparent when using high affinity fluorescent indicators, such as fura-2, indo-1 and fluo-3. Both toxic and non-toxic stimuli result in robust $[Ca^{2+}]_i$ increases that are similar in magnitude and duration (Michaels and Rothman, 1990; Abele et al., 1990; Silver and Erecinska, 1990; Dubinsky and Rothman, 1991; Tymianski et al., 1993b; Rajdev and Reynolds, 1994). Toxic stimuli have been associated with a later loss of $Ca^{2+}$ homeostasis in neurons (Randall and Thayer, 1992; Tymianski et al., 1993a; Kiedrowski et al., 1994a), but this is not a property associated with the acute, injury inducing phase of agonist application.

When $^{45}Ca^{2+}$ or cellular $Ca^{2+}$ content is used to track the movement of $Ca^{2+}$ into neurons during NMDA receptor activation it is possible to estimate the net $Ca^{2+}$ movement rather than the resulting $[Ca^{2+}]_i$. Such studies in cultured neurons showed that significantly more $Ca^{2+}$ is taken up by neurons as the result of stimuli that will ultimately be toxic when compared to agonist exposure conditions that would elevate $[Ca^{2+}]_i$ to the same degree but not result in cell death (Hartley et al., 1993; Eimerl and Schramm, 1994). This important finding indicates that either the $Ca^{2+}$ is going into a compartment to which the dyes do not have access, that the NMDA induced $Ca^{2+}$ load is buffered more effectively so that the $[Ca^{2+}]_i$ is the same even though the load is larger, or that the dyes are under-reporting the $[Ca^{2+}]_i$. Localized high calcium concentrations have been observed in a number of neuronal preparations, and particularly in dendrites following physiological as well as pathophysiological stimulation (Connor et al., 1988; Guthrie et al., 1991; Yuste et al., 1994; Bindokas and Miller, 1995). Typical $[Ca^{2+}]_i$ determinations that are made just in the cell body of cultured neurons may not accurately reflect these local changes, so this represents a form of dye-insensitive compartment. However, there is not strong causative evidence for a "death signal" that is generated in one region of a neuron, and localized high $Ca^{2+}$ concentrations do not appear to predict injury better than routine measurements of $[Ca^{2+}]_i$ in the soma. Most studies that have addressed the issue of dye compartmentalization in organelles using neurons, find that the typical $Ca^{2+}$-sensitive dyes are found largely in the cytoplasm (for example (Brocard et al., 1993)). It is certainly possible that if neurotoxic $Ca^{2+}$ loads are effectively buffered into a compartment such as mitochondria (see below) that does not contain dye, routine fluorescence recordings will not be measuring the correct pool of $Ca^{2+}$, which might also result in the disassociation of $[Ca^{2+}]_i$ changes and injury.

More direct evidence for the third possibility, that the dyes under-report $[Ca^{2+}]_i$, has recently emerged. The effective use of $Ca^{2+}$-sensitive dyes requires attention to a number of important details about their properties. In this case it is necessary to consider the affinity of dyes for $Ca^{2+}$. High affinity indicators bind $Ca^{2+}$ with affinities in the ~200–400 nM range when the affinity is measured in vitro (Grynkiewicz et al., 1985). These dyes include fura-2, indo-1 and fluo-3, which collectively represent the most widely used indicators. In practical terms, the magnitude of the change in the fluorescence signal becomes difficult to detect once the ion concentration exceeds the $k_D$ by about a factor of ten. Assuming that the affinities measured in vitro are correct (an assumption that is not always readily testable), this limits the usefulness of high affinity dyes to $[Ca^{2+}]_i$ changes that are less than ~3 μM. Thus, it would be difficult to detect the difference between a $[Ca^{2+}]_i$ change that was 3 μM and one that reached 300 μM using these agents, because a similar signal would be obtained.

It was widely believed that high affinity dyes were not saturated in experiments that measured $[Ca^{2+}]_i$ changes associated with excitotoxicity. $[Ca^{2+}]_i$ transients triggered by glutamate receptor activation often show initial peaks followed by a decline to a slightly lower plateau. This initial decline was taken as an indicator that the $[Ca^{2+}]_i$ was lower than the initial peak, so that saturation was unlikely (for example Rajdev and Reynolds, 1994). However, this interpretation does not take into account the possibility that the change in the dye signal could be due to alterations in the environment around the dye. Neurons exposed to high glutamate concentrations acutely swell, undergo cytoplasmic acidification and generate reactive oxygen species, any of which could alter the dye signal (Hartley and Dubinsky, 1993; Wang et al., 1993; Irwin et al., 1994; Reynolds and Hastings, 1995; Andrew et al., 1996). There are changes in other ions too, including $Na^+$, $K^+$ and $Mg^{2+}$, which could alter dye properties with a resulting small change in the apparent $[Ca^{2+}]_i$. More direct evidence that the high affinity dyes are under-reporting $[Ca^{2+}]_i$ comes from the use of lower affinity indicators. These include fura-2 dextran, calcium green 5N, BTC and magfura-2, which bind $Ca^{2+}$ with affinities ranging from ~1-50 μM (Regehr and Tank, 1992; Rajdev and Reynolds, 1993; Haugland, 1993; Iatridou et al., 1994; Regehr and Atluri, 1995; Zhao et al., 1996). This gives a practical range for determining $[Ca^{2+}]_i$ up to about 500 μM. Interestingly, when low affinity dyes are used in these experiments it appears that $[Ca^{2+}]_i$ continues to increase during a prolonged exposure to glutamate, during the phase where the high affinity dyes report constant $[Ca^{2+}]_i$ (Hyrc, et al., 1997; Stout and Reynolds, 1996). This is unlikely to be an artefact arising from the selection of different populations of cells between the two experimental groups because it can be observed if cells are loaded with both high and low affinity dyes, and recordings made simultaneously (Stout and Reynolds, 1996). An analysis of the results produced by these studies suggest that stimuli that result in acute, NMDA receptor-mediated excitotoxicity are, in fact, associated with significantly higher $[Ca^{2+}]_i$ concentrations than non-toxic stimuli, and that $[Ca^{2+}]_i$ is elevated to near 10 μM when assessed on a whole cell basis (Stout and Reynolds, 1996; Hyrc et al., 1997). These findings thus correspond to the differential estimates of $Ca^{2+}$ entry made using radioisotope flux measurements (Hartley et al., 1993; Eimerl and Schramm, 1994). It remains possible that a differential localization of $Ca^{2+}$ entry exists that is associated with activation of receptors that result in acute injury, but this does not now appear to be necessary to explain the injury induced by $Ca^{2+}$.

*Where does the calcium go?*

It is widely believed that the injury that results from excessive activation of NMDA receptors occurs when the intracellular $Ca^{2+}$ buffering mechanisms are overwhelmed and the $[Ca^{2+}]_i$ is no longer controlled. The presence of this loss of $Ca^{2+}$ homeostasis has been illustrated by several studies in which the application of an excitotoxic stimulus to cultured neurons ultimately results in very high $[Ca^{2+}]_i$ that fails to recover (Randall and Thayer, 1992; Tymianski et al., 1993a; Schinder et al., 1996). It is debatable whether the failure of $Ca^{2+}$ homeostasis in neurons is the cause of injury, or rather a down-stream consequence of the injury-inducing event (see below). However, in either case the mechanisms that buffer glutamate-induced $Ca^{2+}$ loads are important, either because their failure *is* the cause of injury, or because understanding what causes the loss of homeostasis will provide insight into the target of $Ca^{2+}$ that triggers the injury process.

Neurons clearly have very effective mechanisms for buffering $[Ca^{2+}]_i$ changes. It has been estimated that $<1\%$ of the $Ca^{2+}$ influx that occurs in response to brief activation of VGCC results in an increment in $[Ca^{2+}]_i$ (Helmchen et al., 1996). The remainder is bound by an as yet undefined endogenous buffer capacity that binds $Ca^{2+}$ and is relatively slowly diffusible, if at all (Helmchen et al., 1996). It is possible that this capacity reflects $Ca^{2+}$ binding proteins (CaBPs), such as calbindin, parvalbumin and calmodulin (Neher, 1995). Indeed, it has been suggested that the presence of an abundance of any of these proteins will confer resistance to excitotoxicity on a given population of neurons (Scharfman and Schwartzkroin, 1989; Mattson et al., 1991; Goodman et al., 1993). The introduction of CaBPs into neurons results in smaller increments of $[Ca^{2+}]_i$ in response to a given stimulus, and this effect can be mimicked by incubating cells in the presence of the cell permeant form of the fast $Ca^{2+}$ buffer, BAPTA (Dubinsky, 1993; Tymianski et al., 1993c). It is not clear, however, that chelation of $Ca^{2+}$ in post-synaptic neurons should be neuroprotective because this process does not diminish the total cellular $Ca^{2+}$ load. In addition, CaBPs have a limited binding capacity that relates to the concentration of the protein in the cell. Although there are rather few estimates of the concentration of CaBPs that are normally found in neurons, it seems unlikely that the proteins are present at the high micromolar concentrations that would be necessary to buffer the large $[Ca^{2+}]_i$ changes associated with excitotoxicity. Studies that have investigated the impact of the expression of CaBPs on injury have had mixed results with some finding protection (Mattson et al., 1991; Goodman et al., 1993), and others not (Freund et al., 1990; Möckel and Fischer, 1994). Indeed, the more compelling studies that illustrated the neuroprotective effects of BAPTA ultimately concluded that the efficacy arises from the ability of BAPTA to buffer presynaptic $[Ca^{2+}]_i$ changes, thus reducing glutamate release, rather than a post-synaptic effect on the site of injury (Tymianski et al., 1993c; Abdel-Hamid and Tymianski, 1997). It therefore seems likely that the processes that transport $Ca^{2+}$ out of the cytoplasm are more important in buffering the large $Ca^{2+}$ loads associated with injury.

The major $Ca^{2+}$ transport processes responsible for clearing cytoplasmic $Ca^{2+}$ in neurons are the $Ca^{2+}$ ATPases found on the plasma membrane and the endoplasmic reticulum, the plasma membrane $Na^+/Ca^{2+}$ exchange ($NCE_p$) and the mitochondrial $Ca^{2+}$ uniporter (Miller, 1991). The ATPases are generally believed to have a relatively high affinity but low capacity for $Ca^{2+}$ (Pozzan et al., 1994). This makes them ideally suited to define the resting $[Ca^{2+}]_i$, but limits their ability to transport the large $Ca^{2+}$ loads associated with injury. There is little direct evidence to suggest that these transporters, or their failure, play an important role in excitotoxicity, although it has been suggested that the effects of mitochondrial inhibition on $Ca^{2+}$ homeostasis is, in fact due to the depletion of ATP and the subsequent inhibition of plasmalemmal $Ca^{2+}$ ATPase (Budd and Nicholls, 1996a).

Mitochondria have long been recognized as having the capacity to transport and store $Ca^{2+}$ (Nicholls and Akerman, 1982; Crompton, 1990). Mitochondria use their membrane potential and a transport process called the $Ca^{2+}$ uniporter to accumulate $Ca^{2+}$ from the cytoplasm (Nicholls and Akerman, 1982). This is a relatively low affinity process that shows relatively high capacity. Several studies have demonstrated the ability of acute mitochondrial inhibition to prolong the recovery of $[Ca^{2+}]_i$ to baseline following glutamate stimulation (White and Reynolds, 1995; Schinder et al., 1996), and the relative contribution of this process increases as the intensity of the stimulus is increased (White and Reynolds, 1997). Mitochondria play an additional role in shaping $[Ca^{2+}]_i$ changes in neurons. Mitochondrial $Ca^{2+}$ storage is only transient, and the release of accumulated $Ca^{2+}$ continues for some time after the removal of the stimulus (Thayer and Miller, 1990; Wang and Thayer, 1996; White and Reynolds, 1996). The release of mitochondrial $Ca^{2+}$ in neurons occurs primarily through the mitochondrial $Na^+/Ca^{2+}$ exchange ($NCE_m$), and inhibition of $NCE_m$ increases the rate of recovery of $[Ca^{2+}]_i$ (Wang and Thayer, 1996; White and Reynolds, 1996; White and Reynolds, 1997). Indeed, modifying the rate of recovery of $[Ca^{2+}]_i$ by either inhibiting mitochondrial $Ca^{2+}$ uptake or by decreasing mitochondrial $Ca^{2+}$ release may be the most straightforward way to determine that mitochondria are involved in $Ca^{2+}$ buffering in response to a given stimulus. The accumulation of $Ca^{2+}$ by mitochondria may be an important process whereby the cytoplasm signals cellular activity to the mitochondrion in order to increase the production of ATP, because several elements of glycolysis and respiration are accelerated following mitochondrial $Ca^{2+}$ accumulation (Crompton, 1990; Gunter et al., 1994). However, the accumulation of $Ca^{2+}$ and its subsequent release results in the dissipation of the mitochondrial membrane potential (Nicholls and Akerman, 1982), and mitochondrial $Ca^{2+}$ overload may result in catastrophic failure of respiration (see below).

This leaves the $NCE_p$ as the major mechanism for transporting $Ca^{2+}$ out of neurons. Studies in cultured neurons have established that $NCE_p$ is involved in buffering glutamate-induced $Ca^{2+}$ loads in neurons, because inhibiting $NCE_p$ by removing extracellular $Na^+$ delays recovery of $[Ca^{2+}]_i$ to baseline (White and Reynolds, 1996; White and Reynolds, 1997). It has also been reported that inhibition of $NCE_p$ with an amiloride analogue augments glutamate-induced injury, which is also consistent with the buffering role of $NCE_p$ (Andreeva et al., 1991). Is the exchanger the limiting factor in clearing the cytoplasm of $Ca^{2+}$? This seems unlikely from the studies described above that investigated the role of mitochondria in $Ca^{2+}$ buffering. The rate of recovery from potentially toxic $Ca^{2+}$ loads appears to be dictated by the release of $Ca^{2+}$ from the transiently established mitochondrial $Ca^{2+}$ pool. When inhibitors of the mitochondrial $Ca^{2+}$ release process are applied to neurons the $[Ca^{2+}]_i$ typically recovers much more quickly than in their absence (White and Reynolds, 1996). This suggests that the $NCE_p$ is able to rapidly clear any $Ca^{2+}$ present in the cytoplasm in the absence of on-going $Ca^{2+}$ loading, which may imply that the $NCE_p$ is not rate limiting in this process.

### *What does the calcium do?*

Of the many outstanding questions that surround the mechanism of glutamate induced injury, the nature of the target of $Ca^{2+}$ is perhaps the most important, because identification of the critical action(s) of $Ca^{2+}$ would facilitate the design of more efficacious neuroprotective drugs. There is certainly no shortage of candidate mechanism that have been proposed, many of which are recognized $Ca^{2+}$-dependent processes that are found in the cytoplasm. Such mechanisms would thus be sensitive to the uniquely high $[Ca^{2+}]_i$ found during NMDA receptor activation. However, at this time no single mechanism can fully account for the acute injurious effects of glutamate. This may imply that there are several processes that are each sufficient to lethally injure neurons, or else that each of the proposed mechanisms serve to modulate an as yet unidentified process. It is also possible, given the heterogeneity of neurons in the brain and the well-established phenomenon of the selective vulnerability of certain populations of neurons, that different primary mechanisms operate in different types of neuron.

Several proposed targets bear mentioning. Dawson and colleagues (1991) demonstrated the contribution of nitric oxide (NO) to NMDA mediated by showing that nitric oxide synthase inhibitors block NMDA receptor-mediated neuronal injury. There are a number of possible targets for NO to be cytotoxic in neurons (also see Cancelot and Beal, this volume). NO may modify mitochondrial metabolism by interacting with heme groups in enzymes in the electron transport chain (Stamler, 1994; Bolaños et al., 1994; Brorson et al., 1997), and it has also been proposed that NO generation results in the activation of poly ADP-ribose synthase. This enzyme will consume large quantities of NADH and induce bioenergetic failure

(Zhang et al., 1994; Cosi et al., 1994). The concomitant generation of NO and superoxide (Reynolds and Hastings, 1995; Dugan et al., 1995; Bindokas et al., 1996) results in the formation of peroxynitrite, which is an effective oxidant that may modify proteins and produce DNA strand breaks (Beckman et al., 1996; Smith et al., 1997). There can be no question that NO is an attractive candidate as a $Ca^{2+}$-dependent neurotoxin, although it should be noted that it is possible to express NMDA receptor-mediated injury in neural cell lines that do not express NO synthase (Munir et al., 1995).

Another putative target of $Ca^{2+}$ is calpain. Calpain is a $Ca^{2+}$-activated protease that is expressed in neurons and is activated by the relatively high concentrations of $Ca^{2+}$ that may be associated with injury (Siman et al., 1989; Del Cerro et al., 1994; Wang and Yuen, 1994). Evidence for the activation of calpain comes from studies that have monitored the generation of the breakdown products of calpain-mediated cleavage of spectrin (Siman et al., 1989; Roberts-Lewis et al., 1994), and the efficacy of selective calpain inhibitors in models of ischemic injury (Siman et al., 1989; Arlinghaus et al., 1991; Rami and Krieglstein, 1993; Wang and Yuen, 1994). However calpain activation appears to occur some time after the critical period for inducing injury in these models (Faddis et al., 1997). These findings suggest that calpain may contribute to the destruction of the neuron, perhaps following the loss of $Ca^{2+}$ homeostasis, but is not involved in the initiation of injury. Interestingly, a recent study suggested that calpain activation might even protect neurons by promoting recovery following a sub-lethal stimulation by NMDA (Faddis et al., 1997). There are a number of other $Ca^{2+}$ activated processes that have been proposed to contribute to injury, and these have been reviewed elsewhere (Choi, 1988; Meldrum and Garthwaite, 1990; Siesjö and Smith, 1991; Rothman, 1992; Kristián et al., 1996).

An emerging theme in cell injury research is the role of mitochondria in the injury process (Nakao and Brundin, this volume, Lancelot and Bear this volume). There is ample evidence that mitochondria must play some role in excitotoxicity, but the nature of that role has not been fully defined. It is interesting to note that neurotoxins that target mitochondria produce injury that mimics disease states like Huntingtons disease, and that mitochondrial defects have been associated with Parkinsons disease (Beal et al., 1993; Greene and Greenamyre, 1996; Schapira, 1996). In addition, disease states that are caused by mutations in the mitochondrial genome are characterized by prominent neurological symptoms, including epilepsy and stroke (Johns, 1996). In the simplest sense, neurons depend on oxidative phosphorylation for ATP generation to support critical ionic gradients. Ischemia or hypoxia, the most common cause of excitotoxic injury in vivo, will decrease or eliminate oxidative ATP production, with the result that subsequent exposure to glutamate may prove more injurious in light of the impairment of $Ca^{2+}$ buffering mechanisms.

Recent studies have demonstrated effects of glutamate on mitochondrial function that are independent of ischemia, however. Excitotoxic glutamate exposure results in the $Ca^{2+}$-dependent generation of reactive oxygen species (ROS) that can be detected using electron spin resonance (Lafon-Cazal et al., 1993) or fluorescent dyes (Reynolds and Hastings, 1995; Dugan et al., 1995; Bindokas et al., 1996). It appears that the majority of this signal results from $Ca^{2+}$ transport into mitochondria, and may be the consequence of the $Ca^{2+}$-mediated increase in mitochondrial respiration induced by glutamate (Gunter et al., 1994; White and Reynolds, 1996). Glutamate stimulation also depolarizes neuronal mitochondria. This phenomenon could arise from the cycling of $Ca^{2+}$ through mitochondria, which results in a net positive charge accumulation for each $Ca^{2+}$ cycled (Nicholls and Akerman, 1982). However, there is also evidence for activation of the mitochondrial permeability transition pore (PTP) in response to glutamate. The PTP is a large, non-selective ion channel in the mitochondrial inner membrane. Studies in isolated mitochondria have shown that the PTP is activated by elevated $[Ca^{2+}]_m$, oxidation and depolarization (Bernardi et al., 1994; Zoratti and Szabo, 1995), which all occur during stimulation of neurons by glutamate. Activation of the PTP results in the collapse of the mitochondrial membrane potential and the release of any stored $Ca^{2+}$ (Bernardi et al., 1994; Zoratti and Szabo, 1995). In addition to losing the proton gradient for ATP generation, the cellular energetic status may be further compromised because the $F_0F_1$-ATPase will actually consume ATP in an effort to restore the membrane potential (Budd and Nicholls, 1996a). Thus, induction of transition will result in a prompt bioenergetic collapse. The involvement of the PTP in glutamate induced mitochondrial depolarization is suggested by the ability of cyclosporin A (White and Reynolds, 1996; Schinder et al., 1996; Nieminen et al., 1996) as well as trifluoperazine and dibucaine (Hoyt et al., 1997) to delay or prevent depolarization. Interestingly, cyclosporin A is also neuroprotective (White and Reynolds, 1996; Schinder et al., 1996). However, although it is tempting to conclude that the neuroprotection arises as a result of PTP inhibition, cyclosporin A also inhibits NO synthase activation by virtue of its inhibition of calcineurin (Dawson et al., 1993), so these findings must be interpreted with caution. Nevertheless, PTP activation remains as an attractive possible mechanism whereby NMDA receptor-mediated $Ca^{2+}$ loads could be translated into a lethal injury, and it would be very interesting to test this mechanism with potent and selective PTP inhibitors.

There may be additional mechanisms of neuronal injury associated with mitochondria. Several studies in non-neuronal cells demonstrated that factors released from mitochondria can trigger apoptosis in isolated nuclei (Susin et al., 1996; Yang et al., 1997; Kluck et al., 1997). In one case the factor appears to be cytochrome c, which can be released by an as yet unknown mechanism that apparently does not involve permeability transition (Yang et al., 1997; Kluck et al., 1997). Another laboratory has reported the release of an "apoptosis inducing factor" by a mechanism that does require PTP activation (Susin et al., 1996; Kroemer et al., 1997). Recent studies in neural cells have demonstrated the ability of cytochrome c to reconstitute apoptotic changes in nuclei, suggesting that this process can occur in neurons too (Ellerby et al., 1997). It's not clear that these mechanisms are critically dependent on $Ca^{2+}$, and the contribution of these processes to the necrotic injury that is characteristic of acute glutamate excitotoxicity is not clear because the mechanisms reported thus far relate entirely to apoptotic death.

Mitochondria may also be the critical source of the lethal signal in excitotoxicity. In an intriguing set of experiments Budd and Nicholls (1996b) demonstrated that if $Ca^{2+}$ uptake into mitochondria is prevented under circumstances where the cellular ATP content was preserved then glutamate was no longer toxic. This suggests that the critical event in excitotoxicity is $Ca^{2+}$ uptake into mitochondria, with the implication that mitochondria must generate a "death signal" of some sort. These experiments in cerebellar granule cells are complicated somewhat by the observation that the depolarization of mitochondria is apparently associated with a decrease in $Ca^{2+}$ entry into the cytoplasm (Budd and Nicholls, 1996b), which raises the possibility that it is the mechanism which decreased $Ca^{2+}$ entry that confers protection. However, other studies have illustrated a similar phenomenon in forebrain neurons, where the uncoupler FCCP applied during glutamate exposure protects against toxicity (Stout et al., 1997). In these experiments the $[Ca^{2+}]_i$ rose to very high concentrations during the glutamate exposure even though the neurons did not die. These findings imply that the essential feature of glutamate mediated injury is the entry of $Ca^{2+}$ into mitochondria and the subsequent generation of a lethal signal. Moreover, that neurons do not die even when $[Ca^{2+}]_i$ reaches very high concentrations suggests that the previously described $Ca^{2+}$-activated processes in the cytoplasm only serve a modulatory role in the generation of injury (Stout et al., 1997). The studies that have characterized the ability of cytochrome c to induce apoptosis have suggested that Bcl-2 prevents apoptosis by inhibiting cytochrome c release (Ellerby et al., 1997; Yang et al., 1997; Kluck et al., 1997). It has also been demonstrated that Bcl-2 enhances the ability of mitochondria to take up $Ca^{2+}$ (Murphy et al., 1996), which is perhaps unexpected if mitochondrial $Ca^{2+}$ accumulation is the process that signals cytochrome c release. However, although Bcl-2 and related proteins have been linked to neuronal injury in various paradigms (Martinou et al., 1994; MacGibbon et al., 1997; Chen et al., 1997), no studies have yet demonstrated the impact of Bcl-2 on glutamate mediated $Ca^{2+}$ signaling and mitochondrial function.

## The role of other divalent cations in neurotoxicity

The clear $Ca^{2+}$ dependence of excitotoxic neuronal injury naturally focuses attention on $Ca^{2+}$ as the intracellular mediator of injury. However, there are a number of reasons to consider divalent cations other than $Ca^{2+}$ as possible contributors to neurotoxicity. Recent studies from this laboratory have argued that $Mg^{2+}$ may contribute to glutamate-mediated injury, and it has been widely appreciated that considerable quantities of $Zn^{2+}$ are found in the brain, and that $Zn^{2+}$ may be taken up and released in a neurotransmitter-like fashion. Thus, it is appropriate to consider the circumstances under which these ions may contribute to excitotoxicity.

### *Magnesium*

$Mg^{2+}$ is an abundant intracellular cation. Typical intracellular free $Mg^{2+}$ concentrations ($[Mg^{2+}]_i$) are 0.4–0.6 mM, with 10–20 times as much sequestered within cells (Murphy, 1991; Romani and Scarpa, 1992; Gunther, 1993). This makes $Mg^{2+}$ the second most abundant cation in cells after $K^+$. However, $Mg^{2+}$ transport and $Mg^{2+}$ homeostasis are rather poorly understood compared to the insights into $Ca^{2+}$ homeostasis that have been gained. It is very likely that there are specific mechanisms that control the $[Mg^{2+}]_i$ in cells because resting concentrations are considerably below the equilibrium concentration of $>100$ mM (Murphy, 1991). However, none of the putative $Mg^{2+}$ transport proteins in mammalian cells has been cloned. Many cells express a $Na^+/Mg^{2+}$ exchange mechanism that is located on the plasma membrane (Beyenbach, 1990; Murphy, 1991; Gunther, 1993; Gunzel and Schlue, 1996; Handy et al., 1996), and a phenomenon consistent with $Na^+/Mg^{2+}$ exchange has been reported in neurons (Stout et al., 1996). It is also not clear where in the cell the considerable amount of bound $Mg^{2+}$ is located. Some fraction of the total $Mg^{2+}$ is complexed with ATP, but this fraction is unlikely to account for more than 30% of the total $Mg^{2+}$. The rest may be associated with mitochondria, the nucleus, or else bound to proteins and lipids in the cytoplasm (Kun, 1976; Corkey et al., 1986; Gunther, 1986; Romani et al., 1993).

In most cells the $[Mg^{2+}]_i$ does not appear to vary much from the basal levels. However, using magfura-2 to report $[Mg^{2+}]_i$ we found that glutamate stimulation resulted in rather robust ($>$mM) changes in $[Mg^{2+}]_i$ in cortical neurons (Brocard et al., 1993). The largest part of this phenomenon was mediated by $Ca^{2+}$ entry through NMDA receptors and did not appear to be the result of very high $[Ca^{2+}]_i$, which could produce a magfura-2 signal that is indistinguishable from that produced by $Mg^{2+}$. Interestingly, the stimuli that most effectively increased $[Mg^{2+}]_i$ were those that induced acute excitotoxicity, which raised the possibility that an elevation in $[Mg^{2+}]_i$ might contribute to excitotoxic injury. Moreover, because these apparent increases in $[Mg^{2+}]_i$ were independent of extracellular $Mg^{2+}$ it appeared that $Ca^{2+}$ may be triggering $Mg^{2+}$ release inside neurons. The magnitude of the $[Mg^{2+}]_i$ change would be consistent with a stoichiometric exchange of $Ca^{2+}$ for $Mg^{2+}$ at an intracellular site (Brocard et al., 1993). Subsequent studies, like those described above, have suggested that $[Ca^{2+}]_i$ may be higher than previously appreciated, and have raised doubt about the quantitative interpretation of these findings (Stout and Reynolds, 1996; Hyrc et al., 1997). The resolution of the actual change in $[Mg^{2+}]_i$ awaits more careful determination of the affinity of magfura-2 for $Ca^{2+}$ and $Mg^{2+}$ in situ.

An analysis of the ionic dependence of glutamate triggered increases in $[Mg^{2+}]_i$ revealed a residual component that persisted in the absence of $Ca^{2+}$ (Brocard et al., 1993). This component could be amplified by $Na^+$ removal and by increasing extracellular $Mg^{2+}$ consistent with a $Mg^{2+}$ influx pathway. Subsequent studies demonstrated that $Mg^{2+}$ was entering the cell through the NMDA receptor itself, apparently because the permeability properties of the channel are altered somewhat by the removal of extracellular $Na^+$ and $Ca^{2+}$ (Stout et al., 1996). Although these $[Mg^{2+}]_i$ increases (typically 2–3 mM) were somewhat smaller than those produced by $Ca^{2+}$, they have the advantage that they occur in the complete absence of a change in $[Ca^{2+}]_i$, and thus unequivocally represent an alteration in $[Mg^{2+}]_i$. We exploited this paradigm to test the hypothesis generated by the original observations, that $Mg^{2+}$ can support excitotoxicity. Paradoxically, $Mg^{2+}$ does support glutamate induced injury in the complete absence of $Ca^{2+}$ (Hartnett et al., 1997), although these findings are complicated somewhat because the toxicity of the $[Mg^{2+}]_i$ change appears to depend on which $Na^+$ substitute is used. Nevertheless, this suggests that $[Mg^{2+}]_i$ should be considered as a possible contributor to excitotoxicity.

Several other findings suggest that $Mg^{2+}$ may be an under-appreciated dynamic in brain injury. Several studies measured brain $Mg^{2+}$ content using magnetic resonance spectroscopy and found a decrease in $Mg^{2+}$ after traumatic brain injury (Vink et al., 1991; McIntosh, 1994). Although this initially appears to be at variance with the results from cell culture experiments, the temporal resolution of MRS studies is considerably less than in cultured cells, and it is entirely possible that the decrease in $Mg^{2+}$ content is the result of mobilization of $Mg^{2+}$ stores and the subsequent efflux from neurons. A study with a slightly faster time resolution found a small increase in $Mg^{2+}$ following focal ischemia (Helpern et al., 1993). Interestingly, injury induced alterations in brain $Mg^{2+}$ can be blocked by the NMDA receptor antagonist, ketamine (Shapira et al., 1993).

Clearly, the precise role of $Mg^{2+}$ in neuronal injury is poorly understood. There are a number of potential targets for alterations in $[Mg^{2+}]_i$ that could contribute to injury (Romani and Scarpa, 1992). For example, it has been reported that $Mg^{2+}$ can activate an endonuclease (Sun and Cohen, 1994; Patel et al., 1994). However, in order to approach $Mg^{2+}$ as a potential therapeutic target it will be necessary to first identify the cell signaling mechanisms that $Mg^{2+}$ modifies to cause injury, and also to determine whether it is the acute elevation in $[Mg^{2+}]_i$ or the subsequent loss of cellular $Mg^{2+}$ that is primarily associated with injury.

### *Zinc*

The final divalent cation that merits consideration for a putative role in neuronal injury is $Zn^{2+}$. This trace metal is differentially distributed in the brain and is found in high concentrations in synaptic vesicles, especially in the hippocampus (Crawford and Connor, 1972; Perez and Danscher, 1985). Early studies revealed that $Zn^{2+}$ can be released in a depolarization-induced, $Ca^{2+}$-dependent manner (Assaf and Chung, 1984), and can be taken up into synaptosome preparations (Wensink et al., 1988), thus suggesting a possible neurotransmitter-like role. The subsequent observation that $Zn^{2+}$ can modulate NMDA and $GABA_A$ receptors (Peters et al., 1987; Westbrook and Mayer, 1987) supports this notion. The localization of stores of $Zn^{2+}$ inside vesicles suggests that there are likely to be mechanisms that transport $Zn^{2+}$ both across the plasma membrane and across vesicular membranes too. Palmiter and colleagues (1995; 1996; 1997) have identified three different proteins that are proposed to transport $Zn^{2+}$, namely ZnT1-3. $Zn^{2+}$ transporting activity of ZnT1 has been demonstrated, and this protein is found in plasma membranes (Palmiter and Findley, 1995). ZnT3 has been localized to synaptic vesicular membranes and is highly homologous to ZnT1, but the

$Zn^{2+}$ transport activity of this protein remains to be shown (Palmiter et al., 1997). There is very little specific information available about the regulation of cytoplasmic free $Zn^{2+}$ concentrations ($[Zn^{2+}]_i$). In addition to the putative transporters, $Zn^{2+}$ binds with high affinity to sulfhydryl-containing proteins. In particular, metallothionein appears to be well suited to buffer intracellular $Zn^{2+}$ fluctuations because it is a small molecule that is rich in sulfhydryls, and the expression of metallothionein is induced by exposing cells to Zn (Masters et al., 1994; Aschner, 1996). However, there are no reports that illustrate the ability of any of these processes to actually alter $[Zn^{2+}]_i$ in neurons.

As might be expected from its ability to inhibit NMDA receptor activation, low concentrations of $Zn^{2+}$ are neuroprotective against acute, NMDA-mediated neuronal injury (Peters et al., 1987; Koh and Choi, 1988). However, subsequent studies found that $Zn^{2+}$ could kill neurons in a way that did not require either $Ca^{2+}$ or $Na^+$ when neurons were concentrations of $Zn^{2+}$ above 200 $\mu$M (Choi et al., 1988; Koh and Choi, 1994). In addition, AMPA can potentiate neuronal injury induced by $Zn^{2+}$ (Weiss et al., 1993), and this may be particularly important in neurons that express the $Ca^{2+}$-permeable form of non-NMDA receptors (Weiss et al., 1993; Choi, 1996). Finally, it has been suggested that NMDA receptor activation promotes $Zn^{2+}$ entry into neurons (Koh and Choi, 1994), which may facilitate $Zn^{2+}$-mediated toxicity. The toxicity of $Zn^{2+}$ may be of considerable pathophysiological significance because synaptic $Zn^{2+}$ concentrations have been estimated to reach high micromolar concentrations during intense synaptic activity (Assaf and Chung, 1984). Indeed, the combination of elevated extracellular $Zn^{2+}$ with glutamate receptor activation during synaptic activity provide ideal conditions for $Zn^{2+}$-mediated injury. Direct evidence for the involvement of $Zn^{2+}$ in hippocampal injury in vivo has been provided by the finding that $Zn^{2+}$ translocates from presynaptic terminals to postsynaptic cell bodies following forebrain ischemia, and that $Zn^{2+}$ chelation protects neurons from both $Zn^{2+}$ accumulation and injury (Koh et al., 1997).

An important but unresolved question in this field is the relationship between $Zn^{2+}$ influx and the resulting $[Zn^{2+}]_i$, and also how the toxicity of $Zn^{2+}$ relates to $[Zn^{2+}]_i$. $Zn^{2+}$ homeostasis is poorly understood, and the problem is compounded by the difficulty of measuring $[Zn^{2+}]_i$. The studies that have monitored $Zn^{2+}$ loss and/or accumulation in the above studies have used the fluorescent dye TSQ. This dye is typically used in fixed preparations, which limits its usefulness for the measurement of dynamic changes in $[Zn^{2+}]_i$, and may also recognize $Zn^{2+}$ that is protein-bound. $Zn^{2+}$ chelators that are putatively selective for $Zn^{2+}$ have not proven useful for determining cytoplasmic $Zn^{2+}$ because either a low affinity for $Zn^{2+}$ (Budde et al., 1997) compared to the anticipated free concentrations of 1 nM or less, or because of the lack of selectivity compared to physiological concentrations of other cations. Interestingly, $Ca^{2+}$ and $Mg^{2+}$ sensitive dyes also bind $Zn^{2+}$, and do so with a higher affinity than either ion. Thus fura-2 binds $Zn^{2+}$ with an affinity of 2 nM, compared to ~200 nM for $Ca^{2+}$ (Grynkiewicz et al., 1985). The only difference between the signal produced by $Zn^{2+}$ compared to $Ca^{2+}$ is a slightly smaller maximum 340/380 nm ratio when the dye is saturated with $Zn^{2+}$ (Atar et al., 1995). This raises the interesting possibility that fura-2 and related dyes might, under some circumstances, be reporting a change of $[Zn^{2+}]_i$ rather than $[Ca^{2+}]_i$. This property of $Ca^{2+}$ sensitive dyes has been exploited to estimate $[Zn^{2+}]_i$ in cardiac myocytes (Atar et al., 1995), and, more recently, glutamate-induced $[Zn^{2+}]_i$ changes in neurons (Reynolds and Cheng, 1997). Although this method is far from perfect, in that it is necessary to restrict access of, or at least account for, the other cations that can bind the dye, the method has nevertheless lead to estimates that $[Zn^{2+}]_i$ in the 20–50 nM result in neuronal injury (Reynolds and Cheng, 1997). Adaptations of this approach may be useful for studying neuronal $Zn^{2+}$ homeostasis in the future.

## Conclusions and future directions

Although the blockade of glutamate receptors remains an attractive target for neuroprotective drug action, there remains the need for additional ways to interfere with the mechanisms of neuronal injury. The studies that have centered on the neurotoxic actions of $Ca^{2+}$ have not yet identified the critical mechanisms that link NMDA receptor activation to neuronal death. However, the picture that is beginning to emerge suggests that the injury process is more closely associated with the extent of $Ca^{2+}$ loading than was previously believed, which provides an important parameter by which injury can be inferred. In addition, the identification of the central role of mitochondria in responding to NMDA receptor activation, and as perhaps the source of the "death signal" in excitotoxicity, provides a number of novel targets for neuroprotective drug action. Clearly, it will be important to establish which of the several possible mechanisms does critically influence the survival of neurons before the effectiveness of drugs can be predicted, and it will be interesting to learn how these mechanisms vary between neurons that are selectively vulnerable to injury, compared to those that are not. It is clear that focusing on $Ca^{2+}$ alone is not appropriate, given the good evidence that both $Mg^{2+}$ and $Zn^{2+}$ can support injury to neurons. However, the mechanisms that regulate the cytoplasmic free concentrations of either of these ions are poorly understood, and it will be necessary to gain insight into these processes before meaningful predictions can be made about appropriate targets for drug action.

## Acknowledgements

IJR is an Established Investigator of the American Heart Association. The authors work cited in this paper was supported by NS 34138.

## References

Abdel-Hamid, K.M. and Tymianski, M. (1997) Mechanisms and effects of intracellular calcium buffering on neuronal survival in organotypic hippocampal cultures exposed to anoxia/aglycemia or to excitotoxins. *J. Neurosci.*, 17: 3538–3553.

Abele, A.E., Scholz, K.P., Scholz, W.K. and Miller, R.J. (1990) Excitotoxicity induced by enhanced excitatory neurotransmission in cultured hippocampal primary neurons. *Neuron*, 4: 413–419.

Andreeva, N., Khodorov, B., Stelmashook, E., Cragoe, E., Jr. and Victorov, I. (1991) Inhibition of $Na^{+}/Ca^{2+}$ exchange enhances delayed neuronal death elicited by glutamate in cerebellar granule cell cultures. *Brain Res.*, 548: 322–325.

Andrew, R.D., Adams, J.R. and Polischuk, T.M. (1996) Imaging NMDA- and kainate-induced intrinsic optical signals from the hippocampal slice. *J. Neurophysiol.*, 76: 2707–2717.

Arlinghaus, L., Mehdi, S. and Lee, K.S. (1991) Improved posthypoxic recovery with a membrane-permeable calpain inhibitor. *Eur. J. Pharmacol.*, 209: 123–125.

Aschner, M. (1996) The functional significance of brain metallothioneins. *FASEB J.*, 10: 1129–1136.

Assaf, S.Y. and Chung, S.H. (1984) Release of endogenous $Zn^{2+}$ from brain tissue during activity. *Nature*, 308: 734–736.

Atar, D., Backx, P.H., Appel, M.M., Gao, W.D. and Marban, E. (1995) Excitation-transcription coupling mediated by zinc influx through voltage dependent calcium channels. *J. Biol. Chem.*, 270: 2473–2477.

Beal, M.F., Hyman, B.T. and Koroshetz, W. (1993) Do defects in mitochondrial energy metabolism underlie the pathology of neurodegenerative diseases. *Trends Neurosci.*, 16: 125–131.

Beckman, J.S., Ye, Y.Z., Chen, J. and Conger, K.A. (1996) The interactions of nitric oxide with oxygen radicals and scavengers in cerebral ischemic injury. *Adv. Neurol.*, 71: 339–354.

Bernardi, P., Broekemeier, K.M. and Pfeiffer, D.R. (1994) Recent progress on regulation of the mitochondrial permeability transition pore: A cyclosporin sensitive pore in the inner mitochodrial membrane. *J. Bioenerg. Biomemb.*, 26: 509–517.

Beyenbach, K.W. (1990) Transport of magnesium across biological membranes. *Magnesium Trace Elem.*, 233: 254.

Bindokas, V.P., Jordan, J., Lee, C.C. and Miller, R.J. (1996) Superoxide production in rat hippocampal neurons: Selective imaging with hydroethidine. *J. Neurosci.*, 16: 1324–1336.

Bindokas, V.P. and Miller, R.J. (1995) Excitotoxic degeneration is initiated at non-random sites in cultured rat cerebellar neurons. *J. Neurosci.*, 15: 6999–7011.

Blaustein, M.P., Fontana, G. and Rogowski, R.S. (1996) The $Na^{+}$-$Ca^{2+}$ exchanger in rat brain synaptosomes – Kinetics and regulation. *Ann. NY Acad. Sci.*, 779: 300–317.

Bolaños, J.P., Peuchen, S., Heales, S.J.R., Land, J.M. and Clark, J.B. (1994) Nitric oxide-mediated inhibition of the mitochondrial respiratory chain in cultured astrocytes. *J. Neurochem.*, 63: 910–916.

Brocard, J.B., Rajdev, S. and Reynolds, I.J. (1993) Glutamate induced increases in intracellular free $Mg^{2+}$ in cultured cortical neurons. *Neuron*, 11: 751–757.

Brorson, J.R., Manzolillo, P.A. and Miller, R.J. (1994) $Ca^{2+}$ entry via AMPA/KA receptors and excitotoxicity in cultured cerebellar Purkinje cells. *J. Neurosci.*, 14: 187–197.

Brorson, J.R., Sulit, R.A. and Zhang, H. (1997) Nitric oxide disrupts $Ca^{2+}$ homeostasis in hippocampal neurons. *J. Neurochem.*, 68: 95–105.

Budd, S.L. and Nicholls, D.G. (1996a) A reevaluation of the role of mitochondria in neuronal $Ca^{2+}$ homeostasis. *J. Neurochem.*, 66: 403–411.

Budd, S.L. and Nicholls, D.G. (1996b) Mitochondria, calcium regulation and acute glutamate excitotoxicity in cultured cerebellar granule cells. *J. Neurochem.*, 67: 2282–2291.

Budde, T., Minta, A., White, J.A. and Kay, A.R. (1997) Imaging free zinc in synaptic terminals in live hippocampal slices. *Neuroscience*, 79: 347–358.

Burnashev, N., Khodorova, A., Jonas, P., Helm, P.J., Wisden, W., Monyer, H., Seeburg, P.H. and Sakmann, B. (1992) Calcium-permeable AMPA-kainate receptors in fusiform cerebellar glial cells. *Science*, 256: 1566–1570.

Chen, J., Graham, S.H., Nakayama, M., Zhu, R.L., Jin, K.L., Stetler, R.A. and Simon, R.P. (1997) Apoptosis repressor genes bcl-2 and bcl-x-long are expressed in the rat brain following global ischemia. *J. Cereb. Blood Flow Metab.*, 17: 2–10.

Choi, D.W. (1987) Ionic dependence of glutamate neurotoxicity. *J. Neurosci.*, 7: 369–379.

Choi, D.W. (1988) Glutamate neurotoxicity and diseases of the nervous system. *Neuron*, 1: 623–634.

Choi, D.W. (1996) Zinc neurotoxicity may contribute to selective neuronal death following transient global cerebral ischemia. *Cold Spring Harbor Symp. Quant. Biol.*, 61: 385–387.

Choi, D.W., Maulucci-Gedde, M. and Kriegstein, A.R. (1987) Glutamate neurotoxicity in cortical cell culture. *J. Neurosci.*, 7: 357–368.

Choi, D.W., Yokoyama, M. and Koh, J. (1988) Zinc neurotoxicity in cortical cell culture. *Neuroscience*, 24: 67–79.

Clark, G.D., Clifford, D.B. and Zorumski, C.F. (1990) The effect of agonist concentration, membrane voltage and calcium on *N*-methyl-D-aspartate receptor desensitization. *Neuroscience*, 39: 787–797.

Connor, J.A., Wadman, W.J., Hockberger, P.E. and Wong, R.K.S. (1988) Sustained dendritic gradients of $Ca^{2+}$ induced by excitatory amino acids in CA1 hippocampal neurons. *Science*, 240: 649–653.

Corkey, B.E., Duszynski, J., Rich, T.L., Matschinsky, B. and Williamson, J.R. (1986) Regulation of free and bound magnesium in rat hepatocytes and isolated mitochondria. *J. Biol. Chem.*, 261: 2567–2574.

Cosi, C., Suzuki, H., Milani, D., Facci, L., Menegazzi, M., Vantini, G., Kanai, Y. and Skaper, S.D. (1994) Poly(ADP-ribose) polymerase: Early involvement in glutamate-induced neurotoxicity in cultured cerebellar granule cells. *J. Neurosci. Res.*, 39: 38–46.

Crawford, I.L. and Connor, J.D. (1972) Zinc in maturing rat brain: Hippocampal concentration and localization. *J. Neurochem.*, 19: 1451–1458.

Crompton, M. (1990) Role of mitochondria in intracellular calcium regulation. In: *Intracellular calcium regulation*, edited by F. Bronner. New York: Wiley-Liss, p. 181–209.

Dawson, T.M., Steiner, J.P., Dawson, V.L., Dinerman, J.L., Uhl, G.R. and Snyder, S.H. (1993) Immunosuppressant FK506 enhances phosphorylation of nitric oxide synthase and protects against glutamate neurotoxicity. *Proc. Natl. Acad. Sci. USA*, 90: 9808–9812.

Dawson, V.L., Dawson, T.M., London, E.D., Bredt, D.S. and Snyder, S.H. (1991) Nitric oxide mediates glutamate neurotoxicity in primary cortical cultures. *Proc. Natl. Acad. Sci. USA*, 88: 6368–6371.

Del Cerro, S., Arai, A., Kessler, M., Bahr, B.A., Vanderklish, P., Rivera, S. and Lynch, G. (1994) Stimulation of NMDA receptors activates calpain in cultured hippocampal slices. *Neurosci. Lett.*, 167: 149–152.

Dubinsky, J.M. (1993) Effects of calcium chelators on intracellular calcium and excitotoxicity. *Neurosci. Lett.*, 150: 129–132.

Dubinsky, J.M. and Rothman, S.M. (1991) Intracellular calcium concentrations during "chemical hypoxia" and excitotoxic neuronal injury. *J. Neurosci.*, 11: 2545–2551.

Dugan, L.L., Sensi, S.L., Canzoniero, L.M.T., Handran, S.D., Rothman, S.M., Lin, T.-S., Goldberg, M.P. and Choi, D.W. (1995) Mitochondrial production of reactive oxygen species in cortical neurons following exposure to *N*-methyl-D-aspartate. *J. Neurosci.*, 15: 6377–6388.

Eimerl, S. and Schramm, M. (1994) The quantity of calcium that appears to induce neuronal death. *J. Neurochem.*, 62: 1223–1226.

Eisner, D.A. and Lederer, W.J. (1985) Na-Ca exchange: stoichiometry and electrogenicity. *Am. J. Physiol.*, 248: c189–c202.

Ellerby, H.M., Martin, S.J., Ellerby, L.M., Naiem, S.S., Rabizadeh, S., Salvesen, G. S., Casiano, C.A., Cashman, N.R., Green, D.R. and Bredesen, D.E. (1997) Establishment of a cell-free system of neuronal apoptosis: Comparison of premitochondrial, mitochondrial and postmitochondrial phases. *J. Neurosci.*, 17: 6165–6178.

Faddis, B.T., Hasbani, M.J. and Goldberg, M.P. (1997) Calpain activation contributes to dendritic remodeling after brief excitotoxic injury in vitro. *J. Neurosci.*, 17: 951–959.

Frandsen, A. and Schousboe, A. (1992) Mobilization of dantrolene-sensitive intracellular calcium pools is involved in the cytotoxicity induced by quisqualate and *N*-methyl-D-aspartate but not by 2-amino-3-(3-hydroxy-5-methylisoxazol-4-yl) propionate and kainate in cultured cerebral cortical neurons. *Proc. Natl. Acad. Sci. USA*, 89: 2590–2594.

Freund, T.F., Buzsáki, G., Leon, A., Baimbridge, K.G. and Somogyi, P. (1990) Relationship of neuronal vulnerability and calcium binding protein immunoreactivity in ischemia. *Exp. Brain Res.*, 83: 55–66.

Friel, D.D. and Tsien, R.W. (1992) Phase-dependent contributions from $Ca^{2+}$ entry and $Ca^{2+}$ release to caffeine-induced $[Ca^{2+}]_i$ oscillations in bullfrog sympathetic neurons. *Neuron*, 8: 1109–1125.

Goldberg, M.P. and Choi, D.W. (1993) Combined oxygen and glucose deprivation in cortical cell culture: Calcium-dependent and calcium-independent mechanisms of neuronal injury. *J. Neurosci.*, 13: 3510–3524.

Goodman, J.H., Wasterlain, C.G., Massarweh, W.F., Dean, E., Sollas, A.L. and Sloviter, R.S. (1993) Calbindin-D28k immunoreactivity and selective vulnerability to ischemia in the dentate gyrus of the developing rat. *Brain Res.*, 606: 309–314.

Greene, J.G. and Greenamyre, J.T. (1996) Bioenergetics and glutamate excitotoxicity. *Prog. Neurobiol.*, 48: 613–621.

Grynkiewicz, G., Poenie, M. and Tsien, R.Y. (1985) A new generation of $Ca^{2+}$ indicators with greatly improved fluorescence properties. *J. Biol. Chem.*, 260: 3440–3450.

Gunter, T.E., Gunter, K.K., Sheu, S.-S. and Gavin, C.E. (1994) Mitochondrial calcium transport: Physiological and pathological relevance. *Am. J. Physiol. Cell Physiol.*, 267: C313–C339.

Gunther, T. (1986) Functional compartmentation of intracellular magnesium. *Magnesium*, 5: 53–59.

Gunther, T. (1993) Mechanisms and regulation of $Mg^{2+}$ efflux and $Mg^{2+}$ influx. *Mineral Electrolyte Metab.*, 19: 259–265.

Gunzel, D. and Schlue, W.R. (1996) Sodium-magnesium antiport in Retzius neurones of the leech *Hirudo medicinalis*., *J. Physiol. (Lond.)*, 491: 595–608.

Guthrie, P.B., Segal, M. and Kater, S.B. (1991) Independent regulation of calcium revealed by imaging dendritic spines. *Nature*, 354: 76–80.

Handy, R.D., Gow, I.F., Ellis, D. and Flatman, P.W. (1996) Na-dependent regulation of intracellular free magnesium concentration in isolated rat ventricular myocytes. *J. Mol. Cell. Cardiol.*, 28: 1641–1651.

Hartley, D.M. and Choi, D.W. (1989) Delayed rescue of *N*-methyl-D-aspartate receptor mediated neuronal injury in cortical culture. *J. Pharmacol. Exp. Ther.*, 250: 752–758.

Hartley, D.M., Kurth, M.C., Bjerkness, L., Weiss, J.H. and Choi, D.W. (1993) Glutamate receptor-induced $^{45}Ca^{2+}$ accumulation in cortical culture correlates with subsequent neuronal degeneration. *J. Neurosci.*, 13: 1993–2000.

Hartley, Z. and Dubinsky, J.M. (1993) Changes in intracellular pH associated with glutamate excitotoxicity. *J. Neurosci.*, 13: 4690–4699.

Hartnett, K.A., Stout, A.K., Rajdev, S., Rosenberg, P.A., Reynolds, I.J. and Aizenman, E. (1997) NMDA receptor-mediated neurotoxicity: A paradoxical requirement for extracellular $Mg^{2+}$ in $Na^+/Ca^{2+}$-free solutions in rat cortical neurons in vitro. *J. Neurochem.*, 68: 1836–1845.

Haugland, R.P. (1993) Intracellular ion indicators. In: *Fluorescent and luminescent probes for biological activity*, edited by W.T. Mason. London: Academic Press, pp. 34–43.

Helmchen, F., Imoto, K. and Sakmann, B. (1996) $Ca^{2+}$ buffering and action potential-evoked $Ca^{2+}$ signalling in dendrites of pyramidal neurons. *Biophys. J.*, 70: 1069–1081.

Helpern, J.A., Vande Linde, A.M.Q., Welch, K.M.A., Levine, S.R., Schultz, L.R., Ordidge, R.J., Halvorson, H.R. and Hugg, J.W. (1993) Acute elevation and recovery of intracellular $[Mg^{2+}]$ following human focal cerebral ischemia. *Neurology*, 43: 1577–1581.

Hoyt, K.R., Arden, S.R., Aizenman, E. and Reynolds, I.J. (1998) Reverse $Na+/Ca^{2+}$ exchange contributes to glutamate-induced $[Ca^{2+}]_i$ increases in cultured forebrain neurons. *Mol. Pharmacol.*, 53: 742–749.

Hoyt, K.R., Sharma, T.A. and Reynolds, I.J. (1997) Trifluoperazine and dibucaine inhibit glutamate-induced mitochondrial depolarization in cultured rat forebrain neurones. *Br. J. Pharmacol*, 122: 803–808.

Hua, S.-Y., Tokimasa, T., Takasawa, S., Furuya, Y., Nohmi, M., Okamoto, H. and Kuba, K. (1994) Cyclic ADP-ribose modulates $Ca^{2+}$ release channels for activation by physiological $Ca^{2+}$ entry in bullfrog sympathetic neurons. *Neuron*, 12: 1073–1079.

Hyrc, K., Handran, S.D., Rothman, S.M. and Goldberg, M.P. (1997) Ionized intracellular calcium concentration predicts excitotoxic neuronal death: Observations with low affinity fluorescent calcium indicators. *J. Neurosci.*, 17: 6669–6677.

Iatridou, H., Foukaraki, E., Kuhn, M.A., Marcus, E.M., Haugland, R.P. and Katerinopoulos, H.E. (1994) The development of a new family of intracellular calcium probes. *Cell Calcium*, 15: 190–198

Irwin, R.P., Lin, S.-Z., Long, R.T. and Paul, S.M. (1994) *N*-methyl-D-aspartate induces a rapid, reversible and calcium dependent intracellular acidosis in cultured fetal rat hippocampal neurons. *J. Neurosci.*, 14: 1352–1357.

Iwamoto, T., Watano, T. and Shigekawa, M. (1996) A novel isothiourea derivative selectively inhibits the reverse mode of $Na^+/Ca^{2+}$ exchange in cells expressing NCX1. *J. Biol. Chem.*, 271: 22391–22397.

Jacobs, J.M. and Meyer, T. (1997) Control of action potential-induced $Ca^{2+}$ signaling in the soma of hippocampal neurons by $Ca^{2+}$ release from intracellular stores. *J. Neurosci.*, 17: 4129–4135.

Johns, D.R. (1996) The other human genome: Mitochondrial DNA and disease. *Nature Med.*, 2: 1065–1068.

Kiedrowski, L., Brooker, G., Costa, E. and Wroblewski, J.T. (1994a) Glutamate impairs neuronal calcium extrusion while reducing sodium gradient. *Neuron*, 12: 295–300.

Kiedrowski, L., Wroblewski, J.T. and Costa, E. (1994b) Intracellular sodium concentration in cultured cerebellar granule cells challenged with glutamate. *Mol. Pharmacol.*, 45: 1050–1054.

Kluck, R.M., Bossy-Wetzel, E., Green, D.R. and Newmeyer, D.D. (1997) The release of cytochrome c from mitochondria: A primary site for Bcl-2 regulation of apoptosis. *Science*, 275: 1132–1136.

Koh, J.-Y. and Choi, D.W. (1988) Zinc alters excitatory amino acid neurotoxicity in cortical neurons. *J. Neurosci.*, 8: 2164–2171.

Koh, J.-Y., Suh, S.W., Gwag, B.J., He, Y.Y., Hsu, C.Y. and Choi, D.W. (1997) The role of zinc in selective neuronal death after transient global cerebral ischemia. *Science*, 272: 1013–1016.

Koh, J.-Y. and Choi, D.W. (1994) Zinc toxicity on cultured cortical neurons: Involvement of *N*-methyl-D-aspartate receptors. *Neuroscience*, 60: 1049–1057.

Koh, J.Y., Goldberg, M.P., Hartley, D.M. and Choi, D.W. (1990) Non-NMDA receptor mediated neurotoxicity in cortical culture. *J. Neurosci.*, 10: 693–705.

Kristián, T., Ouyang, Y. and Siesj, B.K. (1996) Calcium-induced neuronal cell death in vivo and in vitro: Are the pathophysiologic mechanisms different? *Adv. Neurol.*, 71: 107–118.

Kroemer, G., Zamzami, N. and Susin, S.A. (1997) Mitochondrial control of apoptosis. *Immunol. Today*, 18: 44–51.

Kudo, Y. and Ogura, A. (1986) Glutamate induced increase in intracellular $Ca^{2+}$ concentration in isolated hippocampal neurones. *Br. J. Pharmacol.*, 89: 191–198.

Kun, E. (1976) Kinetics of ATP-dependent $Mg^{2+}$ flux in mitochondria. *Biochemistry*, 15: 2328–2336.

Lafon-Cazal, M., Pietri, S., Culcasi, M. and Bockaert, J. (1993) NMDA-dependent superoxide production and neurotoxicity. *Nature*, 364: 535–537.

Lederer, W.J., He, S., Luo, S., DuBell, W., Kofuji, P., Kieval, R., Neubauer, C.F., Ruknudin, A., Cheng, H., Cannell, M.B., Rogers, T.B. and Schulze, D.H. (1996) The molecular biology of the $Na^{+}$-$Ca^{2+}$ exchanger and its functional roles in heart, smooth muscle cells, neurons, glia, lymphocytes and nonexcitable cells. *Ann. NY Acad. Sci.*, 779: 7–17.

Lee, H.C., Galione, A. and Walseth, T.F. (1994) Cyclic ADP Ribose: Metabolism and calcium mobilizing function. *Vit. Horm.*, 48: 199–257.

Lei, S.Z., Zhang, D., Abele, A.E. and Lipton, S.A. (1992) Blockade of NMDA receptor-mediated mobilization of intracellular $Ca^{2+}$ prevents neurotoxicity. *Brain Res.*, 598: 196–202.

Llano, I., Dipolo, R. and Marty, A. (1994) Calcium-induced calcium release in cerebellar Purkinje cells. *Neuron*, 12: 663–673.

Llano, I., Dreessen, J., Kano, M. and Konnerth, A. (1991) Intradendritic release of calcium induced by glutamate in cerebellar Purkinje cells. *Neuron*, 7: 577–583.

Lu, Y.M., Yin, H.Z., Chiang, J. and Weiss, J.H. (1996) $Ca^{2+}$-permeable AMPA/kainate and NMDA channels: High rate of $Ca^{2+}$ influx underlies potent induction of injury. *J. Neurosci.*, 16: 5457–5465.

MacDermott, A.B., Mayer, M.L., Westbrook, G.L., Smith, S.J. and Barker, J.L. (1986) NMDA-receptor activation increases cytoplasmic calcium concentration in cultured spinal cord neurones. *Nature*, 321: 519–522.

MacGibbon, G.A., Lawlor, P.A., Sirimanne, E.S., Walton, M.R., Connor, B., Young, D., Williams, C., Gluckman, P., Faull, R.L.M., Hughes, P. and Dragunow, M. (1997) Bax expression in mammalian neurons undergoing apoptosis and in Alzheimer's disease hippocampus. *Brain Res.*, 750: 223–234.

Marcoux, F.W., Probert, A.W., Goodrich, J.E. and Dominick, M.A. (1988) The NMDA antagonist ketamine blocks hypoxia-induced calcium accumulation in cultured cortical neurons and prevents ischemic hippocampal cell injury. In: *Frontiers in excitatory amino acid research*, edited by E.A. Cavalheiro, J. Lehmann and L. Turski. New York: Alan R. Liss, p. 683–686.

Martinou, J.-C., Dubois-Dauphin, M., Staple, J.K., Rodriguez, I., Frankowski, H., Missotten, M., Albertini, P., Talabot, D., Catsicas, S., Pietra, C. and Huarte, J. (1994) Overexpression of BCL-2 in transgenic mice protects neurons from naturally occurring cell death and experimental ischemia. *Neuron*, 13: 1017–1030.

Masters, B.A., Quaife, C.J., Erickson, J.C., Kelly, E.J., Froelick, G.J., Zambrowicz, B.P., Brinster, R.L. and Palmiter, R.D. (1994) Metallothionein III is expressed in neurons that sequester zinc in synaptic vesicles. *J. Neurosci.*, 14: 5844–5857.

Mattson, M.P., Guthrie, P.B. and Kater, S.B. (1989) A role for $Na^{+}$-dependent $Ca^{2+}$ extrusion in protection against neuronal excitotoxicity. *FASEB J.*, 3: 2519–2526.

Mattson, M.P., Rychlik, B., Chu, C. and Christakos, S. (1991) Evidence for calcium-reducing and excito-protective roles for the calcium-binding protein calbindin-$D_{28k}$ in cultured hippocampal neurons. *Neuron*, 6: 41–51.

Mayer, M.L., MacDermott, A.B., Westbrook, G.L., Smith, S.J. and Barker, J.L. (1987) Agonist and voltage gated calcium entry in cultured mouse spinal cord neurons under voltage clamp measured using arsenazo III. *J. Neurosci.*, 7: 3230–3244.

McIntosh, T.K. (1994) Novel pharmacologic therapies in the treatment of experimental traumatic brain injury: A review. *J. Neurotrauma*, 10: 215–261.

Meldrum, B.S. and Garthwaite, J. (1990) Excitatory amino acid neurotoxicity and neurodegenerative disease. *Trends Pharmacol. Sci.*, 11: 379–387.

Michaels, R.L. and Rothman, S.M. (1990) Glutamate neurotoxicity in vitro: Antagonist pharmacology and intracellular calcium concentrations. *J. Neurosci.*, 10: 283–292.

Miller, R.J. (1991) The control of neuronal $Ca^{2+}$ homeostasis. *Prog. Neurobiol.*, 37: 255–285.

Mody, I., Baimbridge, K.G., Shacklock, J.A. and MacDonald, J.F. (1991) Release of intracellularly stored $Ca^{2+}$ in hippocampal neurons by NMDA receptor activation. *Exp. Brain Res.*, 20: 75–80.

Möckel, V. and Fischer, G. (1994) Vulnerability to excitotoxic stimuli of cultured rat hippocampal neurons containing the calcium-binding proteins calretinin and calbindin $D_{28K}$. *Brain Res.*, 648: 109–120.

Munir, M., Lu, L. and McGonigle, P. (1995) Excitotoxic cell death and delayed rescue in human neurons derived from NT2 cells. *J. Neurosci.*, 15: 7847–7860.

Murphy, A.N., Bredesen, D.E., Cortopassi, G., Wang, E. and Fiskum, G. (1996) Bcl-2 potentiates the maximal calcium uptake capacity of neural cell mitochondria. *Proc. Natl. Acad. Sci. USA*, 93: 9893–9898.

Murphy, E. (1991) Cellular magnesium and Na/Mg exchange in heart cells. *Ann. Rev. Physiol.*, 53: 273–287.

Murphy, S. N. and Miller, R. J. (1988) A glutamate receptor regulates $Ca^{2+}$-mobilization in hippocampal neurons. *Proc. Natl. Acad. Sci. (USA)*, 85: 8737–8741.

Murphy, S.N. and Miller, R.J. (1989) Regulation of $Ca^{2+}$ influx into striatal neurons by kainic acid. *J. Pharmacol. Exp. Ther.*, 249: 184–193.

Murphy, S.N., Thayer, S.A. and Miller, R.J. (1987) The effects of excitatory amino acids on intracellular calcium in single mouse striatal neurons in vitro. *J. Neurosci.*, 7: 4145–4158.

Müller, T., Möller, T., Berger, T., Schnitzer, J. and Kettenmann, H. (1992) Calcium entry through kainate receptors and resulting potassium-channel blockade in Bergmann glial cells. *Science*, 256: 1563–1566.

Neher, E. (1995) The use of fura-2 for estimating Ca buffers and Ca fluxes. *Neuropharmacology*, 34: 1423–1442.

Nicholls, D.G. and Akerman, K.E.O. (1982) Mitochondrial calcium transport. *Biochim. Biophys. Acta*, 683: 57–88.

Nieminen, A.-L., Petrie, T.G., Lemasters, J.J. and Selman, W.R. (1996) Cyclosporin A delays mitochondrial depolarization induced by *N*-methyl-D-aspartate in cortical neurons: Evidence of the mitochodrial permeability transition. *Neuroscience*, 75: 993–997.

Palmiter, R.D., Cole, T.B. and Findley, S.D. (1996) ZnT-2, a mammalian protein that confers resistance to zinc by facilitating vesicular sequestration. *EMBO J.*, 15: 1784–1791.

Palmiter, R.D., Cole, T.B., Quaife, C.J. and Findley, S.D. (1997) ZnT-3, a putative transporter of zinc into synaptic vesicles. *Proc. Natl. Acad. Sci. USA*, 93: 14934–14939.

Palmiter, R.D. and Findley, S.D. (1995) Cloning and functional characterization of a mammalian zinc transporter that confers resistance to zinc. *EMBO J.*, 14: 639–649.

Patel, T., Bronk, S.F. and Gores, G.J. (1994) Increases of intracellular magnesium promote glycodeoxycholate-induced apoptosis in rat hepatocytes. *J. Clin. Invest.*, 94: 2183–2192.

Perez, C.J. and Danscher, G. (1985) Intravesicular localization of zinc in rat telencephalic boutons. A histochemical study. *Brain. Res.*, 337: 91–98.

Peters, S., Koh, J. and Choi, D.W. (1987) Zinc selectively blocks the action of *N*-methyl-D-aspartate on cortical neurons. *Science*, 236: 589–593.

Philipson, K.D. (1985) Sodium-calcium exchange in plasma membrane vesicles. *Ann. Rev. Physiol.*, 47: 561–571.

Pinelis, V.G., Segal, M., Greenberger, V. and Khodorov, B.I. (1994) Changes in cytosolic sodium caused by a toxic glutamate treatment of cultured hippocampal neurons. *Biochem. Mol. Biol. Int.*, 32: 475–482.

Pozzan, T., Rizzuto, R., Volpe, P. and Meldolesi, J. (1994) Molecular and cellular physiology of intracellular calcium stores. *Physiol. Rev.*, 74: 595–636.

Rajdev, S. and Reynolds, I.J. (1993) Calcium green 5N, a novel fluorescent probe for monitoring high intracellular free $Ca^{2+}$ concentrations associated with glutamate excitotoxicity in cultured rat brain neurons. *Neurosci. Lett.*, 162: 149–152.

Rajdev, S. and Reynolds, I.J. (1994) Glutamate-induced intracellular calcium changes and neurotoxicity in cortical neurons in vitro: Effect of chemical ischemia. *Neuroscience*, 62: 667–679.

Rami, A. and Krieglstein, J. (1993) Protective effects of calpain inhibitors against neuronal damage caused by cytotoxic hypoxia in vitro and ischemia in vivo. *Brain Res.*, 609: 67–70.

Randall, R.D. and Thayer, S.A. (1992) Glutamate-induced calcium transient triggers delayed calcium overload and neurotoxicity in rat hippocampal neurons. *J. Neurosci.*, 12: 1882–1895.

Regehr, W.G. and Atluri, P.P. (1995) Calcuim transients in cerebellar granule cell presynaptic terminals. *Biophys. J.*, 68: 2156–2170.

Regehr, W.G. and Tank, D.W. (1992) Calcium concentration dynamics produced by synaptic activation of CA1 hippocampal pyramidal cells. *J. Neurosci.*, 12: 4202–4223.

Reynolds, I.J. and Cheng, C. (1997) Glutamate stimulates increases in intracellular $Zn^{2+}$ and $Zn^{2+}$-dependent toxicity in cultured forebrain neurons. *Soc. Neurosci.*, 23: 2292. (Abstract)

Reynolds, I.J. and Hastings, T.G. (1995) Glutamate induces the production of reactive oxygen species in cultured forebrain neurons following NMDA receptor activation. *J. Neurosci.*, 15: 3318–3327.

Roberts-Lewis, J.M., Savage, M.J., Marcy, V.R., Pinsker, L.R. and Siman, R. (1994) Immunolocalization of calpain I-mediated spectrin degradation to vulnerable neurons in the ischemic gerbil brain. *J. Neurosci.*, 14: 3934–3944.

Romani, A., Marfella, C. and Scarpa, A. (1993) Cell magnesium transport and homeostasis: Role of intracellular compartments. *Mineral Electrolyte Metab.*, 19: 282–289.

Romani, A. and Scarpa, A. (1992) Regulation of cell magnesium. *Arch. Biochem. Biophys.*, 298: 1–12.

Rothman, S.M. (1984) Synaptic release of excitatory amino acid neurotransmitter mediates anoxic neuronal death. *J. Neurosci.*, 4: 1884–1891.

Rothman, S.M. (1992) Excitotoxins: Possible mechanisms of action. *Ann. NY Acad. Sci.*, 648: 132–139.

Rothman, S.M. and Olney, J.W. (1987a) Excitotoxicity and the NMDA receptor. *Trends Neurosci.*, 10: 299–302.

Rothman, S.M., Thurston, J.H. and Hauhart, R.E. (1987b) Delayed neurotoxicity of excitatory amino acids in vitro. *Neuroscience*, 22: 471–480.

Schapira, A.H.V. (1996) Neurotoxicity and the mechanisms of cell death in Parkinson's disease. *Adv. Neurol.*, 69: 161–165.

Scharfman, H.E. and Schwartzkroin, P.A. (1989) Protection of dentate hilar cells from prolonged stimulation by intracellular calcium chelation. *Science*, 246: 257–260.

Schinder, A.F., Olson, E.C., Spitzer, N.C. and Montal, M. (1996) Mitochondrial dysfunction is a primary event in glutamate neurotoxicity. *J. Neurosci.*, 16: 6125–6133.

Shapira, Y., Lam, A.M., Artru, A.A., Eng, C. and Soltow, L. (1993) Ketamine alters calcium and magnesium in brain tissue following experimental head trauma in rats. *J. Cereb. Blood Flow Metab.*, 13: 962–968.

Siesjö, B.K. and Smith, M.-L. (1991) The biochemical basis of ischemic brain lesions. *Arzneimittelforschung*, 41: 288–291.

Silver, I.A. and Erecinska, M. (1990) Intracellular and extracellular changes of [$Ca^{2+}$] in hypoxia and ischemia in rat brain in vivo. *J. Gen. Physiol.*, 95: 837–866.

Siman, R., Noszek, J.C. and Kegerise, C. (1989) Calpain I activation is specifically related to excitatory amino acid induction of hippocampal damage. *J. Neurosci.*, 9: 1579–1590.

Simpson, P.B., Challiss, R.A.J. and Nahorski, S.R. (1993) Involvement of intracellular stores in the $Ca^{2+}$ responses to *N*-methyl-D-aspartate and depolarization in cerebellar granule cells. *J. Neurochem.*, 61: 760–763.

Smith, M.A., Harris, P.L., Sayre, L.M., Beckman, J.S. and Perry, G. (1997) Widespread peroxynitrite-mediated damage in Alzheimer's disease. *J. Neurosci.*, 17: 2653–2657.

Stamler, J.S. (1994) Redox signaling: Nitrosylation and related target actions of nitric oxide. *Cell*, 78: 931–936.

Stout, A.K., Li-Smerin, Y., Johnson, J.W. and Reynolds, I.J. (1996) Mechanisms of glutamate-stimulated $Mg^{2+}$ and subsequent $Mg^{2+}$ efflux in rat forebrain neurones in culture. *J. Physiol. (Lond.)*, 492: 641–657.

Stout, A.K., Raphael, H.M., Scanlon, J.M. and Reynolds, I.J. (1997) Glutamate-induced neuronal cell death is blocked by inhibition of mitochondrial calcium uptake with FCCP. *Soc. Neurosci.*, 23: 1666. (Abstract)

Stout, A.K. and Reynolds, I.J. (1996) Simultaneous measurement of intracellular free calcium and intracellular free magnesium in neurons. *Soc. Neurosci.*, 22: 68. (Abstract)

Stys, P.K., Waxman, S.G. and Ransom, B.R. (1991) $Na^{+}$-$Ca^{2+}$ exchanger mediates $Ca^{2+}$ influx during anoxia in mammalian central nervous system white matter. *Ann. Neurol.*, 30: 375–380.

Sugiyama, H., Ito, I. and Hirono, C. (1987) A new type of glutamate receptor linked to inositol phospholipid metabolism. *Nature*, 325: 531–533.

Sun, X.-M. and Cohen, G.M. (1994) $Mg^{2+}$-dependent cleavage of DNA into kilobase pair fragments is responsible for the initial degradation of DNA in apoptosis. *J. Biol. Chem.*, 269: 14857–14860.

Susin, S.A., Zamzami, N., Castedo, M., Hirsch, T., Marchetti, P., Macho, A., Daugas, E., Geuskens, M. and Kroemer, G. (1996) Bcl-2 inhibits the mitochondrial release of an apoptogenic protease. *J. Exp. Med.*, 184: 1331–1341.

Thayer, S.A. and Miller, R.J. (1990) Regulation of the intracellular free calcium concentration in single rat dorsal root ganglion neurones in vitro. *J. Physiol. (Lond.)*, 425: 85–115.

Tymianski, M., Charlton, M.P., Carlen, P.L. and Tator, C.H. (1993a) Secondary $Ca^{2+}$ overload indicates early neuronal injury which precedes staining with viability indicators. *Brain Res.*, 607: 319–323.

Tymianski, M., Charlton, M.P., Carlen, P.L. and Tator, C.H. (1993b) Source specificity of early calcium neurotoxicity in cultured embryonic spinal neurons. *J. Neurosci.*, 13: 2085–2104.

Tymianski, M., Wallace, M.C., Spigelman, I., Uno, M., Carlen, P.L., Tator, C.H. and Charlton, M.P. (1993c) Cell-permeant $Ca^{2+}$ chelators reduce early excitotoxic and ischemic neuronal injury in vitro and in vivo. *Neuron*, 11: 221–235.

Vink, R., McIntosh, T.K. and Faden, A.I. (1991) $Mg^{2+}$ in neurotrauma: Its role and therapeutic implications. In: *$Mg^{2+}$ and excitable membranes*, edited by P. Strata and E. Carbone. Berlin: Springer-Verlag, p. 125–145.

Wang, G.J., Randall, R.D. and Thayer, S.A. (1993) Glutamate-induced intracellular pH changes in hippocampal neurons demonstrate altered energy metabolism resulting from $Ca^{2+}$ and $Na^{+}$ loads. *Soc. Neurosci.*, 19: 1770. (Abstract)

Wang, G.J. and Thayer, S.A. (1996) Sequestration of glutamate-induced $Ca^{2+}$ loads by mitochondria in cultured rat hippocampal neurons. *J. Neurophysiol.*, 76: 1611–1621.

Wang, K.K.W. and Yuen, P.-W. (1994) Calpain inhibition: An overview of its therapeutic potential. *Trends Pharmacol. Sci.*, 15: 412–419.

Watano, T., Kimura, J., Morita, T. and Nakanishi, H. (1996) A novel antagonist, No 7943, of the $Na^{+}/Ca^{2+}$ exchange current in guinea-pig cardiac ventricular cells. *Br. J. Pharmacol.*, 119: 555–563.

Weiss, J., Goldberg, M.P. and Choi, D.W. (1986) Ketamine protects cultured neocortical neurons from hypoxic injury. *Brain Res.*, 380: 186–190.

Weiss, J.H., Hartley, D.M., Koh, J. and Choi, D.W. (1993) AMPA receptor activation potentiates zinc neurotoxicity. *Neuron*, 10: 43–49.

Weiss, J.H., Turetsky, D., Wilke, G. and Choi, D.W. (1994) AMPA/kainate receptor-mediated damage to NADPH-diaphorase-containing neurons is $Ca^{2+}$ dependent. *Neurosci. Lett.*, 167: 93–96.

Wensink, J., Molenaar, A.J., Woroniecka, U.D. and van den Hamer, J.A. (1988) Zinc uptake into synaptosomes. *J. Neurochem.*, 50: 782–789.

Westbrook, G.L. and Mayer, M.L. (1987) Micromolar concentrations of $Zn^{2+}$ antagonize NMDA and GABA responses of hippocampal neurons. *Nature*, 328: 640–643.

White, R.J. and Reynolds, I.J. (1995) Mitochondria and $Na^{+}/Ca^{2+}$ exchange buffer glutamate-induced calcium loads in cultured cortical neurons. *J. Neurosci.*, 15: 1318–1328.

White, R.J. and Reynolds, I.J. (1996) Mitochondrial depolarization in glutamate-stimulated neurons: An early signal specific to excitotoxin exposure. *J. Neurosci.*, 16: 5688–5697.

White, R.J. and Reynolds, I.J. (1997) Mitochondria accumulate $Ca^{2+}$ following intense glutamate stimulation of cultured rat forebrain neurones. *J. Physiol. (Lond.)*, 498: 31–47.

Yang, J., Liu, X., Bhalla, K., Kim, C.N., Ibrado, A.M., Cai, J., Peng, T.I., Jones, D.P. and Wang, X. (1997) Prevention of apoptosis by Bcl-2: Release of cytochrome c from mitochondria blocked. *Science*, 275: 1129–1132.

Yuste, R., Gutnick, M.J., Saar, D., Delaney, K.R. and Tank, D.W. (1994) $Ca^{2+}$ accumulations in dendrites of neocortical pyramidal neurons: An apical band and evidence for two functional compartments. *Neuron*, 13: 23–43.

Zhang, J., Dawson, V.L., Dawson, T.M. and Snyder, S.H. (1994) Nitric oxide activation of poly (ADP-ribose) synthetase in neurotoxicity. *Science*, 263: 687–689.

Zhao, M., Hollingworth, S. and Baylor, S.M. (1996) Properties of tri- and tetracarboxylate $Ca^{2+}$ indicators in frog skeletal muscle fibers. *Biophys. J.*, 70: 896–916.

Zoratti, M. and Szabo, I. (1995) The mitochondrial permeability transition. *Biochim. Biophys. Acta*, 1241: 139–176.

O.P. Ottersen, I.A. Langmoen and L. Gjerstad (Eds.)
*Progress in Brain Research*, Vol 116


CHAPTER 16

# Neurodegeneration and glutamate induced oxidative stress

Naoyuki Nakao[1] and Patrik Brundin[2,*]

[1]*Department of Neurological Surgery, Wakayama Medical College, 7-27 Wakayama, Japan*
[2]*Section for Neuronal Survival, Wallenberg Neuroscience Center, Department of Physiology and Neuroscience, Lund University, Sölvegatan 17, S-223 62 Lund, Sweden*

## Possible mechanisms for neuronal cell death in the basal ganglia

Over the past decade, several molecular mechanisms for neuronal cell death have been suggested, based on a considerable body of in vitro and in vivo experimental work. In this chapter, we will focus on those that are potentially relevant to the pathogenesis of Huntington's and Parkinson's diseases, since this is the main interest of our laboratory. It should be emphasized that pathways leading to neuronal death in these diseases may also be of importance for other neurodegenerative disorders, and various forms of acute neurological injury (Watson and Bullock, 1997; Lancelot and Beal, this volume). Under certain neuropathological conditions, the mechanisms of cell death described below are likely to cooperate in the execution of neuronal death (Fig. 1). Nonetheless, we have chosen to describe them in separate sections, to clarify the distinct role of each mechanism in neurogeneration. Four major mechanisms are proposed: oxidative stress; excitotoxicity; energy failure; and disturbances of intracellular calcium. In the context of the current book, it is noteworthy that oxidative stress and excitoxicity due to glutamate play particularly pivotal roles also in the latter two mechanisms.

### *Oxidative stress*

Oxygen free radicals are defined as atoms or molecules that contain one or more orbitals with single unpaired electron, and typically comprise superoxide radicals ($O_2\cdot^-$), hydrogen peroxide ($H_2O_2$), and hydroxyl radicals ($OH\cdot$). Although $H_2O_2$ itself is not a radical, it is still involved in reactions that produce free radicals. Free radicals readily react with, and damage, various biological molecules, including membrane lipids, cytoskeletal proteins, and deoxyribonucleic acid (DNA). The $OH\cdot$ radical is the most reactive species, and is formed by the spontaneous decomposition of $H_2O_2$. The reaction rate of this process is markedly accelerated in the presence of $Fe^{2+}$ (Halliwell and Gutteridge, 1989). $OH\cdot$ reacts at near diffusion-limited rates within a few Ångstrom of the generation site. The $O_2\cdot^-$ radical is less reactive, but can cross cell membranes and therefore act at a distance. Recently, $O_2\cdot^-$ has been identified as a significant threat to biological molecules. Thus, $O_2\cdot^-$ can interact with the nitric oxide radical ($NO\cdot$) to yield highly reactive oxidizing agents peroxynitrite anion ($ONOO^-$) and $OH\cdot$ (Beckmann et al., 1990). The formation of $ONOO^-$ does not require

*Corresponding author.
Tel.: +46-46-2220525; fax: +46-46-2220531; e-mail: patrik.brundin@neurol.lu.se.

transitional metals, and once formed it can diffuse over several cell diameters to oxidize lipids, protein and nucleic acids. It seems as if neurons are more sensitive to peroxynitrite-induced damage than astrocytes (Bolanos et al., 1995), and the cellular targets of $ONOO^-$ may include mitochodria, leading to energy failure (Bolanos et al., 1995). Recent data also suggest that $ONOO^-$ can release glutamate from presynaptic terminals and cause calcium overload through receptor-mediated excitotoxicity (Leist et al., 1997), as discussed in later sections.

There are significant interactions between oxidative stress and mitochondrial energy metabolism, particularly in the oxidative phosphorylation system (Reynolds, this volume). While mitochondrial oxidative phosphorylation generates most of the free radicals in cells, this system is itself susceptible to oxidative stress (Fig. 1). Free radical-induced mitochondrial damage may be due to oxidative injuries to mitochondrial DNA, proteins and membrane lipids.

Free radical reactions are also a part of normal metabolism. Reactive oxygen species are constantly produced during oxidation-reduction reactions, particularly in the mitochondrial electron transport chain (Halliwell and Gutteridge, 1989). Free radicals cause tissue injury when produced in excess. In this context, "oxidative stress" is an expression for a toxic process which is the consequence of an imbalance between the formation of oxygen free radicals and the antioxidant capacity of the cell (Bast and Bär, 1997). Thus, excessive generation of free radicals and/or defect of antioxidant defenses can lead to oxidative stress. There are naturally occurring antioxidant defense mechanisms which can counteract excessive production of free radicals within cells. Superoxide dismutase (*SOD*) catalyzes the dismutation of $O_2^{\cdot -}$ to $H_2O_2$, which in turn can be reduced to $H_2O$ by glutathione in a reaction catalyzed by *glutathione peroxidase* (see below). $H_2O_2$ can also be reduced by an antioxidant enzyme *catalase*. $H_2O_2$ which is not cleared by glutathione decomposes to form $OH\cdot$. Iron can facilitate this reaction (Haber-Weiss reaction, Fenton reaction).

$$O_2^{\cdot -} \xrightarrow{\text{SOD}} H_2O_2 \xrightarrow{\text{Glutathione peroxidase}} H_2O$$

$$H_2O_2 + Fe^{2+} \longrightarrow OH\cdot + OH^- + Fe^{3+} \quad \text{(Fenton reaction)}$$

$$H_2O_2 + O_2^{\cdot -} \xrightarrow{\text{Fe}} OH\cdot + OH^- + O_2 \quad \text{(Haber-Weiss reaction)}$$

The lipophilic vitamin α-tocopherol can block the chain reaction of lipid peroxidation, and the hydrophilic vitamin ascorbic acid is a reducing agent that can donate electrons directly to free radicals. Ferritin, an iron-binding protein, also acts as an antioxidant by rendering iron relatively unreactive. There are several factors that may render central nervous system (CNS) neurons highly susceptible to oxidative stress (Olanow 1993): (1) brain tissue undergoes a high rate of oxidative phosphorylation and uses a disproportionately large amount of the body's oxygen supply; (2) it is rich in polyunsaturated fatty acids that can act as substrates for lipid peroxidation; (3) enzymatic defense mechanisms such as catalyse and glutathione peroxidase are relatively deficient, compared to other tissues (e.g., liver and erythrocytes); (4) some brain regions (e.g., the substantia nigra) contain a high level of free iron, which facilitates the formation of free radicals.

### *Excitotoxic mechanisms*

The concept of excitotoxicity originated from the experimental studies by Olney and co-workers during early 1970s, who showed that neuronal damage or death results from an excessive activation of excitatory neuronal systems (Olney, 1978, for review see Coyle and Puttfarcken, 1993). As elaborated upon in other chapters in the current volume, glutamate is the principal excitatory neurotransmitter in the mammalian central nervous system, and its interactions with specific membrane receptors play an important role in normal neural functions including cognition, memory, movement, and sensation, as well as in plasticity of synaptic connections. There are two main subtypes of glutamate receptors: the ionotropic receptor that is coupled to membrane ion channels; and the metabotropic receptor that is coupled to G-protein. The ionotropic receptors can be further divided into three subtypes based on their prototype agonists: *N*-methyl-D-aspartate (NMDA), α-amino-3-hydroxy-5-methyl-4-isoxazolepropionate (Ampa), and kainate.

Excitotoxic processes can be started by an *overstimulation* of these receptors. One probable trigger for the excitotoxic pathway is energy failure in cells. The best characterized of the three receptors subtypes, the NMDA receptor, is highly permeable to both sodium and calcium (MacDonald et al., this volume). An *excessive* stimulation of the NMDA receptor leads to a massive influx of calcium into the cell, in turn activating a variety of cellular calcium-dependent enzymes (Lancelot and Beal, this volume). Among those are protease, lipase, and endonuclease, which damage cytoskeleton, cell membrane, and nucleic acids, respectively (Orrenius et al., 1989). Furthermore, elevated levels of intracellular calcium can lead to generation of free radicals by activating phospholipase $A_2$ and NO synthase (NOS) (Coyle and Puttfarcken, 1993; Dawson et al., 1992; Lafon-Cazal et al., 1993; Fig. 1). Oxygen free radicals can also enhance the release of excitatory amino acids (Pellegrini-Giampietro et al., 1990), which might provide a vicious cycle (Fig. 1). Indeed, several studies have indicated that antioxidant treatment can reduce neuronal damage in different paradigms of excitotoxicity (Monyer et al., 1996). An interesting recent study employs water-soluble carboxylic acid derivatives of C60 "Bucky-balls" in cortical cultures treated with a high concentration of NMDA (Dungan et al., 1997). The carboxyafullerences, which are described as particularly potent radical scavengers, were found to block cortical cell death with an exceptional efficacy, even under such conditions of high grade excitotoxicity.

In addition to influx of calcium, an activation of the NMDA receptor causes an influx of sodium ions, which produces a strong depolarization of the neuron (Choi, 1987). This results in an additional massive calcium influx through the opening of the voltage-dependent type-L ion channels.

A dying neuron can essentially fall into one of two different categories of cell death (Ankarcrona, this volume). *Necrotic* cell death typically involves acute damage, cell swelling and rupture of the outer cell membrane, with leakage of cytoplasm and organelle debris causing a local inflammatory response. By contrast, *apoptosis* (or programmed cell death) is slower, involves the activation of well regulated enzyme cascades that break down the cytoskeleton, nuclear envelope and DNA, and elicit cell surface signals that stimulate phagocytosis of the dying cell without an inflammatory response in the surrounding tissue (Nicotera and Leist, 1997). Apoptotic cell death is particularly interesting in the context of neurodegenerative disease, since it offers a time-window for cell rescue and a series of complex molecular events that may be manipulated in therapeutic interventions. An elegant study by Ankarcrona and coworkers (1995) has demonstrated that excitotoxicity can induce both types of cell death in cerebellar granule cells. Following exposure to glutamate, a subpopulation of cultured cerebellar granule neurons exhibit collapse of the mitochondrial membrane potential and undergo rapid necrosis. Other granule cells recover from the initial insult, restore their mitochondrial membrane potential and then undergo apoptosis following a delay of several hours. Emerging data indicates that a family of cysteine proteases named caspases play a pivotal role in apoptosis. Thus, pharmacological caspase inhibition can reduce neuronal death in cortical cultures subjected to high concentrations of glutamate receptor agonists (Nath et al., 1997). The transcription factor NF-κB is also activated in response to glutamate and it has been suggested that NF-κB may participate in the regulation of apoptosis in neurons, although it is still not clear whether it is primarily pro-apoptitic or neuroprotective (O'Neill and Kaltschmidt, 1997).

### *Impairments of energy metabolism*

Energy failure can be caused by a decrease in levels of high energy phosphate compounds which are produced by the mitochondria. Thus, an impairment of mitochondrial metabolism as well as deprivation of energy substrates, i.e. oxygen and/or glucose, can cause energy failure. However, energy failure *per se* is not particularly toxic to neural cells, unless it is protracted, which may cause cell death due to an arrest of all vital processes. It has been hypothesized that impairment of energy metabolism can cause neuronal cell death through the excitotoxic mechanism (Beal et al., 1993a; Lancelot and Beal, this volume).

In the brain, ion gradients across neural membranes, which generate the resting membrane potential, are maintained by energy-dependent ion pumps, such as a $Na^+/K^+$ pump. Energy failure induces an impairment in the repolarization of neural membranes after a depolarizing stimulus. Defective repolarization can relieve a voltage-dependent $Mg^+$ block of NMDA channels, leading to persistent receptor activation by endogenous levels of glutamate (Fig. 1). This hypothesis is derived from an in vitro experiment showing that the inhibition of energy metabolism makes glutamate neurotoxic at concentrations that ordinarily show no toxicity (Novelli et al., 1988). Lending support to this idea, in vitro and in vivo studies have demonstrated that metabolic inhibitors such as mitochondrial toxins reduce the threshold for NMDA-induced neuronal death (Marey-Semper et al., 1995; Zeevalk and Nicklas, 1990; Simpson and Isacson, 1993). In addition, a defect of energy-dependent ion pumps may lead to persistent neuronal depolarization, which would be expected to enhance synaptic release of glutamate. Impaired energy metabolism can also disturb the maintenance of the glial membrane potential. This may impede the re-uptake of glutamate by glial cells, subsequently leading to an excessive accumulation of extracellular glutamate (Lipton and Rosenberg, 1994).

*Disturbance of intracellular calcium homeostasis*

Whereas the calcium ion plays an important role in the function of neural cells, particularly in synaptic transmission, an excessive increase in intracellular calcium levels can lead to several consequences detrimental to cellular structures (see "*Excitotoxic mechanisms*" and chapter by Reynolds, this volume). All types of cells strictly control intracellular levels of calcium ions ($[Ca^{2+}]_i$). In neurons, the resting level of $[Ca^{2+}]_i$ is around 100 nM, which is maintained by the action of energy-dependent $Ca^{2+}$ pumps in the

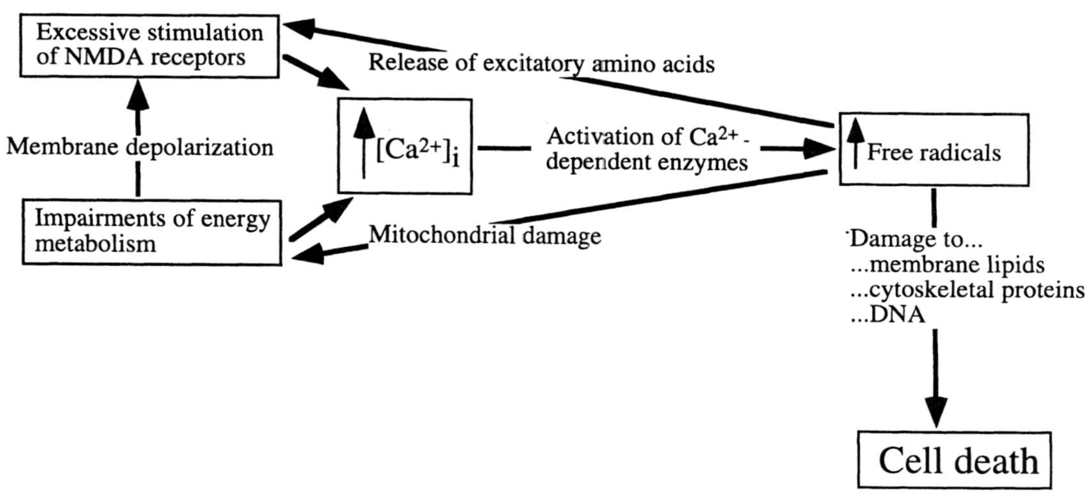


Fig. 1. Interaction between oxidative stress, excitotoxicity, energy failure, and increased $[Ca^{2+}]_i$, illustrating a cascade of events leading to a vicious cycle. See text for details of each mechanism.

plasma membrane and endoplasmic reticulum (Carafoli, 1991). In addition, the $Na^+/Ca^{2+}$ exchanger, localized in the plasma membrane, can rapidly extrude calcium in response to an elevation of $[Ca^{2+}]_i$ (Blaustein, 1988). Under conditions of depolarization of neural membranes, this exchanger system can reverse direction with a resultant increase in $[Ca^{2+}]_i$ (Stys et al., 1992). Another possible buffer system for $[Ca^{2+}]_i$ are calcium-binding proteins, such as calbindin-28 kD and calretinin (Heizmann and Hunziker, 1991).

Since excitotoxic mechanisms involve a massive influx of calcium, and most of the systems for regulating $[Ca^{2+}]_i$ are energy-dependent, the breakdown of intracellular calcium homeostasis may represent an ultimate mechanism leading to cell damage (Fig. 1). As stated earlier, a rise in $[Ca^{2+}]_i$ can facilitate the generation of free radicals secondary to the activation of calcium-dependent enzymes, such as phospholipase A2, xanthine oxidase and NOS (Fig. 1). Budd and Nicholls (1996) have speculated that sequestration of calcium by mitochondria during NMDA receptor stimulation reduces calcium concentration locally in the vicinity of the NMDA receptor and removes a feedback inhibition of calcium on the receptor. Thereby, $[Ca^{2+}]_i$ increases to an even higher level and cell viability is further compromised. In addition, depolarization of the inner mitochondrial membrane can lead to reversal of the membrane bound ATP synthase, causing consumption of ATP and further aggravation of energy failure (Budd and Nicholls, 1996). Thus, the local reduction of calcium concentration through mitochondrial sequestration may ultimately have negative effects on neuronal viability.

## Neurodegeneration involving either the striatum or the substantia nigra, and the proposed pathogenic mechanisms

### *Ischemia and hypoglycemia*

Since neural tissue requires a continuous supply of oxygen and glucose for its energy generation and normal function, prolonged ischemia and severe hypoglycemia both can cause neuronal cells to incur irreversible damage. Brain ischemia that deprives neuronal cells of both oxygen and glucose can be caused either by localized pathological conditions, such as stroke and trauma, or by cardiac arrest and severe shock. Also, severe hypoglycemia can be caused by an overdose of insulin, leading to a comatose state. The neuropathological features of brain damage induced by global ischemia and hypoglycemia are akin to each other, with some variations, in that a similar distribution of neuronal death is observed following these insults. The striatum is one of brain regions which are prone to develop neuronal damage following global ischemia (Pulsinelli et al., 1982; Smith et al., 1984) and hypoglycemia (Auer et al., 1984). Both ischemia and hypoglycemia affect the dorsolateral portion of the striatum, and tend to kill small to medium sized striatal neurons (Francis and Pulsinelli, 1982; Smith et al., 1984; Auer et al., 1984).

Ischemia and hypoglycemia lead to energy failure in neural cells. Impaired energy metabolism causes disturbance of ion homeostasis, including a rise in $[Ca^{2+}]_i$, and membrane depolarization, which could activate the excitotoxic pathway and lead to the generation of free radicals (see "*Disturbance of intracellular calcium homeostasis*"). There is a principal difference in the pathophysiology between the two insults, although the molecular mechanisms for cell death are very similar. During ischemia, high energy phosphate compounds can be produced only by anaerobic glycolysis with a concomitant increase in lactate levels (Siesjö, 1981). During hypoglycemia where cerebral blood flow is maintained, energy substrates other than glucose can be used for oxidative phosphorylation, and there is no lactate accumulation. Thus, in contrast to ischemia, hypoglycemia does not involve acidosis (Pelligrino and Siesjö, 1981).

### *Huntington's disease*

Huntington's disease (HD) is an inherited autosomal dominant neurodegenerative disorder. The major symptoms of the disease are progressive dementia, various psychiatric symptoms, and hyperkinetic, choreiform involuntary movement (Sharp and Ross, 1996). The motor and psychiatric symptoms typically appear between 30–40 years of age, progressively deteriorate and eventually lead to death 15–20 years after diagnosis (for review see Harper, 1991). In 1983, the gene responsible for HD was found to be located on chromosome 4 (Gusella et al., 1983), and recently the gene (IT 15) has been cloned and demonstrated to contain an expansion of CAG trinucleotide repeats from the normal 6–34 to over 36 in HD (The Huntington's disease Collaborative Research Group, 1993). The gene product protein has been named huntingtin and has an, as yet, unknown function.

The most striking neuropathological features of HD are drastic loss of striatal neurons, marked astrogliosis and the resultant severe atrophy of the neostriatum (Sharp and Ross, 1996). The degree of atrophy and cell loss correlates with the severity of clinical symptoms (Myers et al., 1988). In addition to the striatum, striatal projection areas such as, the globus pallidus and the substantia nigra are affected with atrophy and cell loss (Bird, 1980). Within the striatum, the medium-sized densely spiny neurons, which constitute 95% of all striatal neurons are most affected (Graveland et al., 1985). The large cholinergic neurons and the medium-sized aspiny interneurons that contain neuropeptide Y/somatostatin and nicotinamide adenine dinucleotide phosphate-diaphorase (NADPH-d) are relatively spared (Ferrante et al., 1985).

Although the clinical and neuropathological features of HD have been extensively investigated and the responsible gene has been identified, the pathogenic mechanism leading to the characteristic pattern of neuron death in this disorder remains to be elucidated (MacDonald and Gusella, 1996). A central enigma is that the mutant HD gene is widely expressed in the CNS, while the neurodegeneration is strictly localized anatomically. This observation has stimulated a search for and identification of several proteins that bind to huntingtin (MacDonald and Gusella, 1996; Sharp and Ross, 1996). The quest for a pathogenic mechanism involving mutant huntingtin may be aided by the recent construction of a transgenic mouse that expresses Exon 1 of IT15 (Mangiarini et al., 1996) and displays both a neurological phenotype and some neuropathology that is analogous to HD (Davies et al., 1997).

Disregarding precisely how the genetic defect is translated into cell damage, important theories have been proposed to explain the final steps of the cell death process in HD. These include excitotoxicity, oxidative stress, apoptosis, and mitochondrial dysfunction (Sharp and Ross, 1996; Lancelot and Beal, this volume). In 1976, Coyle and Schwarcz (1976), and McGeer and McGeer (1976) provided the first experimental evidence to suggest the implication of excitotoxicity in the pathogenesis of HD. They found that intrastriatal injection of kainic acid (KA) kills most of striatal neurons except for the large-sized cells, and spares axons of passage and nerve terminals. Subsequent studies with quinolinic acid (QUIN), an agonist for the NMDA receptor and an endogenous metabolite of tryptophan, have revealed that it produces striatal lesions that are very similar to those in HD in terms of neurochemical and neuropathological features. Thus, in QUIN-induced striatal lesions the levels of GABA and substance P, both of which are localized in the medium-sized *spiny* neurons, are reduced, and the content of somatostatin present in the medium-sized *aspiny* neurons is preserved (Beal et al., 1986). In accordance with these neurochemical findings, some morphological studies in rats and non-human primates have shown that intrastriatal injections of QUIN preferentially kill medium-sized spiny neurons with relative sparing of medium-sized aspiny neurons and large-sized neurons (Schwarcz et al., 1983; Ferrante et al., 1985, 1993). Other studies, however, have not reproduced the finding of a relative sparing of medium-sized aspiny neurons in QUIN-induced striatal lesions (Davies et al., 1987; Boegman et al., 1987). Some evidence that favours excitotoxicity as a possible pathogenesis of HD. It has been reported that in brains of HD patients the activity of 3-hydroxyanthranilate oxigenase, which is the biosynthetic enzyme for QUIN, is increased (Schwartz et al., 1988). Also, reduced levels of kynurenic acid, an antagonist of NMDA receptors that may modulate QUIN-induced neurotoxicity, have been found in HD brains (Beal et al., 1990).

More recently, an impairment of energy metabolism due to mitochondrial dysfunction has been suggested as a possible mechanism for neurodegeneration in HD (Beal et al., 1993a). In vitro cell culture studies have revealed that defects of energy metabolism make neurons susceptible to excitotoxicity (Marey-Semper, 1995; Novelli et al., 1988; Zeevalk and Nicklas, 1990; Zeevalk et al., 1995). Also, experimental studies in the rat have shown that pharmacological inhibition of the mitochondrial respiratory chain by use of mitochondrial toxins such as 3-nitropropionic acid (3-NP) and 1-methyl-4-phenylpyridinium ($MPP^+$) can cause striatal lesions (Storey et al., 1992; Beal et al., 1993b), and that the neuronal susceptibility to mitochondrial toxins increases with age (Beal et al., 1993c; Brouillet et al., 1993). Nuclear magnetic resonance spectroscopy has disclosed an elevated level of lactate in the basal ganglia of living HD patients, strongly suggesting that a defect of oxidative phosphorylation may occur in HD (Jenkins et al., 1993).

*Parkinson's disease*

The cardinal symptoms of Parkinson's disease (PD) are tremor at rest, rigidity and akinesia and their severity gradually increase with time (Rajput, 1994). In addition to the motor symptoms, some PD patients exhibit autonomic dysfunctions and neuropsychological alterations such as depression and dementia. The mean age of onset is around 55 years, and the incidence of PD increases with age. The neuropathological changes in PD are characterized by neuronal cell death in the pigmented cell nuclei of the brain stem (Jellinger, 1987). The neurodegeneration is associated with morphological changes of cytoplasmic structure, i.e. the appearance of Lewy bodies (Jellinger, 1987). Probably the most important histopathological change is the degeneration of dopamine (DA) neurons in the substantia nigra, leading to a decline of DA levels in the striatum (Agid et al., 1987). The severity of symptoms and duration of disease is correlated with the extent of loss of nigral neurons (McGeer et al., 1988), and DA replacement therapy can ameliorate motor symptoms. The first clinical symptoms are manifest when the striatal content of DA and the number of nigral neurons are reduced by 80% and 50%, respectively (Riederer and Wuketich, 1976). During the presymptomatic period, loss of DA may be compensated by increased activity of the remaining neurons and hypersensitivity of striatal DA receptors (Agid et al., 1987).

The underlying mechanism for the death of DA neurons in the substantia nigra in PD remains unknown. Recent studies on families affected by familial forms of PD with a diffuse distribution of Lewy bodies in several brain regions have identified a missense mutation in the α-synuclein gene (Nussbaum and Polymeropoulos, 1997). The gene codes for a presynaptic protein, part of which is present in plaques in Alzheimer's disease. Albeit very interesting, these findings may only provide us with clues as to why DA neurons die in idiopathic PD. Based on work in animal models and clinical postmortem investigations, several cell death mechanisms have been proposed to be relevant to the pathogenesis of idiopathic PD.

A body of evidence suggests that oxygen free radicals play a role in the pathogenesis of PD (for review see Olanow 1993). First, DA can be oxidized by either monoamine oxidase or auto-oxidation to generate $H_2O_2$. Thus, oxidative degradation of DA itself can be a source of free radicals, and an increased turnover of DA, which has been suggested in the remaining neurons in PD (Agid et al., 1987), may promote the generation of free radicals. Second, levels of glutathione are decreased in the substantia nigra pars compacta (SNc) of patients with PD (Perry et al., 1982), suggesting a reduction in the protective mechanism against OH· formation. Third, increased levels of free iron (Dexter et al., 1989a) and decreased levels of ferritin in the SNc of PD patients (Dexter et al., 1990) both increase the risk that $H_2O_2$ is oxidized to OH·. Fourth, it has been demonstrated that lipid peroxidation, which indicates a state of oxidative stress, is increased in the SNc in PD (Dexter et al., 1989b).

6-Hydroxydopamine (6-OHDA) and 1-methyl-4-phenyl-1,2,3,6-tetrahydropyridine (MPTP) can kill DA neurons, and can be used to produce animal models of PD. Interestingly, the underlying mechanisms by which these toxins can kill DA neurons involve oxidative stress. When injected intracerebrally, 6-OHDA is selectively taken up by DA neurons through the high affinity DA transporter, where it then reacts with molecular oxygen to produce reactive oxygen species (Heikkila and Cohen, 1973). Recently it has been suggested that 6-OHDA also inhibits mitochondrial respiratory chain complexes I and IV, through a mechanism that is independent of free radical generation, although the mitochondrial dysfunction may lead to an increased free radical production in itself (Glinka et al., 1997). MPTP can cause selective degeneration of the nigrostriatal DA system when injected systemically in humans and non-human primates (Langston, 1985). MPTP is catabolized by the action of the B form of monoamine oxidase present in glia to 1-methyl-4-phenyl pyridinium ($MPP^+$). Since $MPP^+$ is a substrate for the presynaptic DA uptake system, it is actively accumulated in DA neurons. $MPP^+$ is a potent inhibitor of mitochondrial ubiquinone oxidoreductase (complex I), and thereby interferes with mitochondrial electron transport. As stated earlier, an impairment of the mitochondrial respiratory chain could lead to the generation of free radicals.

The activity of mitochondrial complex I has been found to be reduced in the substantia nigra, platelets and muscle of PD patients (Mizuno et al., 1989; Schapira et al., 1990). In cells transformed by mitochondria from PD patients there are reduced levels of complex I activity and changes in intracellular calcium handling (Sheehan et al., 1997). It remains unclear whether the reduction in complex I activity in PD results from either abnormalities of gene encoding complex I protein or environmental insults. Complex I comprises 26 peptides, seven of which are encoded by mitochondrial DNA. Mitochondrial DNA undergoes a high rate of spontaneous mutation, and is highly susceptible to oxidative stress (Richter et al., 1988). Whatever the mechanism for the decreased activity of the mitochondrial respiratory chain, it results in bioenergetic failure, and could subsequently trigger the commencement of a cascade comprising detrimental events including excitotoxicity, a rise in cytosolic free calcium levels and oxidative stress (see the previous section). In support of this hypothesis, magnetic resonance spectroscopy has demonstrated an increased level of lactate in the basal ganglia of PD patients (Chen et al., 1994), suggesting that oxidative phosphorylation may be impaired in brains of PD patients. Mesencephalic DA neurons may be particularly sensitive to mitochondrial impairment. Partial inhibition of the Krebs cycle enzyme succinate dehydrogenase is more damaging to DA neurons than their GABAergic counterparts in mesencephalic cultures (Zeevalk et al., 1995). Similarly, inhibition of complex I renders cultured dopaminergic neurons more susceptible to glutamate-induced toxicity compared to GABAergic neurons treated in a similar fashion (Marey-Semper et al., 1995). Indeed, it has been shown that normal metabolism of DA itself can impair the electron transport chain in isolated brain mitochondria (Cohen et al., 1997). Interestingly, studies on cultured forebrain neurons have shown that DA and glutamate can act synergistically to promote neuronal degeneration, and that this death involves oxidative stress (Hoyt et al., 1997).

## Protective actions of antioxidants against environmental insults to cultured striatal neurons

In our laboratory, we have examined several factors that may promote the survival of the neuronal population that displays the most severe pathology in HD. Thus, we recently studied whether the spin-trapping agent PBN (for review see Thomas 1997) and the lazaroid U-83836E (for review see Hall 1997) protect cultured striatal dopamine- and cyclic AMP-regulated phosphoprotein-32 (DARPP-32)-containing neurons against several environmental insults (Nakao et al., 1996). For this purpose, we employed in vitro insult models of hypoglycemia and NMDA exposure. Since a combination of serum deprivation and a low plating cell density is detrimental to the viability of DARPP-32-positive cells in vitro, we also utilized this paradigm to test the neuroprotective effects of antioxidants.

A 24 h period of hypoglycemia reduced the number of surviving DARPP-32-positive cells and the total cell population to around 20% of that in glucose-containing cultures (Fig. 2A). Treatment with 3 mM PBN, but not 0.3 mM, significantly promoted the survival of both DARPP-32-positive cells and the entire neuronal population. U-83836E at a dose of 0.3 $\mu$M, but not 0.03 $\mu$M, also provided similar protection against hypoglycemic damage. Similar treatment with each of the antioxidants also attenuated the neuronal loss induced by a 24 h of NMDA exposure (Fig. 2B). We also explored whether PBN (3 mM) and U-83836E (0.3 $\mu$M) could ameliorate gradual cell death that occurs in serum-free, low-cell density cultures (Fig. 3). In cultures with a high plating density ($1.0 \times 10^5/cm^2$), neither PBN nor U-83836E afforded neuroprotection. In contrast, in cultures with a low plating density of ($0.5 \times 105/cm^2$), protective effects were seen with both the antioxidants. Our data provide evidence that inhibition of free radical processes by PBN or U-83836E can prevent neuronal damage produced by metabolic and excitotoxic insults. Since the underlying mechanisms for hypoglycemic injury are thought to involve an excitotoxic pathway (Wieloch, 1985; Monyer et al., 1989), the results support the notion that the generation of oxygen free radicals may play a role in excitotoxic neuronal damage. Furthermore, the antioxidants afforded neuroprotection in serum-free, low-cell density cultures where trophic support presumably is attenuated. We have also found that lazaroids promote the survival of mesencephalic DA neurons in a similar paradigm (Frodl et al., 1994). Although the molecular mechanisms for neuron death in this culture system remains to be fully determined, it is possible that oxidative stress

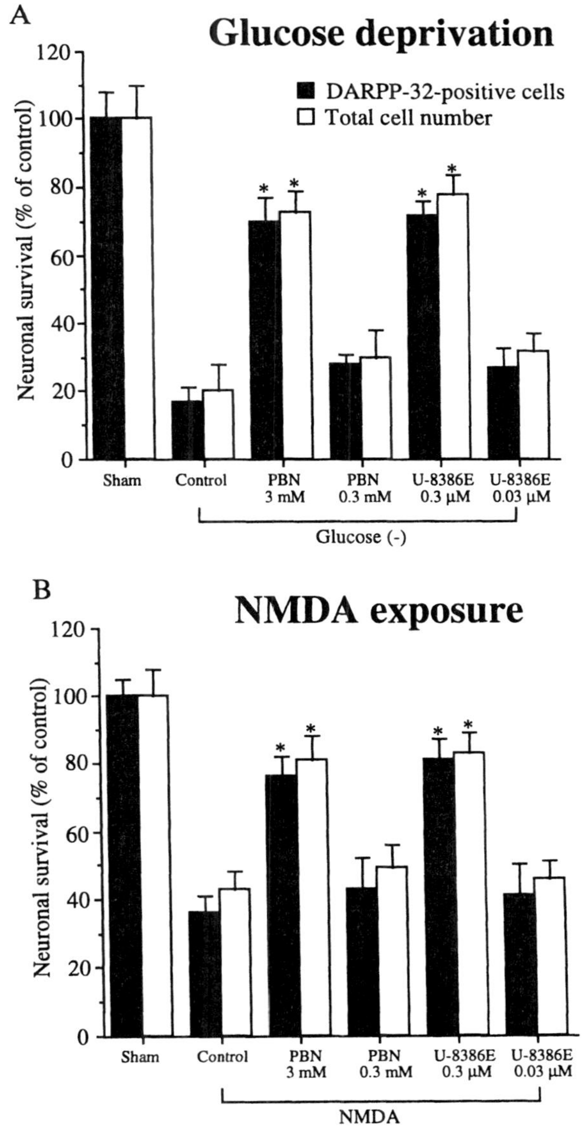


Fig. 2. Effects of PBN (3 mM) and U-83836E (0.3 $\mu$M) on the survival of striatal neurons in cultures subjected to either hypoglycemia (A) or NMDA-induced excitotoxicity (B). Cultures were treated with the antioxidants during the full 24 h of the insults. Cell survival was normalized by setting the number of surviving neurons in sham cultures as 100%. Values are expressed as the mean ± SEM of four independent experiments, each in turn comprising quadruplicate wells. $^*P < 0.0001$, significant difference from the corresponding value in control cultures in the absence of the antioxidants (one-way ANOVA with post-hoc Scheffé' s test). Data from Nakao et al., 1996.

could constitute a final common pathway for cell death. A study on cultured cerebellar granule cells has shown that the removal of serum from the culture medium can trigger an apoptotic form of cell death (Atabay et al., 1996). In that paradigm, there was evidence of increased cellular peroxide after serum deprivation and the fact that NMDA receptor blockade was protective suggests that excitotoxic mechanisms participated too.

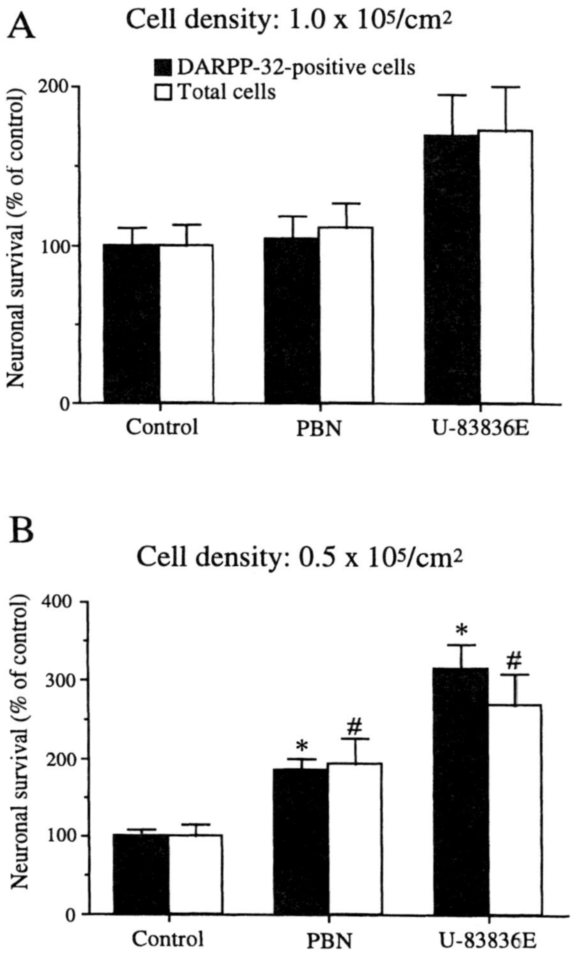


Fig. 3. Cultures were grown at a density of either $1.0 \times 105$ cm$^2$ or $0.5 \times 105$ cm$^2$ under serum-free culture conditions. Treatment with the antioxidants was started on day 1 and cultures were processed for DARPP-32-immunohistochemistry on day 5. The number of surviving neurons is expressed as a percentage of the cell count in untreated control cultures. Data are expressed as the mean ± SEM of four independent experiments with quadruplicate wells. $^{\#}P < 0.0001$, significant difference from the number of surviving DARPP-32-containing neurons in control cultures (one-way ANOVA with post-hoc Scheffé's test). $^{\#}P < 0.02$, significant difference from the total number of surviving neurons in control cultures (one-way ANOVA with post-hoc Scheffé's test). Data from Nakao et al., 1996.

**Protection of striatal neurons by the spin-trapping agent α-phenyl-*tert*-butylnitrone in quinolinic acid lesions in vivo**

We have also studied whether systemically administered PBN affords neuroprotection in a rat model of HD (Nakao et al., 1996). Striatal lesions were produced by intrastriatal injections of the NMDA receptor agonist QUIN. The rationale for using PBN in an in vivo excitotoxic lesion model is that it can readily penetrate the blood-brain barrier (Cheng et al., 1993). Earlier work has shown that PBN treatment can reduce brain protein oxidation and improve cognitive function in aged rats (Carney et al., 1991). In our study, PBN (150 mg/kg) was given at 30 min before lesion surgery, and 12 h and 24 h after the initial treatment. Since PBN has been reported to cause a profound hypothermia, particularly when given at high doses (Yue et al., 1992), we maintained normal body temperature (37–38°C) in rats after lesion surgery in order to preclude the possible influence of hypothermia on the outcome of QUIN lesions.

We examined the survival of two types of striatal neurons, DARPP-32-containing medium-sized *spiny* neurons and NADPH-d-positive medium-sized *aspiny* neurons, at three different levels through the striatum caudal to the QUIN-injection site. In control rats not receiving PBN, the numbers of surviving DARPP-32-positive neurons in the three brain sections were around 10% of the intact side (Fig. 4A). The death of NADPH-d-positive neurons was also significant, although less marked (Fig. 4B). Treatment with PBN significantly prevented the QUIN-induced death of DARPP-32-positive neurons in all the sections monitored. In contrast, a significant protection of NADPH-d-positive neurons by PBN was observed only at the level of the toxin-injection site. We also determined the areas exhibiting MAP-2-immunoreactivity in the lesioned striatum (Fig. 4C). MAP-2 is particularly susceptible to degenerative processes such as ischemia and excitotoxicity (Johnson and Jope, 1992), and therefore could be a sensitive marker for the extent of excitotoxic lesions. PBN significantly attenuated the loss of MAP-2-immunoreactivity throughout all the three sections. In summary, systemically administered PBN ameliorated striatal degeneration produced by intrastriatal injections of QUIN.

### Effects of α-phenyl-*tert*-butyl nitrone on behavioral consequences of striatal lesions produced by quinolinic or 3-nitropropionic acid

This study investigated the neuroprotective efficacy of PBN in unilateral striatal lesions induced by local injections of QUIN or the mitochondrial toxin 3-NP by assessing the functional, behavioural consequences of the lesion (Nakao and Brundin, 1997). We tested lesioned rats for paw-reaching ability, and amphetamine- and apomorphine-induced rotation asymmetry, which are behaviors considered to depend on the functional integrity of striatal circuitry. The protocol for PBN treatment was the same as described above for the previous study.

Both the QUIN and 3-NP-lesions impaired reaching ability of the paw contralateral to the lesion, with the 3-NP lesion producing more severe deficits (Fig. 5). Treatment with PBN significantly improved the paw-reaching deficits induced by the QUIN lesion (Fig. 5). In contrast, the 3-NP-induced deficits were not significantly altered by PBN (Fig. 5). The unilateral striatal lesions caused ipsilateral turning in response to amphetamine and apomorphine. Again, PBN ameliorated drug-induced rotation asymmetry in the QUIN-, but not in the 3-NP-lesioned rats. In agreement with the behavioral outcome, PBN increased the size of the residual spared striatum with preserved DARPP-32-immunoreactivity in the QUIN lesion. There was no significant effect of PBN in the 3-NP lesion. A significant correlation between the behavioural scores and the size of the spared striatum was observed in all the behavioral tests, with the correlation being strongest for the paw-reaching test. In addition to the striatal lesion, the intrastriatal injection of 3-NP produced degeneration of the nigrostriatal DA system as well as of transverse fiber bundles of the internal capsule within the striatum. QUIN killed only intrinsic striatal neurons.

In conclusion, PBN mitigated the behavioral deficits seen in rats with QUIN-induced striatal lesions, which corresponded well with the size of the spared striatum. PBN failed to protect striatal neurons against degeneration produced by intrastriatal injections of 3-NP (Nakao and Brundin, 1997).

### Protective effects of antioxidants against excitotoxic and metabolic injuries to striatal neurons

Strategies for ameliorating neuronal damage associated with excitotoxicity can be targeted to different entry points (Holt, 1997). Stages in the excitotoxic process that could be pharmacologically manipulated are: (1) glutamate release; (2) ligand-glutamate receptors interactions; and (3) intracellular events that follow the receptor activation, i.e. the stimulation of several destructive cascades, including kinase-induced protein phosphorylation, protease-derived protein breakdown, and free radical formation mediated via phospholipase $A_2$ and NOS. Much effort at developing an effective therapy for excitotoxic brain damage has focused on the application of NMDA antagonists (Lipton and Rosenberg, 1994). Nonetheless, since normal NMDA receptor functions are involved in cognition and memory, blockade of NMDA receptors may results in psychotomimetic effects (for review see Holt 1997). Also, it has been reported that neuropathological changes are induced by NMDA antagonists (Olney et al., 1989). In this context, the availability of pharmacological agents that affect processes distal to glutamate receptors are desirable. Among the drugs that can act on the levels downstream from the receptors, antioxidants have been most extensively tested for

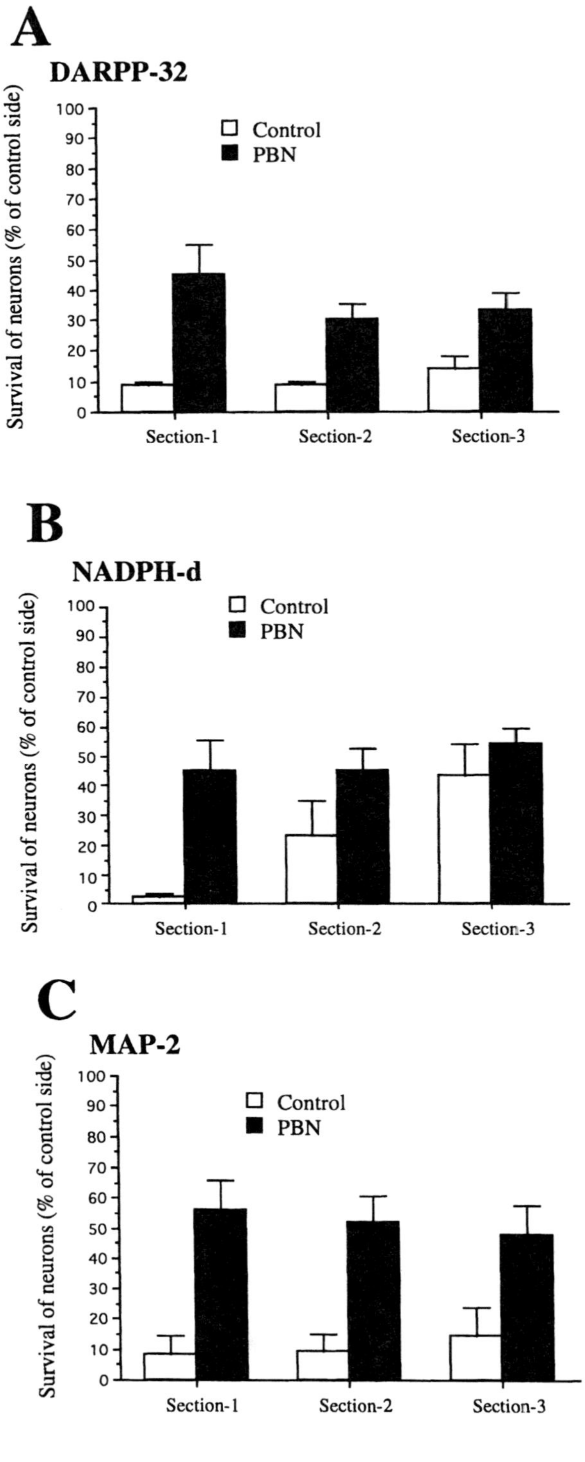


Fig. 4. Effects of systemically administered PBN on QUIN-induced striatal lesion. In PBN-treated rats, PBN (150 mg/kg) was intraperitoneally given on three occasions, 30 min before the striatal lesion, and 12 h and 24 h after the first treatment. Control animals were given physiological saline in an identical fashion. The percent survival of DARPP-32-(A) and NADPH-d-positive (B) neurons relative to the intact side is shown as a function of the distance from the QUIN injection site. The area of the striatum still expressing MAP-2-immunoreactivity is expressed as a percentage of the total cross-sectional area of the striatum in panel (C). Sections 1, 2, and 3 correspond brain levels at 0 mm, 1.0 mm, and 1.5 mm caudal to the QUIN injection site, respectively. Data are depicted as the mean ± SEM. Data from Nakao et al., 1996.

their neuroprotective efficacy in excitotoxicity-related injury in vitro and in vivo.

*In vitro studies*

The idea that excitotoxic neuronal damage is, at least partly, mediated by oxidative stress is supported by growing evidence that the application of NMDA elicits the generation of free radicals, and that antioxidants provide neuroprotection against excitotoxic cell damage (Coyle and Puttfarcken, 1993). Thus, reactive oxygen species are detected in neuronal cultures following exposure to NMDA (Lafon-Cazal et al., 1993; Dugan et al., 1995; Gunasekar et al., 1995; Reynolds and Hastings, 1995). Lending further support to this notion, a complete absence of oxygen (i.e. anoxia), where oxygen-based reactions cannot occur, was found to attenuate glutamate-induced neuronal death in vitro (Dubinsky et al., 1995). Cell culture studies have revealed that free radical scavengers protect against excitotoxic/metabolic injury. Several types of antioxidants, including antioxidant enzymes such as SOD and catalase, allopurinol, and mannitol, prevent kainate-induced death in cerebellar cultures (Dykens et al., 1987; Cheng and Sun, 1994). A free radical scavenger idebenone attenuates the intracellular formation of peroxide following glutamate exposure, and provided protection against excitotoxicity in neuronal cell line cultures (Miyamoto et al., 1989). Cultured cortical neurons overexpressing an antioxidant

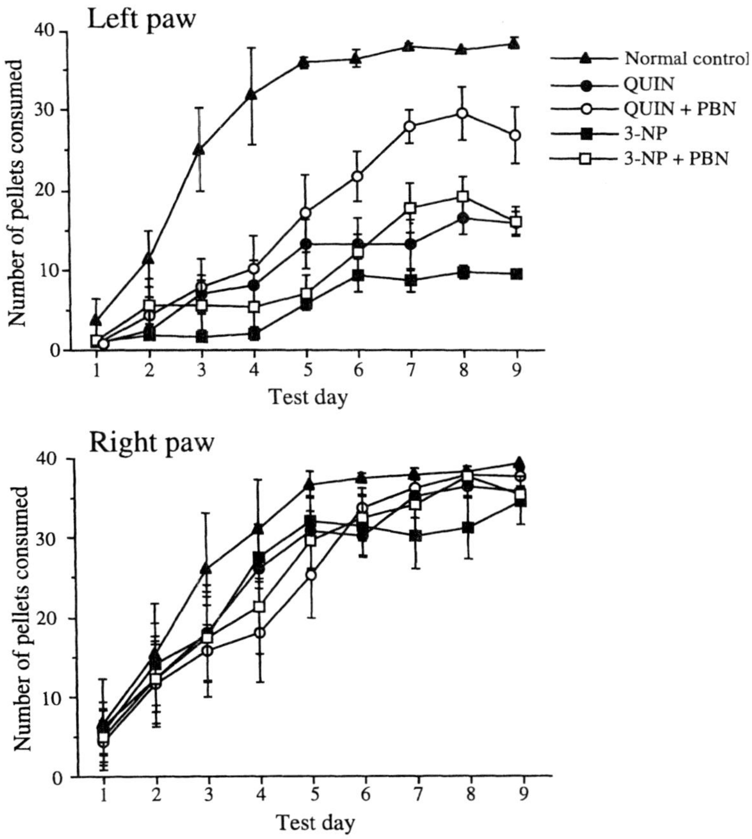


Fig. 5. Paw-reaching ability evaluated over nine consecutive days. Data are expressed as the mean ± SEM. The left paw is contralateral to the striatal lesion, and therefore displays impairment, expressed as a reduced ability to retrieve pellets with that paw. Data from Nakao and Brundin 1997.

enzyme Cu/Zn-SOD exhibit higher resistance to glutamate toxicity compared to non-transgenic control neurons (Chan et al., 1990). It has been demonstrated in vitro that lazaroids attenuate neuronal death induced by combined oxygen and glucose deprivation (Monyer et al., 1990), and that they decrease kainate-induced lipid peroxidation (Puttfarcken et al., 1993). Endogenous antioxidants ubiquinone (Favit et al., 1992) and α-lipoic acid (Müller and Krieglstein, 1995) have been shown to protect cultured neurons against glutamate-induced degeneration. A free radical spin trap agent has proven to be protective in glutamate toxicity in vitro (Lafon-Cazal et al., 1993; Yue et al., 1992). As mentioned earlier, we have shown that U-83836E and PBN both counteract the death of striatal DARPP-32-positive cells in cultures exposed to hypoglycemia and NMDA (Fig. 2) (Nakao et al., 1996).

*In vivo studies*

Systemic antioxidant treatment has been attempted in excitotoxic injury models with varying success. Systemic approaches to prevent QUIN-induced striatal lesions were first made by Beal et al. (1988), who found that a variety of antioxidants, including β-carotene, ascorbic acid, α-tocopherol, and allopurinol, are ineffective in protecting against QUIN toxicity. The lazaroid

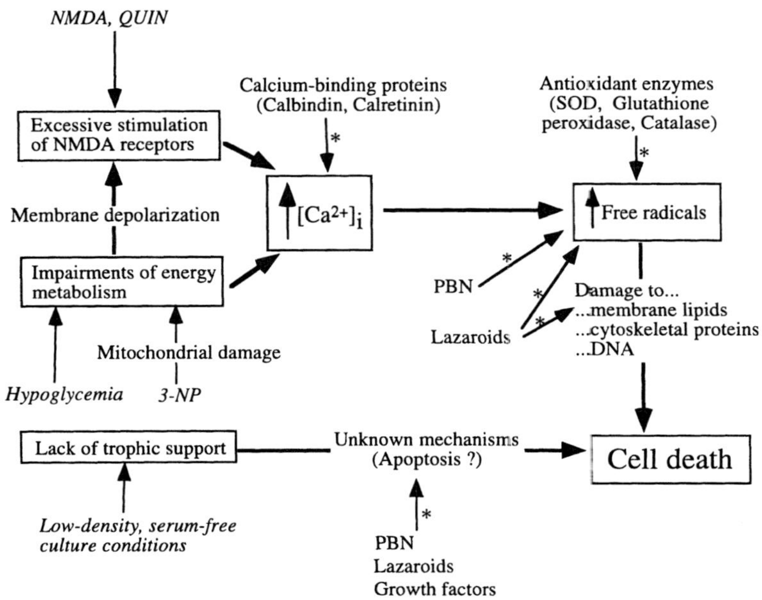


Fig. 6. Summary of the insult models (italics) and neuroprotective factors. This diagram demonstrates possible mechanisms by which different types of insults cause cell death, and sites of actions of neuroprotective agents. *Each detrimental event can be ameliorated by the factors mentioned.

U-78517F also failed to exhibit neuroprotective actions in QUIN-induced lesions of the striatum (Uhler et al., 1994). Since some of the antioxidants tested in those in vivo experiments have previously been shown to protect against excitotoxic/metabolic insults in vitro, the absence of protective effects in vivo may be related to pharmacokinetics. The recent development of potent antioxidant pyrrolopyrimidine lazaroids that readily penetrate the blood-brain barrier (Andrus et al., 1997) opens the possibility that lazaroids may counteract excitotoxicity also in vivo, in the future.

Miyamoto and Coyle (1990) reported that idebenone lessened the decline in the activity of choline acetyltransferase and GAD in striatal lesions produced by intrastriatal injections of KA and quisqualic acid, but not QUIN. Recently, Schulz et al. (1995a) found that PBN can reduce the size of striatal lesions caused by excitotoxins such as NMDA, AMPA and KA, and mitochondrial toxins including malonate (a *reversible* inhibitor of complex II) and $MPP^+$ (complex I inhibitor). In our studies (Nakao et al., 1996; Nakao and Brundin, 1997), PBN significantly ameliorated striatal lesions induced by QUIN, but not by the mitochondrial toxin 3-NP (an *irreversible* inhibitor of complex II). Since our treatment protocol (e.g. dose and frequency of administrations) is similar to that employed by Schulz et al. (1995a), the discrepancy seen in the effect on mitochondrial toxin lesions may be due to the fact that 3-NP exhibits a greater toxicity than $MPP^+$ and malonate. Previous studies have shown that a NMDA antagonist can block striatal lesions induced by local injections of malonate, but not 3-NP (Beal et al., 1993b, 1993c), which supports that 3-NP, particularly when topically administered, causes a more severe energy depletion than malonate.

In our experiments using PBN to protect against striatal QUIN lesions (Nakao et al., 1996; Nakao and Brundin, 1997), we started the treatment with PBN 30 min before intrastriatal injections of QUIN, and did not determine whether administration of PBN after QUIN insult is still protective in our lesion model. Direct injections of either excitotoxins or metabolic inhibitors are poor models of clinical disorders. Nonetheless, it is important to address the issue regarding effects of delayed treatment, which may have significant implications for clinical treatment of excitotoxicity-related brain injuries, particularly brain ischemia and trauma. It has previously been shown that administration of the NMDA antagonist MK-801 at 3 h after toxin injection protects against QUIN lesions (Bakker and Foster, 1991). Recently, PBN was found to protect against malonate-induced striatal lesions even when administered up to 6 h after intrastriatal malonate injections, whereas MK-801 provided protection when given up to 1 h after the lesion (Schulz et al., 1995b). These results indicate that PBN has a wider therapeutic window for protection against this type of metabolic injury, at least compared to MK-801. Also in a model of focal brain ischemia, PBN can still exert neuroprotection when given 3 h (Zhao et al., 1994) or 12 h (Cao and Phillis, 1994) after ischemic insult, whereas the NMDA and non-NMDA antagonists are effective with a delay 1 to 2 h after induction of ischemia (Park et al., 1988; Hatfield et al., 1992).

The PBN-induced protection against excitotoxic and metabolic insults is probably mediated by removal of reactive oxygen species. An electron paramagnetic study showed that PBN prevents increases in the level of free radicals in ischemic brains (Sen and Phillis, 1993). PBN also attenuated malonate-induced increases in hydroxyl radical production, but it affected neither energy metabolism nor spontaneous striatal electrophysiological activity (Schulz et al., 1995a).

## Concluding remarks

We and others have found that antioxidants promote the survival of neurons in several insult models (Fig. 6). Thus, they protected neurons against death caused by exposure both to suboptimal conditions of trophic support (*death by neglect*) and to toxic stimuli (*death by murder*) (Mayer and Noble, 1994). Although these insult models at best only mimic cell death mechanisms occurring in neurodegenerative diseases such as HD and PD, the results indicate that these factors at optimal doses can somehow operate against the cascade of intracellular events leading to cell death (Fig. 6). The molecular mechanism by which the factors protect neurons against the insults remains to be fully explored. An increased understanding of the underlying molecular mechanisms may in the future help to identify protective agents that could be used clinically.

## Acknowledgements

The work from our group described in this review was supported by Grants from the Swedish Medical Research Council, the Hereditary Disease Foundation and the Thorsten and Elsa Segerfalk Foundation.

## References

Agid, Y., Javoy-Agid, F. and Ruberg, M. (1987) Biochemistry of neurotransmitter in Parkinson's disease. In: C.D. Marsden, S. Fahn (Eds) *Movement Disorder*, Vol 2, Butterworths, London, pp. 166–230.

Andrus, P.K., Fleck, T.J., Oostveen, J.A. and Hall, E.D. (1997) Neuroprotective effects of the novel brain-penetrating pyrrolopyrimidine antioxidants U-101033E and U-104067F against post-ischemic degeneration of nigrostriatal neurons. *J., Neurosci. Res.*, 47: 650–654.

Ankarcrona, M., Dypbukt, J.M., Bonfoco, E., Zhivotovsky, B., Orrenius, S., Lipton, S.A. and Nicotera, P. (1995) Glutamate-induced neuronal death: A succession of necrosis and apoptosis depending on mitochondrial function. *Neuron*, 15: 961–973.

Atabay, C., Cagnoli, C.M., Kharlamov, E., Ikonomovic, M.D. and Manev, H. (1996) Removal of serum from primary cultures of cerebellar granule neurons induces oxidative stress and DNA fragmentation: Protection with antioxidants and glutamate receptor antagonists. *J. Neurosci. Res.*, 43: 465–475.

Auer, R., Wieloch, T., Olsson, Y. and Siesjö, B.K. (1984) The distribution of hypoglycemic brain damage. *Acta Neuropathol.*, 64: 117–191.

Bakker, M.H.M. and Foster A.C. (1991) An investigation of the time mechanisms of delayed neurodegeneration caused by direct injection of quinolinate into the rat striatum. *Neuroscience*, 42: 387–395.

Bast, A. and Bär, P.R. (1997) The antioxidant/prooxidant balance in neurodegeneration and neuroprotection. In P.R.

Bär and M.F. Beal (Eds.) *Neuroprotection in CNS diseases*, Marcel Dekker, Inc., pp. 147–160.

Beal, M.F., Kowall, N.W., Ellison, D.W., Mazurek, M.F., Swartz, K.J. and Martin, J.B. (1986) Replication of the neurochemical characteristics of Huntington's disease by quinolinic acid. *Nature*, 321: 168–171.

Beal, M.F., Kowal, N.W., Swartz, K.J., Fernnante, R.J. and Martin, J.B. (1988) Systemic approaches to modifying quinolinic acid striatal lesions in rats. *J. Neurosci.*, 8: 3901–3908.

Beal, M.F., Matson, W.R., Swartz, K., Gamache, P.H. and Bird, E.D. (1990) Kynurenine pathway measurements in Huntington's disease striatum: Evidence for reduced formation of kynurenic acid. *J. Neurochem.*, 55: 1327–1339.

Beal, M.F., Hyman, B.T. and Koroshetz, W. (1993a) Do defects in mitochondrial energy metabolism underlie the pathology of neurodegenerative diseases. *Trends Neurosci.*, 16: 125–131.

Beal, M.F., Brouillet, E., Jenkins, B.G., Ferrante, R.J., Kowall, N.W., Miller, J.M., Storey, E., Srivastava, R., Rosen, B.R. and Hyman, B.T. (1993b) Neurochemical and histologic characterization of striatal excitotoxic lesions produced by the mitochondrial toxin 3-nitropropionic acid. *J. Neurosci.*, 13: 4148–4192.

Beal, M.F., Brouillet, E., Jenkins, B.G., Henshaw, R., Rosen, B.R. and Hyman, B.T. (1993c) Age-dependent striatal excitotoxic lesions produced by the endogenous mitochondrial inhibitor malonate. *J. Neurochem.*, 61: 1147–1154.

Beckmann, J.S., Beckman, T.W., Chen, J., Marshall, P.A. and Freeman, B.A. (1990) Apparent hydroxyl radical production by peroxynitrite: Implications for endothelial injury from nitric oxide and superoxide. *Proc. Natl. Acad. Sci. USA*, 87: 1620–1624.

Bird, E.D. (1980) Chemical pathology of Huntington's disease. *Annu. Rev. Pharmacol. Toxicol.*, 20: 533–551.

Blaustein, M.P. (1988) Calcium transport and buffering in neurons. *Trends Neurosci.*, 11: 438–443.

Boegman, R.J., Smith, Y. and Parent, A. (1987) Quinolinic acid does not spare striatal neuropeptide Y-immunoreactive neurons. *Brain Res.*, 415: 178–182.

Bolanos, J.P., Heales, S.J.R., Land, J.M. and Clark J.B. (1995) Effect of peroxynitrite on the mitochondrial respiratory chain: Differential susceptibility of neurones and astrocytes in primary culture. *J. Neurochem.*, 64: 1965–1972.

Brouillet, E., Jenkins, B.G., Hyman, B.T., Ferrante, R.J., Kowall, N.W., Srivastava, R., Roy, D.S., Rosen, B.R. and Beal, M.F. (1993) Age-dependent vulnerability of the striatum to the mitochondrial toxin 3-nitropropionic acid. *J. Neurochem.*, 60: 356–359.

Budd, S.L. and Nicholls, D.G. (1996) Mitochondria, calcium regulation, and acute glutamate excitotoxicity in cultured cerebellar granule cells. *J. Neurochem.*, 67: 2282–2291.

Cao, X. and Phillis, J.W. (1994) α-Phenyl-*tert*-butyl-nitrone reduces cortical infarct and edema in rats subjected to focal ischemia. *Brain Res.*, 644: 267–272.

Carafoli, E. (1991) Calcium pump of the plasma membrane. *Physiol. Rev.*, 71: 129–153.

Carney, J.M., Starke-Reed, P.E., Oliver, C.N., Landum, R.W., Cheng, M.S., Wu, J.F. and Floyd, R.A. (1991) Reversal of age-related increase in brain oxidation, decrease in enzyme activity, and loss in temporal and spatial memory by chronic administration of the spin-trapping compound *N-tert*-butyl-α-phenylnitrone. *Proc. Natl. Acad. Sci. USA*, 88: 3666–3636.

Chan, P.H., Chu, L., Chen, S.F., Carlson, E.J. and Epstein, C.J. (1990) Reduced glutamate neurotoxicity in transgenic mice expressing human Cu-Zn superoxide dismutase. *Stroke*, 21: 80–82.

Chen, Y.J., Jenkins, B.G. and Rosen, B.R. (1994) Evidence of impairment of energy metabolism in Parkinson's disease using in vivo localized MR spectroscopy. *Proc. Soc. Magn. Res.*, 1: 194.

Cheng, G., Griffin, M., Poyer, J.L. and McCay, P.B. (1993) HPLC procedure for the pharmacokinetics study of the spin-trapping agent, α-phenyl-*n*-nitrone (PBN). *Free Rad. Biol. Med.*, 8: 93–98.

Cheng, Y. and Sun, AY. (1994) Oxidative mechanisms involved in kainate-induced cytotoxicity in cortical neurons. *Neurochem. Res.*, 12: 1557–1564.

Choi, D.W. (1987) Ionic dependence of glutamate neurotoxicity. *J. Neurosci.*, 7: 369–379.

Cohen, G., Farooqui, R. and Kesler, N. (1997) Parkinson's disease: A new link between monoamine oxidase and mitochondrial electron flow. *Proc. Natl. Acad. Sci. USA*, 94: 4890–4894.

Coyle, J.T. and Puttfarcken, P. (1993) Oxidative stress, glutamate, and neurodegenerative disorders. *Science*, 262: 689–695.

Coyle, J.T. and Schwarcz, R. (1976) Lesion of striatal neurones with kainic acid provides a model for Huntington's chorea. *Nature*, 263: 244–246.

Davies, S.W. and Roberts, P.J. (1987) No evidence for preservation of somatostatin-containing neurons after intrastriatal injections of quinolinic acid. *Nature*, 327: 326–329.

Davies, S.W., Turmaine, M., Cozens, B.A., DiFiglia, M., Sharp, A.H., Ross, C.A., Scherzinger, E., Wanker, E.E., Mangiarini, L. and Bates, G. (1997) Formation of neuronal intranuclear inclusions underlies the neurological dysfunction in mice transgenic for the HD mutation. *Cell*, 90: 537–546.

Dawson, T.M., Dawson, V.L. and Snyder, S.H. (1992) A novel neuronal messenger molecule in brain: The free radical, nitric oxide. *Ann. Neurol.*, 32: 297–311.

Dexter, D.T., Wells, F.R., Lees, A.J., Javoy-Agid, F., Agid, Y., Jenner, P. and Marsden, C.D. (1989a) Increased nigral iron content and alterations in other metal ions occurring in brain in Parkinson's disease. *J. Neurochem.*, 52: 1830–1836.

Dexter, D.T., Carter, C.J., Wells, F.R., Javoy-Agid, F., Agid, Y., Lees, A.J., Jenner, P. and Marsden C.D. (1989b) Basal lipid peroxidation in substantia nigra is increased in Parkinson's disease. *J. Neurochem.*, 52: 381–389.

Dexter, D.T., Carayon, A., Vidailhet, M., Ruberg, M., Javoy-Agid, F., Agid, Y., Lees A.J., Wells, F.R., Jenner, P. and Marsden, C.D. (1990) Decreased ferritin levels in brain in Parkinson's disease. *J. Neurochem.*, 55: 16–20.

Dubinski, J.M., Kristal, B.S. and Elizondo-Fournier, M. (1995) An obligate role for oxygen in the early stages of glutamate-induced, delayed neuronal death. *J. Neurosci.*, 15: 7071–7078.

Dugan, L.L., Sensi, S.L., Canzoniero, L.M.T., Handran, S.D., Rothman, S.M., Lin, T.-S., Goldberg, M.P. and Choi, D.W. (1995) Mitochondrial production of reactive oxygen species in cortical neurons following exposure to *N*-methyl-D-aspartate. *J. Neurosci.*, 15: 6377–6388.

Dugan, L.L., Turetsky, D.M., Du, C., Lobner, D., Wheeler, M., Almli, C.R., Shen, C.K.-F., Luh, T.-Y., Choi, D.W. and Lin, T.-S. (1997) Carboxyfullerenes as neuroprotective agents. *Proc. Natl. Acad. Sci. USA*, 94: 9434–9439.

Dykens, J.A., Stern, A. and Trencker, E. (1987) Mechanism of kainate toxicity to cerebellar neurons in vitro is analogous to reperfusion tissue injury. *J. Neurochem.*, 49: 1222–1228.

Favit, A., Nicoletti, F., Scapagnini, U. and Canonico, P.L. (1992) Ubiquinone protects cultured neurons against spontaneous and excitotoxin-induced degeneration. *J. Cereb. Blood Flow Metab.*, 12: 638–645.

Ferrante, R.J., Kowal, N.W., Beal, M.F., Richardson, E.P. Jr., Bird, E.D. and Martin J.B. (1985) Selective sparing of a class of striatal neurons in Huntington's disease. *Science*, 230: 561–563.

Ferrante, R.J., Kowall, N.W., Cipolloni, P.B., Storey, E. and Beal, M.F. (1993) Excitotoxin lesions in primates as a model for Huntington's disease: Histopathologic and neurochemical characterization. *Exp. Neurol.*, 119: 46–71.

Francis, A. and Pulsinelli, W. (1982) The response of GABAergic and cholinergic neurons to transient cerebral ischemia. *Brain Res.*, 243: 271–278.

Frodl, E.M., Nakao, N. and Brundin, P. (1994) Lazaroids improve the survival of cultured rat embryonic mesencephalic neurons. *Neuroreport*, 5: 2393–2396.

Glinka, Y., Gassen, M. and Youdim, M.B.H. (1997) Mechanism of 6-hydroxydoparnine neurotoxicity. *J. Neural Transm. (Suppl)*, 50: 55–66.

Graveland, G.A., Williams, R.S. and DiFiglia, M. (1985) Evidence for degenerative and regenerative changes in neostriatal spiny neurons in Huntington's disease. *Science*, 227: 770–773.

Gunasekar, P.G., Kanthasamy, A.G., Borowitz, J.L. and Isom, G.E. (1995) NMDA receptor activation produces concurrent generation of nitric oxide and reactive oxygen species: Implication for cell death. *J. Neurochem.*, 65: 2016–2021.

Hall, E.D. (1997) Lipid antioxidants as neuroprotectants for acute and chronic neurodegenerative disorders. In P.R. Bär and M.F. Beal (Eds.) *Neuroprotection in CNS diseases*, Marcel Dekker, Inc., pp. 161–182.

Halliwell, B. and Gutteridge, J.M.C. (1989) *Free Radicals in Biology and Medicine.*, Clarendon Press, Oxford.

Harper, P.S. (1996) *Huntington's disease*, 2nd Ed. W.B. Saunders, London. pp. 1438.

Hatfield, R.H., Gill, R. and Brazell, C. (1992) The dose-response relationship and therapeutic window for dizocilpine (MK-801) in a rat focal ischaemia model. *Eur. J. Pharmacol.*, 216: 1–7.

Heikkila, R.E. and Cohen, G. (1973) 6-Hydroxydopamine: Evidence for superoxide radical as an oxidative intermediate. *Science*, 181: 456–457.

Heizmann, C.W. and Hunziker, W. (1991) Intracellular calcium-binding proteins: More site than insights. *Trends Biochem. Sci.*, 16: 98–103.

Holt, W. F. (1997) Glutamate in health and disease: The role of inhibitors. In P.R. Bär and M.F. Beal (Eds.) *Neuroprotection in CNS diseases*, Marcel Dekker, Inc., pp. 87–120.

Hoyt, K.R., Reynolds, I.J. and Hastings, T.G. (1997) Mechanisms of dopamine-induced cell death in cultured rat forebrain neurons: Interactions with and differences from glutamate-induced cell death. *Exp. Neurol.*, 143: 269–281.

Huntington's Disease Collaborative Research Group (1993) A novel gene containing a trinucleotide repeat that is expanded and unstable on Huntington' s disease chromosomes. *Cell*, 72: 971–983.

Jellinger, K. (1987) The pathology of parkinsonism. In: C.D. Marsden, S. Fahn (Eds) *Movement Disorder*, Vol 2, Butterworths, London, pp. 124–165.

Jenkins, B.G., Koroshetz, W.J., Beal, M.F. and Rosen, B.R. (1993) Evidence for impairment of energy metabolism in vivo in Huntington's disease using localized IH NMR spectroscopy. *Neurology*, 43: 2689–2695.

Johnson, G.V.W. and Jope, R.S. (1992) The role of microtubule-associated protein 2 (MAP-2) in neuronal growth, plasticity, and degeneration. *J. Neurosci. Res.*, 33: 505–512.

Lafon-Cazal, M., Pietri, S., Culcasi, M. and Bockaert, J. (1993) NMDA-dependent superoxide production and neurotoxicity. *Nature*, 364: 535–537.

Langston, J.W. (1985) Mechanisms of MPTP toxicity: More answers, more questions. *Trends Pharmacol. Sci.*, 6: 375–378.

Leist, M., Fava, E., Montecucco, C. and Nicotera, P. (1997) Peroxynitrite and nitric oxide donors induce neuronal apoptosis by eliciting autocrine excitotoxicity. *Eur. J. Neurosci.*, 9: 1488–1498.

Lipton, S.A. and Rosenberg, P.A. (1994) Excitatory amino acids as a final common pathway for neurologic disorders. *N. Engl. J. Med.*, 330: 613–622.

MacDonald, M.E. and Gusella, J.F. (1996) Huntington's disease: Translating a CAG repeat into a pathogenetic model. *Curr. Opin. Neurobiol.*, 6: 638–643.

Mangiarini, L., Sathasivam, K., Seller, M., Cozens, B., Harper, A., Hetherington, C., Lawton, M., Trottier, Y., Lehrach, H., Davies, S.W. and Bates, G.P. (1996) Exon 1 of the Huntington's disease gene containing a highly expanded CAG repeat is sufficient to cause a progressive neurological phenotype in transgenic mice. *Cell*, 87: 497–506.

Marey-Semper, I., Gelman, M. and Levi-Strauss, M. (1995) A selective toxicity toward cultured mesencephalic dopaminergic neurons is induced by the synergistic effects of energetic metabolism impairment and NMDA receptor activation. *J. Neurosci.*, 15: 5912–5918.

Mayer, M. and Noble M. (1994) *N*-acetyl-L-cysteine is a pluripotent protector against cell death and enhancer of trophic factor mediated cell survival in vitro. *Proc. Natl. Acad. Sci. USA*, 91: 7496–7500.

McGeer, E.G. and McGeer, P.L. (1976) Duplication of biochemical changes of Huntington's chorea by intrastriatal injection of glutamic acid and kainic acids. *Nature*, 263: 517–519.

McGeer, P.L., Itagaki, S., Akiyama, H. and McGeer, E.G. (1988) Rate of cell death in parkinsonism indicates active neuropathological process. *Ann. Neurol.*, 24: 574–576.

Miyamoto, M., Murphy, T.H., Schnaar, R.L. and Coyle, J.T. (1989) Antioxidants protect against glutamate-induced cytotoxicity in a neuronal cell line. *J. Pharmacol. Exp. Ther.*, 260: 1132–1140.

Miyamoto, M. and Coyle, J.T. (1990) Idebenone attenuates neuronal degeneration induced by intrastriatal injection of excitotoxins. *Exp. Neurol.*, 108: 38–45.

Mizuno, Y., Ohta, S., Tanaka, M., Takiyama, S., Suzuki, K., Sato, T., Oya, H., Ozawa, T. and Kagawa, Y. (1989) Deficiencies in complex I subunits of the respiratory chain in Parkinson's disease. Biochem. *Biophys. Res. Commun.*, 163: 1450–1455.

Monyer, M., Goldberg, M.F. and Choi, D.W. (1989) Glucose deprivation neuronal injury in cortical cell culture. *Brain Res.*, 483: 347–354.

Monyer, M., Hartley, D.M. and Choi, D.W. (1990) 21-Aminosteroids attenuate excitotoxic neuronal injury in cortical cell cultures. *Neuron*, 5: 121–126.

Muller, U. and Krieglstein, J. (1995) Prolonged pretreatment with a-lipoic acid protects cultured neurons against hypoxic, glutamate-, or iron-induced injury. *J. Cereb. Blood Flow Metab.*, 15: 624–630.

Myers, R.H., Vonsattel, J.P., Stevens, T.J., Cupples, L.A., Richardson, E.P., Martin, J.B. and Bird, E.D. (1988) Clinical and neuropathologic assessment of severity in Huntington's disease. *Neurology*, 38: 341–347.

Nakao, N., Grasbon-Frodl, E.M., Widner, H. and Brundin, P. (1996) Antioxidant treatment protects striatal neurons against excitotoxic insults. *Neuroscience*, 73: 185–200.

Nakao, N. and Brundin, P. (1997) Effects of α-phenyl-*tert*-butyl nitrone on neuronal survival and motor function following intrastriatal injections of quinolinate or 3-nitropropionic acid. *Neuroscience*, 76: 749–761.

Nath, R., Probert, A.W., McGinnis, K.M. and Wang, K.K.W. (1997) Activation of apoptosis-linked ICE related caspase(s) in excitotoxin and hypoxia/hypoglycemia induced neuronal death. *Soc. Neurosci. Abstr.*, 567.3.

Nicotera, P. and Leist, M. (1997) Apoptosis and control of cell death. In P.R. Bär and M.F. Beal (Eds.) *Neuroprotection in CNS diseases*, Marcel Dekker, Inc., pp. 319–330.

Novelli, A., Reilly, J.A., Lysko, P.G. and Henneberry, R.C. (1988) Glutamate becomes neurotoxic via the *N*-methyl-D-aspartate receptor when intracellular energy levels are reduced. *Brain Res.*, 451: 205–212.

Nussbaum, R.L. and Polymeropoulos, M.H. (1997) Genetics of Parkinson's disease. *Human Mol. Genet.*, 6: 1987–1691.

Olanow, C.W. (1993) A radical hypothesis for neurodegeneration. *Trends Neurosci.*, 16: 439–444.

Olney, J.W. (1978) Neurotoxicity of excitatory amino acids. In: E.G. McGeer, J.W. Olney, P.L. McGeer (Eds) *Kainic Acid as A Tool in Neurobiology*, Raven Press, New York, pp. 95–121.

Olney, J.W., Labruyera, J. and Price, M.T. (1989) Pathological changes induced in cerebrocortical neurons by phencyclidine and related drugs. *Science*, 244: 1360–1362.

O'Neill, L.A.J. and Kaltschmidt, C. (1997) NF-κB: A crucial transcription factor for glial and neuronal cell function. *Trends Neurosci.*, 20: 252–258.

Orrenius, S., McConkey, D.J., Bellomo, G. and Nicotera, P. (1989) Role of $Ca^{2+}$ in toxic cell killing. *Trends Pharmacol. Sci.*, 10: 281–285.

Park, C.K., Nehls, D.G., Graham, D.I., Teasdale, G.M. and McCullough, J. (1988) Focal cerebral ischaemia in the cat. Treatment with glutamate antagonist MK-801 after induction of ischaemia. *J. Cereb. Blood Flow Metab.*, 8: 757–762.

Patel, M., Day, B.J., Crapo, J.D., Fridovich, I. and McNamara, J.O. (1996) Requirement for superoxide in excitotoxic cell death. *Neuron*, 16: 345–355.

Pellegrini-Giampietro, D.E., Cherici, G., Alesiani, M., Carla, V. and Moroni, F. (1990) Excitatory amino acid release and free radical formation may cooperate in the genesis of ischemia-induced neuronal damage. *J. Neurosci.*, 10: 1035–1041.

Pelligrino, D. and Siesjö, B.K. (1981) Regulation of extra- and intracellular pH in the brain in severe hypoglycemia. *J. Cereb. Blood Flow Metab.*, 4: 85–96.

Perry, T.L., Godin, D.V. and Hansen, S. (1982) Parkinson's disease: A disorder due to nigral glutathione deficiency? *Neurosci. Lett.*, 33: 305–310.

Pulsinelli, W., Brierley, J.B. and Plum, F. (1982) Temporal profile of neuronal damage in a model transient forebrain ischemia. *Ann. Neurol.*, 11: 491–498.

Puttfarcken, P.S., Getz, R.L. and Coyle, J.T. (1993) Kainic acid-induced lipid peroxidation: Protection with butylated hydroxytoluene and U78517F in primary cultures of cerebellar granule cells. *Brain Res.*, 624: 223–232.

Rajput, A.H. (1994) Clinical features and natural history of Parkinson's disease (special consideration of aging). In D.B. Calne (ed) *Neurodegenerative Diseases.*, W.B. Saunders Company, Philadelphia, pp. 555–571.

Reynolds, I.J. and Hastings, T.G. (1995) Glutamate induces the production of reactive oxygen species in cultured forebrain neurons following NMDA receptor activation. *J. Neurosci.*, 15: 3318–3327.

Richter, C., Park, J.W. and Arnes, B.N. (1988) Normal oxidative damage to mitochondrial and nuclear DNA is extensive. *Proc. Natl. Acad. Sci. USA*, 85: 6465–6467.

Riederer, P. and Wuketich, S. (1976) Time course of nigrostriatal degeneration in Parkinson's disease. *J. Neural Transm.*, 38: 277–301.

Schapira, A.H.V., Cooper, J.M., Dexter, D.T., Clark, J.B., Jenner, P. and Marsden, C.D. (1990) Mitochondrial complex I deficiency in Parkinson's disease. *J. Neurochem.*, 54: 823–827.

Schulz, J.B., Henshaw, D.R., Siwek, D., Jenkins, B.G., Ferrante, R.J., Cipolloni, P.B., Kowall, N.W., Rosen, B.R. and Beal, M.F. (1995a) Involvement of free radicals in excitotoxicity in vivo. *J. Neurochem.*, 64: 2239–2247.

Schulz, J.B., Matthews, R.T., Jenkins, B.G., Brar, P. and Beal, M.F. (1995b) Improved therapeutic window for treatment of histotoxic hypoxia with a free radical spin trap. *J. Cereb. Blood Flow Metab.*, 15: 948–952.

Schwarcz, R., Okuno, E., White, R.J., Bird, E.D. and Whetsell, W.O. (1988) 3-Hydroxyanthranilate oxygenase activity is increased in the brains of Huntington's disease victims. *Proc. Natl. Acad. Sci. USA*, 85: 4079–4081.

Schwarcz, R., Whetsell, W.O. Jr. and Mangano, R.M. (1983) Quinolinic acid: An endogenous metabolite that produces axon-sparing lesions in rat brain. *Science*, 219: 316–318.

Sen, S. and Phillis, J.W. (1993) α-phenyl-*tert*-butyl-nitrone (PBN) attenuates hydroxyl radical production during ischemia-reperfusion injury of rat brain – an EPR study. *Free Rad. Res. Commun.*, 19: 255–265.

Sharp, A.H. and Ross, C.A. (1996) Neurobiology of Huntington's disease. *Neurobiol. Dis.*, 3: 3–15.

Sheehan, J.P., Swerdlow, R.H., Parker, W.D., Miller, S.W., Davis, R.E. and Tuttle, J.B. (1997) Altered calcium homeostasis in cells transformed by mitochondria from individuals with Parkinson's disease. *J. Neurochem.*, 68: 1221–1233.

Siesjö, B.K. (1981) Cell damage in the brain: A speculative synthesis. *J. Cereb. Blood Flow Metab.*, 1: 155–185.

Simpson, J.R. and Isacson, O. (1993) Mitochondrial impairment reduces the threshold for in vivo NMDA-mediated neuronal death in the striatum. *Exp. Neurol.*, 121: 57–64.

Smith, M.-L., Auer, R. and Siesjö, B. (1984) The density and distribution of ischemic brain injury in rat following 2–10 min of forebrain ischemia. *Acta Neuropathol.*, 64: 319–332.

Storey, E., Hyman, B.T., Jenkins, B., Brouillet, E., Miller, J.M., Rosen, B.R. and Beal M.F. (1992) 1-Methyl-4-phenylpyridinium produces excitotoxic lesions in rat striatum as a result of impairment of oxidative metabolism. *J. Neurochem.*, 58: 1975–1978.

Stys, P.K., Waxman, S.G. and Ransom, B.R. (1992) Ionic mechanisms of anoxic injury in mammalian CNS white matter: Role of $Na^+$ channels and $Na^+$-$Ca^{2+}$ exchanger. *J. Neurosci.*, 12: 430–439.

Thomas, C.E. (1997) Nitrone spin traps as protectants against oxidative damage. In P.R. Bär and M.F. Beal (Eds.) *Neuroprotection in CNS diseases*, Marcel Dekker, Inc., pp. 183–204.

Uhler, T.A., Frim, D.M., Pakzaban, P. and Isacson, O. (1994) The effects of megadose methylprednisolone and U-78517F on toxicity mediated by glutamate receptors in the rat neostriatum. *Neurosurgery*, 34: 122–128.

Watson, J.C. and Bullock, R. (1997) Neuroprotection in acute neurological disease: Trauma to the brain and spinal cord. In P.R. Bär and M.F. Beal (Eds.) *Neuroprotection in CNS diseases*, Marcel Dekker, Inc., pp. 387–408.

Wieloch, T. (1985) Hypoglycemia-induced neuronal damage prevented by an *N*-methyl-D-aspartate antagonist. *Science*, 230: 681–683.

Yue, T.L., Gu, J.-L., Lysko, P.G., Cheng, H.-Y., Barone, F.C. and Feuerstein, G. (1992) Neuroprotective effects of phenyl-t-butyl-nitrone in gerbil global brain ischemia and in cultured rat cerebellar neurons. *Brain Res.*, 574: 193–197.

Zeevalk, G.D. and Nicklas, W.J. (1990) Chemically induced hypoglycemia and anoxia: Relationship to glutamate receptor-mediated toxicity in retina. *J. Pharmacol. Exp. Ther.*, 253: 1285–1292.

Zeevalk, G.D., Derr-Yellin, E. and Nicklas, W.J. (1995) Relative vulnerability of dopamine and GABA neurons in mesencephalic culture to inhibition of succinate dehydrogenase by malonate and 3-nitropropionic acid and protection by NMDA receptor blockade. *J. Pharmacol. Exp. Therapeut.*, 275: 1124–1130.

Zhao, Q., Pahlmark, K., Smith, M.-L. and Siesjö, B.K. (1994) Delayed treatment with the spin tarp α-phenyl-*N*-*tert*-butyl nitrone (PBN) reduces infarct size following transient middle cerebral artery occlusion in rats. *Acta Physiol. Scand.*, 152: 349–350.

O.P. Ottersen, I.A. Langmoen and L. Gjerstad (Eds.)
*Progress in Brain Research*, Vol 116


CHAPTER 17

# Glutamate induced cell death: Apoptosis or necrosis?

Maria Ankarcrona

*Karolinska Institutet, Department of Clinical Neuroscience and Family Medicine, Division of Geriatric Medicine, KFC, Novum, 4th floor, S-141 86 Huddinge, Sweden*

## Introduction

Glutamate toxicity is involved in the onset of cell death in several pathological conditions. For example, glutamate accumulation during ischemia and ischemic reperfusion often result in widespread cell death in the affected brain regions (Siesjö, 1992). Moreover, glutamate toxicity also seems to play a role in cell deletion in neurodegenerative disorders (Lancelot and Beal, this volume) such as Alzheimer's disease (Copani et al., 1991; Mattson et al., 1992; Weiss et al., 1994), Parkinson's disease (Mitchell et al., 1994), Huntingtońs disease (Portera-Cailliau et al., 1995) and AIDS dementia (Lipton, 1996). Overstimulation of glutamate receptors results in sustained intracellular calcium overload (Choi, 1995) that triggers several lethal processes e.g. DNA-damage, proteolysis, mitochondrial dysfunction and disruption of cytoskeletal organization (Reynolds, this volume). Sudden and massive intracellular calcium overload often lead to necrosis, however disturbances in calcium signalling can also trigger apoptosis (Nicotera et al., 1994). In the present review we discuss whether cell death manifests itself acutely as necrosis or slowly develops as apoptosis after glutamate exposure to a neuronal population. It appears, that to be able to determine which type or types of cell death are involved, it is necessary to use several different techniques to detect cell death and to study the full time-course of degeneration at several time-points after the insult.

## Types of cell death: Necrosis and apoptosis

Two morphologically distinct modes of cell death are in general recognized: necrosis and apoptosis. Necrosis is a passive form of cell death and a result of disorganized breakdown of the cell, often following acute violent insult. It is characterized by the concomitant disruption of several homeostatic processes, rapid energy depletion, organelle swelling and the activation of random catalytic processes. Cells are often destroyed very rapidly and the release of their content into the surroundings evokes inflammatory responses (Wyllie, 1980; Arends and Wyllie, 1991). In contrast, apoptosis is an active and organized cell deletion process, which results in removal of the cells by phagocytosis. Apoptosis occurs physiologically to eliminate unwanted, damaged or unnecessary cells and disturbances in the control of apoptosis can lead to disease (Evan, 1994; Thompson, 1995). Repression of apoptosis may lead to cancer or autoimmune disorders, whereas excess apoptosis may be involved in degenerative diseases. Results from in vitro and in vivo studies show that apoptosis could be the prevalent mode of cell deletion in neurodegenerative processes such as Alzheimer's disease (Cotman and Anderson, 1995), Parkinson's disease (Dispasquale et al., 1991; Mitchell et al., 1994) and Huntingtońs chorea (Portera-Cailliau et al., 1995) as well as neuronal cell death in the penumbra of a stroke (Charriaut-Marlangue et al., 1995).

Cells dying by apoptosis undergo several morphological and biochemical changes, including: cell shrinkage, protease activation, nuclear condensation and DNA-fragmentation (Wyllie, 1980; Arends and Wyllie, 1991). Dead cells are rapidly engulfed and digested by macrophages leaving no trace in the surrounding tissue. In fact, apoptosis is often neglected because of the rapid clearance of dead cells by phagocytosis (Savill, 1993). Recent studies using cell-free systems and enucleated cells suggest that the central components of the cell death machinery are localized in the cytoplasm (Lazebnik et al., 1993; Newmeyer et al., 1994; Jacobson et al., 1994). Several lines of evidence suggest that proteases are good candidates as such cytoplasmic effectors of apoptosis. In the nematode *C. elegans* the exact number of cells dying by apoptosis during development is known. Several genes involved in this cell deletion are identified, among these *ced-3* that must function for apoptosis to proceed (Ellis et al., 1991). *Ced-3* encodes a protein with significant homology with mammalian interleukin-1$\beta$-converting enzyme (ICE) (Yuan et al., 1993). ICE is a cystein protease that cleaves pro-IL-1$\beta$ to IL-1$\beta$ at Asp residues (Thornberry et al., 1992). A whole family of ICE-like proteases (or caspases) has now been described and these seem to be involved in the initiation as well as the execution of apoptosis. Some substrates for caspases have been identified. For example, poly (ADP-ribose) polymerase (PARP) is cleaved and inactivated by caspase-3 (Lazebnik et al., 1994; Fernandez-Alnemri et al., 1995; Tewari et al., 1995) and caspases are also involved in lamin degradation (Lazebnik et al., 1995; Zhivotovsky et al., 1995). Lamin proteolysis appears to play a critical role in nuclear degradation during apoptosis (Zhivotovsky et al., 1997) and precedes chromatin fragmentation in neurons exposed to glutamate (Ankarcrona et al., 1996a). On the other hand, PARP proteolysis is not a prerequisite for nuclear apoptosis and the ability of cells to undergo apoptosis is not affected in PARP knock-out mice (Wang et al., 1995). Still PARP cleavage is a good indicator for nuclear penetration of activated apoptotic proteases (e.g. caspase-3).

### Methods to distinguish necrotic and apoptotic cells

Apoptotic cells stained with a nuclear dye appear with condensed nuclei under a fluorescence microscope and the degraded chromatin, cleaved by proteases and endonucleases into high molecular weight and oligonucleosomal lengthened DNA-fragments, form distinct patterns on agarose gels. On the other hand, necrotic cells swell, lose membrane integrity and their DNA is randomly cleaved forming a smear on agarose gels. Even though DNA-fragmentation into oligonucleosomal lengthed fragments is a classical feature of apoptosis, it does not occur in all situations where apoptosis is involved. Rather, it appears that DNA-cleavage in apoptosis is a multistep process beginning with the formation of high molecular weight DNA-fragments (Filipski et al., 1990; Walker et al., 1995). This type of DNA-cleavage occurs prior to, or in the absence of, internucleosomal fragmentation (Oberhammer et al., 1993; Brown et al., 1993; Ankarcrona et al., 1995) and better correlates with the morphological appearance of apoptotic nuclei. Another method to detect chromatin degradation during apoptosis is in situ nick-end labeling (TUNEL) of fragmentated DNA. This technique has often been used to detect apoptotic cells in tissue slices. However, the labeling is not specific for DNA-fragments formed during apoptosis and it has been shown that also necrotic cells are stained by this technique (Nishiyama et al., 1996). Therefore, the TUNEL-technique should be combined with for example morphological studies of dying cells to be able to determine the type of cell death involved.

While necrotic cells lose membrane permeability, often detected as LDH-leakage or failure to exclude trypan blue, cells dying by apoptosis maintain membrane integrity. Thus necrotic cells can be distinguished from apoptotic cells by using fluorescent nuclear dyes with different properties of cell permeability and emission wavelengths. For example, a membrane permeable dye (e.g. SYTO-13) and a membrane impermeable dye (e.g. propidium iodide) can be used in combination. In our studies, cerebellar granule cells (CGC) grown on cover-slips were loaded with SYTO-13 and then placed under a confocal microscope. Cells were perfused with glutamate and propidium iodide for 30 minutes. During the exposure we observed that some cells turned from green (SYTO-13) to red fluorescence (propidium iodide). In these cells, apparently dying by necrosis, glutamate caused rupture of cell membranes leading to uptake of propidium iodide. In other experiments, CGC were exposed to glutamate, reincubated in the old culture medium for 6 hours, loaded with SYTO-13 and then perfused with propidium iodide under the confocal microscope. Condensed nuclei typical of apoptosis were detected, and all cells excluded propidium iodide. In this case, cells that earlier died by necrosis had been washed away and only apoptotic cells still remaining attached to the cover-slip were observed (Ankarcrona et al., 1995).

Protease activity is detected by immunoblot analysis of substrate cleavage products or assayed in vitro by fluorogenic peptide substrates. In addition, peptide inhibitors of proteases can be used to block apoptosis in many systems. Activation of proteases and cleavage of substrates (e.g. lamin and PARP) often occur early during apoptosis and not in necrosis. For example, caspase-3 processing and activation seems to be specific for apoptosis and was observed only when cerebellar granule cells in culture underwent apoptosis and not necrosis (Armstrong et al., 1997). Finally, cells dying by necrosis rapidly lose mitochondrial membrane potential and cellular energy, while cells undergoing apoptosis restore mitochondrial functions and require energy to complete the death program (see below) (Ankarcrona et al., 1995; Leist et al., 1997). In summary: condensed nuclei, DNA-fragmentation, early protease activation and maintained cell membrane integrity as well as mitochondrial functions can be used as criteria for apoptosis.

By using a combination of the techniques described above it is possible to distinguish apoptotic cells from necrotic cells. To determine the mode of cell death it is also important to study morphological and biochemical changes in cells both at early and late time-points after the insult. This is especially important in the in vivo situation since apoptotic cells are rapidly phagocytised and therefore easily missed if only a few time-points are chosen. A complication in the distinction between apoptosis and necrosis comes from the observation that secondary necrosis of apoptotic cells occurs in vitro or in vivo and reflects an insufficient removal of apoptotic cells by phagocytes (Leist et al., 1995). In this case, secondary processes may cause cell disintegration, mimicking necrosis.

**Glutamate neurotoxicity: A succession of necrosis and apoptosis**

While low concentrations of glutamate exert trophic effects and promote neuronal survival and synapse formation (Zorumski and Thio, 1992), increasing doses can result in neurotoxicity. In glutamate neurotoxicity cell death is mediated by calcium signaling and free radicals and manifests as apoptosis or necrosis (Mattson and Furukawa, 1996). It seems likely that the same fundamental mechanisms of neuronal injury lie behind the two types of cell death which then develop depending on the cell type, severity and duration of the injury and on the state of the cell. Induction of both apoptosis and necrosis in a neuronal cell population exposed to excitatory amino acids has been shown in vitro (Ankarcrona et al., 1995; Bonfoco et al., 1995; Gwag et al., 1995; Cebers et al., 1997) as well as in vivo (Pollard et al., 1994; Ferrer et al., 1995; Portera-Cailliau et al., 1995; van Lookeren Campagne et al., 1995). In vivo models of ischemic brain injury show that necrosis occurs in the focus of the ischemic zone where glutamate accumulation causes severe injury. Cells in the penumbra, also triggered to die, initially survive and later undergo apoptosis (Charriaut-Marlangue et al., 1995). It is of great importance for the treatment of stroke, as well as other neurodegenerative disorders where glutamate toxicity is implicated, to find drugs that block cell death caused by overstimulation of glutamate receptors. Potential targets for such drugs are glutamate receptors themselves and cellular processes upstream and downstream of receptor activation. These cellular processes include release of glutamate, free radical generation, activation of proteases and nitric oxide synthase, opening of mitochondrial transition pores and release of intracellular calcium (Lipton and Rosenberg, 1994; Lancelot and Beal, this volume). Below we discuss the role of glutamate receptors and mitochondria in the onset and progression of cell death.

*Glutamate receptors*

The glutamate receptors are categorized into ionotropic receptors controlling ion channels and metabotropic receptors coupled to G-proteins (Wenthold and Roche, this volume; Bruno et al., this volume). Briefly, the ionotropic receptors are: i) *N*-methyl-D-aspartate (NMDA) receptor sensitive to NMDA and glutamate, ii) α-amino-3-hydoxy-5-methyl-4-isoxazolepropionate (AMPA) receptor sensitive to AMPA, kainic acid and glutamate iii) kainic acid receptor, sensitive to kainic acid and glutamate. When activated by agonists, the NMDA-receptor channel becomes permeable to calcium and sodium. Depending on the exact subunit composition of AMPA and kainic acid receptors their channels are permeable to sodium and sometimes to calcium (Borges and Dingledine, this volume). AMPA and kainic acid receptors mediate fast responses, while NMDA-receptors produce relatively sustained depolarization (Kebabian and Neumeyer, 1994; Lipton and Rosenberg, 1994).

The NMDA-receptor is positively regulated by glycine, polyamines and phosphorylation. Its activity is decreased by oxidation of sulfhydryl groups in a redox-site. Lipton and co-workers have shown that generation of nitrozonium ($NO^+$), produced upon activation of nitric oxide synthases, may result in such downregulation of the NMDA-receptor (Lipton et al., 1993). Oxidation of the redox-site downregulates receptor activity and potentially protects from glutamate induced cell death. Nitroglycerin, a drug used to treat cardiovascular disorders, is active on the NMDA-receptor redox-site and works in a manner resembling $NO^+$. In animal models high concentrations of nitroglycerin have been found to be neuroprotective during various NMDA-receptor mediated insults, including focal ischemia (Sathi et al., 1993).

NMDA-receptor activity is also inhibited by open-channel blockers e.g. dizocilpine (MK-801) and memantine. Our experiments show that the NMDA-receptor antagonist MK-801 blocks both necrosis and apoptosis developing sequentially in cerebellar granule cells exposed to glutamate (Ankarcrona et al., 1995). Moreover, NMDA and quinolinic acid cause striatal apoptosis which was blocked by MK-801 but not by NBQX (AMPA/KA antagonist) (Qin et al., 1996). In other model systems, memantine abolishes neuronal injury in focal ischemia both in vivo and in vitro (Lipton, 1996 and refs therein). The use of memantine for potential treatment of stroke and neurodegenerative disorders has several advantages: i) unlike MK-801, memantine does not remain in the channel for an excessively long time, ii) memantine is clinically tolerated and used in the treatment of Parkinson's disease and spasticity, and iii) memantine mainly blocks effects of pathological concentrations of glutamate and normal NMDA-receptor activity, important in neuronal plasticity, is less affected.

Altogether, these studies indicate a central role for the NMDA-receptor in the onset of glutamate induced cell death. However, by blocking desensitization of AMPA-receptors with cyclothiazide cerebellar granule cells exposed to glutamate were triggered to die also in the presence of NMDA-antagonists (Cebers et al., 1997). These results suggest that activation of AMPA-receptors, permeable to calcium under certain conditions, also contributes to the onset of neuronal apoptosis and necrosis.

*Role of mitochondria in glutamate toxicity*

We used primary cultures of rat cerebellar granule cells (CGC) to study the role of mitochondria in glutamate induced cell death (Ankarcrona et al.,

1995). Differentiated CGC express glutamate receptors predominantly of the NMDA-type and may serve as a model system for studies of glutamate toxicity. CGC are cultured in the presence of cytosine arabinoside to inhibit growth of dividing cells and mature cultures contain more than 95% CGC. In vitro CGC are dependent on constant depolarization for survival and therefore the culture medium is supplemented with 25 mM potassium chloride. In fact a lower concentration of potassium chloride (5 mM) induces cell death by apoptosis (D'Mello et al., 1993). In our studies, CGC were exposed to glutamate for 30 minutes and subsequently reincubated in the old culture medium. Part of the cell population (30–50% depending on glutamate concentration) died rapidly by necrosis up to three hours after exposure. Necrosis was detected as impaired metabolism of 3-(4,5-dimethylthiazole-2-yl)-2,5-diphenyltetrazolium bromide (MTT) to formazan and loss of cell membrane integrity. We also measured a loss of mitochondrial membrane potential and ATP-depletion in the whole cell population immediately after glutamate exposure. Cells initially surviving the insult recovered mitochondrial membrane potential and ATP-levels. This part of the cell population later underwent apoptosis. These experiments show that while necrotic cells undergo rapid depletion of cellular energy, cells dying by apoptosis are dependent on mitochondrial energy production. Other studies have questioned the importance of mitochondrial respiration in apoptosis (Jacobson et al., 1994; Newmeyer et al., 1994), since they observed that cells undergo apoptosis even if deficient in mitochondrial ATP-production. However, artificial ATP-generating systems were included in both studies. We have suggested that cells lacking cellular energy undergo necrosis before the apoptotic program has a chance to develop (Nicotera et al., 1997). This hypothesis is further supported by recent observations in human T-cells triggered to die by apoptosis with staurosporin or CD95 stimulation. Cells preemptied of ATP died by necrosis, while repletion of the extramitochondrial ATP-pool with glucose prevented necrosis and restored the ability of the cells to undergo apoptosis (Leist et al., 1997). The transient loss of mitochondrial membrane potential and cellular energy immediately following glutamate exposure may be an important signal for apoptosis (Reynolds, this volume). Decrease in mitochondrial membrane potential precedes apoptosis in many systems (Deckwerth and Johnson, 1993; Zamzami et al., 1995; Petit et al., 1995; Ankarcrona et al., 1995; Schinder et al., 1996) and leads to opening of permeable transition pores (PT) in the mitochondrial inner membrane. It has been suggested that a molecule active in the apoptotic pathway could be released from mitochondria during PT-opening (Zamzami et al., 1996). Such an apoptosis inducing factor (AIF) has been identified as a soluble protein localized in the intermembrane space of mitochondria and possessing protease activity (Susin et al., 1996). AIF induces typical manifestations of nuclear apoptosis (e.g. nuclear condensation and DNA-fragmentation) and all in vitro activities of AIF are blocked by the cysteine protease inhibitor Z-VAD.fmk, an efficient inhibitor of apoptosis in many cell systems (Zhivotovsky et al., 1995; Pronk et al., 1996; Cain et al., 1996). AIF fails to cleave PARP and lamin, two known substrates for proteases in the caspase family. Its molecular mass and subcellular localization also differs from the proteases in this family and AIF seems to belong to another group of cysteine proteases. Overexpression of Bcl-2, a protein known to inhibit apoptosis in many cells (Hockenbery et al., 1993; Reed, 1994; Newmeyer et al., 1994; Jacobson et al., 1994), blocks PT-opening and release of AIF from mitochondria. This indicates a central role for regulation of PT-opening by Bcl-2 during the initiation of apoptosis (Susin et al., 1996).

One protein released from mitochondria upon apoptotic stimuli has been identified as cytochrome c (Liu et al., 1996). Cytochrome c was required to induce apoptosis in cell-free extracts and caspase-3 activation as well as PARP cleavage were detected. Cytochrome c is encoded by nuclear genes and transported into mitochondria. This supports the notion that all mitochondrial functions critical for apoptosis are encoded by nuclear genes, since cells lacking mitochondrial DNA also release the "death-factor" and die by apoptosis (Zamzami et al., 1996). Whether AIF and cytochrome c are identical or belong to a family of "death-factors" released from mitochondria remains to be elucidated.

We found that CGC exposed to glutamate in the presence of cyclosporin A (CsA), an inhibitor of PT-opening and calcineurin activity, were protected from both necrosis and apoptosis. In addition, the more specific calcineurin inhibitor FK-506 had the same effects (Ankarcrona et al., 1996b). Our studies suggest that both PT-opening and calcineurin activity is important for apoptosis to occur. Hypothetically, a "death-factor" is released from mitochondria into the cytoplasma during transient loss of mitochondrial membrane potential. The "death-factor" is subsequently dephosphorylated by calcineurin and triggers cell death presumably by protease activity. It appears that mitochondria are not only important as an energy source, maybe they also harbour molecules important for initiation of apoptosis.

## Conclusions

Accumulating evidence from in vivo and in vitro studies shows that both necrosis and apoptosis occur in neuronal cell death after glutamate exposure. It is not known why some cells in a neuronal population die by necrosis while others, triggered to die by the same insult, later undergo apoptosis. Parameters that could influence which type of cell death will occur are: type of glutamate receptor activated, cell type, maturity of the cell, effects on mitochondria and cellular energy.

Our studies show that glutamate exposure causes a succession of necrosis and apoptosis in cerebellar granule cells depending on mitochondrial functions. Necrosis associated with extreme energy loss in mitochondria may reflect the failure of neurons to carry out the "default" apoptotic program. The maintenance of mitochondrial function may therefore be a decisive factor in determining the degree and progression of neuronal injury caused by excitotoxins. The recent detection of "death-factors" released from mitochondria upon loss of membrane potential and PT-opening further emphasizes the importance of mitochondria for the onset and progresssion of apoptosis (Susin et al., 1996; Zamzami et al., 1996; Liu et al., 1996; Ankarcrona et al., 1996b). Strong evidence shows a central role for proteases in the apoptotic pathway, and PT-opening might well be an equally important switch in the death program (Marchetti et al., 1996).

## References


Ankarcrona, M., Dypbukt, J.M., Bonfoco, E., Zhivotovsky, B., Orrenius, S., Lipton, S.A. and Nicotera, P. (1995) Glutamate-induced neuronal death: A succession of necrosis or apoptosis depending on mitochondrial function. *Neuron*, 15: 961–973.

Ankarcrona, M., Zhivotovsky, B., Holmström, T., Diana, A., Eriksson, J.E., Orrenius, S. and Nicotera, P. (1996a) Lamin and $\beta$-tubulin fragmentation precede chromatin degradation in glutamate-induced neuronal apoptosis. *NeuroReport*, 7: 2659–2664.

Ankarcrona, M., Dypbukt, J.M., Orrenius, S. and Nicotera, P. (1996b) Calcineurin and mitochondrial function in glutamate-induced neuronal cell death. *FEBS, Lett.*, 394: 321–324.

Arends M.J. and Wyllie A.H. (1991) Apoptosis: Mechanisms and roles in pathology. *Int. Rev. Exp. Pathol.*, 32: 223–254.

Armstrong, R.C., Aja, T.J., Hoang, K.D., Gaur, S., Bai, X., Alnemri, E.S., Litwack, G., Karanewsky, D.S., Lawrence, C.F. and Tomaselli, K.J. (1997) Activation of the CED3/ICE-related protease CPP32 in cerebellar granule neurons undergoing apoptosis but not necrosis. *J. Neurosci.*, 17: 553–562.

Bonfoco, E., Krainc, D., Ankarcrona, M., Nicotera, P. and Lipton S.A. (1995) Apoptosis and necrosis: Two distinct events induced respectively by mild and intense insults with NMDA or nitric oxide/superoxide in cortical cell cultures. *Proc. Natl. Acad. Sci. USA*, 92: 7162–7166.

Brown, D.G., Sun, X. and Cohen, G.M. (1993) Dexamethasone-induced apoptosis involves cleavage of DNA to large fragments prior to internucleosomal fragmentation. *J. Biol. Chem.*, 268: 3037–3039.

Cain, K., Inayathussain, S.H., Couet, C. and Cohen, G.M. (1996) A clevage-site-directed inhibitor of interleukin 1 $\beta$-converting enzyme-like proteases inhibits apoptosis in primary cultures of rat hepatoctes. *Biochem. J.*, 314: 27–32.

Cebers, G., Zhivotovsky, B., Ankarcrona, M. and Liljequist, S. (1997) AMPA neurotoxicity in cultured cerebellar granule neurons: Mode of cell death. *Brain Res. Bull.*, 43: 393–403.

Charriaut-Marlangue, C., Margaill, I., Walsh, R.J., Plotkine, M. and Ben-Ari, Y. (1995) $N^G$-nitro L-arginine methylester

(L-NAME) reduces cortical infarct and necrotic damage but not apoptotic cell loss. *Soc. Neurosci. Abstr.*, 21: 998.

Choi, D.W. (1995) Calcium: Still center-stage in hypoxic-ischemic neuronal death. *TINS*, 18: 58–60.

Copani, A., Koh, J. and Cotman, C.W. (1991) $\beta$-amyloid increases neuronal susceptibility to injury by glucose deprivation. *NeuroReport*, 2: 763–765.

Cotman, C.W. and Anderson, A.J. (1995) A potential role for apoptosis in neurodegeneration and Alzheimers disease. *Mol. Neurobiol.*, 10: 19–45.

Deckwerth, T.L. and Johnson, E.M. (1993) Temporal analysis of events associated with programmed cell death (apoptosis) of sympathetic neurons deprived of nerve growth factor. *J. Cell Biol.*, 123: 1207–1222.

Dispasquale, B., Marini, A.M. and Youle, R.J. (1991) Apoptosis and DNA degradation induced by 1-methyl-4-phenylpyrdinium in neurons. *Biochem. Biophys. Res. Comm.*, 181: 1442–1448.

D'Mello, S.R., Galli, C., Ciotti, T. and Calissano, P. (1993) Induction of apoptosis in cerebellar granule neurons by low potassium: Inhibition of death by insulin-like growth factor I and cAMP. *Proc. Natl. Acad. Sci. USA*, 90: 10989–10993.

Evan, G. (1994) Why we live and why we die. *Chemistry and Biology*, 1: 137–141

Ellis, R.E., Yuan, J. and Horvitz, H.R. (1991) Mechanisms and functions of cell death. *Annu. Rev. Cell Biol.*, 7: 663–698.

Fernandes-Alnemri, T., Litwack, G. and Alnemri, E.S. (1995) Mch2, a new member of the apoptotic Ced-3/Ice cysteine protease gene family. *Cancer Res.*, 55: 2737–2742.

Ferrer, I., Martin, F., Serrano, T., Reiriz, J., Perez-Navarro, E., Alberch, J., Macaya, A. and Planas, A.M. (1995) Both apoptosis and necrosis occur following intrastrialtal administration of excitotoxins. *Acta Neuropathol.*, 90: 504–510.

Filipski, J., Leblanc, J., Youdale, T., Sikorska, M. and Walker, P.R. (1990) Periodicity of DNA folding in higher order chromatin structures. *EMBO J.*, 9: 1319–1327.

Gwag, B.J., Lobner, D., Koh, J.Y., Wie, M.B. and Choi, D.W. (1995) Blockade of glutamate receptors unmasks neuronal apoptosis after oxygen-glucose deprivation in vitro. *Neurosci.*, 68: 615–619.

Hockenbery, D.M., Oltvai, Z.N., Yin, X-M., Milliman, C.L. and Korsmeyer, S.J. (1993) Bcl-2 functions in an antioxidant pathway to prevent apoptosis. *Cell*, 75: 241–251

Jacobson M.D., Burne J.F. and Raff, M.C. (1994) Programmed cell death and Bcl-2 protection in the absence of a nucleus. *EMBO J.*, 13: 1899–1910.

Kebabian, J.W. and Neumeyer, J.L. (1994) *The RBI handbook of receptor classification*, Research Biochemicals Internatinal, Natick.

Lazebnik, Y.A., Cole, S., Cooke, C.A., Nelson, W.G., and Earnshaw, W.C. (1993) Nuclear events of apoptosis in vitro in cell-free mitotic extracts: A model system for analysis of the active phase of apoptosis. *J. Cell. Biol.*, 123: 7–22.

Lazebnik, Y.A., Kaufmann, S.H., Desnoyers, S., Poirier, G.G. and Earnshaw, W.C. (1994) Cleavage of poly(ADP-ribose) polymerase by a proteinase with properties like ICE. *Nature*, 371: 346–347.

Lazebnik, Y.A., Takahashi, A., Moir, R.D., Goldman, R.D., Poirier, G.G., Kaufmann, S.H. and Earnshaw, W.C. (1995) Studies of the lamin proteinase reveal multiple parallel biochemical pathways during apoptotic execution. *Proc. Natl. Acad. Sci. USA*, 92: 9042–9046.

Leist, M.F., Gantner, F., Bohlinger, I., Germann, P.G., Tiegs, G. and Wendel, A. (1995) TNF-induced murine hepatic apoptosis as a pathomechanism os septic liver failure. *Am. J. Pathol.*, 166: 1–15.

Leist, M.F., Single, B., Castoldi, A.F., Kuhnle, S. and Nicotera, P. (1997) Intracellular adenosine triphosphate (ATP) concentration: A switch in the decision between apoptosis and necrosis. *J. Exp. Med.*, 185: 1481–1486.

Lipton, S.A., Choi, Y., Pan, Z., Lei, S.Z., Chen, H.V., Sucher, N.J., Loscalzo, J., Singel, D.J. and Stamler, J.S. (1993) A redox-based mechanism for neuroprotective and neurodestructive effects of nitric oxide and related nitroso-compounds. *Nature*, 364: 626–632.

Lipton, S.A. and Rosenberg, P.A. (1994) Excitatory amino acids as a final common pathway for neurological disorders. *N. Engl. J. Med.*, 330: 613–622.

Lipton, S.A. (1996) Similarity of neuronal cell injury and death in AIDS dementia and focal cerebral ischemia: Potential treatment with NMDA open-channel blockers and nitric oxide-related species. *Brain Pathol.*, 6: 507–517.

Liu, X., Kim C.N., Yang, J., Jemmerson, R. and Wang, X. (1996) Induction of apoptotic program in cell-free extracts: Requirement for dATP and cytochrome c. *Cell*, 86: 147–157.

Marchetti, P., Castedo, M., Susin, S.A., Zamzami, N., Hirsch, T., Macho, A., Haeffner, A., Hirsch, F., Geuskens, M. and Kroemer, G. (1996) Mitochondrial permeability transition is a central coordinating event of apoptosis. *J. Exp. Med.*, 184: 1155–1160.

Mattson, M.P., Cheng, B., Davis, D., Bryant, K., Lieberburg, I. and Rydel, R.E. (1992) $\beta$-amyloid peptides destabilize calcium homeostasis and render human cortical neurons vulnerable to excitotoxicity. *J. Neurosci.*, 12: 379–389.

Mattson, M.P. and Furukawa, K. (1996) Programmed cell life: Anti-apoptotic signaling and theurapeutic strategies for neurodegenerative disorders. *Restorative Neurol. Neurosci.*, 9: 191–205.

Mitchell, I.J., Lawson, S., Moser, B., Laidlaw, S.M., Cooper, A.J., Walkinshaw, G. and Waters, C.M. (1994) Glutamate-induced apoptosis results in a loss of striatal neurons in the Parkinsonian rat. *Neurosci.*, 63: 1–5.

Newmeyer, D.D., Farschon, D.M. and Reed, J.C. (1994) Cell-free apoptosis in Xenopus Egg extracts: Inhibition by Bcl-2 and requirement for an organelle fraction enriched in mitochondria. *Cell*, 79: 353–364.

Nicotera, P., Zhirotorsky, B., Bellozo, G., and Orrenius, S. (1994) Ion signalling in apoptosis. In: Apoptosis (Schimke, R.T. and Mihich, E. eds) pp 97–115, Plenum Press, New York.

Nicotera, P., Ankarcrona, M., Bonfoco, E., Orrenius, S. and Lipton, S.A. (1997) Neuronal necrosis and apoptosis: Two distinct events induced by exposure to glutamate or oxidative stress. *In: Neuronal Regeneration, Reorganization, and Repair*, (Seil FJ ed), pp 95–101. Philadelphia: Lippincott-Raven Publishers.

Nishiyama, K., Kwak, S., Takekoshi, S., Watanabe, K. and Kanazawa, I. (1996) In situ nick end-labeling detects necrosis of hippocampal pyramidal cells induced by kainic acid. *Neurosci. Lett.*, 212: 139–142.

Oberhammer, F., Wilson, J.W., Dive, C., Morris, I.D., Hickman, J.A., Wakeling, A.E., Walker, P.R. and Sikorska, M. (1993) Apoptotic death in epithelial cells: Cleavage of DNA to 300 and/or 50 kb fragments prior to in the absence of internucleosomal fragmentation. *EMBO J.*, 12: 3679–3684.

Petit, P.X., LeCoeur, H., Zorn, E., Dauguet, C., Mignotte, B. and Gougeon, M.L. (1995) Alterations of mitochondrial structure and function are early events of dexamethasone-induced thymocyte apoptosis. *J. Cell Biol.*, 130: 157–167.

Pollard, H., Charriaut-Marlangue, C., Cantagrel, S., Represa, A., Robain, O., Moreau, J. and Ben-Ari, Y. (1994) Kainate-induced apoptotic cell death in hippocampal neurons. *Neurosci.*, 63: 7–18.

Portera-Cailliau, C., Hedreen, J.C., Price, D.L. and Koliatsos, V.E. (1995) Evidence for apoptotic cell death in Huntington disease and excitotoxic animal models. *J. Neurosci*, 15: 3775–3787.

Pronk, G.J. Ramer, K., Amiri, P. and Williams, L.T. (1996) Requirement of an ICE-like protease for induction of apoptosis and ceramide generation by REAPER. *Science (Wash. DC)*, 271: 808–810.

Qin, Z., Wang, Y. and Chase, T.N. (1996) Stimulation of *N*-methyl-D-aspartate receptors induce apoptosis in rat brain. *Brain Res.*, 725: 166–176.

Reed, J.C. (1994) Bcl-2 and the regulation of programmed cell death. *J. Cell Biol.*, 124: 1–6.

Sahti, S., Edgecomb, P., Warach, S., Manchester, K., Donaghey, T. Stieg, P.E., Jensen, F.E. and Lipton, S.A. (1993) Chronic transdermal nitroglycerin (NTG) is neuroprotective in experimental rodent stroke models. *Soc. Neurosci. Abstr.*, 19: 849.

Savill, J.S., Fadok, V., Henson, P. and Haslett, C. (1993) Phagocyte recognition of cells undergoing apoptosis. *Immunol. Today*, 14: 131–136.

Schinder, A.F., Olson, E.C., Spitzer, N.C. and Montal, M. (1996) Mitochondrial dysfunction is a primary event in glutamate neurotoxicity. *J. Neurosci.*, 16: 6125–6133.

Siesjö, B.K. (1992) Pathophysiology and treatment of focal cerebral ischemia. *J. Neurosurg.*, 77: 169–184.

Susin, S.A., Zamzami, N., Castedo, M., Hirsch, T., Marchetti, P., Macho, A., Daugas, E., Geuskens, M. and Kroemer, G. (1996) Bcl-2 inhibits the mitochondrial release of an apoptogenic protease. *J. Exp. Med.*, 184: 1331–1341.

Tewari, M., Quan, L.T., O'Rourke, K., Desnoyers, S., Zeng, Z., Beidler, D.R., Poirier, G.G., Salvesen, G.S. and Dixit, V.M (1995) Yama/CPP32b, a mammalian homolog of CED-3, is a CrmA-inhibitable protease that cleaves the death substrate poly(ADP-ribose) polymearse. *Cell*, 81: 801–809.

Thompson, C.B. (1995) Apoptosis in the pathogenesis and treatment of disease. *Science*, 267: 1456–1462.

Thornberry, N.A., Bull, H.G., Calaycay, J.R., Chapman, K.T. and Howard, A.D. (1992) A novel heterodimeric cysteine protease is required for interleukin-1$\beta$, processing in monocytes. *Nature*, 356: 768–774.

van Lookeren Campagne, M., Lucassen, P.J., Vermeulen, J.P. and Balazs, R. (1995) NMDA and kainate induce internucleosomal DNA cleavage associated with both apoptotic and necrotic cell death in the neonatal brain. *Eur. J. Neurosci.*, 7: 1627–1640.

Walker, P.R., Pandrey, S. and Sikorska, M. (1995) Degradation of chromatin in apoptotic cells. *Cell Death Diff.*, 2: 97–104.

Wang, Z.-Q., Auer, B., Stingl, L., Berghammer, H., Haidacher, D., Schweiger, M. and Wagner, E.F. (1995) Mice lacking ADPRT and poly(ADP-ribosyl)ation develop normally but are susceptible to skin disease. *Genes Dev.*, 9: 509–520.

Weiss, J.H., Pike, C.J. and Cotman, C.W. (1994) $Ca^{2+}$ channel blockers attenuate $\beta$-amyloid peptide toxicity to cortical neurons in culture. *J. Neurochem.*, 62: 372–375.

Wyllie, A.H., Kerr, J.F.R. and Currie, A.R. (1980) Cell death: The significance of apoptosis. *Int. Rev. Cytol.*, 68: 251–306.

Yuan, J., Shaham, S., Ledoux, S., Ellis, H.M. and Horvitz, H.R. (1993) The *c. elegans* cell death gene ced-3 encodes a protein similar to mammalian interleukin-1$\beta$-converting enzyme. *Cell*, 75: 641–652.

Zamzami, N., Marchetti, P., Castedo, M., Zanin, C., Vayssiere, J.-L., Petit, P.X. and Kroemer, G. (1995) Reduction in mitochondrial potential constitutes an early irreversible step of programmed lymphocyte death in vivo. *J. Exp. Med.*, 181: 1661–1672.

Zamzami, N., Susin, S.A., Marchetti, P., Hirsch, T., Gomez-Monterrey, I., Castedo, M. and Kroemer, G. (1996) Mitochondrial control of nuclear apoptosis. *J. Exp. Med.*, 183: 1533–1544.

Zhivotovsky, B., Gahm, A., Ankarcrona, M., Nicotera, P. and Orrenius, S. (1995) Multiple proteases are involved in thymocyte apoptosis. *Exp. Cell Res.*, 221: 404–412.

Zhivotovsky, B., Gahm, A. and Orrenius, S. (1997) Two different proteases are involved in the proteolysis of lamin during apoptosis. *Biochem. Biophys. Res. Com.*, 233: 96–101.

Zorumski, C.F. and Thio, L.L. (1992) Properties of vertebrate glutamate receptors: Calcium mobilization and desensitization. *Prog. Neurobiol.*, 39: 295–336.

O.P. Ottersen, I.A. Langmoen and L. Gjerstad (Eds.)
*Progress in Brain Research*, Vol 116


CHAPTER 18

# Synaptic plasticity in ischemia: Role of NMDA receptors

V. Crépel[1,*] P. Congar[1], L. Aniksztejn[1], H. Gozlan[1], C. Hammond[2] and Y. Ben-Ari[1]

[1]INSERM U 29, 123 Boulevard de Port Royal, 75674 Paris Cedex 14, France
[2]INSERM U 159, Centre Paul Broca, 2 ter, rue d'Alésia, 75014 Paris Cedex, France

## Introduction

In humans and animals, the CA1 region of the hippocampus is one of the most vulnerable areas of the brain to ischemia (Petito and Pulsinelli, 1984; Pulsinelli, 1985; Petito et al., 1987). In this region, a global ischemic episode induces selective neuronal death of the pyramidal neurons, whereas most of the interneurons survive even several months after the ischemic insult (Johansen et al., 1983; Hsu et al., 1994). The death of the CA1 pyramidal neurons leads to the loss of anterograde memory and results in a major impairment of cognitive processes (Zola-Morgan et al., 1986). It is now well accepted that one of the critical steps involved in the loss of CA1 pyramidal cells is the massive release of excitatory amino acids, triggered during and shortly after the ischemic insult (Benveniste et al., 1984; Mitani et al., 1992; Szatkowski and Attwell, 1994), which leads to the over-activation of *N*-methyl-D-aspartate (NMDA) receptors (Zeevalk and Nicklas, 1992), and a massive and toxic influx of $Ca^{2+}$ (Araki et al., 1990; Lobner and Lipton, 1993; Tymianski et al., 1993; Mitani et al., 1994). This physiopathologic mechanism is supported by the following observations: (i) the destruction of the glutamatergic excitatory synaptic inputs of CA1 pyramidal cells protects these neurons from the ischemia-induced degeneration (Wieloch et al., 1985; Johansen et al., 1986; Onodera et al., 1986; Jorgensen et al., 1987; Benveniste et al., 1989; Kaplan et al., 1989; Buchan and Pulsinelli, 1990); (ii) the administration of antagonists of the glutamate receptors α-amino-3-hydroxy-5-methyl-4-isoxazolepropionate (AMPA) and *N*-methyl-D-aspartate (NMDA) (Boast et al., 1987; Gill et al., 1987; Park et al., 1988; Sheardown et al., 1990), before and during the ischemic episode, prevents the neuronal loss in CA1; (iii) finally the administration of a $Ca^{2+}$ chelator also has a protective effect (Tymianski et al., 1993). Interestingly, the cell death of CA1 pyramidal neurons is delayed and occurs 2–3 days after the ischemic insult. Several experiments have been conducted to detect abnormal activities induced after a stroke that might explain the delayed cell death of these neurons. In particular, many laboratories have studied the long-term modification of cellular electrical activities. Along this line, an enhancement of spontaneous high frequency discharges, which can persist for at least 24 h, has been reported following an ischemic episode (Suzuki et al., 1983; Schiff and Somjen, 1985; Chang et al., 1989). However, other groups found neither a rise of spontaneous high frequency discharges (Mitani et al., 1989; Jensen et al., 1991) nor electroencephalographic seizures (Armstrong et al., 1989) during

*Corresponding author.
Tel.: +33 1 53 73 79 32; fax: +33 1 46 34 16 56, e-mail: crepel@u29.cochin.inserm.fr.

the reoxygenation period. In addition, one study clearly showed that there is no correlation between the delayed neuronal death and the hyperexcitability of CA1 pyramidal cells, since their degeneration can occur without increase of spike discharge (Imon et al., 1991). Therefore, hyperexcitability of CA1 pyramidal cells and electroencephalographic seizures fail to fully account for the ischemic cell death. Another possible explanation is that the toxic effects mediated by the activation of glutamatergic receptors and its associated $Ca^{2+}$ influx are not restricted to the ischemic episode but persist several hours after the ischemic insult, at a time when glutamate and aspartate release has returned to basal levels. Supporting this idea, a lesion of the Schaffer collateral pathway (Johansen et al., 1987) or administration of NMDA and AMPA receptor antagonists (Boast et al., 1988; Gill et al., 1988; Urban et al., 1990; Buchan et al., 1991; Nellgard and Wieloch, 1992) performed several hours after the ischemic insults have also protective actions. Therefore, the toxic effects mediated by glutamatergic receptors may result from a persistent change of glutamatergic excitatory transmission. In this review we will focus on the long-term change of glutamatergic synaptic responses, induced by ischemia or ischemic-like conditions, and the importance of the regulatory sites of NMDA receptor-channel complex in this long-term modification. The differences and similarities between long-term changes of synaptic responses, induced by an ischemic-anoxic episode or a tetanic stimulation, will be also discussed.

## An ischemic episode induces a long-term change of NMDA receptor-mediated response

Several studies have proposed that CA1 hippocampal pyramidal cells degenerate as a consequence of the long-term change of synaptic excitation during the reperfusion period. In keeping with this hypothesis, studies performed in vivo and ex vivo have shown that excitatory synaptic transmission is enhanced several hours after ischemic insult (Urban et al., 1989; Andiné et al., 1988; 1992). After the ischemic episode, there is also a persistent increase of NMDA and AMPA receptor-mediated $Ca^{2+}$ influx. The magnitude of the increase in $Ca^{2+}$ influx correlates with the extent of CA1 pyramidal cell death (Andiné et al., 1988; 1992). The increase of excitatory synaptic responses in the CA1 area could result from either a reduction of the inhibitory component of the synaptic responses and/or a change of the glutamatergic responses. We recently reported in vitro, in hippocampal slices, that inhibition is particularly vulnerable to an anoxic episode in the CA1 area. We show that the inhibitory component of the synaptic response is depressed earlier than the excitatory component by anoxia, and this is mainly due to a functional disconnection of the interneurons from their excitatory synaptic inputs during the anoxic episode. We proposed that the early disconnection of CA1 interneurons from glutamatergic inputs may amplify the excitotoxic action of the excitatory synaptic transmission during the ischemic episode (Congar et al., 1994; Khazipov et al., 1995). However, we have demonstrated, as many other laboratories, that GABA receptor-mediated responses fully recover after an anoxic or ischemic episode (Hansen, 1985; Fujiwara et al., 1987; Krnjevic et al., 1991; Zhang and Krnjevic, 1993; Congar et al., 1994; Khazipov et al., 1995). Therefore, a long-term change of the excitatory synaptic responses following the ischemic episode cannot be explained by a persistent reduction of the inhibitory component of the synaptic response. An alternative possibility is that the glutamate receptor-mediated responses are modified following the ischemic episode.

Under physiological conditions, the excitatory synaptic response in CA1 is composed of AMPA and NMDA receptor-mediated components. In these conditions, AMPA receptor-mediated component constitutes the major part of the synaptic transmission at resting membrane potential. Therefore, one important and obvious question was to determine whether AMPA receptor-mediated synaptic response is modified after the stroke.

Studies have reported that administration of AMPA receptors antagonists is strongly protective after the ischemic insult (Buchan et al., 1991; Nellgard and Wieloch, 1992), but it is not clear whether there is a persistent change of the AMPA receptor-mediated response. Indeed, recent observations in hippocampal slices have shown that the AMPA receptor-mediated synaptic response is increased after an aglycemic episode (Tekkok and Krnjevic, 1995; Tekkok and Krnjevic, 1996) but not after an anoxic episode (Crépel et al., 1993b), or an ischemic insult (Hori and Carpenter, 1994) (but see also Tsubokawa et al., 1995a). How can this result be reconciled with the fact that part of the toxic $Ca^{2+}$ influx recorded after ischemia clearly involves the AMPA receptors? One possibility is that the subunit stoichiometry of the AMPA receptor-channel complex has been changed in favor of $Ca^{2+}$-permeable subunits (for example GluRB changed in favor of GluR A, C or D) and this leads to $Ca^{2+}$ accumulation. In keeping with this, it has been shown that there is a reduction of mRNA encoding GluRB subunit (less permeable to $Ca^{2+}$) 12–72h after the ischemic episode (Pellegrini-Giampetro et al., 1992; Pollard et al., 1993; Gorter et al., 1997) and a greater AMPA-induced $Ca^{2+}$ signal (Gorter et al., 1997). An alternative possibility is that AMPA receptors indirectly contribute to the rise of intracellular $Ca^{2+}$. In this scenario, activation of AMPA receptors would be required to relieve the $Mg^{2+}$ block of the NMDA receptor-channel (permeable to $Ca^{2+}$; McDermott et al., 1986; McBain and Mayer, 1994).

In contrast to the AMPA receptors, many studies have reported a long-term change of the NMDA receptor-mediated response after an ischemic episode (Urban et al., 1990; Hori et al., 1991; Hori and Carpenter, 1994) (but see Tsubokawa et al., 1995a). For example, it has been shown that following an ischemic episode the synaptic response includes a late abnormal component (associated with bursting discharges) (Urban et al., 1990; Hori et al., 1991; Tsubokawa et al., 1995b). This late response is mediated by NMDA receptors since a major portion of this response is sensitive to NMDA receptor antagonists (Urban et al., 1990; Hori et al., 1991). The mechanisms triggering and underlying this long-lasting increase of the NMDA receptor-mediated response during the reoxygenation period were not obvious. Part of these questions have been clarified by in vitro and ex vivo experiments.

**The long-term change of the NMDA-receptor mediated response induced by an anoxic episode has a post-synaptic locus and is $Ca^{2+}$-dependent**

To determine the processes involved in the induction and expression of the long-lasting increase of the NMDA receptor-mediated response, we used an in vitro experimental approach. This approach eliminates possible complicating factors, such as damage to cerebral blood vessels, and reduces the influence of other factors, such as acidosis, hyperkalemia, and temperature changes. The ischemic episode was mimicked by an anoxic-aglycemic (AA) episode (in the bath medium glucose was replaced by sucrose and oxygen by nitrogen). The effect of an AA episode was first tested on the pharmacologically isolated NMDA receptor-mediated response recorded in the presence of AMPA and GABA-A receptor antagonists and in the presence of physiological (1.3 mM) or a reduced (0.1–0.6 mM) external concentration of $Mg^{2+}$. The NMDA receptor-mediated excitatory postsynaptic potentials (EPSP) were rapidly blocked by the AA episode (1–3 min). Upon reoxygenation the synaptic transmission recovered and was subsequently and persistently potentiated (by 50% in average) for at least 2 hours. We demonstrated that the persistent enhancement of the NMDA receptor-mediated response is a long-term potentiation since it is due to an increase in the synaptic efficacy (leftward shift of input/output curves) and it was not associated with a change of cell excitability. This event has been called "anoxic long-term potentiation (LTP)" (Crépel et al., 1993a).

The induction of anoxic LTP required: (i) variations of membrane potential (it was not observed when the changes in membrane potential, occurring during and shortly after the AA episode, were prevented by clamping the voltage); (ii) activation of NMDA receptors (it was not observed if the NMDA receptor antagonist D-2-amino-5-phosphonopentanoic acid (D-APV) was applied shortly before and during the anoxic episode); (iii) a rise in $[Ca^{2+}]_i$ (it was prevented by intracellular injection of a $Ca^{2+}$ chelator 1,2-bis (2-aminophenoxy) ethane N,N,N′,N′-tetra-acetic acid (BAPTA); see Fig. 1A) (Crépel et al., 1993a,b; Crépel and Ben-Ari, 1996).

Interestingly, the expression of anoxic LTP only involves a change of the NMDA receptor-mediated response. When the effect of the AA episode was tested on the mixed AMPA and NMDA receptor-mediated responses (i.e. in the absence of an AMPA receptor antagonist and in the presence of 1.3 mM $Mg^{2+}$ in the bath medium), we observed a persistent potentiation of the NMDA receptor-mediated component of the EPSP, and no change of AMPA receptor-mediated response (Crépel et al., 1993b). Finally, we have shown that the locus of the expression of the anoxic LTP is post-synaptic, since currents evoked by a focal application of NMDA (when synaptic activity has been fully blocked) are also persistently enhanced by an AA episode (see Fig. 1B) (Crépel et al., 1993b; for review see Hammond et al., 1994).

In conclusion, we demonstrated that anoxia induces, post-synaptically, a selective long-term change of the NMDA receptor-mediated response which is $Ca^{2+}$-dependent. We propose that the metabolic stress induced during the anoxic episode leads to a persistent modification of the NMDA receptor-channel complex through its modulatory sites (see below).

## Are the regulatory and/or phosphorylation sites of NMDA receptor-channel complex involved in the generation of the anoxic LTP?

The NMDA receptor-channel complex includes many regulatory sites as well as several consensus groups of phosphorylation (for review see McBain and Mayer, 1994 MacDonald et al., this volume) (see Fig. 2A). It has been reported that the NMDA receptor-mediated response can be modulated by glycine (Johnson and Ascher, 1987), polyamines (McGurk et al., 1990; Rock and MacDonald, 1992; Benveniste and Mayer, 1993), protons (Traynelis and Cull-Candy, 1990), nitric oxide (Lei et al., 1992; Manzoni et al., 1992; Lipton et al., 1996), other redox reagents (Aizenman et al., 1990; Tang and Aizenman, 1993), arachidonic acid (Miller et al., 1992), $Mg^{2+}$ (Mayer et al., 1984; Nowak et al., 1984), $Ca^{2+}$ ions (Rosenmund and Westbrook, 1993; Medina et al., 1994; Medina et al., 1995), and by the $Ca^{2+}$- and phospholipid-dependent protein kinase C (PKC) (for review see Ben-Ari et al., 1992). Several of these regulatory sites are highly sensitive to changes that occur during and shortly after an ischemic-anoxic episode. This type of metabolic stress induces: (i) a fall in $[ATP]_i$ and a change in pH (Whittingham et al., 1984); (ii) an extracellular accumulation of glycine (Globus et al., 1991); (iii) an intracellular increase of arachidonic acid (Abe et al., 1987); (iv) an activation of PKC (Onodera et al., 1989; Cardell and Wieloch, 1993) (but see also Cardell et al., 1990); (v) an accumulation of reducing agents (Ginsberg et al., 1976; Tanaka et al., 1986), including cysteines (Slivka and Cohen, 1993), which can shift the equilibrium of the redox site of NMDA receptor toward a reduced state (for review see Lipton, 1993; Gozlan and Ben-Ari, 1995). Some of these changes may persist following the anoxic-ischemic episode and induce long-lasting modifications of the NMDA receptor-mediated response. Along this line, we have shown that the redox sites are instrumental for the induction and the expression of anoxic LTP (Gozlan et al., 1994a). Oxidization of NMDA receptors before and during the AA episode by the thiol-oxidizing reagent 5,5′-dithiobis-2-nitrobenzoic acid (DTNB) prevents anoxic LTP, indicating that this form of LTP involves the reduced form of the NMDA receptors (see Fig. 2B). It has been suggested that disulfide groups on NMDA receptors are persistently reduced by AA episode. The opening of disulfide groups could induce a conformational change of the NMDA receptors and allow a potentiation of the NMDA receptor-mediated response. This process can last for a long time, due to the reducing reagents generated by the AA.

As mentioned above, the persistent increase of the NMDA receptor-mediated response after an anoxic or ischemic episode can also be due to phosphorylation of serine/threonine residues by PKC. We have shown that intracellular injection of inhibitory peptide (PKC 19-36) or bath application of the PKC inhibitor (sphingosine) prevent anoxic LTP (Ben-Ari et al., 1994) (see Fig. 2C). The positive modulation of the NMDA receptor-mediated response by PKC may lead to a reduction of the voltage-dependent block of the NMDA channel by $Mg^{2+}$ (Chen and Huang, 1992). Accordingly, there is an NMDA receptor-mediated component that appears at hyperpolarized membrane potentials after an AA episode and

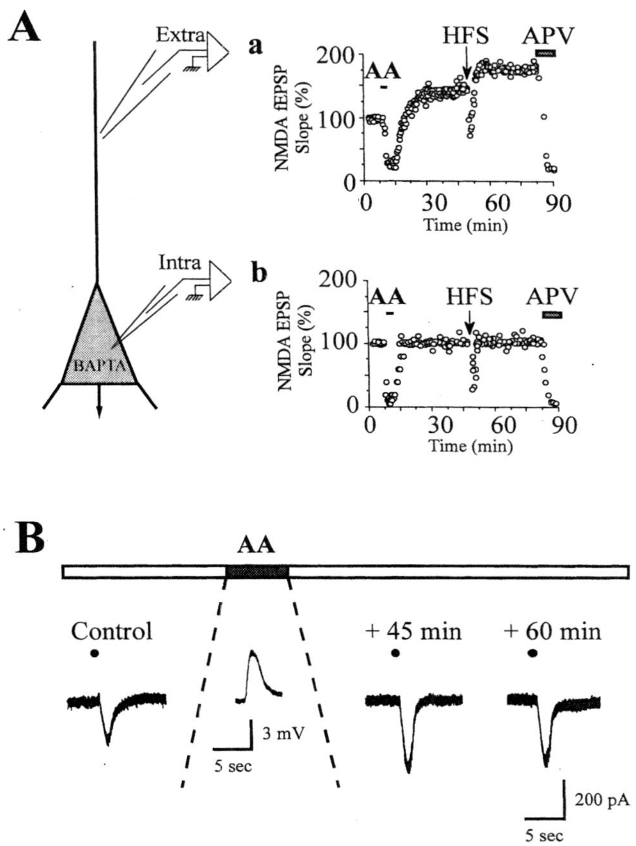


Fig. 1. Anoxic LTP has a post-synaptic locus and is $Ca^{2+}$-dependent. **Aa,b**. Graph showing the slope versus time of the pharmacologically isolated NMDA receptor-mediated field EPSP (fEPSP, a), and intracellular EPSP (b) recorded in CA1 area (in the presence of 10 $\mu$M CNQX, 10 $\mu$M glycine, 10 $\mu$M bicuculline and 0.3 mM $Mg^{2+}$) before, during, and after the anoxic-aglycemic (AA) episode (2.5 min duration). Note that AA episode induces a long-term potentiation of NMDA fEPSP (a). In contrast, in the presence of the $Ca^{2+}$ chelator BAPTA (50 mM) in the pipette solution, there is no change of the initial slope of the EPSP after the AA episode and the high-frequency stimulation (HFS; 500 ms duration, 100 Hz frequency) (b; adapted from Crépel and Ben-Ari, 1996). **B**. NMDA current recorded in voltage-clamp mode ($V_H$ = −60 mV) (in the presence of 1 $\mu$M TTX, 15 $\mu$M CNQX, 10 $\mu$M bicuculline and 1.3 mM $Mg^{2+}$) and evoked by pressure application of NMDA (300 $\mu$M; 350 ms duration; 3.5 bar pressure; ●), before (control), and after (45 min and 60 min) the AA episode (2 min duration). During the AA episode the NMDA responses have been recorded in current-clamp mode (adapted from Crépel et al., 1993b).

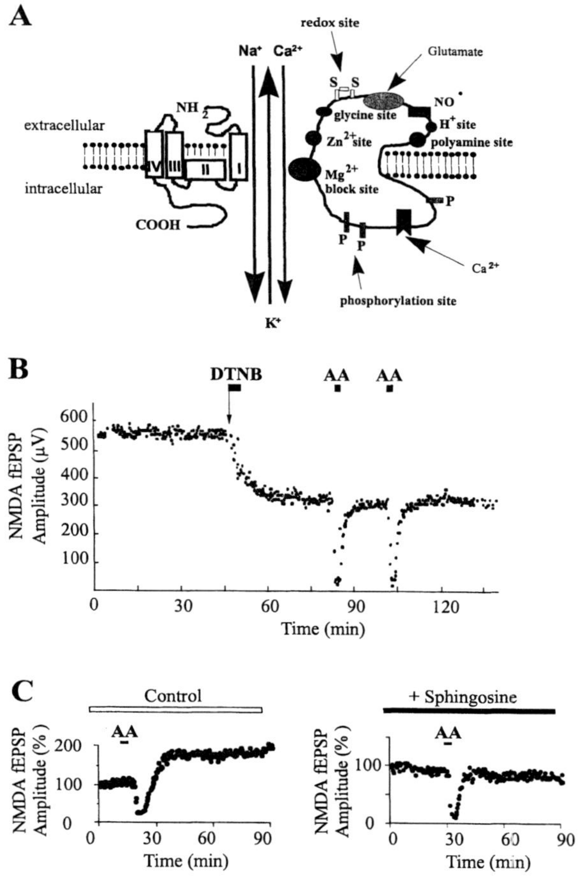


Fig. 2. PKC inhibitor and oxidizing reagent prevent the induction of anoxic LTP. **A.** Schematic representation of NMDA receptor indicating the various regulatory sites (adapted from Hollmann et al., 1994). **B, C.** Graph showing the amplitude versus time of the pharmacologically isolated NMDA receptor-mediated field EPSP (NMDA fEPSP) recorded (in the presence of 10 $\mu$M CNQX, and 0.1 mM $Mg^{2+}$) before, during and after the AA episode (2–3 min duration). In **B**, the AA episodes have been performed after bath application of the thiol-oxidizing reagent, DTNB (200 $\mu$M, 10 min duration) (adapted from Gozlan et al., 1994a). In **C**, the AA episode have been performed in the presence or in the absence of the PKC inhibitor, sphingosine (10 $\mu$M) (adapted from Ben-Ari et al., 1994).

ischemic insult, that was not present in control conditions (Hori et al., 1991; Crépel et al., 1993b; Hori and Carpenter, 1994).

In addition to the rapid changes of the NMDA receptor-mediated responses through their regulatory sites and phosphorylation groups, the increase of NMDA receptor-mediated responses observed several hours after the ischemic insult can be due to a modification of the genetic expression of NMDA receptor-subunits. According to this hypothesis, mRNA encoding NR2C subunit is present at a high level 1 hour after a brief AA episode and persisted at least 3 hours following an AA episode (Perez-Velazquez and Zhang, 1994), in contrast to control tissue where the NR2A and NR2B mRNAs are prominently expressed in hippocampus of adult rat (Monyer et al., 1994). An heteromeric NMDA receptor containing NR2C subunits has been characterized as having a weaker $Mg^{2+}$ block and a prolonged response (Moriyoshi et al., 1991; Monyer et al., 1992). These properties would lead to an increase in $Ca^{2+}$ influx.

In conclusion, protein kinase C and the modulatory sites of NMDA receptors, notably the redox sites, play a crucial role as rapid onset processes involved in the generation of long-term changes of the NMDA receptor-mediated response after an ischemic-anoxic episode. The long-lasting change of the NMDA receptor-mediated response may also involve an alteration of the subunit composition of NMDA receptors.

## Comparison between the anoxic and tetanic LTP

Elucidation of the mechanisms that generate long-term change of synaptic transmission is a major goal of current efforts to clarify the basis of neuronal plasticity, learning, and memory processes (for review see Ben-Ari et al., 1992; McNaughton, 1993; Otani and Ben-Ari, 1993). In normoxic conditions, a brief repetitive stimulation (100 Hz, 1 sec) of hippocampal afferents in the CA1 region of the hippocampus is commonly used to generate a persistent enhancement of synaptic transmission, a change which persists for hours (in vitro) to weeks (in vivo) (for review see McNaughton, 1993). This long-term potentiation has been called tetanic LTP. Interestingly, many studies have reported that the tetanic LTP is primarily mediated by AMPA receptors (Muller and Lynch,

1988; Perkel and Nicoll, 1993), in contrast to the anoxic LTP which is mediated by NMDA receptors (Crépel et al., 1993b). Despite this major difference, the induction of tetanic LTP and anoxic LTP share several common properties. As anoxic LTP, tetanic LTP requires a depolarization of the post-synaptic neuron and the activation of NMDA receptors (for review see Malenka and Nicoll, 1993) and it is prevented by an intracellular injection of a $Ca^{2+}$ chelator (for review see Malenka and Nicoll, 1993). Tetanic LTP is also prevented by intracellular injections of the PKC inhibitory peptide (PKC 19-36) (Malinow et al., 1989), suggesting that phosphorylation cascades are important for induction mechanisms (for review see Ben-Ari et al., 1992). At present, the only difference between the induction of tetanic LTP mediated by AMPA receptors and anoxic LTP mediated by NMDA receptors is their different sensitivity to redox reagents. The thiol-oxidizing reagent DTNB prevents induction of anoxic LTP without affecting tetanic LTP (mediated by AMPA receptors) (Gozlan et al., 1994 a,b). Interestingly, like anoxic LTP, the induction of tetanic LTP, mediated by NMDA receptors, is prevented in the presence of DTNB (Gozlan et al., 1994b). These observations suggest that NMDA receptor-mediated responses cannot be enhanced when they are in a fully oxidized configuration. Our working hypothesis is that the redox site of NMDA receptors provides a selective molecular switch for the amplification of NMDA but not AMPA receptor-mediated responses (for review see Gozlan and Ben-Ari, 1995).

We have shown that the type of LTP, i.e expressed by AMPA (AMPA LTP) or NMDA (NMDA LTP) receptors, can also depend on the amplitude of the NMDA receptor-mediated response evoked during the tetanus (Aniksztejn and Ben-Ari, 1995) (see Fig. 3A,B). A weak tetanus which activates a small NMDA receptor-mediated response or a strong tetanus performed in the presence of 7-Cl-kynurenate (to partially inhibit the NMDA receptor mediated-response), leads to a selective AMPA LTP. In contrast, a strong tetanus which activates a large NMDA receptor-mediated response induces a selective NMDA LTP. Consistent with these observations, other studies have reported that over-activation of the NMDA receptors during the tetanus reduces the probability of generating AMPA LTP (Coan et al., 1989; Huang et al., 1992; Izumi et al., 1992) but not NMDA LTP (Berretta et al., 1991; Xie et al., 1992). Along this line, it is likely that ischemia or anoxia induce NMDA LTP because they lead to a massive release of excitatory amino acids, which in turn strongly activates the NMDA receptors (for review see Choi, 1990; Rothman and Olney, 1995). In keeping with this hypothesis, it was shown that a prolonged application of glutamate in the CA1 region of acute slices triggers a long-term increase in the frequency of single NMDA channel opening (Kovalchuk et al., 1993). Interestingly, the switch between the induction of the AMPA and NMDA LTP is gradual, and in some cases the tetanic stimulation clearly induces both AMPA and NMDA LTP (see Fig. 3). This may explain why some conditions, such as aglycemia, can induce both AMPA and NMDA LTP (Tekkok and Krnjevic, 1995).

In conclusion, in CA1 pyramidal neurons, a persistent increase of NMDA receptor-mediated responses is apparently induced by extreme conditions such as ischemia, anoxia, hypoglycemia, prolonged application of glutamate, strong tetanic stimulation, or tetanic stimulation performed in low $Mg^{2+}$ conditions. This is in contrast to the LTP of the AMPA receptor-mediated response which can be induced in physiological conditions.

**Perspectives**

Our working hypothesis is that the persistent increase of NMDA receptor-mediated responses is one of the key steps involved in the delayed death of the CA1 pyramidal cells induced by a stroke episode. In the patients suffering from a stroke most of them receive medical attention several minutes to several hours after the onset of the stroke episode. Therefore, we propose that acting on NMDA receptors in this therapeutic window (i.e. after the stroke) may open new perspectives to reduce the neuronal infarct induced by ischemia. Along this line, it is crucial to develop pharmacological agents acting on NMDA receptors devoid of the well known side-effect of competitive and noncompetitive NMDA receptor antagonists (for review see Lipton, 1993). The redox site of NMDA receptors could constitute a new target to prevent excitotoxic effects of NMDA receptor-mediated activity without affecting the normal NMDA receptor-dependent synaptic plasticity (for review see Lipton, 1993; Gozlan and Ben-Ari, 1995). Indeed, in the CA1 region, oxidizing the NMDA receptor prevents the persistent increase of NMDA receptor-mediated responses induced by an anoxic-aglycemic episode without modifying the LTP mediated by AMPA receptors.

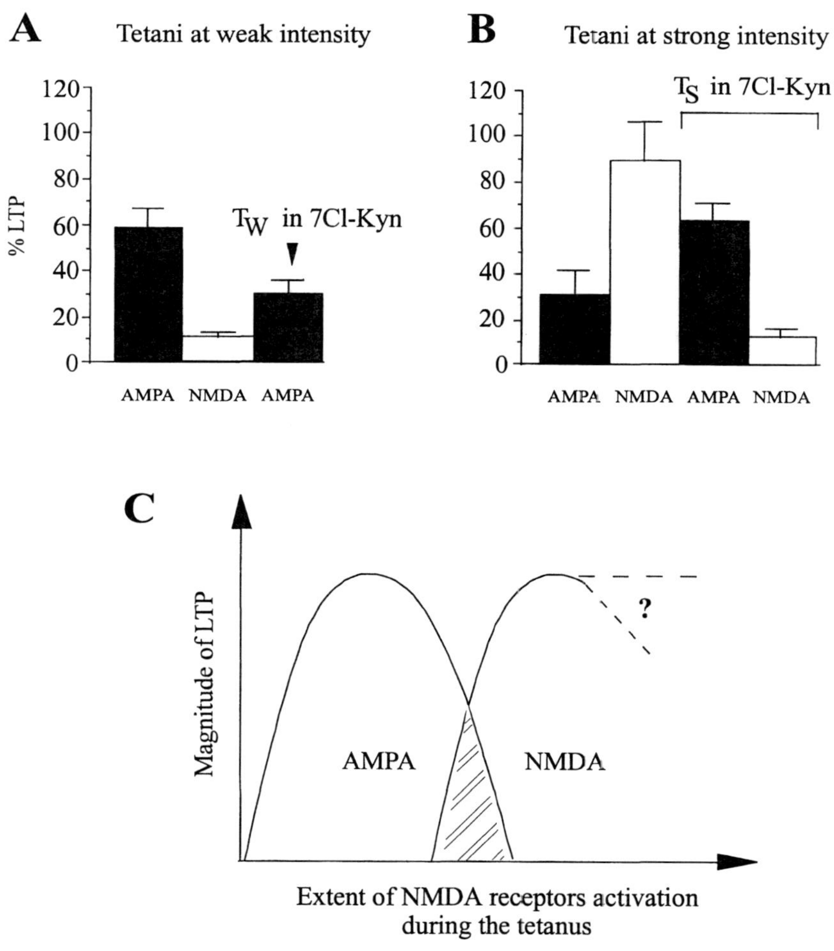


Fig. 3. Expression of LTP by AMPA and/or NMDA receptors is determined by the extent of NMDA receptors activation during the tetanus. **A.** Weak intensity tetanus ($T_W$, which activates small NMDA synaptic response) induces LTP expressed mainly by AMPA receptors. When NMDA response is partially inhibited by 7-Cl-Kynurenate (6 $\mu$M), $T_W$ induces AMPA LTP of a smaller amplitude. **B.** Strong intensity tetanus ($T_S$, which activates large NMDA synaptic response) induces LTP expressed mainly by NMDA receptors. In contrast, when NMDA response is partially inhibited by 7-Cl-Kynurenate (6 $\mu$M), $T_S$ induces LTP expressed mainly by AMPA receptors. **C.** Model representing the relative contribution of the AMPA and NMDA components of the synaptic response to LTP as a function of the extent of NMDA receptors activation during the tetanus (adapted from Aniksztejn and Ben-Ari, 1995).

## Acknowledgements

Financial supports from Institut National de la Santé et de la Recherche Médicale (INSERM), Centre National pour la Recherche Scientifique (CNRS), Direction des Recherches, Etude et Techniques (DRET). The authors thank Drs. S. Williams and D. Doll for helpful comments on the manuscript.

## References

Abe, K., Kogure, K., Yamamoto, H., Imazawa, M. and Miyamaoto, K. (1987) Mechanism of arachidonic acid liberation during ischemia in gerbil cerebral cortex. *J. Neurochem.*, 48: 503–509.

Aizenman, E., Hartnett, K.A. and Reynolds, I.J. (1990) Oxygen free radicals regulate NMDA receptor function via a redox modulatory site. *Neuron*, 5: 841–846.

Andiné, P., Jacobson, I. and Hagberg, H. (1988) Calcium uptake evoked by electrical stimulation is enhanced post-ischemically and precedes delayed neuronaldeath in CA1 of the rat hippocampus: Involvement of the *N*-methyl-D-aspartate receptors. *J. Cerebr. Blood Flow Metab.*, 8: 799–807.

Andiné, P., Jacobson, I. and Hagberg, H. (1992) Enhanced calcium uptake by CA1 pyramidal cell dendrites in the postischemic phase despite subnormal evoked field potentials: Excitatory amino acid receptor dependency and relationship to neuronal damage. *J. Cereb. Blood Flow*, 12: 773–783.

Aniksztejn, L. and Ben-Ari, Y. (1995) Expression of LTP by AMPA and/or NMDA receptors is determined by the extent of NMDA receptors activation during the tetanus. *J. Neurophysiol.*, 74: 2349–2357.

Araki, T., Kato, H. and Kogure, K. (1990) Neuronal damage and calcium accumulation following repeated brief cerebral ischemia in the gerbil. *Brain Res.*, 528: 114–122.

Armstrong, D.R., Neill, K.H., Crain, B.J. and Nadler, J.V. (1989) Absence of electrographic seizures after transient forebrain ischemia in the Mongolian gerbil. *Brain Res.*, 476: 174–178.

Ben-Ari, Y., Aniksztejn, L. and Bregestovski, P. (1992) Protein kinase C modulation of NMDA currents: An important link for LTP induction. *Trends Neurosci.*, 15: 333–339.

Ben-Ari, Y., Congar, P., Diabira, D. and Hammond, C. (1994) Induction of anoxic LTP is prevented by proteine kinase C antagonists. *Soc. Neurosci. Abstr.*, 20: 1517.

Benveniste, H., Drejer, J., Schousboë, A. and Diemer, N.-H. (1984) Elevation of the extracellular concentrations of glutamate and aspartate in rat hippocampus during transient cerebral ischemia monitored by intracerebral microdialysis. *J. Neurochem.*, 43: 1369–1374.

Benveniste, H., Jorgensen, M.B., Sandberg, M., Christensen, T., Hagberg, H. and Diemer, N.H. (1989) Ischemic damage in hippocampal CA1 is dependent on glutamate release and intact innervation from CA3. *J. Cerebr. Blood. Flow Metab.*, 9: 629–639.

Benveniste, M. and Mayer, L. (1993) Multiple effects of spermine on *N*-methyl-D-aspartatic acid receptor responses of rat cultured hippocampal neurones. *J. Physiol. Lond.*, 464: 131–163.

Berretta, N., Berton, F., Bianchi, R., Brunelli, M., Copogna, M. and Francesconi, W. (1991) Long term potentiation of NMDA receptor-mediated EPSP in guinea-pig hippocampal slices. *Eur. J. Neurosci.*, 3: 850–854.

Boast, C.A., Gerhardt, S.C. and Janak, P. (1987) Systemic AP7 reduces ischemic brain damage in gerbils *Excitatory Amino Acid Transmission* Eds. Hicks, T.P., Lodge, D. and McLennan, H., 249–252, New York, Liss.

Boast, C.A., Gerhardt, S.C., Pastor, G., Lehmann, J., Etienne, P.E. and Liebman, J.M. (1988) The *N*-methyl-D-aspartate antagonists CGS19755 and CPP reduce ischemic brain damage in gerbils. *Brain res.*, 442: 345–348.

Buchan, A.M. and Pulsinelli, W. (1990) Septo-hippocampal deafferentation protects CA1 neurons against ischemic injury. *Brain res.*, 512: 7–14.

Buchan, A.M., Li, H., Cho, S. and Pulsinelli, W.A. (1991) Blockade of the AMPA receptor prevents CA1 hippocampal injury following severe but transient forebrain ischemia in adult rats. *Neurosci. lett.*, 132: 255–258.

Cardell, M., Bingren, H., Wieloch, T., Zivin, J. and Saitoh, T. (1990) Protein kinase C is translocated to cell membranes during cerebral ischemia. *Neurosci. Lett.*, 119: 228–232.

Cardell, M. and Wieloch, T. (1993) Time course of the translocation and inhibition of protein kinase C during complete cerebral ischemia in the rat. *J. Neurochem.*, 61: 1308–1314.

Chang, H.S., Sasaki, T. and Kassell, N.F. (1989) Hippocampal unit activity after transient cerebral ischemia in rats. *Stroke*, 20: 1051–1058.

Chen, L. and Huang, L.-Y.M. (1992) Protein Kinase C reduces $Mg^{2+}$ block of NMDA-receptor channels as a mechanism of modulation. *Nature*, 356: 521–523.

Choi, D.W. (1990) The role of glutamate neurotoxicity in Hypoxic-ischemic neuronal death. *Ann. Rev. Neurosci.*, 13: 171–182.

Coan, E.J., Irving, A.J. and Collingridge, G.L. (1989) Low-frequency activation of the NMDA receptor system can prevent the induction of LTP. *Neurosci Lett*, 105: 205–210.

Congar, P., Khazipov, R. and Ben-Ari, Y. (1994) Direct demonstration of functional disconnection by anoxia of inhibitory interneurons from excitatory inputs in rat hippocampus. *J. Neurophysiol.*, 73: 421–426.

Crépel, V., Hammond, C., Krnjevic, K., Chinestra, P. and Ben-Ari, Y. (1993a) Anoxia-induced LTP of isolated NMDA receptors synaptic responses. *J. Neurophysiol.*, 69: 1774–1778.

Crépel, V., Hammond, C., Chinestra, P., Diabira, D. and Ben-Ari, Y. (1993b) A selective LTP of NMDA receptor-mediated currents induced by anoxia in CA1 hippocampal neurons. *J. Neurophysiol.*, 70: 2045–2055.

Crépel, V. and Ben-Ari, Y. (1996) Intracellular injection of a $Ca^{2+}$ chelator prevents generation of anoxic LTP. *J. Neurophysiol.*, 75: 770–779.

Fujiwara, N., Higashi, H., Shimoji, K. and Yoshimura, M. (1987) Effects of hypoxia on rat hippocampal neurones in vitro. *J. Physiol. Lond.*, 384: 131–151.

Gill, R., Foster, A.C. and Woodruff, G.N. (1987) Systemic administration of MK-801 protects against ischemia-induced hippocampal neurodegeneration in the gerbil. *J. Neurosci.*, 7: 3343–3349.

Gill, R., Foster, A.C. and Woodruff, G.N. (1988) MK-801 is neuroprotective in gerbils when administered during the post-ischaemic period. *Neurosci.*, 25: 847–855.

Ginsberg, M.D., Reivich, M., Frinak, S. and Harbig, K. (1976) Pyridine nucleotide redox state and blood flow of cerebral cortex following middle cerebral artery occlusion in the cat. *Stroke*, 7: 125–131.

Globus, M.Y.-T., Busto, R., Martinez, E., Valdés, I., Dietrich, W.D. and Ginsberg, M. (1991) Comparative effect of transient global ischemia on extracellular levels of glutamate, glycine and g-aminobutyric acid in vulnerable and nonvulnerable brain regions in the rat. *J. Neurochem.*, 57: 470–478.

Gorter, J.A., Petrozzino, J.J., Aronica, E.M., Rosenbaum, D.M., Opitz, T., Bennett, M.V., Connor, J.A. and Zukin, R.S. (1997) Global ischemia induces downregulation of Glur2 mRNA and increases AMPA receptor-mediated $Ca^{2+}$ influx in hippocampal CA1 neurons of gerbil. *J. Neurosci.*, 17: 6179–6188.

Gozlan, H., Diabira, D., Chinestra, P. and Ben-Ari, Y. (1994a) Anoxic LTP is mediated by the redox modulatory site of NMDA receptor. *J. Neurophysiol.*, 72: 3017–3022.

Gozlan, H., Chinestra, P., Diabira, D. and Ben-Ari, Y. (1994b) NMDA redox site modulates long-term potentiation of NMDA but not AMPA receptors. *Eur. J. Pharmacol.*, 262: R3–R4.

Gozlan, H. and Ben-Ari, Y. (1995) NMDA receptor redox sites: Are they targets for selective neuronal protection? *Trends Pharmacol. Sci.*, 16: 368–374.

Hammond, C., Crépel, V., Gozlan, H. and Ben-Ari, Y. (1994) Anoxic LTP sheds light on the multiple facets of NMDA receptors. *Trends Neurosci.*, 17: 497–503.

Hansen, A.J. (1985) Effect of anoxia on ion distribution in the brain. *Physiol. Rev.*, 65: 101–148.

Hori, N., Doi, N., Miyahara, S., Shinoda, Y. and Carpenter, D.O. (1991) Appearance of NMDA receptors triggered by anoxia independent of voltage in vivo and in vitro. *Experimental Neurol.*, 112: 304–311.

Hori, N. and Carpenter, D.O. (1994) Transient ischemia causes a reduction of $Mg^{2+}$ blockade of NMDA receptors. *Neurosci. Lett.*, 173: 75–78.

Hsu, M., Sik, A., Gallyas, F., Horvath, Z. and Buzsaki, G. (1994) Short-term and long-term changes in the postischemic hippocampus. *Ann. N. Y. Acad. Sci.*, 743: 121–139.

Huang, Y.Y., Colino, A., Selig, D.K. and Malenka, R.C. (1992) The influence of prior synaptic activity on the induction of long-term potentiation. *Science*, 255: 730–733.

Hollmann, M. and Heinemann, S. (1994) Cloned glutamate receptors. *Ann. Rev. Neurosci.*, 17: 31–108.

Imon, I., Mitani, A., Andou, Y., Arai, T. and Katoaka, K. (1991) Delayed neuronal death is induced without postischemic hyperexcitability: Continuous multiple-unit recording from ischemic CA1 neurons. *J. Cereb. Blood Flow Metabo.*, 11: 819–823.

Izumi, Y., Clifford, D.B. and Zorumski, C.F. (1992) Inhibition of long-term potentiation by NMDA-mediated nitric oxide release. *Science*, 257: 1273–1276.

Jensen, M.S., Lambert, J.D.C. and Johansen, F.F. (1991) Electrophysiological recordings from rat hippocampus slices following in vivo brain ischemia. *Brain Res.*, 554: 166–175.

Johansen, F.F., Jorgensen, M.B. and Diemer, N.H. (1983) Resistance of hippocampal CA-1 interneurons to 20 min of transient cerebral ischemia in the rat. *Acta. Neuropathol.*, 61: 135–140.

Johansen, F.F., Jorgensen, M.B. and Diemer, N.H. (1986) Ischemic CA-1 pyramidal cell loss is prevented by preischemic colchicine destruction of dentate gyrus granule cells. *Brain Res.*, 377: 344–347.

Johansen, F.F., Jorgensen, M.B. and Diemer, N.-H. (1987) Ischemia-induced delayed neuronal death in the CA1 hippocampus is dependent on intact glutamatergic innervation *Excitatory Amino Acid Transmission* Eds. Hicks, T.P., Lodge, D. and McLennan, H. 245–248, New York, Liss.

Johnson, J.W. and Ascher, P. (1987) Glycine potentiates the NMDA response in cultured mouse brain neurons. *Nature (London)*, 325: 529–532.

Jorgensen, M.B., Johansen, F.F. and Diemer, N.-H. (1987) Removal of the entorhinal cortex protects hippocampal CA1 neurons from ischemic damage. *Acta. Neuropathol.*, 73: 189–194.

Kaplan, T.M., Lasner, T.M., Nadler, J.V. and Crain, B.J. (1989) Lesions of excitatory pathways reduce hippocampal cell death after transient forebrain ischemia in the gerbil. *Acta Neuropathol.*, 78: 283–290.

Khazipov, R., Congar, P. and Ben-Ari, Y. (1995) Hippocampal CA1 lacunosum-moleculare interneurons: Modulation of monosynaptic GABAergic IPSCs by presynaptic GABAB receptors. *J. Neurophysiol.*, 74: 2126–2137.

Koroshetz, W.J. and Moskowitz, M.A. (1996) Emerging treatments for stroke in humans. *Trends Pharmacol Sci.*, 17: 227–233.

Kovalchuk, Y., Garaschuk, O. and Krishtal, O.A. (1993) Glutamate induces long-term increase in the frequency of single $N$-methyl-D-aspartate channel openings in hippocampal CA1 neurons examined in situ. *Neuroscience*, 54: 557–559.

Krnjevic, K., Xu, Y.Z. and Zhang, L. (1991) Anoxic block of GABAergic IPSPs. *Neurochem. Res.*, 16: 279–283.

Lei, S.Z., Pan, Z.H., Aggarwal, S.K., Hartman, J., Sucher, N.J. and Lipton, S.A. (1992) Effect of nitric oxide production on the redox modulatory site of the NMDA-receptor-channel complex. *Neuron*, 8: 1087–1099.

Lipton, S.A. (1993) Prospects for clinically tolerated NMDA antagonists: Open-channel blockers and alternative redox states of nitric oxide. *Trends Neurosci.*, 16: 527–532.

Lipton, S.A., Choi, Y.B., Sucher, N.J., Pan, Z.H. and Stamler, J.S. (1996) Redox state, NMDA receptors and NO-related species. *Trends Pharmacol. Sci.*, 17: 186–187.

Lobner, D. and Lipton, P. (1993) Intracellular calcium levels and calcium fluxes in the CA1 region of the rat hippocampal slice during in vitro ischemia: Relationship to electrophysiological cell damage. *J. Neurosci.*, 13: 4861–4871.

Malenka, R.C. and Nicoll, R.A. (1993) NMDA-receptor-dependent synaptic plasticity: Multiple forms and mechanisms. *Trends Neurosci.*, 16: 521–527.

Malinow, R., Schulman, H. and Tsien, R.W. (1989) Inhibition of postsynaptic PKC or CaMKII blocks induction but not expression of LTP. *Science*, 245: 862–865.

Manzoni, O., Prezeau, L., Marin, P., Deshager, S., Bockaert, J. and Fagni, L. (1992) Nitric oxide-induced blockade of NMDA receptors. *Neuron*, 8: 653–662.

Mayer, M.L., Westbrook, G.L. and Guthrie, P.B. (1984) Voltage dependence block by $Mg^{2+}$ of NMDA responses in spinal cord neurones. *Nature*, 309: 262–263.

McBain, C.J. and Mayer, M.L. (1994) $N$-Methyl-D-aspartate acid receptor structure and function. *Physiological reviews*, 74: 723–760.

McDermott, A.B., Mayer, B.C., Westbrook, G.L., Smith, S.J. and Barker, J.L. (1986) NMDA receptor activation increases cytoplasmic calcium concentration in cultured spinal cord neurones. *Nature*, 321: 519–522.

McGurk, J.F., Bennett, M.V. and Zukin, R.S. (1990) Polyamines potentiate responses of $N$-methyl-D-aspartate receptors expressed in Xenopus oocytes. *Proc. Natl. Acad. Sci. USA*, 87: 9971–9974.

McNaughton, B.L. (1993) The mechanism of expression of long-term enhancement of hippocampal synapses: Current issues and theoritical implications *Ann. Rev. Physiol.* Eds. Hoffman, J.F. and DeWeer, P., 55: 375–396, Palo Alto, california, Annual reviews INC.

Medina, I., Filippova, N., Barbin, G., Ben-Ari, Y. and Bregestovski, P. (1994) Kainate-induced inactivation of NMDA currents via an elevation of intracellular $Ca^{2+}$ in hippocampal neurons. *J. Neurophysiol.*, 72: 456–465.

Medina, I., Filippova, N., Charton, G., Rougeole, S., Ben-Ari, Y., Khrestchatisky, M. and Bregestovski, P. (1995) Calcium-dependent inactivation of heteromeric NMDA receptor-channels expressed in human embryonic kidney cells. *J. Physiol. Lond.*, 482.3: 567–573.

Miller, B., Sarantis, M., Traynelis, S.F. and Attwel, D. (1992) Potentiation of NMDA receptor currents by arachidonic acid. *Nature*, 355: 722–725.

Mitani, A., Imon, H. and Kataoka, K. (1989) High frequency discharges of gerbil hippocampal CA1 neurons shortly after ischaemia. *Brain Res. Bull.*, 23: 269–272.

Mitani, A., Andou, Y. and Kataoka, K. (1992) Selective vulnerability of hippocampal CA1 neurons cannot be explained in terms of an increase in glutamate concentration during ischemia in the gerbil: Brain microdialysis study. *Neurosci.*, 48: 307–313.

Mitani, A., Takeyasu, S., Yanase, H., Nakamura, Y. and Kataoka, K. (1994) Changes in intracellular $Ca^{2+}$ and energy levels during in vitro ischemia in the gerbil hippocampal slice. *J. Neurochem.*, 62: 626–634.

Monyer, H., Sprengel, R., Schoepfer, R., Herb, A., Higuchi, M., Lomeli, H., Burnashev, N., Sakmann, B. and Seeburg, P.H. (1992) Heteromeric NMDA receptors: Molecular and functional distinction of subtypes. *Sciences*, 256: 1217–1221.

Monyer, H., Bernashev, N., Laurie, D.J., Sakmann, B. and Seeburg, P.H. (1994) Developmental and regional expression in the rat brain and functional properties of four NMDA receptors. *Neuron*, 12: 529–540.

Moriyoshi, K., Masu, M., Ishii, T.R.S., Mizuno, N. and Nakanishi, S. (1991) Molecular cloning and characterization of the rat NMDA receptor. *Nature*, 354: 31–37.

Muller, D. and Lynch, G. (1988) Long-term potentiation differentially affects two components of synaptic responses in hippocampus. *Proc. Natl. Acad. Sci. USA*, 85: 9346–9350.

Nellgard, B. and Wieloch, T. (1992) Postischemic blockade of AMPA but not NMDA receptors mitigates neuronal damage in the rat brain following transient severe ischemia. *J. Cereb. Blood Flow Metab.*, 12: 2–11.

Nowak, K.L., Bregestovski, P., Ascher, P., Herbert, A. and Prochiantz, A. (1984) Magnesium gates glutamate activated channels in mouse central neurons. *Nature Lond.*, 307: 462–464.

Onodera, H., Sato, G. and Kogure, K. (1986) Lesions of Schaffer collaterals prevent ischemic death of pyramidal cells. *Neurosci. Lett.*, 68: 169–174.

Onodera, H., Araki, T. and Kogure, K. (1989) Protein kinase C activity in the rat hippocampus after forebrain ischemia: autoradiographic anlysis by [3H] phorbol 12, 13-dibutyrate. *Brain Res.*, 481: 1–7.

Otani, S. and Ben-Ari, Y. (1993) Biochemical correlates of long-term potentiation in hippocampal synapses *International Review of Neurobiol.* Eds. Bradley, R.J. and Harris, R.A., 35: 1–41, San Diego, Academic Press, Inc.

Park, C.K., Nehls, D.G., Graham, D.I., Teasdale, G.M. and McCulloch, J. (1988) The glutamate antagonist MK-801 reduces focal ischemic brain damage in the rat. *Ann. Neurol.*, 24: 543–551.

Pellegrini-Giampetro, D.E., Zukin, S.R., Bennett, M.V.L., Cho, S. and Pulsinelli, W.A. (1992) Switch in glutamate receptor subunit gene expression in CA1 subfield of hippocampus following global ischemia in rats. *Proc. Natl. Acad. Sci. USA*, 89: 10499–10503.

Perez-Velazquez, J.L. and Zhang, L. (1994) In vitro hypoxia induces expression of the NR2C subunit of the NMDA receptor in rat cortex and hippocampus. *J. Neurochem.*, 63: 1171–1173.

Perkel, D.J. and Nicoll, R.A. (1993) Evidence for all-or-none regulation of neurotransmitter release: Implications for long-term potentiation. *J. Physiol. Lond.*, 471: 481–500.

Petito, C.K. and Pulsinelli, W.A. (1984) Delayed neuronal recovery and neuronal death in rat hippocampus following severe cerebral ischemia: Possible relationship to abnormalities in neuronal processes. *J. Cereb. Blood Flow metab.*, 4: 194–205.

Petito, C.K., Feldmann, E., Pulsinelli, W.A. and Plum, F. (1987) Delayed hippocampal damage in humans following cardiorespiratory arrest. *Neurology*, 37: 1281–1286.

Pollard, H., Herron, A., Moreau, J., Charriault-Marlangue, C., Khrestchatisky, M., Epelbaum, J. and Ben-Ari, Y. (1993) Alteration of the gluR-B AMPA receptor subunit flip/flop expression and stoichiometry and kainate-induced epilepsy and ischemia. *Soc. Neurosci. Abstr.*

Pulsinelli, W.A. (1985) Selective neuronal vulnerability: Morphological and molecular characteristics. *Progress in brain Res.*, 63: 29–37.

Rock, D.M. and MacDonald, R.L. (1992) Spermine and related polyamines produce a voltage-dependent reduction of *N*-methyl-D-aspartate receptor single-channel conductance. *Mol. Pharmacol.*, 42: 157–164.

Rosenmund, C. and Westbrook, G.L. (1993) Calcium-induced actin depolymerization reduces NMDA channel activity. *Neuron*, 10: 805–814.

Rothman, S.M. and Olney, J.W. (1995) Excitotoxicity and the NMDA receptor – still lethal after eight years. *Trends Neurosci.*, 18: 57–58

Schiff, S.J. and Somjen, G.G. (1985) Hyperexcitability following moderate hypoxia in hippocampal tissue slices. *Brain Res.*, 337: 337–340.

Sheardown, M.J., Nielsen, E.O., Hansen, A.J., Jacobsen, P. and Honoré, T. (1990) 2,3-dihydroxy-6-nitro-7-sulfamoyl-benzo(F)quinoxaline: A neuroprotectant for cerebral ischemia. *Science*, 247: 571–574.

Slivka, A. and Cohen, G. (1993) Brain ischemia markedly elevates levels of the neurotoxic amino acid, cysteine. *Brain Res.*, 608: 33–37.

Suzuki, R., Yamaguchi, T., Li, C.-L. and Klatzo, I. (1983) The effects of 5 min ischemia in Mongolian gerbils: II. Changes of spontaneous neuronal activity in cerebral cortex and CA1 sector of hippocampus. *Acta Neuropathol.*, 60: 217–222.

Szatkowski, M. and Attwell, D. (1994) Triggering and execution of neuronal death in brain ischaemia: Two phases of glutamate release by different mechanisms. *Trends in Neurosci.*, 17: 359–365.

Tanaka, K., Dora, E., Greenberg, J.H. and Reivich, M. (1986) Glucose metabolism during the recovery period after ischemia. Its relationship to NADH-fluorescence, blood flow, ECoG and histology. *Stroke*, 17: 994–1004.

Tang, L. and Aizenman, E. (1993) The modulation of *N*-methyl-D-aspartate receptors by redox and alkylating reagents in rat cortical neurones in vitro. *J. Physiol. Lond.*, 465: 303–323.

Tekkok, S. and Krnjevic, K. (1995) Long-term potentiation in hippocampal slices induced by temporary suppression of glycolysis. *J. Neurophysiol.*, 74: 2763–2766.

Tekkok, S. and Krnjevic, K. (1996) Calcium dependence of LTP induced by 2-deoxyglucose in CA1 neurons. *J. Neurophysiol.*, 76: 2343–2352.

Traynelis, S.F. and Cull-Candy, S.G. (1990) Proton inhibition of *N*-methyl-D-aspartate receptors in cerebellar neurons. *Nature*, 345: 347–350.

Tsubokawa, H., Oguro, K., Robinson, H.P.C., Masuzawa, T. and Kawai, N. (1995a) Single glutamate channels in CA1 pyramidal neurones after transient ischaemia. *Neuroreport*, 6: 527–531.

Tsubokawa, H., Oguro, K., Masuzawa, T. and Kawai, N. (1995b) Spontaneous excitatory postsynaptic currents in hippocampal CA1 pyramidal neurons of the gerbil after transient ischemia. *Neurosci Lett*, 191: 95–98.

Tymianski, M., Wallace, M.C., Spigelman, I., Uno, M., Carlen, P.L., Tator, C.H. and Charlton, M.P. (1993) Cell-permeant $Ca^{2+}$ chelators reduce early excitotoxic and ischemic neuronal injury in vitro and in vivo. *Neuron*, 11: 221–235.

Urban, L., Neill, K.H., Crain, B.J., Nadler, J.V. and Somjen, G.G. (1989) Postischemic synaptic physiology in area CA1 of the gerbil hippocampus studied in vitro. *J. Neurosci.*, 9: 3966–3975.

Urban, L., Neill, K.H., Crain, B.J., Nadler, J.V. and Somjen, G.G. (1990) Postischemic synaptic excitation and *N*-methyl-D-aspartate receptor activation in gerbils. *Stroke*, 21(suppl III): III-23–III-27.

Whittingham, T.S., Lust, W.D. and Passoneau, J.V. (1984) An in vitro model of ischemia: Metabolic and electrical alteration in the hippocampal slice. *J. Neurosci.*, 4: 793–802.

Wieloch, T., Lindvall, O., Blomqvist, P. and Gage, F.H. (1985) Evidence for amelioration of ischemic neuronal damage in the hippocampal formation by lesions of the perforant path. *Neurol. Res.*, 7: 24–26.

Xie, X., Berger, T.W. and Barrionuevo, G. (1992) Isolated NMDA receptor-mediated synaptic responses express both LTP and LTD. *J. Neurophysiol.*, 67: 1009–1013.

Zeevalk, G.D. and Nicklas, W.J. (1992) Evidence that the loss of the voltage-dependent $Mg^{2+}$ block at the *N*-methyl-D-aspartate receptor underlies receptor activation during inhibition of neuronal metabolism. *J. Neurochem.*, 59: 1211–1220.

Zhang, L. and Krnjevic, K. (1993) Whole-cell recording of anoxic effects on hippocampal neurons in slices. *J. Neurophysiol.*, 69: 118–127.

Zola-Morgan, S., Squire, L.R. and Amaral, D.G. (1986) Human amnesia and the medial temporal region: Enduring impairment following a bilateral lesion limited to field CA1 of the hippocampus. *J. Neurosci.*, 6: 2950–2967.

O.P. Ottersen, I.A. Langmoen and L. Gjerstad (Eds.)
*Progress in Brain Research*, Vol 116


CHAPTER 19

# Glutamate in the human brain: Possible roles in synaptic transmission and ischemia

Jon Berg-Johnsen, Tor S. Haugstad and Iver A. Langmoen*

*Institute for Surgical Research, Department of Neurosurgery and Department of Neurology, The National Hospital, Rikshospitalet, University of Oslo, Norway*

Glutamate was early known to be present in the central nervous system in high concentrations (Abderhalden and Weil, 1913). For some time, glutamate was used widely in treatment of epilepsy and mental retardation (Zimmerman et al., 1949; Milliken and Stranden, 1951), until the toxic action of glutamate on nerve cells was discovered (Lucas and Newhouse, 1957).

Three lines of evidence supporting a role of glutamate as a neurotoxic mediator have emerged from studies in experimental animals. First, a massive increase in the extracellular concentration of glutamate occurs during ischemia (Benveniste et al., 1984; Hagberg et al., 1985; Globus et al., 1988). This is closely related to the impairment of neuronal electrical activity (Shimada et al., 1989). The initial release, which is $Ca^{2+}$-dependent and presumably synaptic (Katayama et al., 1991) may contribute to the anoxic depolarisation (Berg-Johnsen et al., 1995) associated with a massive amino acid liberation, of which the bulk is non-synaptic (Kauppinen et al., 1988; Ikeda, 1989; Grøndahl et al., 1991) and mediated by reversal of glutamate transporters in the cell membrane (Attwell et al., 1993; Szatkowski and Attwell, 1994; for a review see Billups et al., 1998, this volume). Cellular swelling may contribute to amino acid release (Kimelberg et al., 1990), but osmotic and ischemic release mainly occur by separate mechanisms (Haugstad et al., 1995, 1996).

Second, glutamate is neurotoxic (Lucas and Newhouse, 1957; Olney, 1969; Olney et al., 1971; Lund Karlsen, 1978; Garthwaite and Garthwaite, 1986, 1989; Meldrum and Garthwaite, 1990) when present in sufficient concentration. This effect does at least in part depend on a large $Ca^{2+}$ (Fig. 1) and $Na^{+}$-influx across the cell membrane (Rothman, 1985; Choi, 1987; Siesjö and Bengtson, 1989; Reynolds, this volume), although with respect to calcium, release from intracellular stores seems to be equally important (Grøndahl and Langmoen, 1996; Grøndahl et al., 1997).

Third, neuronal death may be reduced by severing glutamatergic afferents (Wieloch et al., 1985; Onodera et al., 1986), and by treatment with antagonists against different glutamate receptor subtypes (Simon et al., 1984; Meldrum, 1985; Gill et al., 1988; Swan and Meldrum, 1990; Sheardown et al., 1990; Nellgard and Wieloch, 1992).

## Glutamate in the human brain

It is now generally recognised that glutamate is the major excitatory neurotransmitter in the brain of

*Corresponding author.
Department of Neurosurgery, Karolinska Hospital, 17176 Stockholm, Sweden.

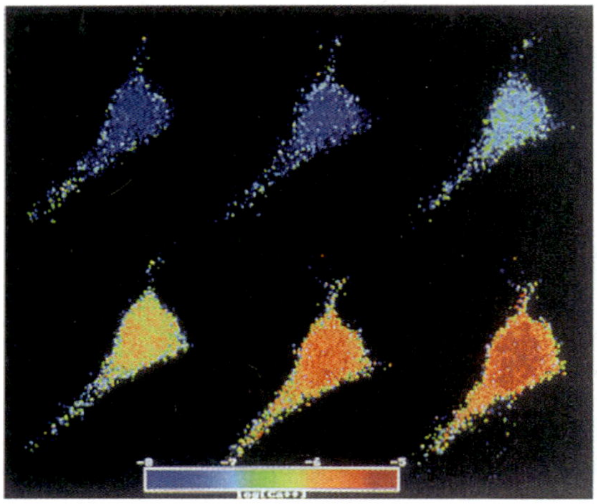

Fig. 1. A pyramidal cell in the CA 1 region of a rat hippocampal slice was loaded with fluo-3/fura red mixture. The neuron was then visualised using confocal laser scanning microscopy, and the ratio of the fluorescence from each probe was used to quantify intracellular calcium concentrations before (image 1 and 2) and during 6, 7, 8 and 9 minutes of ischemia. Intracellular calcium increased from 60 nM to about 30 $\mu$M, i.e. 500 fold (From Grøndahl and Langmoen, 1997 with permission).

lower mammals since it fulfils the criteria for transmitter identification (Storm-Mathisen, 1977; Fonnum, 1984): (1) it is present in the presynaptic terminal, (2) it is released from presynaptic terminals in a calcium dependent manner on depolarisation of the cell membrane, (3) it mimics the action of the naturally occurring transmitter, and (4) its action is rapidly eliminated by reuptake. However, since there are major anatomical, cytoarchitectural, and functional differences between the brain of humans and the brain of laboratory rodents, it is desirable to confirm at least some of the findings from lower mammals by studying human tissue.

*Transmitter identification*

*Presence*

The concentration of glutamate in human cerebral cortex is in a similar order of magnitude as in other mammals, i.e. $\approx$10 mM (Tower, 1955; Van Gelder et al., 1972; Perry and Hansen, 1981; Perry, 1983). The concentration of glutamate in the CSF, which reflects the concentration in the extracellular space, is about 1 $\mu$M (Gjessing et al., 1972; Johnson, 1978). The development of antibodies to amino acids (Storm-Mathisen et al., 1983; Ottersen and Storm-Mathisen, 1984) has provided a mean to analyse the intracellular localisation of amino acids (Storm-Mathisen et al., 1992). Immunhistochemical studies performed in tissue removed from humans during surgery for epilepsy or brain tumors have revealed that glutamate is present in highest concentrations in excitatory terminals (asymmetric synapses on dendritic spines) (Aas et al., 1993).

*Release*

Liberation of glutamate, aspartate and GABA from human neocortex has been investigated in slices prepared from tissue removed during anterior temporal lobectomies for tumor or epilepsy. Potassium induced depolarisation causes a large, mainly calcium-dependent release of glutamate and GABA (Fig. 2) (Haugstad et al., 1992; Hegstad et al., 1992). Immunhistochemical analysis further demonstrate depletion of glutamate from excitatory terminals in the presence, but not in the absence of calcium (Hegstad, 1995).

*Action*

Investigations utilising immunhistochemical methods (Breese and Leonard, 1993; Böckers et al., 1994). binding studies (Quarum et al., 1990; Steele et al., 1990; Dewar et al., 1991; Fukunaga et al., 1991; Steele et al., 1991), and in situ hybridisation and functional expression (Böckers et al., 1994; Garcia-Ladona et al., 1994; Sun et al., 1994) have confirmed that both ionotropic and metabotropic glutamate receptors are expressed in the human brain, with regional differences between the cortex, hippocampus and cerebellum, as well as between different layers and cell types. Fukunaga et al. (1991) have characterised three subtypes of ionotropic glutamate receptors in human cortex. The distribution of NMDA NR1 mRNA and immunoreactivity in human frontal cortex, hippocampus and cerebellum is largely consistent with the expression in the rat brain (Böckers et al., 1994). AMPA/kainate receptor subunits 1 and 2 cloned from human brain show extensive homology with their rat counterparts (GluR1 and GluR2). Coexpression of the two human subunits in *Xenopus* oocytes evokes kainate dependent currents (Sun et al., 1994). High densities of metabotropic glutamate receptors and AMPA receptors are found in the hippocampal CA 1 region (Dewar et al., 1991). Immunhistochemical investigations of the distribution of metabotropic receptor subtypes in the human hippocampus, have revealed regional differences in expression (Blumcke et al., 1996). Thus, while mGluR2/3 is present in all principal neurons in the dentate gyrus as well as in the CA 4, CA 3 and CA 2 regions, pyramidal cells of the CA 1 region exhibited only weak labelling for class II mGluR.

Excitatory postsynaptic potentials evoked in neurones of cortical layers II-IV in human brain are attenuated by both AMPA (CNQX) and NMDA (CPP) receptor antagonists (Avoli and Olivier, 1987; Hwa and Avoli, 1992). Thus, both types of glutamate receptors seem to participate in synaptic transmission.

*Uptake*

Analysis of [$^3$H]L-glutamate uptake in human brain slices has revealed two distinct uptake systems with $K_m$ values of $7.64 \times 10^{-6}$ and $1.38 \times 10^{-3}$ M, and $V_{max}$ of $1.04 \times 10^{-7}$ and $1.22 \times 10^{-5}$ mol $g^{-1}$ $min^{-1}$ (Hegstad, 1995), indicating that uptake is mediated by two kinetically distinct systems of transporters. The $K_m$ values were almost identical to the values found for the uptake of L-glutamate by synaptosomes prepared from frozen post-mortem human brain (Dodd et al., 1989), whereas the $V_{max}$ values were considerably higher. Thus, although tissue disintegration reduces the number of functioning transporters in post

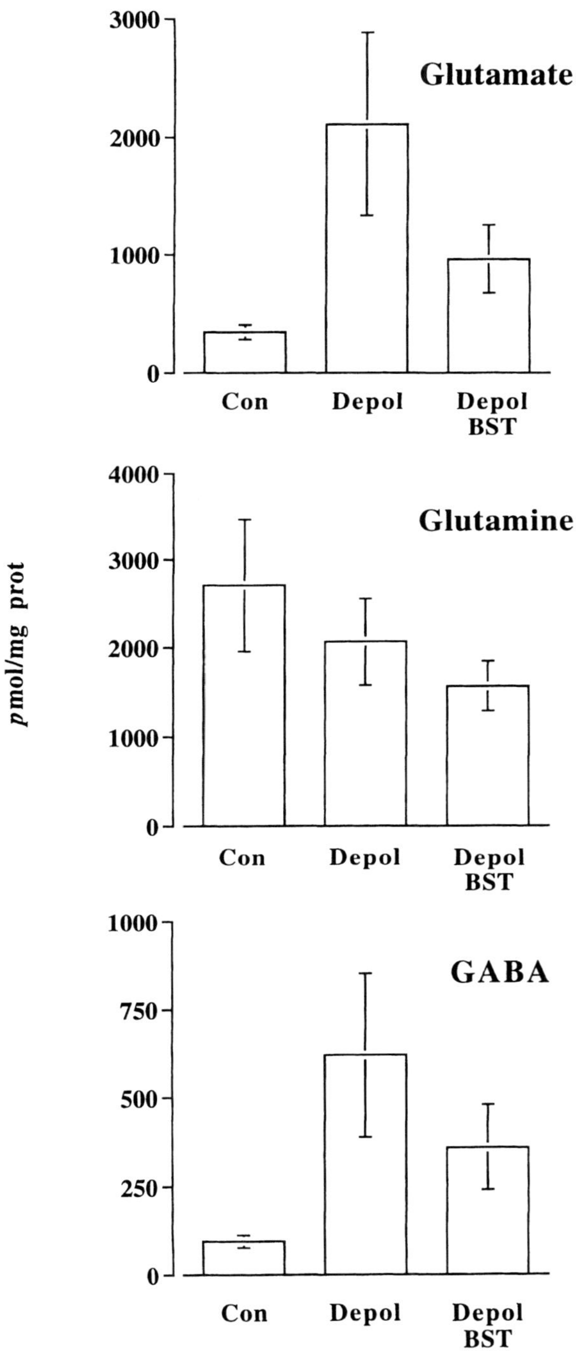


Fig. 2. Human neocortical slices were incubated for 1 minute in artificial cerebrospinal fluid of different compostions: control (Con), 50 mM $K^+$ (Depol) and 50 mM $K^+$, 0 mM $Ca^{2+}$ and 4 mM $Mg^{2+}$ (blocked synaptic transmission, BST). Release of amino acids to the medium was measured by high performance liquid chromatography, and expressed as average of 8 experiments ± SEM. There was a significant calcium dependent release of glutamate and GABA (both $p < 0.05$).

mortem brain, the affinity of preserved transporters to L-glutamate is unchanged. Uptake was reduced by lowering external sodium concentration confirming the $Na^+$-dependency, and by reducing the temperature in the incubation chamber.

The excitatory amino acid transporters EAAT1 (GLAST), EAAT2 (GLT), and EAAT3 (EAAC) have been identified in the human brain by molecular cloning and functional expression (Arriza et al., 1994; see also Danbolt et al., this volume), and electrophoresis and immunoblotting using antibodies against cloned glutamate transporters (Hegstad, 1995). The human transporters appear to have slightly elevated mean molecular masses compared to their rat hippocampal counterparts.

### *The relation of glutamate to ischemia in human brain*

In their pioneering studies of microdialysis in the human brain Hillered and Persson placed microdialysis probes in tumor-free cortical tissue in the frontal lobe of patients undergoing lobectomy as a treatment for brain tumors. Dialysis samples were collected in 10 minutes fractions before and during frontal lobe resection, thus serving as a model of simulated ischemia (Hillered et al., 1990). During the resection a marked release of energy related metabolites and amino acids was observed. They suggested that the use of microdialysis in humans might serve as a method in studies of clinical ischemia and as a clinical method in patient monitoring. Kanthan et al. (1995) used the same method to measure amino acid release in temporal lobes being excised due to epilepsy. During the surgical excision the resected brain undergoes acute ischemia developing from incomplete to complete as the resection proceeds. Extracellular glutamate increased 20-fold during incomplete ischemia, and more than 100-fold when ischemia was complete. Similar trends were also observed for other amino acids.

Hamberger et al. (1995) sampled extracellular fluid from the gyrus rectus and subarachnoid fluid in patients undergoing surgery for intracranial aneurysms. Markedly elevated concentrations of glutamate were observed in only one patient who underwent surgery within 8 hours after the haemorrhage. The total amino acid concentration, however, was inversely correlated to the level of consciousness in the post-operative phase. In a later study they observed specific changes in the concentration of amino acids related to clinical events such as increased intracranial pressure, vasospasms and infarction (Runnerstam et al., 1997). Since then several investigations have shown that glutamate may increase to very high levels during the days following the acute bleed, in the phase where vasospasm, infarction, hydrocephalus and increased intracranial pressure often occur. Further, it has been shown that the concentration of glutamate, and other amino acids in the dialysate correlates well to clinical course, level of consciousness and clinical outcome (Nilsson et al., 1996; Persson et al., 1996; Säveland et al., 1996).

Enblad et al. (1996) studied regional cerebral blood flow and oxygen extraction ratio with simultaneous positron emission tomography (PET). The presence of whole-brain ischemia and/or regional ischemia within the region of the microdialysis probe was associated with increased levels of excitatory amino acids. Increased levels were only occasionally seen when PET did not indicate signs of ischemia or when signs of regional ischemia were found remote from the probe region.

Bullock et al. (1995) inserted a microdialysis probe into the cortex of a man in whom an occlusive stroke with brain stem compression necessitated a partial temporal lobectomy and observed more than 300 times normal glutamate concentration 6 days after infarction. Castillo et al. (1996) have systematically studied CSF samples obtained from stroke patients at admission within 24 hours of symptom start. Cerebrospinal fluid glutamate concentration was 3.46 $\mu$M in control subjects and 6.55 $\mu$M in the patient population. The concentration of glutamate both in plasma and in cerebrospinal fluid was significantly higher in patients with large infarcts, cortical infarcts and with a higher degree of neurological deficits. The presence of glutamate increase in the cerebrospinal fluid cannot be documented for greater than

6 hours in stable ischemic stroke. In contrast, glutamate concentrations measured at any time interval during the first 24 hours from onset in patients with progressing ischemic stroke are significantly higher than among patients with stable stroke and in control groups (Davalos et al., 1997). In stroke patients there is a linear relationship between plasma and cerebrospinal fluid concentrations of glutamate, and glutamate concentrations above 200 $\mu$M in plasma and more than 8.2 $\mu$M in the cerebrospinal fluid are independently and significantly associated with progressive neurological deficits (Castillo et al., 1997).

Investigation of asphyxiated infants to some degree parallels the above mentioned stroke studies as cerebrospinal fluid aspartate level is significantly higher in children with severe compared with mild hypoxic-ischemic encephalopathy. Glutamate is also higher in infants with severe clinical symptoms (Hagberg et al., 1993).

Mendelowitsch et al. (1997) measured pH, lactate and amino acid concentrations in the extracellular fluid by microdialysis in a patient undergoing extra- to intracranial bypass surgery due to chronic cerebral ischemia (sickle cell disease and a moyamoya disease like pattern on angiography). The resting glutamate concentration was relatively high, and showed a short-lived increase during the bypass procedure. Following revascularisation, there was a rapid decrease in the glutamate levels and the neurological deficit improved post-operatively.

Owens et al. (1997) obtained multiple single photon emission tomographic images in five patients with intracerebral haemorrhage 120 minutes after administration of 150 MBq of (+)-3-[123I]Iodo-MK-801. In all patients with cerebral hemorrhages, the initial uptake of (+)-3-[123I]Iodo-MK-801 into the ipsilateral hemisphere was markedly reduced, consistent with a reduced level of cerebral blood flow. In two of five patients, relatively increased tracer retention at later time points (60–120 minutes after tracer administration) could be seen in cortical areas adjacent to the site of the haemorrhage, consistent with activation of NMDA receptors in ischemic brain tissue.

We have investigated amino acid release and glutamate redistribution in human brain slices obtained from tissue removed during temporal lobectomies in epileptic patients. Potassium induced membrane depolarisation causes a release of glutamate, aspartate, GABA and glycine, but not glutamine or leucine. The release of glutamate and GABA is mainly $Ca^{2+}$-dependent. Exposure of the brain tissue to simulated ischemia by combined glucose/oxygen deprivation provokes a $Ca^{2+}$-independent liberation of glutamate, aspartate, GABA, glycine, and taurine which starts after 8 minutes, peaks at the end or shortly after the period of simulated ischemia, and returns to control levels within 11 minutes following its termination (Fig. 3). Preloaded D-[$^3$H]aspartate was released both during potassium stimulation and energy deprivation (Hegstad et al., 1996). Another set of slices was studied by a postembedding immunocytochemical method with polyclonal antibodies raised against glutamate and glutamine (Aas et al., 1993). Combined oxygen and glucose deprivation caused reduced immunolabeling in neuronal somata, while that of glial processes was increased. There was no change in the level of glutamate immunolabeling in axon terminals, consistent with biochemical data indicating that exocytotic glutamate release is energy dependent and therefore inhibited during the larger part of the period of energy deprivation.

## Clinical use of glutamate antagonists in stroke

As noted above, drugs that antagonise glutamatergic neurotransmisson have shown promising effects in preclinical stroke models. Due to intolerable side effects more than lack of efficacy, glutamate antagonists have not been licensed for clinical use so far (for review see Muir and Lees, 1995). New drugs are, however, rapidly developed, and some of these have pharmacological and toxicological profiles that are distinctly different from the older compounds. For a list of glutamate receptor antagonists in different stages of clinical development, see Meldrum (this volume).

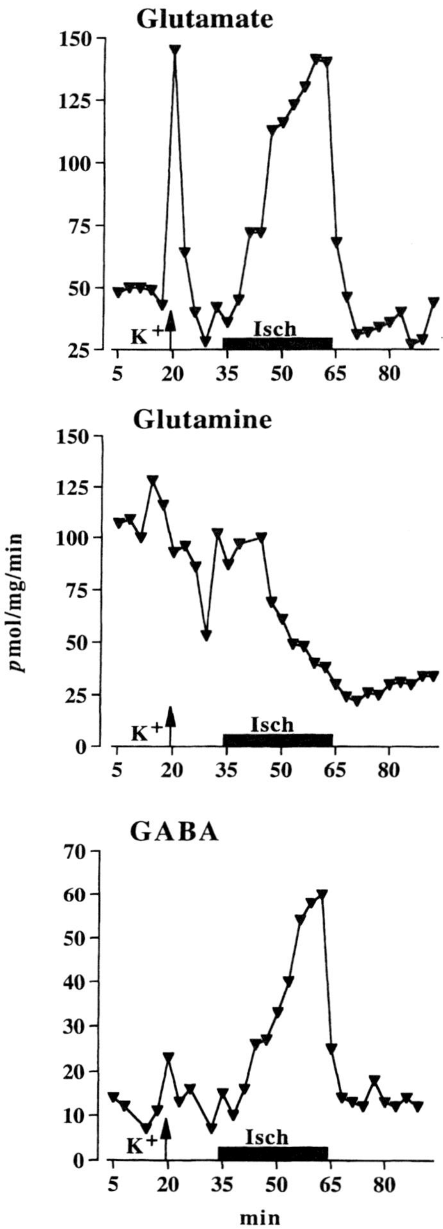


Fig. 3. Slices from human brain were incubated in normal medium for 20 minutes. 50 mM $K^+$ was then added (at the expense of $Na^+$) for 1 minute. After recovery, the slices were exposed for 30 minutes to medium without glucose and oxygen, followed by recovery (30 minutes). The graphs show the release of glutamate, glutamine and GABA sampled from the medium from one experiment.

Cis-4-(phosphonomethyl)-2-piperidine-carboxylic acid (CGS-19755, selfotel), is a potent competitive NMDA receptor antagonist (Lehmann et al., 1988). In the rat middle cerebral artery occlusion model it causes a substantial decrease in infarct size, when administered 5 minutes prior to or 5 minutes following occlusion (Simon and Shiraishi, 1990). The effect is associated with reduction of postischemic cerebral glucose hypermetabolism. A 10 mg/kg bolus dose given immediately after occlusion increases cortical blood flow both in the occluded and in the non-occluded hemisphere, and reduces cortical but not basal ganglia infarct volume (Takizawa et al., 1991)

CGS-19755 is less effective when given 30 minutes after 4-vessel occlusion in rats, as it results in little histological protection of the hippocampus, although it decreases calcium influx into ischemic neurones as evidenced by reduced calcium-calmodulin binding, and significantly improves learning ability when tested after 1 month (Grotta et al., 1990). In the gerbil ischemia model, however, it is effective in reducing ischemia-induced hippocampal brain damage as well as hypermotility (Boast et al., 1988). Postischemic treatment in the latter model also blocks ischemia-induced decreases in brain adenosine and NMDA receptors (Jarvis et al., 1988)

In a phase IIa randomised trial in patients treated within 12 hours of hemispheric ischemic stroke, Grotta et al. (1995) found that CGS 19755 in a single intravenous dose of 1.5 mg/kg CGS 19755 was safe and tolerable in these patients. Adverse experiences (agitation, hallucinations, confusion, paranoia, and delirium) occurred at higher doses from 20 minutes to 22 hours after treatment and lasted 2 to 60 hours. Mortality was 1 of 8 in patients receiving placebo and 3 of 24 in treated patients.

Two randomised, double-blind, placebo-controlled trials testing the effect of 1.5 mg/kg selfotel administered intravenously within 6 hours of stroke onset on functional outcome at 3 months was started in 1994. By December 1995, 509 patients had been enrolled. The mortality rate was 24% in the selfotel group and 19% in the control group.

Although this difference was not significant, review of individual case fatalities by the Data and Safety Monitoring Board raised concern over an imbalance against selfotel in the number of deaths related to brain insults. The studies were therefore terminated (Davis et al., 1997).

The high affinity non-competitive NMDA receptor antagonist N-(1-naphthyl)-N′-(3-ethylphenyl)-

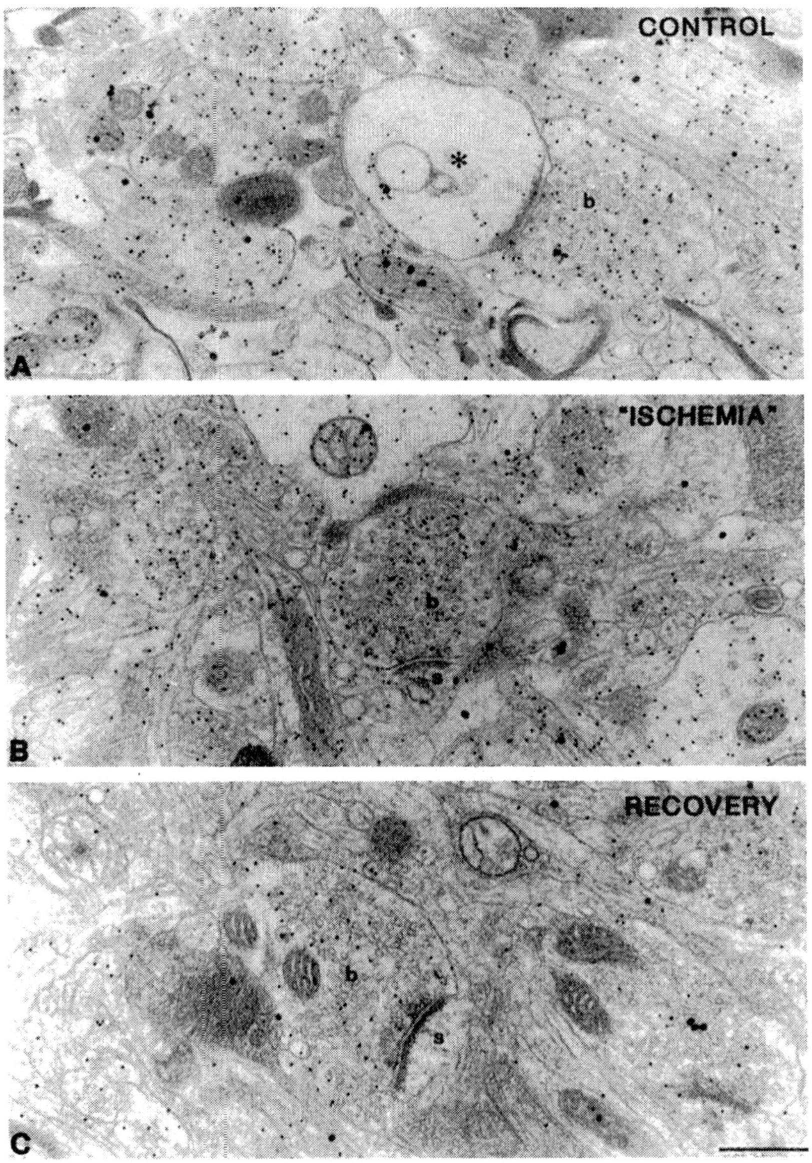


Fig. 4. Electron micrographs showing the effect of 30 minutes of combined hypoxia and glucose deprivation on the levels of glutamate- and glutamine-like immunoreactivity in axon terminals in human neocortex. Synaptic profiles from control (A), "ischemic" (B), and recovery (C) slices are shown. b, boutons; s, spines; asterisk, dendrite. Small gold particles signal glutamate-like immunoreactivity, whereas large gold particles signal glutamine-like immunoreactivity. Except for a slight reduction in the areal density of large gold particles in B, no obvious differences are seen (From Aas et al., 1993 with permission).

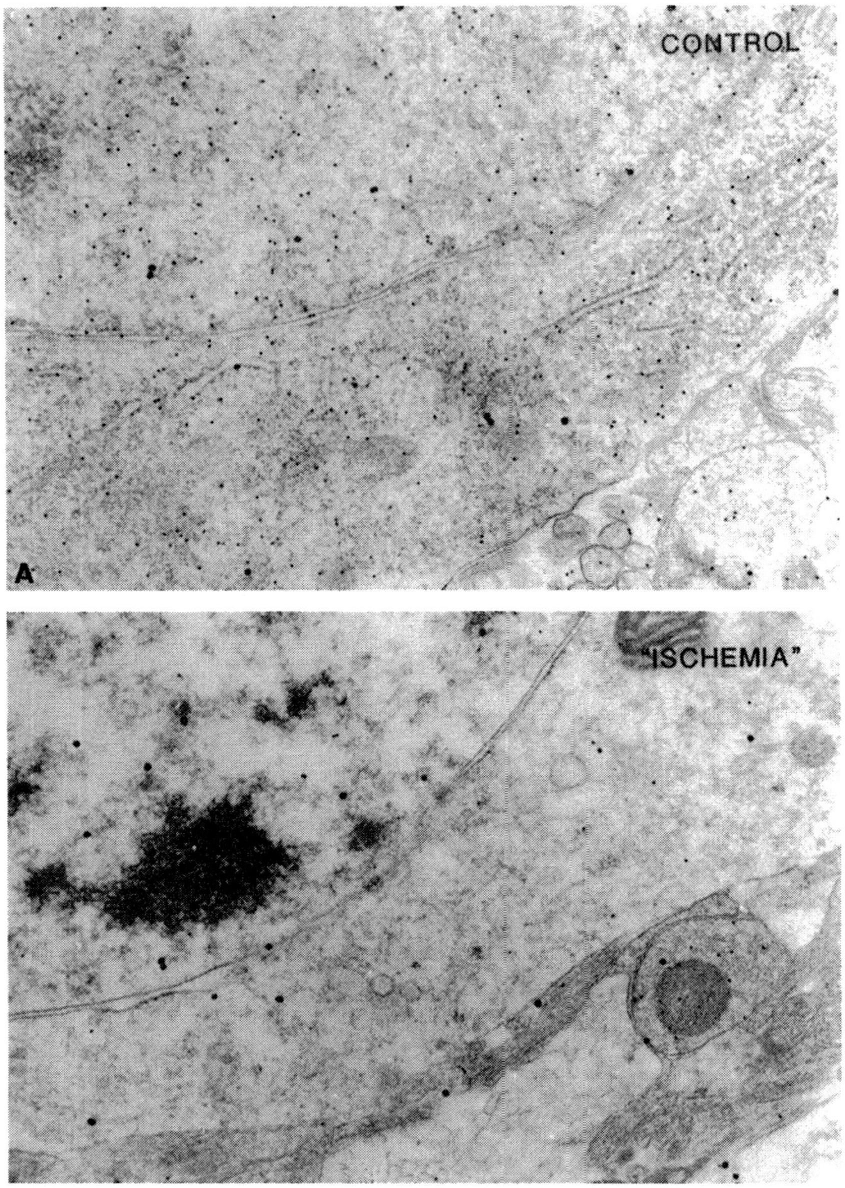


Fig. 5. Electron micrographs showing the effect of 30 minutes of combined hypoxia and glucose deprivation on the levels of glutamate- and glutamine-like immunoreactivity in neuronal somata in human neocortex. Note the marked reduction in the areal density of small gold particles signalling glutamate-like immunoreactivity in the cytoplasmic matrix of the "ischemic" neuron (B) compared to control (A). No significant difference in the areal density of large gold particles signalling glutamine-like immunoreactivity is seen. Also note the condensation of nuclear chromatin and disruption of polyribosomes in the ischemic neuron (B). (From Aas et al., 1993 with permission).

N′-methylguanidinehydrochloride (CNS1 102, aptiganel hydrochloride, cerestat), is neuroprotective in preclinical models of stroke when administered as pretreatment or up to 60 minutes postischemia. In rats undergoing middle cerebral artery occlusion there is a less significant shift to lower frequency EEG activity over the affected hemisphere, and a considerable reduction of post mortem infarct volume at 24 hours of reperfusion (Cohen et al., 1994; Minematsu et al., 1993). At 30 minutes of MCA occlusion the ischemic lesion as quantified by diffusion magnetic resonance imaging is reduced from 50.0 to 10.5% of the hemisphere, and there is additional reduction of the lesion with reperfusion after 3 hours (Minematsu et al., 1993).

In spontaneously hypertensive rats undergoing tandem occlusion of the common carotid/middle cerebral artery (5 to 150 minutes), CNS-1102 extends duration of ischemia before reperfusion producing half-maximal infarct size, but not the maximal infarct volume (Aronowski et al., 1994). It further reduces behavioural dysfunction as measured by different behavioural tests (forearm flex, tape (somatosensory neutralisation), and foot-fault placing) (Aronowski et al., 1996).

In the first administration of CNS 1102 to man Muir et al. (1994) found that the maximum tolerated dose for conscious individuals was 30 $\mu$g/kg given intravenously over 15 minutes. Clinically significant sedation, and increased mean arterial pressure and pulse rate were seen at this and higher doses, whereas sedation and central nervous excitation became unacceptable for conscious individuals at doses of 45 $\mu$g/kg and above. When administered over 4 hours it is well tolerated by healthy males up to total doses of 32 $\mu$g/kg, whereas transient symptoms and signs of peripheral paresthesias, light-headedness, and euphoria are seen with increasing doses and progress to psychomotor retardation, paranoia and catatonia as total administered dose increases. There is a dose-dependent elevation of blood pressure, accompanied by clinical evidence of vasoconstriction. Global cerebral blood flow is maintained, whilst middle cerebral artery flow velocity increases. The plasma half-life is approximately 4.5 hours (Muir et al., 1995; Grosset et al., 1995; Muir et al., 1997).

CNS 1102 is currently in clinical trials both in USA and Europe. A phase III trial for traumatic brain injury was recently stopped due to insufficient evidence of positive clinical impact. In depth analysis is pending, as are the results of a phase III stroke study.

The newly developed halogenated quinoxalinedione derivative ACEA-1021 (5-nitro-6,7-dichloro-quinoxalinedione; licostinel) is a pharmacologic antagonist of the glycine recognition site on the NMDA receptor complex. In contrast to ketamine, ACEA-1021 does not affect the stimulation threshold of spreading depression, but reduces the propagation rate in a dose-dependent fashion (Martin et al., 1994). It reduces CA 1 neuronal damage in hippocampal slice cultures deprived of oxygen and glucose in a dose dependent manner, an effect that is reversed by the addition of glycine (Newell et al., 1995).

ACEA-1021 reduces both cerebral infarct volumes and the incidence of hemiparesis resulting from 90 minutes of middle cerebral artery occlusion in rats (Warner et al., 1995), and further, significantly attenuates involuntary jerking movements (myoclonus) in post-cardiac-arrested rats (Matsumoto et al., 1995). However, it does not effect delayed neuronal necrosis in the hippocampal CA 1 sector, neocortex, or caudoputamen in forebrain ischemia (Warner et al., 1995). Neuroprotection achieved by ACEA-1021 persists when brain temperature is controlled, indicating a potent mechanism of action other than attenuating a hyperthermic response to ischemia (Takaoka et al., 1997).

Although ACEA-1021 causes a reversible alteration in a small proportion of cingulate and retrosplenial cortical neurones it does not cause neuronal necrosis (in contrast to for instance MK-801) and seems to be devoid of lethal neurotoxicity (Auer, 1997). It thus appears that the therapeutic profile is similar, but the toxicologic structural profile is different, from some NMDA receptor antagonists.

The drug is currently in the phase I developmental stage. Clinical studies have shown that it is well tolerated in doses 3–4 times higher than the minimum dose required for efficacy in an animal models of stroke, and considerably below doses that indicate the potential for kidney toxicity. There has not been evidence of the dose-limiting side effects seen with some other experimental stroke drugs, such as hallucinations, delirium, agitation, sedation, and cardiovascular effects (F. Weihmüller, personal communication).

The non-psychotropic cannabinoid (+)-(3S,4S)-7-hydroxy-$\delta$ 6-tetrahydrocannabinol 1,1-dimethylheptyl (HU-211, dexanabinol) is a stereoselective inhibitor of the NMDA receptor and a potent scavenger of peroxy- and hydroxy-radicals in vitro (for review see Biegon and Joseph 1995). Since it combines NMDA receptor antagonistic activity and free radical scavenging abilities in a single molecule it has been suggested that it holds a unique position among putative neuroprotectant agents (Eshhar et al., 1995; Biegon and Joseph, 1995).

In cell culture HU-211 (50 $\mu$M) protects cells against the excitotoxic effect of NMDA and glutamate (Nadler et al., 1993), and significantly attenuates neuronal injury produced by oxygen/glucose deprivation (5 $\mu$M), glial and neuronal cell damage induced by sodium nitroprusside (10 $\mu$M), and reduced protein oxidation initiated by $\gamma$ irradiation (Eshhar et al., 1995). It further protects cultured neurons from toxicity of radical generators.

In rats subjected to 90 minutes of temporary MCA occlusion HU-211 (4 mg/kg i.v. administered 70 minutes after onset of occlusion) significantly improves neurological functioning and reduces both infarct volume and brain swelling (Belayev et al., 1995c).

HU-211 is also effective in models of global ischemia. Thus, rats treated with HU-211 following 20 minutes of 4-vessel occlusion show significant improvement of neurological deficits when compared to vehicle treated controls, and also better preservation of neurones in the hippocampal CA 1 field (Belayev et al., 1995a). The drug was equally effective when given 30 and 60 minutes after ischemia, but neuroprotection was no longer significant after 3 hours. It further conveys significant neuroprotection of the CA 1 subfield of the hippocampus in Mongolian gerbils following 10 minutes bilateral common carotid artery occlusion with a similar therapeutic window (Bar-Joseph et al., 1994). Dose-response studies demonstrate an inverted U curve in which 4 mg/kg induces the best neuroprotection (but see also Vered et al., 1994). It is also effective in protecting against the effects of focal ischemia-induced blood brain-barrier disruption (Belayev et al., 1995b) and reduces $^{45}Ca^{2+}$ accumulation in the weight drop model of closed head injury (Nadler et al., 1995).

A phase I study of HU-211 has shown lack of adverse reactions, and a phase II clinical study has been started in Israel under the auspices of the American Brain Injury Consortium.

Eliprodil (SL 82.0715) is a noncompetitive NMDA receptor antagonist acting at the polyamine modulatory site. The drug has selectivity for the NR1A/NR2B receptor subtype (Avenet et al., 1997). In rat cultured cortical neurones it blocks the inward $Ba^{2+}$ current carried in part by L- and N-type $Ca^{2+}$ channels (Biton et al., 1994) and it also inhibits $Ca^{2+}$ currents in recombinantly expressed human N- and P-type $Ca^{2+}$ channels (Bath et al., 1996).

In a mouse model of focal cerebral ischemia (coagulation of the middle cerebral artery) eliprodil affords substantial protection against cortical infarction when given after the induction of cerebral ischemia (Poignet et al., 1992). Eliprodil provides significant protective effects against ischemia-induced neurodegeneration in the CA 1 region of the hippocampus in gerbils subjected to 5 minutes bilateral carotid artery occlusion (given immediately after occlusion and at 3 and 6 hours post-occlusion) (Bath et al., 1996).

In the rat fluid-percussion brain injury model, eliprodil causes a considerable reduction of the volume of cortical damage, a neuroprotective effect also obtained when the first administration is delayed by up to 12 hours after the brain insult (Toulmond et al., 1993)

The possible effects on memory, psychomotor performance and mood of eliprodil was assessed in a randomised, double-blind, cross-over, placebo-controlled study involving 11 healthy young male volunteers. A single dose of 30 mg was free of any detrimental effect on memory and skilled performance and did not produce either sedation, excitation or psychotomimetic effects in comparison with placebo (Patat et al., 1994). The favourable side effect profile may at least partially be explained by the selectivity of eliprodil for the NR1A/NR2B receptor subtype. The neuroprotective effect may be due to both its NMDA and calcium antagonising properties.

Eliprodil is now evaluated in clinical stroke studies.

[(E)-3[(phenylcarbamoil)ethenyl]-4,6-dichloroindole-2-caboxylic acid sodium salt (GV 150526A) is a recently discovered antagonist at the glycine site of the NMDA receptor (DiFabio et al., 1997 (cited in Bordi et al., 1997)). Postischemia administration of GV150526 up to 6 hours following MCA occlusion in rats (3 mg/kg i.v.) and mice (3 or 10 mg/kg p.o.) results in a significant reduction of the infarct volume measured histologically and improves the functional correlates of the ischemia as evaluated by measurements of the sensory evoked potential (Bordi et al., 1997; Reggiani et al., 1997)

GV150526 has been tested in phase I clinical studies and was recently studied in a double-blind, randomised, placebo-controlled investigation in 25 acute stroke patients. Patients received a loading dose of 800 mg followed by 100, 200 or 400 mg given 12-hourly for five doses, or placebo (Hoke et al., 1997). Three patients experienced serious adverse effects, one of which was fatal. In another study no side effects were observed (Dyker et al., 1997). Patient enrolment in two international multicenter studies for stroke is starting early 1998.

**Perspectives for the future**

Dizocilipine (MK-801) was the first glutamate antagonist that consistently provided neuroprotection in animal models, although some of its effect probably was due to lowering of brain temperature. Clinical testing of this drug was, however, abandoned due to histological changes observed in rat brains. Several compounds acting at glutamate receptors have revealed major side effects when tested in humans. The therapeutic index, i.e. neuroprotective to toxic dose-ratio, has therefore been a major concern. Due to a sharp dose-response curve the difference between the dose giving adequate neuroprotection and doses imposing unacceptable side effects is (too) small for some of these drugs. On the other hand, new antagonists have steadily been discovered and developed ever since the pivotal role of excitatory amino acids in cerebral ischemia was discovered, and some of these have toxicological profiles that are different from the older compounds.

A problem with antagonists at the glycine site of the NMDA receptor is that glycine is present in the concentrations that tend to saturate this site, not least during ischemia when glycine is released to the extracellular space. In contrast to this, open channel blockers would tend to concentrate in ischemic regions due to massive activation of glutamate gated channels.

A promising aspect is the development of drugs that have a combined action at two or more of the processes leading to neuronal injury during ischemia. Thus, as stated above, new antagonists that have a conjoined effect on NMDA receptors and either voltage gated calcium channels or free oxygen radical synthesis are already undergoing clinical evaluation. Drugs with heterogeneous action may turn out to be particularly important as excitotoxicity is only one of several causes of neuronal death consequent to ischemia.

## References


Aas, J.-E., Berg-Johnsen, J., Hegstad, E., Laake, J.H., Langmoen, I.A. and Ottersen O.P. (1993) Redistribution of glutamate and glutamine in slices of human neocortex exposed to combined hypoxia and glucose deprivation in vitro. *J. Cereb. Blood Flow Metab.*, 13: 503–515.

Abderhalden, E. and Weil, A. (1913) Vergleichende Untersuchungen über den Gehalt der verschiedenen Bestandteile des Nervensystems an Aminosäuren. II Mitteilung. Die Aminosäuren der grauen und weissen Substanz des Gehirns. *Zeitschr. Physiol. Chem.*, 83: 425–444.

Aronowski, J., Ostrow, P., Samways, E., Strong, R., Zivin, J.A. and Grotta, J.C. (1994) Graded bioassay for demonstration of brain rescue from experimental acute ischemia in rats. *Stroke*, 25: 2235–2240.

Aronowski, J., Samways, E., Strong, R., Rhoades, H.M. and Grotta, J.C. (1996) An alternative method for the quantitation of neuronal damage after experimental middle cerebral artery occlusion in rats: Analysis of behavioral deficit. *J. Cereb. Blood Flow Metab.*, 16: 705–713.

Arriza, J.L., Fairman, W.A., Wadiche, J.I., Murdoch, G.H., Kavanaugh, M.P. and Amara, S.G. (1994) Functional comparisons of three glutamate transporter subtypes cloned from human motor cortex. *J. Neurosci.*, 14: 5559–5569.

Attwell, D., Barbour, B. and Szatkowski, M. (1993) Nonvesicular release of neurotransmitter. *Neuron*, 11: 401–407.

Auer, R.N. (1997) Structural neurotoxicologic investigation of the glycine antagonist 5-nitro-6,7-dichloroquinoxalinedione (ACEA-1021). *Neurotoxicology*, 18: 53–62.

Avenet, P., Leonardon, J., Besnard, F., Graham, D., Depoortere, H. and Scatton, B. (1997) Antagonist properties of eliprodil and other NMDA receptor antagonists at rat NR1A/NR2A and NR1A/NR2B receptors expressed in Xenopus oocytes. *Neurosci. Lett.*, 223: 133–136.

Avoli, M. and Olivier, A. (1987) Bursting in human epileptogenic neocortex is depressed by an *N*-methyl-D-aspartate antagonist. *Neurosci. Lett.*, 76: 249–254.

Bar-Joseph, A., Berkovitch, Y., Adamchik, J. and Biegon, A. (1994) Neuroprotective activity of HU-211, a novel NMDA antagonist, in global ischemia in gerbils. *Mol. Chem. Neuropathol.*, 23: 125–135.

Bath, C.P., Farrell, L.N., Gilmore, J., Ward, M.A., Hicks, C.A., O'Neill, M.J. and Bleakman, D. (1996) The effects of ifenprodil and eliprodil on voltage-dependent $Ca^{2+}$ channels and in gerbil global cerebral ischaemia. *Eur. J. Pharmacol.*, 299: 103–112.

Belayev, L., Bar-Joseph, A., Adamchik, J. and Biegon, A. (1995a) HU-211, a nonpsychotropic cannabinoid, improves neurological signs and reduces brain damage after severe forebrain ischemia in rats. *Mol. Chem. Neuropathol.*, 25: 19–33.

Belayev, L., Busto, R., Watson, B.D. and Ginsberg, M.D. (1995b) Post-ischemic administration of HU-211, a novel non-competitive NMDA antagonist, protects against blood-brain barrier disruption in photochemical cortical infarction in rats: A quantitative study. *Brain Res.*, 702: 266–270.

Belayev, L., Busto, R., Zhao, W. and Ginsberg, M.D. (1995c) HU-211, a novel noncompetitive *N*-methyl-D-aspartate antagonist, improves neurological deficit and reduces infarct volume after reversible focal cerebral ischemia in the rat. *Stroke*, 26: 2313–9; discussion 2319–2320.

Benveniste, H., Drejer, J., Schousboe, A. and Diemer, N.H. (1984) Elevation of the extracellular concentrations of glutamate and aspartate in rat hippocampus during transient cerebral ischemia monitored by intracerebral microdialysis. *J Neurochem.*, 43: 1369–1374.

Berg-Johnsen, J., Grøndahl, T.Ø., Langmoen, I.A., Haugstad, T.S. and Hegstad, E. (1995) Changes in amino acid release and membrane potential during cerebral hypoxia and glucose deprivation. *Neurol. Res.*, 17: 201–208.

Biegon, A. and Joseph, A.B. (1995) Development of HU-211 as a neuroprotectant for ischemic brain damage. *Neurol Res.*, 17: 275–280.

Biton, B., Granger, P., Carreau, A., Depoortere, H., Scattonm B. and Avenet, P. (1994) The NMDA receptor antagonist eliprodil (SL 82.0715) blocks voltage-operated $Ca^{2+}$ channels in rat cultured cortical neurons. *Eur. J. Pharmacol.*, 257: 297–301.

Blumcke, I., Behle, K., Malitschek, B., Kuhn, R., Knopfel, T., Wolf, H.K. and Wiestler, O.D. (1996) Immunohistochemical distribution of metabotropic glutamate receptor subtypes mGluR1b, mGluR2/3, mGluR4a and mGluR5 in human hippocampus. *Brain. Res.*, 736: 217–226.

Boast, C.A., Gerhardt, S.C., Pastor, G., Lehmann, J., Etienne, P.E.and Liebman, J.M. (1988) The *N*-methyl-D-aspartate antagonists CGS 19755 and CPP reduce ischemic brain damage in gerbils. *Brain. Res.*, 442: 345–348.

Böckers, T.M., Zimmer, M., Müller, A., Bergmann, M., Brose, N. and Kreutz, M.R. (1994) Expression of the NMDA R1 receptor in selected human brain regions. *NeuroReport*, 5: 965–969.

Bordi, F., Pietra, C., Ziviani, L. and Reggiani, A. (1997) The glycine antagonist GV150526 protects somatosensory evoked potentials and reduces the infarct area in the MCAo model of focal ischemia in the rat. *Exp. Neurol.*, 145: 425–433.

Breese, C.R. and Leonard, S.S. (1993) Glutamate receptor subtype expression in human postmortem brain. *J. Mol. Neurosci.*, 4: 263–275.

Bullock, R., Zauner, A., Woodward, J. and Young, H.F. (1995) Massive persistent release of excitatory amino acids following human occlusive stroke. *Stroke*, 26: 2187–2189.

Castillo, J., Davalos, A., Naveiro, J. and Noya, M. (1996) Neuroexcitatory amino acids and their relation to infarct size and neurological deficit in ischemic stroke. *Stroke*, 27: 1060–1065.

Castillo, J., Davalos, A. and Noya, M. (1997) Progression of ischaemic stroke and excitotoxic aminoacids. *Lancet*, 349: 79–83.

Choi, D.W. (1987) Ionic dependence of glutamate neurotoxicity. *J. Neurosci.*, 7: 369–379.

Cohen, R.A., Hasegawa, Y. and Fisher, M. (1994) Effects of a novel NMDA receptor antagonist on experimental stroke quantitatively assessed by spectral EEG and infarct volume. *Neurol. Res.*, 16: 443–448.

Davalos, A., Castillo, J., Serena, J. and Noya, M. (1997) Duration of glutamate release after acute ischemic stroke. *Stroke*, 28: 708–710.

Davis, S.M., Albers, G.W., Diener, H-C., Lees, K.R. and Norris, J. (1997) Termination of acute stroke studies involving selfotel treatment. *Lancet*, 349: 32.

Dewar, D., Chalmers, D.T., Grahan, E.I. and McCulloch, J. (1991) Glutamate metabotropic and AMPA binding sites are reduced in Alzheimer's disease: An autoradiographic study of the hippocampus. *Brain Res.*, 553: 58–64.

Dodd, P.R., Watson, W.E., Morrison, M.M., Johnston, G.A., Bird, E.D., Cowburn, R.F. and Hardy, J.A. (1989) Uptake of g-aminobutyric acid and L-glutamic acid by synaptgosomes from postmortem human cerebral cortex: Multiple sites, sodium dependence and effect of tissue preparation. *Brain Res.*, 490: 320–331.

Dyker, A.D., Lees, K.R., Preston, C. and Caveggion, E. (1997) The safety, tolerability and pharamacokinetics of GV150526A in patients with acute stroke. *Cerebrovasc. Dis.*, 7, Suppl 4: 233.

Enblad, P., Valtysson, J., Andersson, J., Lilja, A., Valind, S., Antoni, G., Langstrom, B., Hillered, L. and Persson, L. (1996) Simultaneous intracerebral microdialysis and positron emission tomography in the detection of ischemia in patients with subarachnoid hemorrhage. *J. Cereb. Blood Flow Metab.*, 16: 637–644.

Eshhar, N., Striem, S., Kohen, R., Tirosh, O. and Biegon, A. (1995) Neuroprotective and antioxidant activities of HU-211, a novel NMDA receptor antagonist. *Eur. J. Pharmacol.*, 283: 19–29.

Fonnum, F. (1984) Glutamate: A neurotransmitter in mammalian brain. *J. Neurochem.*, 42: 1–11.

Fukunaga, R., Ninomiya, H., Shimohama, S., Kimura, J, Kameyama, M. and Taniguchi, T. (1991) Reassessment of [3H]glutamate binding to human brain membrane preparations. *Jap. J. Pharmacol.*, 55: 191–196.

Garcia-Ladona, F.J., Palacios, J.M., Probst, A., Wieser, H.G. and Mengod, G. (1994) Excitatory amino acid AMPA receptor mRNA localization in several regions of normal and neurological disease affected human brain. An in situ hybridization histochemistry study. *Mol. Brain Res.*, 21: 75–84.

Garthwaite, G. and Garthwaite, J. (1986) Amino acid neurotoxicity: Intracellular sites of calcium accumulation associated with the onset of irreversible damage to rat cerebellar neurones in vitro. *Neurosci. Lett.*, 71: 53–58.

Garthwaite, G. and Garthwaite, J. (1989) Neurotoxicity of excitatory amino acid receptor agonists in young rat hippocampal slices. *J. Neurosci. Meth.*, 29: 33–42.

Gill, R., Foster, A.C. and Woodruff, G.N. (1988) MK-801 is neuroprotective in gerbils when administered during the post-ischemic period. *Neuroscience*, 25: 25.

Gjessing, L.R., Gjesdahl, P. and Sjaastad, O. (1972) The free amino acids in human cerebrospinal fluid. *J. Neurochem.*, 19: 1807–1808.

Globus, M.Y-T., Busto, R., Dietrich, W.D., Martinez, E., Valdés, I. and Ginsberg, M.D. (1988) Effect of ischemia on the in vivo release of striatal dopamine, glutamate and $\gamma$-aminobutyric acid studied by intracerebral microdialysis. *J. Neurochem.*, 51: 1455–1464.

Grosset, D.G., Muir, K.W. and Lees, K.R. (1995) Systemic and cerebral hemodynamic responses to the noncompetitive *N*-methyl-D-aspartate (NMDA) antagonist CNS 1102. *J. Cardiovasc. Pharmacol.*, 25: 705–709.

Grotta, J.C., Picone, C.M., Ostrow, P.T., Strong, R.A., Earls, R.M., Yao, L.P., Rhoades, H.M. and Dedman, J.R. (1990) CGS-19755, a competitive NMDA receptor antagonist, reduces calcium-calmodulin binding and improves outcome after global cerebral ischemia. *Ann. Neurol.*, 27: 612–619.

Grotta, J., Clark, W., Coull, B., Pettigrew, L.C., Mackay, B., Goldstein, L.B., Meissner, I., Murphy, D. and LaRue, L. (1995) Safety and tolerability of the glutamate antagonist CGS 19755 (Selfotel) in patients with acute ischemic stroke. Results of a phase IIa randomized trial. *Stroke*, 26: 602–605.

Grøndahl, T.Ø., Langmoen, I.A. and Gregersen, M. (1991) D-[3H]aspartate release from hippocampal slices during hypoxia/glucose deprivation with and without blocked synaptic transmission. *J. Cereb. Blood Flow Metab.*, 11: S102.

Grøndahl, T.Ø. and Langmoen, I.A. (1996) Cytotoxic effect of $Ca^{2+}$ released from intracellular stores during cerebral energy deprivation. *Neurol. Res.*, 18: 499–504.

Grøndahl, T.Ø. and Langmoen, I.A. (1997) Confocal laser scanning microscopy used to monitor intracellular $Ca^{2+}$ changes in hippocampal CA 1 neurons during energy deprivation. *In press.*

Grøndahl, T.Ø., Hablitz, J.J. and Langmoen, I.A. (1997) Depletion of intracellular $Ca^{2+}$ stores or lowering extracellular calcium alters intracellular $Ca^{2+}$ changes during cerebral energy deprivation. *In press.*

Hagberg, H., Lehmann, A., Sandberg, M., Nyström, B., Jacobson, I. and Hamberger, A. (1985) Ischemia-induced shift of inhibitory and excitatory amino acids from intra- to extracellular compartments. *J. Cereb. Blood Flow Metab.*, 5: 413–419.

Hagberg, H., Thornberg, E., Blennow, M., Kjellmer, I., Lagercrantz, H., Thiringer, K., Hamberger, A. and Sandberg, M. (1993) Excitatory amino acids in the cerebrospinal fluid of asphyxiated infants: Relationship to hypoxic-ischemic encephalopathy. *Acta Paediatr.*, 82: 925–929.

Hamberger, A,. Runnerstam, M., Nystrom, B., Starmark, J.E. and von Eccen, (1995) The neuronal environment after subarachnoid haemorrhage – correlation of amino acid and nucleoside levels with post-operative recovery. *Neurol. Res.*, 17: 97–105.

Haugstad, T.S., Hegstad, E. and Langmoen, I.A. (1992) Calcium dependent release of g-aminobutyric acid (GABA) from human cerebral cortex. *Neurosci. Lett.*, 141: 61–64.

Haugstad, T.S., Valø, E.T. and Langmoen, I.A. (1995) Changes in brain amino acid content induced by hyposmolar stress and energy deprivation. *Neurol. Res.*, 17: 402–408.

Haugstad, T.S. and Langmoen, I.A. (1996) Release of brain amino acids during hyposmolar stress and energy deprivation. *J. Neurosurg. Anesthesiol.*, 8: 159–168.

Hegstad, E., Haugstad, T.S., Hauglie-Hanssen, E. and Langmoen, I.A. (1992) Calcium-dependent release of glutamate from human cerebral cortex. *Brain Res.*, 585: 340–342.

Hegstad, E. (1995) *Amino acids in the human brain.*, Ph.D. Thesis, University of Oslo.

Hegstad, E., Berg-Johnsen, J., Haugstad, T.S., Hauglie-Hanssen, E. and Langmoen, I.A. (1996) Amino-acid release from human cerebral cortex during simulated ischaemia in vitro. *Acta Neurochir.*, (Wien) 138: 234–241.

Hillered, L., Persson, L., Ponten, U. and Ungerstedt, U. (1990) Neurometabolic monitoring of the ischaemic human brain using microdialysis. *Acta Neurochir.* (*Wien*), 102: 91–97.

Hoke, J.F., Dyker, A.G., McAllister, A.M. and Lees, K.R. (1997) Pharmacokinetics of GV150526A following multiple intravenous doses in acute stroke patients. *Cerebrovasc. Dis.*, 7, Suppl 4: 29.

Hwa, G.G.C. and Avoli, M. (1992) Excitatory synaptic transmission mediated by NMDA and non-NMDA receptors in the superficial/middle layers of the epileptogenic human neocortex maintained in vitro. *Neurosci. Lett.*, 143: 83–86.

Ikeda, M., Nakazawa, T., Abe, K., Kaneko, T. and Yamatsu, K. (1989) Extracellular accumulation of glutamate in the hippocampus induced by ischemia is not calcium dependent – in vitro and in vivo evidence. *Neurosci. Lett.*, 96: 202–206.

Jarvis, M.F., Murphy, D.E., Williams, M., Gerhardt, S.C. and Boast, C.A. (1988) The novel *N*-methyl-D-aspartate (NMDA) antagonist CGS 19755 prevents ischemia-induced reductions of adenosine A1, NMDA, and PCP receptors in gerbil brain. *Synapse*, 2: 577–584.

Johnson, J.L. (1978) The excitant amino acids glutamic and aspartic acid as transmitter candidates in the vertebrate central nervous system. *Prog. Neurobiol.*, 10: 155–202.

Kanthan, R., Shuaib, A., Griebel, R., Miyashita, H. and Kalra, J. (1996) Glucose-induced decrease in glutamate levels in ischemic human brain by in-vivo microdialysis. *Neurosci. Lett.*, 209: 207–209.

Katayama, Y., Kawamata, T., Tamura, T., Hovda, D.A., Becker, D.P. and Tsubokawa, T. (1991) Calcium-dependent glutamate release concomitant with massive potassium flux during cerebral ischemia in vivo. *Brain Res.*, 558: 136–140.

Kauppinen, R.A., McMahon, H.T. and Nicholls, D.G. (1988) $Ca^{2+}$-dependent and $Ca^{2+}$-independent glutamate release, energy status and cytosolic free $Ca^{2+}$ concentration in isolated nerve terminals following metabolic inhibition: Possible relevance to hypoglycaemia and anoxia. *Neuroscience*, 27: 175–182.

Kimelberg, H.K., Godrie, S.K., Higman, S., Pang, S. and Waniewski, R.A. (1990) Swelling-induced release of glutamate, aspartate, and taurine from astocyte cultures. *J. Neurosci.*, 10: 1583–1591.

Lehmann, J., Hutchison, A.J. and McPherson, S.E. (1988) CGS 19755, a selective an competitive NMDA-type excitatory amino acid receptor antagonist. *J. Pharmacol. Exp. Ther.*, 246: 65–75.

Lucas, D.R. and Newhouse, J.P. (1957) The toxic effect of sodium L-glutamate on the inner layers of the retina. *Arch. Ophthalmol.*, 58: 193–201.

Lund Karlsen, R. (1978) The toxic effect of sodium glutamate and DL-a-aminoadipic acid on the inner layers of the retina: Changes in high affinity uptake of putative neurotransmitters. *J. Neurochem.*, 31: 1055–1061.

Martin, H., Warner, D.S. and Todd, M.M. (1994) Effects of glycine receptor antagonism on spreading depression in the rat. *Neurosci. Lett.*, 180: 285–289.

Matsumoto, R.R., Nguyen, D., Truong, D.D. (1995) Strychnine-insensitive glycine site antagonists attenuate a cardiac arrest-induced movement disorder. *Eur. J. Pharmacol.*, 275: 117–123.

Meldrum, B. (1985) Possible therapeutic applications of antagonists of excitatory amino acid neurotransmitters. *Clin. Sci.*, 68: 113–122.

Meldrum, B. and Garthwaite, J. (1990) Excitatory amino acid neurotoxicity and neurodegenerative disease. *Trends Pharmacol. Sci.*, 11: 379–387.

Mendelowitsch, A., Sekhar, L.N., Clemente, R. and Shuaib, A. (1997) EC-IC bypass improves chronic ischemia in a patient with moyamoya disease secondary to sickle cell disease: An in vivo microdialysis study. *Neurol. Res.*, 19: 66–70.

Milliken, J.R. and Standen, J.L. (1951) An investigation into the effects of glutamic acid on human intelligence. *J. Neurol. Neurosurg. Psych.*, 14: 47–54.

Minematsu, K., Fisher, M., Li, L. and Sotak, C.H. (1993) Diffusion and perfusion magnetic resonance imaging studies to evaluate a noncompetitive *N*-methyl-D-aspartate antagonist and reperfusion in experimental stroke in rats. *Stroke*, 24: 2074–2081.

Muir, K.W. and Lees, K.R. (1995) Clinical experience with excitatory amino acid antagonist drugs. *Stroke*, 26: 503–513.

Muir, K.W., Grosset, D.G. and Lees, K.R. (1995) Clinical pharmacology of CNS 1102 in volunteers. *Ann. N.Y. Acad. Sci.*, 765: 279–289.

Muir, K.W., Grosset, D.G. and Lees, K.R. (1997) Effects of prolonged infusions of the NMDA antagonist aptiganel hydrochloride (CNS 1102) in normal volunteers. *Clin. Neuropharmacol.*, 20: 311–321.

Nadler, V., Mechoulam, R. and Sokolovsky, M. (1993) The non-psychotropic cannabinoid (+)-(3S,4S)-7-hydroxy-δ

6-tetrahydrocannabinol 1,1-dimethylheptyl (HU-211) attenuates *N*-methyl-D-aspartate receptor-mediated neurotoxicity in primary cultures of rat forebrain. *Neurosci. Lett.*, 162: 43–45.

Nadler, V., Biegon, A., Beit-Yannai, E., Adamchik, J. and Shohami, E. (1995) 45Ca accumulation in rat brain after closed head injury; attenuation by the novel neuroprotective agent HU-211. *Brain Res.*, 685: 1–11.

Nellgard, B. and Wieloch, T. (1992) Postischemic blockade of AMPA but not NMDA receptors mitigates neuronal damage in the rat brain following transient severe cerebral ischemia. *J. Cereb. Blood Flow Metab.*, 12: 2–11.

Newell, D.W., Barth, A. and Malouf, A.T. (1995) Glycine site NMDA receptor antagonists provide protection against ischemia-induced neuronal damage in hippocampal slice cultures. *Brain Res.*, 675: 38–44.

Nilsson, O.G., Saveland, H., Boris-Moller, F., Brandt, L. and Wieloch, T. (1996) Increased levels of glutamate in patients with subarachnoid haemorrhage as measured by intracerebral microdialysis. *Acta Neurochir. Suppl.*, (Wien) 67: 45–47.

Olney, J.W. (1969) Brain lesions, obesity, and other disturbances in mice treated with monosodium glutamate. *Science*, 164: 719–721.

Olney, J.W., Ho, O.L. and Rhee, V. (1971) Cytotoxic effects of acidic and sulphur containing amino acids on the infant mouse central nervous system. Experimental *Brain Res.*, 14: 61–76.

Onodera, H., Sato, G. and Kogure, K. (1986) Lesions to Schaffer collaterals prevent ischemic death of CA1 pyramidal cells. *Neurosci. Lett.*, 68: 169–174.

Ottersen, O.P. and Storm-Mathisen, J. (1984) Glutamate- and GABA-containing neurons in the mouse and rat brain, as demonstrated with a new immunocytochemical technique. *J. Comp. Neurol.*, 229: 57–66.

Owens, J., Wyper, D.J., Patterson, J., Brown, D.R., Elliott, A.T., Teasdale, G.M. and McCulloch, J. (1997) First SPET images of glutamate (NMDA) receptor activation in vivo in cerebral ischaemia. *Nucl. Med. Commun.*, 18: 149–158.

Patat, A., Molinier, P., Hergueta, T., Brohier, S., Zieleniuk, I., Danjou, P., Warot, D. and Puech, A. (1994) Lack of amnestic, psychotomimetic or impairing effect on psychomotor performance of eliprodil, a new NMDA antagonist. *Int. Clin. Psychopharmacol.*, 9: 155–162.

Perry, T.L. and Hansen, S. (1981) Amino acid abnormalities in epileptogenic foci. *Neurology*, 31: 872–876.

Perry, T. (1983) Levels of glutamine, glutamate and GABA in CSF and brain under pathological conditions. In L. Hertz, E. Kvamme, E. G. McGeer, and A. Schousboe (Eds.), *Glutamine, Glutamate, and GABA in the Central Nervous System*, New York, Alan R. Liss, pp. 581–594.

Persson, L., Valtysson, J., Enblad, P., Warme, P.E., Cesarini, K., Lewen, A. and Hillered, L. (1996) Neurochemical monitoring using intracerebral microdialysis in patients with subarachnoid hemorrhage. *J. Neurosurg.*, 84: 606–616.

Poignet, H., Nowicki, J.P. and Scatton, B. (1992) Lack of neuroprotective effect of some sigma ligands in a model of focal cerebral ischemia in the mouse. *Brain Res.*, 596: 320–324.

Quarum, M.L., Parker, J.D., Keana, J.F. and Weber, E. (1990) (+)-[3H]MK-801 binding sites in postmortem human brain. *J. Neurochem.*, 54: 1163–1168.

Reggiani, A., Costa, S., Pietra, C., Ratti, E., Trist, D., Zivani, L. and Gaviraghi, G. (1995) Neuroprotective activity of GV150526A, a novel poetent and selective glycine antagonist. *Eur. J. Neurol.*, 2, Suppl 2: 6–7.

Rothman, S.M. (1985) The neurotoxicity of excitatory amino acids is produced by passive chloride influx. *J. Neurosci.*, 5: 1483–1489.

Runnerstam, M., von Essen, C., Nystrom, B., Rosengren, L. and Hamberger, A. (1997) Extracellular glial fibrillary acidic protein and amino acids in brain regions of patients with subarachnoid hemorrhage – correlation with level of consciousness and site of bleeding. *Neurol. Res.*, 19: 361–368.

Säveland, H., Nilsson, O.G., Boris-Møller, F., Wieloch, T. and Brandt, L. (1996) Intracerebral microdialysis of glutamate and aspartate in two vascular territories after aneurysmal subarachnoid hemorrhage. *Neurosurgery*, 38: 12–19.

Sheardown, M.J., Nielsen, E.O., Hansen, A.J., Jacobsen, P. and Honoré, T. (1990) 2,3-dihydro-6-nitro-7-sulfamoyl-benzo(F)quinoxaline: A neuroprotectant for cerebral ischemia. *Science*, 247: 571–574.

Shimada, N., Graf, R., Rosner, G., Wakayama, A., George, C.P. and Heiss, W.-D. (1989) Ischemic flow threshold for extracellular glutamate increase in cat cortex. *J. Cereb. Blood Flow Metab.*, 9: 603–606.

Siesjö, B.K. and Bengtsson, F. (1989) Calcium fluxes, calcium antagonists, and calcium-related pathology in brain ischemia, hypoglycemia, and spreading depression: A unifying hypothesis. *J. Cereb. Blood Flow Metab.*, 9: 127–140.

Simon, R.P., Swan, J.H., Griffiths, T. and Meldrum, B.S. (1984) Blockade of *N*-methyl-D-aspartate receptors may protect against ischemic damage in the brain. *Science*, 226: 850–852.

Simon, R. and Shiraishi, K. (1990) *N*-methyl-D-aspartate antagonist reduces stroke size and regional glucose metabolism. *Ann. Neurol.*, 27: 606–611.

Steele, J.E., Bowen, D.M., Francis, P.T., Green, A.R. and Cross, A.J. (1990) Spermidine enchancement of [3H]MK-801 binding to the NMDA receptor complex in human cortical membranes. *Eur. J. Pharmacol.*, 189: 195–200.

Steele, J.E., Robinson, T.N., Cross, A.J., Bowen, D.M. and Green, A.R. (1991) A comparison of [3H]MK-801 and N-[1-(2-thienyl)cyclohexyl]-3,4-[3H]piperidine binding to the *N*-methyl-D-aspartate receptor complex in human brain. *J. Neurochem.*, 56: 1248–1254.

Storm-Mathisen, J. (1977) Localization of transmitter candidates in the brain: The hippocampal formation as a model. *Prog. Neurobiol.*, 8: 119–181.

Storm-Mathisen, J., Leknes, A.K., Bore, A.T., Vaaland, J.L., Edminson, P., Haug, F.-M.S. and Ottersen, O.P. (1983) First visualization of glutamate and GABA in neurones by immunocytochemistry. *Nature*, 301: 517–520.

Storm-Mathisen, J., Danbolt, N.C., Rothe, F., Torp, R., Zhang, N., Aas, J.-E., Kanner, B.I., Langmoen, I. and Ottersen, O.P. (1992) Ultrastructural immunocytochemical observations on the localization, metabolism and transport of glutamate in normal and ischemic brain tissue. *Prog. Brain Res.*, 94: 225–241.

Sun, W., Ferrer-Montiel, A.V. and Montal, M. (1994) Primary structure and functional expression of the AMPA/kainate receptor subunit 2 from human brain. *NeuroReport*, 5: 441–444.

Swan, J.H. and Meldrum, B.S. (1990) Protection by NMDA antagonist against selective cell loss following transient ischemia. *J. Cereb. Blood Flow Metab.*, 10: 343–351.

Szatkowski, M. and Attwell, D. (1994) Triggering and execution of neuronal death in brain ischemia: Two phases of glutamate release by different mechanisms. *Trends Neurosci.*, 17: 359–365.

Takaoka, S., Bart, R.D., Pearlstein, R., Brinkhous, A. and Warner, D.S. (1997) Neuroprotective effect of NMDA receptor glycine recognition site antagonism persists when brain temperature is controlled. *J. Cereb. Blood Flow Metab.*, 17: 161–167.

Takizawa, S., Hogan, M. and Hakim, A.M. (1991) The effects of a competitive NMDA receptor antagonist (CGS-19755) on cerebral blood flow and pH in focal ischemia. *J. Cereb. Blood Flow Metab.*, 11: 786–793.

Toulmond, S., Serrano, A., Benavides, J. and Scatton, B. (1993) Prevention by eliprodil (SL 82.0715) of traumatic brain damage in the rat. Existence of a large (18 h) therapeutic window. *Brain Res.*, 620: 32–41.

Tower, D.B. (1955) Nature and extent of the biochemical lesion in human epileptogenic cerebral cortex. *Neurology*, 5: 113–130.

Van Gelder, N.M., Sherwin, A.L. and Rasmussen, T. (1972) Amino acid content of epileptogenic human brain: Focal versus surrounding regions. *Brain Res.*, 40: 385–393.

Vered, M., Bar-Joseph, A., Belayev, L., Berkovich, Y. and Biegon, A. (1994) Anti-ischemia activity of HU-211, a non-psychotropic synthetic cannabinoid. *Acta Neurochir. Suppl.*, (Wien) 60: 335–337.

Warner, D.S., Martin, H., Ludwig, P., McAllister, A., Keana, J.F. and Weber, E. (1995) In vivo models of cerebral ischemia: Effects of parenterally administered NMDA receptor glycine site antagonists. *J. Cereb. Blood Flow Metab.*, 15: 188–196.

Wieloch, T., Lindvall, O., Blomquist, P. and Gage, F.H. (1985) Evidence for amelioration of ischaemic neuronal damage in the hippocampal formation by lesions of the perforant path. *Neurol. Res.*, 7: 24–26.

Zimmermann, F.T., Burgermeister, B.B. and Putnam, T.J. (1949) Effect of glutamic acid on the intelligence of patients with Mongolism. *Arch. Neurol. Psych.*, 61: 275–287.

O.P. Ottersen, I.A. Langmoen and L. Gjerstad (Eds.)
*Progress in Brain Research*, Vol 116


CHAPTER 20

# Glutamate and its receptors in the pathophysiology of brain and spinal cord injuries

B. Alessandri* and R. Bullock

*Medical College of Virginia, Department of Neurosurgery, P. O. Station Box 693, 1225 E. Marshall St., Richmond, VA 23298, USA*

## Human head and spinal cord injury

Traumatic brain injury (TBI) is one of the leading causes for morbidity and mortality in both developing and fully industrialized countries, and more than 350 000 individuals are hospitalized each year as a result of TBI in the USA alone. TBI is responsible for 56 000 lives lost annually (Kraus and McArthur, 1996) and almost 100 000 persons sustain moderate to severe brain injuries resulting in lifelong disability (Kraus and McArthur, 1996; Kraus et al., 1994). The cost of TBI has been estimated to be about 37–38 billion US dollars per year in the USA alone (Max et al., 1990). Although less frequent than TBI, almost 10 000 spinal cord injuries (SCI) occur each year in the USA. In about 45% of all cases, SCI result in a total loss of sensation and motor function below the injured cord (NSCIA, 1996), most of these spinal injury patients survive, and create a major financial and social burden for society, and their caregivers. The percentage of deaths caused by neurotrauma declined during 1979 and 1992 by about 21%, mainly due to reduced motor vehicle-related deaths (Sosin et al., 1995). However, there is no clinically available pharmacological treatment to reduce the extent of cerebral damage following neural injury at present. Over recent years, several clinical trials with 'neuroprotectants' have failed to show beneficial effects in head injured patients (Doppenberg and Bullock, 1997). Thus, demand for new pharmacological treatments of traumatic brain and spinal cord injury is higher than ever, at present, and at the time of writing, glutamate antagonists are currently the most studied neuroprotectant category.

Although many aspects of the pathophysiology of TBI and SCI remain poorly understood, research has provided new insights into pathophysiological mechanisms over recent years, providing promising avenues for new therapies (for recent review see (McIntosh et al., 1996a)). Studies have focused on the prevention of secondary damage and the regeneration of damaged neuronal tissue TBI or SCI injury. Targets for the prevention of secondary damage are glutamatergic, cholinergic and catecholaminergic neurotransmission systems, free radical production, lipid peroxidation, calcium channels, growth factors, inflammation processes, endogenous opioid receptors, enzymes (e.g. calpain), apoptotic cell death, repair mechanisms and others. However, this chapter reviews the contribution of the excitatory neurotransmitter glutamate to CNS damage fol-

* Corresponding author.
Tel.: 1 804 828 1318; fax: 1 804 828 3276; e-mail: balessandri@gems.vcu.edu.

lowing TBI and SCI, and in particular evidence for glutamate-induced neuronal damage in head-injured patients.

## Glutamate and head injury

Glutamate is the most widely distributed excitatory neurotransmitter in the mammalian brain. Under physiological conditions, glutamate is removed from the extracellular space (ECS) by high-affinity glutamate transporters in neurons (EAAC1, EAAT4) and glial cells (GLAST, GLT-1 (Tanaka et al., 1997; see also Danbolt et al., and Billups et al. this volume). These transporters work against a massive extracellular-intracellular concentration gradient, (for vesicles: 1:10 000) and are highly energy dependent. Once in glial cells, glutamate is transformed to glutamine which is subsequently transported to neurons and transformed back to glutamate. In addition to its important physiological role as a neurotransmitter, glutamate is of particular interest because of its involvement in both acute and chronic neuronal damage. The hypothesis that glutamate contributes significantly to brain damage after neurotrauma depends on the following facts:

(1) Glutamate is toxic to neurons and mixed cell cultures and normal brain tissue *in vivo* in concentrations of 5 $\mu$M, 100–200 $\mu$M and 1000–5000 $\mu$M, respectively, depending on duration of application, the status of transporters and tissue defenses;
(2) Glutamate is released into the ECF, after impact in both animal models, and humans, in concentrations putatively as high as 10–200 $\mu$M and remaining high for up to 4–6 days;
(3) Glutamate receptor antagonists, and presynaptic release inhibitors, given before impact (and onset of secondary ischemia) can prevent histological, functional and behavioral consequences of neurotrauma.

Many studies using animal models of traumatic brain injury (Panter and Faden, 1992) spinal cord injury (Faden et al., 1988a), ischemia (Arvin et al., 1994), hypoxia (Schurr et al., 1995) and subdural hematoma (Bullock et al., 1990) have shown this ‘neuroprotection’ by glutamate receptor antagonists. We and others speculate that factors such as secondary energy depletion (hypoxia, ischemia) or pyrexia, seizures, or ionic imbalance may potentiate the toxicity of glutamate after neurotrauma, such that the concentrations shown in cerobrospinal fluid (CSF) and microdialysis studies, maybe harmful. These synergistic mechanisms remain poorly understood (Bullock and Di, 1997; Ishige et al., 1987; Jenkins et al., 1988; Koizumi et al., 1997; Nawashiro et al., 1995; Palmer et al., 1993b; Simpson RK et al., 1990).

## Pathophysiology of traumatic brain injury

### *Human traumatic brain injury (TBI)*

Traumatic brain or spinal cord injury results in neurological dysfunction (e.g. motor, sensory and memory impairment) and death through both direct (mechanical damage) and indirect (secondary or delayed) mechanisms. Primary traumatic brain damage may directly damage blood vessels, axons, neurons and glia in a focal, multifocal or diffuse pattern of involvement. The mechanical injury subsequently initiates dynamic and evolving processes which may differ for each component part. In patients, several types of injury are distinguished, namely diffuse axonal injury, focal contusions and intracranial hematomas caused by subdural, epidural, or intracerebral hemorrhage. The pathophysiology is in many cases worsened by secondary (delayed) events including ischemic and hypoxic damage, cerebral swelling (vasogenic and cytotoxic edema), increased ICP, hypotension, seizures, infections, pyrexia, and respiratory failure.

Severe diffuse axonal injury (DAI) without mass lesions occurs in about 25% of patients who die with severe TBI. This type of injury is a major cause of the persistent vegetative or severely disabled state which occurs in 4% and 15% of all patients, respectively. The most prominent morphological feature of this type of TBI is the white matter axonal damage demonstrated by

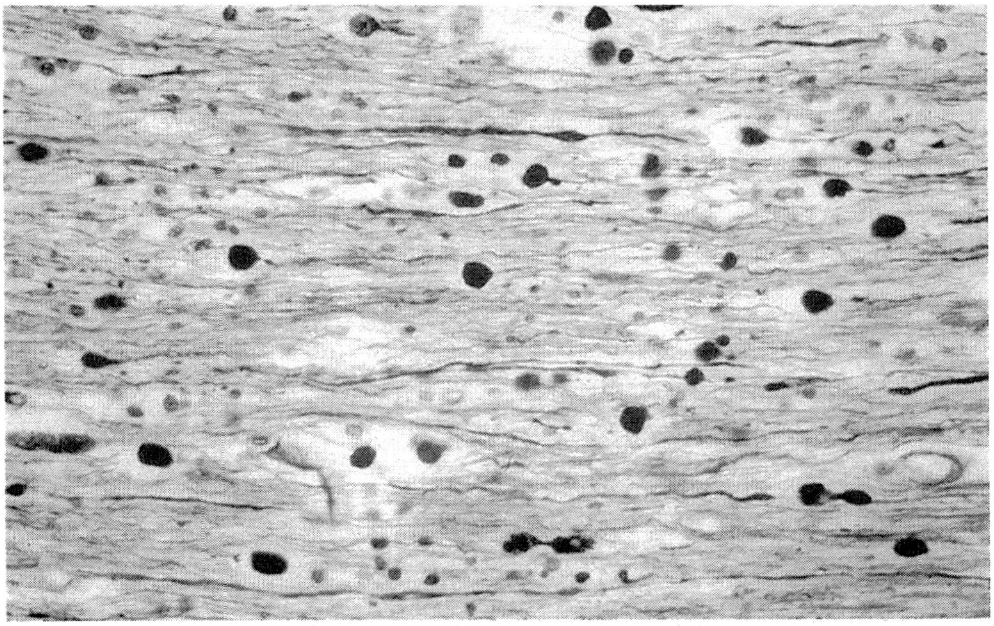

Fig. 1. Human diffuse axonal injury with retraction balls (silver stain, original magnification x800).

axonal swelling and the development of 'retraction balls' (Fig. 1), leading to permanent interruption of axons 24 to 72 hours after injury (Povlishock, 1992). DAI also leads to small hemorrhages in brain regions such as corpus callosum and brain stem which may be seen on CT and MRI scans.

Massive calcium entry into axons via voltage-dependent ion channels or mechanoporation, may play a pivotal role in the neurodegeneration which follows (Fig. 2) (Maxwell et al., 1997). Although the triggering mechanism for the cytoskeletal changes which then ensue, are not known, calpains (non-lysosomal cysteine protease) are probably important (see Reynolds, this volume). Animal models such as the weight-drop injury model (impact acceleration injury model; (Foda and Marmarou, 1994; Marmarou et al., 1994), the inertial acceleration injury model (Gennarelli et al., 1981), and fluid percussion injury model (Dixon et al., 1987), all produce such DAI, histologically. There is, as yet, no clear evidence implicating glutamate in the pathogenesis of DAI itself, and glutamatergic ion channel receptors are not believed to be present on white matter axons.

Glutamate seems to have a more important role in the pathogenesis of non-hemorrhagic/hemorrhagic focal contusions. Cerebral contusions are focal injuries which result when localized mechanical forces damage blood vessels (e.g. capillaries), neurons, neurophils and astrocytes. Micro-hemorrhages, 0.1–5 mm in size, on the surface or in deeper layers of the gyri are usually the most obvious features of contused tissue. Contusions often increase in size over hours to days due to the evolving events related to the interplay of hemorrhage, early ischemic necrosis, cytotoxic swelling, and delayed vasogenic edema (Fig. 3). Thus, patients may show progressive or sudden neurological deterioration after a lucid interval (Blumbergs, 1997). Studies in monkeys have confirmed that the depth of a contusion lesion increases with time, and this may induce secondary ischemic damage in surrounding normal brain, due to compression, or harmful mediators (Gennarelli et al., 1979). Clinical studies show that cerebral blood flow (CBF) around contusions and hematomas is almost always reduced to ischemic levels ($\pm$18 ml/100 g/min) (Schroder et al., 1995;

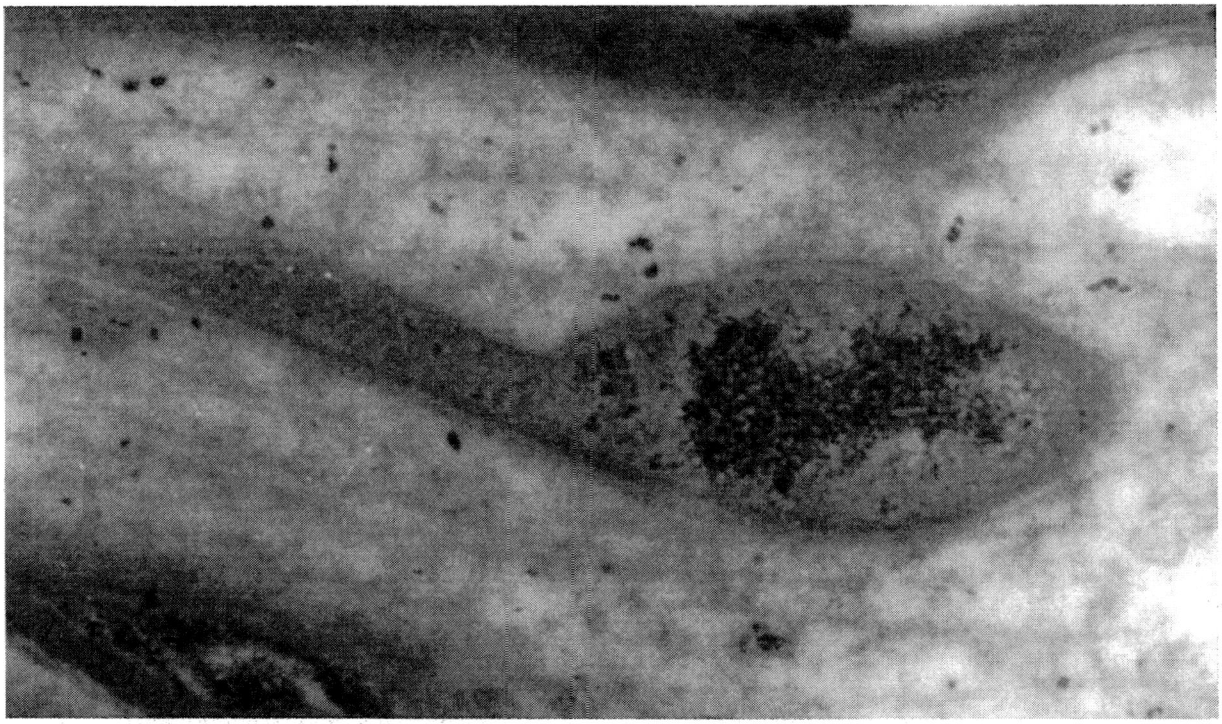

Fig. 2. A longitudinal, thin section of a swollen mitochondrion after experimental DAI in material processed for the pyroantimonate technique for calcium localization. The middle of the organelle contains an electron dense precipitate. Original magnification x600 000 (reproduced from Maxwell et al., 1997; with permission).

Schroder et al., 1994), and concomitantly glucose utilization is increased during the first few hours or days (Bergsneider et al., 1997; Hovda et al., 1990; Kuroda et al., 1992; Yoshino et al., 1991). In addition, microdialysis studies in patients indicate massive elevation of extracellular glutamate in these injury types (Bullock et al., 1995a). This leads to a disturbance of ion homeostasis, massive calcium influx and cytotoxic edema which causes increased intracranial pressure (ICP). Animal models such as fluid percussion, controlled cortical impact, weight drop injury, and acute subdural hematoma seem to depict certain of these pathological and biochemical events as seen in patients (Dixon et al., 1987; Gennarelli et al., 1981; Marmarou et al., 1994; Miller et al., 1990a)

Secondary ischemic brain damage is extremely common after severe head injury, especially superimposed upon hematomas, and 80–90% of all patients who die, show ischemia upon post-mortem histo-pathological examination of the brain (Table 1; Graham, 1985). In about one-third of severely head-injured patients the cerebral blood flow is reduced to ischemic level 2–6 hours after injury This may result in a reduced substrate delivery to the brain (mainly oxygen, glucose), increased metabolic demands, and loss of normal $pCO_2$ vasoconstriction and autoregulation. A persistently ischemic blood flow subsequently leads to a shift from aerobic to anaerobic metabolism, further glutamate release and loss of ion homeostasis (massive calcium influx), brain swelling (increased ICP) and necrotic cell death. In animal models of focal and global ischemia a massive immediate increase of glutamate was found persisting up to several hours after ischemia (Shimada et al., 1990; Shimada et al., 1989). Shimada et al. (Shimada et al., 1989), in a global

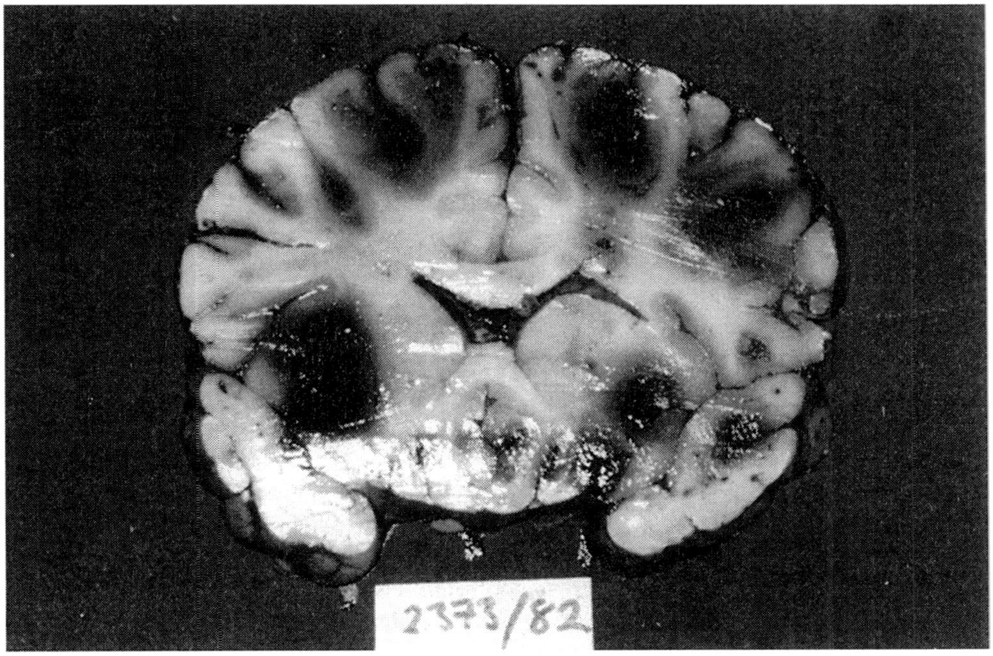


Fig. 3. Coronal human brain section in a patient who died with multiple focal cortical and subcortical contusions, and brain swelling.

model of brain ischemia in cats, have demonstrated a 30-fold increase in glutamate when the CBF fell below a blood flow threshold of 20 ml/100 g/min (Fig. 4). In human head-injury patients, a close relationship between CBF and glutamate release has been found (Zauner et al., 1996) (Fig. 4).

TABLE 1

Incidence, cause, localization and pattern of ischemic brain damage after severe head injury (modified from Graham, 1985)

| | | |
|---|---|---|
| Incidence | Overall | 88% |
| Cause/Localisation | Hematomas | 46% |
| | Hippocampus | 84% |
| | High ICP | 49% |
| Pattern | Focal, contusional | 59% |
| | "Boundary zone" | 8% |
| | Diffuse/whole brain | 24% |

In addition to ischemic-hypoxic secondary brain damage, cerebral edema is a common, but variable secondary response to trauma. However, the causes and consequences of this brain swelling are poorly understood. Clinically, five types of cerebral swelling have been described, namely vasogenic, cytotoxic, hydrostatic, osmotic and interstitial brain edema. Swelling occurs almost always around cerebral contusions, intracerebral hemorrhages, and under subdural hematomas, less often one hemisphere or even the entire brain is involved ≤10–15%. Edema, chiefly cytotoxic, causes an increase of ICP and consequently decrease of cerebral perfusion pressure (CPP) which may lead to further global ischemia. With respect to glutamate, the cytotoxic edema is of particular interest, since this earliest form of swelling, is caused by disturbance of ionic gradients, with accumulation of intracellular fluid, and is especially associated with secondary ischemia/hypoxia following TBI. Data from animal models of fluid percussion injury (FPI) (Dixon et al., 1987), acute subdural

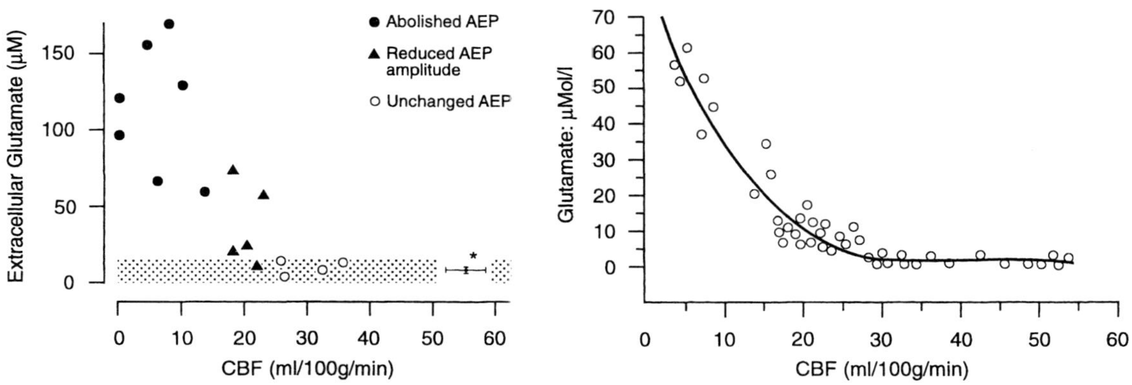


Fig. 4. Relationship between extracellular glutamate and cerebral blood flow in cats following global ischemia (left graph) (Shimada et al., 1989) and in severely head-injured patients (right graph) (Zauner et al., 1996). Note that a threshold for glutamate release exists, at about 18–20 ml/100 g/min (reproduction with permission).

hematoma (ASDH) (Fujisawa et al., 1994) and infusion of glutamate in vivo (Landolt et al., 1998; Maxwell et al., 1994) indicate that agonist-gated ion channels such as the NMDA (*N*-methyl-D-aspartate) glutamate receptors are important contributors to this type of edema. We have shown a significant relationship between extracellular $K^+$ (an indicator of trans-membrane ionic flux) and extracellular glutamate, after severe human head injury, using microdialysis.

*Human spinal cord injury (SCI)*

As with head injury, the severity of impairment of neurological function is dependent upon the magnitude of shear force imposed, but this may be modified by the presence or absence of osteophytes which may narrow the spinal canal and predispose the cord to damage with trauma. In those patients who die central hematomyelia is usually seen within the cord acutely. This is usually surrounded by a zone of contusional damage with petechial perivascular hemorrhage extending outward from the center. Such lesions are always accompanied by swelling of the cord and compression of the cord may be produced by a displaced bony fragment. Magnetic resonance imaging (MRI) has made a major contribution to the diagnosis and management of patients with acute spinal cord injuries, because MRI can visualize the zones of hemorrhage and peri-hemorrhagic oedema. Even in the absence of bony injury, MRI will frequently reveal a 'bright lesion' within the center of the cord in those patients who have sustained a hyperextension injury producing microvascular damage and cord ischemia together with stretching of axons with both functional and delayed structural axotomy (Bullock, 1993).

The relationship between the size and nature of an acute spinal cord lesion and the degree of consequent functional impairment which is produced is poorly understood. Patients with apparently severe lesions on MRI imaging may subsequently recover considerably, while others with small or even absent lesions on MRI may remain paraplegic or quadriplegic permanently. There is also evidence, both from animal models of SCI, and in vitro excitoxicity models, that glutamate plays a major role in neuronal damage after SCI (Agrawal and Fehlings, 1997; Holtz and Gerdin, 1991; Liu et al., 1991; Martinez-Arizala et al., 1990; Panter and Faden, 1992; Panter et al., 1990; Regan and Goi, 1991; Sun and Faden, 1994).

*Animal models of TBI*

Over the last years systematic efforts to develop animal models that reproduce specific features of TBI and SCI in humans have yielded a variety of models, which reproduce different aspects of TBI, but no single model adequately depicts the situation in humans (Gennarelli, 1994). The pathophysiology of head injury has been studied predominately in rodents, but also in gyrencephalic animals such as primates (Gennarelli et al., 1981), dogs (Griffiths, 1976), pigs (Smith et al., 1997), sheeps (Lewis et al., 1996) and cats (DeWitt et al., 1986; Katayama et al., 1985; Lewelt et al., 1980; Shima and Marmarou, 1991). Several models are widely used to study certain aspects of human head injury to evaluate drug effects and pathophysiological mechanisms. Many of these models have recently been reviewed in relation to the neuropathology in head-injured patients (Gennarelli, 1994). The most frequently used animal models are the central or lateral fluid percussion injury model (FPI) (Dixon et al., 1987), the central or lateral controlled cortical impact injury model (CCII; also called: rigid percussion model) (Lighthall, 1988), the weight-drop injury model (WDI; also called: impact acceleration model) (Marmarou et al., 1994) and several injection models such as acute subdural hematoma model (ASDH) (Miller et al., 1990a) or intracerebral hematoma models (ICH) (Jenkins et al., 1989). To evaluate the isolated effects of 'pure' agonist/receptor effects, excitotoxicity models have been developed (Alessandri et al., 1996; Boisvert and Schreiber, 1992; Fujisawa et al., 1993; Van Harreveld and Fifkova, 1971). These models cover different aspects of head injury in patients such as diffuse axonal injury, contusion, fracture, acute subdural, epidural and intracerebral hematoma and brain swelling. It has been shown that they model morphological changes after TBI relatively well, whereas neurological changes (cognition, memory, long-term outcome), physiology (e.g. cerebral metabolism) and biochemistry (e.g. gene expression, ionic changes) are less well characterized by animal models (partially due to a lack of data from patients) (Gennarelli, 1994). Almost no data are available about the relationship between subtle cognitive and behavioral effects, after moderate injury, in humans, and the relationship between these effects and functional sequelae in animal models, such as FPI.

*Animal models of SCI*

A variety of different methods have been developed to evaluate spinal cord injury in animals, such as spinal cord ischemia models (Faden et al., 1984; Marsala and Yaksh, 1994b; Robertson et al., 1986), spinal cord contusion (Panter et al., 1990; Wrathall et al., 1985). Similar to the weight-drop TBI model, a contusion of the spinal cord is produced by either dropping a weight (e.g. 10 g) on a impounder which rests on the dura, or by mechanical compression using clips (Khan and Griebel, 1983). These models have been characterized in terms of biomechanics, somatosensory evoked potentials, blood-brain barrier disruption, histopathology and functional deficits (Gale et al., 1985; Khan and Griebel, 1983; Noble and Wrathall, 1985; Noble and Wrathall, 1987).

*Glutamate- and excitotoxin-infusion models*

Although excitatory amino acid (EAA) release has been shown in human TBI and animal models of TBI and SCI (see below), it is not clear whether this is a significant pathological factor, or merely a epiphenomenon. To address this more clearly, several authors, including ourselves, have infused glutamate (Fig. 5a), NMDA, AMPA ($\alpha$-amino-3-hydroxy-5-methylisoxazole-4-propionic acid), kainate etc. into the intact brain (Alessandri et al., 1996; Boisvert and Schreiber, 1992; Foster et al., 1987; Fujisawa et al., 1993; Landolt et al., 1998), and after TBI (Bullock and Di, 1997), at different concentrations, to try to determine whether the concentrations measured in animal models and patients after TBI are in fact toxic to neurons.

Contrary to the injection of kainic acid, NMDA, or AMPA, glutamate infusion activates all receptor subtypes and is taken up by astrocytes. Van Harreveld and Fifkova (Van Harreveld and Fifkova, 1971) showed for the first time that glutamate, electrophoretically applied to the brain, causes morphological changes, indicating the potential danger of glutamate in the CNS. Fujisawa and coworkers (Fujisawa et al., 1993) reported dose-dependent, ischemia-like cortical lesions after

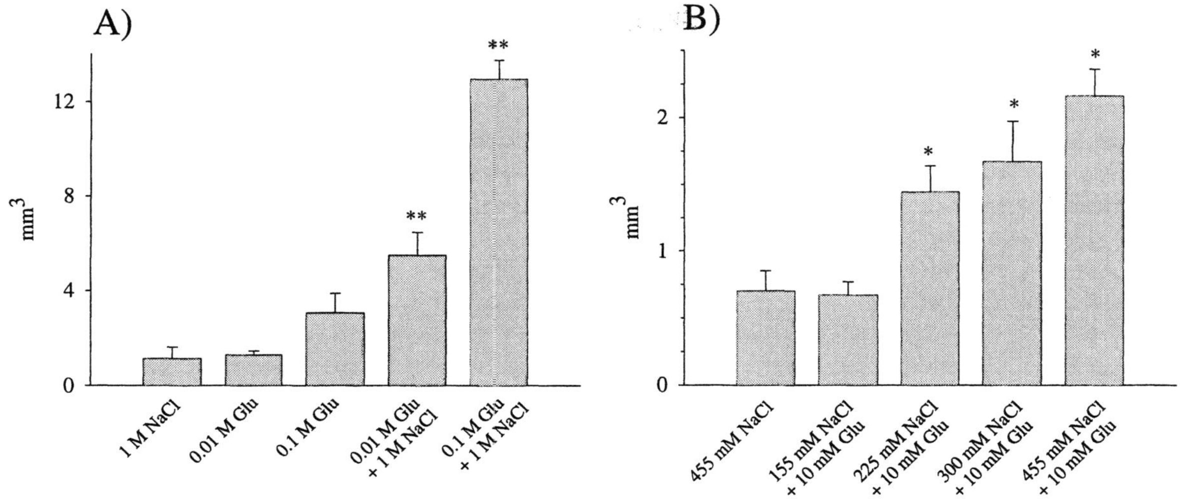


Fig. 5. Acute dose-dependent glutamate-induced cortical lesion and potentiation by sodium ions. (A) Glutamate infusion (10, 100 $\mu$M) for 90 min through microdialysis into the cortex of rats through a microdialysis probe for 90 min and (B) Potentiation of glutamate neurotoxicity (500 $\mu$M in perfusate for 30 min) by increasing concentrations of sodium in the perfusate (Alessandri, unpublished data).

90-min infusion of glutamate through a microdialysis probe into the cortex of rats (Fig. 5a). Similar to ischemia, the infusion of glutamate produced hypermetabolism, measured as increased 2-deoxyglucose uptake, in the boundary zone around the glutamate-induced cortical damage (Fujisawa et al., 1996). In this model, neuroprotection was achieved by NMDA and AMPA/KA receptor antagonists (D-CPPene, MK801, NBQX), sodium-channel blocker (BW103c87), and the free radical and lipid peroxidase inhibitor U74006F, supporting the excitotoxic nature of glutamate in vivo and the involvement of free radicals (Alessandri et al., 1995; Boisvert and Schreiber, 1992; Di et al., 1996b; Fujisawa et al., 1993). A recently performed study also indicates that increased extracellular sodium concentrations may potentiate the excitotoxicity of glutamate (unpublished data, Fig. 5b). A synergistic damaging effect of glutamate on cortical tissue has been found if glutamate was superimposed upon traumatic brain damage, such as FPI, indicating increased susceptibility of neurons to glutamate after an initial brain injury (Bullock and Di, 1997).

Unlike glutamate, NMDA, AMPA, kainic acid, or other excitotoxins can give specific information about which glutamate receptor subtype is most involved in glutamate-induced neuronal death. Studies have shown, that NMDA, AMPA, kainic acid as well as glutamate have excitotoxic properties, in vitro, as well as in vivo (Regan and Choi, 1991). The concentrations needed to kill murine cortical neurons in cultures is lowest for NMDA ($EC_{50} = 10$ $\mu$M), followed by AMPA ($EC_{50} = 12$ $\mu$M) and kainic acid ($EC_{50} = 50$ $\mu$M), indicating that the strongest excitotoxic effect of glutamate is transmitted through the NMDA receptor subtype (Regan, 1996).

Spinal cord damage and behavioral deficits due to microinjection of NMDA (Douglas et al., 1993; Liu, 1994; Liu et al., 1997), AMPA (Liu et al., 1997), quisqualic acid (Douglas et al., 1993; Yezierski et al., 1993), kainic acid (Douglas et al., 1993; Liu, 1994), indicate a major role of the glutamate receptors in SCI. In vitro experiments also showed neuronal, but not glial cell death after 5-min exposure to 10–1000 $\mu$M glutamate ($EC_{50}$ 100–200 $\mu$M) (Regan and Choi, 1991). The $EC_{50}$ for AMPA (5 $\mu$M; 24 h-exposure) was much lower than for kainate (20 $\mu$M) and NMDA (30 $\mu$M), suggesting massive non-NMDA mediated excitotoxicity following SCI (Regan, 1996). Contrary to massive NMDA receptor involvement in TBI, the target for glutamate receptor antagonist therapy in patients after SCI may therefore focus more on non-NMDA receptor modulation. This is supported by in vitro results from Regan (1996), demonstrating that cortical neuronal tissue is much more vulnerable to NMDA than spinal cord tissue, with $EC_{50}$ of 10 $\mu$M and of 30 $\mu$M, respectively (Regan, 1996).

*Ionic and metabolic changes*

Ion channels are by far the most frequent structures embedded in neuronal and glial membranes. To date over 15 varieties of neuronal ion channels have been identified. The most common of these, voltage gated channels, are closely linked functionally with the sodium/potassium ATPase pump. Many types of ion channel are linked to the specific agonist-gated receptors, and others are linked within the cell with second messenger systems, such as adenylate cyclase and G-proteins (Sachs, 1991; Siegelbaum and Koester, 1991).

The effects of traumatic mechanical shearing upon neuronal and glial membranes is not well understood in brain injury. Membranes themselves appear to be 'stretchy' and relatively resistant to deformation and movement which might be explained by the constant force on membranes by arterial pulse pressure waves and extremes of sagittal plane neck movements which stretches the spinal cord and medulla several centimeters (Bullock, 1997). However, mechanical deformation at certain shear magnitudes and rates have been shown to radically alter ion channel function in neurons cultured on a deformable surface (Zhang et al., 1996). Specific classes of 'mechanotransducing' ion channels have been identified using patch clamp techniques in both neurons and glia (Bowman et al., 1992; Sachs, 1991). Ion channels may remain 'leaky' for several hours after mechanical deformation (Tavalin et al., 1995). The majority of voltage-sensitive and agonist-gated ion channels could probably be affected by transient mechanical deformation by shearing forces. In vitro experiments have shown that monolayer neuron and astrocyte cultures growing on a flexible plastic membrane, which are rapidly deformed by a brief air jet impulse, undergo rapid calcium entry and subsequent neuronal death, with efflux of lactate and potassium into the culture medium. Recently, it could be demonstrated that this deformation leads to a release of the

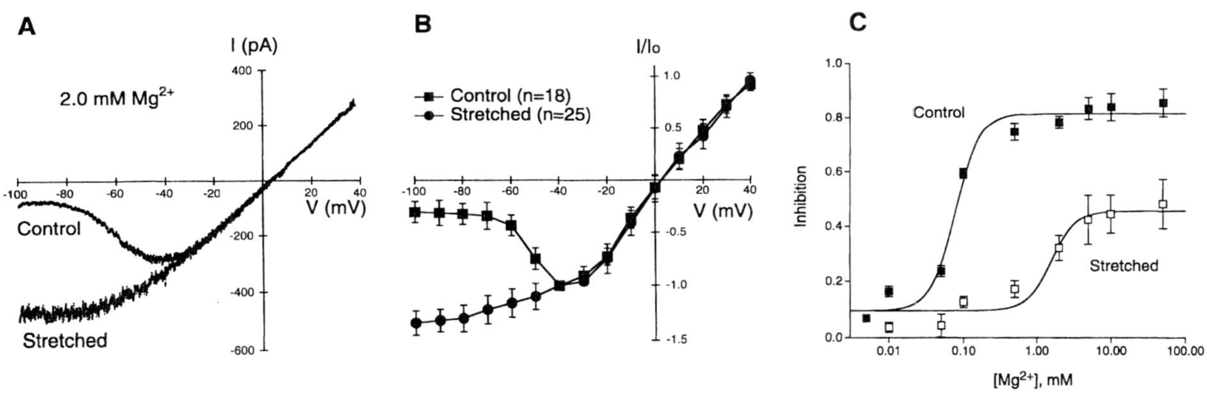


Fig. 6. Stretch-induced neuronal injury reduced voltage-dependent $Mg^{2+}$ blockade of NMDA currents in vitro. (A) Whole cell NMDA currents recorded from control and stretched neurons; (B) Averaged current voltage relationship of normalized NMDA currents from 18 control and 25 stretched neurons; (C) The NMDA currents of control and stretched neurons differed in their sensitivity to extracellular $Mg^{2+}$ (reproduced with permission from Zhang et al., 1996).

magnesium block of ionotropic NMDA receptor channels (Zhang et al., 1996), which may increase the vulnerability of neurons to glutamate (Fig. 6).

Data from in vivo trauma models such as FPI and CCI models have shown massive rapid transient efflux of potassium into the ECF, associated with a fall in sodium content in the ECF (Bullock et al., 1995b; Di et al., 1996c; Katayama et al., 1990; Nilsson et al., 1993). These changes may be explained by rapid alterations in the relationship between ion fluxes through voltage-dependent ion channels, $Na^+/K^+$ dependent ATPase pumps, and agonist-gated ion channels. Blockade of voltage-gated sodium channels prior to FPI also fails to ameliorate the negative neurological and behavioral effects and had only moderate effect on potassium fluxes, suggesting that greater pathological effects of TBI maybe upon agonist operated ion channels such as the NMDA receptor. Contrary to TBI, the sodium blocker tetrodotoxin (TTX) has been found to improve behavioral outcome after spinal cord contusion in rats, showing an important role for voltage-gated sodium channels in SCI (Teng et al., 1997). This increased 'leakiness' for potassium may be buffered by astrocytic uptake which results in enormous astrocytic swelling. This astrocyte swelling is a hallmark of acute cerebral ischemia (Katayama et al., 1992), cerebral contusion (Ito et al., 1996; Van den Brink et al., 1990) acute subdural hematoma (Fujisawa et al., 1994; Kinoshita et al., 1994) and glutamate excitotoxicity (Landolt et al., 1998; Maxwell et al., 1994) (Fig. 7). Such cytotoxic brain swelling is mainly responsible for early post-traumatic ICP and concomitant reduced CPP, producing secondary focal ischemia

Massive ionic fluxes and glutamate release as found after TBI, also produce an intense energy

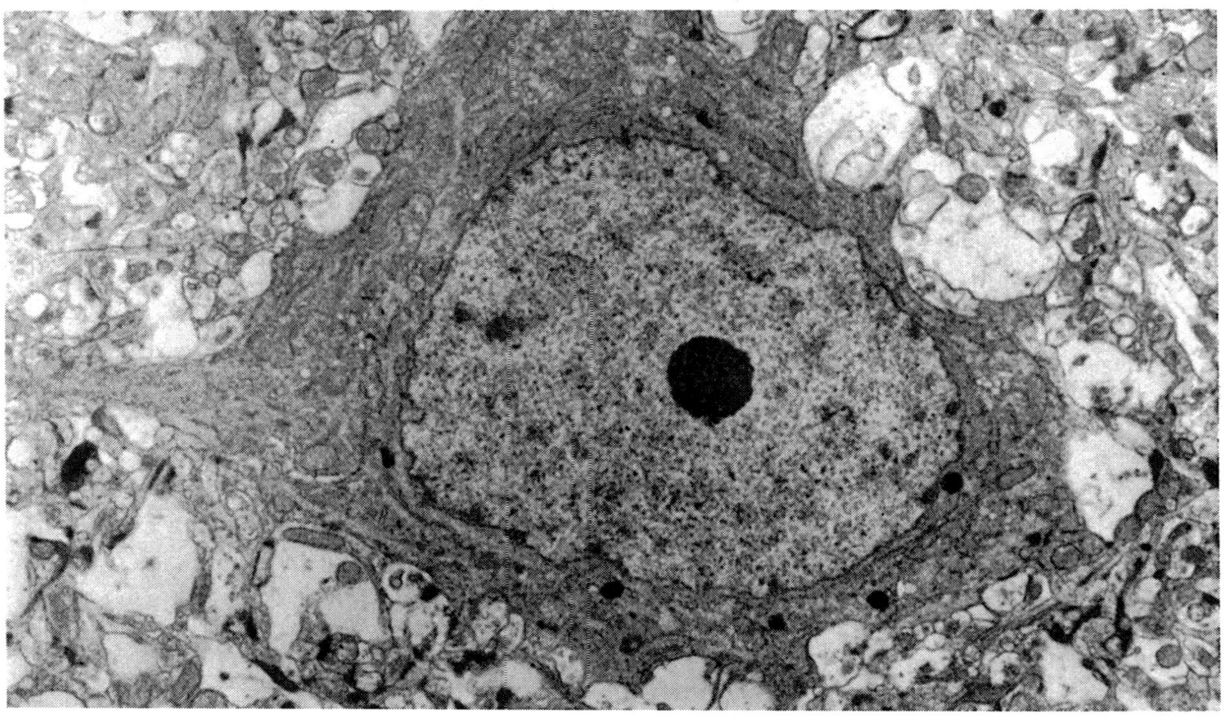

Fig. 7. Swollen peri-neuronal astrocytes processes after glutamate infusion into the cortex of a rat via microdialysis probe (reproduction with permission from Maxwell et al., 1994).

demand, resulting in an abrupt increase in glucose utilization. Glucose measurements in the ECF and 2-deoxyglucose studies using animal models of glutamate excitotoxicity (Alessandri et al., 1996; Fujisawa et al., 1996), FPI (Andersen and Marmarou, 1992; Hayes et al., 1988b; Hovda et al., 1990; Unterberg et al., 1988), WDI (Dimlich et al., 1990) and ASDH (Inglis et al., 1992; Kuroda et al., 1992) have revealed a massive increase in glucose utilization following injury. This hypermetabolism, especially found in the 'penumbral' zone of focal damage, can even persist up to 7 days. Normally, this hypermetabolism is followed by a period of hypometabolism, lasting days up to weeks after impact (Bergsneider et al., 1997; Hovda et al., 1991; Hovda et al., 1990; Kuroda et al., 1992; Yoshino et al., 1991). The effect of hypermetabolism on cell survival might be worsened by an early uncoupling of CBF and substrate demand ($O_2$, glucose). Compromised substrate delivery may directly affect mitochondrial functions, enzyme functions, and ion channel functioning, leading to processes such as free radical production, lipid peroxidation, accumulation of lactate and glutamine and ionic imbalances. Lactate production in the brain has recently been linked directly to glutamate uptake into astrocytes (Pellerin and Magistretti, 1996). In vitro studies showed that the uptake of glutamate stimulates glycolysis in astrocytes, leading to increased lactate. Lactate is then transported to neurons and is used there as an energy substrate for oxidative metabolism (Pellerin and Magistretti, 1994).

A combined stable Xenon CT and microdialysis study from our center revealed that in more than 50% of patients with elevated dialysate lactate, CBF was higher than the ischemic threshold (Doppenberg et al., 1996). Positive correlations between glutamate and lactate levels in human head injured patients, as measured by microdialysis, support glutamate-induced glycolysis after trauma (Alessandri et al., 1998; Alessandri et al., 1997). This might indicate that massive ischemic processes early after trauma, which lead to massive anaerobic lactate production, overshadow this more subtle glutamate-lactate interrelationship.

## Glutamate and amino acid changes after TBI and SCI

### *Glutamate release in animals*

Although histological studies in TBI and SCI models show many similarities to the neuropathology seen in head-injured patients who die, the neurochemical basis for these morphological changes are less well understood. Glutamate may be a major factor involved in the pathogenesis of these brain lesions. The major effect of glutamate is the *activation of ionotropic* (NMDA, AMPA/KA) and *metabotropic glutamate receptors*, inducing the *opening of ion channels* with subsequent massive influx of calcium and sodium ions; the *release of calcium from intracellular stores* via second messenger pathways; the *initiation of down-stream processes* such as gene expression; and the *activation of high-affinity glutamate carriers* in neurons and astrocytes leading to increased energy demand, upon the tissue.

Following FPI, increased levels of glutamate and other amino acids have been reported by several authors (Katayama et al., 1990; Panter and Faden, 1992). For example, a 18-fold increase in glutamate was found following severe FPI in rats (Faden et al., 1989). Peak levels seems to be reached within minutes, and although glutamate elevations may persist up to 1 hour after severe injury, baseline levels are usually reached after 30 minutes after impact (Fig. 8a). In a model of concussive brain injury, similar transient (20–30 min) extracellular increases (8–13 fold) were reported with peak levels after 10 minutes (Nilsson et al., 1990).

Using the CCII model, the amplitude and time-course of extracellular glutamate elevation (30–60 min) were directly related to the depth of contusion (Palmer et al., 1993a; Palmer et al., 1993b). Glutamate release and the subsequent massive inflow of calcium into cells have been linked to mitochondrial dysfunctions in this model (Xiong et al., 1997). Histological examination revealed very similar pathological changes in the CCII model, when compared to NMDA and AMPA injection into the normal brain (Palmer et al., 1993a). In this model, hippocampal damage was also seen distant from the contusion, and produces enduring cognitive deficits (Hamm et al., 1992).

In the rodent model of ASDH, the acute cortical ischemic lesion produced by subdural injection of 400 $\mu$l of autologeous blood ($\pm$20% of intracranial volume) is associated with long-lasting release of glutamate (Bullock et al., 1991; Miller et al., 1990a). The dialysate levels of glutamate increased 7–8 fold over baseline value in the cortex and 3–4 fold in the hippocampus underlying the hematoma (Fig. 8c).

Microdialysis studies have shown only modest post-impact glutamate levels in a recently devised weight drop injury model (Marmarou et al., 1994). However, when a delayed secondary ischemic insult was superimposed in this model, dialysate glutamate increased about 10-fold.

In contrast to cerebral damage, the involvement of glutamate receptors in the development of spinal cord damage has been less thoroughly investigated. Over recent years, EAAs as a factor in the development of secondary pathology following traumatic SCI have been studied (Haghighi et al., 1996; Simpson RK et al., 1990; Wrathall et al., 1992b; Wrathall et al., 1994). EAAs are elevated in models of traumatic (Farooque et al., 1996; Liu et al., 1991; Panter et al., 1990), and ischemic SCI (Marsala et al., 1994a; Simpson RK

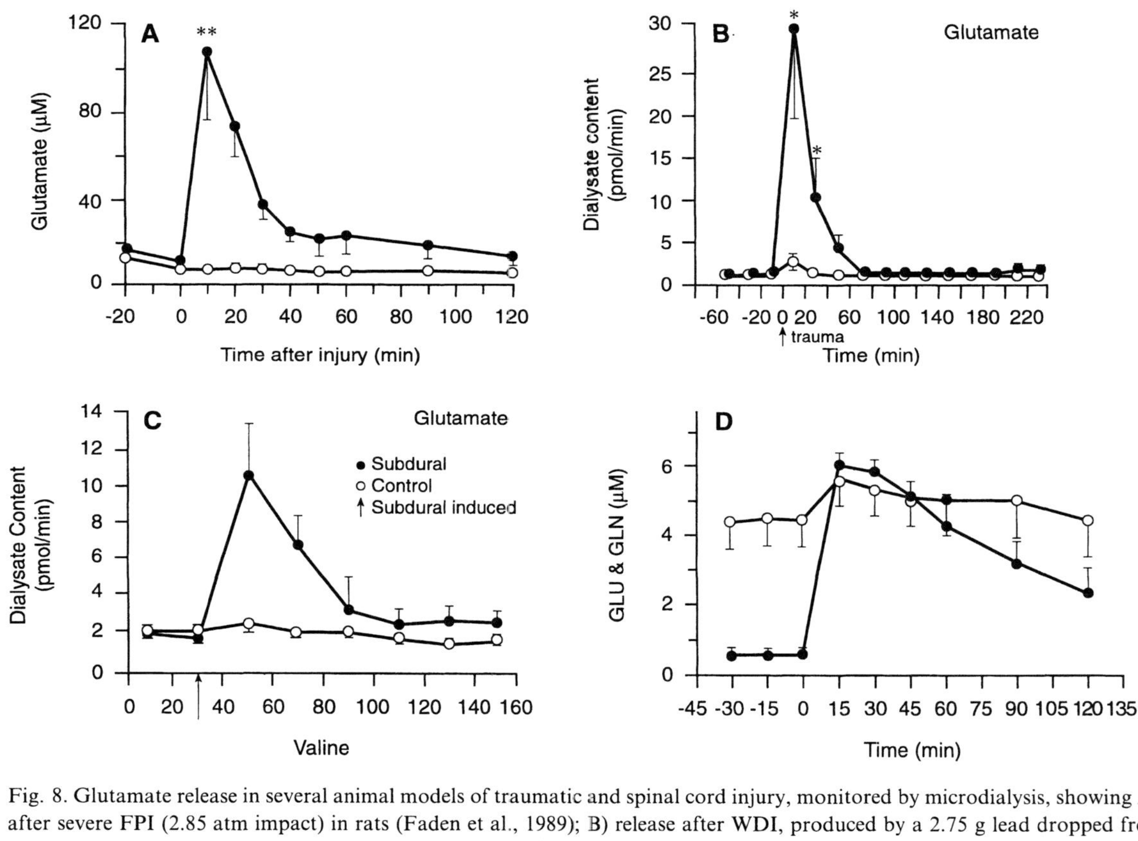


Fig. 8. Glutamate release in several animal models of traumatic and spinal cord injury, monitored by microdialysis, showing A) release after severe FPI (2.85 atm impact) in rats (Faden et al., 1989); B) release after WDI, produced by a 2.75 g lead dropped from 90 cm height, in rats (Koizumi et al., 1997); C) release following ASDH in rats, induced by subdural injection of 400 $\mu$l autologeous blood (Bullock et al., 1991); and D) glutamate and glutamine O release after spinal cord WDI in rats, produced by a 10 g weight dropped from 5 cm height (Liu et al., 1991). Note: The concentrations found in the different models cannot be directly compared due to differences in microdialysis recovery rates (all figures reproduced with permission).

et al., 1990; Yum and Faden, 1990). Using a WDI model, Panter et al. (Panter et al., 1990) found that increased glutamate and aspartate release was directly related to the severity of spinal cord injury, with 2–4 fold elevation following moderate, and 4–6 fold elevation after severe injury. After severe SCI, glutamate was increased up to 60 min. A similar increase was found by Farooque et al. (Farooque et al., 1996) with a 5–6 fold peak increase after severe weight-drop injury. Another study using a spinal cord WDI model, reported higher (10-fold) and longer lasting (up to 120 min) increases of extracellular glutamate with a peak elevation as soon as 15 min post-injury (Liu et al., 1991) (Fig. 8d). Ischemia (20 min) and reperfusion (60 min), produced by means of a inflatable balloon, caused an immediate increase of glutamate (about 4-fold), which lasted for more than 60 min. after reperfusion (Simpson RK et al., 1990). In a similarly designed experiment, Marsala et al. (1994) found a shorter time-course and less massive glutamate increase (2.5-fold) using 20-min ischemia with 120-min reperfusion, whereas the increase was about 4.5-fold after complete ischemia (cardiac arrest) (Marsala et al., 1994a).

*Human EAA and structural amino-acid changes*

Glutamate concentrations in humans have been reported to be about 600 μmoles/100 g in brain tissue and 0.8 μmole/100 ml in CSF (Siegel et al., 1989). The measurement of glutamate in CSF revealed markedly increased levels after traumatic brain injury, predominantly in patients with hematoma, contusions, and secondary ischemic events (for references see: Doppenberg and Bullock, 1997). Recently, the use of microdialysis in patients made it possible to measure local concentrations of glutamate and other excitatory amino acids in the brain. Glutamate was measured following severe head-injury (Alessandri et al., 1998; Alessandri et al., 1997; Bullock et al., 1995a; Bullock et al., 1994; Bullock et al., 1995b; Doppenberg et al., 1996; Goodman et al., 1996a; Goodman et al., 1996b; Kanthan and Shuaib, 1995; Kanthan et al., 1996; Persson et al., 1996; Zauner et al., 1996), subarachnoid hemorrhage (Enblad et al., 1996; Persson et al., 1992), stroke (Bullock et al., 1995c; see also Berg-Johnsen et al., this volume), during epilepsy surgery (During and Spencer, 1993), cerebrovascular surgery (Mendelowitsch et al., 1997, 1998), and in Parkinson's disease patients (Meyerson et al., 1990). Using this sampling technique, prolonged glutamate release has been found in head-injured patients (Bullock et al., 1995a; Bullock et al., 1994; Bullock et al., 1995b) with peak concentrations up to 250 μM glutamate in the dialysate. Our analysis of microdialysis data showed that glutamate levels are prolonged and markedly increased (50–60 fold) in patients with secondary ischemic insults and focal contusions, whereas patients with diffuse axonal injury showed only transient increases (Bullock et al., 1995a; Bullock et al., 1995b). The strong positive correlation of mean glutamate levels with the patient outcome indicates a major potential impact of glutamate on the fate of neurons following injury. However, the dynamic process of release and immediate uptake by astrocytes and pre-synaptic vessels, makes it difficult to calculate true extracellular levels from these dialysates values, and to draw conclusions about the quantity and duration of EAA exposure, needed to damage brain tissue after human TBI (Fig. 9).

Glutamate release in human patients has been shown to be closely related to CBF (Zauner et al., 1996), and lactate (Alessandri et al., 1998; Alessandri et al., 1997; Doppenberg et al., 1996). The highest glutamate levels in patients were found at ischemic CBF below 20 ml/100 mg/min (Fig. 4). Lactate was positively correlated with the release of glutamate, and negatively with CBF in trauma patients. However, there were many patients with non-ischemic CBF and high lactate levels, indicating dynamic and complex regulation mechanisms following TBI. Since there is evidence that glutamate 'drives' astrocytic glycolysis and subsequently production of lactate, which serves as an energy substrate for neurons (Alessandri et al., 1998; Alessandri et al., 1997; Bullock et al., 1995c; Doppenberg et al., 1996; Magistretti and Pellerin, 1997; Pellerin and Magistretti, 1994; Pellerin and

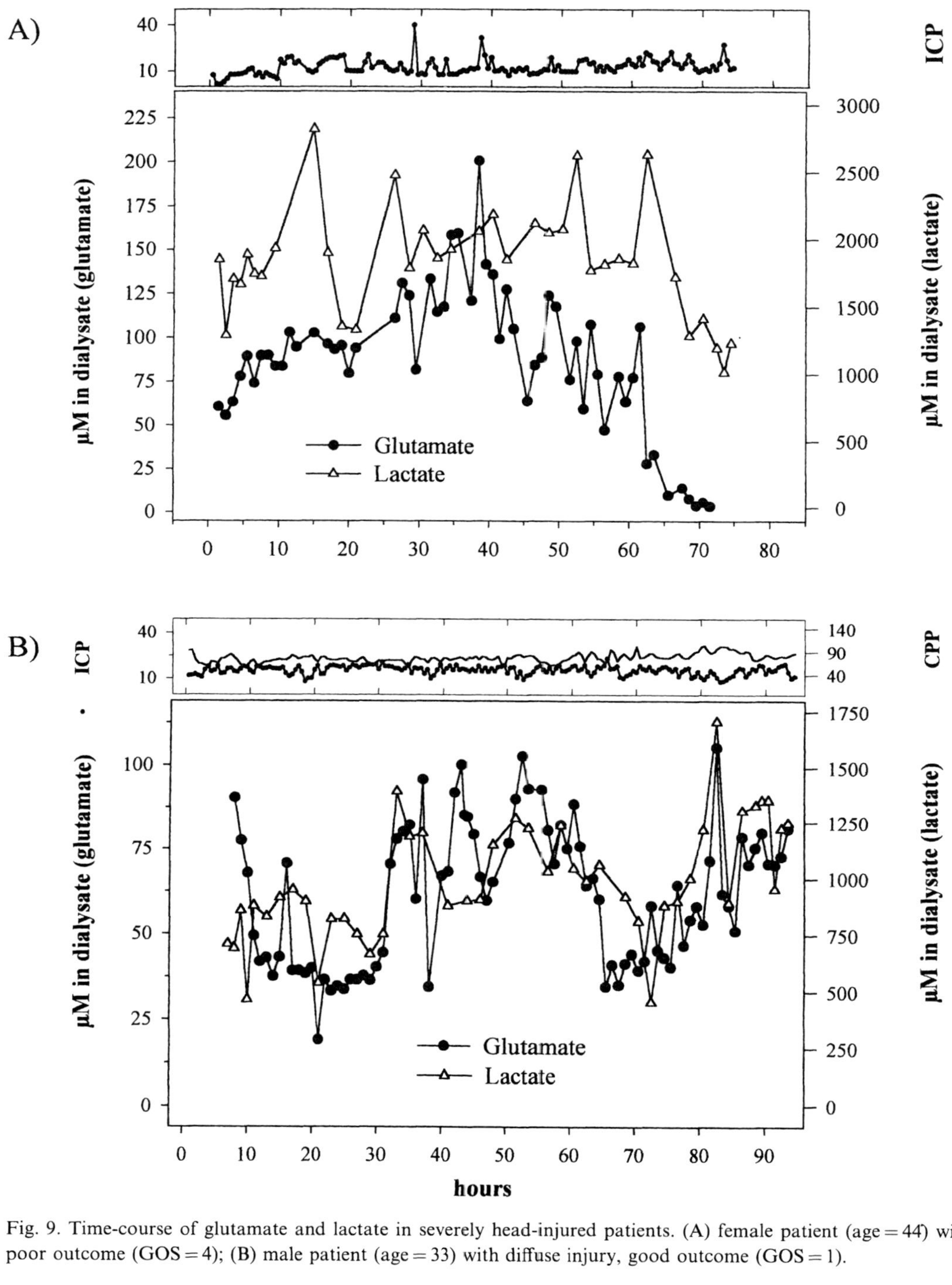


Fig. 9. Time-course of glutamate and lactate in severely head-injured patients. (A) female patient (age = 44) with severe contusion, poor outcome (GOS = 4); (B) male patient (age = 33) with diffuse injury, good outcome (GOS = 1).

Magistretti, 1996), a close relationship between glutamate and lactate may also indicate a unique mechanism for the brain to support neurons with energy following head-injury, when EAA's are making metabolic demands upon the tissue.

Structural amino acids have been investigated less thoroughly than EAA. However, structural amino acids may give an indication of disturbed membrane integrity (e.g. threonine) or compromised enzyme systems following trauma. A recent analysis of our amino-acid database in human neurotrauma revealed that high structural amino acid levels, as compared to glutamate, indicate a more favorable outcome, whereas low levels point towards poor outcome with more ischemic events. The relationship between EAA and structural amino acids, an 'amino acid tissue signature' concept, has to be investigated more thoroughly to make outcome predictions (Bullock, Birgfeld, Alessandri, 1997: unpublished data).

In human SCI, no such microdialysis studies have been reported as yet. It is therefore uncertain whether or not glutamate increases to levels which might be considered to be excitotoxic after SCI. Future studies will help to elucidate the role of glutamate in this type of human CNS injury.

*Effect of glutamate antagonists on brain or spinal cord damage*

*NMDA receptor antagonists* were the first drugs investigated for their neuroprotective properties in models of TBI. Histology, behavior, and metabolism were studied following traumatic brain and spinal cord injury. It was shown that the competitive NMDA receptor antagonist CGS19755 (selfotel; *cis*-4-(phosphomethyl)-2-piperidine carboxylic acid) and dextrorphan (a weak non-competitive NMDA receptor antagonist) attenuated trauma-induced glutamate release (Panter and Faden, 1992). Other antagonists such as 2-amino-5-phosphovaleric acid (APV) and kynurenic acid, were found to reduce trauma-induced glucose-hypermetabolism (Kawamata et al., 1992). 3-(2-carboxypiperizin-4yl)-propyl-1-phosphonic acid (CPP), phencyclidine (PCP) and dizocilopine (MK-801) also improved neurological motor outcome after TBI (Faden et al., 1989; Hayes et al., 1988a). High doses of MK-801 (3 mg/kg) improved outcome and brain edema even if injected up to 1 hour after injury (Shapira et al., 1990). Similarly, dextromethorphan and HU-211 (7-hydroxy-tetrahydrocannabinol 1,1-dimethyl heptyl; a cannabinoid with affinity to the NMDA receptor) have been shown to improve neurological motor and memory function and decrease regional edema formation when administered post-injury (Faden et al., 1989; Shohami et al., 1995; Shohami et al., 1993). However, a low dose of MK801, which was effective if injected pre-injury, did not show effects on FPI-induced cognitive deficits when administered 15 min post-injury (Hamm et al., 1993).

Other ways to affect glutamate excitotoxicity include the modulation of the NMDA receptor by *magnesium* (voltage-dependent magnesium block), or by compounds binding to the NMDA-associated glycine and polyamine sites. In vitro and in vivo experiments have demonstrated that traumatic injury disrupts magnesium homeostasis (Vink et al., 1987; Vink et al., 1988), and the magnesium block of the NMDA receptor (Zhang et al., 1996). Magnesium therapy was associated with improved cognitive (Smith et al., 1993a) and neurological motor outcome (McIntosh et al., 1988), and reduced brain swelling in various studies (e.g. Okiyama et al., 1995).

The modulation of NMDA receptor activation is an important neuroprotective mechanism. Chronic injection of D-cycloserine, an antibiotic and partial *NMDA glycine-site antagonist*, after lateral FPI attenuated trauma-induced cognitive deficits in rats (Temple and Hamm, 1996), further indicating the potency of glycine-site antagonists. Modulation of NMDA receptor activation through *polyamine sites* by ifenprodil or eliprodil (SL82.0715), and recently CP101-606 have been reported to be neuroprotective in TBI (Toulmond et al., 1993), and ASDH (Tsuchida et al., 1997), as well as in reducing astrocyte swelling following a cryo-lesion (Trout et al., 1995).

The *AMPA receptor* is another possible site of glutamate-mediated neuroprotection following

TBI or SCI. AMPA receptor antagonists, such as indole-2-carboxylic acid (I2CA) (Smith et al., 1993b), kynurenic acid (Hicks et al., 1994; Kawamata et al., 1992; Smith et al., 1993b) and 6-cyano-7-nitroquinaxaline (CNQX) (Kawamata et al., 1992) have been shown to improve neurological motor and cognitive function, and to restore regional ion homeostasis, cerebral edema and post-traumatic hyperglycolysis.

The *AMPA/KA receptor* is an ionotropic receptor which is permeable to sodium, potassium and, depending on its subunit composition, also to calcium. 2,3-Dihydroxy-6-nitro-7-sulfamoylbenzo[*f*]quinoxaline (NBQX), a potent AMPA antagonist, is effective in reducing glutamate-induced brain swelling in rats (Westergren and Johansson, 1992). Recently, it has been shown that NBQX, given before or up to 7 hours after TBI reduced cortical and hippocampal damage in adult rats (Ikonomidou and Turski, 1996). Delayed neuroprotection (cognitive outcome) was also found with 1-(4-aminophenyl)-methyl-7,8-methylenedioxy-5H-2,3-benzodiazepine HCl (GYKI-52466) when administered at 15 min or 3 hours following experimental brain injury (Hylton et al., 1995), indicating a potentially longer time-window for therapy with AMPA/KA receptor antagonists, than NMDA antagonists.

Another approach to modulate glutamate-induced excitotoxicity following TBI or SCI is the *presynaptic inhibition of glutamate release*. Compounds such as 5-[2,3,5-trichlorophenyl] pyrimidine-2,4-diamine ethylane sulphonate (BW1003c87), and 4-amino-2-[4-methyl-1-piperazinyl]-5-[2,3,5-trichlorophenyl]-pyrimidine mesylate monohydrate (619c89) inhibit veratrine- but not potassium-stimultated glutamate release, potentially by reducing ion flux through voltage-gated $Na^+$ channels. BW1003c87 significantly reduced glutamate-induced cortical lesion in the rat (Fujisawa et al., 1993). After ASDH, 619c89 significantly reduced ischemic brain damage as well

TABLE 2

Neuroprotective effects of glutamate receptor antagonist in animal models of FPI

| | Reference | Drug | Time | Effects |
|---|---|---|---|---|
| TBI | Hayes et al., 1988 | PCP | pre-trauma | PCP sign. reduced behavioral deficits |
| TBI | McIntosh et al., 1988 | Magnesium | pre-trauma | Sign. improved cellular bioenergetic state (inorg. phosphate, phospho-creatine, ATP) and neurological outcome |
| TBI | Smith et al., 1990 | Dextromethorphan CGS19755 | post-trauma | Sign. improved neurological outcome by dextromethorphan, but not CGS19755; no effects on edema formation |
| TBI | Katayama et al., 1992 | APV, CNQX, Kynurinic acid | pre-trauma | Reduction of increased glucose utilization following injury, measured using labelled 2-deoxyglycose. |
| TBI | Hamm et al., 1993 | MK-801 | pre- or post | Pre-, but not post-treatment reduced sign. cognitive deficits |
| TBI | Toulmond et al., 1993 | Eliprodil | post-trauma | 60% reduction of lesion volume |
| | | MK-801 | post-trauma | Non-sign. reduction of lesion volume |
| TBI | Hicks et al., 1994 | Kynurenic acid | post-trauma | Sign. neuroprotection in CA3 region of hippocampus |
| TBI | Ikonomidou and Turski, 1996 | CPP, NBQX | pre- or post | Pre-treatment reduced cortical and hippocampal lesion; NBQX, but not CPP post-treatment was neuroprotective |
| TBI | Temple and Hamm, 1996 | D-cycloserine | post-trauma | partial NMDA receptor agonist improved cognitive performance |
| SCI | Faden, 1993 | Dextrorphan | post-trauma | improved sign. behavioral recovery at 2 weeks post-trauma |

as glutamate release (Tsuchida et al., 1996). The glutamate release inhibitor 619c89 attenuated cognitive deficits (Voddi et al., 1995), whereas BW1003c87 reduced cerebral edema (Okiyama et al., 1995). Recent studies also reported neuroprotective properties for the glutamate release inhibitor riluzole, which has also shown clinical efficacy for motor neurone disease. Although no histological neuroprotection was found, the compound improved neurological motor and cognitive outcome after TBI (McIntosh et al., 1996b; Wahl et al., 1997).

Recently, there is increasing evidence that the metabotropic glutamate receptor family (mGlu) is also involved in TBI. In vitro and in vivo it has been shown that antagonism of the phosphoinositol-coupled subtype (mGluR1) protects against cell death following traumatic injury (Mukhin et al., 1996). In a model of FPI, blocking Group I metabotropic receptors by (+)-MCPG (α-methyl-4-carboxyphenylglycine) reduced neurological motor and cognitive deficits (Gong et al., 1995), indicating the involvement of certain μGlu receptor subtypes in TBI induced neuronal damage. However, the development of new, highly selective (against subtypes) metabotropic receptor agonists and antagonists is necessary to role out the specific involvement in the positive and negative effects on TBI and SCI in vivo.

NMDA receptor antagonists such as MK-801 and CPP also improved histological and behavioral outcome following traumatic (Faden et al., 1988a; Wrathall et al., 1992a; Wrathall et al., 1992b) and ischemic SCI (Hao et al., 1992; Martinez-Arizala et al., 1990). Recent studies, however, indicate that non-NMDA ionotropic as well as metabotropic receptors play an important part in glutamate excitotoxicity after SCI. The AMPA/KA antagonist NBQX potently decreased lesion size and improved recovery in models of spinal cord ischemia (von Euler et al., 1994) and contusion (Wrathall et al., 1992b; Wrathall et al., 1994). The

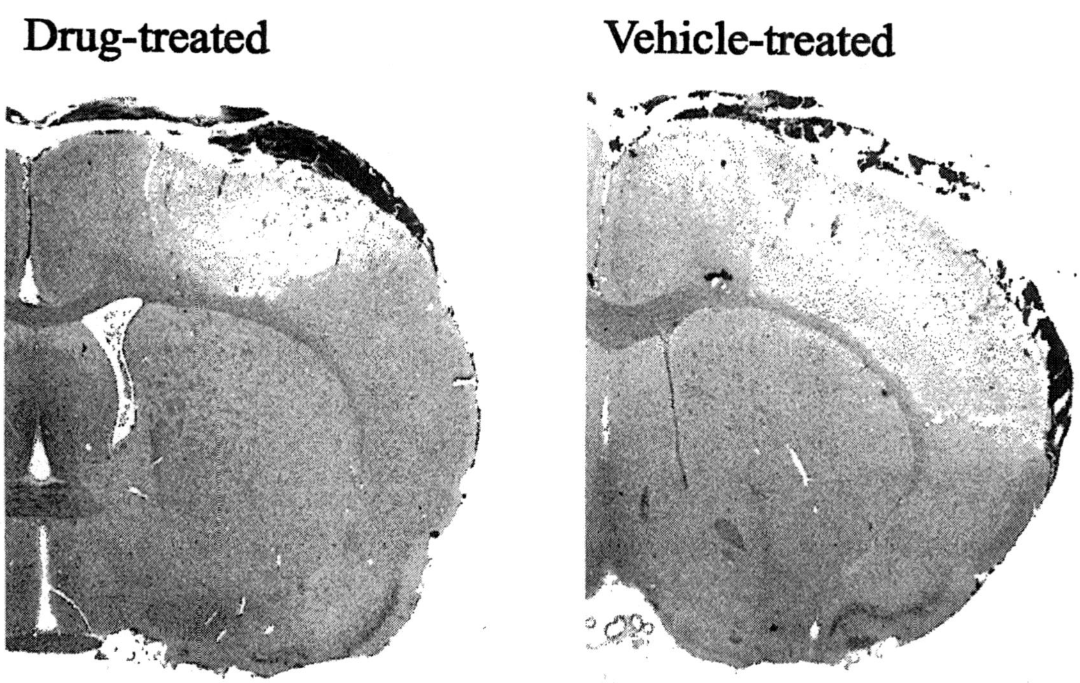


Fig. 10. Effect of CP101-606 (left), a glycine-site NMDA receptor antagonist on the acute ischemic lesion in a model of ASDH in rats (hematoxylin-eosin stained); right: section from control animal injected with vehicle following FPI.

TABLE 3

Neuroprotective effects of glutamate receptor antagonist in animal models of WDI and CCII

| | Reference | Drug | Time | Effects |
|---|---|---|---|---|
| TBI | Shapira et al., 1990 | MK-801 | post-trauma | Sign. reduced edema formation and improved neurological score; non-sign. effect on pathological score |
| TBI | Katoh et al., 1997 | MK-801 | pre-trauma | Glutamate increase lasted < 30 min, no effect of MK-801 on peak and time-course of glutamate release |
| SCI | Faden and Simon, 1988b | MK-801 | post-trauma | Improved neurological recovery at 4 weeks post-trauma |
| SCI | Wrathall et al., 1992b | NBQX | pre-trauma | Sign. reduction of behavioral deficits at 4 weeks post-injury |
| SCI | Wrathall et al., 1992a | MK-801 Dextromethorphan Kynurenic acid | pre- and post-trauma | Kynurenate, but not MK-801 or dextromethorphan reduced significantly injury-induced behavioral deficits at 4 weeks post-trauma |
| SCI | Wrathall et al., 1994 | NBQX | post-trauma | Sign. reduction of average lesion length, behavioral deficits, and increase of residual white matter tissue at 4 weeks post-injury |
| SCI | Wrathall et al., 1996 | NBQX | | Amelioration of trauma-induced functional deficits |
| SCI | Faden et al., 1988a | MK-801 | post-trauma | Improved neurological recovery and histological outcome at 4 weeks post-injury; effects on ion balance, but not on water content |
| CCI | Shohami et al., 1993 | HU-211 | post-trauma | Reduction of edema and BBB breakdown |
| CCI | Shohami et al., 1995 | HU-211 | post-trauma | Improved trauma-induced neurological and cognitive deficits at 4 weeks after TBI |

neuroprotective effect of NBQX exceeded that of CPP in the same model (Wrathall et al., 1992b and von Euler et al., 1997) found no protective effect for memantine, an NMDA receptor antagonist, which has a much lower NMDA receptor affinity in spinal cord than in the cortex of rats (von Euler et al., 1997). Recently, Mukhin and Faden (1996) reported neuroprotection by (+)MCPG using in vitro and in vivo models of SCI and they found an exacerbation of LDH release in vitro following

TABLE 4

Neuroprotection by glutamate receptor antagonist in animal models of ASDH in rats

| Reference | Drug | Time | Effects |
|---|---|---|---|
| Tsuchida and Bullock, 1995 | ACEA-1021 | pre- or post-subdural | Pre- (5, 15 mg/kg) and post-treatment (15 mg/kg) reduced ischemic lesion at 4 hours post-subdural (26–39%) |
| Inglis et al., 1992 | D-CPPene | pre-subdural | reduced sign. 2-deoxyglucose hypermetabolism |
| Di and Bullock, 1996a | ACEA-1021 MK-801 | pre-subdural | Sign. (up to 5-fold) increased 125iodo-MK801 binding in penumbra; sign. reduction by drugs |
| Kuroda et al., 1994 | MK801, D-CPPene CGS 19755 | post-subdural | about 30–35% reduction of ischemic lesion in all groups; no effect on ICP or CPP |
| Tsuchida et al., 1997 | CP101, 606 | post-subdural | 37% reduction of ischemic lesion size when given 20 min. after ASDH (2 or 5 mg/kg i.v. + continue infusion) |

injury by DHPG, an selective mGluR1 receptor agonist (Mukhin et al., 1996). Metabotropic receptors might also have some effects on regenerative membrane properties (Morisset and Nagy, 1996). However, the development of new, highly selective (against subtypes) metabotropic receptor agonists and antagonists is necessary to role out the specific involvement in the positive and negative effects on TBI and SCI in vivo.

*Glutamate receptor changes*

In a rat model of ASDH, our group showed neuroprotective effects of ACEA-1021, a glycine-site NMDA receptor antagonist (Tsuchida et al., 1995) and binding of $^{125}$iodo-MK-801 binding in the penumbra of ASDH-induced cortical lesion (Fig. 11), which could be blocked by ACEA-1021

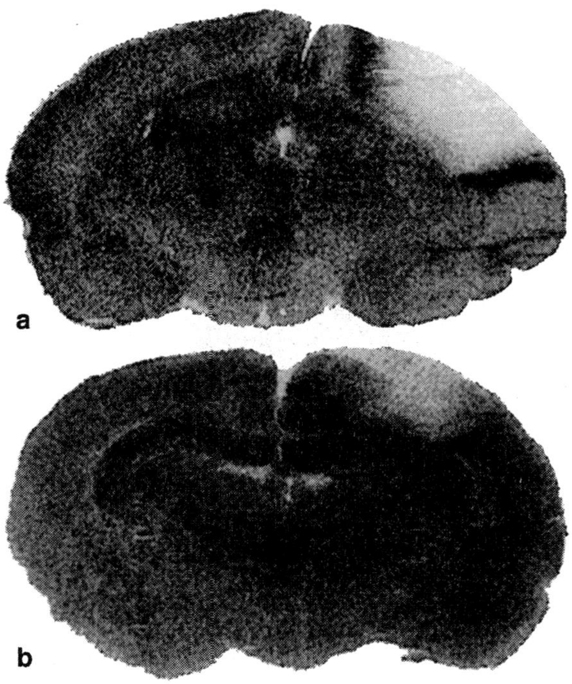


Fig. 11. upper photo: Autoradiogram of $^{125}$iodo-MK-801 binding following ASDH in a rat, with marked increased binding in the penumbral region of the ischemic lesion; lower photo: Autoradiogram after pretreatment of a rat with ACEA-1021, with markedly reduced binding in the penumbra (with permission from Di and Bullock, 1996a).

(5-nitro-6,7-dichloro-1,4-dihydro-2,3-quinoxaline dione) (Di and Bullock, 1996a). These results indicate massive involvement of NMDA receptor channel opening in traumatic brain injury in rats.

A recent study, investigating changes in glutamate receptor subtypes following FPI, showed that [$^3$H]glutamate binding was reduced in the hippocampus (up to 3 hrs after injury) and neocortex (up to 24 hrs), whereas [$^3$H]KA binding was decreased only in the neocortex, 24 hrs after injury and [$^3$H]AMPA was not affected by FPI (Miller et al., 1990b). These results showed for the first time, that brain regions are selectively vulnerable to moderate FPI. M. Temple and coworkers (Temple et al., 1997) reported a reduction in staining intensity for both NR1 and NR2A/B subunits in the hippocampus 24 hours, 8 and 16 days following lateral FPI (Fig. 12). The effect was more pronounced for the NR1 subunit and seems not to be related to cell loss, indicating a downregulation of NR1 receptors after TBI. This is supported by a study by Reeves and coworkers showing that central fluid percusssion (no cell loss) had no effect on hippocampal immunobinding of NR1 antibodies, whereas FPI and an entorhinal cortex deafferentiation lesion increased the intensity of staining, particular in stratum lacunosum-moleculare. In the combined injury group, however, less marked effects were found, supporting a downregulative effect of TBI upon the NMDAR (Reeves et al., 1997). Recently, a study using a weight-drop SCI model, demonstrated a reduction in labelled MK801 binding 4 hours but not 24 hours after spinal cord injury (Sun and Faden, 1994). It might be speculated that downregulation of glutamate receptors following TBI or SCI helps the tissue to prevent excitotoxic lesions by glutamate. In head or spinal cord injured patients, changes in glutamate receptors have not been reported yet.

*Glutamate transporters*

Beside the breakdown of ion homeostasis and electrochemical ion gradients necessary for glutamate uptake, reductions of glutamate transporters may also contribute to glutamate neurotoxicity

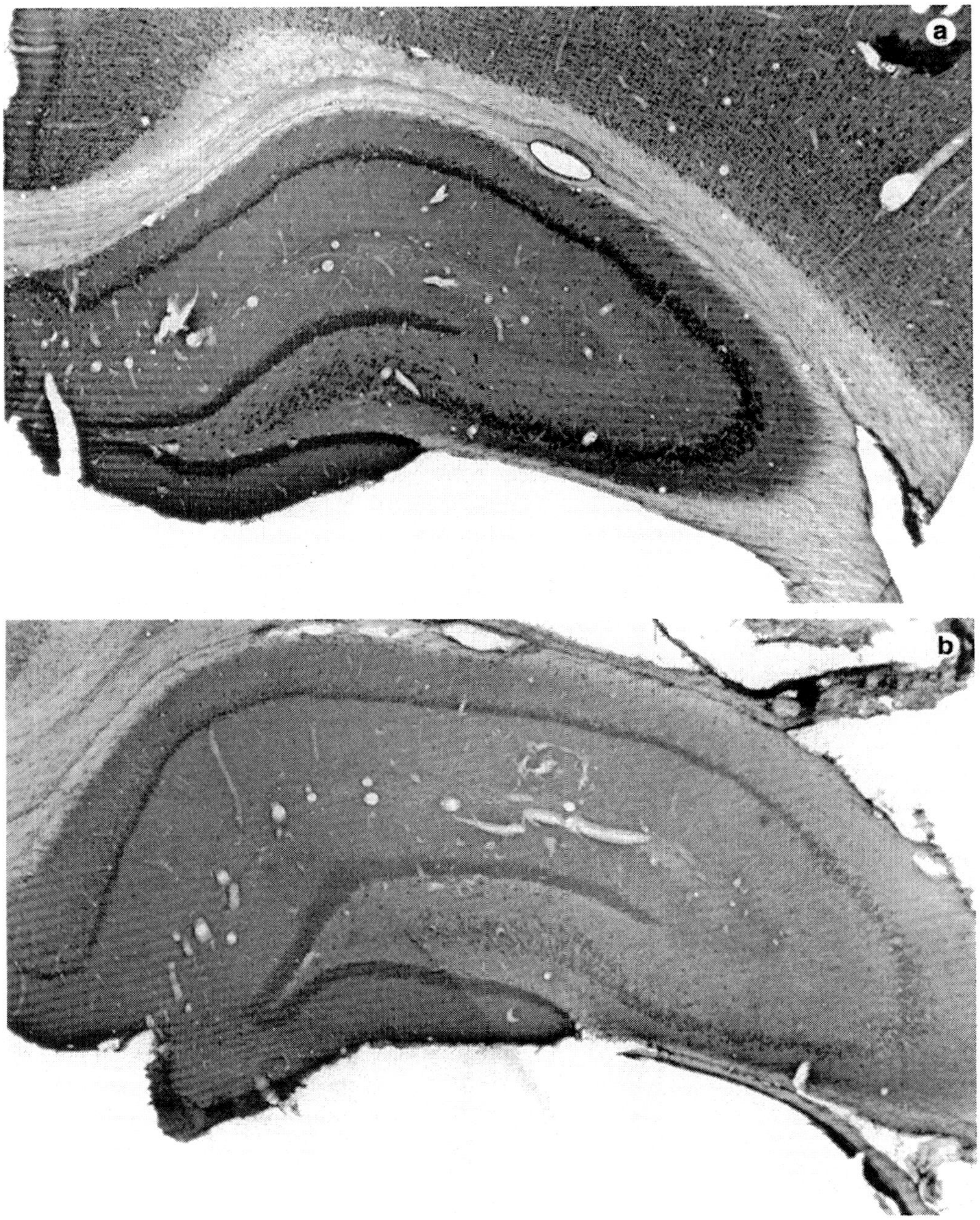


Fig. 12. Immunohistochemical staining of NMDAR1 (NR1) receptors subtypes in rats receiving sham operation (upper graph) and FPI (lower graph). Staining of sections 24 hours after sham operation or injury indicates a decreased hippocampal staining intensity following FPI (with permission from M. Temple, 1997).

following TBI. There are at least four types of glutamate transporters known (Danbolt et al., this volume), which are located in neurons (EAAC1, EAAT4) and glial cells (GLAST, GLT-1) (Rothstein et al., 1994). Knock-out mice lacking the astrocytic glutamate transporter GLT-1 were more susceptible to acute epileptic cortical injury than wild-type animals (Tanaka et al., 1997), showing the key role of glutamate uptake. As shown by quantitative immunoblotting, GLT-1 is reduced up to 20% in the hippocampus following global ischemia, whereas mRNA content was modestly reduced in hippocampus (especially CA1) 3 hours and partially reduced 24 hours after an ischemic event (Torp et al., 1995). The involvement of glutamate transporters in animal models of TBI or head-injured patients has not been shown to date. The characterization of alterations of uptake transporters and glutamate receptors (subtypes) following TBI will be essential for a better understanding of glutamate neurotoxicity in humans and therapeutical use of glutamate receptor antagonists in humans.

## Conclusions

Numerous animal studies have demonstrated transient high levels of glutamate release, after TBI and SCI, and this has been confirmed in humans, and shown to persist for several days, after injury, in certain injury types, in some cortical areas. The concentration of ECF glutamate appears to reach levels at least as high as 100–400 $\mu$M, in the worst affected patients. The persistence of this glutamate, for hours to days, suggests inactivation of uptake mechanisms. In vitro and in vivo studies in animals suggest that exposure to glutamate at this magnitude and duration, is probably directly neurotoxic, but this is not yet conclusively shown. Secondary exacerbating factors, such as poor substrate delivery, EAA induced increased tissue metabolic demand, inactivation of uptake mechanisms, receptor upregulation, receptor/ion-channel 'sensitization' to effect of glutamate, and EAA-induced seizures may all magnify this neurotoxic effect, which maybe mediated via all the receptor subtypes.

The wide spectrum of beneficial effects shown by various pharmacological antagonists to glutamate, and its co-agonists in numerous TBI-relevant animal models, has been a powerful impetus to the pharmaceutical industry, to evaluate glutamate antagonists in TBI clinical trials. Although, to date, two major efficacy trials have been terminated prematurely in human severe TBI, because of lack of evidence of benefit for the patients. At least 6 other trials have shown early evidence of benefit, in phase II studies, and several major phase III studies are currently in progress. To date, antagonists of the AMPA/KA and metabotropic glutamate receptor sites have not been evaluated in human TBI, or SCI. For the next 3–5 years at least, the evaluation of the role of NMDA receptor antagonists in human TBI, and other indications, remains one of the most exciting, and important areas in neuroscience.

## Acknowledgements

We would like to thank Dr. Bruce Lyeth for his comments which helped to improve the value of this manuscript. Many of the studies performed at the Medical College of Virginia were, in part, supported by NIH Grant NS12587, and by the Lind Lawrence Foundation.

## References

Agrawal, S. and Fehlings, M. (1997) Role of NMDA and non-NMDA ionotropic glutamate receptors in traumatic spinal cord axonal injury. *J. Neurosci.*, 17: 1055–1063.

Alessandri, B., Bullock, R. and Woodward, J. (1997) Evidence for time-dependent glutamate-mediated glycolysis in head-injured patients: A microdialysis study. In: Bullock et al. (Eds), *Neurochemical Monitoring, Satellite Symposium of the 10th International Symposium on Intracranial Pressure*, Williamsburg, VA, USA, pp. OS-6

Alessandri, B., Doppenberg, E., Koura, J., Woodward, J. and Bullock, R. (1998) Excitatory neurotransmitters influence energy metabolism after severe human head injury: Effect of injury type. *Acta Neurochir. Suppl.* (Wien), 71 400p

Alessandri, B., Landolt, H., Langemann, H. and Gratzl, O. (1995) Peracute effects of microdialytically applied glutamate in the cortex of rats. *J. Cereb. Blood Flow Metab.*, 15: S420.

Alessandri, B., Landolt, H., Langemann, H., Gregorin, J., Hall, J. and Gratzl, O. (1996) Application of glutamate in the cortex of rats: A microdialysis study. *Acta Neurochir. Suppl. (Wien)*, 67: 6–12.

Andersen, B.J. and Marmarou, A. (1992) Post-traumatic selective stimulation of glycolysis. *Brain Res.*, 585: 184–189.

Arvin, B., Lekieffre, D., Graham, J.L., Moncada, C., Chapman, A.G. and Meldrum, B.S. (1994) Effect of the non-NMDA receptor antagonist GYKI 52466 on the microdialysate and tissue concentrations of amino acids following transient forebrain ischaemia. *J. Neurochem.*, 62: 1458–1467.

Bergsneider, M., Hovda, D., Shalmon, E., Kelly, D., Vespa, P., Martin, N., Phelps, M., McArthur, D., Caron, M., Kraus, J. and Becker, D. (1997) Cerebral hyperglycolysis following severe traumatic brain injury in humans: A positron emission tomography study. *J. Neurosurg.*, 86: 241–51.

Blumbergs, P.C. (1997) Pathology. In: P. Reilly and R. Bullock (Eds), *Head injury: Pathophysiology and management of severe closed injury*, Chapman & Hall Medical, London, pp. 39–70.

Boisvert, D.P.J. and Schreiber, C. (1992) Interrelationship of excitotoxic and free radical mechanisms. In: J. Krieglstein and H. Oberpichler-Schwenk (Eds), *Pharmacology of cerebral ischemia*, Marburg, Germany: Wissenschaftliche Verlagsgesellschaft mbH, Stuttgart, pp. 311–320.

Bowman, C.L., Ding, J.P., Sachs, F. and Sokabe, M. (1992) Mechanotransducing ion channels in astrocytes. *Brain Res.*, 584: 272–286.

Bullock, R. (1993) Pathophysiological alterations in the central nervous system due to trauma. *Schweiz. Med. Wochenschr.*, 123: 449–458.

Bullock, R. (1997) Injury and cell function. In: P. Reilly and R. Bullock (Eds), *Head injury: Pathophysiology and management of severe closed injury*, Chapman & Hall Medical, London, pp. 121–141.

Bullock, R., Butcher, S., Chen, M., Kendall, L. and McCulloch, J. (1991) Correlation of the extracellular glutamate concentration with extent of blood flow reduction after subdural hematoma in the rat. *J. Neurosurg.*, 74: 794–802.

Bullock, R. and Di, X. (1997) Synergistic effect of glutamate on secondary brain damage following fluid percussion brain injury in rats. *J. Neurotrauma*, 14: 765.

Bullock, R., McCulloch, J., Graham, D., Lowe, D., Chen, M. and Teasdale, G. (1990) Focal ischemic damage is reduced by CPP-ene studies in two animal models. *Stroke*, 21: III32–6.

Bullock, R., Zauner, A., Myseros, J.S., Marmarou, A., Woodward, J.J. and Young, H.F. (1995a) Evidence for prolonged release of exitatory amino acids in severe human head trauma. *Ann. N.Y. Acad. Sci.*, 765: 290–297.

Bullock, R., Zauner, A., Tsuji, O., Woodward, J., Young, H. and Marmarou, A. (1994) Excitatory amino acid releases after severe head trauma: Effect of intracranial pressure and cerebral perfusion pressure changes. In: H. Nagai, K. Kamiya, and S. Ishii (Eds), *Ninth International Symposium on Intracranial Pressure*, Nagoya, Japan, pp. 264–267.

Bullock, R., Zauner, A., Tsuji, O., Woodward, J.J., Marmarou, A.T. and Young, H.F. (1995b) Patterns of excitatory amino acid release and ionic flux after severe human head trauma. In: T. Tsubokawa, et al. (Eds), *Neurochemical monitoring in the intensive care unit.*, Springer-Verlag, Tokyo, pp. 64–71.

Bullock, R., Zauner, A., Woodward, J. and Young, H.F. (1995c) Massive persistent release of excitatory amino acids following human occlusive stroke. *Stroke*, 26: 2187–2189.

DeWitt, D., Jenkins, L., Wei, E., Lutz, H., Becker, D. and Kontos, H. (1986) Effects of fluid-percussion brain injury on regional cerebral blood flow and pial arteriolar diameter. *J. Neurosurg.*, 64: 787–794.

Di, X. and Bullock, R. (1996a) Effect of the novel high-affinity glycine-site *N*-methyl-D-aspartate antagonist ACEA-1021 on 125I-MK-801 binding after subdural hematoma in the rat: An in vivo autoradiographic study. *J. Neurosurg.*, 85: 655–661.

Di, X., Harpold, T., Watson, J.C. and Bullock, M.R. (1996b) Excitotoxic damage in neurotrauma: Fact or fiction. *Restor. Neurol. Neurosci.*, 9: 231–241.

Di, X., Lyeth, B.G., Hamm, R.J. and Bullock, M.R. (1996c) Voltage dependent $Na^+/K^+$ ion channel blockade fails to ameliorate behavioral deficits after traumatic brain injury in the rat. *J. Neurotrauma*, 13: 497–504.

Dimlich, R.V., Tornheim, P.A., Kindel, R.M., Hall, E.D., Braughler, J.M. and McCall, J.M. (1990) Effects of a 21-aminosteroid (U-74006F) on cerebral metabolites and edema after severe experimental head trauma. *Adv. Neurol.*, 52: 365–375.

Dixon, C.E., Lyeth, B.G., Povlishock, J.T., Findling, R.L., Hamm, R.J., Marmarou, A., Young, H.F. and Hayes, R.L. (1987) A fluid percussion model of experimental brain injury in the rat. *J. Neurosurg.*, 67: 110–119.

Doppenberg, E. and Bullock, R. (1997) Clinical neuro-protection trials in severe traumatic brain injury: Lessons from previous studies. *J. Neurotrauma*, 14: 71–80.

Doppenberg, E.M.R., Zauner, A., Bullock, R., Woodward, J.L. and Young, H.F. (1996) Evidence for glutamate-mediated anerobic glycolysis, after human head injury? A microdialysis study. *J. Neurotrauma*, 13: 597.

Douglas, J., Noga, B., Dai, X. and Jordan, L. (1993) The effects of intrathecal administration of excitatory amino acid agonists and antagonists on the initiation of locomotion in the adult cat. *J. Neurosci.*, 13: 990–1000.

During, M.J. and Spencer, D.D. (1993) Extracellular hippocampal glutamate and spontaneous seizure in the conscious human brain [see comments]. *Lancet*, 341: 1607–1610.

Enblad, P., Valtysson, J., Andersson, J., Lilja, A., Valind, S., Antoni, G., Langstrom, B., Hillered, L. and Persson, L. (1996) Simultaneous intracerebral microdialysis and positron emission tomography in the detection of ischemia in patients with subarachnoid hemorrhage. *J. Cereb. Blood Flow Metab.*, 16: 637–644.

Faden, A., Jacobs, T. and Smith, M. (1984) Evaluation of the calcium channel antagonist nimodipine in experimental spinal cord ischemia. *J. Neurosurg.*, 60: 796–799.

Faden, A., Lemke, M., Simon, R. and Noble, L. (1988a) *N*-methyl-D-aspartate antagonist MK801 improves outcome following traumatic spinal cord injury in rats: Behavioral, anatomic, and neurochemical studies. *J. Neurotrauma*, 5: 33–45.

Faden, A.I. (1993) Comparison of single and combination drug treatment strategies in experimental brain trauma. *J. Neurotrauma*, 10: 91–100.

Faden, A.I., Demediuk, P., Panter, S.S. and Vink, R. (1989) The role of excitatroy amino acids and NMDA receptors in traumatic brain injury. *Science*, 244: 798–800.

Faden, A.I. and Simon, R.P. (1988b) A potential role for excitotoxins in the pathophysiology of spinal cord injury. *Ann. Neurol.*, 23: 623–626.

Farooque, M., Hillered, L., Holtz, A. and Olsson, Y. (1996) Changes of extracellular levels of amino acids after graded compression trauma to the spinal cord: An experimental study in the rat using microdialysis. *J. Neurotrauma*, 13: 537–548.

Foda, M. and Marmarou, A. (1994) A new model of diffuse brain injury in rats. Part II: Morphological characterization. *J. Neurosurg.*, 80: 301–313.

Foster, A.C., Gill, R., Kemp, J.A. and Woodruff, G.N. (1987) Systemic administration of MK-801 prevents *N*-methyl-D-aspartate-induced neuronal degeneration in rat brain. *Neurosci. Lett.*, 76: 307–311.

Fujisawa, H., Dawson, D., Browne, S.E., MacKay, K.B., Bullock, R. and McCulloch, J. (1993) Pharmacological modification of glutamate neurotoxicity in vivo. *Brain Res.*, 629: 73–78.

Fujisawa, H., Landolt, H. and Bullock, R. (1996) Patterns of increased glucose use following extracellular infusion of glutamate: An autoradiographic study. *J. Neurotrauma*, 13: 245–254.

Fujisawa, H., Maxwell, W.L., Graham, D.I., Teasdale, G.M. and Bullock, R. (1994) Focal microvascular occlusion after acute subdural haematoma in the rat: A mechanism for ischaemic damage and brain swelling? *Acta Neurochir. Suppl. (Wien)*, 60: 193–196.

Gale, K., Kerasidis, H. and Wrathall, J. (1985) Spinal cord contusion in the rat: Behavioral analysis of functional neurologic impairment. *Exp. Neurol.*, 88: 123–34.

Gennarelli, T.A. (1994) Animal models of human head injury. *J. Neurotrauma*, 11: 357–368.

Gennarelli, T.A., Abel, J.M., Adams, H. and Graham, D. (1979) Differential tolerance of frontal and temporal lobes to contusion induced by angular acceleration. In: *Proceedings of the 23rd Stapp Car Crash Conference*, Society of Automotive Engineers, New York, pp. 561–586.

Gennarelli, T.A., Adams, J.H. and Graham, D.I. (1981) Acceleration induced head injury in the monkey. I. The model, its mechanical and physiological correlates. *Acta Neuropathol Suppl (Berl)*, 7: 23–25.

Gong, Q.Z., Delahunty, T.M., Hamm, R.J. and Lyeth, B.G. (1995) Metabotropic glutamate antagonist, MCPG, treatment of traumatic brain injury in rats. *Brain Res.*, 700: 299–302.

Goodman, J.C., Gopinath, S.P., Valadka, A.B., Narayan, R.K., Grossman, R.G., Simpson RK, Jr. and Robertson, C.S. (1996a) Lactic acid and amino acid fluctuations measured using microdialysis reflect physiological derangements in head injury. *Acta Neurochir. Suppl. (Wien)*, 67: 37–39.

Goodman, J.C., Valadka, A B., Gopinath, S.P., Cormio, M. and Robertson, C.S. (1996b) Lactate and excitatory amino acids measured by microdialysis are decreased by pentobarbital coma in head-injured patients. *J. Neurotrauma*, 13: 549–556.

Graham, D. (1985) The pathology of brain ischaemia and possibilities for therapeutic intervention. *Br. J. Anaesth.*, 57: 3–17.

Griffiths, I.R. (1976) Spinal cord blood flow after acute experimental cord injury in dogs. *J. Neurol. Sci.*, 27: 247–259.

Haghighi, S., Johnson, G., de Vergel, C. and Vergel-Rivas, B. (1996) Pretreatment with NMDA receptor antagonist MK801 improves neurophysiological outcome after an acute spinal cord injury. *Neurol. Res.*, 18: 509–515.

Hamm, R., Dixon, C., Gbadebo, D., Singha, A., Jenkins, L., Lyeth, B. and Hayes, R. (1992) Cognitive deficits following traumatic brain injury produced by controlled cortical impact. *J. Neurotrauma*, 9: 11–20.

Hamm, R.J., O'Dell, D.M., Pike, B.R. and Lyeth, B.G. (1993) Cognitive impairment following traumatic brain injury: The effect of pre- and post-injury administration of scopolamine and MK-801. *Brain Res. Cogn. Brain Res.*, 1: 223–226.

Hao, J., Watson, B., Xu, X., Wiesenfeld-Hallin, Z., Seiger, A. and Sundstrom, E. (1992) Protective effect of the NMDA antagonist MK-801 on photochemically induced spinal lesions in the rat. *Exp. Neurol.*, 118: 143–152.

Hayes, R., Jenkins, L., Lyeth, B., Balster, R., Robinson, S., Clifton, G., Stubbins, J. and Young, H. (1988a) Pretreatment with phencyclidine, an *N*-methyl-D-aspartate antagonist, attenuates long-term behavioral deficits in the rat produced by traumatic brain injury. *J. Neurotrauma*, 5: 259–274.

Hayes, R., Katayama, Y., Jenkins, L., Lyeth, B., Clifton, G., Gunter, J., Povlishock, J. and Young, H. (1988b) Regional rates of glucose utilization in the cat following concussive head injury. *J. Neurotrauma*, 5: 121–137.

Hicks, R.R., Smith, D.H., Gennarelli, T.A. and McIntosh, T. (1994) Kynurenate is neuroprotective following experimental brain injury in the rat. *Brain Res.*, 655: 91–96.

Holtz, A. and Gerdin, B. (1991) MK 801, an OBS *N*-methyl-D-aspartate channel blocker, does not improve the functional recovery nor spinal cord blood flow after spinal cord compression in rats. *Acta Neurol. Scand.*, 84: 334–338.

Hovda, D.A., Yoshino, A., Kawamata, T., Katayama, Y. and Becker, D.P. (1991) Diffuse prolonged depression of cerebral oxidative metabolism following concussive brain injury in the rat: A cytochrome oxidase histochemistry study. *Brain Res.*, 567: 1–10.

Hovda, D.A., Yoshino, A., Kawamata, T., Katayama, Y., Fineman, I. and Becker, D.P. (1990) The increase in local cerebral glucose utilization following fluid percussion brain injury is prevented with kynurenic acid and is associated with an increase in calcium. *Acta Neurochir. Suppl. (Wien)*, 51: 331–333.

Hylton, C., Perri, B.R. and Voddi, M.D. (1995) Non-NMDA antagonist GYKI-52466 enhances spatial memory after experimental brain injury. *J. Neurotrauma*, 12: 124.

Ikonomidou, C. and Turski, L. (1996) Prevention of trauma-induced neurodegeneration in infant and adult rat brain: Glutamate antagonists. *Metab. Brain Dis.*, 11: 125–141.

Inglis, F., Kuroda, Y. and Bullock, R. (1992) Glucose hypermetabolism after acute subdural hematoma is ameliorated by a competitive NMDA antagonist. *J. Neurotrauma*, 9: 75–84.

Ishige, N., Pitts, L.H., Hashimoto, T., Nishimura, M.C. and Bartkowski, H.M. (1987) Effect of hypoxia on traumatic brain injury in rats: Part 1. Changes in neurological function, electroencephalograms, and histopathology. *Neurosurgery*, 20: 848–853.

Ito, J., Marmarou, A., Barzo, P., Fatouros, P. and Corwin, F. (1996) Characterization of edema by diffusion-weighted imaging in experimental traumatic brain injury. *J. Neurosurg.*, 84: 97–103.

Jenkins, A., Maxwell, W.L. and Graham, D.I. (1989) Experimental intracerebral haematoma in the rat: Sequential light microscopical changes. *Neuropathol Appl Neurobiol*, 15: 477–486.

Jenkins, L., Lyeth, B., Lewelt, W., Moszynski, K., Dewitt, D., Balster, R., Miller, L., Clifton, G., Young, H. and Hayes, R. (1988) Combined pretrauma scopolamine and phencyclidine attenuate posttraumatic increased sensitivity to delayed secondary ischemia. *J. Neurotrauma*, 5: 275–287.

Kanthan, R. and Shuaib, A. (1995) Clinical evaluation of extracellular amino acids in severe head trauma by intracerebral in vivo microdialysis. *J. Neurol. Neurosurg. Psychiatry*, 59: 326–327.

Kanthan, R., Shuaib, A., Griebel, R., Miyashita, H. and Kalra, J. (1996) Glucose-induced decrease in glutamate levels in ischemic human brain by in-vivo microdialysis. *Neurosci. Lett.*, 209: 207–209.

Katayama, Y., Becker, D.P., Tamura, T. and Hovda, D.A. (1990) Massive increases in extracellular potassium and the indiscriminate release of glutamate following concussive brain injury. *J. Neurosurgery*, 73: 889–900.

Katayama, Y., Glisson, J.D., Becker, D.P. and Hayes, R.L. (1985) Concussive head injury producing suppression of sensory transmission within the lumbar spinal cord in cats. *J. Neurosurg.*, 63: 97–105.

Katayama, Y., Tamura, T., Becker, D.P. and Tsubokawa, T. (1992) Early cellular swelling during cerebral ischemia in vivo is mediated by excitatory amino acids released from nerve terminals. *Brain Res.*, 577: 121–126.

Katoh, H., Sima, K., Nawashiro, H., Wada, K. and Chigasaki, H. (1997) The effect of MK-801 on extracellular neuroactive amino acids in hippocampus after closed head injury followed by hypoxia in rats. *Brain Res.*, 758: 153–162.

Kawamata, T., Katayama, Y., Hovda, D.A., Yoshino, A. and Becker, D.P. (1992) Administration of excitatory amino acid antagonists via microdialysis attenuates the increase in glucose utilization seen following concussive brain injury. *J. Cereb. Blood Flow Metab.*, 12: 12–24.

Khan, M. and Griebel, R. (1983) Acute spinal cord injury in the rat: Comparison of three experimental techniques. *Can. J. Neurol. Sci.*, 10: 161–165.

Kinoshita, K., Katayama, Y., Kano, T., Hirayama, T. and Tsubokawa, T. (1994) Attenuation of hemispheric swelling associated with acute subdural hematomas by excitatory amino acid antagonist in rats. *Acta Neurochir. Suppl. (Wien)*, 60: 505–507.

Koizumi, H., Fujisawa, H., Ito, H., Maekawa, T., Di, X. and Bullock, R. (1997) Effects of mild hypothermia on cerebral blood flow-independent changes in cortical extracellular levels of amino acids following contusion trauma in the rat. *Brain Res.*, 747: 304–312.

Kraus, J. and McArthur, D. (1996) Epidemiologic aspects of brain injury. *Neurol. Clin.*, 14: 435–450.

Kraus, J., McArthur, D. and Silberman, T. (1994) Epidemiology of mild brain injury. *Semin. Neurol.*, 14: 1–7.

Kuroda, Y., Fujisawa, H., Strebel, S., Graham, D.I. and Bullock, R. (1994) Effect of neuroprotective *N*-methyl-D-aspartate antagonists on increased intracranial pressure: studies in the rat acute subdural hematoma model. *Neurosurgery*, 35: 106–112.

Kuroda, Y., Inglis, F., Miller, J., McCulloch, J., Graham, D. and Bullock, R. (1992) Transient glucose hypermetabolism after acute subdural hematoma in the rat. *J. Neurosurg.*, 76: 471–477.

Landolt, H., Fujisawa H., Graham, D.I., Maxwell, W.L. and Bullock, R. (1998) Reproducible peracute glutamate-induced focal lesion of the normal rat brain using microdialysis. *J. Clin. Neurosci.*, 5: 193–202.

Lewelt, W., Jenkins, L. and Miller, J. (1980) Autoregulation of cerebral blood flow after experimental fluid percussion injury of the brain. *J. Neurosurg.*, 53: 500–511.

Lewis, S.B., Finnie, J.W., Blumbergs, P.C., Scott, G., Manavis, J., Brown, C., Reilly, P.L., Jones, N.R. and McLean, A.J.

(1996) A head impact model of early axonal injury in the sheep. *J. Neurotrauma*, 13: 505–514.

Lighthall, J. (1988) Controlled cortical impact: A new experimental brain injury model. *J. Neurotrauma*, 5: 1–15.

Liu, D. (1994) An experimental model combining microdialysis with electrophysiology, histology, and neürochemistry for studying excitotoxicity in spinal cord injury. Effect of NMDA and kainate. *Mol. Chem. Neuropathol.*, 23: 77–92.

Liu, D., Thangnipon, W. and McAdoo, D.J. (1991) Excitatory amino acids rise to toxic levels upon impact injury to the rat spinal cord. *Brain Res.*, 547: 344–348.

Liu, S., Ruenes, G.L. and Yezierski, R.P. (1997) NMDA and non-NMDA receptor antagonists protect against excitotoxic injury in the rat spinal cord. *Brain Res.*, 756: 160–167.

Magistretti, P.J. and Pellerin, L. (1997) Metabolic coupling during activation. A cellular view. *Adv. Exp. Med. Biol.*, 413: 161–166.

Marmarou, A., Foda, M.A., van den Brink, W., Campbell, J., Kita, H. and Demetriadou, K. (1994) A new model of diffuse brain injury in rats. Part I: Pathophysiology and biomechanics. *J. Neurosurg.*, 80: 291–300.

Marsala, M., Sorkin, L. and Yaksh, T. (1994a) Transient spinal ischemia in rat: Characterization of spinal cord blood flow, extracellular amino acid release, and concurrent histopathological damage. *J. Cereb. Blood Flow Metab.*, 14: 604–514.

Marsala, M. and Yaksh, T. (1994b) Transient spinal ischemia in the rat: Characterization of behavioral and histopathological consequences as a function of the duration of aortic occlusion. *J. Cereb. Blood Flow Metab.*, 14: 526–535.

Martinez-Arizala, A., Rigamonti, D.D., Long, J.B., Kraimer, J.M. and Holaday, J.W. (1990) Effects of NMDA receptor antagonists following spinal ischemia in the rabbit. *Exp. Neurol.*, 108: 232–240.

Max, W., Rice, D. and MacKenzie, E. (1990) The lifetime cost of injury. *Inquiry*, 27: 332–343.

Maxwell, W.L., Bullock, R., Landholt, H. and Fujisawa, H. (1994) Massive astrocytic swelling in response to extracellular glutamate – A possible mechanism for post-traumatic brain swelling? *Acta Neurochir. Suppl. (Wien)*, 60: 465–467.

Maxwell, W.L., Povlishock, J.T. and Graham, D.L. (1997) A mechanistic analysis of nondisruptive axonal injury: A review. *J. Neurotrauma*, 14: 419–440.

McIntosh, T.K., Faden, A.I., Yamakami, I. and Vink, R. (1988) Magnesium deficiency exacerbates and pretreatment improves outcome following traumatic brain injury in rats: 31P magnetic resonance spectroscopy and behavioral studies. *J. Neurotrauma*, 5: 17–31.

McIntosh, T.K., Smith, D.H. and Garde, E. (1996a) Therapeutic approaches for the prevention of secondary brain injury. *Eur. J. Anaesth.*, 13: 291–309.

McIntosh, T.K., Smith, D.H., Voddi, M., Perri, B.R. and Stutzmann, J.M. (1996b) Riluzole, a novel neuroprotective agent, attenuates both neurologic motor and cognitive dysfunction following experimental brain injury in the rat. *J. Neurotrauma*, 13: 767–780.

Mendelowitsch, A., Sekhar, L.N., Clemente, R. and Shuaib, A. (1997) EC-IC bypass improves chronic ischemia in a patient with moyamoya disease secondary to sickle cell disease: An in vivo microdialysis study. *Neurol. Res.*, 19: 66–70.

Mendelowitsch, A., Sekhar, L.N., Wright, D.C., Nadel, A., Miyashita, H., Richardson, R., Kent, M. and Shuaib, A. (1998) An increase in extracellular glutamate is a sensitive method of detecting ischemic neuronal damage during cranial base and cerebrovascular surgery: An in vivo microdialysis study. *Acta Neurochir.*, in press.

Meyerson, B.A., Linderoth, B., Karlsson, H. and Ungerstedt, U. (1990) Microdialysis in the human brain: Extracellular measurements in the thalamus of parkinsonian patients. *Life Sci.*, 46: 301–308.

Miller, J., Bullock, R., Graham, D., Chen, M. and Teasdale, G. (1990a) Ischemic brain damage in a model of acute subdural hematoma. *Neurosurgery*, 27: 433–439.

Miller, L.P., Lyeth, B.G., Jenkins, L.W., Oleniak, L., Panchision, D., Hamm, R.J., Phillips, L.L., Dixon, C.E., Clifton, G.L. and Hayes, R.L. (1990b) Excitatory amino acid receptor subtype binding following traumatic brain injury. *Brain Res.*, 526: 103–107.

Morisset, V. and Nagy, F. (1996) Modulation of regenerative membrane properties by stimulation of metabotropic glutamate receptors in rat deep dorsal horn neurons. *J. Neurophysiol.*, 76: 2794–2798.

Mukhin, A., Fan, L. and Faden, A.I. (1996) Activation of metabotropic glutamate receptor subtype mGluR1 contributes to post-traumatic neuronal injury. *J. Neurosci.*, 16: 6012–6020.

Nawashiro, H., Shima, K. and Chigasaki, H. (1995) Selective vulnerability of hippocampal CA3 neurons to hypoxia after mild concussion in the rat. *Neurol. Res.*, 17: 455–460.

Nilsson, P., Hillered, L., Olsson, Y., Sheardown, M.J. and Hansen, A.J. (1993) Regional changes in interstitial $K^+$ and $Ca^{2+}$ levels following cortical compression contusion trauma in rats. *J. Cereb. Blood Flow Metab.*, 13: 183–192.

Nilsson, P., Hillered, L., Ponten, U. and Ungerstedt, U. (1990) Changes in cortical extracellular levels of energy-related metabolites and amino acids following concussive brain injury in rats. *J. Cereb. Blood Flow Metab.*, 10: 631–637.

Noble, L. and Wrathall, J. (1985) Spinal cord contusion in the rat: Morphometric analyses of alterations in the spinal cord. *Exp. Neurol.*, 88: 135–149.

Noble, L. and Wrathall, J. (1987) The blood-spinal cord barrier after injury: Pattern of vascular events proximal and distal to a transection in the rat. *Brain Res.*, 424: 177–188.

NSCIA (1996) Factsheet #2: Spinal cord injury statistical information. National spinal cord injury association, Cambridge, MA.

Okiyama, K., Smith, D.H., Gennarelli, T.A., Simon, R.P., Leach, M. and McIntosh, T.K. (1995) The sodium channel blocker and glutamate release inhibitor BW1003C87 and magnesium attenuate regional cerebral edema following experimental brain injury in the rat. *J. Neurochem.*, 64: 802–809.

Palmer, A., Marion, D., Botscheller, M., Swedlow, P., Styren, S. and DeKosky, S. (1993a) Traumatic brain injury-induced excitotoxicity assessed in a controlled cortical impact model. *J. Neurochem.*, 61: 2015–2024.

Palmer, A.M., Marion, D.W., Botscheller, M.L. and Redd, E.E. (1993b) Therapeutic hypothermia is cytoprotective without attenuating the traumatic brain injury-induced elevations in interstitial concentrations of aspartate and glutamate. *J. Neurotrauma*, 10: 363–372.

Panter, S. and Faden, A. (1992) Pretreatment with NMDA antagonists limits release of excitatory amino acids following traumatic brain injury. *Neurosci. Lett.*, 136: 165–168.

Panter, S., Yum, S. and Faden, A. (1990) Alteration in extracellular amino acids after traumatic spinal cord injury. *Ann. Neurol.*, 27: 96–99.

Pellerin, L. and Magistretti, P.J. (1994) Glutamate uptake into astrocytes stimulates aerobic glycolysis: A mechanism coupling neuronal activity to glucose utilization. *Proc. Natl. Acad. Sci. USA*, 91: 10625–10629.

Pellerin, L. and Magistretti, P.J. (1996) Excitatory amino acids stimulate aerobic glycolysis in astrocytes via an activation of the $Na^+/K^+$ ATPase. *Dev. Neurosci.*, 18: 336–342.

Persson, L. and Hillered, L. (1992) Chemical monitoring of neurosurgical intensive care patients using intracerebral microdialysis. *J. Neurosurg.*, 76: 72–80.

Persson, L., Valtysson, J., Enblad, P., Warme, P.E., Cesarini, K., Lewen, A. and Hillered, L. (1996) Neurochemical monitoring using intracerebral microdialysis in patients with subarachnoid hemorrhage. *J. Neurosurg.*, 84: 606–616.

Povlishock, J.T. (1992) Traumatically induced axonal injury: Pathogenesis and pathobiological implications. *Brain Pathol.*, 2: 1–12.

Reeves, T.M., Zhu, J., Povlishock, J.T. and Phillips, L.L. (1997) The effect of combined fluid percussion and entorhinal cortical lesions on long term potentiation. *Neuroscience*, 77: 431–444.

Regan, F.R. (1996) The vulnerability of spinal cord neurons to excitotoxic injury: Comparison with cortical neurons. *Neurosci. Lett.*, 213: 9–12.

Regan, R. and Choi, D. (1991) Glutamate neurotoxicity in spinal cord cell culture. *Neuroscience*, 43: 585–591.

Robertson, C., Foltz, R., Grossman, R. and Goodman, J. (1986) Protection against experimental ischemic spinal cord injury. *J. Neurosurg.*, 64: 633–642.

Rothstein, J.D., Martin, L., Levey, A.I., Dykes-Hoberg, M., Jin, L., Wu, D., Nash, N. and Kuncl, R.W. (1994) Localization of neuronal and glial glutamate transporters. *Neuron*, 13: 713–725.

Sachs, F. (1991) Mechanical transduction by membrane ion channels: A mini review. *Mol. Cell Biochem.*, 104: 57–60.

Schroder, M.L., Muizelaar, J.P., Bullock, M.R., Salvant, J.B. and Povlishock, J.T. (1995) Focal ischemia due to traumatic contusions documented by stable xenon-CT and ultrastructural studies. *J. Neurosurg.*, 82: 966–971.

Schroder, M.L., Muizelaar, J.P. and Kuta, A.J. (1994) Documented reversal of global ischemia immediately after removal of an acute subdural hematoma. Report of two cases. *J. Neurosurg.*, 80: 324–327.

Schurr, A., Payne, R.S., Heine, M.F. and Rigor, B.M. (1995) Hypoxia, excitotoxicity, and neuroprotection in the hippocampal slice preparation. *J. Neurosci. Methods*, 59: 129–138.

Schwartz, J.H. and Kandell, E.R. (1991) Synaptic transmission mediated by second messangers. In: E.R. Kandell, J.H. Schwartz, and T.M. Jessel (Eds), *Principles of Neuroscience*, Elsevier Science, New York, pp. 173–194.

Shapira, Y., Yadid, G., Cotev, S., Niska, A. and Shohami, E. (1990) Protective effect of MK801 in experimental brain injury. *J. Neurotrauma*, 7: 131–139.

Shima, K. and Marmarou, A. (1991) Evaluation of brain-stem dysfunction following severe fluid-percussion head injury to the cat. *J. Neurosurg.*, 74: 270–277.

Shimada, N., Graf, R., Rosner, G. and Heiss, W.D. (1990) Differences in ischemia-induced accumulation of amino acids in the cat cortex. *Stroke*, 21: 1445–1451.

Shimada, N., Graf, R., Rosner, G., Wakayama, A., George, C.P. and Heiss, W.D. (1989) Ischemic flow threshold for extracellular glutamate increase in cat cortex. *J. Cereb. Blood Flow Metab.*, 9: 603–606.

Shohami, E., Novikov, M. and Bass, R. (1995) Long-term effect of HU-211, a novel non-competitive NMDA antagonist, on motor and memory functions after closed head injury in the rat. *Brain Res.*, 674: 55–62.

Shohami, E., Novikov, M. and Mechoulam, R. (1993) A nonpsychotropic cannabinoid, HU-211, has cerebroprotective effects after closed head injury in the rat. *J. Neurotrauma*, 10: 109–119.

Siegel, G.J., Agranoff, B.W., Albers, R.W. and Molinoff, P.B. (1989) Basic neurochemistry: *Molecular, cellular, and medical aspects*. 4. ed., New York: Raven Press. 984 p.

Siegelbaum, S.A. and Koester, J. (1991) Ion channels. In: E.R. Kandel, J.H. Schwartz, and T.M. Jessel (Eds), *Principles of Neuroscience*, Elsevier Science, New York, pp. 66–79.

Simpson, R.K.J., Robertson, C. and Goodman, J. (1990) Spinal cord ischemia-induced elevation of amino acids: Extracellular measurement with microdialysis. *Neurochem. Res.*, 15: 635–639.

Smith, D.H., Chen, X.H., Xu, B.N., McIntosh, T.K., Gennarelli, T.A. and Meaney, D.F. (1997) Characterization of diffuse axonal pathology and selective hippocampal damage following inertial brain trauma in the pig. *J. Neuropathol. Exp. Neurol.*, 56: 822–834.

Smith, D.H., Okiyama, K., Gennarelli, T.A. and McIntosh, T.K. (1993a) Magnesium and ketamine attenuate cognitive dysfunction following experimental brain injury. *Neurosci. Lett.*, 157: 211–214.

Smith, D.H., Okiyama, K., Thomas, M.J. and McIntosh, T.K. (1993b) Effects of the excitatory amino acid receptor antagonists kynurenate and indole-2-carboxylic acid on behavioral and neurochemical outcome following experimental brain injury. *J. Neurosci.*, 13: 5383–5392.

Smith, D.H., Thomas, M.J., Soares, H.D. and McIntosh, T.K. (1990) Differential effects of competitive and non-competitive *N*-methyl-D-aspartate (NMDA) receptor antagonists in experimental brain injury. *FASEB*, 4: 771.

Sosin, D., Sniezek, J. and Waxweiler, R. (1995) Trends in death associated with traumatic brain injury, 1979 through 1992: Success and failure. *JAMA*, 273: 1778–1780.

Sun, F.Y. and Faden, A.I. (1994) High- and low-affinity NMDA receptor-binding sites in rat spinal cord: Effects of traumatic injury. *Brain Res.*, 666: 88–92.

Tanaka, K., Watase, K., Manabe, T., Yamada, K., Watanabe, M., Takahashi, K., Iwama, H., Nishikawa, T., Ichihara, N., Kikuchi, T., Okuyama, S., Kawashima, N., Hori, S., Takimoto, M. and Wada, K. (1997) Epilepsy and exacerbation of brain injury in mice lacking the glutamate transporter GLT-1. *Science*, 276: 1699–1702.

Tavalin, S.J., Ellis, E.F. and Satin, L.S. (1995) Mechanical perturbation of cultured cortical neurons reveals a stretch-induced delayed depolarization. *J. Neurophysiol.*, 74: 2767–2773.

Temple, M.D. and Hamm, R.J. (1996) Chronic, post injury administration of D-cycloserine, an NMDA partial agonist, enhances cognitive performance following experimental brain injury. *Brain Res.*, 741: 246–251.

Temple, M.D., Phillips, L.L., Hamm, R.J., Delahunty, T.M., Zhu, J.P., Floyd, C.L. and Lyeth, B.G. (1997) Changes in *N*-methyl-D-aspartate receptor (NMDAR) 16 days after traumatic brain injury (TBI). *J. Neurotrauma*, 14: 766.

Teng, Y. and Wrathall, J. (1997) Local blockade of sodium channels by tetrodotoxin ameliorates tissue loss and long-term functional deficits resulting from experimental spinal cord injury. *J. Neurosci.*, 17: 4359–4366.

Torp, R., Lekieffre, D., Levy, L.M., Haug, F.M., Danbolt, N.C., Meldrum, B.S. and Ottersen, O.P. (1995) Reduced post-ischemic expression of a glial glutamate transporter, GLT1, in the rat hippocampus. *Exp. Brain Res.*, 103: 51–58.

Toulmond, S., Serrano, A., Benavides, J. and Scatton, B. (1993) Prevention by eliprodil (SL 82.0715) of traumatic brain damage in the rat. Existence of a large (18 h) therapeutic window. *Brain Res.*, 620: 32–41.

Trout, J.J., Lu, C.Y., Goldstone, A.D. and Sahgal, S. (1995) Polyamines and NMDA receptors modulate pericapillary astrocyte swelling following cerebral cryo-injury in the rat. *J. Neuroscytol.*, 24: 341–346.

Tsuchida, E. and Bullock, R. (1995) The effect of the glycine site-specific *N*-methyl-D-aspartate antagonist ACEA1021 on ischemic brain damage caused by acute subdural hematoma in the rat. *J. Neurotrauma*, 12: 279–88.

Tsuchida, E., Harms, J., Woodward, J. and Bullock, R. (1996) A use-dependent sodium channel antagonist, 619C89, in reduction of ischemic brain damage and glutamate release after acute subdural hematoma in the rat. *J. Neurosurg.*, 85: 104–111.

Tsuchida, E., Rice, M. and Bullock, R. (1997) The neuroprotective effect of the forebrain-selective NMDA antagonist CP101,606 upon focal ischemic brain damage caused by acute subdural hematoma in the rat. *J. Neurotrauma*, 14: 409–417.

Unterberg, A.W., Andersen, B.J., Clarke, G.D. and Marmarou, A. (1988) Cerebral energy metabolism following fluid-percussion brain injury in cats. *J. Neurosurg.*, 68: 594–600.

Van den Brink, W.A., Marmarou, A. and Avezaat, C.J. (1990) Brain oedema in experimental closed head injury in the rat. *Acta Neurochir. Suppl. (Wien)*, 51: 261–262.

Van Harreveld, A. and Fifkova, E. (1971) Light- and electron-microscopic changes in central nervous tissue after electrophoretic injection of glutamate. *Exp. Mol. Pathol.*, 15: 61–81.

Vink, R., McIntosh, T.K., Demediuk, P. and Faden, A.I. (1987) Decrease in total and free magnesium concentration following traumatic brain injury in rats. *Biochem. Biophys. Res. Commun.*, 149: 594–599.

Vink, R., McIntosh, T.K., Demediuk, P., Weiner, M.W. and Faden, A.I. (1988) Decline in intracellular free $Mg^{2+}$ is associated with irreversible tissue injury after brain trauma. *J. Biol. Chem.*, 263: 757–761.

Voddi, M.D., Perri, B.R., Perlman, K.G., Smith, D.H., Leach, M. and McIntosh, T.K. (1995) The pre-synaptic glutamate release inhibitor BW619c89 attenuates memory dysfunction following experimental brain injury. *J. Neurotrauma*, 12: 146.

von Euler, M., Li-Li, M., Whittemore, S., Seiger, A. and Sundstrom, E. (1997) No protective effect of the NMDA antagonist memantine in experimental spinal cord injuries. *J. Neurotrauma*, 14: 53–61.

von Euler, M., Seiger, A., Holmberg, L. and Sundstrom, E. (1994) NBQX, a competitive non-NMDA receptor antagonist, reduces degeneration due to focal spinal cord ischemia. *Exp. Neurol.*, 129: 163–168.

Wahl, F., Renou, E., Mary, V. and Stutzmann, J.M. (1997) Riluzole reduces brain lesions and improves neurological function in rats after a traumatic brain injury. *Brain Res.*, 756: 247–255.

Westergren, I. and Johansson, B.B. (1992) NBQX, an AMPA antagonist, reduces glutamate-mediated brain edema. *Brain Res.*, 573: 324–326.

Wrathall, J., Bouzoukis, J. and Choiniere, D. (1992a) Effect of kynurenate on functional deficits resulting from traumatic spinal cord injury. *Eur. J. Pharmacol.*, 218: 273–281.

Wrathall, J., Pettegrew, R. and Harvey, F. (1985) Spinal cord contusion in the rat: Production of graded, reproducible, injury groups. *Exp. Neurol.*, 88: 108–122.

Wrathall, J., Teng, Y., Choiniere, D. and Mundt, D. (1992b) Evidence that local non-NMDA receptors contribute to functional deficits in contusive spinal cord injury. *Brain Res.*, 586: 140–143.

Wrathall, J.R., Choiniere, D. and Teng, Y.D. (1994) Dose-dependent reduction of tissue loss and functional impairment after spinal cord trauma with the AMPA/kainate antagonist NBQX. *J. Neurosci.*, 14: 6598–6607.

Wrathall, J.R., Teng, Y.D. and Choiniere, D. (1996) Amelioration of functional deficits from spinal cord trauma with systemically administered NBQX, an antagonist of non-*N*-methyl-D-aspartate receptors. *Exp. Neurol.*, 137: 119–126.

Xiong, Y., Gu, Q., Peterson, P., Muizelaar, J. and Lee, C. (1997) Mitochondrial dysfunction and calcium perturbation induced by traumatic brain injury. *J. Neurotrauma*, 14: 23–34.

Yezierski, R., Santana, M., Park, S. and Madsen, P. (1993) Neuronal degeneration and spinal cavitation following intraspinal injections of quisqualic acid in the rat. *J. Neurotrauma*, 10: 445–456.

Yoshino, A., Hovda, D.A., Kawamata, T., Katayama, Y. and Becker, D.P. (1991) Dynamic changes in local cerebral glucose utilization following cerebral conclusion in rats: Evidence of a hyper- and subsequent hypometabolic state. *Brain Res.*, 561: 106–119.

Yum, S. and Faden, A. (1990) Comparison of the neuroprotective effects of the *N*-methyl-D-aspartate antagonist MK-801 and the opiate-receptor antagonist nalmefene in experimental spinal cord ischemia. *Arch. Neurol.*, 47: 277–281.

Zauner, A., Bullock, R., Kuta, A.J., Woodward, J. and Young, H.F. (1996) Glutamate release and cerebral blood flow after severe human head injury. *Acta Neurochir. Suppl. (Wien)*, 67: 40–44.

Zhang, L., Rzigalinski, B., Ellis, E. and Satin, L. (1996) Reduction of voltage-dependent $Mg^{2+}$ blockade of NMDA current in mechanically injured neurons. *Science*, 274: 1921–1923.

O.P. Ottersen, I.A. Langmoen and L. Gjerstad (Eds.)
*Progress in Brain Research*, Vol 116


CHAPTER 21

# Glutamate toxicity in chronic neurodegenerative disease

Eric Lancelot and M. Flint Beal*

*Department of Neurology, Warren 408, Massachusetts General Hospital, 32 Fruit Street, Boston, MA 02114, USA*

## Introduction

The majority of synapses in the vertebrate central nervous system use L-glutamate and L-aspartate as excitatory neurotransmitters (Cotman et al., 1981). Evidence indicating a role for these amino acids in mediating excitation in the brain was first provided by Hayashi (1954) who observed that topical application of glutamate to the rodent cerebral cortex triggered severe cortical seizures. Glutamate is involved in various physiological neuronal processes including execution of motor acts, nociception, learning, and memory. However, the observation of Lucas and Newhouse (1957) that systemic administration of glutamate to mice resulted in retinal degeneration suggested that, beside their involvement in normal brain functions, excitatory amino acids (EAA) could also act as endogenous neurotoxins. Subsequently, the work by Olney and colleagues linked the activation of EAA receptors to neurotoxicity and led to the general concept of excitotoxicity (Olney et al., 1971). Important advances in clarifying mechanisms of excitotoxicity were those of Rothman (1984) demonstrating that neuronal cell death in anoxic hippocampal cultures was due to the release of EAA, and of Choi (1987) linking delayed glutamate neurotoxicity to $Ca^{2+}$ influx. A substantial body of evidence has implicated excitotoxicity as a mechanism of cell death in acute neurologic diseases such as stroke, head trauma, and hypoglycemia. A possible role in neurodegenerative diseases has also been supported by circumstantial evidence. Improved knowledge of the mechanisms of excitotoxicity may therefore lead to the development of new therapeutic strategies for treatment of these illnesses.

## Structure and classification of excitatory amino acid receptors

EAA receptors are expressed at the membranes of neurons and glial cells. They have been classified into ionotropic receptors (coupled to ion fluxes through membrane ion channels) and metabotropic receptors (coupled *via* G proteins to phospholipase C activation or adenylate cyclase inhibition). The ionotropic receptors have been further subdivided into the *N*-methyl-D-aspartate (NMDA), α-amino-3-hydroxy-5-methyl-4-isoxazole propionate (AMPA), and kainate receptors according to their preferential agonists (Watkins et al., 1990). Activated NMDA receptors allow the influx of $Na^{+}$ and $Ca^{2+}$ whereas AMPA and kainate receptors mainly allow fluxes of monovalent ions. Several recognition sites have been characterized at the NMDA receptor (Lipton and Rosenberg,

*Corresponding author.
Tel.: +1 617 726 8463; fax: +1 617 724 1480; e-mail: beal@helix.mgh.harvard.edu.

1994; MacDonald et al., this volume): (1) a binding site for endogenous (glutamate and aspartate) or exogenous (NMDA) agonists; (2) a co-agonist recognition site for glycine, the activation of which potentiates NMDA responses; (3) a facilitatory binding site sensitive to polyamines; (4) three intra-ion channel sites at which $Mg^{2+}$, $Zn^{2+}$, and phencyclidine (PCP)-related drugs bind to block ion fluxes through the channel; and (5) two extra-ion channel modulatory sites sensitive to the redox state or the pH of the milieu. A binding site for benzodiazepines was also described on the AMPA receptor. NMDA, AMPA, and metabotropic sites are widely distributed in the mammalian central nervous system whereas high-affinity kainate sites are concentrated in a few specific brain areas (Young and Fagg, 1990). The localization of EAA receptors is largely postsynaptic, except for some subtypes of the metabotropic receptor family which are located presynaptically (Watkins et al., 1990; Sánchez-Prieto et al., 1996 Takumi et al., this volume).

Ionotropic receptors for EAA are heteromeric complexes composed of distinct subunits that can assemble in various combinations. A subunit for the NMDA receptor was initially cloned and sequenced by Moriyoshi and colleagues (1991). This protein, termed NMDAR1 (or NR1), possesses all properties characteristic of the NMDA receptor and is widely and uniformly distributed in the brain (Nakanishi, 1992). Subsequently, four additional NMDA receptor subunits, termed NMDAR2A through NMDAR2D (or NR2A through NR2D), were cloned (Monyer et al., 1992; Meguro et al., 1992; Kutsuwada et al., 1992). They share less than 20% amino acid sequence homology to NR1 and demonstrate a more restricted distribution. They evoke no electrophysiological responses after agonist application but can combine with NR1 subunits to form functional NMDA receptors with distinct physiological and pharmacological properties (Sucher et al., 1996). For instance, heterodimers composed of NR1/NR2C subunits are more resistant to $Mg^{2+}$ block and more sensitive to glycine than those generated from NR1/NR2A and from NR1/NR2B. The high $Ca^{2+}$ permeability of NR1 is imparted by an asparagine located on the second transmembrane segment of the subunit. An asparagine to glutamine mutation results in decreased $Ca^{2+}$ permeability (Burnashev et al., 1992; Sakurada et al., 1993). Non-neuronal kidney cells transfected with NMDA receptors containing the mutated NR1 subunit show less vulnerability to cell death than those which do not express the mutation (Anegawa et al., 1995).

Four AMPA subunits were cloned and termed GluR1 through GluR4 (or GluRA through GluRD) (Hollmann et al., 1989; Keinänen et al., 1990). They can assemble in various combinations that determine the functional characteristics of the receptor Borges and Dingledine, this volume. For example, although GluR1 and GluR3 can form homomeric or heteromeric ion channels permeable to $Ca^{2+}$, the inclusion of a GluR2 subunit abolishes their $Ca^{2+}$ permeability (Hollmann et al., 1991). However, an arginine to glutamine mutation confers $Ca^{2+}$ permeability to GluR2 (Hume et al., 1991; Verdoorn et al., 1991).

Further studies led to the identification of five kainate subunits called GluR5 through GluR7, KA-1, and KA-2 (Egebjerg et al., 1991; Herb et al., 1992; Werner et al., 1991). Eight subtypes of the metabotropic receptor were also reported (Nicoletti et al., 1996). They were termed mGluR1 through mGluR8 and were classified into three groups according to their sequence similarities, transduction pathways, and agonist selectivities Bruno et al., this volume. Receptors from group I are coupled to phospholipase C and lead to mobilization of intracellular $Ca^{2+}$. Receptors from other groups are negatively coupled to adenylate cyclase and voltage-sensitive $Ca^{2+}$ channels, and are generally associated with reduced neuronal excitation.

These findings suggest that excitotoxicity is highly dependent on the diversity and anatomical distribution of EAA receptors. The amino acid sequences of receptor subunits are also particularly important since a modification of a single amino acid may alter $Ca^{2+}$ permeability or other properties, which may then lead to excitotoxicity.

## Cellular mechanisms of excitotoxicity

### *Elevation of intracellular calcium concentration*

The disruption of intracellular $Ca^{2+}$ homeostasis is a key mechanism in the neurotoxicity of EAA (Reynolds, this volume). A large body of evidence has actually suggested that intracellular $Ca^{2+}$ accumulation is responsible for the early neuronal degeneration arising after a brief exposure to EAA, as well as for the delayed neurotoxicity occurring up to many hours following the initial insult. When neurons are lethally challenged with glutamate, they exhibit a biphasic elevation of intracellular $Ca^{2+}$ concentration characterized by a primary, transient rise that decays within minutes, followed by a secondary, sustained, and irreversible increase that indicates imminent cell death (Randall and Thayer, 1992; Tymianski et al., 1993a). The accumulation of $Ca^{2+}$ may be attributed to influx through NMDA receptor- and, to a lesser extent, non-NMDA receptor-gated channels, as well as the release from intracellular $Ca^{2+}$ stores (Frandsen and Schousboe, 1993; Hartley et al., 1993).

Many studies have failed in establishing a link between $Ca^{2+}$ overload and glutamate-induced neurodegeneration: while one report suggested a correlation between the magnitude of the rise in free intracellular $Ca^{2+}$ concentration and neurotoxicity (Milani et al., 1991), others could not demonstrate such associations (Michaels and Rothman, 1990; Dubinsky and Rothman, 1991; Witt et al., 1994). More recently, two studies have shown a close correlation between the absolute amount of $Ca^{2+}$ taken up *via* the NMDA receptor and the amount of cell death (Hartley et al., 1993; Eimerl and Schramm, 1994), thus demonstrating that $Ca^{2+}$ influx through NMDA receptor-gated channels is a key step in excitotoxicity. In accordance with this hypothesis, previous reports showed that the delayed neuronal loss in cerebellar slices and cultured hippocampal neurons exposed to glutamate could be prevented by removal of $Ca^{2+}$ from the extracellular medium (Garthwaite et al., 1986; Rothman et al., 1987), while retinal ganglion neurons, which are normally resistant to glutamate toxicity, were made sensitive by increasing the extracellular $Ca^{2+}$ concentration (Hahn et al., 1988). Finally, one of the most direct pieces of evidence that $Ca^{2+}$ overload triggers excitotoxic injury of the brain was the observation that cell-permeant $Ca^{2+}$ chelators attenuated both glutamate-induced intracellular $Ca^{2+}$ increases and neurotoxicity (Tymianski et al., 1993b).

It appears that neuronal death depends much more on the source of $Ca^{2+}$ than on the intracellular concentration itself since the $Ca^{2+}$ load entering through NMDA receptor-associated channels is much more toxic than that entering through other sources (Dubinsky and Rothman, 1991; Tymianski et al., 1993a). Therefore, it has been suggested that $Ca^{2+}$ influx by the NMDA receptor may have access to a sensitive compartment which makes this $Ca^{2+}$ more efficacious in triggering cell death. One of these compartments could be the mitochondria in which $Ca^{2+}$ rapidly accumulates when cytoplasmic levels rise abnormally (Garthwaite and Garthwaite, 1986; White and Reynolds, 1995). $Ca^{2+}$ sequestration by mitochondria leads to intracellular acidification (Wang et al., 1994) and free radical generation (Reynolds and Hastings, 1995), which may contribute to the mitochondrial damage associated with excitotoxicity.

Elevation in intracellular $Ca^{2+}$ activates a number of $Ca^{2+}$-dependent enzymes, which may trigger cell death through different pathways. The relative contribution of these pathways is still unclear. For example, activated phospholipase $A_2$ ($PLA_2$), proteases, and endonucleases mediate respectively extensive membrane breakdown, cytoskeleton alterations, and DNA fragmentation (Orrenius et al., 1989). Activation of $PLA_2$ following NMDA receptor stimulation also results in arachidonic acid release from membrane phospholipids (Dumuis et al., 1988), leading to the generation of toxic free radicals (Lafon-Cazal et al., 1993). $PLA_2$ inhibitors were shown to attenuate both the production of free radicals and neurotoxicity induced by glutamate (Cazevieille et al., 1994; Gunasekar et al., 1995). Also, in some neurons,

NMDA receptor-mediated $Ca^{2+}$ influx serves as a signal for activation of nitric oxide synthase (NOS) (Bredt and Snyder, 1992). This $Ca^{2+}$/calmodulin-dependent enzyme generates nitric oxide (NO), which is lethal to surrounding neurons (Dawson et al., 1991). NOS inhibitors were shown to block NO formation and to markedly reduce excitotoxic injury in cultured neurons (Dawson et al., 1993; Gunasekar et al., 1995).

### Generation of free radicals

An increasing number of studies have suggested that a causal relationship exists between free radicals and excitotoxicity (Bondy and LeBel, 1993; Coyle and Puttfarcken, 1993; Nakao and Brundin, this volume). An initial study showed that the free radical scavengers superoxide dismutase and mannitol could protect cerebellar neurons from kainate-induced death (Dykens et al., 1987). Subsequent studies showed that antioxidants such as ascorbic acid, ubiquinone, and 21-aminosteroids also attenuated glutamate neurotoxicity in vitro (Majewska and Bell, 1990; Monyer et al., 1990; Favit et al., 1992). The spin trapping agent α-phenyl-*N-tert*-butylnitrone (PBN) protected cultured neurons from NMDA-induced toxicity (Nakao et al., 1996; Nakao and Brundin, this volume), while the vitamin E analogue trolox exerted neuroprotection against AMPA (Chow et al., 1994).

Recently, several techniques have become available to study the mechanisms of free radical formation. Using electron paramagnetic resonance, Bockaert and colleagues demonstrated that cultured neurons exposed to NMDA generate superoxide by a mechanism involving $PLA_2$ and NOS (Lafon-Cazal et al., 1993; Culcasi et al., 1994). Measurements of hydroxyl radicals in vivo by the means of the salicylate trapping technique confirmed the hypothesis that NOS contributes to the generation of free radicals following NMDA receptor activation (Hammer et al., 1993). Gunasekar and colleagues (1995) used the fluorescent dye dichlorofluorescein to show an increased production of oxidants in cerebellar granule cells treated with glutamate. Intracellular oxidant generation and neurotoxicity were attenuated by inhibitors of $PLA_2$ and NOS. Uncouplers of mitochondrial electron transport also reduced the rise in fluorescence occurring in cultured neurons after NMDA receptor activation (Reynolds and Hastings, 1995; Dugan et al., 1995). A critical role for mitochondria in the production of oxygen radicals associated with excitoxicity has therefore been suggested. Finally, inhibitors of xanthine oxidase reduced the production of free radicals in a model of cerebral ischemia (Phillis et al., 1994) and protected cerebellar neurons from kainate toxicity (Dykens et al., 1987).

Beal and co-workers examined the relationship of excitotoxicity to free radical production in vivo (Schulz et al., 1995a). Neuronal injury was induced in rat striatum by local injection of malonate, a neurotoxin that impairs mitochondrial energy metabolism and produces secondary excitotoxic damage. The free radical spin trap *N-tert*-butyl-α-(2-sulfophenyl)-nitrone (S-PBN) attenuated both the generation of hydroxyl radicals and the striatal lesion induced by malonate. It also reduced the volumes of the lesions produced by local injections of NMDA, kainate, and AMPA. Similarly, PBN was shown to alleviate the neurotoxicity of glutamate in rat striatum by preventing the accumulation of hydroxyl radicals (Lancelot et al., 1997). These findings provide direct in vivo evidence that free radicals are involved in EAA neurotoxicity.

### Generation of nitric oxide

A role for NO in excitotoxicity was initially suggested by Dawson and colleagues (1991) who reported that NMDA toxicity in cultured neurons was blocked by NOS inhibitors, by the NO scavenger hemoglobin, and by depleting the culture medium of L-arginine, which is the precursor of NO. They subsequently showed that the selective killing of NOS-containing neurons by quisqualate made other neurons resistant to glutamate toxicity (Dawson et al., 1993). Further studies confirmed the idea that NO mediates NMDA toxicity in neuronal cultures (Cazevieille et al., 1993; Gunasekar et al., 1995), in brain slices

(Izumi et al., 1992; Kollegger et al., 1993), and in vivo (Moncada et al., 1992; Fujisawa et al., 1993). The neuronal degeneration resulting from stimulation of non-NMDA receptors was not prevented by NOS inhibitors.

Several mechanisms have been proposed for NO-mediated neurotoxicity, one of which is based on the combination of NO with superoxide to form peroxynitrite (Beckman, 1996). Peroxynitrite is responsible for sulfhydryl oxidation and tyrosine nitration, subsequently reducing the levels of free radical scavengers (cysteine, glutathione) and altering the structure of various proteins (Radi et al., 1991; Ischiropoulos et al., 1992). It was also reported to spontaneously decompose in vitro and to produce an oxidant species with a hydroxyl radical-like activity (Beckman et al., 1990). Other studies suggested that NO may directly damage the DNA (Nguyen et al., 1992) or inhibit its replication (Kwon et al, 1991; Lepoivre et al., 1991). It may activate poly(ADP-ribose) synthetase (PARS), an enzyme responsible for the repair of DNA, and subsequently produce energy depletion. PARS inhibitors were reported to attenuate the toxicity of NO-donors and NMDA in hippocampal slices and cortical cultures (Wallis et al., 1993; Zhang et al., 1994). Finally, NO was shown to inhibit mitochondrial enzymes (aconitase, NADH dehydrogenase, and succinate dehydrogenase) containing an iron-sulfur group in their catalytic site (Drapier and Hibbs, 1988; Stadler et al., 1991). It was therefore suggested that inhibition of mitochondrial respiration by NO may be one mechanism leading to excitotoxic neuronal death.

Contrary to the initial reports of Dawson and colleagues (1991), several studies could not prevent glutamate-induced neurotoxicity by inhibiting NOS activity (Demerlé-Pallardy et al., 1991; Pauwels and Lysen, 1992; Zinkand et al., 1993; Hewett et al., 1993). This discrepancy was attributed to differences in origin or composition of the cell cultures, as well as the composition of the culture medium. Conflicting results were also obtained in vivo, some studies reporting either no effect or even a potentiation of EAA neurotoxicity by NOS inhibitors (Lerner-Natoli et al., 1992; Haberny et al., 1992; Galasso et al., 1995). This appeared to be due to the nonspecificity of the inhibitors tested, which affected both the neuronal and the endothelial NOS isoforms, leading to vascular effects (Globus et al., 1995). Recently, Beal and co-workers showed that the selective neuronal NOS inhibitor 7-nitroindazole attenuated the lesions produced by intrastriatal injections of NMDA or malonate, but not kainate or AMPA (Schulz et al., 1995b). It also blocked the toxicity of the dopaminergic neurotoxin 1-methyl-4-phenyl-1,2,3,6-tetrahydropyridine (MPTP) (Schulz et al., 1995c). S-methylthiocitrulline, another neuronal NOS inhibitor, protected against malonate and MPTP (Matthews et al., 1997). Altogether these findings strongly favor a role for neuronal NOS in excitotoxicity.

**Excitatory amino acids and neurodegenerative diseases**

*Concept of slow excitotoxicity*

A contribution of excitotoxicity to the pathophysiology of chronic neurodegenerative diseases is being increasingly accepted. Yet, contrary to cerebral ischemia and brain trauma, there is no consistent evidence of increased extracellular glutamate concentration. An excess of other endogenous excitotoxins, such as quinolinic acid and homocysteic acid, or a lack of the endogenous NMDA receptor antagonist kynurenic acid were proposed to explain chronic neurodegeneration (Schwarcz et al., 1983; Do et al., 1986; Beal et al., 1990), but further research in these directions has proved unsuccessful. Subsequently, the theory of excitotoxicity has been modified and the expressions "slow" or "weak" excitotoxicity were coined (Albin and Greenamyre, 1992; Beal, 1992). The slow excitotoxic hypothesis is based on the concept that some pathological changes may increase the vulnerability of specific populations of neurons toward glutamate. Thus, excitotoxicity could occur in sensitized neurons even in the absence of abnormally elevated glutamate levels.

Genetic mutations leading to changes in the configuration of glutamate receptors and resulting in increased $Ca^{2+}$ influx represent one possible disturbance causing enhanced cell vulnerability to glutamate. Another possibility is that slow excitotoxicity results from an impairment in energy metabolism. Evidence in support of this hypothesis was first provided by Novelli and colleagues (1988) who reported that inhibition of oxidative phosphorylation by cyanide or $Na^+/K^+$ ATPase by ouabain in neuronal cultures caused innocuous concentrations of glutamate to become toxic. Another study showed that impairment of energy metabolism with cyanide in cultured chick retina triggered excitotoxic lesions sensitive to NMDA antagonists (Zeevalk and Nicklas, 1990). Therefore, it is believed that impaired mitochondrial function depletes ATP, disrupts $Na^+/K^+$ ATPase activity, and causes neuronal depolarization. This may alleviate the voltage-dependent $Mg^{2+}$ block of NMDA receptor-gated channels, and lead to persistent receptor activation by ambient levels of glutamate. Consistent with this hypothesis was the observation that toxicity induced by metabolic inhibition could be mimicked by membrane depolarization with $K^+$ (Zeevalk and Nicklas, 1991).

The initial disturbance leading to mitochondrial dysfunction could be genetic mutations in components of the electron transport chain or in Krebs cycle enzymes. Oxidative damage to mitochondria could also account for the bioenergetic defects. Moreover, the buffering of the $Ca^{2+}$ load by mitochondria is believed to cause further impairment in energy metabolism. It appears therefore that a mild disturbance of mitochondrial function will lead to excitotoxicity which will, in turn, further damage the mitochondria. This complex relationship between excitotoxicity and bioenergetic defects may play a role in the pathophysiology of chronic neurodegenerative diseases (Beal, 1992).

**Involvement of excitatory amino acids in disease**

Several lines of evidence favor a role for excitotoxicity in amyotrophic lateral sclerosis (ALS). First, lathyrism and western Pacific ALS were linked respectively to ingestion of the AMPA agonist *β-N*-oxalylamino-L-alanine and the NMDA agonist *β-N*-methylamino-L-alanine (Spencer et al., 1991). Second, glutamate concentrations are reduced in brain and spinal cord of ALS patients (Plaitakis et al., 1988), but increased in plasma and cerebrospinal fluid (Plaitakis and Caroscio, 1987; Rothstein et al., 1990). This could be the consequence of an increased synthesis and release (Malessa et al., 1991), or an impairment of glutamate uptake (Rothstein et al., 1992). Third, there is a decrease in the number of NMDA receptors in ALS spinal cord, suggesting that degeneration affects preferentially the neurons which possess this type of receptors (Allaoua et al., 1992; Krieger et al., 1993). NMDA receptor activation is known to induce free radical generation, and mutations in $Cu^{2+}/Zn^{2+}$ superoxide dismutase have been linked to familial ALS (Rosen et al., 1993). Furthermore, there is an increase in kainate receptors (Shaw et al., 1994), and the gene encoding the kainate receptor subunit GluR5 is located on human chromosome 21 in the vicinity of the gene for familial ALS (Eubanks et al., 1993).

The neurotoxicity of MPTP and its active metabolite 1-methyl-4-phenylpyridinium ($MPP^+$) has been extensively used to model the selective degeneration of the dopaminergic neurons of the substantia nigra seen in Parkinson's disease (PD). A role of EAA in PD was initially suggested by Turski and his associates (1991), who demonstrated that $MPP^+$ neurotoxicity could be prevented by NMDA antagonists. Subsequently, conflicting results were obtained by Sonsalla and colleagues (1992), but two other studies reported neuroprotective effects of NMDA antagonists in primates (Zuddas et al., 1992; Lange et al., 1993). Beal and co-workers showed that $MPP^+$ injections in rat striatum produced a depletion of GABA and substance P, with relative sparing of somatostatin and neuropeptide Y, consistent with an NMDA receptor-mediated excitotoxic process (Storey et al., 1992). The lesion was attenuated either by decortication or by treatment with the non-competitive NMDA receptor antagonist MK-801. Further evidence are the potentiation of the antiparkinsonian action of levodopa by NMDA receptor blockers (Klockgether and Turski, 1990), and the improved survival of PD patients treated with amantadine, a low-affinity NMDA receptor antagonist (Uitti et al., 1993). Lastly, a loss of NMDA and AMPA sites was described in post-mortem PD substantia nigra (Difazio et al., 1992), and NMDA and AMPA were shown to induce parkinsonism in rats when injected in the anterior striatum or the basal ganglia output nuclei respectively (Klockgether and Turski, 1993).

Various pieces of evidence support the idea of the involvement of excitotoxicity in the pathogenesis of Alzheimer's disease (AD; see Farber et al., this volume). Lesions of the rat nucleus basalis with EAA induce degeneration of cholinergic projections to the cortex that are also lost in AD (Arendash et al., 1987). Neurofibrillary tangles appear predominantly in neurons involved in cortico-cortical pathways (Rogers and Morrison, 1985; Pearson et al., 1985), which are known to use glutamate as a neurotransmitter. Furthermore, a loss of NMDA receptors was described in the hippocampus and neocortex of patients with AD (Chalmers et al., 1990; Ulas et al., 1992). In vitro, $\beta$-amyloid was shown to sensitize cultured neurons to impaired energy metabolism (Copani et al., 1991) and to EAA toxicity (Koh et al., 1990; Mattson et al., 1992). It is therefore possible that a bioenergetic defect in AD makes neurons vulnerable to slow excitotoxicity, which may be exacerbated by $\beta$-amyloid.

The possibility that excitotoxicity may play a role in Huntington's disease (HD) was first suggested by observations that kainic acid striatal lesions reproduced many of the neuropathological features of HD (Coyle and Schwarcz, 1976; McGeer and McGeer, 1976). Subsequently, it was found that lesions induced by the NMDA agonist quinolinic acid more closely resembled those in HD since there was a relative sparing of NADPH-diaphorase and parvalbumin interneurons (Beal et al., 1986; Waldvogel et al., 1991). The modelling of the disease was further improved by chronic administration of quinolinic acid to rats, which resulted in shrinkage of the striatal matrix compartment (Beal et al., 1991). Administration of the NMDA agonist ibotenic acid to primates produced an apomorphine-inducible movement disorder (Hantraye et al., 1990). Furthermore, there is evidence of a depletion of NMDA receptors in HD striatum (Young et al., 1988). It was recently demonstrated that genotype variations of the GluR6 subunit of kainate receptors may partly account for the variance in the age of onset of HD (Rubinsztein et al., 1997). It has been therefore postulated that GluR6-mediated excitotoxicity may contribute to the pathogenesis of HD.

The hypothesis that slow excitotoxicity may play a role in chronic neurodegenerative diseases has been strongly supported by experimental studies using mitochondrial toxins. As we previously mentioned, the mitochondrial complex I inhibitor $MPP^+$ replicates the neurochemical and neuropathological features of PD, that can be blocked by NMDA antagonists. Beal and co-workers and others have intensively used inhibitors of succinate dehydrogenase to model HD. Succinate dehydrogenase is a component of mitochondrial complex II and the tricarboxylic acid cycle. Intrastriatal injections of the reversible inhibitor malonate were shown to produce ATP depletions and excitotoxic lesions which spared NADPH-diaphorase neurons. The lesions were attenuated by NMDA antagonists and glutamate release inhibitors (Beal et al., 1993a; Greene et al., 1993; Henshaw et al., 1994). Moreover, subtoxic concentrations of malonate caused large lesions when co-injected with non-toxic concentrations of glutamate, NMDA, or AMPA (Greene and Greenamyre, 1995a). This indicates that bioenergetic defects may exacerbate excitotoxicity, and that a mild metabolic disturbance may induce severe brain damage by rendering neurons vulnerable to ambient levels of glutamate.

3-nitropropionic acid (3-NP) is an irreversible inhibitor of succinate dehydrogenase. In cultured cerebellar neurons, it produced neurotoxicity which was delayed but not prevented by NMDA antagonists (Weller and Paul, 1993). Intrastriatal injections of 3-NP in rats resulted in age-dependent lesions with neuronal loss and gliosis as well as ATP depletions and increases in lactate levels (Brouillet et al., 1993). Systemic administration produced selective striatal lesions which were attenuated by the glutamate release inhibitor lamotrigine or by prior decortication (Beal et al., 1993b; Schulz et al., 1996a). Consistent with 3-NP inducing secondary excitotoxicity, microdialysis studies showed no significant increase in extracellular glutamate concentrations after 3-NP administration, and in vivo [$^3$H]MK-801 receptor autoradiography showed an activation of NMDA receptors (Beal et al., 1993b; Wüllner et al., 1994). Chronic low-dose administration to rats was associated with sparing of NADPH-diaphorase neurons and proliferative changes in dendrites of spiny neurons, similar to findings in HD (Beal et al., 1993b). In non-human primates, it produced selective striatal degeneration and abnormal choreiform movements (Brouillet et al., 1995). Using magnetic resonance imaging in vivo, Beal and co-workers reported marked increases in lactate concentrations in the basal ganglia and cerebral cortex of HD patients (Jenkins et al., 1993), thus providing direct evidence of bioenergetic defects in HD. These findings strengthen the contention that 3-NP replicates the neurochemical and histologic features of HD, and suggest that impairment of energy metabolism and secondary excitotoxicity may be a relevant mechanism for the disease process.

## Therapeutic directions

### *Excitatory amino acid antagonists*

Experimental data suggest that blockade of excitotoxicity may prolong the life-span of cells undergoing energy failure, which may occur in several neurodegenerative diseases. This can be achieved by decreasing presynaptic glutamate release with $Na^+$ channel blockers. One such compound, lamotrigine, was found to be neuroprotective against kainate striatal lesions in rats and against MPTP dopaminergic neurotoxicity in mice (McGeer and Zhu, 1990; Jones-Humble et al., 1994). It attenuated lesions produced by intrastriatal administrations of $MPP^+$ and malonate, or by systemic administration of 3-NP (Schulz et al., 1996a). It alleviated the striatal toxicity of 3-acetylpyridine, a neurotoxin which impairs energy metabolism by substituting itself for nicotinamide in the formation of NAD and NADP (Schulz et al., 1994). Lamotrigine has recently been approved for use as an antiepileptic in man (Pellock, 1994), and has been reported to exert enhanced symptomatic benefits in PD when used in combination with carbidopa-levodopa (Zipp et al., 1993). Riluzole is another glutamate release inhibitor that prevented the depletion of dopamine in mice treated with MPTP (Boireau et al., 1994). It also prevented the occurrence of MPTP-induced parkinsonism in rhesus monkeys (Benazzouz et al., 1995). It attenuated decreases in dopamine and it improved behavioral abnormalities induced by the toxin 6-hydroxydopamine in rats (Barneoud et al., 1996). Furthermore, riluzole prolongs survival and slows the progression of patients with ALS (Bensimon et al., 1994).

Another means to reduce secondary excitotoxicity is to use glutamate receptor antagonists. The non-competitive NMDA antagonist MK-801 showed inconsistent efficacy against $MPP^+$ neurotoxicity in the substantia nigra (Turski et al., 1991; Sonsalla et al., 1992). Yet, two studies using either MK-801 or the competitive NMDA antagonist CPP reported neuroprotection against MPTP in primates (Zuddas et al., 1992; Lange et al., 1993). Beal and co-workers observed significant attenuation of malonate and 3-acetylpyridine toxicity in rat striatum after treatment with MK-801 (Schulz et al., 1995a; 1996a). These results suggest that NMDA antagonists may be useful in the treatment of chronic neurodegenerative diseases. Non-competitive antagonists have the advantage that their blocking action is use-dependent. However, their clinical use as neuroprotective agents has largely been hampered by the fact that they exhibit a high affinity for the PCP-binding site and therefore frequently show strong psychotomimetic side effects. The recent finding that the well-tolerated antiparkinsonian drugs memantine and budipine are actually low-affinity blockers of the PCP site (Parsons et al., 1993; Klockgether et al., 1993), suggests that low-affinity PCP-like NMDA receptor antagonists may be valuable in preventing slow excitotoxicity associated with chronic neurodegenerative diseases. Consistent with this notion was the protective effect of memantine against malonate toxicity in vivo (Schulz et al., 1996a). Inhibitors of NMDA receptor modulatory sites may also be of interest since malonate toxicity was blocked by the glycine site antagonist 7-chlorokynurenate (Greene and Greenamyre, 1995b). Similarly, the therapeutic potential of AMPA antagonists has been questioned. In vitro, AMPA antagonists showed more efficacy than NMDA antagonists in blocking delayed toxicity mediated by glutamate (Prehn et al., 1995). Rothstein and colleagues (1993) demonstrated that motor neuron toxicity triggered by chronic inhibition of glutamate uptake in spinal cord slices could be attenuated by the AMPA antagonists CNQX and GYKI 52466. NBQX also prevented the striatal toxicity of 3-NP (Turski and Ikonomidou, 1994), but not that of malonate (Greene and Greenamyre, 1995b). These results suggest that AMPA antagonists may represent an alternative approach to NMDA antagonists in preventing slow excitotoxicity. Yet, experimental data are not sufficient to make predictions about possible success of these drugs in the clinic.

### Modulators of intracellular calcium concentration

A growing body of evidence has suggested that neurotrophic factors can alleviate excitotoxicity. Protection against quinolinic acid toxicity was reported in adult rat striatum following local implantation of fibroblasts secreting nerve growth factor (NGF) or basic fibroblast growth factor (bFGF) (Frim et al., 1993a). NGF protected against glutamate-induced neurotoxicity in cultured cortical neurons (Shimohama et al., 1993), and bFGF attenuated NMDA striatal lesions in neonatal rats (Nozaki et al., 1993). In addition, these factors are effective against mitochondrial toxins. bFGF and glial cell line-derived neurotrophic factor (GDNF) prevented MPTP-induced dopaminergic neurotoxicity in mice (Otto and Unsicker, 1990; Tomac et al., 1995). Systemic administration of bFGF also protected against malonate lesions in adult rats, and $MPP^+$ toxicity in neonatal animals (Kirschner et al., 1995). Furthermore, NGF-secreting fibroblasts attenuated striatal 3-NP lesions (Frim et al., 1993b). The neuroprotective effect of these neurotrophins is postulated to be due to their ability to maintain $Ca^{2+}$ homeostasis in cells undergoing energy impairment. In vitro, NGF and bFGF prevented increases in intracellular $Ca^{2+}$ concentrations induced by hypoglycemia (Cheng and Mattson, 1991). bFGF also protected neurons against $Ca^{2+}$ ionophore neurotoxicity, suggesting that it may enhance $Ca^{2+}$ buffering or extrusion (Mattson and Rychlik, 1990). bFGF and NGF were recently reported to attenuate glutamate-induced accumulation of reactive oxygen species in cultured hippocampal neurons (Mattson et al., 1995), as well as $MPP^+$-induced generation of hydroxyl radicals in vivo (Kirschner et al., 1996). NGF was shown to reduce 3-nitrotyrosine levels in 3-NP-treated animals (Galpern et al., 1996). Altogether, these findings suggest that growth factors may be useful for treatment of neurologic diseases mediated by excitotoxicity.

### Free radical scavengers

Since EAA toxicity is associated with increased generation of free radicals, a number of studies have investigated the potential therapeutic effects of free radical scavengers against both direct and indirect excitotoxicity (also see Nakao and Brundin, this volume). The antioxidants coenzyme $Q_{10}$ and idebenone protected against glutamate toxicity in neuronal cultures (Murphy et al., 1990; Favit et al., 1992), and improved the neurological deficits in patients with mitochondrial encephalomyopathy MELAS (Ihara et al., 1989). Dihydrolipoate reduced NMDA- and malonate-induced striatal lesions in rats (Greenamyre et al., 1994), and the free radical scavenger bromocriptine prevented 6-hydroxydopamine neurotoxicity in mice (Ogawa et al., 1994). 21-aminosteroids exhibited a striking neuroprotective effect against NMDA toxicity in cortical cultures (Monyer et al., 1990), and were found effective in models of cerebral ischemia and head trauma (Hall et al., 1988; Park and Hall, 1994).

Among the most promising antioxidants are the free radical spin traps PBN, S-PBN, and 5,5-dimethyl-1-pyrroline-*N*-oxide (DMPO). PBN was shown to attenuate hydroxyl radical production and striatal lesion induced by glutamate in rats (Lancelot et al., 1997). S-PBN reduced lesion volumes caused by intrastriatal injections of NMDA, AMPA, and kainate, as well as those produced by the mitochondrial toxins 3-acetylpyridine, $MPP^+$, and malonate (Schulz et al., 1995a). It also attenuated malonate-induced increases in hydroxyl radicals. Similarly, DMPO protected against systemic administration of 3-NP in rats (Schulz et al., 1996b). The neuroprotective effect of S-PBN against malonate was potentiated when the drug was given in combination with MK-801 or with the putative enhancer of mitochondrial function nicotinamide (Schulz et al., 1995a; 1996a). These results suggest that combination of free radical scavengers with agents acting at different steps in the neurodegenerative process can produce additive neuroprotective effects. Another advantage of free radical spin traps is their exceptionally long therapeutic window. Beal and co-workers actually found that S-PBN was still effective when administered up to 6 hours after malonate injection in rat striatum (Schulz et al., 1995d).

## Nitric oxide synthase inhibitors

Excitotoxicity has been linked to activation of NOS (Dawson et al., 1991; 1993). The effects of NOS inhibitors in animal models of neurodegenerative diseases were therefore investigated. The non-specific NOS inhibior $N^G$-nitro-L-arginine was shown to attenuate malonate-induced striatal lesions in rats (Maragos and Silverstein, 1995). Similarly, Beal and co-workers demonstrated that the relatively selective NOS inhibitors 7-nitroindazole and *S*-methylthiocitrulline protected against malonate toxicity in vivo (Schulz et al., 1995b; Matthews et al., 1997). They found that the concentration of 3-nitrotyrosine, which is a marker of tyrosine nitration by peroxynitrite, was increased after malonate injection. Similarly, since peroxynitrite generates a hydroxyl radical-like intermediate, they found that the hydroxyl radical markers 2,3- and 2,5-dihydroxybenzoic acids (2,3- and 2,5-DHBA) were increased. 7-nitroindazole and *S*-methylthiocitrulline reduced both 3-nitrotyrosine and 2,3-and 2,5-DHBA concentrations in malonate-treated animals. Both 7-nitroindazole and *S*-methylthiocitrulline also protected against MPTP-induced neurotoxicity in mice (Schulz et al., 1995c; Matthews et al., 1997). Consistent with these results was the recent report that mice deficient in neuronal NOS showed significant protection against MPTP dopamine depletions (Przedborski et al., 1996). Inbaboons,7-nitroindazole almost completely prevented the biochemical, histologic, and behavioral deficits induced by MPTP (Hantraye et al., 1996). These findings demonstrate that NO, whose toxicity involves generation of peroxynitrite, plays a role in animal models of HD and PD. Selective inhibitors of neuronal NOS may therefore be useful in treating neurodegenerative illnesses.

## Conclusions

Unequivocal demonstration has been obtained of the involvement of EAA in the pathogenesis of ischemic and traumatic brain injury. It has been more difficult, however, to link neuronal degeneration that occurs in chronic neurodegenerative diseases to an excitotoxic mechanism. Studies in rodents and non-human primates using mitochondrial toxins have suggested that slow excitotoxicity may occur as a consequence of impaired energy metabolism. EAA antagonists, modulators of intracellular $Ca^{2+}$ concentration, free radical scavengers, and NOS inhibitors showed efficacy in these experimental models. The development of well tolerated NMDA receptor antagonists, potent free radical spin traps, and relatively selective

NOS inhibitors has resulted in new perspectives for treatment of chronic neurodegenerative diseases.

## References

Albin, R.L. and Greenamyre, J.T. (1992) Alternative excitotoxic hypotheses. *Neurology*, 42: 733–738.

Allaoua, H., Chaudieu, I., Krieger, C., Boska, P., Privat, A. and Quirion, R. (1992) Alterations in spinal cord excitatory amino acid receptors in amyotrophic lateral sclerosis patients. *Brain Res.*, 579: 169–172.

Anegawa, N.J., Lynch, D.R., Verdoorn, T.A. and Pritchett, D.B. (1995) Transfection of *N*-methyl-D-aspartate receptors in a nonneuronal cell line leads to cell death. *J. Neurochem*, 64: 2004–2012.

Arendash, G.W., Millard, W.J., Dunn, A.J. and Meyer, E.M. (1987) Long-term neuropathological and neurochemical effects of nucleus basalis lesions in the rat. *Science*, 238: 952–956.

Barneoud, P., Mazadier, M., Miquet, J.M., Parmentier, S., Dubedat, P., Doble, A. and Boireau, A. (1996) Neuroprotective effects of riluzole on a model of Parkinson's disease in the rat. *Neuroscience*, 74: 971–983.

Beal, M.F. (1992) Does impairment of energy metabolism result in excitotoxic neuronal death in neurodegenerative illnesses? *Ann. Neurol.*, 31: 119–130.

Beal, M.F., Kowall, N.W., Ellison, D.W., Mazurek, M.F., Swartz, K.J. and Martin, J.B. (1986) Replication of the neurochemical characteristics of Huntington's disease by quinolinic acid. *Nature*, 321: 168–171.

Beal, M.F., Matson, W.R., Swartz, K.J., Gamache, P.H. and Bird, E.D. (1990) Kynurenine pathway measurements in Huntington's disease striatum: Evidence for reduced formation of kynurenic acid. *J. Neurochem.*, 55: 1327–1339.

Beal, M.F., Ferrante, R.J., Swartz, K.J. and Kowall, N.W. (1991) Chronic quinolinic acid lesions in rats closely resemble Huntington's disease. *J. Neurosci.*, 11: 1649–1659.

Beal, M.F., Brouillet, E., Jenkins, B., Henshaw, R., Rosen, B. and Hyman, B.T. (1993a) Age-dependent striatal excitotoxic lesions produced by the endogenous mitochondrial inhibitor malonate. *J. Neurochem.*, 61: 1147–1150.

Beal, M.F., Brouillet, E., Jenkins, B.G., Ferrante, R.J., Kowall, N.W., Miller, J.M., Storey, E., Srivastava, R., Rosen, B.R. and Hyman, B.T. (1993b) Neurochemical and histologic characterization of striatal excitotoxic lesions produced by the mitochondrial toxin 3-nitropropionic acid. *J. Neurosci.*, 13: 4181–4192.

Beckman, J.S. (1996) Oxidative damage and tyrosine nitration from peroxynitrite. *Chem. Res. Toxicol.*, 9: 836–844.

Beckman, J.S., Beckman, T.W., Chen, J., Marshall, P.A. and Freeman, B.A. (1990) Apparent hydroxyl radical production by peroxynitrite: Implications for endothelial injury from nitric oxide and superoxide. *Proc. Natl. Acad. Sci. USA*, 87: 1620–1624.

Benazzouz, A., Boraud, T., Dubedat, P., Boireau, A., Stutzmann, J.M. and Gross, C. (1995) Riluzole prevents MPTP-induced parkinsonism in the rhesus monkey: A pilot study. *Eur. J. Pharmacol.*, 284: 299–307.

Bensimon, G., Lacomblez, L., Meiniger, V and the ALS/Riluzole Study Group. (1994) A controlled trial of riluzole in amyotrophic lateral sclerosis. *N. Engl. J. Med.*, 330: 585–591.

Boireau, A., Dubedat, P., Bordier, F., Peny, C., Miquet, J.M., Durand, G., Meunier, M. and Doble, A. (1994) Riluzole and experimental parkinsonism: Antagonism of MPTP-induced decrease in central dopamine levels in mice. *Neuroreport*, 5: 2657–2660.

Bondy, S.C. and LeBel, C.P. (1993) The relationship between excitotoxicity and oxidative stress in the central nervous system. *Free Rad. Biol. Med.*, 14: 633–642.

Bredt, D.S. and Snyder, S.H. (1992) Nitric oxide, a novel neuronal messenger. *Neuron*, 8: 3–11.

Brouillet, E., Jenkins, B.G., Hyman, B.T., Ferrante, R.J., Kowall, N.W., Srivastava, R., Samanta Roy, D., Rosen, B.R. and Beal, M.F. (1993) Age-dependent vulnerability of the striatum to the mitochondrial toxin 3-nitropropionic acid. *J. Neurochem.*, 60: 356–359.

Brouillet, E., Hantraye, P., Ferrante, R.J., Dolan, R., Leroy-Willig, A., Kowall, N.W. and Beal, M.F. (1995) Chronic mitochondrial energy impairment produces selective striatal degeneration and abnormal choreiform movements in primates. *Proc. Natl. Acad. Sci. USA*, 92: 7105–7109.

Burnashev, N., Schoepfer, R., Monyer, H., Ruppersberg, J.P., Günther, W., Seeburg, P.H. and Sakmann, B. (1992) Control by asparagine residues of calcium permeability and magnesium blockade in the NMDA receptor. *Science*, 257: 1415–1419.

Cazevieille, C., Muller, A., Meynier, F. and Bonne, C. (1993) Superoxide and nitric oxide cooperation in hypoxia/reoxygenation-induced neuron injury. *Free Rad. Biol. Med.*, 14: 389–395.

Cazevieille, C., Muller, A., Meynier, F., Dutrait, N. and Bonne, C. (1994) Protection by prostaglandins from glutamate toxicity in cortical neurons. *Neurochem. Int.*, 24: 395–398.

Chalmers, D.T., Dewar, D., Graham, D.I., Brooks, D.N. and McCulloch, J. (1990) Differential alterations of cortical glutamatergic binding sites in senile dementia of the Alzheimer type. *Proc. Natl. Acad. Sci. USA*, 87: 1352–1356.

Cheng, B. and Mattson, M.P. (1991) NGF and bFGF protect rat hippocampal and human cortical neurons against hypoglycemic damage by stabilizing calcium homeostasis. *Neuron*, 7: 1031–1041.

Choi, D.W. (1987) Ionic dependence of glutamate neurotoxicity. *J. Neurosci.*, 7: 369–379.

Chow, H.S., Lynch III, J.J., Rose, K. and Choi, D.W. (1994) Trolox attenuates cortical neuronal injury induced by iron, ultraviolet light, glucose deprivation, or AMPA. *Brain Res.*, 639: 102–108.

Copani, A., Koh, J.Y. and Cotman, C.W. (1991) β-amyloid increases neuronal susceptibility to injury by glucose deprivation. *Neuroreport*, 2: 763–765.

Cotman, C.W., Foster, A.C. and Lanthorn, T.H. (1981) Overview of glutamate as a neurotransmitter. *Adv. Biochem. Psychopharmacol.*, 27: 1–27.

Coyle, J.T. and Schwarcz, R. (1976) Lesion of striatal neurones with kainic acid provides a model for Huntington's chorea. *Nature*, 263: 244–246.

Coyle, J.T. and Puttfarcken, P. (1993) Oxidative stress, glutamate, and neurodegenerative disorders. *Science*, 262: 689–695.

Culcasi, M., Lafon-Cazal, M., Pietri, S. and Bockaert, J. (1994) Glutamate receptors induce a burst of superoxide *via* activation of nitric oxide synthase in arginine-depleted neurons. *J. Biol. Chem.*, 269: 12589–12593.

Dawson, V.L., Dawson, T.M., London, E.D., Bredt, D.S. and Snyder, S.H. (1991) Nitric oxide mediates glutamate neurotoxicity in primary cortical cultures. *Proc. Natl. Acad. Sci. USA*, 88: 6368–6371.

Dawson, V.L., Dawson, T.M., Bartley, D.A., Uhl, G.R. and Snyder, S.H. (1993) Mechanisms of nitric oxide-mediated neurotoxicity in primary brain cultures. *J. Neurosci.*, 13: 2651–2661.

Demerlé-Pallardy, C., Lonchampt, M.O., Chabrier, P.E. and Braquet, P. (1991) Absence of implication of L-arginine/nitric oxide pathway on neuronal cell injury induced by L-glutamate or hypoxia. *Biochem. Biophys. Res. Commun.*, 181: 456–464.

Difazio, M.C., Hollingsworth, Z., Young, A.B. and Penney Jr., J.B. (1992) Glutamate receptors in the substantia nigra of Parkinson's disease brains. *Neurology*, 42: 402–406.

Do, K.Q., Herrling, P.L., Streit, P., Turski, W.A. and Cuenod, M. (1986) In vitro release and electrophysiological effects *in situ* of homocysteic acid, an endogenous *N*-methyl-(D)-aspartic acid agonist, in the mammalian striatum. *J. Neurosci.*, 6: 2226–2234.

Drapier, J.C. and Hibbs Jr., J.B. (1988) Differentiation of murine macrophages to express nonspecific cytotoxicity for tumor cells results in L-arginine-dependent inhibition of mitochondrial iron-sulfur enzymes in the macrophage effector cells. *J. Immunol.*, 140: 2829–2838.

Dubinsky, J.M. and Rothman, S.M. (1991) Intracellular calcium concentrations during "chemical hypoxia" and excitotoxic neuronal injury. *J. Neurosci.*, 11: 2545–2551.

Dugan, L.L., Sensi, S.L., Canzoniero, L.M.T., Handran, S.D., Rothman, S.M., Lin, T.S., Goldberg, M.P. and Choi, D.W. (1995) Mitochondrial production of reactive oxygen species in cortical neurons following exposure to *N*-methyl-D-aspartate. *J. Neurosci.*, 15: 6377–6388.

Dumuis, A., Sebben, M., Haynes, L., Pin, J.P. and Bockaert, J (1988) NMDA receptors activate the arachidonic acid cascade system in striatal neurons. *Nature*, 336: 68–70.

Dykens, J.A., Stern, A. and Trenkner, E. (1987) Mechanism of kainate toxicity to cerebellar neurons in vitro is analogous to reperfusion tissue injury. *J. Neurochem.*, 49: 1222–1228.

Egebjerg, J., Bettler, B., Hermans-Borgmeyer, I. and Heinemann, S. (1991) Cloning of a cDNA for a glutamate receptor subunit activated by kainate but not AMPA. *Nature*, 351: 745–748.

Eimerl, S. and Schramm, M. (1994) The quantity of calcium that appears to induce neuronal death. *J. Neurochem.*, 62: 1223–1226.

Eubanks, J.H., Puranam, R.S., Kleckner, N.W., Bettler, B., Heinemann, S.F. and McNamara, J.O. (1993) The gene encoding the glutamate receptor subunit GluR5 is located on human chromosome 21q21.1–22.1 in the vicinity of the gene for familial amyotrophic lateral sclerosis. *Proc. Natl. Acad. Sci. USA*, 90: 178–182.

Favit, A., Nicoletti, F., Scapagnini, U. and Canonico, P.L. (1992) Ubiquinone protects cultured neurons against spontaneous and excitotoxin-induced degeneration. *J. Cereb. Blood Flow Metab.*, 12: 638–645.

Frandsen, A. and Schousboe, A. (1993) Excitatory amino acid-mediated cytotoxicity and calcium homeostasis in cultured neurons. *J. Neurochem.*, 60: 1202–1211.

Frim, D.M., Uhler, T.A., Short, M.P., Ezzedine, Z.D., Klagsbrun, M., Breakefield, X.O. and Isacson, O. (1993a) Effect of biologically delivered NGF, BDNF and bFGF on striatal excitotoxic lesions. *Neuroreport*, 4: 367–370.

Frim, D.M., Simpson, J., Uhler, T.A., Short, M.P., Bossi, S.R., Breakefield, X.O., Isacson, O. (1993b) Striatal degeneration induced by mitochondrial blockade is prevented by biologically delivered NGF. *J. Neurosci. Res.*, 35: 452–458.

Fujisawa, H., Dawson, D., Browne, S.E., MacKay, K.B., Bullock, R. and McCulloch, J. (1993) Pharmacological modification of glutamate neurotoxicity in vivo. *Brain Res.* 629: 73–78.

Galasso, J.M., Bazzett, T.J., Becker, J.B. and Albin, R.L. (1995) Synergistic effect of intrastriatal co-administration of L-NAME and quinolinic acid. *Neuroreport*, 6: 1505–1508.

Galpern, W.R., Matthews, R.T., Beal, M.F. and Isacson, O. (1996) NGF attenuates 3-nitrotyrosine formation in a 3-NP model of Huntington's disease. *Neuroreport*, 7: 2639–2642.

Garthwaite, G. and Garthwaite, J. (1986) Amino acid neurotoxicity: Intracellular sites of calcium accumulation associated with the onset of irreversible damage to rat cerebellar neurones in vitro. *Neurosci. Lett.*, 71: 53–58.

Garthwaite, G., Hajós, F. and Garthwaite, J. (1986) Ionic requirements for neurotoxic effects of excitatory amino acid analogues in rat cerebellar slices. *Neuroscience*, 18: 437–447.

Globus, M.Y.T., Prado, R., Sanchez-Ramos, J., Zhao, W., Dietrich, W.D., Busto, R. and Ginsberg, M.D. (1995) Dual role for nitric oxide in NMDA-mediated toxicity in vivo. *J. Cereb. Blood Flow Metab.*, 15: 904–913.

Greenamyre, J.T., Garcia-Osuna, M. and Greene, J.G. (1994) The endogenous cofactors, thioctic acid and dihydrolipoic acid, are neuroprotective against NMDA and malonic acid lesions of striatum. *Neurosci. Lett.*, 171: 17–20.

Greene, J.G., Porter, R.H.P., Eller, R.V. and Greenamyre, J.T. (1993) Inhibition of succinate dehydrogenase by malonic acid produces an "excitotoxic" lesion in rat striatum. *J. Neurochem.*, 61: 1151–1154.

Greene, J.G. and Greenamyre, J.T. (1995a) Exacerbation of NMDA, AMPA, and L-glutamate excitotoxicity by the succinate dehydrogenase inhibitor malonate. *J. Neurochem.*, 64: 2332–2338.

Greene, J.G. and Greenamyre, J.T. (1995b) Characterization of the excitotoxic potential of the reversible succinate dehydrogenase inhibitor malonate. *J. Neurochem.*, 64: 430–436.

Gunasekar, P.G., Kanthasamy, A.G., Borowitz, J.L. and Isom, G.E. (1995) NMDA receptor activation produces concurrent generation of nitric oxide and reactive oxygen species: Implication for cell death. *J. Neurochem.*, 65: 2016–2021.

Haberny, K.A., Pou, S. and Eccles, C.U. (1992) Potentiation of quinolinate-induced hippocampal lesions by inhibition of NO synthesis. *Neurosci. Lett.*, 146: 187–190.

Hahn, J.S., Aizenman, E. and Lipton, S.A. (1988) Central mammalian neurons normally resistant to glutamate toxicity are made sensitive by elevated extracellular $Ca^{2+}$: Toxicity is blocked by the *N*-methyl-D-aspartate antagonist MK-801. *Proc. Natl. Acad. Sci. USA*, 85: 6556–6560.

Hall, E.D., Yonkers, P.A., McCall, J.M. and Braughler, J.M. (1988) Effects of the 21-aminosteroid U74006F on experimental head injury in mice. *J. Neurosurg.*, 68: 456–461.

Hammer, B., Parker Jr., W.D. and Bennett Jr., J.P. (1993) NMDA receptors increase OH radicals in vivo by using nitric oxide synthase and protein kinase C. *Neuroreport*, 5: 72–74.

Hantraye, P., Riche, D., Mazire, M. and Isacson, O. (1990) A primate model of Huntington's disease: Behavioral and anatomical studies of unilateral excitotoxic lesions of the caudate-putamen in the baboon. *Exp. Neurol.*, 108: 91–104.

Hantraye, P., Brouillet, E., Ferrante, R., Palfi, S., Dolan, R., Matthews, R.T. and Beal, M.F. (1996) Inhibition of neuronal nitric oxide synthase prevents MPTP-induced parkinsonism in baboons. *Nat. Med.*, 2: 1017–1021.

Hartley, D.M., Kurth, M.C., Bjerkness, L., Weiss, J.H. and Choi, D.W. (1993) Glutamate receptor-induced $^{45}Ca^{2+}$ accumulation in cortical cell culture correlates with subsequent neuronal degeneration. *J. Neurosci.*, 13: 1993–2000.

Hayashi, T. (1954) Effects of sodium glutamate on the nervous system. *Keio J. Med.*, 3: 183–192.

Henshaw, R., Jenkins, B.G., Schulz, J.B., Ferrante, R.J., Kowall, N.W., Rosen, B.R. and Beal, M.F. (1994) Malonate produces striatal lesions by indirect NMDA receptor activation. *Brain Res.*, 647: 161–166.

Herb, A., Burnashev, N., Werner, P., Sakmann, B., Wisden, W. and Seeburg, P.H. (1992) The KA-2 subunit of excitatory amino acid receptors shows widespread expression in brain and forms ion channels with distantly related subunits. *Neuron*, 8: 775–785.

Hewett, S.J., Corbett, J.A., McDaniel, M.L. and Choi, D.W. (1993) Inhibition of nitric oxide formation does not protect murine cortical cell cultures from *N*-methyl-D-aspartate neurotoxicity. *Brain Res.*, 625: 337–341.

Hollmann, M., O'Shea-Greenfield, A., Rogers, S.W. and Heinemann, S. (1989) Cloning by functional expression of a member of the glutamate receptor family. *Nature*, 342: 643–648.

Hollmann, M., Hartley, M. and Heinemann, S. (1991) $Ca^{2+}$ permeability of KA-AMPA-gated glutamate receptor channels depends on subunit composition. *Science*, 252: 851–853.

Hume, R.I., Dingledine, R. and Heinemann, S.F. (1991) Identification of a site in glutamate receptor subunits that controls calcium permeability. *Science*, 253: 1028–1031.

Ihara, Y., Namba, R., Kuroda, S., Sato, T. and Shirabe, T. (1989) Mitochondrial encephalomyopathy (MELAS): Pathological study and successful therapy with coenzyme $Q_{10}$ and idebenone. *J. Neurol. Sci.*, 90: 263–271.

Ischiropoulos, H., Zhu, L., Chen, J., Tsai, M., Martin, J.C., Smith, C.D. and Beckman, J.S. (1992) Peroxynitrite-mediated tyrosine nitration catalyzed by superoxide dismutase. *Arch. Biochem. Biophys.*, 298: 431–437.

Izumi, Y., Benz, A.M., Clifford, D.B. and Zorumski, C.F. (1992) Nitric oxide inhibitors attenuate *N*-methyl-D-aspartate excitotoxicity in rat hippocampal slices. *Neurosci. Lett.*, 135: 227–230.

Jenkins, B.G., Koroshetz, W.J., Beal, M.F. and Rosen, B.R. (1993) Evidence for impairment of energy metabolism in vivo in Huntington's disease using localized $^{1}H$ NMR spectroscopy. *Neurology*, 43: 2689–2695.

Jones-Humble, S.A., Morgan, P.F. and Cooper, B.R. (1994) The novel anticonvulsant lamotrigine prevents dopamine depletion in C57 black mice in the MPTP animal model of Parkinson's disease. *Life Sci.*, 54: 245–252.

Keinänen, K., Wisden, W., Sommer, B., Werner, P., Herb, A., Verdoorn, T.A., Sakmann, B. and Seeburg, P.H. (1990) A family of AMPA-selective glutamate receptors. *Science*, 249: 556–560.

Kirschner, P.B., Henshaw, R., Weise, J., Trubetskoy, V., Finkelstein, S., Schulz, J.B. and Beal, M.F. (1995) Basic fibroblast growth factor protects against excitotoxicity and chemical hypoxia in both neonatal and adult rats. *J. Cereb. Blood Flow Metab.*, 15: 619–623.

Kirschner, P.B., Jenkins, B.G., Schulz, J.B., Finkelstein, S.P., Matthews, R.T., Rosen, B.R. and Beal, M.F. (1996) NGF, BDNF and NT-5, but not NT-3 protect against $MPP^{+}$ toxicity and oxidative stress in neonatal animals. *Brain Res.*, 713: 178–185.

Klockgether, T. and Turski, L. (1990) NMDA antagonists potentiate antiparkinsonian actions of L-dopa in monoamine-depleted rats. *Ann. Neurol.*, 28: 539–546.

Klockgether, T. and Turski, L. (1993) Toward an understanding of the role of glutamate in experimental parkinsonism:

agonist-sensitive sites in the basal ganglia. *Ann. Neurol.*, 34: 585–593.

Klockgether, T., Jacobson, P., Löschmann, P.A. and Turski, L. (1993) The antiparkinsonian agent budipine is an *N*-methyl-D-aspartate antagonist. *J. Neural Transm. Park. Dis. Dement. Sect.*, 5: 101–106.

Koh, J.Y., Yang, L.L. and Cotman, C.W. (1990) β-amyloid protein increases the vulnerability of cultured cortical neurons to excitotoxic damage. *Brain Res.*, 533: 315–320.

Kollegger, H., McBean, G.J. and Tipton, K.F. (1993) Reduction of striatal *N*-methyl-D-aspartate toxicity by inhibition of nitric oxide synthase. *Biochem. Pharmacol.*, 45: 260–264.

Krieger, C., Wagey, R. and Shaw, C. (1993) Amyotrophic lateral sclerosis: Quantitative autoradiography of [$^3$H]MK-801/NMDA binding sites in spinal cord. *Neurosci. Lett.*, 159: 191–194.

Kutsuwada, T., Kashiwabuchi, N., Mori, H., Sakimura, K., Kushiya, E., Araki, K., Meguro, H., Masaki, H., Kumanishi, T., Arakawa, M. and Mishina, M. (1992) Molecular diversity of the NMDA receptor channel. *Nature*, 358: 36–41.

Kwon, N.S., Stuehr, D.J. and Nathan, C.F. (1991) Inhibition of tumor cell ribonucleotide reductase by macrophage-derived nitric oxide. *J. Exp. Med.*, 174: 761–767.

Lafon-Cazal, M., Pietri, S., Culcasi, M. and Bockaert, J. (1993) NMDA-dependent superoxide production and neurotoxicity. *Nature*, 364: 535–537.

Lancelot, E., Revaud, M.L., Boulu, R.G., Plotkine, M. and Callebert, J. (1997) α-phenyl-*N*-*tert*-butylnitrone attenuates excitotoxicity in rat striatum by preventing hydroxyl radical accumulation. *Free Rad. Biol. Med.*, in press.

Lange, K.W., Löschmann, P.A., Sofic E., Burg, M., Horowski, R., Kalveram, K.T., Wachtel, H. and Riederer, P. (1993) The competitive NMDA antagonist CPP protects substantia nigra neurons from MPTP-induced degeneration in primates. *Naunyn-Schmiedeberg's Arch. Pharmacol.*, 348: 586–592.

Lepoivre, M., Fieschi, F., Coves, J., Thelander, L. and Fontecave, M. (1991) Inactivation of ribonucleotide reductase by nitric oxide. *Biochem. Biophys. Res. Commun.*, 179: 442–448.

Lerner-Natoli, M., Rondouin, G., de Bock, F. and Bockaert, J. (1992) Chronic NO synthase inhibition fails to protect hippocampal neurones against NMDA toxicity. *Neuroreport*, 3: 1109–1112.

Lipton, S.A. and Rosenberg, P.A. (1994) Excitatory amino acids as a final common pathway for neurologic disorders. *N. Engl. J. Med.*, 330: 613–622.

Lucas, D.R. and Newhouse, J.P. (1957) The toxic effect of sodium L-glutamate on the inner layers of the retina. *AMA Arch. Ophtalmol.*, 58: 193–201.

Majewska, M.D. and Bell, J.A. (1990) Ascorbic acid protects neurons from injury induced by glutamate and NMDA. *Neuroreport*, 1: 194–196.

Malessa, S., Leigh, P.N., Bertel, O., Sluga, E. and Hornykiewicz, O. (1991) Amyotrophic lateral sclerosis: Glutamate dehydrogenase and transmitter amino acids in the spinal cord. *J. Neuro. Neurosurg. Psych.*, 54: 984–988.

Maragos, W.F. and Silverstein, F.S. (1995) Inhibition of nitric oxide synthase activity attenuates malonate lesions in rats. *J. Neurochem.*, 64: 2362–2365.

Matthews, R.T., Yang, L. and Beal, M.F. (1997) *S*-methylthiocitrulline, a neuronal nitric oxide synthase inhibitor, protects against malonate and MPTP neurotoxicity. *Exp. Neurol.*, 143: 282–286.

Mattson, M.P. and Rychlik, B. (1990) Glia protect hippocampal neurons against excitatory amino acid-induced degeneration: Involvement of fibroblast growth factor. *Int. J. Dev. Neurosci.*, 8: 399–415.

Mattson, M.P., Cheng, B., Davis, D., Bryant, K., Lieberburg, I. and Rydel, R.E. (1992) β-amyloid peptides destabilize calcium homeostasis and render human cortical neurons vulnerable to excitotoxicity. *J. Neurosci.*, 12: 376–389.

Mattson, M.P., Lovell, M.A., Furukawa, K. and Markesbery, W.R. (1995) Neurotrophic factors attenuate glutamate-induced accumulation of peroxides, elevation of intracellular $Ca^{2+}$ concentration, and neurotoxicity and increase antioxidant enzyme activities in hippocampal neurons. *J. Neurochem.*, 65: 1740–1751.

McGeer, E.G. and McGeer, P.L. (1976) Duplication of biochemical changes of Huntington's chorea by intrastriatal injections of glutamic and kainic acids. *Nature*, 263: 517–519.

McGeer, E.G. and Zhu, S.G. (1990) Lamotrigine protects against kainate but not ibotenate lesions in rat striatum. *Neurosci. Lett.*, 112: 348–351.

Meguro, H., Mori, H., Araki, K., Kushiya, E., Kutsuwada, T., Yamazaki, M., Kumanishi, T., Arakawa, M., Sakimura, K. and Mishina, M. (1992) Functional characterization of a heteromeric NMDA receptor channel expressed from cloned cDNAs. *Nature*, 357: 70–74.

Michaels, R.L. and Rothman, S.M. (1990) Glutamate neurotoxicity in vitro: Antagonist pharmacology and intracellular calcium concentrations. *J. Neurosci.*, 10: 283–292.

Milani, D., Guidolin, D., Facci, L., Pozzan, T., Buso, M., Leon, A. and Skaper, S.D. (1991) Excitatory amino acid-induced alterations of cytoplasmic free $Ca^{2+}$ in individual cerebellar granule neurons: Role in neurotoxicity. *J. Neurosci. Res.*, 28: 434–441.

Miller, B., Sarantis, M., Traynelis, S.F. and Attwell, D. (1992) Potentiation of NMDA receptor currents by arachidonic acid. *Nature*, 355: 722–725.

Moncada, C., Lekieffre, D., Arvin, B. and Meldrum, B. (1992) Effect of NO synthase inhibition on NMDA- and ischaemia-induced hippocampal lesions. *Neuroreport*, 3: 530–532.

Monyer, H., Hartley, D.M. and Choi, D.W. (1990) 21-aminosteroids attenuate excitotoxic neuronal injury in cortical cell cultures. *Neuron*, 5: 121–126.

Monyer, H., Sprengel, R., Schoepfer, R., Herb, A., Higuchi, M., Lomeli, H., Burnashev, N., Sakmann, B. and Seeburg, P.H. (1992) Heteromeric NMDA receptors:

Molecular and functional distinction of subtypes. *Science*, 256: 1217–1221.

Moriyoshi, K., Masu, M., Ishii, T., Shigemoto, R., Mizuno, N. and Nakanishi, S. (1991) Molecular cloning and characterization of the rat NMDA receptor. *Nature*, 354: 31–37.

Murphy, T.H., Schnaar, R.L. and Coyle, J.T. (1990) Immature cortical neurons are uniquely sensitive to glutamate toxicity by inhibition of cystine uptake. *FASEB J.* 4: 1624–1633.

Nakanishi, S. (1992) Molecular diversity of glutamate receptors and implications for brain function. *Science*, 258: 597–603.

Nakao, N., Grasbon-Frodl, E.M., Widner, H. and Brundin, P. (1996) Antioxidant treatment protects striatal neurons against excitotoxic insults. *Neuroscience*, 73: 185–200.

Nguyen, T., Brunson, D., Crespi, C.L., Penman, B.W., Wishnok, J.S. and Tannenbaum, S.R. (1992) DNA damage and mutation in human cells exposed to nitric oxide in vitro. *Proc. Natl. Acad. Sci. USA*, 89: 3030–3034.

Nicoletti, F., Bruno, V., Copani, A., Casabona, G. and Knöpfel, T. (1996) Metabotropic glutamate receptors: A new target for the therapy of neurodegenerative disorders? *Trends Neurol. Sci.*, 19: 267–271.

Novelli, A., Reilly, J.A., Lysko, P.G. and Henneberry, R.C. (1988) Glutamate becomes neurotoxic *via* the *N*-methyl-D-aspartate receptor when intracellular energy levels are reduced. *Brain Res.*, 451: 205–212.

Nozaki, K., Finklestein, S.P. and Beal, M.F. (1993) Delayed administration of basic fibroblast growth factor protects against *N*-methyl-D-aspartate neurotoxicity in neonatal rats. *Eur. J. Pharmacol.*, 232: 295–297.

Ogawa, N., Tanaka, K.I., Asanuma, M., Kawai, M., Masumizu, T., Kohno, M. and Mori, A. (1994) Bromocriptine protects mice against 6-hydroxydopamine and scavenges hydroxyl free radicals in vitro. *Brain Res.*, 657: 207–213.

Olney, J.W., Ho, O.L. and Rhee, V. (1971) Cytotoxic effects of acidic and sulphur containing amino acids on the infant mouse central nervous system. *Exp. Brain Res.*, 14: 61–76.

Orrenius, S., McConkey, D.J., Bellomo, G. and Nicotera, P. (1989) Role of $Ca^{2+}$ in toxic cell killing. *Trends Pharmacol. Sci.*, 10: 281–285.

Otto, D. and Unsicker, K. (1990) Basic FGF reverses chemical and morphological deficits in the nigrostriatal system of MPTP-treated mice. *J. Neurosci.*, 10: 1912–1921.

Park, C.K. and Hall, E.D. (1994) Dose-response analysis of the effect of 21-aminosteroid tirilazad mesylate (U-74006F) upon neurological outcome and ischemic brain damage in permanent focal cerebral ischemia. *Brain Res.*, 645: 157–163.

Parsons, C.G., Gruner, R., Rozental, J., Millard J. and Lodge, D. (1993) Patch clamp studies on the kinetics and selectivity of *N*-methyl-D-aspartate receptor antagonism by memantine (1-amino-3,5-dimethyladamantan). *Neuropharmacology*, 32: 1337–1350.

Pauwels, P.J. and Leysen, J.E. (1992) Blockade of nitric oxide formation does not prevent glutamate-induced neurotoxicity in neuronal cultures from rat hippocampus. *Neurosci. Lett.*, 143: 27–30.

Pearson, R.C.A., Esiri, M.M., Hiorns, R.W., Wilcock, G.K. and Powell, T.P.S. (1985) Anatomical correlates of the distribution of the pathological changes in the neocortex in Alzheimer disease. *Proc. Natl. Acad. Sci. USA*, 82: 4531–4534.

Pellock, J.M. (1994) Clinical efficacy of lamotrigine as an antiepileptic drug. *Neurology*, (Suppl. 8): 29–35.

Phillis, J.W., Sen, S. and Cao, X. (1994) Amflutizole, a xanthine oxidase inhibitor, inhibits free radical generation in the ischemic/reperfused rat cerebral cortex. *Neurosci. Lett.*, 169: 188–190.

Plaitakis, A. and Caroscio, J.T. (1987) Abnormal glutamate metabolism in amyotrophic lateral sclerosis. *Ann. Neurol.*, 22: 575–579.

Plaitakis, A., Constantakakis, E. and Smith, J. (1988) The neuroexcitotoxic amino acids glutamate and aspartate are altered in the spinal cord and brain in amyotrophic lateral sclerosis. *Ann. Neurol.*, 24: 446–449.

Prehn, J.H.M., Lippert, K. and Krieglstein, J. (1995) Are NMDA or AMPA/kainate receptor antagonists more efficacious in the delayed treatment of excitotoxic neuronal injury? *Eur. J. Pharmacol.*, 292: 179–189.

Przedborski, S., Jackson-Lewis, V., Yokoyama, R., Shibata, T., Dawson, V.L. and Dawson, T.M. (1996) Role of neuronal nitric oxide in 1-methyl-4-phenyl-1,2,3,6-tetrahydropyridine (MPTP)-induced dopaminergic neurotoxicity. *Proc. Natl. Acad. Sci. USA*, 93: 4565–4571.

Radi, R., Beckman, J.S., Bush, K.M. and Freeman, B.A. (1991) Peroxynitrite oxidation of sulfhydryls: The cytotoxic potential of superoxide and nitric oxide. *J. Biol. Chem.*, 266: 4244–4250.

Randall, R.D. and Thayer, S.A. (1992) Glutamate-induced calcium transient triggers delayed calcium overload and neurotoxicity in rat hippocampal neurons. *J. Neurosci.*, 12: 1882–1895.

Reynolds, I.J. and Hastings, T.G. (1995) Glutamate induces the production of reactive oxygen species in cultured forebrain neurons following NMDA receptor activation. *J. Neurosci.*, 15: 3318–3327.

Rogers, J. and Morrison, J.H. (1985) Quantitative morphology and regional and laminar distributions of senile plaques in Alzheimer's disease. *J. Neurosci.*, 5: 2801–2808.

Rosen, D.R., Siddique, T., Patterson, D., Figlewicz, D.A., Sapp, P., Hentati, A., Donaldson, D., Goto, J., O'Regan, J.P., Deng, H.X., Rahmani, Z., Krizus, A., McKenna-Yasek, D., Cayabyab, A., Gaston, S.M., Berger, R., Tanzi, R.E., Halperin, J.J., Herzfeldt, B., Van der Bergh, R., Hung, W.Y., Bird, T., Deng, G., Mulder, D.W., Smyth, C., Laing, N.G., Soriano, E., Pericak-Vance, M.A., Haines, J., Rouleau, G.A., Gusella, J.S., Horvitz, H.R. and Brown Jr., R.H. (1993) Mutations in Cu/Zn superoxide dismutase gene are associated with familial amyotrophic lateral sclerosis. *Nature*, 362: 59–62.

Rothman, S. (1984) Synaptic release of excitatory amino acid neurotransmitter mediates anoxic neuronal death. *J. Neurosci.*, 4: 1884–1891.

Rothman, S.M., Thurston, J.H. and Hauhart, R.E. (1987) Delayed neurotoxicity of excitatory amino acids in vitro. *Neuroscience*, 22: 471–480.

Rothstein, J.D., Tsai, G., Kuncl, R.W., Clawson, L., Cornblath, D.R., Drachman, D.B., Pestronk, A., Stauch, B.L. and Coyle, J.T. (1990) Abnormal excitatory amino acid metabolism in amyotrophic lateral sclerosis. *Ann. Neurol.*, 28: 18–25.

Rothstein, J.D., Martin, L.J. and Kuncl, R.W. (1992) Decreased glutamate transport by the brain and spinal cord in amyotrophic lateral sclerosis. *N. Engl. J. Med.*, 326: 1464–1468.

Rothstein, J.D., Jin, L., Dykes-Hoberg, M. and Kuncl, R.W. (1993) Chronic inhibition of glutamate uptake produces a model of slow neurotoxicity. *Proc. Natl. Acad. Sci. USA*, 90: 6591–6595.

Rubinsztein, D.C., Leggo, J., Chiano, M., Dodge, A., Norbury, G., Rosser, E. and Craufurd, D. (1997) Genotypes at the GluR6 kainate receptor locus are associated with variation in the age of onset of Huntington disease. *Proc. Natl. Acad. Sci. USA*, 94: 3872–3876.

Sakurada, K., Masu, M. and Nakanishi, S. (1993) Alteration of $Ca^{2+}$ permeability and sensitivity to $Mg^{2+}$ and channel blockers by a single amino acid substitution in the *N*-methyl-D-aspartate receptor. *J. Biol. Chem.*, 268: 410–415.

Sánchez-Prieto, J., Budd, D.C., Herrero, I., Vázquez, E. and Nicholls, D.G. (1996) Presynaptic receptors and the control of glutamate exocytosis. *Trends Neurol. Sci.*, 19: 235–239.

Schulz, J.B., Henshaw, D.R., Jenkins, B.G., Ferrante, R.J., Kowall, N.W., Rosen, B.R. and Beal, M.F. (1994) 3-acetylpyridine produces age-dependent excitotoxic lesions in rat striatum. *J. Cereb. Blood Flow Metab.*, 14: 1024–1029.

Schulz, J.B., Henshaw, D.R., Siwek, D., Jenkins, B.G., Ferrante, R.J., Cipolloni, P.B., Kowall, N.W., Rosen, B.R. and Beal, M.F. (1995a) Involvement of free radicals in excitotoxicity in vivo. *J. Neurochem.*, 64: 2239–2247.

Schulz, J.B., Matthews, R.T., Jenkins, B.G., Ferrante, R.J., Siwek, D., Henshaw, D.R., Cipolloni, P.B., Mecocci, P., Kowall, N.W., Rosen, B.R. and Beal, M.F. (1995b) Blockade of neuronal nitric oxide synthase protects against excitotoxicity in vivo. *J. Neurosci.*, 15: 8419–8429.

Schulz, J.B., Matthews, R.T., Muqit, M.M.K., Browne, S.E. and Beal, M.F. (1995c) Inhibition of neuronal nitric oxide synthase by 7-nitroindazole protects against MPTP-induced neurotoxicity in mice. *J. Neurochem.*, 64: 936–939.

Schulz, J.B., Matthews, R.T., Jenkins, B.G., Brar, P. and Beal, M.F. (1995d) Improved therapeutic window for treatment of histotoxic hypoxia with a free radical spin trap. *J. Cereb. Blood Flow Metab.*, 15: 948–952.

Schulz, J.B., Matthews, R.T., Henshaw, D.R. and Beal, M.F. (1996a) Neuroprotective strategies for treatment of lesions produced by mitochondrial toxins: Implications for neurodegenerative diseases. *Neuroscience*, 71: 1043–1048.

Schulz, J.B., Henshaw, D.R., MacGarvey, U. and Beal, M.F. (1996b) Involvement of oxidative stress in 3-nitropropionic acid neurotoxicity. *Neurochem. Int.*, 29: 167–171.

Schwarcz, R., Whetsell Jr., W.O. and Mangano, R.M. (1983) Quinolinic acid: An endogenous metabolite that produces axon-sparing lesions in rat brain. *Science*, 219: 316–318.

Shaw, P.J., Chinnery, R.M. and Ince, P.G. (1994) Non-NMDA receptors in motor neuron disease (MND): A quantitative autoradiographic study in spinal cord and motor cortex using [$^3$H]CNQX and [$^3$H]kainate. *Brain Res.*, 655: 186–194.

Shimohama, S., Ogawa, N., Tamura, Y., Akaike, A., Tsukahara, T., Iwata, H. and Kimura, J. (1993) Protective effect of nerve growth factor against glutamate-induced neurotoxicity in cultured cortical neurons. *Brain Res.*, 632: 296–302.

Sonsalla, P.K., Zeevalk, G.D., Manzino, L., Giovanni, A. and Nicklas, W.J. (1992) MK-801 fails to protect against the dopaminergic neuropathology produced by systemic 1-methyl-4-phenyl-1,2,3,6-tetrahydropyridine in mice or intranigral 1-methyl-4-phenylpyridinium in rats. *J. Neurochem.*, 58: 1979–1982.

Spencer, P.S., Allen, C.N., Kisby, G.E., Ludolph, A.C., Ross, S.M. and Roy, D.N. (1991) Lathyrism and western Pacific amyotrophic lateral sclerosis: Etiology of short and long latency motor system disorders. *Adv. Neurol.*, 56: 287–299.

Stadler, J., Billiar, T.R., Curran, R.D., Stuehr, D.J., Ochoa, J.B. and Simmons, R.L. (1991) Effect of exogenous and endogenous nitric oxide on mitochondrial respiration of rat hepatocytes. *Am. J. Physiol.*, 260: C910–C916.

Storey, E., Hyman, B.T., Jenkins, B.G., Brouillet, E., Miller, J.M., Rosen, B.R. and Beal, M.F. (1992) 1-methyl-4-phenylpyridinium produces excitotoxic lesions in rat striatum as a result of impairment of oxidative metabolism. *J. Neurochem.*, 58: 1975–1978.

Sucher, N.J., Awobuluyi, M., Choi, Y.B. and Lipton, S.A. (1996) NMDA receptors: From genes to channels. *Trends Pharmacol. Sci.*, 17: 348–355.

Tomac, A., Lindqvist, E., Lin, L.F.H., Ögren, S.O., Young, D., Hoffer, B.J. and Olson, L. (1995) Protection and repair of the nigrostriatal dopaminergic system by GDNF in vivo. *Nature*, 373: 335–339.

Turski, L., Bressler, K., Rettig, K.J., Löschmann, P.A. and Wachtel, H. (1991) Protection of substantia nigra from $MPP^+$ neurotoxicity by *N*-methyl-D-aspartate antagonists. *Nature*, 349: 414–418.

Turski, L. and Ikonomidou, H. (1994) Striatal toxicity of 3-nitropropionic acid prevented by the AMPA antagonist NBQX. *Soc. Neurosci. Abst.*, 20: 1677.

Tymianski, M., Charlton, M.P., Carlen, P.L. and Tator, C.H. (1993a) Source specificity of early calcium neurotoxicity in cultured embryonic spinal neurons. *J. Neurosci.*, 13: 2085–2104.

Tymianski, M., Wallace, M.C., Spigelman, I., Uno, M., Carlen, P.L., Tator, C.H. and Charlton, M.P. (1993b) Cell-permeant $Ca^{2+}$ chelators reduce early excitotoxic and ischemic neuronal injury in vitro and in vivo. *Neuron*, 11: 221–235.

Uitti, R.J., Rajput, A.H., Ahlskog, J.E., Offord, K.P., Ho, M.M., Prasad, M., Rajput, A. and Basran, P. (1993) Amantadine treatment is an independent predictor of improved survival in parkinsonism. *Can. J. Neurol. Sci.*, 20: S235.

Ulas, J., Brunner, L.C., Geddes, J.W., Choe, W. and Cotman, C.W. (1992) *N*-methyl-D-aspartate receptor complex in the hippocampus of elderly, normal individuals and those with Alzheimer's disease. *Neuroscience*, 49: 45–61.

Verdoorn, T.A., Burnashev, N., Monyer, H., Seeburg, P.H. and Sakmann, B. (1991) Structural determinants of ion flow through recombinant glutamate receptor channels. *Science* 252: 1715–1718.

Waldvogel, H.J., Faull, R.L.M., Williams, M.N. and Dragunow, M. (1991) Differential sensitivity of calbindin and parvalbumin immunoreactive cells in the striatum to excitotoxins. *Brain Res.*, 546: 329–335.

Wallis, R.A., Panizzon, K.L., Henry, D. and Wasterlain, C.G. (1993) Neuroprotection against nitric oxide injury with inhibitors of ADP-ribosylation. *Neuroreport*, 5: 245–248.

Wang, G.J., Randall, R.D. and Thayer, S.A. (1994) Glutamate-induced intracellular acidification of cultured hippocampal neurons demonstrates altered energy metabolism resulting from $Ca^{2+}$ loads. *J. Neurophysiol.*, 72: 2563–2569.

Watkins, J.C., Krogsgaard-Larsen, P. and Honoré, T. (1990) Structure-activity relationships in the development of excitatory amino acid receptor agonists and competitive antagonists. *Trends Pharmacol. Sci.*, 11: 25–33.

Weller, M. and Paul, S.M. (1993) 3-nitropropionic acid is an indirect excitotoxin to cultured cerebellar granule neurons. *Eur. J. Pharmacol.*, 248: 223–228.

Werner, P., Voight, M., Keinänen, K., Wisden, W. and Seeburg, P.H. (1991) Cloning of a putative high-affinity kainate receptor expressed predominantly in hippocampal CA3 cells. *Nature*, 351: 742–744.

White, R.J. and Reynolds, I.J. (1995) Mitochondria and $Na^{+}/Ca^{2+}$ exchange buffer glutamate-induced calcium loads in cultured cortical neurons. *J. Neurosci.*, 15: 1318–1328.

Witt, M.R., Dekermendjian, K., Frandsen, A., Schousboe, A. and Nielsen, M. (1994) Complex correlation between excitatory amino acid-induced increase in the intracellular $Ca^{2+}$ concentration and subsequent loss of neuronal function in individual neocortical neurons in culture. *Proc. Natl. Acad. Sci. USA*, 91: 12303–12307.

Wüllner, U., Young, A.B., Penney, J.B. and Beal, M.F. (1994) 3-nitropropionic acid toxicity in the striatum. *J. Neurochem.*, 63: 1772–1781.

Young, A.B., Greenamyre, J.T., Hollingsworth, Z., Albin, R., D'Amato, C., Shoulson, I. and Penney, J.B. (1988) NMDA receptor losses in putamen from patients with Huntington's disease. *Science*, 241: 981–983.

Young, A.B. and Fagg, G.E. (1990) Excitatory amino acid receptors in the brain: Membrane binding and receptor autoradiographic approaches. *Trends Pharmacol. Sci.*, 11: 126–133.

Zeevalk, G.D. and Nicklas, W.J. (1990) Chemically induced hypoglycemia and anoxia: Relationship to glutamate receptor-mediated toxicity in retina. *J. Pharmacol. Exp. Ther.*, 253: 1285–1292.

Zeevalk, G.D. and Nicklas, W.J. (1991) Mechanisms underlying initiation of excitotoxicity associated with metabolic inhibition. *J. Pharmacol. Exp. Ther.*, 257: 870–878.

Zhang, J., Dawson, V.L., Dawson, T.M. and Snyder, S.H. (1994) Nitric oxide activation of poly(ADP-ribose) synthetase in neurotoxicity. *Science*, 263: 687–689.

Zinkand, W.C., Stumpo, R.J., Thompson, C., Patel, J. and Pullan, L.M. (1993) Lack of involvement of nitric oxide in NMDA-induced neuronal cell death in cortical culture. *Neuroreport*, 5: 148–150.

Zipp, F., Baas, H. and Fischer, P.A. (1993) Lamotrigine-antiparkinsonian activity by blockade of glutamate release? *J. Neural Transm. Park. Dis. Dement. Sect.*, 5: 67–75.

Zuddas, A., Oberto, G., Vaglini, F. Fascetti, F., Fornai, F. and Corsini, G.U. (1992) MK-801 prevents 1-methyl-4-phenyl-1,2,3,6-tetrahydropyridine-induced parkinsonism in primates. *J. Neurochem.*, 59: 733–739.

SECTION V

# Conditions of perturbed glutamatergic transmission

O.P. Ottersen, I.A. Langmoen and L. Gjerstad (Eds.)
*Progress in Brain Research*, Vol 116


CHAPTER 22

# Electrophysiological substrates for focal epilepsies

John G.R Jefferys* and Roger D. Traub

*Neuroscience Unit, Department of Physiology, The Medical School, University of Birmingham, Vincent Drive, Edgbaston, Birmingham B15 2TT, UK*

Glutamate plays a central role in the neuronal circuitry that recruits neurons into epileptic discharges. In the hippocampal CA3 region glutamatergic synapses between pyramidal cells provide the substrate for a chain reaction which leads to the synchronous epileptic discharge. Both AMPA and NMDA receptors are present at these recurrent synapses; AMPA receptors are responsible for the synchronization of epileptic bursts in the bicuculline/picrotoxin models and NMDA in the low-$Mg^{2+}$ model. The slower kinetics of NMDA receptors can prolong discharges from the ~100 ms of the interictal spike to the first few hundred ms of an afterdischarge. Events lasting as long as a full seizure ( > 10's of seconds) can occur in several experimental epilepsies in the slice. Seizure-like events in the bicuculline model are not sensitive to NMDA receptor antagonists; they may be due to spontaneous release of glutamate from presynaptic terminals and/or activation of metabotropic glutamate receptors.

## Glutamate's role in epilepsy

Amongst the earliest uses of the selective glutamate antagonists that became available during the 1970's was the dissection of their anticonvulsant actions. Around the same time it became clear that glutamate could kill neurons through "excitotoxicity" (Choi, 1995) (Mody and MacDonald, 1995) (Olney et al., 1986) (Simon et al., 1986). More recently it has been shown explicitly that glutamate is released during epileptic discharges (During and Spencer, 1993). We will briefly outline evidence from these sources that glutamate receptors play a crucial role in epilepsy, before discussing in more detail what that role might be in the context of the synchronization of local networks of neurons.

A critical step forward in this story was the development of agonists and antagonists to the various glutamate receptors, largely by the group of J.C. Watkins. As is made clear elsewhere in this volume, (see Borges and Dingledine; Monaghan, this volume) the major ligand gated channels that concern us here are called AMPA and NMDA after their selective agonists. The third major class of glutamate-gated ion channel is the kainate receptor. We will also mention metabotropic glutamate receptors (mGluRs) which are the G-protein coupled class of glutamate receptor (see Bruno et al., this volume). It quickly became clear that NMDA and kainate agonists resembled glutamate itself in being convulsant and also toxic to neurons (Ben-Ari et al., 1980). NMDA receptor antagonists proved effective anticonvulsants in several animal models in vivo, and protected against neuronal loss induced by seizures (and

*Corresponding author.
Tel.: +44 121 414 7629. 7525; fax: +44 121 414 7625; e-mail: j.g.r.jefferys@bham.ac.uk.

ischaemia) (Coutinho-Netto et al., 1981; Croucher et al., 1982; Meldrum, 1994). NMDA receptor antagonists were also able to block the development of kindling, but not its expression in animals kindled previously (Callaghan and Schwark, 1980; Croucher et al., 1988). Kindling is a chronic model of epilepsy in which repetition of initially subconvulsive stimulation lead to a reduction in seizure threshold and increase in seizure severity which lasts indefinitely. Analogous results have been found in models in vitro too (Stasheff et al., 1989). The fact that it was the induction phase that was selectively blocked by NMDA antagonists suggested that the kindling process itself depended on plastic synaptic changes analogous to those in long term potentiation (LTP) (Bliss and Collingridge, 1993), but that the final expression more likely depended on AMPA receptors (again this has parallels with LTP).

The release of glutamate during seizures can kill neurons, mainly as a result of $Ca^{2+}$ entering with the $Na^+$ and $K^+$ through the NMDA receptor and through certain kinds of AMPA receptor (those which lack the GluR-B subunit) (Bochet et al., 1994; Jonas et al., 1994). The anticonvulsant, antiepileptic and neuroprotective roles of NMDA receptor antagonists made them promising novel drugs for the treatment of epilepsy and some kinds of brain injury (see Chapman; Taubøll and Gjerstad, this volume). Unfortunately adverse side effects of these drugs have so far prevented their translation into the clinic.

## Basic mechanisms of epilepsies

While glutamatergic drugs so far have failed to live up to their initial promise, they have been a major asset in attempting to unravel the basic mechanisms of epileptic activity. The central question is why should large numbers of neurons abruptly start to discharge synchronously to produce the epileptic seizure. Much of the work in this area has concentrated on brief synchronous events which last about 100 ms, and which resemble interictal spikes on EEGs from patients with focal epilepsies. Early electrophysiological studies of a number of experimental models had shown that neurons appeared electrically normal until they suddenly became recruited into an epileptic discharge; they experienced a so-called "paroxysmal depolarization shift" (PDS) synchronously with each other (and consequently with the extracellular field potential discharge).

A key step forward was the adoption of the brain slice preparation by electrophysiologists, and the recognition that several epilepsy models worked in vitro. Early on it was recognised that drugs that blocked inhibition (e.g. penicillin (Dingledine and Gjerstad, 1979), bicuculline and picrotoxin) produced epileptic discharges in hippocampal and neocortical slices (Dingledine and Gjerstad, 1979; Wong and Traub, 1983; Gutnick et al., 1982). In the hippocampus most kinds of epileptic discharge start amongst the pyramidal cells of the CA3 region (Gjerstad et al., 1981; Wong and Traub, 1983; Gutnick et al., 1982). In a critical experiment Johnston and Brown (1981) showed that the PDS in CA3 pyramidal neurons had the properties of a giant EPSP, in that both PDS and EPSP reversed at the same membrane potential. They also showed that the PDS continued at the same rate when the impaled neuron had been hyperpolarized sufficiently to stop its firing, which lent strong support to the idea that the PDS was a property of the network rather than of individual neurons.

## Neuronal circuits for epileptogenesis

At that stage it was not clear how the giant EPSP could come about. Two properties of CA3 pyramidal neurons seemed potentially important. Firstly, they could fire complex spikes due to the presence of voltage-sensitive calcium channels (Wong and Prince, 1981; Schwartzkroin and Slawsky, 1977) (for a more detailed model of CA3 pyramidal cells see (Traub et al., 1994a)). Secondly, they were joined by monosynaptic excitatory connections, albeit at a relatively low probability of ~2% (MacVicar and Dudek, 1980; Miles et al., 1984; Miles and Wong, 1987). It took another technical advance to show whether this low level of connectivity was necessary and sufficient to explain the epileptic synchronization and the PDS. That advance was the development of realistic computer models of real CA3 pyramidal neurons and their assembly into synaptic networks. These models showed that the network of excitatory synapses between pyramidal neurons provides the substrate for the build-up of excitation and synchrony (Traub and Wong, 1982; Traub and Miles, 1991). Epileptic discharges occur: (a) if neurons are connected by divergent, excitatory synaptic connections; (b) if these synapses a strong enough to excite the postsynaptic neurons with a reasonable probability (Miles and Wong, 1987); and (c) if the population of neurons is large enough (the "minimum aggregate", ~1000–2000 CA3 pyramidal neurons in disinhibited slice and in computer models with spatially structured synaptic connectivity (Miles et al., 1984)). This scheme works for a diverse range of experimental models. Conditions a and c are determined by anatomy, but may change in chronic epilepsies. Condition b depends on: the properties of the synapses between pyramidal cells, the operation of inhibitory neurons, and the intrinsic properties of pyramidal cells. Voltage sensitive calcium currents play a key role in amplifying excitation through the CA3 pyramidal cell network in several models (Traub and Wong, 1982; Traub and Miles, 1991). Persistent sodium currents result from a non-inactivating state of the tetrodotoxin sensitive sodium channel. Its low threshold and persistence allows it to amplify EPSPs (Crill, 1996; Schwindt and Crill, 1995) and thus can contribute to epileptogenesis by amplifying condition (b) above (Segal and Douglas, 1997). Under some conditions persistent sodium currents can underlie intrinsic neuronal bursts, for instance in pyramidal cells of the neocortex and subiculum (Tian et al., 1995), much as calcium currents do in CA3 (Traub and Wong, 1982; Traub and Miles, 1991).

The vast majority of fast excitatory synapses in the brain are glutamatergic. Those made by CA3 pyramidal neurons are no exception. They appear to use both AMPA (non-NMDA) and NMDA receptors. The presence of AMPA receptors at the recurrent synapses explains why antagonists to AMPA, and not to NMDA, receptors block epileptic discharges induced by blocking synaptic inhibition (Traub et al., 1993) (e.g. with bicuculline or picrotoxin), or by blocking certain potassium channels (with 4-aminopyridine) (Traub et al., 1995). In contrast, antagonists at NMDA, and not AMPA, receptors block epileptic activity induced by removing magnesium ions from the bathing medium. This model-specific pharmacology, and our computer simulation studies argue that the recurrent synapses between CA3 pyramidal cells must use both AMPA and NMDA receptors, although we cannot distinguish whether these receptors are colocalized at the same synapses or whether they occur in different subsets of these synapses. Selective kainate receptor ligands are a relatively recent development. They suggest that kainate receptors in the CA3 region are activated by the mossy fibre (granule cell axon) input from the dentate area and not by synapses made by pyramidal cells (Vignes and Collingridge, 1997). Therefore kainate receptors will not play a role in the initiation of epileptic activity in CA3, but they could play a role in vivo when seizures spread through the limbic system, and CA3 will receive excitation from the entorhinal cortex through the mossy fibres.

Magnesium ions block, in a voltage-dependent manner, NMDA receptors under physiological conditions (and also in the bicuculline and 4-aminopyridine models). The removal of $Mg^{2+}$ leads to a considerable strengthening of the recurrent synapses in CA3, because of the increase in the NMDA receptor current, and this leads to epileptic synchronisation (Traub et al., 1994b). (NMDA receptor desensitization contributes to the termination of the epileptic discharge.) This could be important in the rare instances of seizures induced by $Mg^{2+}$ deficiency, but also has wider significance, given the prominent role of NMDA receptors in chronically epileptic tissue, e.g. in kainate-lesioned and kindled hippocampus (Martin et al., 1992; Kraus et al., 1994; Köhr and Mody, 1994; Mody and Heinemann, 1987; Mody et al., 1988) and in human epileptogenic tissue (Masukawa et al., 1991).

While the main content of this review relates to the hippocampus, we note that similar principles apply in other brain regions prone to focal epilepsies, including the neocortex, entorhinal cortex and the piriform cortex (Wadman and Gutnick, 1993; Haberly and Sutula, 1992; Lee and Hablitz, 1991; Jones and Lambert, 1990; Chagnac-Amitai and Connors, 1989; Gutnick et al., 1982). The detailed connectivity and intrinsic neuronal properties differ between these different regions, but in each the key factor in epileptogenesis is the recruitment of neurons into synchronous discharges through glutamatergic synaptic networks.

**Glutamate in prolonged epileptic bursts**

NMDA receptors have slower kinetics than AMPA. In the picrotoxin/bicuculline model they are required to extend the brief (~100 ms) interictal burst to the several hundreds of ms of an afterdischarge or polyspike (Traub et al., 1993; Traub et al., 1993) (Fig. 1). We developed a theory of these longer bursts of epileptic activity (Traub et al., 1993). It depends on the following. An initial AMPA receptor-dependent "primary" burst starts the event ①. The glutamate released also activates NMDA receptors. If the activation of NMDA receptors is strong enough, the inward current they generate ② can eventually counterbalance the outward potassium currents of the intrinsic afterhyperpolarization or AHP that normally terminate the interictal burst. When the inward current through the NMDA receptor exceeds the outward current through the potassium channels of the AHP pyramidal cells start to fire and initiate a "secondary burst". The activity of neuronal population is synchronized into the secondary burst through AMPA receptors ④, so that if some pyramidal cells start to fire a secondary bursts before the others, AMPA receptor mediated EPSPs from the leading cells will advance the phase of secondary bursts in connected cells. The glutamate released during this process also re-activates the NMDA receptors and prolongs the slow depolarization ②. Calcium channels open as a result of the depolarization and contribute to the depolarization of the secondary burst ③. Calcium entry and depolarization activate the AHP which terminates the secondary burst, and the cycle continues. Another way of looking at this process is that the persistent NMDA receptor-mediated excitation, in neurons throughout the population, activates a system of coupled neuronal oscillators dependent on intrinsic dendritic membrane currents which determines the shape and frequency of secondary bursts which comprise the late phase of this discharge. Currents through AMPA-type glutamate receptors at recurrent synapses synchronize both the initiation of the discharge, when they must be effective enough to initiate firing in connected neurons, and the phase locking of the secondary bursts across the population of neurons, when their effect is only to influence the timing of each other's firing (Traub et al., 1993). In principle other slow excitatory mechanisms could operate to play the role of NMDA receptors, such as accumulation of extracellular potassium ions or the activation of slow excitatory synapses. Pharmacological experiments show that NMDA receptors play this role in models where GABA-ergic synaptic inhibition is blocked.

Focal epileptic seizures last tens of seconds up to a few minutes. Epilepsy models in slices mostly produce much briefer synchronous discharges. This supported the contention that seizures needed larger amounts of brain tissue, and perhaps several discrete regions joined by long-range projections (Lothman, 1994). While it is clear that many structures become involved in limbic seizures, evidence that seizures depend on activity cycling around a long "re-entrant" loop remains weak. In particular, where phase lags have been measured during seizures, most often there is no consistent lag between regions that might be expected to take part in such loops (De Curtis et al., 1994; Bragin et al., 1997). Evidence from brain slices now suggests that several brain regions can sustain events lasting as long as a seizure, and we believe that seizures in vivo are much more likely to represent the interaction of separate regions each of which can sustain seizure-like discharges. In this model seizures do not have to circulate through long loops, but rather they synchronize between regions with low average phase lag, because the "lead" region can fluctuate back and forth from cycle to cycle.

Several groups have identified seizure-like or "ictaform" events in brain slices (Korn et al., 1987) (Swartzwelder et al., 1987) (Anderson et al., 1986). Some of these resemble low-$Ca^{2+}$ field bursts, in that they have negative fields and they preferentially start in CA1 within the hippocampus. These kinds of prolonged burst depend in large part on non-synaptic synchronization (through field effects, fluctuations in extracellular ion concentrations and electrotonic junctions) (Jefferys, 1995).

Other kinds of seizure-like event more closely resemble discharges we and others have recorded in vivo. For instance, slices of ventral hippocampus exposed to bicuculline produced epileptic discharges lasting up to 1.5 minutes (Traub et al., 1996), especially if the extracellular $K^+$ concentration is increased from 3 to 6 mM. These bursts start in the same way as the afterdischarge described above – a primary, AMPA receptor-dependent burst followed by a train of NMDA receptor-dependent secondary bursts at ~10 per second. These are followed by tertiary bursts at 1–3 per second. The tertiary bursts are not blocked by NMDA receptor antagonists. Rather they look like a very rapid train of interictal discharges joined by an increase in the spontaneous release of glutamate from terminals. This asynchronous release has the effect of shifting the network into a more excitable state for the duration of the seizure-like event (Traub et al., 1996). Spontaneous synaptic release certainly occurs in the brain, and can be modulated by neuronal activity (Strowbridge and Schwartzkroin, 1996). In at least some cases the asynchronous synaptic activity is evoked by "ectopic" action potentials that can arise spontaneously from axons. Ectopic action potentials were first described in the thalamocortical relay neurons during cortical focal discharges by Gutnick and Prince (1972). Subsequently they have also been reported in other cortical models and in hippocampus during epileptic activity induced by 4-aminopyridine and by tetanic stimulation (Stasheff et al., 1993; Traub et al., 1995; Pinault, 1995).

Finally, "metabotropic" glutamate receptors (mGluRs) have recently been shown to modify

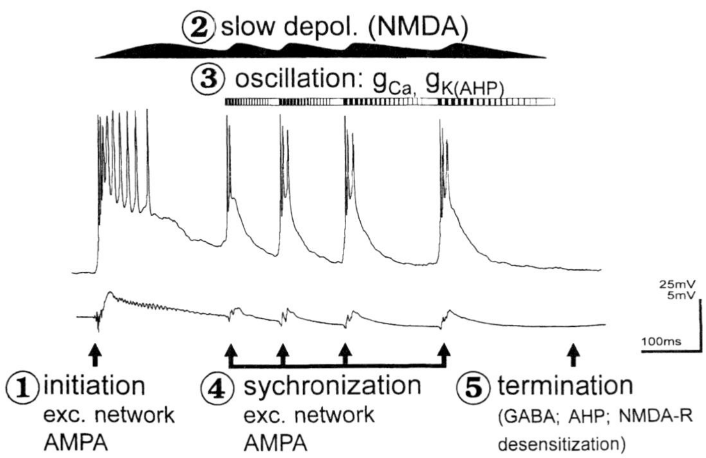


Fig. 1. Intracellular recording from a CA3 pyramidal cell (upper trace) and a simultaneous field potential recording from an adjacent part of the pyramidal cell layer (lower trace) made during a spontaneous epileptic discharge in a rat hippocampal slice exposed to 20 $\mu$M bicuculline. Labels cross reference with the text.

the frequency of epileptic discharges in hippocampal slices in which synaptic inhibition had been blocked by picrotoxin (Merlin et al., 1995). In particular, mGluR agonists can transform interictal bursts into synchronous epileptic discharges, at rates of ~10–20 Hz, and lasting many seconds (Taylor et al., 1995). With group I mGluR agonists such as (*S*)-3-hydroxyphenylglycine, this effect lasts for many hours after the removal of the agonist, which suggests a long lasting potentiation of mGluRs that can then be activated by glutamate released during the epileptic discharges.

### Human epileptic foci

The value of the work described above on animal tissues, mostly studied in vitro, depends largely on what it can tell us about the human disease. Brain slices can be prepared from tissue removed from humans in the course of surgery for medically intractable epilepsy (Berg-Johnson et al., this volume). On the whole the results from such work have not shown that any one experimental model embodies the human condition. Overall these studies show that there are: modest changes in the kinetics of membrane currents in neurons from these slices, a reduction in synaptic inhibition, and, of particular interest for this review, an increase in the strength and duration of synaptic excitation (see review (Schwartzkroin, 1994)). Given that these changes mostly are small in relation with those found in acute models of epilepsy, it is likely that epileptogenesis results from several underlying biophysical and structural changes, including changes in EPSPs that tend to make them larger, more prolonged, or more likely to be sustained during prolonged discharges. Thus it is likely that glutamatergic transmission, in combination with other factors, plays a significant role in human epileptogenesis.

### Conclusion

Glutamate plays keys roles in several distinct aspects of epileptic activity. What these roles are depend on the kind of experimental (and perhaps clinical) epilepsy. Discriminating them from the physiological functions of glutamatergic transmission in order to develop novel therapies remains a major challenge.

### Acknowledgements

The authors' work was supported by the Wellcome Trust, IBM and Human Frontiers.

### References

Anderson, W.W., Lewis, D.V., Swartzwelder, H.S. and Wilson, W.A. (1986) Magnesium-free medium activates seizure-like events in the rat hippocampal slice. *Brain Res.*, 398: 215–219.

Ben-Ari, Y., Tremblay, E., Ottersen, O.P. and Meldrum, B.S. (1980) The role of epileptic activity in hippocampal and "remote" cerebral lesions induced by kainic acid. *Brain Res.*, 191: 79–97.

Bliss, T.V.P. and Collingridge, G.L. (1993) A synaptic model of memory: Long-term potentiation in the hippocampus. *Nature*, 361: 31–39.

Bochet, P., Audinat, E., Lambolez, B., Crépel, F., Rossier, J., Iino, M., Tsuzuki, K. and Ozawa, S. (1994) Subunit composition at the single-cell level explains functional properties of a glutamate-gated channel. *Neuron*, 12: 383–388.

Bragin, A., Csisvari, J., Penttonen, M. and Buzsáki, G. (1997) Epileptic afterdischarge in the hippocampal-entorhinal system: Current source density and unit studies. *Neuroscience*, 76: 1187–1203.

Callaghan, D.A. and Schwark, W.S. (1980) Pharmacological modification of amygdaloid-kindled seizures. *Neuropharm.*, 19: 1131–1136.

Chagnac-Amitai, Y. and Connors, B.W. (1989) Synchronized excitation and inhibition driven by intrinsically bursting neurons in neocortex. *J. Neurophysiol.*, 62: 1149–1162.

Choi, D.W. (1995) Calcium: Still center-stage in hypoxic-ischemic neuronal death. *Trends Neurosci.*, 18: 58–60.

Coutinho-Netto, J., Abdul-Ghani, A.S., Collins, J.F. and Bradford, H.F. (1981) Is glutamate a trigger factor in epileptic hyperactivity. *Epilepsia*, 22: 289–296.

Crill, W.E. (1996) Persistent sodium current in mammalian central neurons. *Annu. Rev. Physiol.*, 58: 349–362.

Croucher, M.J., Bradford, H.F., Sunter, D.C. and Watkins, J.C. (1988) Inhibition of the development of electrical kindling of the prepyriform cortex by daily focal injections of excitatory amino acid antagonists. *Eur. J. Pharmacol.*, 152: 29–38.

Croucher, M.J., Collins, J.F. and Meldrum, B.S. (1982) Anticonvulsant action of excitatory amino acid antagonists. *Science*, 216: 899–901.

De Curtis, M., Biella, G., Forti, M. and Panzica, F. (1994) Multifocal spontaneous epileptic activity induced by restricted bicuculline ejection in the piriform cortex of the isolated guinea pig brain. *J. Neurophysiol.*, 71: 2463–2476.

Dingledine, R. and Gjerstad, L. (1979) Penicillin blocks hippocampal IPSPs unmasking prolonged EPSPs. *Brain Res.*, 168: 205–209.

During, M.J. and Spencer, D.D. (1993) Extracellular hippocampal glutamate and spontaneous seizure in the conscious human brain. *Lancet*, 341: 1607–1610.

Gjerstad, L., Andersen, P., Langmoen, I.A., Lundervold, A. and Hablitz, J. (1981) Synaptic triggering of epileptiform discharges in CA1 pyramidal cells in vitro. *Acta Physiol. Scand.*, 113: 245–252.

Gutnick, M.J., Connors, B.W. and Prince, D.A. (1982) Mechanisms of neocortical epileptogenesis in vitro. *J. Neurophysiol.*, 48: 1321–1335.

Gutnick, M.J. and Prince, D.A. (1972) Cortical relay neurons: Antidromic invasion of spikes from a cortical epileptogenic focus. *Science*, 176: 424–426.

Haberly, L.B. and Sutula, T.P. (1992) Neuronal processes that underlie expression of kindled epileptiform events in the piriform cortex in vivo. *J. Neurosci.*, 12: 2211–2224.

Jefferys, J.G.R. (1995) Non-synaptic modulation of neuronal activity in the brain: Electric currents and extracellular ions. *Physiol. Rev.*, 75: 689–723.

Johnston, D. and Brown, T.H. (1981) Giant synaptic potential hypothesis for epileptiform activity. *Science*, 211: 294–297.

Jonas, P., Racca, C., Sakmann, B., Seeburg, P.H. and Monyer, H. (1994) Differences in $Ca^{2+}$ permeability of AMPA-type glutamate receptor channels in neocortical neurons caused by differential GluR-B subunit expression. *Neuron*, 12: 1281–1289.

Jones, R.S.G. and Lambert, J.D.C. (1990) Synchronous discharges in the rat entorhinal cortex in vitro: Site of initiation and the role of excitatory amino acid receptors. *Neuroscience*, 34: 657–670.

Korn, S.J., Giacchino, J.L., Chamberlin, N.L. and Dingledine, R. (1987) Epileptiform burst activity induced by potassium in the hippocampus and its regulation by GABA-mediated inhibition. *J. Neurophysiol.*, 57: 325–340.

Köhr, G. and Mody, I. (1994) Kindling increases *N*-methyl-D-aspartate potency at single *N*-methyl-D-aspartate channels in dentate gyrus granule cells. *Neuroscience*, 62: 975–981.

Kraus, J.E., Yeh, G.-C., Bonhaus, D.W., Nadler, J.V. and McNamara, J.O. (1994) Kindling induces the long-lasting expression of a novel population of NMDA receptors in hippocampal region CA3. *J. Neurosci.*, 14: 4196–4205.

Lee, W.-L. and Hablitz, J.J. (1991) Excitatory synaptic involvement in epileptiform bursting in the immature rat neocortex. *J. Neurophysiol.*, 66: 1894–1901.

Lothman, E.W. (1994) Seizure circuits in the hippocampus and associated structures. *Hippocampus*, 4: 286–290.

MacVicar, B.A. and Dudek, F.E. (1980) Local synaptic circuits in rat hippocampus: Interaction between pyramidal cells. *Brain Res.*, 184: 220–223.

Martin, D., McNamara, J.O. and Nadler, J.V. (1992) Kindling enhances sensitivity of CA3 hippocampal pyramidal cells to NMDA. *J. Neurosci.*, 12: 1928–1935.

Masukawa, L.M., Higashima, M., Hart, G.J., Spencer, D.D. and O'Connor, M.J. (1991) NMDA receptor activation during epileptiform responses in the dentate gyrus of epileptic patients. *Brain Res.*, 562: 176–180.

Meldrum, B.S. (1994) The role of glutamate in epilepsy and other CNS disorders. *Neurology*, 44: S14–23.

Merlin, L.R., Taylor, G.W. and Wong, R.K.S. (1995) Role of metabotropic glutamate receptor subtypes in the patterning of epileptiform activities in vitro. *J. Neurophysiol.*, 74: 896–900.

Miles, R. and Wong, R.K.S. (1987) Latent pathways revealed after tetanic stimulation in the hippocampus. *Nature*, 329: 724–726.

Miles, R., Wong, R.K.S. and Traub, R.D. (1984) Synchronized afterdischarges in the hippocampus: Contribution of local synaptic interactions. *Neuroscience*, 12: 1179–1189.

Mody, I. and Heinemann, U. (1987) NMDA receptors of dentate gyrus granule cells participate in synaptic transmission following kindling. *Nature*, 326: 701–704.

Mody, I. and MacDonald, J.F. (1995) NMDA receptor-dependent excitotoxicity: The role of intracellular $Ca^{2+}$ release. *Trends Pharmacol. Sci.*, 16: 356–359.

Mody, I., Stanton, P.K. and Heinemann, U. (1988) Activation of *N*-methyl-D-aspartate receptors parallels changes in cellular and synaptic properties of dentate gyrus granule cells after kindling. *J. Neurophysiol.*, 59: 1033–1054.

Olney, J.W., Collins, R.C. and Sloviter, R.S. (1986) Excitotoxic mechanisms of epileptic brain damage. *Adv. Neurol.*, 44: 857–877.

Pinault, D. (1995) Backpropagation of action potentials generated at ectopic axonal loci: Hypothesis that axon terminals integrate local environmental signals. *Brain Res. Rev.*, 21: 42–92.

Schwartzkroin, P.A. (1994) Cellular electrophysiology of human epilepsy. *Epilepsy Res.*, 17: 185–192.

Schwartzkroin, P.A. and Slawsky, M. (1977) Probable calcium spikes in hippocampal neurons. *Brain Res.*, 133: 157–161.

Segal, M.M. and Douglas, A.F. (1997) Late sodium channel openings underlying epileptiform activity are preferentially diminished by the anticonvulsant phenytoin. *J. Neurophysiol.*, 77: 3021–3034.

Simon, R.P., Schmidley, J.W., Meldrum, B.S., Swan, J.H. and Chapman, A.G. (1986) Excitotoxic mechanisms in hypoglycaemic hippocampal injury. *Neuropathol. Appl. Neurobiol.*, 12: 567–576.

Stasheff, S.F., Anderson, W.W., Clark, S. and Wilson, W.A. (1989) NMDA antagonists differentiate epileptogenesis from seizure expression in an in vitro model. *Science*, 245: 648–651.

Stasheff, S.F., Hines, M. and Wilson, W.A. (1993) Axon terminal hyperexcitability associated with epileptogenesis in vitro. I. Origin of ectopic spikes. *J. Neurophysiol.*, 70: 961–975.

Strowbridge, B.W. and Schwartzkroin, P.A. (1996) Transient potentiation of spontaneous EPSPs in rat mossy cells induced by depolarization of a single neurone. *J. Physiol. (Lond.)*, 494: 493–510.

Swartzwelder, H.S., Lewis, D.V., Anderson, W.W. and Wilson, W.A. (1987) Seizure-like events in brain slices: Suppression by interictal activity. *Brain Res.*, 410: 362–366.

Taylor, G.W., Merlin, L.R. and Wong, R.K.S. (1995) Synchronized oscillations in hippocampal CA3 neurons induced by metabotropic glutamate receptor activation. *J. Neurosci.*, 15: 8039–8052.

Tian, L.M., Otoom, S. and Alkadhi, K.A. (1995) Endogenous bursting due to altered sodium channel function in rat hippocampal CA1 neurons. *Brain Res.*, 680: 164–172.

Traub, R.D., Borck, C., Colling, S.B. and Jefferys, J.G.R. (1996) On the structure of ictal events in vitro. *Epilepsia*, 37: 879–891.

Traub, R.D., Colling, S.B. and Jefferys, J.G.R. (1995) Cellular mechanisms of 4-aminopyridine-induced synchronized after-discharges in the rat hippocampal slice. *J. Physiol. (Lond.)* 489: 127–140.

Traub, R.D., Jefferys, J.G.R. and Miles, R. (1993) Analysis of the propagation of disinhibition-induced after-discharges along the guinea-pig hippocampal slice in vitro. *J. Physiol. (Lond.)* 472: 267–287.

Traub, R.D., Jefferys, J.G.R., Miles, R., Whittington, M.A. and Tóth, K. (1994a) A branching dendritic model of a rodent CA3 pyramidal neurone. *J. Physiol. (Lond.)* 481: 79–95.

Traub, R.D., Jefferys, J.G.R. and Whittington, M.A. (1994b) Enhanced NMDA conductance can account for epileptiform activity induced by low $Mg^{2+}$ in the rat hippocampal slice. *J. Physiol. (Lond.)* 478: 379–393.

Traub, R.D. and Miles, R. (1991) Neuronal networks in the hippocampus. Cambridge: Cambridge University Press.

Traub, R.D., Miles, R. and Jefferys, J.G.R. (1993) Synaptic and intrinsic conductances shape picrotoxin-induced synchronized after-discharges in the guinea-pig hippocampal slice. *J. Physiol. (Lond.)*, 461: 525–547.

Traub, R.D. and Wong, R.K.S. (1982) Cellular mechanism of neuronal synchronization in epilepsy. *Science*, 216: 745–747.

Vignes, M. and Collingridge, G.L. (1997) The synaptic activation of kainate receptors. *Nature*, 388: 179–182.

Wadman, W.J. and Gutnick, M.J. (1993) Non-uniform propagation of epileptiform discharge in brain slices of rat neocortex. *Neuroscience*, 52: 255–262.

Wong, R.K.S. and Prince, D.A. (1981) Afterpotential generation in hippocampal cells. *J. Neurophysiol.*, 45: 86–97.

Wong, R.K.S. and Traub, R.D. (1983) Synchronized burst discharge in the disinhibited hippocampal slice. I. Initiation in the CA2-CA3 region. *J. Neurophysiol.*, 49: 459–471

O.P. Ottersen, I.A. Langmoen and L. Gjerstad (Eds.)
*Progress in Brain Research*, Vol 116


CHAPTER 23

# Involvement of glutamate in human epileptic activities

Richard D. Scheyer

*Department of Neurology, Yale University School of Medicine, New Haven, CT, USA*

## Introduction

Epilepsy may be described as a disruption of the normal balance between excitation and inhibition. Much effort has been devoted toward understanding the role of perturbations of the excitatory neurotransmitter glutamate. Traditional studies of brain biopsies and autopsy samples have been expanded by the application of newer genetic techniques allowing identification of receptor subtypes. In vivo studies, previously limited primarily to analysis of plasma and cerebrospinal fluid (CSF), have been supplemented by newer techniques of microdialysis and magnetic resonance spectroscopy (MRS).

The role of glutamate in human epilepsy will be presented. Related information from animal studies will be mentioned, but the reader is directed to the appropriate chapters for more details of animal models of epilepsy, as well as additional information on glutamate receptors, in vitro electrophysiology, and antiepileptic medications.

Current address:
Global Clinical Pharmacology, Hoechst Marion Roussel, Inc., P.O. Box 6800, Bridgewater, NJ 08807-0800, USA.

## Indirect assessments

Early studies attempted to look at glutamate concentrations in readily accessible locations, the plasma and CSF.

Van Gelder et al. (1980) reported elevated plasma levels of glutamate not only in patients with three per second spike-wave epilepsy, but also in their first degree relatives. Janjua et al. (1992), also reported elevations in plasma glutamate in patients, and to a lesser extent, family members, of patients with partial epilepsy. Glutamate is an intermediate of energy metabolism throughout the body, and it is doubtful if plasma glutamate reflects brain neurotransmitter activity (Fillenz, 1995). Other studies (e.g., Haines et al., 1985; Mutani et al., 1975) have not seen glutamate elevations even in epilepsy patients.

Studies measuring glutamate in the CSF have also reported variable findings. Kälviäinen et al. (1993), found elevated glutamate concentrations in patients with untreated partial seizures, but Mutani et al. (1974) reported markedly diminished CSF glutamate concentrations in patients with severe epilepsy. Engelsen and Elsayed (1984) found normal levels of glutamate but elevated aspartic acid and glutamine. Following release from neurons, glutamate is largely taken up into astrocytes, converted to glutamine, and transported back to the presynaptic neurons, where it is converted back to glutamate for reuse. Hence increased glutamine levels are indirect evidence of increased glutamatergic activity (Westergaard et al., 1995).

**Tissue**

Much of our knowledge has been derived from direct study of epileptic brain. Most patients with epilepsy are managed with antiepileptic drugs (AEDs), but perhaps one third continue to have seizures despite optimal AED therapy. In some of these patients, seizures may be controlled by removal of the epileptogenic zone. The most common pathology seen in resected tissue is hippocampal sclerosis (neuronal loss in the CA1 region of the hippocampus with some loss in CA3/CA4). Depending on the extent of pathology, the terms mesial temporal sclerosis (indicating the additional involvement of the amygdala), Ammon's Horn Sclerosis (implying sparing of dentate gyrus), or (least precisely) temporal lobe epilepsy, may be used (Swanson, 1995). Criteria vary from center to center, and terms are sometimes used interchangeably. The "epileptogenic zone" is defined as the area of the brain that is necessary and sufficient for habitual seizure generation (Lüders et al., 1993). This may in practice only be defined by successful surgical resection, but results of long term outcome are usually not available. The term "irritative zone" may be used to refer to areas of cortex generating interictal spikes, and "ictal onset zone" to more specifically describe areas of cortex where seizures originate.

An early study by van Gelder et al. (1972) found a tendency toward decreased glutamate and elevated glutamine concentrations in actively spiking cortex ("irritative zone") as compared to surrounding tissue. Perry and Hansen (1981) found elevations of both glycine and glutamate in actively spiking cortex.

Transmitter assays in biopsy samples analyze a homogenate of tissue including neuronal and astrocytic elements along with myelin and interstitial fluid. Neurotransmitters may also be affected by anesthesia, time since the last seizure, and by the trauma of surgery. An assessment of the enzymes involved in formation and degradation of neurotransmitters might be less affected by acute events. Lloyd et al. (1986) found a diminution in glutamic acid decarboxylase (GAD) in the irritative zone. Low activity of this enzyme, which converts glutamate to GABA, might be expected to elevate glutamate and diminish GABA concentrations. Other groups, however, (Babb et al., 1989; Sherwin et al., 1984; Tursky et al., 1976) have reported normal GAD activity. Glutamic acid dehydrogenase (GDH), involved in glutamate formation, and tyrosine hydroxylase, involved in catecholamine synthesis, have also been reported to be elevated in seizure foci (Sherwin et al., 1984; Sherwin and Van Gelder, 1986).

Peeling and Sutherland (1993), using magnetic resonance spectroscopy (MRS) on resected tissue, found glutamate and aspartate levels in the resected hippocampus were depressed relative to neocortical control cortex. This depression of transmitters in the epileptogenic hippocampus may be a reflection of neuronal loss since MRS, like tissue homogenates, looks at total, rather than extracellular concentrations. Glutamate levels tended to be higher in spiking than non-spiking neocortex.

Recent advances in MRS technology have provided sufficient sensitivity to assay neurotransmitter concentrations in vivo, although current anatomical resolution of neurotransmitter spectroscopy has largely limited the technique to regions of healthy brain. The non-invasiveness of MRS permits inclusion of healthy controls, and serial measurements in epilepsy patients. Petroff et al. (1995), using in vivo MRS of the occipital cortex of normal and epileptic subjects, found elevated levels of glutamate, and decreased levels of GABA, remote from the seizure focus, and a tendency for these findings to become more pronounced with increasing seizure frequency (Petroff et al., 1996). Fazekas et al. (1995) reported elevations of either glutamate or glutamine (their technique did not permit specific identification) after focal status epilepticus.

Localization and quantification of glutamate receptors have been studied in brain tissue removed at the time of epilepsy surgery. More detail on glutamate receptors may be found in other chapters (see section III). In brief, glutamate receptors are classified as ionotropic or metabotropic. Ionotropic receptors open a channel permeable to calcium (or in some instances, sodium) in response to stimulation. These receptors are subdivided into those which respond to the glutamate analogue *N*-methyl-D-aspartate (NMDA receptors), those which respond to L-α-amino-3-hydroxy-5-methyl-4-isoxazole propionate (AMPA receptors), and those which resond to kainic acid stimulation (kainate receptors). AMPA and kainate receptors have overlapping affinity and are often referred to as AMPA/kainate or "non-NMDA" receptors. Metabotropic receptors link receptors to response via G-protein activation.

Geddes et al. (1990), used receptor autoradiography on tissue samples from five patients who had undergone temporal lobectomy for refractory seizures, and compared them to post-mortem controls. They found a loss of NMDA and kainate receptors in the sclerotic CA3, dentate hilus, and CA1, but preserved or even increased binding in the subiculum and molecular layer of the dentate gyrus, and increased NMDA and kainate receptors in the parahippocampal gyrus. Hosford et al. (1991), also compared epileptogenic hippocampi to post mortem controls, and reported increased AMPA binding in the molecular layer of the dentate gyrus. Ben-Ari et al. (1992) noted increased kainic acid binding in the stratum lucidum-pyramidal of CA3 and the supragranular layer of the dentate gyrus in five children with temporal lobe epilepsy. McDonald et al. (1991) showed increased binding to the NMDA recognition site in CA1 and the dentate gyrus, but not in CA4. Phencyclidine (PCP) binding (associated with the NMDA modulated calcium channel), was decreased in all regions, suggesting an alteration in the normal NMDA receptor complex. AMPA receptor binding was increased in the molecular layer of CA1, but otherwise was unchanged from autopsy controls.

Newer techniques have allowed more precise characterizations of receptor subtypes. García-Ladona et al. (1994), using in situ hybridization, identified elevated concentrations of mRNA coding for AMPA receptor subtypes one and two in the granule cell layer of the dentate gyrus, and possibly a patchy increase of the same subtypes in the hilus of surgically resected epileptogenic hippocampi. Blümcke et al. (1996), using immunohistochemical techniques, found an increase in the GluR2 subunit of the AMPA receptor in the dentate gyrus of patients with Ammon's Horn Sclerosis (compared to autopsy and lesional epilepsy controls), with no alteration in concentrations of the NR1 subunit of the NMDA receptor. De Lanerolle et al. (1998) reported a punctate increased expression of the GluR1 AMPA receptor subunit in the hilar neurons and CA3 regions of patients with mesial temporal sclerosis, but not in epilepsy patients whose hippocampi were removed in association with the resection of mass lesions ("mass-associated epilepsy"). Similiar findings were reported by Babb et al. (1996). Mathern et al. (1996) noted increased NMDA NR2 staining in the dentate gyrus inner molecular layer correlated with inner molecular layer mossy fiber sprouting in patients with hippocampal sclerosis. Bayer et al. (1995), comparing epilepsy surgery patients with autopsy controls, found a decrease in NR1 binding in sclerotic hippocampi, but noted that it appeared to correspond to regions of neuronal dropout.

Very little information is available regarding metabotropic receptors in human epilepsy. Activation of metabotropic receptors in audiogenic seizure (DBA/2) mice have both proconvulsant and anticonvulsant effects (Ghauri et al., 1996). Glazier et al. (1997) recently reported decreases in expression of mGluR1 and mGluR2 levels in the hippocampi of epileptic patients.

Neocortical epilepsy has not been studied as thoroughly as hippocampal disease. Most early studies lumped neocortical and mesial onset seizures, and more recent studies have tended to focus on the hippocampus. Sherwin et al. (1988) studied the neocortex of 21 patients undergoing temporal lobectomy. In patients showing neocortical spikes during intraoperative electrocorticography (presumably neocortical onset epilepsy), they found a 26% elevation in glutamate and 55% elevation of glycine, a co-agonist of the NMDA glutamate receptor, compared to patients who did not have neocortical spikes (presumably hippocampal or amygdala onset seizures). The effect of the interictal spiking itself, versus the effect of the focus, could not be separated. Wolf et al. (1996) studied 30, primarily neocortical, resections, which had sufficient morphologically normal peri-lesional and remote tissue. While there was considerable inter-individual variability, immunohistochemistry studies showed a tendency for decreased expression of the NR1 subunit of the NMDA receptor as well as decreased GABA receptor expression in the peri-lesional cortex.

Glutamatergic synaptic activity is terminated by reuptake into neurons or glia. It would be anticipated that abnormalities of glutamate reuptake cause seizures, and experimental knockdown of the EAAC-1 neuronal glutamate transporter has been associated with epilepsy in the rat (Rothstein et al., 1996). Knockdowns of the astroglial GLAST and GLT-1 glutamate transporters were associated with elevated extracellular glutamate, neurodegeneration, and progressive motor impairments (but not epilepsy). In amygdala kindled rats, EAAC-1 transporter is increased (but the GLAST transporter is decreased) (Miller et al., 1997). O'Connor et al. (1996) recently reported evidence of impaired astrocytic reuptake of glutamate in patients with temporal lobe epilepsy.

## Physiology – in vitro

Tissue studies provide a picture of the brain at a single point in time, usually during a neurosurgical procedure in which the manipulation of the brain and the use of anesthetics may affect results. Interictal spiking may be present on electrocorticography, but seizures are rare. It is difficult to separate the neurochemical abnormalities that may be causing seizures or neuronal injury from those which are a result of chronic epilepsy. More interesting than merely recording glutamate, receptor, and transporter concentrations, would be to relate this information to brain function.

Electrophysiological findings in hippocampi resected from patients with mesial temporal sclerosis include a loss of inhibitory postsynaptic potentials and a propensity to fire evoked epileptiform bursts (Williamson et al., 1995). These bursts are not seen in hippocampi resected from patients with seizures related to extra-hippocampal abnormalities (Strowbridge et al., 1992), and may be blocked by the addition of the NMDA antagonist 2-amino-5-phosphonovaleric acid (APV) (Avoli and Olivier, 1987; Isokawa et al., 1991; Masukawa et al., 1991; Williamson and Spencer, 1995).

## Physiology – in vivo

Devinsky et al. (1993) studied CSF glutamate in relation to seizures, and found elevations of post-seizure, but not interictal, CSF glutamate concentrations. While not providing anatomic information, nor information regarding whether glutamate elevations were cause or effect of seizures, the study did associate glutamate with seizures, rather than just with the epileptogenic substrate.

CSF is readily available, but has a number of limitations. CSF is produced primarily by the choroid plexus. Only 10% is derived from the brain extracellular fluid (Segal, 1993). Even to the extent that CSF is sampling the brain, the time lag from formation until sampling (especially in the lumbar space), may be measured in hours, and sampling of specific brain regions is not possible.

Other techniques have permitted more direct sampling of the brain extracellular fluid. Do et al. (1991) using push-pull cannulas, reported increases in excitatory amino acids during brief intraoperative recordings in anesthetized humans undergoing electrical stimulation. Microdialysis allows the brain's neurochemistry to be sampled with minimal disruption of the normal physiology (Hamani et al., 1997). This technique has now been used in several human studies. Hamberger et al. (1991), combining microdialysis of the brains surface with intraoperative electrocorticography, found a suggestion of increased glutamate when the catheter was placed over the seizure focus. Carlson et al. (1992), sampling the extracellular space with an intraparenchymal catheter placed during intraoperative electrocorticography, studied one patient who had multiple intraoperative seizures, and found a sixteen fold ictal elevation in glutamate, as well as major increases in aspartate, glycine, and serine. Ronne-Engström et al. (1992) confirmed these findings with intraoperative seizures recorded in an additional three patients.

Techniques used to identify the epileptogenic zone prior to resective surgery include the neurologic exam, electroencephalography (EEG), and neuroimaging; but localization and/or confirmation of the focus occasionally requires recording of the intracranial EEG, from electrodes placed either in the subdural space or into the brain parenchyma.

Such electrodes, once implanted, may be left in place for days or even weeks. These extended recordings offer a greater chance to record typical spontaneous seizures. Microdialysis catheters, attached to intraparenchymal "depth" EEG electrodes, permit continuous sampling of the extracellular space. In patients with suspected mesial temporal lobe epilepsy, electrodes may be implanted through occipital burr holes, traversing the length of the hippocampi.

Catheters are continuously perfused with artificial ECF, typically at 2.5 $\mu$l/min. Samples are routinely collected every 1/2 hour, but the deadspace in the outflow tubing provides a 10 minute delay, and the sampling rate is changed to every three minutes when seizures are detected. During and Spencer (1993) reported changes in glutamate concentrations associated with spontaneous seizures in six patients with medically intractable epilepsy of unilateral mesial temporal lobe onset. Interictal glutamate concentrations were 11% higher in the epileptogenic hippocampus than in the contralateral hippocampus. During spontaneous seizures, glutamate levels in the epileptogenic hippocampus rose to 600% of baseline. Of particular interest, initial glutamate elevations were seen in the dialysate sample immediately preceding the onset of the electrographic (and clinical) seizure. Contralateral elevations in glutamate were more modest (approximately 300% of baseline), and were not seen until after seizure onset. Delayed elevations in GABA were seen in both hippocampi, and were maximal contralateral to the side of seizure onset.

Confirmatory data were provided by Wilson et al. (1994, 1996), who also noted seven fold ictal elevations of glutamate, five fold elevations of GABA, and ten fold elevations of aspartate. In other studies, they showed modest elevations glutamate and aspartate during cognitive tasks (Fried et al., 1996) or electrical stimulation (Wilson et al., 1997).

Dialysis is most useful for determining relative changes in the extracellular environment. The system is a non-equilibrium dialysis, and calibration is difficult (Morrison et al., 1991) although not impossible (Bungay et al., 1989; Miele et al., 1996; Scheyer et al., 1994). By reducing perfusion rates to allow greater equilibration with the extracellular space, it was possible to estimate that dialysate concentrations are approximately 12% of the ECF concentrations (During and Spencer, 1993). Seizure-related peak dialysate concentrations of 65 $\mu$M might be expected to be correspond to neurotoxic ECF glutamate concentrations.

The extracellular concentrations measured by microdialysis represent a balance of release and reuptake. While most studies have ignored this distinction, it seems unlikely that abnormalities of glutamate uptake could fully explain the changes in extracellular concentrations seen in association with seizures (Herrera-Marschitz et al., 1996).

Microdialysis is not limited to passive recording. Increasing the potassium in the dialysate evokes the release of glutamate, with similar increases found in the epileptogenic and non-epileptogenic hippocampi. This chemically induced release supports the concept that microdialysis can detect neurotransmitter as well as metabolic glutamate. This has been further tested by perfusing with a zero calcium perfusate, depleting the local environment of calcium, and blocking vesicular release of glutamate (M. During, personal communication).

In other studies, During et al. (1995) demonstrated that perfusion of glutamate was able to trigger the release of GABA, and that this effect was attenuated in the epileptogenic hippocampus. By depleting calcium in the immediate vicinity of the dialysis probe, he demonstrated that this release was calcium independent and therefore non-vesicular in nature. This glutamate-induced release of GABA might be partially responsible for the delayed GABA elevation seen in the epileptogenic hippocampus.

**Glutamate related syndromes**

Glutamatergic abnormalities are well established for certain human epilepsies. Lott et al. (1978) found elevated brain glutamate (and diminished GABA) in the brain of a child who died as a result of status epilepticus related to pyridoxine dependency, a genetic condition in which glutamic acid decarboxylase (GAD), the enzyme which converts glutamate to GABA, does not function normally. In this patient, the effect of dysfunctional GAD could not be isolated from the effects of status epilepticus. Baumeister et al. (1994) found a 200 fold elevation of CSF glutamate after a five minute seizure in a patient with untreated pyridoxine dependency, which reverted to normal with pyridoxine therapy.

Rogers et al. (1994) identified autoantibodies to the GluR3 subunit of the AMPA receptor in the serum of four patients with Rasmussen's encephalitis, a progressive focal epilepsy. Preliminary data indicates that this condition can be favorably affected by immune modulation with plasma exchange (Rogers et al., 1994) or immunoglobulin administration (Hart et al., 1994).

Elevated CSF glutamate levels have been reported in Rett syndrome, and implicated in its pathophysiology (Lappalainen and Riikonen, 1996). Uldall and Hansen (1993) reported promising results in control of seizures in Rett syndrome using lamotrigine, a drug which appears to inhibit glutamate release.

Intoxication with the kainic acid analogue domoic acid has been associated with both acute seizures and with the development of hippocampal atrophy and temporal lobe epilepsy (Cendes et al., 1995; Perl et al., 1990; Teitelbaum et al., 1990), supporting a causative role of glutamate in epileptogenesis.

**Effects of antiepileptic drugs on glutamatergic neurotransmission**

Several antiepileptic drugs have anti-glutamatergic actions that may account for a part of their antiepileptic effects (see Gjerstad, this volume). Felbamate inhibits binding of glycine at the NMDA receptor (Calabresi et al., 1996; Coffin et al., 1994). Topiramate blocks the AMPA/kainate receptor (Hart et al., 1994). Lamotrigine and perhaps carbamazepine may decrease presynaptic glutamate release by blocking sodium and perhaps calcium channels (Waldmeier et al., 1995; Wang et al., 1996). Gabapentin may decrease glutamate formation (Goldlust et al., 1995). Remacemide, currently in clinical trials for both epilepsy and neuroprotection, has a metabolite with NMDA antagonist properties (Srinivasan et al., 1995).

Glutamate measurements may be useful in selecting antiepileptic drug therapy. Kälviäinen et al. (1993) studied 42 patients with newly diagnosed epilepsy, prior to randomization to either vigabatrin or carbamazepine. Patients whose seizures responded to vigabatrin had pretreatment CSF glutamate concentrations 32% higher than controls, and 26% higher than non-responders. Glutamate decreased 14%, and glutamine and GABA both increased, in patients treated with vigabatrin; no changes in amino acid concentrations were seen in patients treated with carbamazepine. Petroff et al. (1995), using in vivo MRS, also found elevated glutamine and GABA, and decreased glutamate, in patients treated with vigabatrin, but not in patients given standard antiepileptic drugs.

Given the role of glutamate in epilepsy, and the apparent neurotoxicity of NMDA receptor activation, a number of NMDA antagonists have been evaluated for the treatment of epilepsy (Gjerstad and Tauboll, this volume, and Meldrum, this volume). Animal studies of the non-competitive NMDA antagonist ketamine did not show a clear effect on seizures, but did suggest a protection against central nervous system (CNS) injury (Fujikawa, 1995). The non-competitive NMDA antagonist MK-801 (dizocilpine) was quite potent in a variety of animal models, but only transient improvements in seizures were observed in a human trial (Rogawski and Porter, 1990; Troupin et al., 1986). The competitive NMDA antagonist d-CPPene, given at doses causing impaired concentration, showed no evidence of reducing seizures (Sveinbjornsdottir et al., 1993; Herrling et al., 1997).

## Discussion

There is compelling evidence that glutamate is involved in human focal seizures and in seizure-related neuronal injury. Limited evidence also suggests a role in generalized seizures. Familial elevations of plasma glutamate, reported by van Gelder et al. (1980), suggest the existence of genetic predispositions to epilepsy associated with systemic abnormalities in glutamate metabolism.

CA3 and dentate gyrus, commonly affected in human mesial temporal sclerosis, have high densities for both NMDA and AMPA/kainate receptors. There is evidence for increased AMPA receptor density in mesial temporal sclerosis.

A causal role of glutamate in human epilepsy has been more difficult to establish. Tissue, CSF, and systemic measures do not separate neurotransmitter from metabolic glutamate, and receptor changes may be showing the effect, rather than the etiology, of recurrent seizures. A causal relationship is perhaps better shown by inadvertent domoic acid intoxications, and supported by the early rise in microdialysate glutamate in the epileptogenic hippocampus.

Receptor binding data, and efficacy of drugs in animal models of epilepsy, have not always translated into effective treatments of human epilepsy. Most drugs targeted at glutamatergic mechanisms have been directed at the NMDA receptor. Human trials have generally been disappointing, with marked neurotoxicity and limited efficacy (Meldrum, this volume). The NMDA receptor is but one of many sites of glutamate activity. The role of NMDA receptors may be age related, with a decrease in NMDA relative to AMPA binding as age increased from neonate to in an autopsy series of humans without neurologic disease (Johnson et al., 1996), and decreased antiepileptic efficacy of MK-801 was noted in adult rats versus rat pups (Moshé, 1993).

Data which may reconcile the prominent neuroprotective and limited antiepileptic effects of NMDA antagonists come from the psychiatric literature. Moghaddam et al. (1997) recently reported that sub-anesthetic doses of ketamine resulted in increased glutamate outflow in the rat pre-frontal cortex. Ketamine-induced dopamine release in the pre-frontal cortex could be blocked by administration of the AMPA/kainate receptor antagonist CNQX, suggesting that NMDA antagonists could cause increased glutamatergic activity at non-NMDA receptors, which may be responsible for seizure generation.

Federico and Macvicar (1996), among others, have found AMPA/kainate receptor activation to be pro-convulsant, whereas NMDA activation had little effect on seizures in their bicuculline model. Recent development of CNS penetrant blockers of non-NMDA receptors will allow further delineation of the role of these receptors in initiation and propagation of seizures (Lees, 1996). While the role of NMDA blockade in preventing human seizures may prove to be limited, NMDA antagonists appear able to prevent seizure-related excitotoxic injury and epileptogenesis (Löscher, 1998), and a variety of NMDA antagonists have been tried as neuroprotectants (Meldrum, this volume).

## Conclusion

The concept of epilepsy as an excess of excitatory activity has made glutamate an appealing target for study. Most evidence of its involvement is indirect, as elevated glutamate levels in brain or CSF of individuals with frequent seizures could be a result of metabolic, rather than neurotransmitter, glutamate. There is nevertheless a growing body of evidence supporting the role of glutamate in causing both seizures and seizure-related CNS injury, and there is evidence for glutamatergic mechanisms of action for current and investigational antiepileptic drugs.

## References


Avoli, M. and Olivier, A. (1987) Bursting in human epileptogenic neocortex is depressed by an *N*-methyl-D-aspartate antagonist. *Neurosci. Lett.*, 76: 249–254.

Babb, T.L., Pretorius, J.K., Kupfer, W.R. and Crandall, P.H. (1989) Glutamate decarboxylase-immunoreactive neurons are preserved in human epileptic hippocampus. *J. Neurosci.*, 9: 2562–2574.

Babb, T.L., Mathern, G.W., Leite, J.P., Pretorius, J.K., Yeoman, K.M. and Kuhlman, P.A. (1996) Glutamate AMPA receptors in the fascia dentata of human and kainate rat hippocampal epilepsy. *Epilepsy Res.*, 26: 193–205.

Baumeister, F.A., Gsell, W., Shin, Y.S. and Egger, J. (1994) Glutamate in pyridoxine-dependent epilepsy: Neurotoxic glutamate concentration in the cerebrospinal fluid and its normalization by pyridoxine. *Pediatrics*, 94: 318–321.

Bayer, T.A., Wiestler, O.D. and Wolf, H.K. (1995) Hippocampal loss of *N*-methyl-D-aspartate receptor subunit 1 mRNA in chronic temporal lobe epilepsy. *Acta Neuropathologica*, 89: 446–450.

Ben-Ari, Y., Represa, A., Tremblay, E., Robain, O., LeGal LaSalle, G., Rovira, C., Gho, M. and Cherubini, E. (1992) Epileptogenesis and neuronal plasticity: Studies on kainate receptor in the human and rat hippocampus. *Epilepsy Res.*, (Suppl. 8): 369–373.

Blümcke, I., Beck, H., Scheffler, B., Hof, P.R., Morrison, J.H., Wolf, H.K., Schramm, J., Elger, C.E. and Wiestler, O.D. (1996) Altered distribution of the α-amino-3-hydroxy-5-methyl-4-isoxazole propionate receptor subunit glur2(4) and the *N*-methyl-D-aspartate receptor subunit NMDAr1 in the hippocampus of patients with temporal lobe epilepsy. *Acta Neuropathologica*, 92: 576–587.

Bungay, P.M., Morrison, P.F. and Dedrick, R.L. (1989) Steady-state theory for quantitative microdialysis of solutes and water in vivo and in vitro. *Life Sci.*, 46: 105–119.

Calabresi, P., Siniscalchi, A., Pisani, A., Stefani, A., Mercuri, N.B. and Bernardi, G. (1996) A field potential analysis on the effects of lamotrigine, GP 47779, and felbamate in neocortical slices. *Neurology*, 47: 557–562.

Carlson, H., Ronne-Engström, E., Ungerstedt, U. and Hillered, L. (1992) Seizure related elevations of extracellular amino acids in human focal epilepsy. *Neurosci. Lett.*, 140: 30–32.

Cendes, F., Andermann, F., Carpenter, S., Zatorre, R.J. and Cashman, N.R. (1995) Temporal lobe epilepsy caused by domoic acid intoxication: Evidence for glutamate receptor-mediated excitotoxicity in humans. *Ann. Neurol.*, 37: 123–126.

Coffin, V., CohenWilliams, M. and Barnett, A. (1994) Selective antagonism of the anticonvulsant effects of felbamate by glycine. *Eur. J. Pharmacol*, 256: R9-R10.

de Lanerolle, N.C., Eid, T., Von Campe, G., Kovacs, I., Spencer D.O. and Brines, M. (1998) Glutamate receptor subumits GluR1 and GluR2/3 distribution shows reorganization in the human epileptogenic hippocampus *Eur. J. Neurosci.*, 10: 1687–1703.

Devinsky, O., Emoto, S., Nadi, N.S. and Theodore, W.H. (1993) Cerebrospinal fluid levels of neuropeptides, cortisol, and amino acids in patients with epilepsy. *Epilepsia*, 34: 255–261.

Do, K.Q., Klancnik, J.M., Gahwiler, B., Perschak, H., Weiser, H.G. and Cuenod, M. (1991) Release of excitatory amino acids: Animal studies and epileptic foci studies in humans. In B.S. Meldrum, F. Moroni, R.P. Simon and J.S. Woods (Eds.), Fidia Research Foundation Symposium: Excitatory Amino Acids, Vol. 5, Raven Press, New York, pp. 677–685.

During, M.J. and Spencer, D.D. (1993) Extracellular hippocampal glutamate and spontaneous seizure in the conscious human brain. *Lancet*, 341: 1607–1610.

During, M.J., Ryder, K.M. and Spencer, D.D. (1995) Hippocampal gaba transporter function in temporal-lobe epilepsy. *Nature*, 376: 174–177.

Engelsen, B. and Elsayed, S. (1984) Increased concentrations of aspartic acid in the cerebrospinal fluid of patients with epilepsy and trigeminal neuralgia: An effect of medication? *Acta Neurol.* Scand., 70: 70–76.

Fazekas, F., Kapeller, P., Schmidt, R., Stollberger, R., Varosanec, S., Offenbacher, H., Fazekas, G. and Lechner, H. (1995) Magnetic resonance imaging and spectroscopy findings after focal status epilepticus. *Epilepsia*, 36: 946–949.

Federico, P. and Macvicar, B.A. (1996) Imaging the induction and spread of seizure activity in the isolated brain of the guinea pig – the roles of gaba and glutamate receptors. *J. Neurophysiol.*, 76: 3471–3492.

Fillenz, M. (1995) Physiological release of excitatory amimo acids. *Behav. Brain. Res.*, 71: 51–67.

Fried, I., Maidment, N., Behnke, E., MacDonald, K. and Wilson, C. (1996) In vivo microdialysis in human hippocampus and amygdala during seizures and cognitive tasks (abstract). *Epilepsia*, 37(Suppl. 5): 81.

Fujikawa, D.G. (1995) Neuroprotective effect of ketamine administered after status epilepticus onset. *Epilepsia*, 36: 186–195.

Garcia-Ladona, F.J., Palacios, J.M., Probst, A., Wieser, H.G. and Mengod, G. (1994) Excitatory amino acid AMPA receptor mRNA localization in several regions of normal and neurological disease affected human brain: An in situ hybridization histochemistry study. *Molecular Brain Research*, 21: 75–84.

Geddes, J.W., Cahan, L.D., Cooper, S.M., Kim, R.C., Choi, B.H. and Cotman, C.W. (1990) Altered distribution of excitatory amino acid receptors in temporal lobe epilepsy. *Exp. Neurol.*, 108: 214–220.

Ghauri, M., Chapman, A.G. and Meldrum, B.S. (1996) Convulsant and anticonvulsant actions of agonists and antagonists of group III mGluRs. *Neuroreport*, 7: 1469–1474.

Glazier, S., Bell, W., Nosel, E., Pons, T., Tytell, M., Deadwyler, S. and Grigorenko, E.V. (1997) Changes in glutamate receptor subunit composition in hippocampal and cortex of patients with refractory epilepsy (abstract). *Epilepsia*, 38 (Suppl. 3): 191.

Goldlust, A., Su, T.Z., Welty, D.F., Taylor, C.P. and Oxender, D.L. (1995) Effects of anticonvulsant drug gabapentin on the enzymes in metabolic pathways of glutamate and GABA. *Epilepsy Res.*, 22: 1–11.

Haines, J.L., Rich, S.S., Tsai, M.Y. and Anderson, V.E. (1985) Altered amino acid levels in multiply affected sibships with seizures. *Epilepsia*, 26: 642–648.

Hamani, C., Luer, M.S. and Dujovny, M. (1997) Microdialysis in the human brain – review of its applications. *Neurological Research*, 19: 281–288.

Hamberger, A., Nystrom, B., Larsson, S., Silfvenius, H. and Nordborg, C. (1991) Amino acids in the neuronal microenvironment of focal human epileptic lesions. *Epilepsy Res.*, 9: 32–43.

Hart, Y.M., Cortez, M., Andermann, F., Hwang, P., Fish, D.R., Dulac, O., Silver, K., Fejerman, N., Cross, H. and Sherwin, A., et al. (1994) Medical treatment of Rasmussen's syndrome (chronic encephalitis and epilepsy): Effect of high-dose steroids or immunoglobulins in 19 patients. *Neurology*, 44: 1030–1036.

Herrera-Marschitz, M., You, Z.B., Goiny, M., Meana, J.J., Silveira, R., Godukhin, O.V., Chen, Y., Espinoza, S., Pettersson, E., Loidl, C.F., Lubec, G., Andersson, K., Nylander, I., Terenius, L. and Ungerstedt, U. (1996) On the origin of extracellular glutamate levels monitored in the basal ganglia of the rat by in vivo microdialysis. *J. Neurochem.*, 66: 1726–1735.

Herrling, P.L., Emre, M. and Watkins, J.C. (1997) D-CPPene (SDZ EAA-494) – A competative NMDA antagonist pharmacology and results in humans. Excitatory amino acids – clinical results with antagonists. Academic press pp. 7–21.

Hosford, D.A., Crain, B.J., Cao, Z., Bonhaus, D.W., Friedman, A.H., Okazaki, M.M., Nadler, J.V. and McNamara, J.O. (1991) Increased AMPA-sensitive quisqualate receptor binding and reduced NMDA receptor binding in epileptic human hippocampus. *J. Neurosci.*, 11: 428–434.

Isokawa, M., Avanzini, G., Finch, D.M., Babb, T.L. and Levesque, M.F. (1991) Physiologic properties of human dentate granule cells in slices prepared from epileptic patients. *Epilepsy Res.*, 9: 242–250.

Janjua, N.A., Itano, T., Kugoh, T., Hosokawa, K., Nakano, M., Matsui, H. and Hatase, O. (1992) Familial increase in plasma glutamic acid in epilepsy. *Epilepsy Res.*, 11: 37–44.

Johnson, M., Perry, R.H., Piggott, M.A., Court, J.A., Spurden, D., Lloyd, S., Ince, P.G. and Perry, E.K. (1996) Glutamate receptor binding in the human hippocampus and adjacent cortex during development and aging. *Neurobiol. Aging*, 17: 639–651.

Kälviäinen, R., Halonen, T., Pitkänen, A. and Riekkinen, P.J. (1993) Amino acid levels in the cerebrospinal fluid of newly diagnosed epileptic patients – Effect of vigabatrin and carbamazepine monotherapies. *J. Neurochem.*, 60: 1244–1250.

Lappalainen, R. and Riikonen, R.S. (1996) High levels of cerebrospinal fluid glutamate in Rett syndrome. *Pediatr. Neurol.*, 15: 213–216.

Lees, G.J. (1996) Therapeutic potential of AMPA receptor ligands in neurological disorders. *CNS Drugs* 5: 51–74.

Lloyd, K.G., Bossi, L., Morselli, P.L., Munari, C., Rougier, M. and Loiseau, H. (1986) Alterations of GABA-mediated synaptic transmission in human epilepsy. *Adv. Neurol.*, 44: 1033–1044.

Löscher, W. (1998) Pharmacology of glutamate receptor in the kindling model of epilepsy. *Prog. Neurobiol.*, 54: 721–741.

Lott, I.T., Coulombe, T., Di Paolo, R.V., Richardson, E.P., Jr. and Levy, H.L. (1978) Vitamin B6-dependent seizures: Pathology and chemical findings in brain. *Neurology*, 28: 47–54.

Lüders, H.O., Engel, J., Jr. and Munari, C. (1993) General Principles. In J. Engel Jr. (Ed.), Surgical Treatment of the Epilepsies, Second Ed. edn., Raven Press, New York pp. 137–153.

Masukawa, L.M., Higashima, M., Hart, G.J., Spencer, D.D. and O'Connor, M.J. (1991) NMDA receptor activation during epileptiform responses in the dentate gyrus of epileptic patients. *Brain Res.*, 562: 176–180.

Mathern, G.W., Leite, J.P., Babb, T.L., Pretorius, J.K., Kuhlman, P.A., Mendoza, D., Fried, I., Sakamoto, A.C., Assirati, J.A., Adelson, P.D. and Peacock, W.J. (1996) Aberrant hippocampal mossy fiber sprouting correlates with greater NMDAR2 receptor staining. *Neuroreport*, 7: 1029–1035.

McDonald, J.W., Garofalo, E.A., Hood, T., Sackellares, J.C., Gilman, S., McKeever, P.E., Troncoso, J.C. and Johnston, M.V. (1991) Altered excitatory and inhibitory amino acid receptor binding in hippocampus of patients with temporal lobe epilepsy. *Ann. Neurol.*, 29: 529–541.

Miele, M., Berners, M., Boutelle, M.G., Kusakabe, H. and Fillenz, M. (1996) The determination of the extracellular concentration of brain glutamate using quantitative microdialysis. *Brain Res.*, 707: 131–133.

Miller, H.P., Levey, A.I., Rothstein, J.D., Tzingounis, A.V. and Conn, P.J. (1997) Alterations in glutamate transporter protein levels in kindling-induced epilepsy. *J. Neurochem.*, 68: 1564–1570.

Moghaddam, B., Adams, B., Verma, A. and Daly, D. (1997) Activation of glutamatergic neurotransmission by ketamine –

A novel step in the pathway from nmda receptor blockade to dopaminergic and cognitive disruptions associated with the prefrontal cortex. *J. Neurosci.*, 17: 2921–2927.

Morrison, P.F., Bungay, P.M., Hsiao, J.K., Mefford, I.N., Dykstra, K.H. and Dedrick, R.L. (1991) Quantitative microdialysis. In T.E. Robinson and J.B. Justice Jr. (Eds.), Microdialysis in the Neurosciences, Elsevier, New York pp. 47–80.

Mosh, S.L. (1993) Seizures in the developing brain. *Neurology*, 43(Suppl. 5): S3-S7.

Mutani, R., Monaco, F., Durelli, L. and Delsedime, M. (1974) Free amino acids in the cerebrospinal fluid of epileptic subjects. *Epilepsia*, 15: 593–597.

Mutani, R., Monaco, F., Durelli, L. and Delsedime, M. (1975) Levels of free amino acids in serum and cerebrospinal fluid after administration of taurine to epileptic and normal subjects. *Epilepsia*, 16: 765–769.

O'Connor, E.R., Pizzonia, J.H., Spencer, D.D. and de Lanerolle, N.C. (1996) Glutamate, aspartate, and GABA transport in astrocytes cultured from human epileptic foci (abstract). *Epilepsia*, 37(Suppl. 5): 51.

Peeling, J. and Sutherland, G. (1993) Proton magnetic resonance spectroscopy of extracts of human epileptic neocortex and hippocampus. *Neurology*, 43: 589–594.

Perl, T.M., Bedard, L., Kosatsky, T., Hockin, J.C., Todd, E.C. and Remis, R.S. (1990) An outbreak of toxic encephalopathy caused by eating mussels contaminated with domoic acid. N. Engl. *J. Med.*, 322: 1775–1780.

Perry, T.L. and Hansen, S. (1981) Amino acid abnormalities in epileptogenic foci., *Neurology* 31: 872–876.

Petroff, O.A.C., Rothman, D.L., Behar, K.L. and Mattson, R.H. (1995) Initial observations on effect of vigabatrin on in vivo h-1 spectroscopic measurements of gamma-aminobutyric acid, glutamate, and glutamine in human brain. *Epilepsia*, 36: 457–464.

Petroff, O.A.C., Rothman, D.L., Behar, K.L. and Mattson, R.H. (1996) Low brain GABA and high glutamate correlate with poor seizure control in patients with complex partial epilepsy. Proceedings of the 3rd International Symposium on Basic Mechanisms of Epilepsy.

Rogawski, M.A. and Porter, R.J. (1990) Antiepileptic drugs: pharmacological mechanisms and clinical efficacy with consideration of promising developmental stage compounds. *Pharmacol. Rev.*, 42: 223–286.

Rogers, S.W., Andrews, P.I., Gahring, L.C., Whisenand, T., Cauley, K., Crain, B., Hughes, T.E., Heinemann, S.F. and McNamara, J.O. (1994) Autoantibodies to glutamate receptor GluR3 in Rasmussen's encephalitis. *Science*, 265: 648–651.

Ronne-Engström, E., Hillered, L., Flink, R., Spännare, B., Ungerstedt, U. and Carlson, H. (1992) Intracerebral microdialysis of extracellular amino acids in the human epileptic focus. *J. Cereb. Blood Flow Metab.*, 12: 873–876.

Rothstein, J.D., Dykes-Hoberg, M., Pardo, C.A., Bristol, L.A., Jin, L., Kuncl, R.W., Kanai, Y., Hediger, M.A., Wang, Y., Schielke, J.P. and Welty, D.F. (1996) Knockout of glutamate transporters reveals a major role for astroglial transport in excitotoxicity and clearance of glutamate. *Neuron*, 16: 675–686.

Scheyer, R.D., During, M.J., Hochholzer, J.M., Spencer, D.D., Cramer, J.A. and Mattson, R.H. (1994) Phenytoin concentrations in the human brain – An in vivo microdialysis study. *Epilepsy Res.*, 18: 227–232.

Segal, M.B. (1993) Extracellular and cerebrospinal fluids. *Journal of Inherited Metabolic Disease*, 16: 617–638.

Sherwin, A.L. and Van Gelder, N.M. (1986) Amino acid and catecholamine markers of metabolic abnormalities in human focal epilepsy. *Adv. Neurol.*, 44: 1011–1032.

Sherwin, A., Quesney, F., Gauthier, S., Olivier, A., Robitaille, Y., McQuaid, P., Harvey, C. and Van Gelder, N. (1984) Enzyme changes in actively spiking areas of human epileptic cerebral cortex. *Neurology*, 34: 927–933.

Sherwin, A., Robitaille, Y., Quesney, F., Olivier, A., Villemure, J., Leblanc, R., Feindel, W., Andermann, E., Gotman, J., Andermann, F., Ethier, R. and Kish, S. (1988) Excitatory amino acids are elevated in human epileptic cerebral cortex. *Neurology*, 38: 920–923.

Srinivasan, J., Richens, A. and Davies, J.A. (1995) The effect of the desglycinyl metabolite of remacemide hydrochloride (FPL 12495aa) and dizocilpine (MK-801) on endogenous amino acid release from mouse cortex. *Br. J. Pharmacol.*, 116: 3087–3092.

Strowbridge, B.W., Masukawa, L.M., Spencer, D.D. and Shepherd, G.M. (1992) Hyperexcitability associated with localizable lesions in epileptic patients. *Brain Res.*, 587: 158–163.

Sveinbjornsdottir, S., Sander, J.W.A.S., Upton, D., Thompson, P.J., Patsalos, P.N., Hirt, D., Emre, M., Lowe, D. and Duncan, J.S. (1993) The excitatory amino acid antagonist D-CPP-ene (SDZ EAA-494) in patients with epilepsy. *Epilepsy Res.*, 16: 165–174.

Swanson, T.H. (1995) The pathophysiology of human mesial temporal lobe epilepsy. *J. Clin. Neurophysiol.*, 12: 2–22.

Teitelbaum, J.S., Zatorre, R.J., Carpenter, S., Gendron, D., Evans, A.C., Gjedde, A. and Cashman, N.R. (1990) Neurologic sequelae of domoic acid intoxication due to the ingestion of contaminated mussels. *N. Engl. J. Med.*, 322: 1781–1787.

Troupin, A.S., Mendius, J.R., Cheng, F. and Risinger, M.W. (1986) MK 801. In B. Meldrum and R. Porter (Eds.), New anticonvulsant drugs, John Libbey, London, pp. 191–201.

Tursky, T., Lassánová, M., Sramka, M. and Nádvornák, P. (1976) Formation of glutamate and GABA in epileptogenic tissue from human hippocampus in vitro. Acta *Neurochir.*, (Suppl. 23): 111–118.

Uldall, P. and Hansen, F.J. (1993) Lamotrigine in Rett syndrome (abstract). *Epilepsia*, 34 (Suppl. 2): 158.

Van Gelder, N.M., Sherwin, A.L. and Rasmussen, T. (1972) Amino acid content of epileptogenic human brain: Focal versus surrounding regions. *Brain Res.*, 40: 385–393.

Van Gelder, N.M., Janjua, N.A., Metrakos, K., MacGibbon, B. and Metrakos, J.D. (1980) Plasma amino acids in 3/sec spike-wave epilepsy. *Neurochem. Res.*, 5: 659–671.

Waldmeier, P.C., Baumann, P.A., Wicki, P., Feldtrauer, J.J., Stierlin, C. and Schmutz, M. (1995) Similar potency of carbamazepine, oxcarbazepine, and lamotrigine in inhibiting the release of glutamate and other neurotransmitters. *Neurology*, 45: 1907–1913.

Wang, S.J., Huang, C.C., Hsu, K.S., Tsai, J.J. and Gean, P.W. (1996) Inhibition of n-type calcium currents by lamotrigine in rat amygdalar neurones. *Neuroreport*, 7: 3037–3040.

Westergaard, N., Sonnewald, U. and Schousboe, A. (1995) Metabolic trafficking between neurons and astrocytes – The glutamate glutamine cycle revisited. *Dev. Neurosci.*, 17: 203–211.

Williamson, A. and Spencer, D. (1995) Zinc reduces dentate granule cell hyperexcitability in epileptic humans. *Neuroreport*, 6: 1562–1564.

Williamson, A., Spencer, S.S. and Spencer, D.D. (1995) Depth electrode studies and intracellular dentate granule cell recordings in temporal lobe epilepsy. *Ann. Neurol.*, 38: 778–787.

Wilson, C.L., Maidment, N.T., Behnke, E.J., Shomer, M.H., Velasco, A.L., Engel, J., Jr. and Fried, I. (1994) Ictal and interictal correlation of in vivo microdialysis and neuronal activity of hippocampal formation and amygdala in complex partial epilepsy. *Epilepsia*, 35 (Suppl. 8): 6.

Wilson, C.L., Maidment, N.T., Shomer, M.H., Behnke, E.J., Ackerson, L., Fried, I. and Engel, J. (1996) Comparison of seizure related amino acid release in human epileptic hippocampus versus a chronic, kainate rat model of hippocampal epilepsy. *Epilepsy Res.*, 26: 245–254.

Wilson, C.L., Maidment, N.T., Behnke, E.J., Fried, I., Mathern, G. and Engel, J., Jr. (1997) Analysis of variables influencing temporal lobe neurotransmitter level in epilepsy patients (abstract). *Epilepsia*, 38(Suppl. 3): 178.

Wolf, H.K., Roos, D., Blümcke, I., Pietsch, T. and Wiestler, O.D. (1996) Perilesional neurochemical changes in focal epilepsies. *Acta Neuropathologica*, 91: 376–384.

O.P. Ottersen, I.A. Langmoen and L. Gjerstad (Eds.)
*Progress in Brain Research*, Vol 116


CHAPTER 24

# Glutamate receptors in epilepsy

Astrid G. Chapman

*Department of Clinical Neuroscience, Institute of Psychiatry, De Crespigny Park, Denmark Hill, London SE5 8AF, UK*

## Convulsant and anticonvulsant activities of glutamatergic ligands

### *Convulsant action of glutamatergic agonists*

Glutamate plays a key role in the initiation and spread of seizure activity. Focal administration of glutamate into different brain structures (Hayaishi, 1952), and focal or systemic administration of more potent glutamatergic agonists, such as NMDA, AMPA or kainic acid, cause convulsions in experimental animals *in vivo* and epileptiform activity in *in vitro* cell- and slice-preparations.

Glutamate exerts its excitatory action via ligand-gated ion channels (NMDA- and non-NMDA receptors) to increase sodium and calcium conductance as discussed elsewhere in this volume (Section III). Glutamate also acts on neuronal and glial metabotropic receptors. Metabotropic receptors can be categorized into 3 classes of receptors depending on pharmacological diversity and coupling to second messenger systems (Bruno et al., this volume). Activation of different classes of metabotropic receptors can result in both convulsant and anticonvulsant action as discussed below.

### *Anticonvulsant action of ionotropic glutamatergic antagonists*

The anticonvulsant properties of all classes of ionotropic glutamate receptor antagonists have been comprehensively reviewed recently (Rogawski, 1992; Chapman, 1995; Meldrum, 1995; Meldrum & Chapman, 1998b) and will be briefly summarized below.

### *NMDA receptor antagonists*

*i) Competitive NMDA antagonists*
Shortly after 2-amino-7-phosphonoheptanoic acid (AP7) and 2-amino-5-phosphonovaleric acid (AP5) were identified as potent selective NMDA receptor antagonists it was shown that these compounds were potent antiepileptic agents when given intracerebroventricularly (icv) in mice (Croucher et al., 1982; Meldrum et al., 1983). Competitive NMDA antagonists are particularly effective against models of reflex epilepsy. A single oral dose of D-CPPene can give 24–48 hours protection against photically-induced epileptic responses in the photosensitive baboon, *Papio papio*, without any apparent motor or sedative side effects (Patel et al., 1990). Competitive NMDA antagonists are also effective against chemically-induced seizures and maximal electroshock. They are rather less effective against fully-kindled seizures in rats, and the motor side effects are relatively more marked in kindled versus non-kindled animals (Hönack & Löscher, 1995) presumably reflecting the altered properties of NMDA receptors.

*ii) Channel site antagonists*

High affinity open-channel NMDA receptor blockers such as phencyclidine and dizocilpine (MK-801) are also potent anticonvulsants in reflex epilepsy and chemically-induced seizures. However, they produce abnormal patterns of motor activity including headweaving and other stereotypes and enhanced motor activity at doses less than the median anticonvulsant dose.

Low affinity open-channel NMDA receptor blockers include memantine, remacemide and its desglycinated metabolite, FPL 12495, and ADCI, ((±)-5-aminocarbonyl-10,11-dihydro-5H-dibenzo[a,d] cyclohepten-5,10-imine). These compounds all display anticonvulsant activity with a much better therapeutic ratio than the high-affinity uncompetitive antagonists (Rogawski et al, 1991; Rogawski, 1996).

*iii) Glycine site antagonists*

A large number of compounds that act as NMDA receptor antagonists by competing with glycine or D-serine at its recognition site on the NMDAR1 subunit have been described. These include L-687,414, MDL 104,653, and ACEA 1021 (5-nitro-6,7-dichloroquinoxalinedione). These compounds show a similar spectrum of anticonvulsant activity to competitive NMDA antagonists and a similar therapeutic ratio (Smith & Meldrum, 1992; Chapman et al., 1995; Singh et al., 1990).

*iv) Polyamine site/NR1/NR2B selective antagonists*

Ifenprodil and eliprodil can both potentiate and block NMDA mediated responses through action at a site sensitive to the polyamines, spermine and spermidine. Similar properties are shown by Ro 25,6981 and CP 101,606. These compounds act preferentially on recombinant NMDA receptors containing NR1/ NR2B (Avenet et al., 1997). They are relatively free from cognitive or motor side effects, but their anticonvulsant effects are variable.

*v) Downregulation of NMDA receptor subunits*

Antisense probes against the NMDAR1 subunit have been given icv to DBA/2 mice to test the effect of modest reductions in NMDA receptor density on sound-induced seizures. Protection is seen 7–24 hours after probe administration when protein expression as judged by [3H]MK-801 binding is reduced by approximately 20% (Chapman et al., 1996). In a separate study, the icv administration of an antisense probe to NMDAR1 caused a significantly higher latency to NMDA-induced seizures in Swiss mice and a 25% reduction in the binding of the competitive NMDA antagonists, CGS 19755 and CGP 39653, but no alteration in the binding at the glycine site (Zapata et al., 1997).

*Clinical experience with glutamatergic antagonists*

In spite of the broad range of positive preclinical data with NMDA antagonists, to date no clinical trial with a pure NMDA receptor antagonist has given evidence of therapeutic benefit (Taubøll and Gjerstad, this volume). This applies to trials of dextromethorphan, dizocilpine and D-CPPene. The latter drug given as add-on therapy to drug resistant candidates for neurosurgery yielded no therapeutic benefit, but exacerbation of neurological and cognitive side effects was seen (Sveinbjornsdottir et al., 1993).

*AMPA/Kainate antagonists*

As AMPA receptors are responsible for fast excitatory transmission throughout the brain it might be expected that their use as anticonvulsants would be associated with unacceptable motor and cognitive side effects. Surprisingly, the preclinical data indicates that this is not the case (Chapman et al., 1991; Rogawski, 1996). Two principal classes of compounds have been tested. These are either quinoxalinedione derivatives (that are competitive glutamate antagonists at AMPA and kainate receptors, such as NBQX and YM90K) or 2,3-benzodiazepines (such as GYKI 52466 and GYKI 53655, that are allosteric antagonists with significant selectivity for AMPA versus kainate receptors). Both groups of compounds are active against reflex epilepsies in rodents and primates with a therapeutic index comparable to that of some established anticonvulsant drugs (Smith et al., 1991; Chapman et al., 1991). They are also effective against electroshock and chemically-induced seizures (Yamaguchi et al., 1993). Against fully-kindled seizures they have modest activity (Dürmüller et al., 1994). No pure AMPA antagonist has entered clinical trial for epilepsy. Barbiturates, however, show AMPA antagonist activity at concentrations similar to those enhancing GABA-mediated inhibition (Simmonds & Horne, 1987). Topiramate also shows both AMPA antagonism and effects on sodium channels and GABA receptors (White et al., 1997; see also Taubøll and Gjerstad, this volume).

*Metabotropic glutamate receptors*

The functions of glutamate metabotropic receptors (mGluRs) are clearly modified in chronic epilepsy (see Kindling section below). It is likely that they play a significant role in seizures and in epileptogenesis. The pharmacological tools for analyzing these effects have been very inadequate, but progress has been made in the last 2–3 years (Bedingfield et al., 1995; Conn & Pin, 1997).

*i) Classification of mGluRs*

The 8 mGluRs so far cloned and sequenced, fall into three groups on the basis of their sequence homology, second messenger effectors and pharmacology (There may be a fourth group related to phospholipase D activation) (Pin & Duvoisin, 1995; Conn & Pin, 1997). Thus Group I comprises mGluR1 and mGluR5, which are G-protein linked to activation of phospholipase C with diacylglycerol and IP3 as the intracellular second messengers. Group II comprises mGluR2 and mGluR3 which are negatively linked to adenylyl cyclase activation. Group III comprises mGluR4, mGluR6, mGluR7, mGluR8, and are also negatively coupled to adenylyl cyclase.

*ii) Convulsant action of agonists for Group I mGluRs*

The icv or focal cerebral injection of agonists acting on Group I receptors leads to epileptic activity with clinical manifestations that depend on the site of injection (Tizzano et al., 1995; Ghauri et al., 1996). This can be confidently linked to Group I activation in the case of 3,5-dihydroxyphenylglycine, which is selective for this Group. Group I agonists enhance neuronal excitability by several mechanisms, including blockade of accommodation to a steady current, potentiation of the effects of NMDA and AMPA and depolarization (resulting from a decrease in a calcium activated potassium current and an activation of sodium/calcium exchange).

*iii) Anticonvulsant action of Group I antagonists*

The potent Group I antagonist (S)4-carboxy-3-hydroxyphenylglycine is anticonvulsant against sound-induced and chemically-induced seizures in mice (Thomsen et al., 1994). Some other group I antagonists are also anticonvulsant in rodent models (e.g. AIDA = (RS)-1-aminoindan-1,5-dicarboxylic acid and LY 367385; Chapman, Meldrum, Tang, Yip, unpublished).

*iv) Group II agonists and antagonists*

Several anticonvulsant effects of Group II agonists have been described (Tizzano et al., 1995), but in some test systems transient proconvulsant effects precede prolonged anticonvulsant effects (Tang et al., 1997). The available Group II antagonists have failed to yield a consistent picture.

*v) Group III agonists and antagonists*

Two group III selective agonists, L-AP4 (L(+)-2-amino-4-phosphonobutyric acid) and L-SOP (L-serine-O-phosphate) have been tested in a variety of models and shown to possess anticonvulsant properties (Ghauri et al., 1996; Abdul-Ghani et al., 1997) although here also transient excitatory effects may be observed. Group III antagonists also fail to give a consistent picture, but one antagonist that is active in biochemical assays, rather than in electrophysiological ones, MCPA ((S)-α-methyl-3-carboxyphenylalanine) is anticonvulsant (Ghauri et al., 1996).

## Glutamatergic parameters in the kindling model of limbic seizures

Development of limbic seizes in rats by periodic electrical stimulation or 'kindling' of the amygdala is probably the most thoroughly documented animal seizure model. Both the kindling process (epileptogenesis) itself and the chronic, fully kindled limbic seizures (considered a close model of complex partial seizures in man, with which it shares many morphological, electrophysiological and pharmacological features) are suppressed by different classes of NMDA- and AMPA antagonists as summarized below. This section will be devoted to a description of glutamatergic parameters (levels and release of excitatory amino acid transmitters, ionotropic and metabotropic glutamatergic receptor responses, levels, subunit composition and editing, and levels of glutamate transporters) that are altered during the kindling process. Whenever similar glutamatergic alterations have been described in other chronic or genetic experimental seizure models, or in patients with epilepsy (mainly temporal lobe epilepsy) these will be addressed in the appropriate connection.

### *Inhibition of the kindling process by NMDA antagonists and metabotropic ligands*

NMDA antagonists are well known to block LTP in hippocampal slice preparations and in vivo. They also block the kindling process when given focally or systemically just prior to each kindling stimulus. All classes of NMDA-receptor antagonists (competitive-, channel-site-, and glycine-site antagonist) have this effect, whereas AMPA/kainate antagonists (both competitive, NBQX, or non-competitive, GYKI 52466, antagonists) provide a much weaker inhibition of the kindling process (see Chapman, 1995).

Metabotropic ligands can also affect the kindling process. The Group I antagonist, MCPG, the Group II agonist, 1S,3S-ACPD, and the Group III agonists, L-AP4, shorten ADD and delay kindling, (Beach et al., 1995; Attwell et al., 1995; Abdul-Ghani et al., 1997) in agreement with the anti-epileptic action of Group I antagonists and Groups II and III agonists demonstrated above for other seizure models.

### *Inhibition of fully kindling seizures by NMDA and non-NMDA-antagonists and metabotropic ligands*

Both competitive-(AP5, AP7, CGP 37849, CPPene) and channel-site-NMDA-antagonists (MK 801) are relatively weak anticonvulsants against fully kindled seizures in rats. The doses required to reduce ADD and suppress motor-seizures are approximately 50 to 150 fold higher than those required to block generalized reflex-induced seizures in rodents (see Chapman, 1995). Furthermore, kindled rats appear much more sensitive to the behavioral side-effects of the competitive NMDA antagonists, such as CPPene (Hönack & Löscher, 1995) perhaps related to the similar enhanced sensitivity of patients with drug-resistant complex partial seizures to CPPene-induced cognitive impairment relative to the effects seen in healthy volunteers (Sveinbjornsdottir et al., 1993).

Glycine-site NMDA-antagonists, such as MDL 104,653) have a better therapeutic profile against fully kindled seizures, and both competitive and non-competitive AMPA/kainate antagonists are relatively potent against fully kindled seizures (see Chapman, 1995).

The non-selective metabotropic agonist 1S,3R-ACPD has a delayed (24 hours) protection against kindled seizures, and the Group III agonist, L-AP4, shows a very prolonged (1–3 days) protection against fully kindled seizures (Attwell et al., 1995; Suzuki et al., 1996).

### *Glutamatergic responses in kindled rats*

Mody and Heinemann (1987) provided the earliest demonstration of an enhanced excitation of the hippocampal NMDA-system in kindled rats by showing that NMDA-induced calcium influx had extended from the pyramidal cell layer in control rats to include the dendritic layer of CA1 in kindled rats. A similar enlargement of the area showing enhanced susceptibility to NMDA-induced calcium influx has been demonstrated in human cortical epileptic tissue (Louvel & Pumain, 1992) and in sensory motor cortex of a genetic strain of rats with absence-like epilepsy (GAERS; Pumain et al., 1992).

There is also a decrease in the channel blocking effect of magnesium on NMDA receptors in dentate granule cells isolated from hippocampi of kindled rats (Köhr et al., 1993; Köhr & Mody, 1994).

There is a long-lasting (>4 weeks) functional enhancement of group I metabotropic receptors (quisqualate or 1S,3R-ACPD-evoked phosphoinositide hydrolysis) in the amygdala and pyriform cortex of kindled rats, as well as a corresponding transient enhancement of hippocampal Group I metabotropic activity (Akiyama et al., 1992a). Consistent with this increases in depolarizing responses to 1S,3R-ACPD have been described in the contralateral amygdala in amygdala-kindled rats (Holmes et al., 1996).

The presynaptic inhibitory action of Group III agonists (L-SOP, LAP4) on the lateral perforant path input to dentate granule cells is markedly reduced both in hippocampal slices from kindled rats (Klapstein et al., 1997) and in temporal lobectomy specimens from patients with complex partial seizures with Ammon's horn sclerosis. (Kral et al., 1997). These changes could contribute to the epileptogenic process. In contrast an enhanced sensitivity of inhibitory presynaptic metabotropic receptors (Groups II and III) has been described in the contralateral amygdala in amygdala kindled rats (Neugebauer et al., 1997).

### *Enhanced release of glutamate and aspartate in kindled rats*

There has been a general expectation – partly based on parallel observations following brain ischemia, partly based on the implications of an excitatory mechanism of seizure induction – that there should be an increased release (and a concomitant rise in extracellular levels) of glutamate and aspartate during epileptic seizures. Indeed, this has been shown to be the case during spontaneous seizures in ambulatory patients who are candidates for temporal lobectomy for their drug-resistant seizures and are undergoing a telemetry evaluation of their seizure focus. In this group of patients microdialysis probes can be implanted in the hippocampus in association with depth EEG-electrodes. Several centers utilizing such a setup have shown that spontaneous seizures in ambulatory patients, and evoked seizures in intra-operative patients, are associated with marked, transient, bilateral increases (larger in the epileptic than in the contralateral hemisphere) in the extracellular hippocampal levels of aspartate and glutamate (Ronne-Engström et al., 1992; During & Spencer, 1993; Wilson et al., 1996), as discussed in more detail by Scheyer in this volume.

In experimental animal seizure models, it has been more difficult to establish a consistent link between seizures and elevated extracellular glutamate (and aspartate) levels. There are many examples of acute, evoked experimental seizures occurring in the absence of any detectable rise in extracellular levles of glutamate, even when the experimental conditions are optimized to detect very transient changes and reuptake has been suppressed with uptake inhibitors (see Chapman, 1997), and a correlation between high extracellular glutamate and seizure activity has been disputed (Obrenovitch et al., 1996). However, in chronic seizure models (e.g. kindled seizures, spontaneous, recurrent seizures following an incident of kainate-induced status epilepticus, and in genetically epilepsy prone rats (GEPRs) there are consistent enhancements of extracellular levels of glutamate and aspartate observed in the hippocampus, amygdala and substantia nigra during seizures (Lothman et al., 1987; Lehmann, 1989; Jarvie et al., 1990; Zhang et al., 1991; Doretto et al., 1994; Kaura et al., 1995; Ueda et al., 1995; Wilson et al., 1996; Rowley et al., 1997; Liu et al., 1997). On the other hand, the baseline hippocampal and cortical tissue levels of glutamate and aspartate have been reported as either normal in kindled brains assessed 1 week (Leach et al., 1985) or 5 weeks (Löscher et al., 1993) after the last kindled seizure, or elevated (Lothman et al., 1987) 1 week after the last kindled seizure.

*Glutamate transporter changes in kindled rats*

The neuronal (EAAC-1) and glial (GLT-1 and GLAST) glutamate (and aspartate) transporters facilitate excitatory transmitter reuptake. A down-regulation of glutamate transporter levels would be expected to contribute to enhanced excitatory activity in kindled rats or other seizure models. Transgenic mice lacking the glutamate transporter, GLT-1, exhibit spontaneous epileptic activity in agreement with such a hypothesis (Tanaka et al., 1997), and an early study (Leach et al., 1987) reported a marked reduction in in vitro $[H]^3$-D-aspartate uptake in slices from kindled rats. No change was found in the hippocampal levels of GLT-1 and GLAST mRNA in amygdala-kindled rats at 4 weeks after the last evoked seizure (Akbar et al., 1997). Studies from Rothstein and coworkers (Miller et al., 1997) confirm that GLT-1 levels are unchanged in multiple brain regions when examined 24 hours or 1 month following the last evoked seizure in fully kindled rats, while GLAST levels are significantly reduced in the piriform cortex/amygdala when examined 24 hours after the last evoked seizure, but recover to control levels 1 month later. Unexpectedly, the levels of EAAC-1 are transiently (24 hours) *increased* in the piriform cortex, amygdala and hippocampus of kindled rats (Miller et al., 1997). Previous work from the same laboratory has shown that prolonged (1 week) in vivo icv administration of antisense oligodeoxynucleotide (ODN) probes to GLT-1, GLAST or EAAC-1 has diverging effects on transporter protein levels and transporter activity, extracellular glutamate levels and motor function and epilepsy in the striatum of rats, with only EAAC-1 antisense treatment causing epilepsy in the rats (Rothstein et al., 1996). The epileptogenic effect of chronic (1 week) EAAC-1 antisense treatment was associated with a 78% reduction in EAAC-1 protein level, a 22% decrease in functional striatal transport and no change in extracellular striatal glutamate levels. Antisense probes to GLT-1 and GLAST caused hindlimb paralysis and other motor impairment in the rats, but only 5–10% incidence of seizures. This was associated with marked decreases in transporter protein level (58% decrease in GLT-1; 84% decrease in GLAST) and functional transport (56% decrease with GLT-1 antisense; 35% decrease with GLAST antisense) and 13 to 32 fold increases in extracellular striatal glutamate levels (Rothstein et al., 1996).

Recently, it has been shown that 3 models of chemically induced *status epilepticus* in rats (induced by kainic acid, pentylenetetrazol or pilocarpine) are associated with decreased levels of hippocampal GLT-1, but not of GLAST. (Sutherland et al., 1997). GLT-1 mRNA expression is transiently reduced, followed by more prolonged decreases in GLT-1 protein level.

*Altered expression of NMDA receptors in kindled rats*

Attempts at quantifying glutamatergic receptor levels in control and kindled rat brain from the mid 1980s to the early 1990s usually employed radioactively labelled receptor ligand binding to regional membrane preparations or to slices subjected to autoradiography. The majority of studies detected no differences (Jones & Johnson, 1989; Chaudieu et al., 1991; Akiyama et al., 1992b) between control and kindled rat brains, or reduced binding in kindled brains (Savage et al., 1984; Sircar et al., 1987; Okazaki et al., 1989) regarding overall NMDA and AMPA/kainate receptor levels in assorted limbic areas when assessed 24 hours to 6 weeks after the last kindled seizure. A unilateral upregulation of NMDA- and AMPA-receptors in ipsilateral hippocampus and cortex of kindled rats was also reported 5 days after the last kindled seizure (Cincotta et al., 1991).

The corresponding glutamatergic receptor level data in resected human temporal lobe tissue is likewise diverse with reports of increased AMPA receptor levels in dentate gyrus, and increased levels of NMDA and kainate binding in entorhinal cortex, and relative preservation of NMDA receptors in dentate gyrus (Geddes et al., 1990; McDonald et al., 1991; Hosford et al., 1991; Baird et al., 1997).

There have been numerous recent reports of alterations in NMDA receptor levels and composition in kindled rat brain or other forms of acute or chronic seizures. So far, we do not have sufficient information about the exact effect of induced gene expression, the reciprocal interaction of the NMDA system with an abundance of regulatory gene products, NMDA subunit composition, slice variants or subunit turnover, receptor editing, phosphorylation, desensitization, and membrane anchoring, to incorporate all the reported changes into a coherent model of an activated NMDA receptor system, or indeed if the alterations contribute to, or are a consequence of the epileptogenesis. However, a picture is gradually emerging concerning the sensitivity of different heteromeric recombinant NMDA receptors to modulation by ligand binding, ionic gradients, pH and redox changes and to direct interaction with other modulatory proteins.

The prominent effect of direct and indirect calcium effects on the glutamatergic receptor systems illustrates the reciprocal complexity of the system. It is known that a mutant mouse strain lacking the α-subunit of a calcium dependent calmodulin kinase II has limbic seizures (Butler et al., 1995). A substantial inactivation of NMDA by direct interaction of calmodulin with NR1 subunit of the NMDA receptor has been reported (Ehlers et al., 1996). Furthermore there is a phosphorylation-dependent regulation of NMDA by calmodulin (Hisatsune et al., 1997), calcium-modulation of NMDA is affected by alternate splicing in NR1 (Koltchine et al., 1996; Kraus et al., 1996), and calcium-dependent inactivation of recombinant NMDA is NR2 specific (Krupp et al., 1996). Therefore, the number of alterations in NMDA expression and composition observed in kindling and summarized below may contribute to the enhanced excitability during kindling, even though the precise mechanism is not yet known.

Alternate splicing at C-terminal of NMDAR1 receptor is altered in kindled rats (Vezzani et al., 1995) and kindling induces a transient reduction of specific splice isoforms of NR1 (Kraus et al., 1996). There are altered NMDA subunit mRNA in rat supraoptic nucleus (Al-Ghoul et al., 1997) and in rat cortex (Kikuchi et al., 1996) during kindling.

Although there was no change in hippocampal mRNA for NR1, NR2A, NR2B, NR2C, NR2D when assessed at 24 hours or 4 weeks after the last kindled seizure, there was, however, a 2.8 fold increase in CPP binding, but not CGS19755, at 4 weeks (Kraus et al., 1994).

In fully kindled rats there is a transient, marked reduction of NR1 mRNA 0–4 hours post kindled seizure, and an increases at 4 hours in NR2A and NR2B mRNA levels, which have returned to normal by 12 hours (Pratt et al., 1993).

Related information in other seizure models include a rapid (1–2 hours) upregulation of NMDAR1 (mRNA and protein) following pentylenetetrazol seizures in rats, followed by a depression (6–24 hours) in receptor protein levels. The rapid upregulation occurred at a time when NGFI-A (that has the capability of binding to the promoter region of NMDAR1) is expected to be induced (Jensen et al., 1997). Repeated electroshocks in rats had no effect on NMDAR1 mRNA in rat hippocampus, when assessed 24 hours after the last electroshock (Naylor et al., 1996), and there is a reported overexpression of NMDAR1 subunit in inferior colliculus of a strain of mice susceptible to sound-induced seizures (Marianowski et al., 1995).

There is an altered distribution and loss of NMDAR1 subunit mRNA in resected hippocampal tissue from chronic temporal lobe epilepsy, but only as a consequence of cell loss (Bayer et al., 1995; Blümcke et al., 1996).

*Altered expression of AMPA/kainate receptors in kindled rats*

Most of the interest in possible AMPA receptor modification in connection with epileptogenisis has been focused on the presence of GluR2 subunit in heteromeric AMPA receptors, since its presence produces a calcium impermeable receptor assembly (see Pellegrini-Giampietro et al., 1997). Kainate-induced status epilepticus causes a down-regulation of hippocampal GluR2 (the 'expected' result during hippocampal excitation), and not in GluR1 expression (Friedman et al., 1994).

Amygdala-kindling, likewise, causes a selective, transient (at 24 hours; return to control levels by 4 weeks) down-regulation of GluR2 receptor protein in the limbic forebrain, piriform cortex and amygdala (but not in hippocampus), with no effect on the levels of GluR1 and GluR4 in any of the regions studied (Prince et al., 1995). In the hippocampus a transient decrease in mRNA for GluR2 has been observed, but only in connection with similar decreases in the mRNA levels for the other AMPA subunits (Lee et al., 1994). There is no change in mRNA editing for any of the AMPA/kainate subunits 24 hours after last kindled seizure (Kamphuis et al., 1995), while hippocampal GluR1 flip expression is enhanced 4 weeks after last kindled seizure (Kamphuis et al., 1994). The functional significance of these latter changes is not established.

*Altered expression of metabotropic glutamate receptors in kindled rats*

mGluR1 receptor levels are transiently enhanced 24 hours after the last fully kindled seizure, while mGluR5 receptor levels are decreased at the same time point (Akbar et al., 1996).

## Transgenic mouse models with seizure disorders

Glutamatergic receptors and transporters have been considered likely 'candidate genes' for mutations responsible for epileptic seizures in man or genetically seizure susceptible rodents. However, none of the established mutations underlying naturally occurring epileptic syndroms in rodents or man directly involve the glutamatergic system. Rather, mutations in ion channels (voltage-gated calcium-, potassium-, and sodium/hydrogen exchange-channels), nicotine acetylcholine receptor α4 subunit, and the protease inhibitor, cystatin B, represent known genetic defects in epilepsy (see Allen & Walsh, 1996; Meldrum & Chapman, 1998a).

Transgenic mice lacking genes for ionotropic or metabotropic glutamatergic receptor subunits have been studied, and some of these strains exhibit seizures, as briefly summarized below.

*Transgenic mice with NMDA receptor deficiencies*

The global elimination of the NMDA NR1 subunits (NMDAR1 knockouts) causes neonatal death in mice (Forrest et al., 1994). However, specific gene knockout can be restricted to a particular anatomical subregion, for instance NMDAR1 knockouts in the hippocampal CA1 region. These mice grow apparently normally, but display impaired plasticity and learning, but no seizures, at least not in this particular hippocampal domain (Wilson & Tonegawa, 1997). Combined gene disruption of NMDA-NR2A and NR2C causes motor discoordination, but no seizures (no effect of NR2A or NR2C knockout alone) (Nakanishi et al., 1996). Mice with a synapse-selective impairment of NMDA subunits NR2A or NR2B likewise have no seizures (Ito et al., 1997)

*Transgenic mice with AMPA/kainate receptor deficiencies*

Transgenic mice with an editing-deficient GluR2 display early onset of epilepsy (Brusa et al., 1995).

Over-expression by direct hippocampal injection of a HSV-1 vector expressing the kainate receptor GluR6 into hippocampus causes immdiate and prolonged seizure activity in rats (During et al., 1993). A similar over-expression of the AMPA subunit GluR1, but not GluR2, into deep prepiriform cortex of rats facilitates seizures (Kondratyev et al., 1997).

*Transgenic mice with glutamate transporter deficiencies*

Glutamate transporter GLT-1 knockout mice displays epilepsy (Tanaka et al., 1997).

*Transgenic mice with metabotropic glutamate receptor deficiencies*

There are a number of transgenic mice lacking genes for different metabotropic glutamate receptors, especially mGluR1 and mGluR5 knockout. On the whole, these strains display impaired plasticity, long term potentiation and learning, but not seizure disorders (Bordi et al., 1997; Ferraguti et al., 1997; Lu et al., 1997).

## Summary

Glutamatergic synapses play a critical role in all epileptic phenomena. Broadly enhanced activation of post-synaptic glutamate receptors (ionotropic and metabotropic) is proconvulsant.

Antagonists of NMDA receptors and AMPA receptors are powerful anticonvulsants in many animal models of epilepsy. A clinical application of pure specific glutamate antagonists has not yet been established.

Many different alterations in glutamate receptors or transporters can potentially contribute to epileptogenesis. Several genetic alterations have been shown to be epileptogenic in animal models but no specific mutation relating to glutamatergic function has yet been linked to a human epilepsy syndrome. There is clear evidence for altered NMDA receptor function in acquired epilepsy in animal models and in man. Changes in metabotropic receptor function may also play a key role in epileptogenesis.

## References

Abdul-Ghani, A.S., Attwell, P.J.E., Kent, N.S., Bradford, H.F., Croucher, M.J. and Jane, D.E. (1997) Anti-epileptogenic and anticonvulsant activity of L-2-amino-4-phosphonobutyrate, a presynaptic glutamate receptor agonist. *Brain Res.*, 755: 202–212.

Akbar, M.T., Rattray, M., Powell, J.F. and Meldrum, B.S. (1996) Altered expression of group I metabotropic glutamate receptors in the hippocampus of amygdala-kindled rats. *Mol. Brain Res.*, 43: 105–116.

Akbar, M.T., Torp, R., Danbolt, N.C., Levy, L.M., Meldrum, B.S. and Ottersen, O.P. (1997) Expression of glial glutamate transporters GLT-1 and GLAST is unchanged in the hippocampus in fully kindled rats. *Neuroscience*, 78: 351–359.

Akiyama, K., Daigen, A., Yamada, N., Itoh, T., Kohira, I., Ujike, H. and Otsuki, S. (1992a) Long-lasting enhancement of metabotropic excitatory amino acid receptor-mediated polyphosphoinositide hydrolysis in the amygdala/pyriform cortex of deep prepiriform cortical kindled rats. *Brain Res.*, 569: 71–77.

Akiyama, K., Yoneda, Y., Ogita, K., Itoh, T., Daigen, A., Sora, I., Kohira, I., Ujike, H. and Otsuki, S. (1992b) Ionotropic excitatory amino acid receptors in discrete brain regions of kindled rats. *Brain Res.*, 587: 73–82.

Al-Ghoul, W.M., Meeker, R.B. and Greenwood, R.S. (1997) Amygdala kindling alters *N*-methyl-D-aspartate receptor subunit messenger RNA expression in the rat supraoptic nucleus. *Neuroscience*, 77: 985–992.

Allen, K.M. and Walsh, C. (1996) Shaking down new epilepsy genes. *Nature Med.*, 2: 516–518.

Attwell, P.J.E., Kaura, S., Sigala, G., Bradford, H.F., Croucher, M.J., Jane, D.E. and Watkins, J.C. (1995) Blockade of both epileptogenesis and glutamate release by (1S,3S)-ACPD, a presynaptic glutamate receptor agonist. *Brain Res.*, 698: 155–162.

Avenet, P., Léonardon, J., Besnard, F., Graham, D., Depoortere, H. and Scatton, B. (1997) Antagonist properties of eliprodil and other NMDA receptor antagonists at rat NR1A/NR2A and NR1A/NR2B receptors expressed in Xenopus oocytes. *Neurosci. Lett.*, 223: 133–136.

Baird, V.H., Bowery, N.G., Van Paesschen, W., Hand, K.S., Ravesz, T., Harkness, W. and Duncan, J.S. (1997) NMDA-receptor binding and neuron densities in hippocampus of patients with temporal-lobe epilepsy. *Epilepsia*, 38 (Suppl. 8): 4. (Abstract)

Bayer, T.A., Wiestler, O.D. and Wolf, H.K. (1995) Hippocampal loss of *N*-methyl-D-aspartate receptor subunit 1 mRNA in chronic temporal lobe epilepsy. *Acta Neuropathol.* (*Berl.*), 89: 446–450.

Bedingfield, J.S., Kemp, M.C., Jane, D.E., Tse, H.W., Roberts, P.J. and Watkins, J.C. (1995) Structure-activity relationships for a series of phenylglycine derivatives acting at metabotropic glutamate receptors (mGluRs). *Br. J. Pharmacol.*, 116: 3323–3329.

Blümcke, I., Beck, H., Scheffler, B., Hof, P.R., Morrison, J.H., Wolf, H.K., Schramm, J., Elger, C.E. and Wiestler, O.D. (1996) Altered distribution of the α-amino-3-hydroxy-5-methyl-4-isoxazole propionate receptor subunit GluR2(4) and the *N*-methyl-D-aspartate receptor subunit NMDAR1 in the hippocampus of patients with temporal lobe epilepsy. *Acta Neuropathol.* (*Berl.*), 92: 576–587.

Bordi, F., Reggiani, A. and Conquet, F. (1997) Regulation of synaptic plasticity by mGluR1 studied in vivo in mGluR1 mutant mice. *Brain Res.*, 761: 121–126.

Bruno, V., Battaglia, G., Copani, A., Casabona, G., Storto, M., Di Giorgio Gerevini, V., Ngomba, R. and Nicoletti, F. (1998) Chapter 14 of this volume, 209–222.

Brusa, R., Zimmermann, F., Koh, D.S., Feldmeyer, D., Gass, P., Seeburg, P.H. and Sprengel, R. (1995) Early-onset epilepsy and postnatal lethality associated with an editing-deficient GluR-B allele in mice. *Science*, 270: 1677–1680.

Butler, L.S., Silva, A.J., Abeliovich, A., Watanabe, Y., Tonegawa, S. and McNamara, J.O. (1995) Limbic epilepsy in transgenic mice carrying a $Ca^{2+}$/calmodulin-dependent kinase II (α-subunit mutation. *Proc. Natl. Acad. Sci. USA*, 92: 6852–6855.

Chapman, A.G. (1995) Excitatory neurotransmission and antiepileptic drug development: A status report. In: Recent Advances in Epilepsy, Vol. 6. T.A. Pedley and B.S. Meldrum, eds. Churchill Livingstone, Edinburgh, pp. 1–21.

Chapman, A.G. (1997) Microdialysis. In: Epilepsy – A comprehensive textbook. J.Jr. Engel and T.A. Pedley, eds. Raven Press, New York, pp. 1067–1072.

Chapman, A.G., Dürmüller, N., Harrison, B.L., Baron, B.M., Parvez, N. and Meldrum, B.S. (1995) Anticonvulsant activity of a novel NMDA/glycine site antagonist, MDL 104,653, against kindled and sound-induced seizures. *Eur. J. Pharmacol.*, 274: 83–88.

Chapman, A.G., Smith, S.E. and Meldrum, B.S. (1991) The anticonvulsant effect of the non-NMDA antagonists, NBQX and GYKI 52466, in mice. *Epilepsy Res.*, 9: 92–96.

Chapman, A.G., Woodburn, V.L., Woodruff, G.N. and Meldrum, B.S. (1996) Anticonvulsant effect of reduced NMDA receptor expression in audiogenic DBA/2 mice. *Epilepsy Res.*, 26: 25–35.

Chaudieu, I., Allaoua, H., Bellaidi, M., Rondouin, G., Kamenka, J.M. and Chicheportiche, R. (1991) Effect of amygdaloid kindling on the high- and low-affinity [3H]TCP binding sites of the rat CNS. *Neurosci. Lett.*, 131: 263–266.

Cincotta, M., Young, N.A. and Beart, P.M. (1991) Unilateral up-regulation of glutamate receptors in limbic regions of amygdaloid-kindled rats. *Exp. Brain Res.*, 85: 650–658.

Conn, P.J. and Pin, J.P. (1997) Pharmacology and function of metabotropic glutamate receptors. *Annu. Rev. Pharmacol. Toxicol.*, 37: 205–237.

Croucher, M.J., Collins, J.F. and Meldrum, B.S. (1982) Anticonvulsant action of excitatory amino acid antagonists. *Science*, 216: 899–901.

Doretto, M.C., Burger, R.L., Mishra, P.K., Garcia-Cairasco, N., Dailey, J.W. and Jobe, P.C. (1994) A microdialysis study of amino acid concentrations in the extracellular fluid of the substantia nigra of freely behaving GEPR-9s: Relationship to seizure predisposition. *Epilepsy Res.*, 17: 157–165.

During, M.J., Mirchandani, G.R., Leone, P., Williamson, A., de Lanerolle, N.C., Geschwind, M.D., Bergold, P.J. and Federoff, H.J. (1993) Direct hippocampal injection of a HSV-1 vector expressing GLUR6 results in spontaneous seizures, hyperexcitability in CA1 cells and loss of CA1, hilar and CA3 neurons. *Soc. Neurosci. Abstr.*, 19: 21 (Abstract).

During, M.J. and Spencer, D.D. (1993) Extracellular hippocampal glutamate and spontaneous seizure in the conscious human brain. *Lancet*, 341: 1607–1610.

Dürmüller, N., Craggs, M. and Meldrum, B. (1994) The effect of the non-NMDA antagonists GYKI 52466 and NBQX and the competitive NMDA receptor antagonist D-CPPene on the development of amygdala kindling and on amygdala-kindled seizures. *Epilepsy Res.*, 17: 167–174.

Ehlers, M.D., Zhang, S., Bernhardt, J.P. and Huganir, R.L. (1996) Inactivation of NMDA receptors by direct interaction of calmodulin with the NR1 subunit. *Cell*, 84: 745–755.

Ferraguti, F., Pietra, C., Valerio, E., Corti, C., Chiamulera, C. and Conquet, F. (1997) Evidence against a permissive role of the metabotropic glutamate receptor 1 in acute excitotoxicity. *Neuroscience*, 79: 1–5.

Forrest, D., Yuzaki, M., Soares, H.D., Ng, L., Luk, D.C., Sheng, M., Stewart, C.L., Morgan, J.I., Connor, J.A. and Curran, T. (1994) Targeted disruption of NMDA receptor 1 gene abolishes NMDA response and results in neonatal death. *Neuron*, 13: 325–338.

Friedman, L.K., Pellegrini-Giampietro, D.E., Sperber, E.F., Bennett, M.V.L., Moshé, S.L. and Zukin, R.S. (1994) Kainate-induced status epilepticus alters glutamate and GABAA receptor gene expression in adult rat hippocampus: An in situ hybridization study. *J. Neurosci.*, 14: 2697–2707.

Geddes, J.W., Cahan, L.D., Cooper, S.M., Kim, R.C., Choi, B.H. and Cotman, C.W. (1990) Altered distribution of excitatory amino acid receptors in temporal lobe epilepsy. *Exp. Neurol.*, 108: 214–220.

Ghauri, M., Chapman, A.G. and Meldrum, B.S. (1996) Convulsant and anticonvulsant actions of agonists and antagonists of group III mGluRs. *Neuroreport* 7: 1469–1474.

Hayashi, T. (1952) A physiological study of epileptic seizures following cortical stimulation in animals and its application to human clinics. *Jap. J. Physiol.*, 3: 46–64.

Hisatsune, C., Umemori, H., Inoue, T., Michikawa, T., Kohda, K., Mikoshiba, K. and Yamamoto, T. (1997) Phosphorylation-dependent regulation of *N*-methyl-D-aspartate receptors by calmodulin. *J. Biol. Chem.*, 272: 20805–20810.

Holmes, K.H., Keele, N.B. and Shinnick-Gallager, P. (1996) Loss of metabotropic glutamate receptor (mGluR)-mediated hyperpolarizations and increase in mGluR depolarizations in basolateral amygdala neurons in kindling-induced epilepsy. *J. Neurophysiol.*, 76: 1208–1212.

Hosford, D.A., Crain, B.J., Cao, Z., Bonhaus, D.W., Friedman, A.H., Okazaki, M.M., Nadler, J.V. and McNamara, J.O. (1991) Increased AMPA-sensitive quisqualate receptor binding and reduced NMDA receptor binding in epileptic human hippocampus. *J. Neurosci.*, 11: 428–434.

Hönack, D. and Löscher, W. (1995) Kindling increases the sensitivity of rats to adverse effects of certain antiepileptic drugs. *Epilepsia.*, 36: 763–771.

Ito, I., Futai, K., Katagiri, H., Watanabe, M., Sakimura, K., Mishina, M. and Sugiyama, H. (1997) Synapse-selective impairment of NMDA receptor functions in mice lacking

NMDA receptor e1 or e2 subunit. *J. Physiol.* (*Lond.*), 500: 401–408.

Jarvie, P.A., Logan, T.C., Geula, C. and Slevin, J.T. (1990) Entorhinal kindling permanently enhances $Ca^{2+}$-dependent L-glutamate release in regio inferior of rat hippocampus. *Brain Res.*, 508: 188–193.

Jensen, P.J., Millan, N. and Mack, K.J. (1997) Cortical NMDAR-1 gene expression is rapidly upregulated after seizure. *Mol. Brain Res.*, 44: 157–162.

Jones, S.M. and Johnson, K.M. (1989) Effects of amygdaloid kindling on NMDA receptor function and regulation. *Exp. Neurol.*, 106: 52–60.

Kadotani, H., Hirano, T., Masugi, M., Nakamura, K., Nakao, K., Katsuki, M. and Nakanishi, S. (1996) Motor discoordination results from combined gene disruption of the NMDA receptor NR2A and NR2C subunits, but not from single disruption of the NR2A or NR2C subunit. *J. Neurosci.*, 16: 7859–7867.

Kamphuis, W., De Rijk, T.C., Talamini, L.M. and Lopes da Silva, F.H. (1994) Rat hippocampal kindling induces changes in the glutamate receptor mRNA expression patterns in dentate granule neurons. *Eur. J. Neurosci.*, 6: 1119–1127.

Kamphuis, W. and Lopes da Silva, F.H. (1995) Editing status at the Q/R site of glutamate receptor-A, -B, -5 and -6 subunit mRNA in the hippocampal kindling model of epilepsy. *Mol. Brain Res.*, 29: 35–42.

Kaura, S., Bradford, H.F., Young, A.M.J., Croucher, M.J. and Hughes, P.D. (1995) Effect of amygdaloid kindling on the content and release of amino acids from the amygdaloid complex: In vivo and in vitro studies. *J. Neurochem.*, 65: 1240–1249.

Kikuchi, S., Iwasa, H., Inagaki, N., Suzuki, K., Sato, T. and Seino, S. (1996) Alterations in *N*-methyl-D-aspartate receptor subunit mRNA expression levels in cerebral cortex in kindling-induced epileptogenesis. *Epilepsia*, 37 Suppl. 3: 113–114.

Klapstein, G.J., Meldrum, B.S. and Mody, I. (1997) Decreased inhibition of glutamate release through Group III metabotropic glutamate receptors in the rat dentate gyrus following kindling. *Soc. Neurosci.*, Abstr. (Part 1) 23: 2421.

Koltchine, V.V., Anantharam, V., Bayley, H. and Treistman, S.N. (1996) Alternative splicing of the NMDAR1 subunit affects modulation by calcium. *Mol. Brain Res.*, 39: 99–108.

Kondratyev, A., Neve, R., Rogers, S., Wolfe, B. and Gale, K. (1997) Overexpression of AMPA receptors modulates seizure activity evoked from area tempestas. *Epilepsia*, 38 (Suppl. 8): 6. (Abstract)

Köhr, G., De Koninck, Y. and Mody, I. (1993) Properties of NMDA receptor channels in neurons acutely isolated from epileptic (kindled) rats. *J. Neurosci.*, 13: 3612–3627.

Köhr, G. and Mody, I. (1994) Kindling increases *N*-methyl-D-aspartate potency at single *N*-methyl-D-aspartate channels in dentate gyrus granule cells. *Neuroscience*, 62: 975–981.

Krai, T., Dietrich, D., Clusmann, H. and Schramm, J. (1997) Mossy fiber sprouting correlates with loss of synaptic depression by Group III metabotropic glutamate receptors (mGluRs) in the area dentata of patients with temporal lobe epilepsy (TLE). *Soc. Neurosci., Abstr.* (Part 1) 23: 817.

Kraus, J.E., Nadler, J.V. and McNamara, J.O. (1996) Regulation of alternative splicing of NMDAR1 in the kindling model. *Mol. Brain Res.*, 41: 97–104.

Kraus, J.E., Yeh, G.-C., Bonhaus, D.W., Nadler, J.V. and McNamara, J.O. (1994) Kindling induces the long-lasting expression of a novel population of NMDA receptors in hippocampal region CA3. *J. Neurosci.*, 14: 4196–4205.

Krupp, J.J., Vissel, B., Heinemann, S.F. and Westbrook, G.L. (1996) Calcium-dependent inactivation of recombinant *N*-methyl-D-aspartate receptors is NR2 subunit specific. *Mol. Pharmacol.*, 50: 1680–1688.

Leach, M.J., Marden, C.M., Miller, A.A., O'Donnell, R.A. and Weston, S.B. (1985) Changes in cortical amino acids during electrical kindling in rats. *Neuropharmacology*, 24: 937–940.

Leach, M.J., O'Donnell, R.A., Collins, K.J., Marden, C.M. and Miller, A.A. (1987) Effect of cortical kindling on [3H]D-aspartate uptake and glutamate metabolism in rats. *Epilepsy Res.*, 1: 145–148.

Lee, S., Miskovsky, J., Williamson, J., Howells, R., Devinsky, O., Lothman, E. and Christakos, S. (1994) Changes in glutamate receptor and pro-enkephalin gene expression after kindled seizures. *Mol. Brain Res.*, 24: 34–42.

Lehmann, A. (1989) Abnormalities in the levels of extracellular and tissue amino acids in the brain of the seizure-susceptible rat. *Epilepsy Res.*, 3: 130–137.

Liu, Z., Stafstrom, C.E., Sarkisian, M.R., Yang, Y.L., Hori, A., Tandon, P. and Holmes, G.L. (1997) Seizure-induced glutamate release in mature and immature animals: An in vivo microdialysis study. *Neuroreport*, 8: 2019–2023.

Lothman, E.W., Bennett, J.P. and Perlin, J.B. (1987) Alterations in neurotransmitter amino acids in hippocampal kindled seizures. *Epilepsy Res.*, 1: 313–320.

Louvel, J. and Pumain, R. (1992) *N*-methyl-D-aspartate-mediated responses in epileptic cortex in man: An in vitro study. In: Neurotransmitters, Seizures and Epilepsy IV. J. Engel, ed. Elsevier, Amsterdam, pp. 487–495.

Löscher, W., Hörstermann, D., Hönack, D., Rundfeldt, C. and Wahnschaffe, U. (1993) Transmitter amino acid levels in rat brain regions after amygdala-kindling or chronic electrode implantation without kindling: Evidence for a pro-kindling effect of prolonged electrode implantation. *Neurochem. Res.*, 18: 775–781.

Lu, Y.M., Jia, Z.P., Janus, C., Henderson, J.T., Gerlai, R., Wojtowicz, J.M. and Roder, J.C. (1997) Mice lacking metabotropic glutamate receptor 5 show impaired learning and reduced CA1 long-term potentiation (LTP) but normal CA3 LTP. *J. Neurosci.*, 17: 5196–5205.

Marianowski, R., Pollard, H., Moreau, J., Després, G., Ben Ari, Y., Tran Ba Huy, P. and Romand, R. (1995) *N*-methyl-

D-aspartate receptor subunits NR1 and NR2C are overexpressed in the inferior colliculus of audiogenic mice. *Neurosci. Lett.*, 189: 190–194.

McDonald, J.W., Garofalo, E.A., Hood, T., Sackellares, J.C., Gilman, S., McKeever, P.E., Troncoso, J.C. and Johnston, M.V. (1991) Altered excitatory and inhibitory amino acid receptor binding in hippocampus of patients with temporal lobe epilepsy. *Ann. Neurol.*, 29: 529–541.

Meldrum, B.S. (1995) Excitatory amino acid receptors and their role in epilepsy and cerebral ischemia. *Ann. NY Acad. Sci.*, 757: 492–505.

Meldrum, B.S. and Chapman, A.G. (1998a) Epileptic seizures. In: Basic Neurochemistry. G. Siegel, B. Agranoff, R.W. Albers, S.K. Fisher and M.D. Uhler, eds. Lippincott-Raven Publishers, Philadelphia, Vol. 6th, (in press).

Meldrum, B.S. and Chapman, A.G. (1998b) Excitatory amino acids and antiepileptic drug development. In: Basic Mechanisms of the Epilepsies. R.J. Porter, ed. Lippincott-Raven Publishers, Philadelphia (in press).

Meldrum, B.S., Croucher, M.J., Badman, G. and Collins, J.F. (1983) Antiepileptic action of excitatory amino acid antagonists in the photosensitive baboon, Papio Papio. *Neurosci. Lett.*, 39: 101–104.

Miller, H.P., Levey, A.I., Rothstein, J.D., Tzingounis, A.V. and Conn, P.J. (1997) Alterations in glutamate transporter protein levels in kindling-induced epilepsy. *J. Neurochem.*, 68: 1564–1570.

Minamoto, Y., Itano, T., Tokuda, M., Matsui, H., Janjua, N.A., Hosokawa, K., Okada, Y., Murakami, T.H., Negi, T. and Hatase, O. (1992) In vivo microdialysis of amino acid neurotransmitters in the hippocampus in amygdaloid kindled rat. *Brain Res.*, 573: 345–348.

Mody, I. and Heinemann, U. (1987) NMDA receptors of dentate gyrus granule cells participate in synaptic transmission following kindling. *Nature*, 326: 701–704.

Naylor, P., Stewart, C.A., Wright, S.R., Pearson, R.C.A. and Reid, I.C. (1996) Repeated ECS induces GluR1 mRNA but not NMDAR1A-G mRNA in the rat hippocampus. *Mol. Brain Res.*, 35: 349–353.

Neugebauer, V., Keele, N.B. and Shinnick-Gallagher, P. (1997) Epileptogenesis in vivo enhances the sensitivity of inhibitory presynaptic metabotropic glutamate receptors in basolateral amygdala neurons in vitro. *J. Neurosci.*, 17: 983–995.

Obrenovitch, T.P., Urenjak, J. and Zilkha, E. (1996) Evidence disputing the link between seizure activity and high extracellular glutamate. *J. Neurochem.*, 66: 2446–2454.

Okazaki, M.M., McNamara, J.O. and Nadler, J.V. (1989) *N*-methyl-D-aspartate receptor autoradiography in rat brain after angular bundle kindling. *Brain Res.*, 482: 359–363.

Patel, S., Chapman, A.G., Graham, J.L., Meldrum, B.S. and Frey, P. (1990) Anticonvulsant activity of the NMDA antagonists, D(-)4-(3-phosphonopropyl)piperazine-2-carboxylic acid (D-CPP) and D(-) (E)-4-(3-phosphonoprop-2-enyl)piperazine-2-carboxylic acid (D-CPPene) in a rodent and a primate model of reflex epilepsy. *Epilepsy Res.*, 7: 3–10.

Pellegrini-Giampietro, D.E., Gorter, J.A., Bennett, M.V.L. and Zukin, R.S. (1997) The GluR2 (GluR-B) hypothesis: $Ca^{2+}$-permeable AMPA receptors in neurological disorders. *TINS*, 20: 464–470.

Pin, J.-P. and Duvoisin, R. (1995) The metabotropic glutamate receptors: Structure and functions. *Neuropharmacology*, 34: 1–26.

Pratt, G.D., Kokaia, M., Bengzon, J., Kokaia, Z., Fritschy, J.-M., Möhler, H. and Lindvall, O. (1993) Differential regulation of *N*-methyl-D-aspartate receptor subunit messenger RNAs in kindling-induced epileptogenesis. *Neuroscience*, 57: 307–318.

Prince, H.K., Conn, P.J., Blackstone, C.D., Huganir, R.L. and Levey, A.I. (1995) Down-regulation of AMPA receptor subunit GluR2 in amygdaloid kindling. *J. Neurochem.*, 64: 462–465.

Pumain, R., Louvel, J., Gastard, M., Kurcewicz, I. and Vergnes, M. (1992) Responses to *N*-methyl-D-aspartate are enhanced in rats with Petit Mal-like seizures. *J. Neural Transm.*, 35 [Suppl]: 97–108.

Rogawski, M.A. (1992) The NMDA receptor, NMDA antagonists and epilepsy therapy: A status report. *Drugs*, 44: 279–292.

Rogawski, M.A. (1996) Epilepsy. In: Neurotherapeutics: Emerging strategies. L. Pullan and J. Patel, eds. Humana Press Inc., Totowa, N.J. pp. 193–273.

Rogawski, M.A., Yamaguchi, S.-I., Jones, S.M., Rice, K.C., Thurkauf, A. and Monn, J.A. (1991) Anticonvulsant activity of the low-affinity uncompetitive *N*-methyl-D-aspartate antagonist (±)-5-aminocarbonyl-10,11-dihydro-5H-dibenzo [a,d]cyclohepten-5,10-imine (ADCI): Comparison with the structural analogs dizocilpine (MK- 801) and carbamazepine. *J. Pharmacol. Exp. Ther.*, 259: 30–37.

Ronne-Engström, E., Hillered, L., Flink, R., Spännare, B., Ungerstedt, U. and Carlson, H. (1992) Intracerebral microdialysis of extracellular amino acids in the human epileptic focus. *J. Cereb. Blood Flow Metab.*, 12: 873–876.

Rothstein, J.D., Dykes-Hoberg, M., Pardo, C.A., Bristol, L.A., Jin, L., Kuncl, R.W., Kanai, Y., Hediger, M.A., Wang, Y.F., Schielke, J.P. and Welty, D.F. (1996) Knockout of glutamate transporters reveals a major role for astroglial transport in excitotoxicity and clearance of glutamate. *Neuron*, 16: 675–686.

Rowley, H.L., Marsden, C.A. and Martin, K.F. (1997) Generalised seizure-induced changes in rat hippocampal glutamate but not GABA release are potentiated by repeated seizures. *Neurosci. Lett.*, 234: 143–146.

Savage, D.D., Nadler, J.V. and McNamara, J.O. (1984) Reduced kainic acid binding in rat hippocampal formation after limbic kindling. *Brain Res.*, 323: 128–131.

Scheyer, R.D. (1998) Chapter 23 of this volume, 359–370.

Simmonds, M.A. and Horne, A.L. (1987) Antagonism of Excitatory amino acids by barbiturates. In: Excitatory amino acid transmission. T.P. Hicks, D. Lodge and H. McLennan, eds. Alan R.Liss, New York, pp. 99–105.

Singh, L., Oles, R.J. and Tricklebank, M.D. (1990) Modulation of seizure susceptibility in the mouse by the strychnine-insensitive glycine recognition site of the NMDA receptor/ion channel complex. *Br. J. Pharmacol.*, 99: 285–288.

Sircar, R., Ludvig, N., Zukin, S.R. and Moshe, S.L. (1987) Down-regulation of hippocampal phencyclidine (PCP) receptors following amygdala kindling. *Eur. J. Pharmacol.*, 141: 167–168.

Smith, S.E., Dürmüller, N. and Meldrum, B.S. (1991) The non-*N*-methyl-D-aspartate antagonists, GYKI 52466 and NBQX are anticonvulsant in two animal models of reflex epilepsy. *Eur. J. Pharmacol.*, 201: 179–183.

Smith, S.E. and Meldrum, B.S. (1992) The glycine-site NMDA receptor antagonist, R-(+)-cis-β-methyl-3-amino-1-hydroxy-pyrrolid-2-one, L-687,414 is anticonvulsant in baboons. *Eur. J. Pharmacol.*, 211: 109–111.

Sutherland, M.L., Delaney, T.A. and Noebels, J.L. (1997) Subtype-specific downregulation of glutamate transporter gene expression in three models of temporal lobe epilepsy. *Epilepsia*, 38 (Suppl. 8): 5–6. (Abstract)

Suzuki, K., Mori, N., Kittaka, H., Iwata, Y., Yamada, Y., Osonoe, K. and Niwa, S. (1996) Anticonvulsant action of metabotropic glutamate receptor agonists in kindled amygdala of rats. *Neurosci. Lett.*, 204: 41–44.

Sveinbjornsdottir, S., Sander, J.W.A.S., Upton, D., Thompson, P.J., Patsalos, P.N., Hirt, D., Emre, M., Lowe, D. and Duncan, J.S. (1993) The excitatory amino acid antagonist D-CPP-ene (SDZ EAA-494) in patients with epilepsy. *Epilepsy Res.*, 16: 165–174.

Tanaka, K., Watase, K., Manabe, T., Yamada, K., Watanabe, M., Takahashi, K., Iwama, H., Nishikawa, T., Ichihara, N., Hori, S., Takimoto, M. and Wada, K. (1997) Epilepsy and exacerbation of brain injury in mice lacking the glutamate transporter GLT-1. *Science*, 276: 1699–1702.

Tang, E., Yip, P.K., Chapman, A.G., Jane, D.E. and Meldrum, B.S. (1997) Prolonged anticonvulsant action of glutamate metabotropic receptor agonists in inferior colliculus of genetically epilepsy-prone rats. *Eur. J. Pharmacol.*, 327: 109–115.

Taubøll, E. and Gjerstad, L. (1998) Chapter 25 of this volume, 385–394.

Thomsen, C., Klitgaard, H., Sheardown, M., Jackson, H.C., Eskesen, K., Jacobsen, P., Treppendahl, S. and Suzdak, P.D. (1994) (S)-4-carboxy-3-hydroxyphenylglycine, an antagonist of metabotropic glutamate receptor (mGluR)1a and an agonist of mGluR2, protects against audiogenic seizures in DBA/2 mice. *J. Neurochem.*, 62: 2492–2495.

Tizzano, J.P., Griffey, K.I. and Schoepp, D.D. (1995) Induction or protection of limbic seizures in mice by mGluR subtype selective agonists. *Neuropharmacology*, 34: 1063–1067.

Ueda, Y. and Tsuru, N. (1995) Simultaneous monitoring of the seizure-related changes in extracellular glutamate and gamma-aminobutyric acid concentration in bilateral hippocampi following development of amygdaloid kindling. *Epilepsy Res.*, 20: 213–219.

Vezzani, A., Speciale, C., Della Vedova, F., Tamburin, M. and Benatti, L. (1995) Alternative splicing at the C-terminal but not at the N-terminal domain of the NMDA receptor NR1 is altered in the kindled hippocampus. *Eur. J. Neurosci.*, 7: 2513–2517.

White, H.S., Brown, S.D., Woodhead, J.H., Skeen, G.A. and Wolf, H.H. (1997) Topiramate enhances GABA-mediated chloride flux and GABA-evoked chloride currents in murine brain neurons and increases seizure threshold. *Epilepsy Res.*, 28: 167–179.

Wilson, C.L., Maidment, N.T., Shomer, M.H., Behnke, E.J., Ackerson, L., Fried, I. and Engel, J.,Jr. (1996) Comparison of seizure related amino acid release in human epileptic hippocampus versus a chronic, kainate rat model of hippocampal epilepsy. *Epilepsy Res.*, 265: 245–254.

Wilson, M.A. and Tonegawa, S. (1997) Synaptic plasticity, place cells and spatial memory: Study with second generation knockouts. *TINS*, 20: 102–106.

Yamaguchi, S., Donevan, S.D. and Rogawski, M.A. (1993) Anticonvulsant activity of AMPA/kainate antagonists: Comparison of GYKI 52466 and NBQX in maximal electroshock and chemoconvulsant seizure models. *Epilepsy Res.*, 15: 179–184.

Zapata, A., Capdevila, J.L., Tarrason, G., Adan, J., Martínez, J.M., Piulats, J. and Trullas, R. (1997) Effects of NMDA-R1 antisense oligodeoxynucleotide administration: Behavioral and radioligand binding studies. *Brain Res.*, 745: 114–120.

Zhang, W.Q., Hudson, P.M., Sobotka, T.J., Hong, J.S. and Tilson, H.A. (1991) Extracellular concentrations of amino acid transmitters in ventral hippocampus during and after the development of kindling. *Brain Res.*, 540: 315–318.

O.P. Ottersen, I.A. Langmoen and L. Gjerstad (Eds.)
*Progress in Brain Research*, Vol 116


CHAPTER 25

# Effects of antiepileptic drugs on the activation of glutamate receptors

Erik Tauboll and Leif Gjerstad*

*Department of Neurology, Rikshospitalet, The National Hospital, University of Oslo, N-0027 Oslo, Norway*

Epilepsy can be caused by a variety of brain disorders. Seizures, usually generalized tonic-clonic seizures, can also occur during pathophysiological aberrations induced by systemic disorders such as thyreotoxicosis. There are in addition, several form of seizures (see Brodtkorp, this volume) which most likely involve different mechanisms in the generation of the abnormal neuronal discharges.

The hallmark of *partial* seizures is the paroxysmal depolarizing shift (PDS) occurring synchronously in a large cell population when the epileptic focus is activated (see Jefferys and Traub, this volume). The PDS is most often generated when activation of AMPA and NMDA receptors opens postsynaptic ion channels which depolarize the cell membrane sufficiently to also trigger voltage-dependent depolarizing currents carried by calcium and sodium (Dingledine et al., 1990).

The development of an ictal activity can be related to a repetitive and strong glutamatergic activation of a large population of neurones (see Jefferys and Traub, this volume). Other factors like reduced inhibition, accumulation of extracellular potassium, restricted extracellular space and reduced extracellular calcium may also contribute to the transition from interictal to ictal activity. A partial seizure and in particular a secondary generalized seizure necessarily also include activation of other neuronal populations than the original one. This widespread pathological activation of neuronal populations can be obtained by non-synaptic mechanisms. However, the release of glutamate and activation of glutamatergic receptors, both ionotropic and metabotropic, most likely play a crucial role.

Thus, limiting glutamatergic activation either by reducing the release of glutamate or by blocking the glutamate receptors may be an important approach in controlling partial seizures including the secondary generalized ones.

Glutamatergic activation can be decreased in four different ways: (1) Reducing the release of glutamate by blocking the repetitive activation of the terminals; (2) Blocking specifically the release of glutamate from the pre-synaptic terminals; (3) Blocking the activation of the postsynaptic glutamate receptors; (4) Increase the removal of free glutamate from the synaptic cleft.

## Blocking the sustained repetitive firing of action potentials may reduce the release of glutamate

The bursts of actions potentials underlying the interictal and ictal epileptiform discharges may be suppressed by blocking the fast Na-current of

*Corresponding author.
Tel.: +47 228 68902; fax: +47 228 8890; e-mail: leif.gjerstad@-klinmed.uio.no.

the action potentials by drugs like phenytoin and carbamazepine (McLean and Macdonald 1983, 1986; Matsuki et al., 1984, Willow et al., 1985). The block is voltage-dependent, use-dependent and time-dependent, leaving the initial action potentials intact while suppressing subsequent actions potentials for several hundred milliseconds.

The blockade of phenytoin has been related to a prolonged inactivation but does not depend upon a completed Na inactivation as block may occur even with small depolarizations that do not trigger inactivation, thus inactivation and drug binding may be independent processes (Kuo and Bean, 1994). In addition, phenytoin diminish the late sodium channel openings responsible for the persistent Na-current that do not easily inactivate (Chao and Alzheimer, 1995). This persistent sodium current may be an essential component of the sustained depolarizations underlying epileptiform activity (Taylor, 1993; Segal, 1994) and it has recently been shown to be more sensitive to phenytoin than the peak current (Segal and Douglas, 1997). While phenytoin has a higher affinity for the sodium channel than carbamazepine, the binding rate constant for the latter drug is five times faster (Kuo et al., 1997). These differences may be related to a difference in action on short versus longer depolarizations.

The well documented effect of drugs like phenytoin, carbamazepine and valproate on sodium channels can be directly related to their effect on seizure models (Rogawski and Porter, 1990). Blocking the repetitive firing of neurons obviously also causes a reduced hyperactivity and reduced temporal summation of activation. However, the anticonvulsant properties of these and similar sodium channel blocking drugs like lamotrigine (see below), may in fact also be related to a reduced release of glutamate secondary to their strong ability to block sustained repetitive firing (Lees and Leach, 1993; Taylor and Meldrum, 1995; Xie et al., 1995). New drugs like topiramate, felbamate, remacemide and many other compounds with anticonvulsant properties also seem to have a main action on voltage-dependent sodium channels (for a review, see Taylor and Narasimhan, 1997).

**Effects of drugs on the release of glutamate**

An inhibitory effect of barbiturates on the presynaptic release of glutamate is probably related to inhibition of calcium influx (Sohn and Ferrendelli, 1976; French-Mullen et al., 1993). In hippocampal slices the potassium induced depletion of glutamate from nerve terminals was inhibited both by phenobarbital and by 5α-pregnan-3α-ol-20-one suggesting a pre-synaptic barbiturate-like activity of this progesterone metabolite (Taubøll, et al., 1993; see Fig. 1). A presynaptic modulation of transmitter release may also possibly contribute to the effect of phenytoin and carbamazepine when these drugs are applied at very high concentrations.

Lamotrigine inhibited veratrine-evoked release of glutamate and aspartate but had no effect on potassium evoked release suggesting an effect caused by blockade of sodium channels (Leach et al., 1986). The effect on GABA release was less pronounced. Like phenytoin, lamotrigine has been shown to inhibit burst firing in cultured cortical neurones in an activity dependent way and in higher concentration also inhibits calcium currents (Lees and Leach, 1993). Lamotrigine was found to prolong the inactivation of recombinant sodium channels (Xie et al., 1995) suggesting a similar main mechanism as phenytoin and carbamazepine. However, newer studies indicate an additional effect on high-voltage calcium currents which may be of importance for a presynaptic effect on transmitter release (Stefani et al., 1996). On the other hand, a specific effect on glutamate release has also been questioned (Waldmeier, Martin, Stocklin, Portet and Schmutz, 1996).

Most likely the main mechanism for the effect on partial seizures for drugs like phenyotin, carbamazepine, valproate and lamotrigine (and some newer drugs) is a sodium channel blockade causing a decreased transmitter release. More specific drugs acting on different calcium channels may be potentially useful in blocking the release of

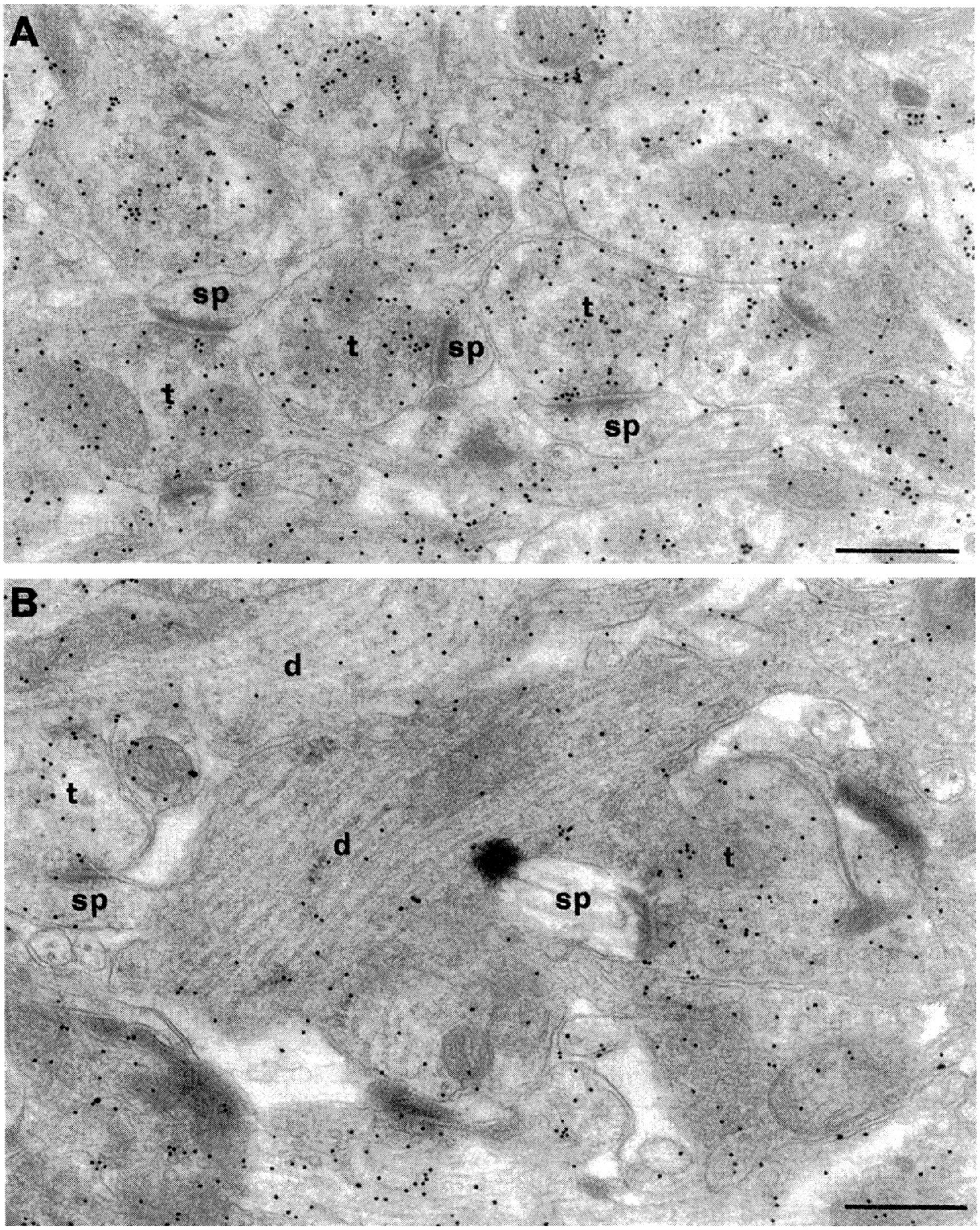


Fig. 1. *Reduced release of glutamate from pre-synaptic terminals by phenobarbital.* Electron micrographs from rat hippocampal CA1 region treated with a highly selective antiserum against glutamate. The gold particle density (i.e. the level of glutamate-like immunoreactivity) is significantly higher in sections exposed to 8.6 mM phenobarbital plus 35 mM potassium (A) than in sections exposed to potassium alone (B). t = terminal; sp = spine; d = dendrite; bar = 0.4 $\mu$m.

glutamate (Gross et al., 1997; see Meldrum, this volume).

Glutamate release also is controlled by metabotropic glutamatergic receptors located pre-synaptically (see Bruno et al., this volume; Nicholls, this volume). A new group of compounds ((1S,3S)-ACPD; L-AP4; LY354740) with anticonvulsant properties exerts their effect by acting as agonists on such receptors (Attwell et al., 1995; Monn et al., 1997; Abdul-Ghani et al., 1997; see also Chapman, this volume). A promising property of these agonists is their effect on the seizures as well as the kindling process. The effect after oral administration of a group II mGluR agonist in mice (Monn et al., 1997) makes it more likely that this compound may also be used in humans.

**Effects of antiepileptic drugs on the post-synaptic glutamate receptors**

The post-synaptic ionotropic glutamate receptors can be divided on pharmacological grounds in the three major receptor subtypes NMDA, kainate and AMPA (see Takumi et al., this volume). Drugs acting on either of these receptor subtypes are putative anticonvulsants and the mechanism of action of some of the currently used antiepileptic drugs seems to be related to their effects on these receptor subtypes.

*The NMDA receptors*

Based upon some experimental studies (see Jefferys and Traub, this volume) a blockade of NMDA receptors was supposed to be crucial in stopping seizure activity. Different specific binding sites within the NMDA-receptor channel complex exist (see MacDonald et al., this volume), and the most important sites of action for putative anticonvulsants are: (1) the transmitter recognition site, (2) the ion channel site, and (3) modulatory sites like the glycine site.

The first type of organic compounds to be developed as NMDA receptor antagonists was those that acted competitively at the transmitter recognition site. In the search for possible therapeutic agents acting on this site, several conformationally restricted analogues of AP5 and AP7 have been synthesized. The development of a rigid piperazine analogue of AP7 gave rise to 3-(2-carboxy-piperazin-4-yl)propyl-l-phosphonic acid (CPP) (Davies et al., 1986; Harris et al., 1986; Wong and Kemp, 1991). CPP-ene appeared to possess antiepileptic properties in electroshock tests, pentylenetetrazol induced seizures, in sound sensitive DBA/2 mice and in photosensitive baboons. After these promising results in animal studies, the drug was tested in humans, but the number of side-effects was unacceptably high (Sveinbjornsdottir et al., 1993). In a trial of 8 patients with intractable complex partial seizures, all had to withdraw from the study due to side-effects while seizures were unchanged in four and exacerbated in three patients.

A diverse group of substances can act as blockers of the NMDA receptor channel. Probably the best known are the dissociative anaesthetics ketamine and phencyclidine. Since these compounds act at a site distinct from the NMDA recognition site (Martin and Lodge, 1985; Harrison and Simmonds, 1985), they have been classified as noncompetitive NMDA antagonists. Their action is voltage- and use-dependent (Honey et al., 1985; Huettner and Bean, 1988) and involves an open-channel blocking mechanism. These drugs are effective anticonvulsants, but their obvious intolerable sedative side-effects limit their use in epilepsy except for possible benefit in the treatment of status epilepticus. Another drug in this class of noncompetitive antagonists at the NMDA receptors is dizocilpine (MK-801). This drug has been used in clinical trials for epilepsy in humans. However, because of unimpressive efficacy and psychological side-effects MK-801 is no longer considerd as a putative antiepileptic drug (Porter, 1989).

Glycine acts as a co-agonist on specific sites at the NMDA receptor ionophore complex and plays a permissive role in NMDA-mediated effects (Kleckner and Dingledine, 1988; Wroblewski and Danysz, 1989; Monaghan et al., 1989). In fact, occupancy of the glycine binding site appears to be necessary to elicit NMDA receptor responses.

Glycine binding to the NMDA receptor is strychnine-insensitive, whereas the classically defined glycine receptors, found mainly in the brainstem and spinal cord, are inhibitory in nature and sensitive to strychnine.

Glycine increases NMDA-evoked currents by increasing the opening frequency of the NMDA channel (Johnson and Ascher, 1987; Kleckner and Dingledine, 1988). Since activation of the NMDA channel requires occupation of the strychnine-insensitive binding site by glycine, antagonists acting at this receptor may afford protection against pathologies mediated by NMDA-preferring glutamate receptors.

The new anticonvulsant felbamate (2-phenyl-1,3-propanediol dicarbamate) has been shown to be effective in preventing NMDA-induced clonus in mice, suggesting that the NMDA receptor is a site of anticonvulsant action (Sofia et al., 1994). The observation that felbamate could inhibit the binding of the antagonist 5,7-dichlorokynurenic acid (5,7-DCKA) to the strychnine-insenitive glycine receptor suggests that the anticonvulsant effect of the drug could be mediated by its ability to modulate this receptor site (McCabe et al., 1993). This has been replicated in human tissue using temporal cortical slices from 5 males. Addition of felbamate to the incubation media inhibited specific 5,7-DCKA binding (Wamsley et al., 1994). It was concluded from these studies that felbamate probably acted as a functional antagonist at the strychnine-insensitive glycine receptor site. In addition, White et al. (1995) found that in a model with audiogenic seizure-susceptible mice, the glycine agonist D-serine completely blocked the anticonvulsant activity of a maximally effective dose of felbamate. This effect of D-serine could be reversed by increasing the administered dose of felbamate.

Also, felbamate decreased the magnitude of glycine-enhanced NMDA-induced intracellular calcium transients in mouse cerebellar granule cells (White et al., 1995). In addition to the effect on the glycine-site, felbamate might, however, also block directly the NMDA channel (Rho et al., 1994). In whole-cell voltage clamp recordings from cultured rat hippocampal neurons, felbamate only in very high concentration inhibited NMDA responses. Single-channel recordings indicated that this effect on NMDA responses occurred via an open channel blocking mechanism.

Another region for neuromodulation of NMDA-mediated channel activity is the polyamine site (Wong and Kemp, 1991; Williams et al., 1989). Agents such as spermine and spermidine are able to open the NMDA receptor channel and increase the binding of compounds like MK-801. In contrast, occupancy of ion-specific sites such as those for $Mg^{2+}$ and $Zn^{2+}$ appears to block NMDA receptor-mediated responses (Wong and Kemp, 1991). To date, no antiepileptic drugs in current use seem to act on this site.

Another way to reduce glutamate mediated excitation is to use drugs targeting specific NMDA receptor subtypes. This is exemplified by Conantokin-R (Con-R), a 27 aminoacid peptide recently isolated from the venom of the fish-hunting snail Conus radiatus, which is a potent and selective antagonist of the NMDA-preferring glutamate receptor. Con-R displays a unique selectivity for the NMDAR1-1B/NR2B subunit. In one study, Con-R was administered intracerebroventricularly to Frings audiogenic seizure-susceptible mice and evaluated for its ability to block sound-induced seizures (McCabe et al., 1997). It was also tested in mice for its ability to block maximal electroshock seizures. Con-R produced a time-and dose-dependent inhibition of audiogenic and electrically induced tonic extension seizures. Con-R was 49 times more potent that MK801 and 1900 times more potent than ifenprodil, another selective NMDA receptor antagonist at the NR2B receptor subtype. When effective dose for Con-R was evaluated in relationship to the toxic dose ($TD_{50}$), Con-R was superior. These results indicate that the subtype selective Con-R affords a distinct advantage over the non-competitive NMDA antagonists MK-801 and also ifenprodil. It further suggests that subtype selectivity may play an important role in differentiating anticonvulsant and toxic effects of NMDA antagonists and may represent a new class of antiepileptic drugs.

*The kainate/AMPA receptors*

The mechanism of action of the new antiepileptic drug topiramate (2,3:4,5-bis O-(1-metylethylidene)-$\beta$-D-fructopyranose sulfamate) is, at least in part, due to an effect on kainate receptors. The drug is now available for clinical use in several countries. Pretreatment with topiramate (50–90 mg/kg) before intracerebroventricular injections of NMDA, kainate and AMPA in mice significantly reduced the percentage of mice with seizures in response to kainate, but not NMDA or AMPA (Piña-Garza and McLean, 1996). Topiramate also blocked neuronal responses to kainate at 2 $\mu$g/ml in a use-dependent manner, but did not block NMDA or AMPA responses at 10 $\mu$g/ml in cultured mouse central neurons (Piña-Garza and McLean, 1996).

The effect of topiramate on excitatory amino-acid-evoked responses has also been studied in cultured hippocampal neurons using perforated-patch whole-cell voltage-clamp techniques (Coulter et al., 1995; Severt et al., 1995). Topiramate blocked kainate-evoked inward currents in hippocampal neurons in a concentration dependent manner. The rising phase of the concentration-response relationship was steepest between 10 and 100 $\mu$M, a concentration range overlapping free serum levels of topiramate in patients. At 100 $\mu$M, topiramate had no effect on NMDA-evoked inward currents. It is therefore likely that the anticonvulsant effect of topiramate is at least in part related to the effect of the drug on the kainate/AMPA receptor.

Antagonists of AMPA receptor subtypes (see Borges and Dingledine, this volume) of the glutamate receptor have powerful anticonvulsant activities. A key difference between NMDA and AMPA receptor antagonists is that NMDA receptor antagonists can prevent the development of kindled seizures, while AMPA receptor antagonists do not (Dürmüller et al., 1994). No currently used antiepileptic drug has been shown to act at the AMPA receptor subtype although it has been proposed that phenobarbital might affect AMPA-EPSP with no concomitant effect on the NMDA-EPSP (Ko et al., 1997).

The competitive AMPA receptor antagonists YM90K, LY293558, NS 229, S 17625 and the noncompetitive antagonists of the 2,3-benzodiazepine type exert potent anticonvulsant activity in various seizure models (Yamaguchi et al., 1993; Donevan et al., 1994; Chapman et al., 1991; Löscher and Hönack, 1994; Chimirri et al., 1997). Studies in normal human volunteers with LY300164, the active (R) enantiomer of the 2,3-benzodiazepine GYKI 52466, indicated that high plasma concentrations were associated with side effects such as ataxia, diplopia, sedation and euphoria (Donevan et al., 1994). Whether lower plasma concentrations can confer anticonvulsant and neuroprotective activity without toxicity remains to be determined.

In animals, it has been demonstrated that low, non-toxic doses of AMPA receptor antagonists can potentiate the anticonvulsant activity of conventional antiepileptic agents (Zarnowski et al., 1993) and NMDA antagonists (Löscher et al., 1993; Czuczwar et al., 1995). Based on these findings the most appropiate use of AMPA receptor antagonists may be in combination therapy.

## Removal of free glutamate from the synaptic cleft

So far no drugs seem to facilitate the processes involved in removal of glutamate from the synaptic cleft. However, a sodium channel blockade may potentially limit non-synaptic glutamate release and indirectly support removal during pathological conditions (Taylor and Narasimhan, 1997).

## Concluding remarks

Increased glutamate activation is of paramount importance in epilepsy. Several of the presently used antiepileptic drugs affect the glutamatergic activation. However, so far no clinically useful drug has been developed that selectively target the glutamate receptors. The variety and different distribution of the receptors may be one reason for the present situation. The importance of glutamatergic activation for normal brain function may also limit the usefulness of blockers of glutamate receptors. Promising results may come from a further development of agonist of presynaptic metabotropic receptors and antagonists of AMPA receptors (see Meldrum, this volume). A main challenge will be to identify the specific pathophysiological changes of the receptors in different kind of seizures and epilepsies so as to make it possible to attack the receptors that give rise rise to the abnormal response. So far, the application of drugs blocking sodium currents has proved to be the most useful clinical approach for reducing glutamatergic activation.

## References


Abdul-Ghani, A.-S., Attwell, P.J.E., Kent, N.S., Bradford, H.F., Croucher, M.J. and Jane, D.E. (1997) Anti-epileptogenic and anticonvulsant activity of L-2-amino-4-phosphonobutyrate, a presynaptic glutamate receptor agonist. *Brain. Res.*, 755: 202–212.

Attwell, P.J.E., Kaura, S., Sigala, G., Bradford, H.F., Croucher, M.J., Jane, D.E. and Watkin, J.C., Blockade of both epileptogenesis and glutamate release by (1S,3S)-ACPD, a presynaptic glutamate receptor agonist. *Brain Res.*, 698: 155–162.

Chao, T.I. and Alzheimer, C. (1995) Effects of phenytoin on the persistent $Na^+$ current of mammalian CNS neurones. *NeuroReport*, 6: 1778–1780.

Chapman, A.G., Smith, S.E. and Meldrum, B.S. (1991) The anticonvulsant effect of the non-NMDA antagonists, NBQX and GYK1 52466, in mice. *Epilepsy Res.*, 9: 92–96.

Chimirri, A., De Sarro, G., De Sarro, A., Gitto, R., Grasso, S., Quartarone, S., Zappala, M., Giusti, P., Libri, V., Constanti, A. and Chapman. A.G. (1997) 1-aryl-3,5-dihydro-4H-2,3-benzodiazepin-4-ones: Novel AMPA receptor antagonists. *J. Med. Chem.*, 40: 1258–1269.

Coulter, D.A., Sombati, S. and DeLorenzo, R.J. (1995) Topiramate effects on excitatory amino acid-mediated responses in cultured hippocampal neurons: Selective blockade of kainate currents. *Epilepsia*, 37(suppl 3): S40.

Czuczwar, S.J., Borowicz, K.K., Kleinrok, Z., Tutka, P., Zarnowski, T. and Turski, W.A. (1995) Influence of combined treatment with NMDA and non-NMDA receptor antagonists on electroconvulsions in mice. *Eur. J. Pharmacol.*, 281: 327–333.

Davies, J., Evans, R.H., Herrling, P.L., Jones, A.W., Olverman, H.J., Pook, P. and Watkins, J.C. (1986) CPP, a new potent and selective NMDA antagonist. Depression of central neuron responses, affinity for 3H-D-AP5 binding sites on brain membranes and anticonvulsant activity. *Brain Res.*, 382: 169–173.

Dingledine, R., Mcbain, C.J. and Mcnamara, J.O. (1990) Excitatory amino acids receptors in epilepsy. *Trends Physiol. Sci.*, 11: 334–338.

Donevan, S.,D., Yamaguchi, S. and Rogawski, M.A. (1994) Non-*N*-methyl-D-aspartate receptor antagonism by 3-N-substituted 2,3-benzodiazepines: Relationship to anticonvulsant activity. *J. Pharmacol. Exp. Ther.*, 271: 25–29.

Dürmüller, N., Craggs, M. and Meldrum, B.S. (1994) The effect of the non-NMDA receptor antagonist GYK1 52466 and NBQX and the competitive NMDA receptor antagonist D-CPPene on the development of amygdala kindling and on amygdala-kindled seizures. *Epilepsy Res.*, 17: 167–174.

French-Mullen, J.M.H., Barker J.L. and Rogawski, M.A. (1993) Calcium current block by (−)-pentobarbital, phenobarbital, and CHEB but not (+)-pentobarbital in acutely isolated hippocampal CA1 neurons: Comparison with effects on GABA-activated CL-current. *J. Neurosci.*, 13: 3211–3221.

Ghauri, M., Chapman, A.G., Jane, D.E. and Meldrum, B. (1997) Anticonvulsant and proconvulsant effects of metabotropic glutamate receptor agonists and antagonists. *Epilepsia*, 38(suppl 3): 223.

Gross, R.A., Covey, D.F. and Ferrendelli, J.A. (1997) Voltage-dependent calcium channels as targets for convulsant and anticonvulsant alkyl-substituted thiobutyrolactones. *J. Pharmacol. Exp. Ther.*, 280: 686–694.

Harris, E.W., Ganong, A.H., Monaghan, D.T., Watkins, J.C. and Cotman, C.W. (1986) Action of 3-((+/−)-2-carboxypiperazin-4-yl)-propyl-1-phosphonic acid (CPP): A new and highly potent antagonist of *N*-methyl-D-aspartate receptors in hippocampus. *Brain Res.*, 382: 174–177.

Harrison, N.L. and Simmonds, M.A. (1985) Quantitative studies on some antagonists of *N*-methyl-D-aspartate in slices of rat cerebral cortex. *Br. J. Pharmacol.*, 84: 381–391.

Honey, C.R., Miljkovic, Z. and MacDonald, J.F. (1985) Ketamine and phencyclidine cause a voltage-dependent block of responses to L-aspartic acid. *Neurosci. Lett.*, 61: 135–139.

Huettner, J.E. and Bean B.P. (1988) Block of *N*-methyl-D-aspartate activated current by the anticonvulsant MK-801: selective binding to open channels. *Proc. Natl. Acad. Sci., USA* 85: 1307–1311.

Johnson, J.W. and Ascher, P. (1987) Glycine potentiates the NMDA response in cultured mouse brain neurons. *Nature*, 325: 529–531.

Kleckner, N.W. and Dingledine, R.D. (1988) Requirement for glycine in activation of NMDA receptors expressed in Xenopus oocytes. *Science*, 241: 835–837.

Leach, M.J., Marden, C.M. and Miller, A.A. (1986) Pharmacological studies on lamotrigine, a novel potential antielipeptic drug. *Epilepsia*, 27: 490–497

Lees, G. and Leach, M.J. (1993) Studies on the mechamism of action of the novel anticonvulsant lamotrigine (Lamictal) using primary neuroglial cultures from rat cortex. *Brain Res.*, 190–199.

Ko, G.Y.L., Brown-Croyts, L.M. and Teyler, T.J. (1997) The effects of anticonvulsant drugs on NMDA-EPSP, AMPA-EPSP, and GABA-IPSP in the rat hippocampus. *Brain Res. Bull.*, 42: 297–302.

Kuo, C.-C. and Bean, B.P. (1994) Slow binding of phenytoin to inactivated sodium channels in rat hippocampal neurons. *Mol. Pharmacol.*, 46: 716–725.

Kuo, C.-C, Chen, R.-S, Lu, L. and Chen, R.-C. (1997) Carbamazepine inhibition of neuronal Na currents: Quantitative distinction from phenytoin and possible therapeutic implications. *Mol. Pharmacol.*, 51: 1077–1083.

Löscher, W. and Hönack, D. (1994) Effects of the non-NMDA antagonists NBQX and the 2,3-benzodiazepine GYK1 52466 on different seizure types in mice: Comparison with diazepam and interactions with flumazenil. *Br. J. Pharmacol.*, 113: 1349–1357.

Löscher, W., Rundfeldt, C. and Hönack, D. (1993) Low doses of NMDA receptor antagonists synergistically increase the anticonvulsant effect of the AMPA receptor antagonist NBQX in the kindling model of epilepsy. *Eur. J. Neurosci.*, 5: 1545–1550.

Martin, D. and Lodge, D. (1985) Ketamine acts as a non-competitive *N*-methyl-D-aspartate antagonist on frog spinal cord in vitro. *Neuropharmacology*, 24: 999–1003.

Matsuki, N.F.N., Quandt, R.E., Ten Eick and Yeh, J.Z. (1984) Characterization of the block of sodium channels by phenytoin in mouse neuroblastoma cells. *J. Pharmacol. Exp. Ther.*, 228: 523–530.

McCabe, R.T., Olivera, B.M., Jimenez, E., Cruz, L.J., Zhou, L.M. and White, H.S. (1997) Anticonvulsant profile on Conantokin-R: A novel and selective antagonist at the NMDAR1-1b/NR2b subunit. *Epilepsia*, 38(suppl 3): S230.

McCabe, R.T., Wasterlain C.G., Kucharczyk, N., Sofia, R.D. and Vogel, J.R. (1993) Evidence for anticonvulsant and neuroprotectant action of felbamate mediated by strychnine-insensitive glycine receptors. *J. Pharmacol. Exp. Ther.*, 264: 1248–1252.

McLean, M.J. and Macdonald, R.L. (1983) Multiple actions of phenytoin on mouse spinal cord neurons in cell cultures. *J. Pharmacol. Exp. Ther.*, 227: 779–789.

McLean, M.L. and Macdonald, R.L. (1986) Carbamazepine and 10.11-epoxycarbamazeoine produce use- and voltage-dependent limitation of rapidly firing action potentials of mouse central neurons in cell culture. *J. Pharmacol. Exp. Ther.*, 238: 727–732.

Monaghan, D.T., Bridges, R.J. and Cotman, C.W. (1989) The excitatory amino acid receptors: Their classes, pharmacology, and distinct properties in the function of the central nervous system. *Annu. Rev. Pharmacol. Toxicol.*, 29: 365–402.

Monn, J.A., Valli, M.J., Massey, S.M., Wright, R.A., Salhoff, C.R., Johnson, B.G., Howe, T., Alt, C.A., Rhodes, G.A., Robey, R.L., Griffey, K.R., Tizzano, J.P., Kallmann, M.J., Helton, D.R. and Schoepp, D.D. (1997) Design, synthesis, and pharmacological characterization of (+)-2-amino-biocyclo(3.1.0.)hexane-2,6-dicarboxylic acid (LY354740): A potent, selective, and orally active group 2 metabotropic glutamate receptor agonist possesing anticonvulsant and anxiolytic properties. *J. Med. Chem.*, 40: 528–537.

Piña-Garza, J.E. and McLean, M.J. (1996) Different effects of topiramate and phenytoin on mouse seizures and responses of cultured neurons to excitatory amino acids. *Epilepsia*, 37(suppl 5): S26.

Porter, R.J. (1989) Mechanisms of action of new antiepileptic drugs. *Epilepsia*, 30(suppl. 1): S29–S34.

Rho, J.M., Donevan S.D. and Rogawski, M.A. (1994) Mechanism of action of the anticonvulsant felbamate: Opposing effects on *N*-methyl-D-aspartate and $\gamma$-aminobutyric acid a receptors. *Ann. Neurol.*, 35: 229–234.

Rogawski, M.A. and Porter, R. (1990) Antiepileptic drugs: Pharmacological mechanisms and clinical efficacy with consideration of promising developmental stage compounds. *Pharmacol. Rev.*, 42: 223–286.

Segel, M.M. (1994) Endogenous bursts underlie seizurelike activity insolitary excitatory hippocampal neurons in microcultures. *J. Neurphysiol.*, 72: 1874–1884.

Segal, M.M. and Douglas, A.F. Late sodium channel openings underlying epileptiform activity are preferentially diminished by the anticonvulsant phenytoin, *J. Neurophysiol.*, 77: 3021–3034.

Sofia, R.D., Gordon, R., Gels, M. and Diamantis, W. (1994) Comparative effects of felbamate and other compounds on *N*-methyl-D-aspartic acid-induced convulsions and lethality in mice. *Pharmacol. Res.*, 29: 139–144.

Sveinbjornsdottir, S., Sander, J.W.A.S., Upton, D., Thompson, P.J., Patsalos, P.N., Hirt, D., Emre, M., Lowe, D. and Duncan, J.S. (1993) The excitatory amino acid antagonist D-CPP-ene (SDZ EAA-494) in patients with epilepsy. *Epilepsy Res*, 16: 165–174.

Severt, L., Coulter, D.A., Sombati, S. and DeLorenzo, J. (1995) Topiramate selectively blocks kainate currents in cultured hippocampal neurons. *Epilepsia*, 36(suppl 4): S38.

Sohn, R.S. and Ferendelli, J.A. (1976) Anticonvulsant drug mechanisms: Phenytoin, phenobarbital and anf ethosuximide and calcium flux in isolated presynaptic endings. *Arch. Neurol.*, 33: 626–629.

Stefani, A., Spadoni, F., Sinicalchi., A. and Bernardi, G. (1996) Lamotrigine inhibits $Ca^{+}$ currents in cortical neurons; Functional implications. *Eur. J. Pharmacol.*, 307: 113–116.

Taubøll, E., Ottersen, O.P. and Gjerstad, L. (1993) The progesterone metabolite 5$\alpha$-pregnan-3$\alpha$.ol-20-one reduces $K^{+}$-induced GABA and glutamate release: A semiquantitative immunocytochemical study. *Brain. Res.*, 623: 329–333.

Taylor, C.P. (1993) $Na^{+}$ currents that fail to inactivate. *Trends Neurosci.*, 16: 455–460.

Taylor, C.P. and Meldrum, B.S. (1995) $Na^+$ channels as targets for neuroprotective drugs. *Trends Pharmacol. Sci.*, 16: 309–316.

Taylor, C.P. and Narasimhan, L.S. (1997) Sodium channels and therapy of central nervous system diseases. *Adv. Pharmacol.*, 39: 47–98.

Waldmeier, P.C., Martin, P., Stocklin, K., Portet, C. and Schmutz, M. (1996) Effect of carbamazepine, oxcarbazepine and lamotrigine on the indcrease in extracellular glutamate elicited by veratridine in rat cortex and striatum. *Naunyn-Schmiedebergs Arch. Pharmacol.*, 354: 164–172.

Wamsley, J.K., Sofia, R.D., Faull, R.L.M., Narang, N., Ary, T., Gan, X-D. and McCabe, R.T. (1994) Felbamate: Interaction with glycine receptors in human cerebral cortex. *Proc. West. Pharmacol. Soc.*, 37: 81–83.

White, H.S., Harmsworth, W.L., Sofia, R.D. and Wolf, H.H. (1995) Felbamate modulates the strychnine-insensitive glycine receptor. *Epilepsy Res.*, 20: 41–48.

Williams, K., Roman, C. and Molinoff, P.B. (1989) Effects of polyamines on the binding of [3H]MK-801 to the *N*-methyl-D-aspartate receptor: Pharmacological evidence for the existence of a polyamine recognition site. *Mol. Pharmacol.*, 36: 575–581.

Willow, M., Gonoi, T. and Caterall, W.A. (1985) Voltage clamp analysis of the inhibitory actions of diphenylhydantoin and carbamazepine on voltage-sensitive sodium channels in neuroblastoma cells. *Mol. Pharmacol.*, 27: 549–548.

Wong, E.H.F. and Kemp, J.A. (1991) Sites for antagonism on the *N*-methyl-D-aspartate receptor channel complex. *Annu. Rev. Pharmacol. Toxicol.*, 31: 401–425.

Wroblewski, J.T. and Danysz, W. (1989) Modulation of glutamate receptors: Molecular mechanisms and functional implications. *Annu. Rev. Pharmacol. Toxicol.*, 29: 441–474.

Xie, X., Lancaster, B., Peakman, Y. and Garthwaite, J. (1995) Interaction of the antiepileptic drug lamotrigine with the recombinant rat brain type IIA $Na^+$ channels and with native $Na^+$ channels in rat hippocampal neurones. *Pfluegers Arch.*, 430: 437–446.

Yamaguchi, S., Donevan, S.D. and Rogawski, M.A. (1993) Anticonvulsant activity of AMPA/kainate antagonists: Comparison of GYK1 52466 and NBQX in maximal electroshock and chemoconvulsant seizure models. *Epilepsy Res.*, 15: 179–184.

Zarnowski, T., Kleinrok, Z., Turski, W.A. and Czuczwar, S.J. (1993) 2,3-dihydroxy-6-nitro-7-sulfamoylbenzo[F]quinoxaline enhances the protective activity of common antiepileptic drugs against maximal electroshock-induced seizures in mice. *Neuropharmacology*, 32: 895–900.

O.P. Ottersen, I.A. Langmoen and L. Gjerstad (Eds.)
*Progress in Brain Research*, Vol 116


CHAPTER 26

# Antiepileptic drug treatment: Clinical considerations and concerns

Eylert Brodtkorb

*Department of Neurology, University Hospital, Trondheim, Norway*

Epileptology is a multifaceted discipline spanning from basic neurobiology to the psychosocial problems of the individual patient. A good communication between the different professions in this wide and complex field is required to achieve new developments in the comprehensive treatment of epilepsy patients. The ongoing close co-operation between neuroscientists and clinicians has accelerated the creation of new antiepileptic drugs (AEDs) (Table 1) and is generating a time of considerable activity and promise in the field of clinical epileptology.

Among the diseases for which the glutamate synapse is a potential therapeutic target, the epilepsies are rather unique since many drugs with other mechanisms of action are already available. The present chapter will give a general overview of AEDs to provide some insight in how current and future drugs affecting glutamatergic transmission will find their place among the existing therapies. The chapter will also highlight some of the general concerns that must be kept in mind when introducing new drugs for the treatment of neurological disease.

## Antiepileptic drug development

Effective epilepsy treatment dates back to the introduction of bromides in 1857. A breakthrough was made in AED treatment with *phenobarbital* (PB) in 1912 and a further major advance was made in 1937 with the discovery of *phenytoin* (PHT) by Putnam and Merritt. These scientists developed the capability to test AEDs in experimental animal models. By then it was understood that the sedative and the anticonvulsive effects of AEDs could be separated (Goldensohn et al., 1997).

For several decades PB and PHT were the major drugs (Table 1). In the 1980s, *carbamazepine* (CBZ) and *valproate* (VPA) gained an increasing part of the market. The consumption of PB and PHT has been decreasing in large parts of the world. PB is still among the most consumed AEDs in the epilepsy population. However, at present, few new patients receive PB in Europe and North America.

Over the last few years several novel AEDs have been discovered worldwide, partly through serendipity, partly by empirical screening and by introducing minimal changes in the structure of established drugs, and now increasingly based on what is known about synaptic transmission and neurotransmittor activity and metabolism. New information on the mechanisms of action, the efficacy and the safety of AEDs has been reported.

It has been believed that the traditional AEDs primarily have three major mechanisms of action. Drugs against generalized tonic-clonic seizures and partial seizures appear to delay the recovery of sodium channels from activation. Drugs against absence seizures appear to reduce low-threshold

TABLE 1

Antiepileptic drugs, their international abbreviations (Editorial, 1993) and year of introduction*

| | | |
|---|---|---|
| Phenobarbital | PB | 1912** |
| Phenytoin | PHT | 1944 |
| Primidone | PRM | 1953 |
| Ethosuximide | ESM | 1961 |
| Carbamazepine | CBZ | 1965 |
| Clonazepam | CZP | 1974 |
| Valproate | VPA | 1981 |
| Vigabatrin | VGB | 1993 |
| Lamotrigin | LTG | 1994 |
| Felbamate | FBM | 1996 |
| Gabapentin | GBP | 1996 |
| Topiramate | TPM | 1997 |
| Clobazam | CLB | *** |
| Oxcarbazepine | OCBZ | *** |
| Tiagabine | TGB | *** |

*) Year of introduction refers to year of licensing in Norway
**) Introduced as a pharmacy product before license regulations
***) Licensed in several countries, but not yet in Norway

calcium currents. Finally, drugs that are effective against myoclonic seizures generally seem to enhance $GABA_A$-receptor inhibition (Macdonald, 1997).

The chemical structure of many new drugs has been designed to affect specific mechanisms involved in the balance between neuronal excitation and inhibition (Dreifuss, 1994). *Vigabatrin* (VGB) is an inhibitor of GABA-transaminase and enhances GABAergic neuronal inhibition (Grant and Heel, 1991). Its clinical effects are primarily limited to partial seizures and infantile spasms. The increased brain GABA content may contribute to adverse effects on mood and cognitive function in some patients, and may occasionally lead to psychosis. The narrow clinical spectrum of efficacy of VGB was at first a surprise, and it was also soon realized that VGB may facilitate spike and wave discharges, contrary to the benzodiazepines which have a broadspectered effect. This diversity may be explained by $GABA_A$ and $GABA_B$ receptor differences (Meldrum, 1997).

*Lamotrigine* (LTG) inhibits presynaptic voltage-dependent sodium channels leading to decreased release of excitatory amino acids (Fitton and Goa, 1995). LTG has been shown to inhibit calcium currents in cultured rat cortical neurons (Binnie, 1997). However, its mechanistic profile is not completely known (see Tauboll and Gjerstad, this volume). LTG has a hitherto unexplained clinical effect in absences. It is questioned whether this is due to some form of additional GABA-ergic effect or if it suppresses voltage-dependent calcium currents in the thalamus, like ethosuximide (ESM).

*Felbamate* (FBM) also has an effect on sodium voltage channels, and probably several more actions. It has an NMDA blocking effect and a facilitating effect on GABA-ergic responses. The mechanism of action of *gabapentin* (GBP) is also still partly not understood in spite of its structural similarity to GABA. Nevertheless, it appears to increase GABA turnover in various regions of the rat brain (Meldrum, 1996). At least three pharmacological properties of *topiramate* (TPM) seem to contribute to its antiepileptic action. It has an effect on sodium channels, potentiates the effect of GABA and exhibits a weak glutamate antagonistic effect on AMPA/kainate receptors. *Tiagabine* (TGB) inhibits GABA re-uptake and augments and prolongs the inhibitory effect of GABA (Macdonald and Greenfield, 1997).

*Oxcarbazepine* (OCBZ) and *clobazam* (CLB) are on the market in a limited number of countries. OCBZ is the keto-derivative of CBZ. It was designed to improve the pharmacokinetic properties of this substance and its antiepileptic effect is similar to that of CBZ. CLB is a 1,5-benzodiazepine and is supposed to have a greater therapeutic potential than the 1,4-benzodiazepines, such as clonazepam (CZP) (Johannessen and Sundqvist, 1998).

As already mentioned, the relationship of the mechanisms of action and the clinical utility may be difficult to predict from preclinical data. Molecular mechanisms are often multiple and may be concentration-dependent. Such drugs represent polypharmacology in themselves. Different actions may be responsible for various effects and side-effects, which paradoxically may occasionally even cause seizure exacerbation in some patients. (Perucca et al., 1998)

The various mechanisms of FBM and TPM include glutamate antagonistic effects. The mechanistic complexity of these drugs suggests a broad spectrum of clinical utility, which is now supported by clinical experience. Another good example of a multi-mechanistic drug is the oldest AED, PB. The main mechanisms of action which produce its antiepileptic effect, are a combination of depressing glutamate excitability and enhancing postsynaptic GABA-ergic inhibition. The role of the effect on potassium and calcium conductance is not clear. (O'Connell and Patsalos, 1996). The sedative effects may depend on dissimilar mechanisms of action. A crucial question is whether we will ever have "clean" AEDs with one single, predictable mode of action at the molecular level.

Modern drug design will further widen our antiepileptic armamentarium. New compounds that interact with excitatory amino acid receptors are thought to have a great potential as AEDs (see Tauboll and Gjerstad, this volume). However, they may have pronounced neurologic and cognitive side effects. Differential effects at different receptor subunits may be relevant for specific syndromes involving altered glutamatergic receptor function (Meldrum, 1997).

In the clinic, new AEDs are first tested as add-on treatment in randomised, placebo-controlled clinical trials in patients with refractory partial epilepsy. Subsequently, they are usually compared with standard treatments, predominantly in monotherapy studies in patients with newly diagnosed epilepsy (French, 1997).

## Epileptic seizures and epilepsy syndromes

The selection of an AED is based primarily on its efficacy for specific types of seizures and epilepsies. The current classification of seizure types (Table 2) and epileptic syndromes (Table 3) has been proposed by the International League against epilepsy (Commission on Classification and Terminology, 1981 and 1989). Epileptic syndromes can be diagnosed on the basis of a constellation of characteristics, including not only seizure types but also factors such as age at onset, EEG findings, and etiology. The course and prognosis of the disorders and the precipitants and timing of the seizures may be dependent on the epilepsy syndrome. Thus, the comprehensive treatment of epilepsy is based on the correct classification of both seizures and epilepsy syndromes.

TABLE 2

International classification of epileptic seizures

I. Partial seizures (seizures beginning locally)
- A. Simple partial seizures (consciousness not impaired)
- B. Complex partial seizures (with impairment of consciousness)
- C. Partial seizures secondarily generalized

II. Generalized seizures (Bilaterally symmetrical and without local onset)
- A. Absence seizures
- B. Myoclonic seizures
- C. Clonic seizures
- D. Tonic seizures
- E. Tonic-clonic seizures
- F. Atonic seizures

III. Unclassified epileptic seizures (inadequate or incomplete data)

Abstracted from the Commission on Classification and Terminology, 1981

## General strategy

Monotherapy is the key principle in antiepileptic treatment (Reynolds and Shorvon, 1981; Genton and Roger, 1997) and unneccessary polytherapy should be avoided. Approximately 70 to 80% of new patients achieve satisfactory control with a single drug. One drug implies less side-effects, no interactions, less cost, and probably a better compliance. For the remaining patients rational combinations should be explored. Synergisms in the form of favourable pharmacodynamic interactions are desired. One way to enhance efficacy is to combine drugs with complementary modes of

TABLE 3

International classification of epilepsies and epileptic syndromes

1. Localization-related
    - 1.1 Idiopathic, e.g.
        - Benign childhood epilepsy with centro-temporal spikes
    - 1.2 Cryptogenic
        - Defined by seizure type/ clinical features/ etiology/ anatomical localization
    - 1.3 Symptomatic
        - Temporal/ frontal/ parietal/ occipital
        - Chronic progressive epilepsia partialis of childhood
2. Generalized
    - 2.1. Idiopathic, e.g.
        - Childhood absence epilepsy
        - Juvenile absence epilepsy
        - Juvenile myoclonic epilepsy
        - Epilepsy with GTC on awakening
    - 2.2. Cryptogenic or symptomatic, e.g.
        - West syndrome
        - Lennox-Gastaut syndrome
3. Undetermined epilepsies
    - 3.1. With both generalized and focal seizures, e.g.
        - Neonatal seizures
        - Epilepsy with continuous spike-waves during sleep
        - Landau-Kleffner syndrome
    - 3.2. Without unequivocal generalized or focal seizures

Abstracted from the Commission on Classification and Terminology, 1989

action (sodium channel drugs, GABAergic compounds, anti-glutaminergic drugs). Another approach may be to add a drug which manipulates the same transmitter target. The tolerability, therapeutic index and pharmacokinetic interaction potentials of such drugs need consideration. The concept of rational polypharmacy is now a challenge to clinical research (Richens, 1995).

Several studies have shown that after a period of complete seizure control of 2–5 years, medication can be stopped without recurrence (at least within several years) in more than 60% of patients. The type of epilepsy is an important predictor of the outcome (Shinnar and Berg, 1995); e.g., juvenile myoclonic epilepsy conveys a high probability of relapse, whereas absence epilepsy often remits. Benign epilepsy of childhood with rolandic spikes disappears by definition in adolescence, but even temporal lobe epilepsy with complex partial seizures may remit after a long period of control (Brodie and Dichter, 1997)

**Therapeutic drug monitoring**

Assessments of plasma levels may be of great help in guiding the treatment in patients with uncontrolled seizures and may also help the clinician to gain experience of the kinetics of the prescribed drugs. When starting AED therapy, the dose is gradully increased towards the anticipant maintainance level. The lowest effective dose for each patient is aimed at. Adjustments should be performed first of all on clinical grounds. However, since the treatment is prophylactic, the exact level of clinical response is often difficult to evaluate. Due to marked inter-individual differences in pharmacokinetics, the dosages of most AEDs need to be individualized. In some patients, it may be necessary to increase the dose until the plasma levels reaches the upper target level (for drugs with a defined therapeutic range) or the patients get symptoms or signs of toxicity. It should always be borne in mind that the estimated therapeutic ranges are statistical values referring to the average population. In reality, the "therapeutic" ranges are patient-specific. Some patients require higher levels and, nevertheless, are free from manifestations of unwanted effects. Others may do well on levels below the recommended range and their doses should not be increased (Commission on Antiepileptic Drugs, 1993; Brodie and Dichter, 1997).

Frequent routine assessments in well controlled patients are of little value. Usually, plasma levels should be limited to evaluate the cause of either seizure break-throughs or symptoms that may represent dose-related toxicity. However, it may also be useful in patients who may exhibit changes in drug kinetics in the course of development; e.g. in pregnancy and especially during concomitant treatment with other drugs with interaction potentials. Only morning drug fasting levels are meaningful to compare.

It is of utmost importance to identify seizure break-throughs due to lapses in drug compliance. Dose increment in place of appropriate information and education, may lead to toxicity and a vicious circle of increased mutual mistrust and communication problems between the patient and the treating physician.

## Drug selection

Table 4 shows a recommendation of the choice of AEDs in relation to seizure type. The AED of first choice for localization-related epilepsies with partial and secondarily generalized tonic clonic seizures is for most patients still CBZ. VPA may be a good alternative, but PHT and the newer drugs are listed as third choice or add-on therapy (Tomson et al., 1998).

For generalized epilepsies, the first choice is usually VPA. In most cases of idiopathic generalized epilepsies with absence, tonic clonic, and myoclonic seizures, the response to treatment is excellent (Osservatorio Regionale per L'Epilessia, 1996). For absence seizures, the second drug is ESM. Symptomatic and cryptogenic generalized epilepsies with multiple seizure types (tonic, atonic etc.) are difficult to treat. It should be noted that PB is no longer on this list of recommended drugs due to its tendency to give central nervous side effects.

TABLE 4

Recommended choice of antiepileptic drugs in relation to seizure type (After Tomson et al., 1998)

| Seizure type | 1. choice | 2. choice | 3. choice or add-on* |
|---|---|---|---|
| Partial seizures | CBZ | VPA | GBP, LTG, PHT, VGB, TPM |
| Generalized seizures | | | |
| Myoclonic | VPA | CZP | |
| Absences | VPA | ESM | CZP, LTG |
| Tonic-clonic | VPA | CBZ | LTG, PHT |

*Drugs in this group are listed in alphabetical order due to absence of an evidence-based ranking at the present.

## Patient groups with special requirements

Such a general recommendation (Table 4), should not be interpreted as a rigid direction. In several subpopulations of epilepsy patients, special considerations are required. Some categories may be more prone to particular side effects and complications. Vulnerable groups are fertile women, persons with mental retardation, and the elderly. There has recently been an increased tendency to a more individual approach in epilepsy treatment.

Women with epilepsy may face difficult decisions about their reproductive health. It may be essential to choose drugs with low or absent teratogenic potentials. It may also be of importance to avoid drugs that interact with hormonal contraception (Ben-Menachem, 1997).

Cognitive side-effects are of particular concern in children and students, and also in other people exposed to intellectual challenges. Patients with organic brain lesions, including demented and mentally retarded individuals may show a lowered threshold for central nervous side effects (Helmstaedter et al., 1993; Mattson, 1996). In multihandicapped patients, investigation and treatment are often hampered by communication problems. Difficulties in achieving a satisfactory balance between seizure control and adverse reactions may be pronounced. The unwanted effects may be masked by the pre-existing intellectual deficits or sometimes manifest themselves as behavioural problems (Brodtkorb, 1998). Drugs with simple intake regimens may improve compliance in patients with memory or concentration difficulties (Willmore, 1996). Elderly patients often suffer from underlying concomitant disease. When multiple therapies are needed, drugs with low interaction potentials may be preferred (Pedersen, 1997). Seizures in sick and medically complex patients, e.g., with hepatic or renal impairment require special knowledge and management (Boggs, 1997). In some patients with undetermined seizure types and in epilepsy encephalopathies with multiple seizure types of both partial and generalized origin, broad-spectred agents may be the most appropriate primary drugs.

**New drugs**

Currently, the new drugs are recommended in seizure disorders in which traditional drugs are not effective or cause intolerable side effects (Dichter and Brodie, 1996). For VGB, the clinical efficacy appears to be best in partial seizures with and without generalization and in symptomatic infantile spasms. VGB seems to have a negative effect on myoclonic and absence seizures (Grant and Heel, 1991). LTG has an apparently broader spectrum of action, permitting its use also in patients with multiple or unknown seizure types. Recent findings point to a useful role in the Lennox-Gastaut syndrome (Motte et al., 1997). In children with intractable seizures, uncontrolled studies have shown a better efficacy in generalized than in localization-related epilepsies, including the more difficult to treat cryptogenic and symptomatic forms (Schlumberger et al., 1994). GBP and TGB are effective in partial seizures, whereas FBM and TPM seem to be more broad-spectred drugs. The use of FBM is restricted due to rare but severe idiosyncratic side-effects in the form of aplastic anaemia and liver toxicity (Bebin et al., 1995). With the exception of FBM and recently also LTG, the new drugs are at present only licensed for use in partial seizures.

Each new drug represents progress in the treatment of epilepsy (e.g., mode of action, side-effect profile and/or pharmacokinetic properties) which, as already outlined, may be advantageous to many patients. However, there is currently no scientific evidence that these drugs are more potent in monotherapy in newly diagnosed patients than the traditional AEDs. The effect of the new drugs are initially documented in the heterogeneous group of refractory localization-related epilepsy. Many of these compounds have not yet found their proper place in epilepsy treatment in spite of widespread use. More controlled studies in selective epilepsy syndromes and groups of patients are called for.

Patients with uncontrolled seizures or unwanted effects with traditional drugs should be offered to try the new drugs, but already successful treatment should not be changed. The new drugs are expensive. Possible long term adverse reactions and rare complications are not yet known for many of them. The FBM story has told the clinicians a lesson. The severe, but rare adverse reactions from the bone marrow and the liver was a post-marketing experience. If no definite gain is achieved with novel treatment, a return to traditional drugs is again advocated. However, only a small improvement in the balance between seizure control and unwanted effects may be important in long term treatment. The cumulated gain for patients with chronic epilepsy may be considerable. Many of the new compounds seem to have an improved side effect profile, but drug tolerance is a highly individual matter.

**Pharmacokinetic interaction potentials**

Correct utilization of AEDs involves knowledge of pharmacokinetics and drug interactions. The pharmacokinetic properties are the factors that determine how optimal plasma concentrations can be achieved and maintained. Most physicians tend to choose drugs that are more easy to use and combine, if they are otherwise equally safe and potent. In clinical practice, the two most important pharmacokinetic properties are linear elimination kinetics and lack of pharmacokinetic interactions with other drugs.

The drugs that are metabolized by the hepatic P 450 isoenzyme system are prone to drug-drug interactions. The enzyme inducers (PB, PHT, and CBZ) have the most pronounced interaction potentials, while VPA may act as a potent enzyme inhibitor. The metabolism of LTG may be induced as well as inhibited, whereas the proportion of TPM that is metabolized in the liver may be induced. VGB, GBP, and TGB are excreted unchanged, but VGB still has an unexplained tendency to lower the plasma levels of PHT somewhat (approximately 20%) (Bourgeois, 1995).

Recently a great step forward has taken place in the understanding of inhibition-based drug interactions. Apparently, a limited number of isoforms of the cytochrome P450 are involved in the hepatic metabolism of most drugs. It is posssible to correlate such isoforms with specific metabolic pathways and the formation of certain metabolites. Specific interactions may now be mapped in vitro and such studies will probably have to be included in the developmental programmes of new drugs (Levy, 1995)

**Comparisons between drugs**

The prescribing doctors are faced with an increasing number of new treatments. The important clinical question is how a new drug compares with the standard or first choice treatment currently available (Marson and Chadwick, 1996). In the patient with seizures refractory to traditional treatment, the choice of further options should preferably be based upon evidence-based facts in the form of comparative randomised controlled trials.

**Potency**

Several studies have compared the efficacy of the traditional drugs. In the first of these studies, CBZ, PHT, PB, and primidone (PRM) were compared (Mattson et al., 1985). When all seizure types were considered together there was no difference between seizure treatment groups with respect to seizure frequency, but when partial seizures were considered separately, CBZ was more likely to render patients seizure-free. In another study CBZ and VPA were compared in partial epilepsy (Mattson et al., 1992). No significant difference was found concerning GTC, but for CPS there was a significant reduction in frequency in favour of CBZ, but not between the numbers of completely seizure-free patients. However, other studies have not supported that CBZ is more effective than VPA in partial seizures (Richens et al., 1994; Verity, et al., 1995). A recent study comparing CBZ and LTG in newly diagnosed patients indicated that these two drugs were similarly efficacious in that setting (Brodie et al., 1995).

A meta-analysis of monotherapy randomized studies using the proportion of patients who showed 50% or greater reduction in seizure frequency has recently been reported (Marson et al., 1996). All six included drugs (VGB, LTG, GBP, TPM. TGB, and Zonisamide (ZNS)) were significantly better than placebo, but the 95% confidence interval for all drugs overlapped, indicating that their efficacy were not significantly different. Nevertheless, the authors felt that TPM, VGB and TGB had the best apparent evidence of potency and would seem the best choice for patients with poor seizure control with standard drugs.

However, in these studies, the dose-effect relationship could not be adequately taken into account. The optimal doses are difficult to explore and may be different in different sub-populations of patients and also in each individual. The doses of LTG and GBP were probably not optimal in these studies. Further controlled clinical trials comparing different drugs in individual dose regimens are needed in order to optimize the current approach to refractory epilepsy in different patient categories and epilepsy syndromes.

**Tolerability**

The goal of drug therapy in epilepsy is not to stop seizures at any cost. The goal is to achieve an optimal therapeutic effect with a minimum of undesirable effects. All effective drugs have non-therapeutic effects that need careful clinical consideration.

In clinical practice, the side effects of AEDs may be classified into four main categories (Table 5): (1) dose-related central nervous side effects, (2) unpredictable (idiosyncratic) reactions, (3) complications, and (4) late effects of long term treatment.

The most common adverse events are dose-related (or more precisely drug-concentration-

TABLE 5

Adverse reactions of antiepileptic drugs

1. Drug concentration-related central nervous side effects: sedative, cognitive and behavioural symptoms
2. Unpredictable (idiosyncratic) reactions
   –immunological, e.g. rash
   –metabolic, e.g. valproate hepatotoxicity
3. Complications
   e.g. teratogenic effect, hormonal contraceptive failure
4. Long term treatment effects
   e.g. weight changes, hormonal problems, osteomalacia, hirsutism, gingival hyperplasia, Dupuytren's contracture, phenytoin encephalopathy etc.

related) CNS symptoms that usually appear early in treatment and often subside with continuing therapy and adequate dose adjustment. The tendency to such side effects may be substantially reduced when the start dose is low and subsequently carefully adjusted to appropriate clinical response. All AEDs probably can compromise cognitive functioning, particularly when plasma concentrations are high. Knowledge of the comparative tolerability of AEDs is essential when making a choice between the different drugs. The central nervous side effects are usually subtle and sneaking and may be difficult to evaluate, particularly in the large group of patients with intellectual disabilities who do not communicate well (Brodtkorb, 1998). A variety of psychometric tools have been used to investigate this issue, but current studies are difficult to compare (Marson and Chadwick, 1996). Some studies have suggested that CBZ and VPA have less potential for negative effects than other medications, particularly barbiturates. Recent studies show findings that may reflect a subtle compromise in cognitive functioning even with these drugs in standard doses (Prevey et al., 1996).

In the study comparing CBZ, PHT, PB, and PRM (Mattson et al., 1985), there was little difference in the retention of patients with either CBZ and PHT, but it was better than in the group randomised to PB, which in turn was better than in the PRM group. In two studies comparing CBZ and VPA, one in adults (Richens et al., 1994) and one in children (Verity et al., 1995), the overall prevalence of side effects was higher in the CBZ-treated adult group (mainly rashes), but similar in children with a different side-effect profile for each drug. In the above mentioned meta-analysis of the efficacy and tolerability of the new antiepileptic drugs (VGB, LTG, GBP, TPM, ZNS) the 95% confidence intervals for withdrawals also overlapped, indicating that the tolerability was not significantly different. However, the tendencies indicate that GBP and LTG are better tolerated than TGB, TPM and VGB (Marson et al., 1996). The results may again be biased by dose inequivalency, but clinical experience is consistent with a wide therapeutic index of LTG and GBP. For patients who exhibit low thresholds for dose dependent side effects and have a less severe epilepsy, these two may be the best options among the novel drugs.

These data are derived form large groups of patients. It is always important for the clinician to bear in mind that individuals are treated. The variable adverse-effect profiles of AEDs between different patients are enormous and must never be lost in means and p-values. Some patients are sensitive to one or several AEDs wheras others are remarkably tolerant (Devinsky, 1995).

Some aspects of drug tolerability currently cause particular concern and need further comments:

*Rash* is a hypersensitivity reaction that occurs in a few percent of patients exposed to the traditional drugs, such as PB, PHT, and CBZ. A rash is very rare in VPA treatment and probably does not occur with VGB. Initial high plasma levels predispose to skin rash in LTG treatment. This occurs particularly when used as adjunctive treatment to VPA. The reactions may be severe and the incidence seems higher in children than in adults. When the dose is escalated very slowly, the frequency of rash approximates that of other drugs (Shorvon and Stefan, 1997).

*Blood idiosyncrasies* may be caused by several AEDs. In FBM therapy, the incidence of aplastic anemia has been estimated as high as 1:4–5000. This was not detected before extensive post-marketing use (Bebin et al., 1995). The potential risk of using this drug should be carefully weighed against its benefit.

*VPA hepatotoxicity*, histologically a microvesicular steatosis similar to that found in Reye's syndrome, predominantly affects children. The formation of the toxic compound, 4-en VPA, is dependent on individual factors in the liver such as maturity and the load of other drugs. Enzyme inducing drugs may enhance its formation and some children may have a pre-existing metabolic defect. The presenting signs are vomiting, lethargy and seizure exacerbation (Bryant and Dreifuss, 1996).

*Offspring malformations* are increased 2 to 3 times in women using AEDs (for a review, see Samrén and Lindhout, 1997). However, more than 90% of women with epilepsy give birth to normal babies. It is impossible to rank the teratogenicity of the traditional drugs, but their teratogenic profiles are somewhat different (Table 6). Several theories about the mechanisms have been suggested, e.g., disturbed folate metabolism, the formation of endogenous toxic, reactive radicals and there has also been speculations about a disturbed Vitamin A metabolism (Danielson, 1997; Nau, 1997).

None of the newer AEDs are licensed for use in pregnancy and extensive experience is lacking with these drugs. LTG, GBP and TGB are promising in these respects. No teratogenic potential has been shown in animal studies, but species specificities may be high and caution is recommended (Morell, 1996; Ben-Menachem, 1997; Tomson et al., 1997)

TABLE 6

Teratogenic profiles of traditional antiepileptic drugs

| | |
|---|---|
| Valproate: | 1–2% Neural tube defects |
| | Radial aplasia |
| Phenytoin/Phenobarbital: | Heart defects |
| | Orofacial clefts |
| Carbamazepine: | Hypospadia, |
| | 0,9% Neural tube defects |

According to Lindhout et al., 1992

*Constricted visual fields* have recently been detected in some patients during VGB treatment (Krauss et al., 1998).

*Weight gain* may limit the utilization of VPA. VPA may affect fatty acid beta-oxidation (Breum et al., 1992). The availability of fatty acids for deposition may be increased. Weight increase may also be a problem in VGB and GBP treatment, and sometimes also with CBZ. It is an interesting fact that weight loss is a recognized side effect of some of the new drugs, TPM and FBM.

*Late metabolic effects* are at present a focus of increasing clinical attention and concern. It is important to bear in mind that we do not exclusively treat the sick part of the brain with AEDs. We expose the entire body to a pharmacologically active compound. Enzyme inducers may interact with other drugs as well as with the intrinsic body milieu. The homeostasis of e.g. hormones, carrier proteins and lipids may be altered by them. Metabolism of thyroid and steroid hormones may be influenced. An increased production of sex hormone binding globulin may also occur, resulting in less free and active testosterone and estrogen (Isojärvi et al., 1994, 1995a,b). Enzyme inhibition by e.g. valproate may possibly also have harmful effects. In VPA treatment, an increased occurrence of polycystic ovaries and hyperandrogenism have been reported in obese women (Isojärvi et al., 1993). For some patients a modified lipoprotein metabolism may alter cholesterol levels and potentially influence cardiovascular morbidity.

In spite of these facts, the use of traditional AEDs is generally well tolerated and beneficial for most patients. However, awareness of the risks is warranted. We welcome the new AEDs that do not affect liver enzyme action.

## Future perspectives

Neuroprotection is an important field in modern neurology (Trembley, 1995). An unanswered question is if a neuroprotective effect is important in epilepsy. Clinical studies that correlate the possible effects of seizures and cognitive functioning is difficult to perform, as they usually will be confounded by AED treatment as well as underlying neuropathologies (Lynch et al., 1996). It is believed that a neuroprotective effect may block further seizure-induced damage in the brain and associated functional alterations, which possibly may include seizure promoting factors. In prolonged febrile seizures, exitotoxic cell death in epileptogenic regions like the hippocampus, seems to be responsible for future spontaneous seizures (Maher and McLachlan, 1995).

The present AEDs have a proven symptomatic anti-seizure effect. There is no scientific evidence to support that they have an effect on the development or on the prognosis of the underlying disorders in epilepsy patients (Shinnar and Berg, 1996). However, animal studies have shown that FBM and the GABA-enhancing agents VGB and TGB may act as neuroprotective agents. In spite of the side effects elicited by many NMDA antagonists, it is hoped that differential effects may be detected among these compounds that may render them suitable for clinical application in this field (Sagratella, 1995). These speculations currently indicate novel directions for research. Agents that not only prevent symptomatic seizures, but also lower the likelihood of developing epilepsy after a brain injury may prove to be very valuable. Compounds with antiepileptogenic and neuroprotective properties could possibly alter the natural course of the disorder and prevent the development of intractable epilepsy and hopefully also the progressive cognitive decline, particularly associated with some forms of childhood-onset epilepsies (e.g., West and Lennox-Gastaut syndromes). The arrival of such drugs will have a major influence on the future AED choice.

Continued and growing clinical research is needed to obtain further knowledge about the optimal use of the expanding antiepileptic armamentarium and how to tailor the treatment in each individual patient according to clinical circumstances. Epileptologists have many tools and must continuously learn how to use them better to improve the quality of life of their patients.

## References


Bebin, E.M., Sofia, R.D. and Dreifuss, F.E. (1995) Felbamate. Toxicity. In Levy, R.H., Mattson, R.H. and Meldrum B.S. (Eds.), *Antiepileptic Drugs*, Raven Press, New York, pp. 823–827.

Ben-Menachem, E. (1997) Women with epilepsy: Must Contraception and Pregnancy add to the burden of epilepsy? *Epilepsia*, 38 (suppl 2): 13–18.

Binnie, C.D. (1997) Lamotrigine. In Engel, J., Pedley, T.A. (Eds.) *Epilepsy: A comprehensive textbook, Vol. 2*, Lippincott-Raven Publishers, Philadelphia pp. 1531–1540.

Boggs, J.G. (1997) Seizures in medically complex patients. *Epilepsia*, 38(suppl 4): 55–59.

Bourgeois, B.F.D. (1995) Important pharmacokinetic properties of antiepileptic drugs. *Epilepsia*, 36 (suppl 5): 1–7.

Breum, L., Astrup, A., Gram, L., Andersen, T., Stokholm, K.H., Christensen, N.J., Werdelin, L. and Madsen, J. (1992) Metabolic changes during treatment with valproate in humans: Implications for untoward weight gain. *Metabolism*, 41: 666–670.

Brodtkorb, E. (1998) Treatment of epilepsy in the mentally retarded. In *Treatment of Epilepsy Workshop 1998, 1*, Medical Product Agency, Uppsala, Sweden. 189–200.

Brodie, M.J., Richens, A. and Yuen, A.W.C. (1995) Double blind comparison of lamotrigine and carbamazepine in newly diagnosed epilepsy. *Lancet*, 345: 476–479.

Brodie, M.J. and Dichter, M.A. (1997) Established antiepileptic drugs. *Seizure*, 6: 159–174.

Bryant, A.E. and Dreifuss, F.E. (1996) Valproic acid hepatic fatalities. III. U.S. experience since 1986. *Neurology*, 46: 465–469.

Commission on Antiepileptic Drugs, International League against Epilepsy. (1993) Guidelines for therapeutic monitoring of antiepileptic drugs. *Epilepsia*, 34: 585–587.

Commission on Classification and Terminology of the International League against Epilepsy. (1981) Proposal for revised clinical and electroencephalographic classification of epileptic seizures. *Epilepsia*, 22: 489–501.

Commission on Classification and Terminology of the International League against Epilepsy. (1989) Proposal for revised classification of epilepsies and epileptic syndromes. *Epilepsia*, 30: 389–399.

Danielson, B.R. (1997). Mechanisms of teratogenesis of antiepileptic drugs. In Tomson, T., Gram, L., Sillanpää, M., Johannessen, S.I. (Eds.) *Epilepsy and Pregnancy*, Wrightson Biomedical Publishing Ltd., Petersfield, Bristol pp. 17–33.

Devinsky, O. (1995) Cognitive and behavioral effects of antiepileptic drugs. *Epilepsia*, 36(suppl 2): 46–65.

Dichter, M.A. and Brodie, M.J. (1996) New antiepileptic drugs. *N. Engl. J. Med.*, 334: 1583–1590.

Dreifuss, F.E. (1994) New antiepileptic drug development. *Epilepsia*, 35(suppl 5): 6–9.

Editorial. (1993) Official names of antiepileptic drugs. *Epilepsia*, 34: 115.

Fitton, A. and Goa, K.L. (1995) Lamotrigine. An update of its pharmacology and therapeutic use in epilepsy. *Drugs*, 50: 691–713.

French, J.A. (1997) What trials, which designs? *Epilepsia*, 38: 263–265.

Genton, P. and Roger, J. (1997) Antiepileptic drug monotherapy versus polytherapy: A historical perspective. *Epilepsia*, 38(suppl 5): 2–5.

Goldensohn, E.S., Porter, R.J. and Schwartzkroin, P.A. (1997) The American Epilepsy Society: An historic perspective on 50 years of advances in research. *Epilepsia*, 38: 124–150.

Grant, S.M. and Heel, R.C. (1991) Vigabatrin. A review of its pharmacodynamic and pharmacokinetic properties, and therapeutic potential in epilepsy and disorders of motor control. *Drugs*, 41: 889–926.

Helmstaedter, C., Wagner, G. and Elger, C.E. (1993). Differential effects of first antiepileptic drug application on cognition in lesional and non-lesional patients with epilepsy. *Seizure*, 2: 125–130.

Isojärvi, J.I.T., Laatikainen, T.J., Pakarinen, A.J., Juntunen, K.T.S. and Myllylä, V.V. (1993) Polycystic ovaries and hyperandrogenism in women taking valproate for epilepsy. *N. Engl. J. Med.*, 329: 1383–1388.

Isojärvi, J.I.T., Pakarinen, A.J., Rautio, A., Pelkonen, O. and Myllylä, V.V. (1994) Liver enzyme induction and serum lipid levels after replacement of carbamazepine with oxcarbazepine. *Epilepsia*, 35: 1217–1220.

Isojärvi, J.I.T., Airaksinen, K.E.J., Mustonen, J.N., Pakarinen, A.J., Rautio, A., Pelkonen, O. and Myllylä, V.V. (1995a) Thyroid and myocardial function after replacement of carbamazepine with oxcarbazepine. *Epilepsia*, 36: 810–816.

Isojärvi, J.I.T., Pakarinen, A.J., Rautio, A., Pelkonen, O. and Myllylä, V.V. (1995b) Serum sex hormone levels after replacement of carbamazepine with oxcarbazepine. *Eur. J. Clin. Pharmacol.*, 47: 461–464.

Johannessen, S.I. and Sundqvist, A. (1998) Review of present and potential antiepileptic drugs (including side-effects). In *Treatment of Epilepsy Workshop 1998, 1,* Medical Product Agency, Uppsala, Sweden. 105–153.

Krauss, G.L., Johnson, M.A. and Miller, N.R. (1998) Vigabatrin-associated retinal cone system dysfunction. Electroretinogram and ophthalmologic findings. *Neurology*, 50: 614–618.

Levy, R.H. (1995) Cytochrome P450 isozymes and antiepileptic drug interactions. *Epilepsia*, 36(suppl 5): 8–13.

Lindhout, D., Meinardi, H., Meijer, J.W.A. and Nau, H. (1992) Antiepileptic drugs and teratogenesis in two consecutive cohorts: Changes in prescription policy paralleled by changes in pattern of malformations. *Neurology*, 42 (suppl 5): 94–110.

Lynch, M.W., Rutecki, P.A. and Sutula, T.P. (1996) The effects of seizures on the brain. *Curr. Opin. Neurol.*, 9: 97–102.

Macdonald, R.L (1997) Cellular effects of antiepileptic drugs. In Engel, J., Pedley, T.A. (Eds.) *Epilepsy: A comprehensive textbook, Vol.2,* Lippincott-Raven Publishers, Philadelphia pp. 1383–1391.

Macdonald, R.L. and Greenfield, L.J. (1997) Mechanisms of action of new antiepileptic drugs. *Curr. Opin. Neurol.*, 10: 121–128.

Maher, J. and McLachlan, R.S. (1995) Febrile convulsions. Is seizure duration the most important predictor of temporal lobe epilepsy? *Brain*, 118: 1521–1528.

Marson, A.G. and Chadwick, D. (1996) Comparing antiepileptic drugs. *Curr. Opin. Neurol.*, 9: 103–106.

Marson, A.G., Kadir, Z.A. and Chadwick, D.W. (1996) New antiepileptic drugs: A systematic review of their efficacy and tolerability. *BMJ*, 313: 1169–1174.

Mattson, R.H. (1996) The role of the old and the new drugs in special populations: Mental and multiple handicaps. *Epilepsia*, 37 (suppl 6): 45–53.

Mattson, R.H., Cramer, J.A. and Collins, J.F. (1992) A comparison of valproate with carbamazepine for the treatment of complex partial seizures and secondarily generalized tonic clonic seizures in adults. *N. Engl. J. Med.*, 327: 765–771.

Mattson, R.H., Cramer, J.A., Collins, J.F., Smith, D.B., Delgado-Escueta, A.V., Browne, T.R., Williamson, T.R., Treiman, D.M., McNamara, J.O. and McCutchen, C.B. (1985) Comparison of carbamazepine phenobarbital phenytoin and primidone in partial and secondarily generalized tonic-clonic seizures. *N. Engl. J. Med.*, 313: 145–151.

Meldrum, B.S. (1996) Update on the mechanism of action of antiepileptic drugs. *Epilepsia*, 37 (suppl 6): 4–11.

Meldrum, B.S. (1997) Current strategies for designing and identifying new antiepileptic drugs. In Engel, J., Pedley, T.A. (Eds.) *Epilepsy: A comprehensive textbook, Vol. 2,* Lippincott-Raven Publishers, Philadelphia pp. 1405–1416..

Morell, M. (1996) New antiepileptic drugs and women: Efficacy, reproductive health, pregnancy, and fetal outcome. *Epilepsia*, (suppl 6): 34–44.

Motte, J., Trevathan, E., Arvidsson, J.F.V., Barrera, M.N., Mullens, E.L., Manasco, P. and the Lamictal Lennox-Gastant Study Group (1997) Lamotrigine for generalized seizures associated with the Lennox-Gastant syndrome. *N. Engl. J. Med.*, 337: 1807–12.

Nau, H. (1997) Towards the mechanism of valproic acid induced neural tube defects. In Tomson T., Gram L., Sillanpää M., Johannessen S.I. (Eds.) *Epilepsy and Pregnancy*, Wrightson Biomedical Publishing Ltd., Petersfield, Bristol, pp. 35–42.

O'Connell, M.T. and Patsalos, P.N. Phenobarbitone. In Shorvon S, Dreifuss F, Fish D, Thomas D. (Eds.) *The treatment of epilepsy*, Blackwell Science, Oxford, pp. 446–453.

Osservatorio Regionale per L'Epilessia (OREp), Lombardy. (1996) ILAE classification of epilepsies: Its applicability and practical value of different diagnostic categories. *Epilepsia*, 37: 1051–1059.

Pedersen, B. (1997) Elderly patients with epilepsy: Specific requirements. *Epilepsia*, 38 (suppl 2): 19–22.

Perucca, E., Gram, L., Avanzini, G. and Dulac, O. (1998) Antiepileptic drugs as a cause of worsening seizures. *Epilepsia*, 39: 5–17.

Prevey, M.L., Delaney, R.C., Cramer, J.A., Cattanach, L., Collins, J.F. and Mattson, R.H. (1996) Effect of valproate on cognitive functioning. Comparison with carbamazepine. *Arch. Neurol.*, 53: 1008–1016.

Reynolds, E.H. and Shorvon, S.D. (1981) Monotherapy or polytherapy for epilepsy. *Epilepsia*, 22: 1–10.

Richens, A. (1995) Rational monotherapy. *Seizure*, 4: 211–214.

Richens, A., Davidson, D.L.W., Cartlidge, N.E.F. and Easter, D.J. (1994) A multicenter comparative trial of sodium valproate and carbamazepine in adult onset epilepsy. *J. Neurol. Neurosurg. Psychiatry*, 57: 682–687.

Sagratella, S. (1995) NMDA antagonists: Antiepileptic-neuroprotective drugs with diversified neuropharmacological profiles. *Pharmacol Res*, 32: 1–13.

Samrén, E.B. and Lindhout, D. (1997) Major malformations associated with maternal use of antiepileptic drugs. In Tomson, T., Gram, L., Sillanpää, M., Johannessen, S.I. (Eds). *Epilepsy and Pregnancy* Wrightson Biomedical Publishing Ltd., Petersfield, Bristol, pp. 43–61.

Schlumberger, E., Chavez, F., Palacios, L., Rey, E., Pajot, N. and Dulac, O. (1994) Lamotrigine in 120 children with epilepsy. *Epilepsia*, 35: 359–367.

Shinnar, S. and Berg, A.T. (1996) Does antiepileptic drug therapy prevent the development of "chronic" epilepsy? *Epilepsia*, 37: 701–708.

Shinnar, S. and Berg, A.T. (1995) Withdrawal of antiepileptic drugs. *Curr. Opin. Neurol.*, 8: 103–106.

Shorvon, S. and Stefan, H. (1997) Overview of the safety of antiepileptic drugs. *Epilepsia*, 38 (suppl 1): 45–51.

Tomson, T., Gram, L., Sillanpää, M. and Johannessen, S.I. (1997) Recommendations for the management and care of pregnant women with epilepsy. In Tomson, T., Gram, L., Sillanpää, M., Johannessen, S.I. (Eds.) *Epilepsy and Pregnancy* Wrightson Biomedical Publishing Ltd., Petersfield, Bristol, pp. 201–208.

Tomson, T., Johannessen, S.I. and Olsson, I. (1998) Basic principles of prophylactic antiepileptic drug treatment. In *Treatment of Epilepsy Workshop 1998, 1,* Medical Product Agency, Uppsala, Sweden. 155–175.

Trembley, B. (1995). Clinical potential for the use of neuroprotective agents. A brief overview. *Ann. N.Y. Acad. Sci.*, 765: 1– 20.

Verity, C.M., Hosking, G. and Easter, D.J. (1995) A multicenter comparative trial of sodium valproate and carbamazepine in paediatric epilepsy. *Dev. Med. Child. Neurol.*, 37: 97–108.

Willmore, L.J. (1996) Management of epilepsy in the elderly. *Epilepsia*, 37 (suppl 6): 23–33.

O.P. Ottersen, I.A. Langmoen and L. Gjerstad (Eds.)
*Progress in Brain Research*, Vol 116


CHAPTER 27

# The glutamate synapse: A target in the pharmacological management of hyperalgesic pain states

M.O. Urban and G.F. Gebhart*

*Department of Pharmacology, University of Iowa College of Medicine, Iowa City, IA 52242-1109, USA*

## Introduction

Over the past two decades, it has been become increasingly evident that glutamate and its receptors play key roles in the transmission and production of nociception (pain). Particular attention has focused on the role of glutamate in the production of hyperalgesia (enhanced response to a noxious stimulus) that is observed in models of persistent pain. A variety of such models, including cutaneous neurogenic, neuropathic, and visceral, have been extensively studied in recent years. It has been proposed in persistent pain models that the hyperalgesia, enhanced excitability of spinal dorsal horn nociceptive neurons (termed "central sensitization"), and expanded peripheral receptive fields are the result of a neuroplasticity that is largely dependent on glutamate activation of NMDA receptors in the spinal cord.

## Role of spinal NMDA receptors and NO˙ in neuroplasticity and pain

A role for EAAs in the spinal cord in nociception was suggested by the early observation that iontophoretic application of glutamate excites spinal dorsal horn neurons (Curtis et al., 1960). Glutamate was subsequently localized in the spinal cord, including the central terminals of primary afferent nociceptors (Miller et al., 1988), as have been EAA receptors, including ionotropic NMDA and AMPA/kainate receptors and metabotropic glutamate (mGlu) receptors (Greenamyre et al., 1985; Li et al., 1997). Each of these receptors has been implicated in spinal nociceptive transmission because iontophoretic application of NMDA or AMPA receptor agonists enhances spinal dorsal horn neuron responses to noxious stimulation (Aanonsen et al., 1990), and spinal administration of NMDA, AMPA, or mGlu receptor agonists produce spontaneous pain behaviors and hyperalgesia (Aanonsen and Wilcox, 1987; Brambilla et al., 1996; Coderre et al., 1997; Meller et al., 1992a).

That NMDA receptors in the spinal dorsal horn are an integral component of a plasticity which results in central sensitization and hyperalgesia following prolonged noxious stimulation has been the focus of numerous studies (for reviews see Coderre et al., 1993; Wilcox, 1991). Repetitive, high frequency electrical stimulation of C-fiber nociceptors produces a "wind-up," in which dorsal horn nociceptive neurons are characterized as having increased spontaneous activity, enhanced response magnitudes to subsequent C-fiber input, prolonged afterdischarges, and expanded receptive fields (Mendell, 1966; Cook et al., 1987). This

*Corresponding author.
Tel.: +1 319 335 7946; fax: +1 319 335 8930; e-mail: gf-gebhart@uiowa.edu.

increased excitability of spinal dorsal horn nociceptive neurons and the behavioral hyperalgesia that occurs following high frequency stimulation of nociceptors (via direct electrical stimulation or following peripheral injury) is largely dependent on a central plasticity because there is often an expansion of receptive fields and hyperalgesia to areas outside, and even distant to the site of injury/stimulation. Centrally mediated "wind-up" and plasticity following high frequency C-fiber stimulation is NMDA receptor-mediated because spinal administration of NMDA receptor antagonists block this effect (e.g., Davies and Lodge, 1987; Dickenson and Sullivan, 1987).

A model has been proposed in which prior glutamate activation of spinal ionotropic AMPA receptors and mGlu receptors (in addition to neuropeptide activation of other receptors; e.g., neurokinin receptors) produces a depolarization that removes the voltage dependent $Mg^{2+}$ block of the NMDA receptor channel and allows enhanced NMDA receptor-mediated $Ca^{2+}$ currents through the channel (for reviews see Coderre et al., 1993; Wilcox, 1991). Activation of protein kinase C via mGlu receptors and the subsequent increase in $[Ca^{2+}]i$ results in enhanced NMDA receptor $Ca^{2+}$ currents and glutamate release. Activation of the NMDA receptor, located on the terminals of substance P-containing primary afferent neurons (Shigemoto et al., 1992; Liu et al., 1994) as well as post-synaptic to those terminals, has recently been reported to cause the release of substance P (Liu et al., 1997), completing a positive feedback loop which likely contributes to central sensitization.

Following NMDA receptor activation, the production of nitric oxide (NO$^•$) in the spinal cord has been identified as an important component of plasticity and central sensitization (for review see Meller and Gebhart, 1993). It is proposed that NMDA receptor activation and increased $[Ca^{2+}]_i$ in spinal dorsal horn neurons results in increased activity of the constitutive, neuronal nitric oxide synthase (NOS) and increased NO$^•$ production. Production of NO$^•$ and subsequent activation of soluble guanylate cyclase in the spinal cord appear to be important mechanisms of central sensitization and hyperalgesia following spinal NMDA administration (Meller et al., 1992a).

Animal models of peripheral inflammation involving carrageenan, formalin, or mustard oil treatment all sensitize nociceptors, resulting in central sensitization and hyperalgesia. Peripheral inflammation following carrageenan results in enhanced glutamate release in the spinal cord (Sorkin et al., 1992). Accordingly, the expansion of receptive fields, enhanced dorsal horn nociceptive neuron responses, and behavioral hyperalgesia that develop after carrageenan administration are blocked following spinal administration of NMDA receptor antagonists (Ren et al., 1992a, b). Furthermore, inhibition of spinal NOS has been shown to block carrageenan-induced hyperalgesia (Meller et al., 1994). Similar to carrageenan, application of mustard oil (allyl isothiocyanate; a C-fiber excitant) has been shown to produce a central sensitization that is blocked by NMDA receptor antagonists (Wolf and Thompson, 1991). Lastly, injection of formalin is a well characterized model of nociception in which rodents exhibit spontaneous pain behaviors in two distinct phases: an early phase involving activation of peripheral nociceptors (Puig and Sorkin, 1995) and a late phase presumably involving central sensitization (Coderre et al., 1990). Accordingly, spinal administration of NMDA receptor antagonists and NOS inhibitors are found to be most effective in blocking the later phase nociceptive behaviors in the formalin test (Coderre and Van Empel, 1994b; Malmberg and Yaksh, 1993).

Animal models of neuropathic pain involving loose ligation of peripheral nerves have improved our understanding of neuropathic pain in humans. Loosely constrictive ligatures around peripheral nerves produces hyperalgesia, spontaneous pain behaviors, and enhanced responses of spinal dorsal horn nociceptive neurons (for review see Bennett, 1993). Both systemic and spinal administration of NMDA receptor antagonists have been found to effectively reduce this hyperalgesia, while non-NMDA receptor antagonists are less effective (Mao et al., 1992; Qian et al., 1996). A role for spinal NO˙ in neuropathic pain has also been demonstrated by increased spinal NOS activity (Choi et al., 1996) and the observation that spinal administration of NOS inhibitors blocks nerve injury-induced hyperalgesia (Meller et al., 1992b).

While mechanisms of cutaneous pain have received a great deal of attention, pain originating from the viscera is less well understood. Pain associated with the gastrointestinal tract, for example, occurs in subjects with inflammatory or functional bowel disorders and is likely a consequence of altered colonic reflexes and hypersensitivity. Colorectal distension (CRD) has been well characterized as a noxious visceral stimulus that results in pain in man (Ness and Gebhart, 1990; Ness et al., 1990) and produces reliable physiological and behavioral responses in animals, including a pseudaffective visceromotor response (Ness and Gebhart, 1988). A role for spinal glutamate in visceral pain is supported by the observation that spinal administration of NMDA enhances the visceromotor response, increases responses of spinal dorsal horn neurons to noxious CRD, and results in an expansion of convergent cutaneous receptive fields (Kolhekar and Gebhart, 1994, 1996). The enhanced behavioral responses and dorsal horn neuron responses to noxious CRD following spinal NMDA treatment were all blocked by NMDA receptor antagonists, suggesting a contribution of spinal NMDA receptors to visceral pain and hyperalgesia. A contribution of spinal NMDA receptors to visceral hyperalgesia was demonstrated more directly by several studies in which enhanced responses to noxious CRD following inflammation of the colon were inhibited by administration of spinal NMDA receptor antagonists (Coutinho et al., 1996; Ide et al., 1997).

It has been generally proposed that while spinal NMDA receptors contribute to central sensitization and facilitation of nociceptive responses produced by high frequency nociceptive input, responses to acute, phasic noxious stimuli do not involve spinal NMDA receptors. This conclusion is based on results from a number of studies in which doses of spinal NMDA receptor antagonists effective in blocking the facilitation of nociceptive withdrawal reflexes of an affected hindlimb have no effect on nociceptive withdrawal reflexes of the unaffected, contralateral hindlimb (Ren et al., 1992 a,b; Yamamoto and Yaksh, 1992). Results from studies such as these suggest that low frequency input from acute, phasic noxious stimulation likely involves fast synaptic transmission mediated by spinal non-NMDA (e.g., AMPA) receptors. This notion is supported by a recent study which found that spinal non-NMDA receptor antagonists were significantly more antinociceptive in an acute pain model than NMDA receptor antagonists (Lutfy et al., 1997). Other studies, however, do report an inhibition of acute nociceptive withdrawal reflexes following spinal administration of NMDA receptor antagonists, but only at doses that produce a concomitant hindlimb motor paralysis (Aanonsen and Wilcox, 1987; Cahusac et al., 1984; Coderre and Van Empel, 1994a). These results are somewhat confounded by a report that spinal administration of CPP, a potent NMDA receptor antagonist, was found to produce antinociception in tests involving phasic noxious stimulation at doses that did not impair motor function (Kristensen et al., 1994).

Given the conflicting nature of the results obtained with spinal NMDA receptor antagonists, we recently reexamined the effects of a commonly used NMDA receptor antagonist, AP-5, on two different behavioral nociceptive tests, as well as hindlimb motor function, following spinal administration. Acute responses were measured in separate experiments using two commonly used tests of nociception, the thermal tail-flick reflex and thermal paw withdrawal response. Hindlimb motor function was tested using an inclined plane. We found spinal administration of AP-5 to dose-dependently inhibit the tail-flick reflex at relatively low doses (3–30 nmol) that had no apparent effect on motor function. Greater doses (100–200 nmol), however, produced a maximal inhibition of the tail-flick reflex and impaired hindlimb motor function. In contrast, the paw withdrawal response was found to be inhibited only at greater doses (100–200 nmol) that also produced a concomitant hindlimb paralysis. Thus, spinal NMDA receptor antagonists can produce an antinociception in tests involving an acute, phasic noxious stimulus (tail-flick reflex) at doses that do not impair motor function. That spinal AP-5 was found to have no effect on the nociceptive paw withdrawal response is also consistent with previous reports, and may be explained by activation of different nociceptors, perhaps due to the dramatically different rates of heating inherent in these models (Yeomans and Proudfit, 1996).

**Role of peripheral EAA receptors in nociceptive processing**

In addition to modulation of nociceptive transmission in the spinal cord, peripheral EAAs and their receptors have been implicated in nociceptive mechanisms. A recent study localized NMDA and non-NMDA receptors on small diameter, unmyelinated axons in the periphery, and glutamate injection into the hindpaw produced hyperalgesia (Carlton et al., 1995). Consistent with this finding, peripheral injection of selective agonists for NMDA or AMPA receptors were found to produce a similar hyperalgesia, and these effects were selectively blocked by peripheral injection of the appropriate receptor antagonists (Zhuo et al., 1996). A role for peripheral EAA receptors in nociceptive modulation was further demonstrated by the finding that peripheral injection of selective NMDA or AMPA receptor antagonists into an inflamed hindpaw attenuated the hyperalgesia produced by the inflammation (Jackson et al., 1995). These results support a role for peripheral NMDA and non-NMDA receptors in the activation/sensitization of primary afferent nociceptors and development of hyperalgesia following peripheral tissue injury.

**Role of supraspinal EAAs in nociceptive modulation**

While the majority of studies investigating the role of EAAs in the CNS in the modulation of nociception have focused on mechanisms intrinsic to the spinal cord, an involvement of EAAs in supraspinal sites has received relatively little attention. It is well established that spinal nociceptive input is subject to modulation from a number of supraspinal sites, including the midbrain periaqueductal gray (PAG) and rostral ventromedial medulla (RVM) (for reviews see Basbaum and Fields, 1984; Fields et al., 1991; Gebhart and Randich, 1990). The majority of efferent projections from the PAG terminate in the RVM (Abols and Basbaum, 1981) which, in turn, sends descending projections that terminate in laminae I, II, and V of the trigeminal n. caudalis and laminae I–III and V–VII of the spinal cord dorsal horn (e.g., Basbaum et al., 1978; Holstege and Kuypers, 1982).

Although descending pain modulation was originally thought to exert primarily an inhibitory influence, recent studies have demonstrated that descending systems from the PAG and RVM can facilitate spinal nociceptive transmission as well (Light et al., 1986; Zhuo and Gebhart, 1992, 1997; Urban and Gebhart, 1997). The contribution of supraspinal sites, and descending pain facilitatory systems, to the hyperalgesia that develops after peripheral inflammation/injury has received little attention, although a role for supraspinal sites has been proposed (Wiertelak et al., 1994a; Herrero and Cervero, 1996). We have recently demonstrated an involvement of descending pain facilitatory systems in a model of hyperalgesia involving topical application of the C-fiber excitant mustard oil to the hind leg and measurement of the nociceptive tail-flick reflex (Urban et al., 1996). It was found that mustard oil-induced hyperalgesia is blocked in spinal transected rats as well as those in which the RVM had been destroyed by an electrolytic lesion. Similarly, in a model of acute arthritis involving carrageenan/kaolin injection into the hind leg knee joint cavity, we found that facilitation of the paw withdrawal response to noxious heat is significantly attenuated in animals with an RVM lesion. Because the tail-flick reflex and paw withdrawal response are at least partially spinally organized (both responses remain intact in spinal transected animals), it is reasonable to conclude that the facilitation of these responses following peripheral inflammation is, at least in part, the result of activation of a centrifugal descending pain facilitatory system involving a spino-bulbar-spinal loop.

Activation of a spino-bulbar-spinal loop following noxious stimulation is supported by studies using retrograde tracing methods and antidromic stimulation of identified dorsal horn nociceptive neurons. These studies identified ascending projections of dorsal horn nociceptive neurons in the spinoreticular, spinomesencephalic, and spinothalamic tracts which terminate in the RVM, PAG, and thalamus, respectively (for review see Willis, 1986). Consistent with a role for these sites in nociceptive processing, electrical stimulation in these sites has been shown to produce a characteristic behavioral response similar to that observed following peripheral noxious stimulation. This reaction, termed spontaneous pain state, involves characteristic motor responses, including aversive escape responses and jumping, in addition to vocalization and autonomic responses (Casey, 1971; Waldbillig, 1975). A role for supraspinal glutamate in nociceptive processing has been suggested because glutamate is localized in many of these ascending projections (DiBiasi et al., 1994; Magnusson et al., 1987) and microinjection of glutamate into these brainstem sites has been shown to elicit this spontaneous pain state response (Jensen and Yaksh, 1992).

After documenting an involvement of supraspinal sites (RVM) in two different models of hyperalgesia following peripheral inflammation, we examined the potential contribution of glutamate in the RVM to the hyperalgesia observed in these models. Topical application of mustard oil to the lateral surface of the left hind leg produces a neurogenic inflammation and facilitation of the tail-flick reflex (Urban et al., 1996). Microinjection of the competitive NMDA receptor antagonist AP-5 (1–1000 fmol/1 μl) into the RVM before topical mustard oil application dose-dependently inhibited this hyperalgesia (Fig. 1a). Additionally, the greatest dose of AP-5 tested (1000 fmol) produced a significant inhibition of the tail-flick reflex. In contrast, injection of the non-NMDA receptor antagonist DNQX before mustard oil application further enhanced the facilitation of the tail-flick reflex. Consistent with a role for NMDA receptors in the RVM in descending facilitation of nociception, microinjection of NMDA into the RVM of naïve animals produced a dose-dependent, short-lived facilitation of the tail-flick reflex. The mechanisms of NMDA receptor-mediated descending facilitation from the RVM were further investigated in a series of experiments designed to examine the potential contribution of NO$^•$ in the RVM. Similar to AP-5, microinjection of the NOS inhibitor L-NAME (100–1000 nmol/1 μl) into the RVM before topical mustard oil application dose-dependently blocked the mustard oil-induced facilitation of the tail-flick reflex (Fig. 1b) and produced an inhibition of the tail-flick reflex at the greatest dose tested. Microinjection of the inactive stereoisomer D-NAME into the RVM before mustard oil application was without effect. To support the notion that NMDA receptor mediated facilitation of the tail-flick reflex occurs via a NO$^•$ mechanism in the RVM, we found that microinjection of GEA 5024, a NO$^•$ donor, into the RVM also produced a facilitation of the tail-flick reflex. Finally, induction of NOS activity in the RVM following topical mustard oil application was examined using NADPH-diaphorase histochemistry. Rats that had received topical mustard oil showed increased numbers of NADPH-diaphorase positive cells in the RVM at the time of maximal hyperalgesia (15 min following application) compared with rats treated with mineral oil as a control. Consistent with the behavioral data, these results demonstrate an increase in NOS activity in the RVM in consequence of mustard oil-induced hyperalgesia, and further support a role for NMDA receptors and NO$^•$ in the RVM in the development of hyperalgesia following peripheral tissue injury.

We have recently begun to examine a potential contribution of supraspinal glutamate to the hyperalgesia in two additional models of peripheral tissue inflammation: cutaneous hyperalgesia

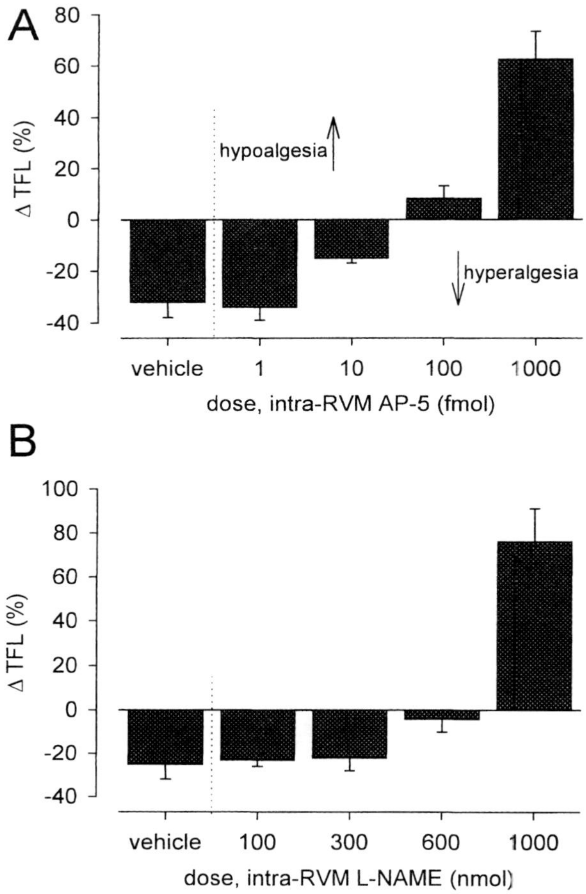


Fig. 1. Effects on mustard oil-induced hyperalgesia by AP-5 (A) or L-NAME (B) microinjection (1 μl) into the RVM 5 min prior to topical mustard oil application. Data expressed as percentage change in tail flick latency from baseline (ΔTFL (%)) at the time of maximal drug effect (10 min following AP-5 or L-NAME or vehicle [saline] injection). Both AP-5 and L-NAME dose-dependently reversed the hyperalgesia and, at the greatest doses tested, produce a hypoalgesia. D-NAME, the inactive stereoisomer of L-NAME, was without effect (not shown).

following carrageenan/kaolin injection into the knee joint, and visceral hyperalgesia following intracolonic administration of zymosan. Injection of carrageenan/kaolin into the hind leg knee joint cavity has been characterized as a model of acute arthritis in which rats exhibit localized edema, decreased weight bearing and guarding of the affected limb, and a thermal hyperalgesia four hours after intra-articular injection (Schaible and Schmidt, 1985; Sluka and Westlund, 1993). Microinjection of AP-5 (1 pmol/1 μl) into the RVM at the time of maximal hyperalgesia was found to significantly reverse the hyperalgesia in the ipsilateral limb compared with animals that received saline (vehicle) injection into the RVM. In contrast, AP-5 injection into the RVM was found to have no effect on the withdrawal latency of the contralateral (unaffected) limb.

In a model of visceral hyperalgesia in which the colon is inflamed with zymosan and the visceromotor response to CRD is enhanced (Coutinho et al., 1996), microinjection of AP-5 (10–100 fmol/1 μl) into the RVM at the time of maximal hyperalgesia (3 hours following intracolonic zymosan) was found to attenuate the enhanced visceromotor response to CRD in a dose-dependent manner (Fig. 2a) while having no effect on responses in control animals that had received intracolonic saline. Similar to results observed with mustard oil, injection of the non-NMDA receptor antagonist DNQX into the RVM at the time of maximal hyperalgesia produced a further enhancement of the already exaggerated visceromotor response. Additionally, injection of the NOS inhibitor L-NAME (100–1000 nmol/μl) into the RVM attenuated the enhanced visceromotor response to CRD in rats that had received intracolonic zymosan (Fig. 2b), while having no effect in control animals. Increases in NADPH-diaphorase positive cells were also observed in the RVM three hours after intracolonic instillation of zymosan compared with controls. Thus, similar to cutaneous hyperalgesia following peripheral inflammation, it appears that supraspinal NMDA receptors and $NO^{\bullet}$ in the RVM similarly contribute to enhanced behavioral responses in a model of visceral hyperalgesia following colonic inflammation.

These data demonstrating a contribution of supraspinal NMDA receptors to the hyperalgesia following peripheral inflammation are consistent with previous studies implicating supraspinal EAAs in the modulation of nociception. Glutamate and aspartate are localized throughout supraspinal sites that modulate nociception such

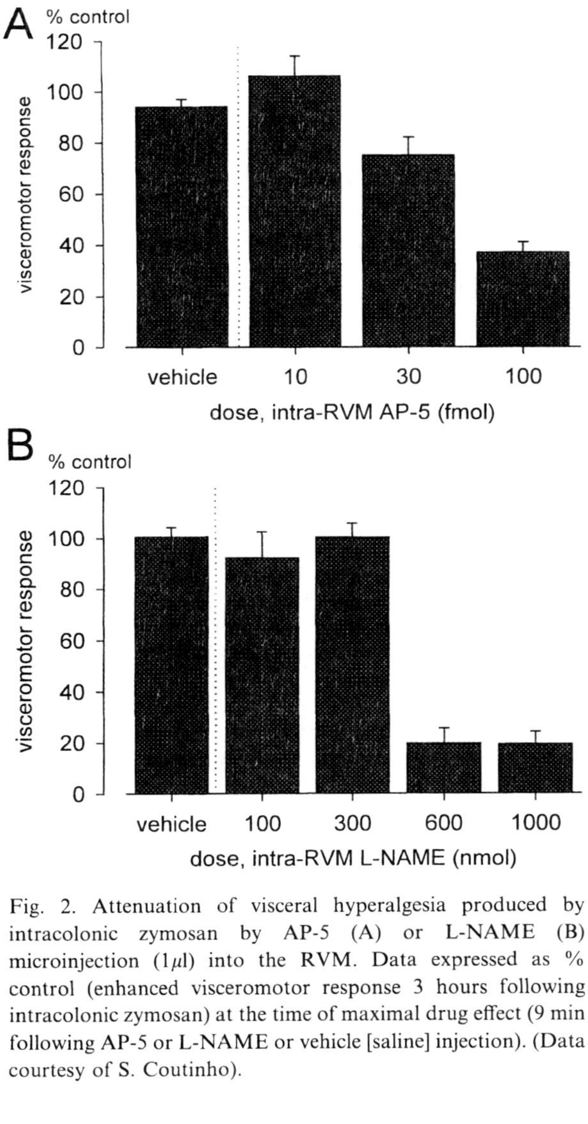


Fig. 2. Attenuation of visceral hyperalgesia produced by intracolonic zymosan by AP-5 (A) or L-NAME (B) microinjection (1$\mu$l) into the RVM. Data expressed as % control (enhanced visceromotor response 3 hours following intracolonic zymosan) at the time of maximal drug effect (9 min following AP-5 or L-NAME or vehicle [saline] injection). (Data courtesy of S. Coutinho).

as the PAG, RVM, and thalamus (Beitz, 1990; Clements et al., 1987; Dibiasi et al., 1994). A role for EEAs in the thalamus in the modulation of hyperalgesia was recently reported by Kolhekar et al. (1997) who showed that the hyperalgesia produced by intraplantar carrageenan injection was blocked by thalamic microinjection of either AP-5 or antisense oligodeoxynucleotides to the NMDA receptor. In addition to the thalamus, glutamate/aspartate is found in descending projections from the PAG to the RVM, and thus a role for glutamate in descending pain modulation from the PAG has been proposed. Van Praag and Frenk (1990) found that the antinociception produced by morphine injection into the PAG was blocked by injection of the non-NMDA receptor antagonist PCB into the RVM; injection of AP-5 into the RVM was less effective. These data suggest that non-NMDA receptors in the RVM mediate descending pain inhibition and are consistent with a dual role for EAAs in the RVM in the modulation of spinal nociceptive transmission via different receptors.

Glutamate microinjection into the RVM has been shown to both facilitate and inhibit spinal nociceptive transmission at lesser and greater doses, respectively (Zhuo and Gebhart, 1992; 1997), and the block of hyperalgesia by AP-5 injection into the RVM following peripheral inflammation supports a role for NMDA receptors in the RVM in descending pain facilitation. That NMDA injection into the RVM was found to facilitate spinal nociceptive transmission further supports this notion. Interestingly, we found that injection of the non-NMDA receptor antagonist DNQX into the RVM enhances both cutaneous and visceral hyperalgesia following topical application of mustard oil or intracolonic instillation of zymosan, respectively. Because DNQX was without effect in the absence of inflammation, it would appear that DNQX is selectively blocking a descending pain *inhibitory* component, resulting in an enhanced hyperalgesia. It is well established that noxious stimuli can activate descending pain inhibitory systems, and the notion that a noxious peripheral insult can activate both descending pain inhibitory and facilitatory systems has received recent attention (for reviews see Fields, 1992; Maier et al., 1992) and has been demonstrated in several reports (Wiertelak, et al., 1994a,b). Additionally, we found that the greatest doses of AP-5 and L-NAME injected into the RVM produced a significant inhibition of the tail-flick reflex following mustard oil application. Because these doses were without effect in the absence of mustard oil, it appears that the selective block of facilitation by AP-5 and L-NAME revealed a masked descending inhibitory component. These results further support the hypothesis that descending pain facilitation and inhibition involves NMDA and non-NMDA receptors in the RVM, respectively.

Although our results suggest that NMDA and non-NMDA receptors in the RVM are involved in descending facilitation and inhibition, respectively, somewhat conflicting results have been reported in other studies (Aimone and Gebhart, 1986; Spinella et al., 1996). The differences observed are likely due to several factors, including the means by which descending systems of inhibition were activated (i.e., intra-PAG electrical stimulation and morphine microinjection). More importantly, it is highly likely that persistent nociceptor input such as produced by peripheral inflammation engages systems different than those activated by short-lived, phasic inputs.

In addition to involvement of supraspinal NMDA receptors, results suggesting a contribution of NO$^\bullet$ in the RVM to the behavioral hyperalgesia following peripheral inflammation is supported by a number of studies. NOS has been found to be widely distributed throughout the rat CNS, including the RVM, and supraspinal NO$^\bullet$ has been implicated in pain facilitation (Kawabata et al., 1993; Moore et al., 1991; Semos and Headley, 1994; Shibuta et al., 1995; Vincent and Kimura, 1992). Again, somewhat conflicting results have been obtained regarding the role of supraspinal NO$^\bullet$ in nociceptive modulation, likely due to differences in the nature of the stimulus (Iwamoto and Marion, 1994; Kawabata et al., 1993).

## Implications for clinical pain management

The targeting of the NMDA receptor for the treatment of chronic pain through the use of NMDA receptor antagonists has received recent clinical application. The use of non-competitive NMDA receptor channel blockers such as ketamine and magnesium chloride have been found to be effective in the treatment of a variety of pain disorders, including neuropathic (Felsby et al., 1996), phantom limb (Stannard and Porter, 1993) and ischemic pain (Maurset et al., 1989). The clinical efficacy of non-competitive NMDA receptor antagonists is likely due to a selective effect on central sensitization because these antagonists

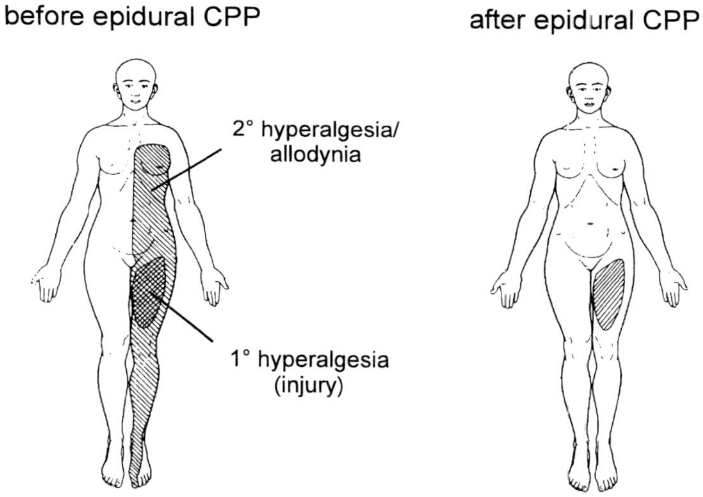


Fig. 3. Anatomical mapping of the area of pain sensation after low-threshold mechanical stimulation of the left thigh before (left) and after (right) intrathecal treatment with the NMDA receptor antagonist CPP. The area of pain sensation after treatment with CPP was identical to the territory of the injured nerve (double hatched area). Intrathecal administration of CPP completely reversed the expanded area of dysesthetic sensation. (Reproduced with permission from Kristensen, J.D., Svensson, B., Gordh, T.: Pain 51: 249–253, 1992.)

block temporal summation of second pain following repeated electrical stimulation (a human correlate to "wind-up") while having no effect on pain evoked by the initial stimulus (Price et al., 1994). The use of competitive NMDA receptor antagonists in the treatment of pain disorders has received somewhat less attention, likely due to the potential for undesirable side effects such as motor disturbances and dysphoria. Kristensen et al. (1992) reported that spinal administration of a potent, competitive NMDA receptor antagonist, CPP, was effective in eliminating spontaneous pain, touch-evoked pain, and the expansion of pain outside the area of nerve injury in a patient suffering from neuropathic pain (Fig. 3). Additionally, this treatment was found to have no effect on normal pain sensations.

The use of NMDA receptor antagonists as analgesics has been limited by undesirable psychotomimetic and motor side effects, which may be overcome by pharmacological manipulation of glycine and polyamine modulatory sites on the NMDA receptor. Glycine, by a specific interaction with an allosteric site on the NMDA receptor, enhances NMDA receptor function by enhancing competitive binding at the NMDA receptor (Johnson and Ascher, 1987), while spermine enhances NMDA receptor $Ca^{2+}$ currents and the binding of NMDA receptor channel blockers such as MK-801 (Williams et al., 1989). Thus, antagonists at the glycine modulatory site on the NMDA receptor complex have been shown to prevent C-fiber wind-up and pain behaviors following peripheral injury (Dickenson and Aydar, 1991; Lutfy and Weber, 1996). Interestingly, the combination of glycine with competitive NMDA receptor antagonists increases the antinociceptive potency of NMDA receptor antagonists, presumably by enhancing the binding of the competitive antagonist (Coderre et al., 1997). A similar enhanced antinociceptive potency of NMDA receptor channel blockers is observed by coadministration with spermine. These results suggest that combination therapy involving lesser doses of NMDA receptor antagonists that lack potential side effects may be a useful option in clinical pain management.

Finally, NMDA receptors have been shown to play a prominent role in the development of morphine tolerance and dependence that occurs with chronic morphine use. Tolerance and dependence to morphine during the treatment of chronic pain is a concern, and numerous studies have shown that NMDA receptor antagonists enhance the antinociceptive potency of morphine and prevent the development of morphine tolerance and dependence (Ben-Eliyahu et al., 1992; Chapman and Dickenson, 1992; Trujillo and Akil, 1991; Wong et al., 1996). A prolonged activation of NMDA receptors by chronic morphine use, resulting in a reduced antinociceptive potency and tolerance, is supported by studies demonstrating NMDA receptor down-regulation in such sites as hypothalamus, midbrain, and spinal cord (Bhargava et al., 1995). Thus, in addition to direct pain control, NMDA receptor antagonists may find increased utilization as adjunct therapy in cases involving chronic opioid treatment.

## Conclusions

A large body of evidence supports the involvement of glutamate in the modulation of acute and particularly persistent pain. It has become increasingly clear that glutamate both contributes to and modulates pain by a variety of mechanisms at the level of the peripheral nociceptor, in the spinal cord, and recently, supraspinally, including the RVM (Fig. 4). The hyperalgesia and central sensitization that develop following peripheral injury/inflammation of either cutaneous or visceral tissue involves sensitization of peripheral nociceptors presumably, in part, via glutamate release from peripheral nociceptor terminals. Sensitization of nociceptors results in enhanced responses and increased excitability of spinal dorsal horn nociceptive neurons (central sensitization), an effect that is largely dependent on spinal NMDA receptors and the production of NO˙ Additionally, sustained afferent input associated with peripheral tissue injury results in the activation of centrifugal descending pain modulatory

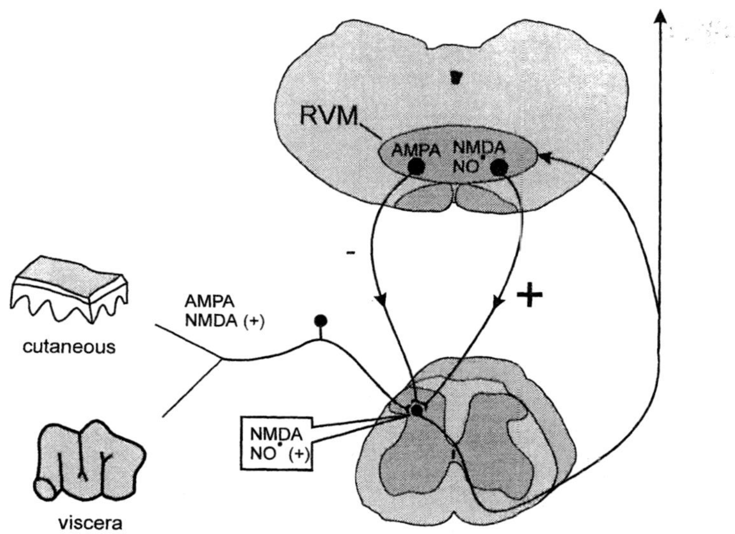


Fig. 4. Summary diagram illustrating peripheral and central EAA involvement in pain and hyperalgesia following peripheral cutaneous and/or visceral injury/inflammation. Hyperalgesia and central sensitization following peripheral injury is likely a consequence of sensitization/activation (+) of peripheral nociceptors, in part, via release in the periphery of EAAs and activation of NMDA and AMPA receptors. Prolonged activation of nociceptors results in an increased spinal release of EAAs (and peptides such as substance P, not shown), activation of spinal NMDA receptors and production of NO•, and an increase in the excitability of dorsal horn neurons. Additionally, activation of descending pain modulatory systems via a centrifugal spino-bulbar-spinal loop involving the RVM both inhibits (-) and facilitates (+) spinal nociceptive transmission, with a dominant facilitatory influence. Concurrent activation of descending facilitatory and inhibitory systems following peripheral tissue insult involves NMDA receptors and NO•, and non-NMDA receptors in the RVM, respectively.

systems from the RVM via a spino-bulbar-spinal loop, with a dominant descending facilitatory influence. Both descending facilitatory and inhibitory systems are activated in the RVM involving NMDA receptors and generation of NO• and non-NMDA receptors, respectively.

The use of NMDA receptor antagonists in clinical pain management has been rather limited due to undesirable psychotomimetic and motor effects that accompany pain relief. Administration of NMDA receptor antagonists to selective sites, and/or pharmacological modulation of these receptors and their modulatory sites, may be useful strategies to reduce these effects and deserves further study. Another consideration relates to the subunit composition of the NMDA receptor and the possibility that it may change in the face of sustained afferent input arising from injured tissue. If such a change did occur, it may be restricted, for example, to the spinal segments of nociceptor termination (and to supraspinal sites of termination of spinal second order neurons) and present a novel target for drug development. There are no data currently in the literature that address this possiblilty, but it may be a fruitful avenue of investigation. Clearly, the significant role that NMDA receptors play throughout the neuraxis in the development and maintenance of hyperalgesia emphasizes its importance as a therapeutic target in pain management warranting continued investigation.

**Acknowledgements**

The authors thank Sue Birely for excellent secretarial assistance, Michael Burcham for production of the graphics, Santosh Coutinho for sharing data and results, and Frank Porreca, University of

Arizona, for suggesting that alterations in NMDA receptor subunit composition may change in the presence of tissue injury. The authors are supported by F32 DA 05673, NS 19912, and DA 02879.

## Abbreviations

AMPA, α-amino-3-hydroxy-5-methylisoxazole-4-propionic acid; AP-5, 2-amino-5-phosphonopentanoate; CNS, central nervous system; CRD, colorectal distension; CPP, 3-(2-carboxypiperazin-4-yl)propyl-1-phosphonic acid; D-NAME, $N^G$-nitro-D-arginine methyl ester; DNQX, 6,7-dinitroquinoxaline-2,3-dione; EAAs, excitatory amino acids; L-NAME, $N^G$-nitro-L-arginine methyl ester; mGlu, metabotropic glutamate; MK-801, dizocilpine; NADPH, nicotinamide adenine dinucleotide phosphate; NMDA, *N*-methyl-D-asparate; NO, nitric oxide; NOS, nitric oxide synthase; PAG, periaqueductal gray; PCB, 1-(*p*-chlorobenzoyl)-piperazine-2,3-dicarboxylate; RVM, rostral ventromedial medulla

## Figure legend abbreviations

AMPA, α-amino-3-hydroxy-5-methylisoxazole-4-propionic acid; AP-5, 2-amino-5-phosphonopentanoate; CPP, 3-(2-carboxypiperazin-4-yl)propyl-1-phosphonic acid; D-NAME, $N^G$-nitro-D-arginine methyl ester; EAAs, excitatory amino acids; L-NAME, NG-nitro-L-arginine methyl ester; NMDA, *N*-methyl-D-asparate; NO, nitric oxide; RVM, rostral ventromedial medulla; TFL, tail-flick latency

## References


Aanonsen, L.M., Lei, S. and Wilcox, G.L. (1990) Excitatory amino acid receptors and nociceptive neurotransmission in rat spinal cord. *Pain*, 41: 309–321.

Aanonsen, L.M. and Wilcox, G.L. (1987) Nociceptive action of excitatory amino acids in the mouse: Effects of spinally administered opioids, phencyclidine and sigma agonists. *J. Pharmacol. Exp. Ther.*, 243: 9–19.

Abols, I.A. and Basbaum, A.I. (1981) Afferent connections of the rostral medulla of the cat: A neural substrate for midbrain-medullary interactions in the modulation of pain. *J. Comp. Neurol.*, 201: 285–297.

Aimone, L.D. and Gebhart, G.F. (1986) Stimulation-produced spinal inhibition from the midbrain in the rat is mediated by an excitatory amino acid neurotransmitter in the medial medulla. *J. Neurosci.*, 6: 1803–1813.

Basbaum, A.I., Clanton, C.H. and Fields, H.L. (1978) Three bulbospinal pathways from the rostral medulla of the cat: An autoradiographic study of pain modulating systems. *J. Comp. Neurol.*, 178: 209–224.

Basbaum, A.I. and Fields, H.L. (1984) Endogenous pain control systems: Brainstem spinal pathways and endorphin circuitry. *Annu. Rev. Neurosci.*, 7: 309–338.

Beitz, A.J. (1990) Relationship of glutamate and aspartate to the periaqueductal gray-raphe magnus projection: Analysis using immunocytochemistry and microdialysis. *J. Histochem. Cytochem.*, 38: 1755–1765.

Ben-Eliyahu, S., Marek, P., Vaccarino, A.L., Mogil, J.S., Sternberg, W.F. and Liebeskind, J.C. (1992) The NMDA receptor antagonist MK-801 prevents long-lasting non-associative morphine tolerance in the rat. *Brain Res.*, 575: 304–308.

Bennett, G.J. (1993) An animal model of neuropathic pain: A review. *Muscle and Nerve.*, 16: 1040–1048.

Bhargava, H.N., Reddy, P.L. and Gedehithlu, K.P. (1995) Down-regulation of *N*-methyl-D-aspartate (NMDA) receptors of brain regions and spinal cord of rats treated chronically with morphine. *Gen. Pharmacol.*, 26: 131–136.

Brambilla, A., Prudentino, A., Grippa, N. and Borsini,, F. (1996) Pharmacological characterization of AMPA-induced biting behaviour in mice. *Eur. J. Pharmacol.*, 305: 115–117.

Cahusac, P.M., Evans, R.H., Hill, R.G., Rodriguez, R.E. and Smith, D.A. (1984) The behavioural effects of an *N*-methylaspartate receptor antagonist following application to the lumbar spinal cord of conscious rats. *Neuropharmacol.*, 23: 719–724.

Carlton, S.M., Hargett, G.L. and Coggeshall, R.E. (1995) Localization and activation of glutamate receptors in unmyelinated axons of rat glabrous skin. *Neurosci. Lett.*, 197: 25–28.

Casey, K.L. (1971) Escape elicted by bulboreticular stimulation in the cat. *Int. J. Neurosci.*, 2: 29–34.

Chapman, V. and Dickenson, A.H. (1992) The combination of NMDA antagonism and morphine produces profound antinociception in the rat dorsal horn. *Brain Res.*, 573: 321–323.

Choi, Y., Raja, S.N., Moore, L.C. and Tobin, J.R. (1996) Neuropathic pain in rats is associated with altered nitric oxide synthase activity in neural tissue. *J. of Neurol. Sci.*, 138: 14–20.

Clements, J.R., Magl, J.E., Johnson, R.L., Larson, A.A. and Beitz, A.J. (1987) Localization of glutamate, glutaminase, aspartate, and aspartate aminotransferase in the rat midbrain periaqueductal gray. *Exp. Brain Res.*, 67: 594–602.

Coderre, T.J., Fisher, K. and Fundytus, M.E. The role of ionotropic and metabotropic glutamate receptors in persistent nociception. In: T.S. Jensen, J.A. Turner and Z. Wiesenfeld-Hallin (Eds.). Progress in Pain Research and Management, Vol. 8, Proc. VIII World Congress on Pain, IASP Press, Seattle, 1997, pp. 259–275.

Coderre, T.J., Katz, J., Vaccarino, A.L. and Melzack, R. (1993) Contribution of central neuroplasticity to pathological pain: Review of clinical and experimental evidence. *Pain*, 52: 259–285.

Coderre, T.J., Vaccarino, A.L. and Melzack, R. (1990) Central nervous system plasticity in the tonic pain response to subcutaneous formalin injection. *Brain Res.*, 535: 155–158.

Coderre, T.J. and Van Empel, I. (1994a) The utility of excitatory amino acid (EAA) antagonists an analgesic agents. I. Comparison of the antinociceptive activity of various classes of EAA antagonists in mechanical, thermal and chemical nociceptive tests. *Pain*, 59: 345–352.

Coderre, T.J. and Van Empel, I. (1994b) The utility of excitatory amino acid (EAA) antagonists as analgesic agents. II. Assessment of the antinociceptive activity of combinations of competitive and non-competitive NMDA antagonists with agents acting at allosteric-glycine and polyamine receptor sites. *Pain*, 59: 353–359.

Cook, A.J., Woolf, C.J., Wall P.D. and McMahon, S.B. (1987) Dynamic receptive field plasticity in rat spinal dorsal horn following C-primary afferent input. *Nature*, 325: 151–153.

Coutinho, S.V., Meller, S.T. and Gebhart, G.F. (1996) Intracolonic zymosan produces visceral hyperalgesia in the rat that is mediated by spinal NMDA and non-NMDA receptors. *Brain Res.*, 736: 7–15.

Curtis, D.R., Phillips, J.W. and Watkins, J.C. (1960) Chemical excitation of spinal neurons. *Nature*, 183: 611–612.

Davies, S.N. and Lodge, D. (1987) Evidence for involvement of *N*-methyl-D-aspartate receptors in "wind-up" of class 2 neurones in the dorsal horn of the rat. *Brain Res.*, 424: 402–406.

Dickenson, A.H. and Aydar, E. (1991) Antagonism at the glycine site on the NMDA receptor reduces spinal nociception in the rat. *Neurosci. Lett.*, 121: 263–266.

Dickenson, A.H. and Sullivan, A.F. (1987) Evidence for a role of the NMDA receptor in the frequency dependent potentiation of deep rat dorsal horn nociceptive neurones following C fiber stimulation. *Neuropharmacol.*, 26: 1235–1238.

DiBiasi, S., Amadeo, A., Spreafico, R. and Rustioni, A. (1994) Enrichment of glutamate immunoreactivity in lemniscal terminals in the ventropostero-lateral thalamic nucleus of the rat: An immunogold and WGA-HRP study. *Anat. Record*, 240: 131–140.

Felsby, S., Nielsen, J., Arendt-Nielsen, L. and Jensen, T.S. (1996) NMDA receptor blockade in chronic neuropathic pain: A comparison of ketamine and magnesium chloride. *Pain*, 64: 283–291.

Fields, H.L. (1992) Is there a facilitating component to central pain modulation, *Am. Pain Soc. J.*, 1: 72–78.

Fields, H.L., Heinricher, M.M. and Mason, P. (1991) Neurotransmitters in nociceptive modulatory circuits. *Annu. Rev. Neurosci.*, 7: 309–338.

Gebhart, G.F. and Randich, A. Brainstem modulation of nociception. In: *Brainstem Mechanisms of Behavior*, edited by W.R. Klemm and R.P. Vertes. New York: Wiley, 1990, p.315–352.

Greenamyre, J.T., Olson, J.M., Penney, J.B. and Young, A.B. (1985) Autoradiographic characterization of *N*-methyl-D-aspartate-, quisqualate- and kainate-sensitive glutamate binding sites. *J. Pharmacol. Exp. Ther.*, 233: 254–263.

Herrero, J.F. and Cervero, F. (1996) Supraspinal influences on the facilitation of rat nociceptive reflexes induced by carrageenan monoarthritis. *Neurosci. Lett.*, 209: 21–24.

Holstege, G. and Kuypers, H.G.J.M. (1982) The anatomy of brain stem pathways to the spinal cord in cat. A labeled amino acid tracing study. *Prog. Brain Res.*, 57: 145–175.

Ide, Y., Maehara, Y., Tsukahara, S., Kitahata, L.M. and Collins, J.G. (1997) The effects of an intrathecal NMDA antagonist (AP5) on the behavioral changes induced by colorectal inflammation with turpentine in rats. *Life Sci.*, 60: 1359–1363.

Iwamoto, E.T. and Marion, L. (1994) Pharmacological evidence that nitric oxide mediates the antinociception produced by muscarinic agonists in the rostral ventral medulla of rats. *J. Pharmacol. Exp. Ther.*, 269: 699–708.

Jackson, D.L., Graff, C.B., Richardson, J.D. and Hargreaves, K.M. (1995) Glutamate participates in the peripheral modulation of thermal hyperalgesia in rats. *Eur. J. Pharmacol.*, 284: 321–325.

Johnson, J.W. and Ascher, P. (1987) Glycine potentiates the NMDA responses in cultured mouse brain neurons. *Nature*, 325: 529–532.

Kawabata, A., Umeda, N. and Takagi, H. (1993) L-arginine exerts a dual role in nociceptive processing in the brain: Involvement of the kyotorphin-Met-enkephalin pathway and NO-cyclic GMP pathway. *Br. J. Pharmacol.*, 109: 73–79.

Kolhekar, R. and Gebhart, G.F. (1994) NMDA and quisqualate modulation of visceral nociception in the rat. *Brain Res.*, 651: 215–226.

Kolhekar, R. and Gebhart, G.F. (1996) Modulation of spinal visceral nociceptive transmission by NMDA receptor activation in the rat. *J. Neurophysiol.*, 75: 2344–2353.

Kolhekar, R., Murphy, S. and Gebhart, G.F. (1997) Thalamic NMDA receptors modulate inflammation-produced hyperalgesia in the rat. *Pain*, 71: 31–40.

Kristensen, J.D., Karlsten, R., Gordh, T. and Berg, O. (1994) The NMDA antagonist 3-(2-carboxypiperazin-4-yl)propyl-1-phosphonic acid (CPP) has antinociceptive effect after intrathecal injection in the rat. *Pain*, 56: 59–67.

Kristensen, J.D., Svensson, B.A. and Gordh, T. (1992) The NMDA-receptor antagonist CPP abolishes neurogenic "wind-up pain" after intrathecal administration in humans. *Pain*, 51: 249–253.

Li, H., Ohishi, H., Kinoshita, A., Shigemoto, R., Nomura, S. and Mizuno, N. (1997) Localization of a metabotropic glutamate receptor, mGluR7, in axon terminals of presumed nociceptive, primary afferent fibers in the superficial layers of the spinal dorsal horn: An electron microscope study in the rat. *Neurosci. Lett.*, 223: 153–156.

Light, A.R., Casale, E.J. and Menetrey, D.M. (1986) The effects of focal stimulation in nucleus raphe magnus and periaqueductal gray on intracellularly recorded neurons in spinal laminae I and II. *J. Neurophysiol.*, 56: 555–571.

Liu, H., Mantyh, P.W. and Basbaum, A.I. (1997) NMDA-receptor regulation of substance P release from primary afferent nociceptors. *Nature*, 386: 721–724.

Liu, H., Wang, H., Sheng, M., Jan, L.Y., Jang, Y.N., Basbaum, A.I. (1994) Evidence for presynaptic *N*-methyl-D-aspartate receptors in the spinal cord dorsal horn. *Proc. Nat. Acad. Sci. USA*, 92: 8383–8387.

Lutfy, K., Cai, S.X., Woodward, R.M. and Weber, E. (1997) Antinociceptive effects of NMDA and non-NMDA receptor antagonists in tail flick test in mice. *Pain*, 70: 31–40.

Lutfy, K. and Weber, E. (1996) Attenuation of nociceptive responses by ACEA-1021, a competitive NMDA receptor/glycine site antagonist, in mice. *Brain Res.*, 743: 17–23.

Magnusson, K.R., Clements, J.R., Larson, A.A., Madl J.E. and Beitz, A.J. (1987) Localization of glutamate in trigeminothalamic projection neurons: A combined retrograde transport-immunohistochemical study. *Somatosens. Res.*, 4: 177–190.

Maier, S.F., Wiertelak, E.P. and Watkins, L.R. (1992) Endogenous pain facilitory systems – Antianalgesia and hyperalgesia, *Am. Pain Soc.*, J. 1: 191–198.

Malmberg, A.B. and Yaksh, T.L. (1993) Spinal nitric oxide synthesis inhibition blocks NMDA-induced thermal hyperalgesia and produces antinociception in the formalin test in rats. *Pain*, 54: 291–300.

Mao, J., Price, D.D., Hayes, R.L., Lu, J. and Mayer, D.J. (1992) Differential roles of NMDA and non-NMDA receptor activation in induction and maintenance of thermal hyperalgesia in rats with painful peripheral mononeuropathy. *Brain Res.*, 598: 271–278.

Maurset, A., Skoglund, L.A., Justveit, O. and Oye, I. (1989) Comparison of ketamine and pethidine in experimental and postoperative pain. *Pain*, 36: 37–41.

Mendell, L.M. (1966) Physiological properties of unmyelinated fiber projections to the spinal cord. *Exp. Neurol.*, 16: 316–332.

Meller, S.T., Cummings, C.P., Traub, R.J. and Gebhart, G.F. (1994) The role of nitric oxide in the development and maintenance of the hyperalgesia produced by intraplantar injection of carrageenan in the rat. *Neurosci.*, 60: 367–374.

Meller, S.T., Dykstra, C. and Gebhart, G.F. (1992a) Production of endogenous nitric oxide and activation of soluble guanylate cyclase are required for *N*-methyl-D-aspartate-produced facilitation of the nociceptive tail-flick reflex. *Eur. J. Pharmacol.*, 214: 93–96.

Meller, S.T. and Gebhart, G.F. (1993) Nitric oxide (NO) and nociceptive processing in the spinal cord. *Pain*, 52: 127–136.

Meller, S.T., Pechman, P.S., Gebhart, G.F. and Maves, T.J. (1992b) Nitric oxide mediates the thermal hyperalgesia produced in a model of neuropathic pain in the rat. *Neurosci.*, 50: 7–10.

Miller, K.E., Clements, J.R., Larson, A.A. and Beitz, A.J. (1988) Organization of glutamate-like immunoreactivity in the rat superficial dorsal horn: Light and electron microscopic observations. *Synapse*, 2: 28–36

Moore, P.K., Oluyomi, A.O., Babbedge, R.C., Wallace, P. and Hart, S.L. (1991) L-NG-nitro arginine methyl ester exhibits antinociceptive activity in the mouse. *Br. J. Pharmacol.*, 102: 198–202.

Ness, T.J. and Gebhart, G.F. (1988) Colorectal distension as a noxious visceral stimulus: Physiologic and pharmacologic characterization of pseudaffective reflexes in the rat. *Brain Res.*, 450: 153–169.

Ness, T.J. and Gebhart, G.F. (1990) Visceral pain: A review of experimental studies. *Pain*, 41: 167–234.

Ness, T.J., Metcalf, A.M. and Gebhart, G.F. (1990) A psychophysical study in humans using phasic colonic distension as a noxious visceral stimulus. *Pain*, 43: 377–386.

Price, D.D., Mao, J., Frenk, H. and Mayer, D.J. (1994) The *N*-methyl-D-aspartate receptor antagonist dextromethorphan selectively reduces temporal summation of second pain in man. *Pain*, 59: 165–174.

Puig, S. and Sorkin, L.S. (1995) Formalin-evoked activity in identified primary afferent fibers: Systemic lidocaine suppresses phase-2 activity. *Pain*, 64: 345–355.

Qian, J., Brown, S.D. and Carlton, S.M. (1996) Systemic ketamine attenuates nociceptive behaviors in a rat model of peripheral neuropathy. *Brain Res.*, 715: 51–62.

Ren, K., Hylden, J.K.L., Williams, G.M., Ruda, M.A. and Dubner, R. (1992a) The effects of a non-competitive NMDA receptor antagonist, MK-801, on behavioral hyperalgesia and dorsal horn neuronal activity in rats with unilateral inflammation. *Pain*, 50: 331–344.

Ren, K., Williams, G.M., Hylden, J.K.L., Ruda, M.A. and Dubner, R. (1992b) The intrathecal administration of excitatory amino acid receptor antagonists selectively attenuate carrageenan-induced behavioral hyperalgesia in rats. *Eur. J. Pharmacol.*, 219: 235–243.

Schaible, H.-G. and Schmidt, R.F. (1985) Effects of an experimental arthritis on the sensory properties of fine articular afferent units. *J. Neurophysiol.*, 54: 1109–1122.

Semos, M.L. and Headley, P.M. (1994) The role of nitric oxide in spinal nociceptive reflexes in rats with neurogenic and non-neurogenic peripheral inflammation. *Neuropharm.*, 33: 1487–1497.

Shibuta, S., Mashimo, T., Ohara, A., Zhang, P. and Yoshiya, I. (1995) Intracerebroventricular administration of a nitric oxide-releasing compound, NOC-18, produces thermal hyperalgesia in rats. *Neurosci. Lett.*, 187: 103–106

Shigemoto, R., Ohishi, H., Nakanishi, S. and Mizuno, N. (1992) Expression of the mRNA for the rat NMDA receptor (NMDAR1) in the sensory and autonomic ganglion neurons. *Neurosci. Lett.*, 144: 229–232.

Sluka, K.A. and Westlund, K.N. (1993) Behavioral and immunohistochemical changes in an experimental arthritis model in rats. *Pain*, 55: 367–377.

Sorkin, L.S., Westlund, K.N., Sluka, K.A., Dougherty, P.M. and Willis, W.D. (1992) Neural changes in acute arthritis in monkeys. IV. Time-course of amino acid release into the lumbar dorsal horn. *Brain Res. Rev.*, 17: 39–50.

Spinella, M., Cooper, M.L. and Bodnar, R.J. (1996) Excitatory amino acid antagonists in the rostral ventromedial medulla inhibit mesencephalic morphine analgesia in rats. *Pain*, 64: 545–552.

Stannard, C.F. and Porter, G.E. (1993) Ketamine hydrochloride in the treatment of phantom limb pain. *Pain*, 54: 227–230.

Trujillo, K.A. and Akil, H. (1991) Inhibition of morphine tolerance and dependence by the NMDA receptor antagonist MK-801. *Science*, 251: 85–87.

Urban, M.O. and Gebhart G.F.(1997) Characterization of biphasic modulation of spinal nociceptive transmission by neurotensin in the rat rostral ventromedial medulla. *J. Neurophysiol.*, 78: 1550–1562.

Urban, M.O., Jiang, M.C. and Gebhart, G.F. (1996) Participation of central descending nociceptive facilitatory systems in secondary hyperalgesia produced by mustard oil. *Brain Res.*, 737: 83–91.

Van Praag, H. and Frenk H. (1990) The role of glutamate in opiate descending inhibition of nociceptive spinal reflexes. *Brain Res.*, 524: 101–105.

Vincent, S.R. and Kimura, H. (1992) Histochemical mapping of nitric oxide synthase in the rat brain. *Neurosci.*, 46: 755–784.

Waldbillig, R.J. (1975) Attack, eating, drinking and gnawing elicited by electrical stimulation of rat mesencephalon and pons. *J. Comp. Physiol. Psychol.*, 89: 200–212.

Wiertelak, E.P., Furness, L.E., Horan, R., Martinez, J., Maier, S.F. and Watkins, L.R. (1994a) Subcutaneous formalin produces centrifugal hyperalgesia at a non-injected site via the NMDA-nitric oxide cascade. *Brain Res.*, 649: 19–26.

Wiertelak, E.P., Furness, L.E., Watkins, L.R. and Maier, S.F. (1994b) Illness-induced hyperalgesia is mediated by a spinal NMDA-nitric oxide cascade. *Brain Res.*, 664: 9–16.

Wilcox, G.L. (1991) Excitatory neurotransmitters and pain. In: M.R. Bond, J.E. Charlton and C.J. Woolf (Eds.) Pain Research and Clinical Management, Vol. 4 Proc. VIth World Congress on Pain, Elsevier, Amsterdam, pp. 97–117

Williams, K., Romano, C. and Molinoff, P.B. (1989) Effects of polyamines on the binding of [3H] MK-801 to the *N*-methyl-D-aspartate receptor: Pharmacological evidence for the existence of a polyamine recognition site. *Mol. Pharmacol.*, 36: 575–581.

Willis, W.D. (1986) Ascending somatosensory systems. In: Spinal afferent processing, edited by T.L. Yaksh. New York: *Plenum*, p. 243–274.

Wong, C.S., Cherng, C.H., Luk, H.N., Ho, S.T. and Tung, C.S. (1996) Effects of NMDA receptor antagonists on inhibition of morphine tolerance in rats: Binding at mu-opioid receptors. *Eur. J. Pharmacol.*, 297: 27–33.

Woolf, C.J. and Thompson, S.W.N. (1991) The induction and maintenance of central sensitization is dependent on *N*-methyl-D-aspartic acid receptor activation: Implications for post-injury pain hypersensitivity states. *Pain*, 44: 293–299.

Yamamoto, T. and Yaksh, T.L. (1992) Spinal pharmacology of thermal hyperesthesia induced by contriction injury of sciatic nerve. Excitatory amino acid antagonists. *Pain*, 49: 121–128.

Yeomans, D.C. and Proudfit, H.K. (1996) Nociceptive responses to high and low rates of noxious cutaneous heating are mediated by different nociceptors in the rat: electrophysiological evidence. *Pain*, 68: 141–150.

Zhuo, M. and Gebhart, G.F. (1992) Characterization of descending facilitation and inhibition of spinal nociceptive transmission from the nuclei reticularis gigantocellularis and gigantocellularis pars alpha in the rat. *J. Neurophysiol.*, 67: 1599–1614.

Zhuo, M. and Gebhart, G.F. (1997) Biphasic modulation of spinal nociceptive transmission from the medullary raphe nuclei in the rat. *J. Neurophysiol.*, 78: 746–758, 1997.

Zhuo, S., Bonasera, L. and Carlton, S.M. (1996) Peripheral administration of NMDA, AMPA, or KA results in pain behaviors in rats. *Neuroreport*, 7: 895–900.

O.P. Ottersen, I.A. Langmoen and L. Gjerstad (Eds.)
*Progress in Brain Research*, Vol 116


CHAPTER 28

# The glutamate synapse in neuropsychiatric disorders

## Focus on schizophrenia and Alzheimer's disease

Nuri B. Farber*, John W. Newcomer and John W. Olney

*Department of Psychiatry, Washington University School of Medicine, St. Louis, MO 6310-1093, USA*

### Introduction

Extensive research focusing on the amino acid, glutatmate (Glu), has documented the central role played by this compound in both the normal and abnormal functioning of the central nervous system (CNS). Glu is now recognized to be the main excitatory neurotransmitter in the CNS, estimated to be released at up to half of the synapses in the brain. In addition, Glu is also an excitotoxin that can destroy CNS neurons by excessive activation of excitatory receptors on dendritic and somal surfaces.

Two major classes of Glu receptors – ionotropic and metabotropic – have been identified. Glu exerts excitotoxic activity through three receptor subtypes which belong to the ionotropic family. These three receptors are named after agonists to which they are differentially sensitive – *N*-methyl-D-aspartate (NMDA), amino-3-hydroxy-5-methyl-4-isoxazole propionic acid (AMPA), and kainic acid (KA). Of these three, the NMDA receptor (depicted schematically in Fig. 1) has been the most extensively studied and the most frequently implicated in CNS diseases (Olney, 1990). Excessive activation of NMDA receptors (NMDA receptor hyperfunction [NRHyper]) plays an important role in the pathophysiology of acute CNS injury syndromes such as hypoxia-ischemia, trauma and status epilepticus (Olney, 1990; Choi, 1992). Recently, hyperstimulation of AMPA/KA receptors and consequent excitotoxicity has been proposed to underlie neurodegeneration in amyotrophic lateral sclerosis (ALS, Lou Gerhig's Disease, [Rothstein et al., 1995; Ikonomidou et al., 1996]). The role of Glu excitotoxicity in the pathology seen in several other neuropsychiatric disorders has been extensively reviewed elsewhere (Choi, 1988; Olney, 1990; Lancelot and Beal, this volume) and will not be the focus of this chapter. Instead, we will focus on a recently discovered unconventional form of neurotoxicity that arises from the *under*excitation of NMDA receptors. We postulate that this underexcitation of NMDA receptors is operative in two major neuropsychiatric disorders – schizophrenia and Alzheimer's Disease (AD). In the following sections we will describe initially the type of neuronal damage produced by hypoactivation of the NMDA receptor and the complex neural circuitry that is postulated to be perturbed as a consequence of NMDA receptor hypofunction (NRHypo). Then we will discuss the proposal that NRHypo may be involved in both schizophrenia and AD.

*Corresponding author.
Tel.: +1 314 362 2459; fax: +1 314 362 9902; e-mail: FarberN@psychiatry.wustl.edu.

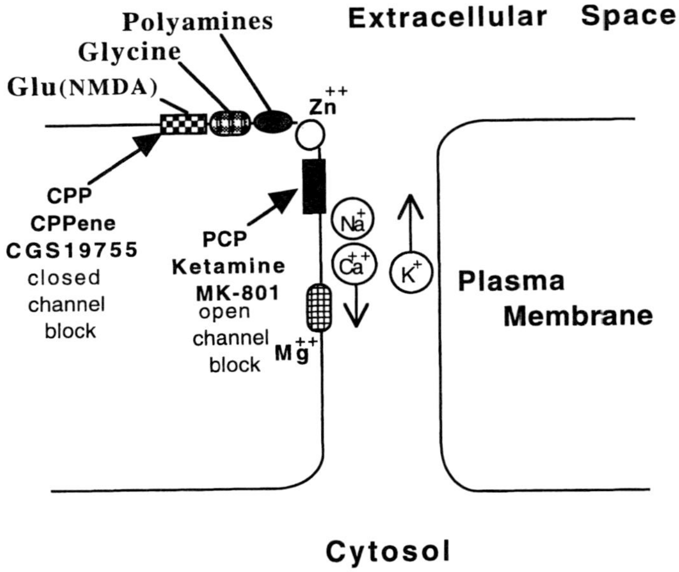


Fig. 1. Associated with the NMDA receptor are multiple recognition sites through which receptor channel function is modulated. Currently it is recognized that NMDA receptor hypofunction (NRHypo) can be induced by agents (PCP, ketamine, MK-801) acting at the PCP site to perform an open channel block or agents (CPP, CPPene, CGS 19755) acting at the NMDA site to perform a closed channel block.

**NRHypo produces brain damage in the rat**

*Descriptive characterization*

Low doses of NMDA receptor antagonist (NRA) drugs such as ketamine, MK-801, tiletamine, phencyclidine [PCP], CPP, and CPP-ene reliably injure certain cerebrocortical neurons (Olney et al., 1989; Olney et al., 1991; Hargreaves et al., 1993). At these doses the injury is confined to the posterior cingulate and retrosplenial (PC/RS) cortex and consists of the formation of intracytoplasmic vacuoles in layer III-IV pyramidal neurons. These changes are transient and resolve by 24 hours (Olney et al., 1989). While these neurons will continue to express the 72 kDa form of heat shock protein (HSP-72) for up to two weeks (Sharp et al., 1991; Olney et al., 1991), they do not become argyrophilic (de Olmos cupric silver method) or die.

Administering an NRA in high dosage or by continuous infusion for several days induces a prolonged NRHypo state which causes irreversible injury involving the death of neurons in many cerebrocortical and limbic brain regions (Allen and Iversen, 1990; Ellison, 1994; Corso et al., 1997; Horvath et al., 1997). Large- to medium-sized pyramidal and multipolar neurons are preferentially affected, although smaller neurons are also involved. The full pattern of damage (Fig. 2) includes the PC/RS, frontal, temporal, entorhinal, perirhinal, piriform and prefrontal cortices, the amygdala and hippocampus (Corso et al., 1997). At 4 hours, the reaction in PC/RS cortex consists of intracytoplasmic vacuole formation, but in other brain regions a spongiform reaction featuring edematous swelling of spines on proximal dendrites is the most prominent cytopathological change. At 24 to 48 hours, the affected neurons become argyrophilic and immunopositive for HSP-72 and begin to display cytoskeletal abnormalities, including a conspicuous corkscrew deformity of their apical dendrites. In the 72 to 96 hour interval many of the degenerating neurons display conspicuous fragmentation, but cytoplasmic organelles and cytoskeletal elements within the cell body and mainstem dendrites of some cells con-

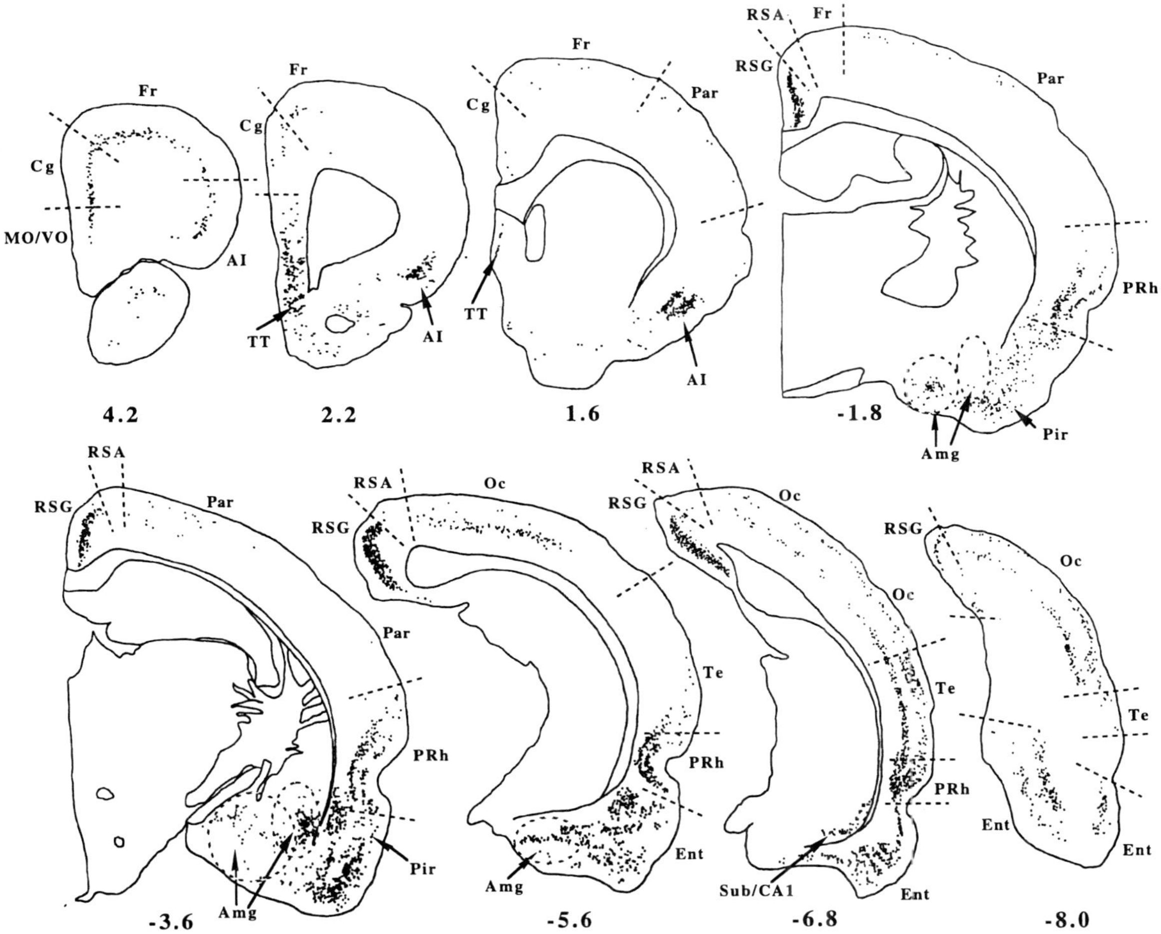


Fig. 2. Schematic representation of the distribution of degenerating neurons at eight rostrocaudal levels of the adult rat brain 4 days following treatment with a high dose of an NMDA antagonist. These cell plots illustrate the neuronal populations that are primarily at risk to degenerate when the NMDA receptor system is rendered profoundly hypofunctional. At risk populations include pyramidal and multipolar neurons in the prefrontal, PC/RS, occipital, temporal, parietal, entorhinal, perirhinal and piriform cortices and in the anterior olfactory nucleus, taenia tecta, amygdala and hippocampus. Rostro-caudal level and anatomical nomenclature are from Paxinos and Watson (1986). The number under each cell plot is the distance in mm from Bregma. AI: agranular insular cortex, Amg: amygdala, Cg: anterior cingulate, Ent: entorhinal cortex, Fr: frontal cortex, MO/VO: medial orbital/ventral orbital cortex, Oc: occipital cortex, Par: parietal cortex, Pir: piriform cortex, PRh: perirhinal cortex, RSA: retrosplenial agranular cortex, RSG: retrosplenial granular cortex, Sub/CA1: subiculum/CA1 field of hippocampus, Te: temporal cortex, TT: taenia tecta. Note: The area designated as RSG is the equivalent to the region referred to herein as PC/RS. See Corso et al. (1997) for histological documentation of this degeneration pattern.

tinue to show mixed signs of viability and degeneration for at least 10 days. Over this period the degenerative reaction does not elicit a robust glial or phagocytic response and the overall appearance is one of a subacute protracted neurodegenerative process.

*Receptor mechanisms and neural circuitry involved*

In a series of recent studies, our group (Olney et al., 1991; Farber et al., 1993; Price et al., 1994; Farber and Olney, 1995; Farber et al., 1998) and others (Sharp et al., 1992; Sharp et al., 1994; Sharp et al.,

1995) have found that several different classes of drugs (Table 1) effectively block the PC/RS neurotoxic action of NRA drugs. These findings, in conjuction with other work (Price et al., 1994; Farber et al., 1997; Kim et al., 1997), indicate that a surprisingly complex neural circuitry is involved in NRHypo neurotoxicity (Fig. 3).

*–Main excitatory limbs*

The key feature of this neural circuitry is that Glu, acting at NMDA receptors, functions in this circuit as a regulator of inhibitory tone. This regulation is accomplished through tonic stimulation of NMDA receptor-bearing GABAergic interneurons. Each of these interneurons, in turn, inhibits excitatory projections that converge onto the vulnerable PC/RS neuron. One of these excitatory projections consists of a cholinergic neuron that resides in the basal forebrain and projects to the PC/RS cortex where it synapses on a $m_3$ muscarinic receptor located on the vulnerable pyramidal neuron (Olney et al., 1991; Price et al., 1994; Farber et al., 1997; Kim et al., 1997). A second limb consists of a glutamatergic neuron that originates in the anterior thalamus and synapses on a KA receptor on the PC/RS pyramidal neuron (Price et al., 1994; Sharp et al., 1995; Farber et al., 1997). A third projection is less well defined and appears to consist of a sigmaergic neuron intrinsic to the PC/RS cortex (Farber et al., 1993; Price et al., 1994). Based on reports that neuropeptide Y (NPY) may activate the sigma receptor (Riviere et al., 1990; Monnet et al., 1990; Gue et al., 1992; Monnet et al., 1992), we propose that the sigmaergic neuron releases NPY onto a sigma receptor located on the PC/RS pyramidal neuron. By blocking Glu's tonic stimulation of GABAergic interneurons, NRA drugs remove inhibitory control from the three excitatory neurons. The resulting disinhibition of these three excitatory projections, which all converge on the vulnerable PC/RS pyramidal neuron, leads to excessive stimulation of the pyramidal neuron. Through this excessive stimulation the vulnerable pyramidal neuron is injured and develops histological changes (Price et al., 1994; Farber et al., 1997). Thus, blockade of one of the most common excitatory receptors in the CNS paradoxically leads to a hyperexcitatory state that can cause widespread corticolimbic neurodegeneration.

TABLE 1

Efficacy of various agents in blocking NRHypo neurotoxicity

| Test Compound | $ED_{50}$ (mg/kg ip) | Confidence limits (25th & 75th percentiles) |
|---|---|---|
| *Novel Antipsychotics* | | |
| Olanzapine | 0.6 | (0.2–1.9) |
| Clozapine | 2.8 | (1.5–5.1) |
| $\alpha_2$ Adrenergic Agonists | | |
| Clonidine | 0.044 | (0.0088–0.22) |
| Guanabenz | 0.21 | (0.13–0.33) |
| *$5HT_{2A}$ Serotonergic Agonists* | | |
| Lisuride | 0.17 | (0.03–0.85) |
| *Barbiturates* | | |
| Pentobarbital | 13.1 | (9.5–16.6) |
| Thiopental | 27.4 | (16.1–38.6) |
| *Benzodiazepines* | | |
| Diazepam | 1.0 | * |
| *Antimuscarinics* | | |
| Scopolamine | 0.13 | (0.80–0.18) |
| Benztropine | 1.29 | (0.92–1.65) |
| Trihexyphenidyl | 2.41 | (1.26–3.57) |
| *Sigma Agents* | | |
| Trihexyphenidyl | 2.41 | (1.26–3.57) |
| Haloperidol | 5.1 | (2.4–10.9) |
| Rimcazole | 17.8 | (8.9–35.6) |

*Diazepam did not show protection greater than 50% and thus confidence limits could not be calculated.

*–Accessory control elements*

In addition to the primary control provided by the NMDA receptor-bearing GABAergic neurons, there are at least two secondary control elements, which we have termed "accessory" because they provide additional control over the main excitatory projections. One accessory control element consists of an NMDA receptor-bearing noradrenergic (NE) neuron that, via stimulation of an $\alpha_2$ adrenergic receptor, controls the release of ACh by the basal forebrain cholinergic neuron (Farber and Olney, 1995; Kim et al., 1997). The second accessory control element consists of an NMDA receptor-bearing serotonergic (5HT) neuron that exerts agonist activity at a $5HT_{2A}$ receptor (Farber et al., 1998) that we postulate is located on a GABAergic neuron. Stimulation of either the $\alpha_2$ receptor or the $5HT_{2A}$ receptor restores inhibitory control over one or more of the excitatory projections and thus prevents the neurotoxic reaction from occuring (Farber and Olney, 1995; Farber et al., 1998). These two control elements can be thought of as back-up or accessory braking systems that help to maintain inhibition within the circuit. When generalized NRHypo is produced by the systemic administration of an NRA drug, the three excitatory limbs become disinhibited while at the same time Glu is blocked from activating these accessory control elements. Control over the disinhibited excitatory projections is lost, resulting in excessive excitation of the pyramidal neuron and consequent histological damage.

*–Output control element*

In addition to the primary and accessory control elements described above, the circuit also contains an inhibitory feedback loop that directly controls output of the pyramidal neuron. The loop consists of a recurrent collateral given off by the long projecting axon of the PC/RS pyramidal neuron; this collateral terminates on an NMDA receptor-bearing GABAergic neuron which synapses back onto the PC/RS pyramidal neuron (Farber et al., unpublished observations). Increased firing of the PC/RS neuron automatically causes increased activity in this inhibitory loop and provides a means by which the pyramidal neuron can regulate its own firing. Inactivation or under-functioning of this inhibitory feedback loop would compromise this control, causing the pyramidal neuron to fire excessively and in a poorly modulated manner, thereby disturbing mental functioning subserved by the PC/RS neuron and its projections. While inactivation of this loop by NRA drugs would lead to reduced control over the firing of the PC/RS neuron, it is unclear whether this inactivation contributes to the injury of the PC/RS neuron.

*–Other comments*

On-going work in our laboratory suggests that at least 5 different populations of NMDA receptors, one for each of the NMDA receptor-bearing GABAergic neurons in the three different excitatory limbs and one for each of the two accessory control arms of the circuit, need to be hypofunctional in order to induce the acute neurotoxic reaction. In other words, it requires that all three of the excitatory input channels be disinhibited simultaneously (and thereby simultaneously hyperactivated) for acute physical injury of the pyramidal neuron to occur. Preventing hyperactivation of any one of the excitatory input receptors (KA-glutamatergic, $m_3$-muscarinic, sigmaergic receptors), either by blocking one of these three receptors or by restoring inhibitory tone over one of the excitatory limbs, prevents the neurotoxic reaction from occuring (Table 1).

Inspection of the circuit reveals that Glu plays two key but opposing roles in this neurotoxic process. By stimulating the NMDA receptors Glu prevents the circuit from wreaking self destruction on itself. However, by stimulating the KA subtype of non-NMDA receptor on the vulnerable pyramidal neuron, Glu also is partially responsible for causing degeneration of the vulnerable pyramidal neurons. Studies aiming to confirm the role of Glu and NRHypo in neuropsychiatric disorders will have to be conducted by methods that separately evaluate the status of Glu at NMDA and KA receptors.

## NRHypo hypothesis of schizophrenia

*NRHypo causes psychosis*

A variety of NRA drugs (e.g., ketamine, PCP, CPP, CPP-ene, CGS19755, CNS 1102), when administered to humans, cause a psychotic state (Domino and Luby, 1981; Javitt and Zukin, 1991; Kristensen et al., 1992; Herrling, 1994; Krystal et al., 1994; Grotta et al., 1995; Malhotra et al., 1996; Muir et al., 1997; Newcomer et al., 1998). In depth studies of two of these agents, PCP and ketamine, indicate that these drugs, more faithfully than other psychotomimetics (e.g., amphetamines, LSD), can mimic a broad range of symptoms seen in schizophrenia, including impairments in cognitive function (Domino and Luby, 1981; Krystal et al., 1994; Malhotra et al., 1996; Newcomer et al., 1998). Because of these differences among the different psychosis-producing agents, NRA drugs have been described as "schizophrenomimetic" and are being intensively studied in hope of gaining insight into the pathophysiology of schizophrenia.

*NRHypo-induced disinhibition: A common denominator of the neurotoxic and psychotic effects of NRA drugs*

Several classes of GABAergic agents – benzodiazepines (Olney et al., 1991), barbiturates (Olney et al., 1991), and GABAergic anesthetics, including halothane, isoflurane, and propofol (Ishimaru et al., 1995; Jevtovic-Todorovic et al., 1997) – block the neurotoxic action of NRA drugs, presumably by restoring GABAergic inhibition to the excitatory limbs in the relevant neural circuitry (Fig. 3, see discussion above). Similarly anesthesiologists have long reported that the psychotomimetic effects of NRA drugs, which they term "emergence reactions," can be prevented by benzodiazepines (White et al., 1982; Reich and Silvay, 1989) and barbiturates (Magbagbeola and Thomas, 1974). The ability of GABAergic agents to block both the neurotoxic and psychotomimetic effects of NRA drugs has lead us to propose that the complex polysynaptic disinhibition mechanism that underlies the neurotoxic action of NRA drugs also underlies their psychotomimetic effects (Olney and Farber, 1995).

In addition, we have found that pre-pubertal rats are insensitive to the neurotoxic effects of NRA drugs, even following very high doses, but during adolescence they become increasingly sensitive to the neurotoxic action so that by late adolescence even low doses of an NRA drug can cause neurotoxicity (Farber et al., 1995). Similarly, anesthesiologists have found that the ability of ketamine to produce emergence reactions is dependent on the age of the patient. Infants and children prior to puberty rarely develop an emergence reaction when exposed to these agents, whereas up to fifty percent of post-pubertal teenagers and adults develop these symptoms (Reich and Silvay, 1989; Marshall and Longnecker, 1990). The similar age dependency profile of these two NRA drug-induced phenomena reinforces the proposal that a common pathophysiology underlies the neurotoxic and psychotomimetic effects.

*NRHypo-induced disinhibition: A common denominator of NRA drug effects and schizophrenia*

Given the similarity of the mental disturbances seen with NRA drugs and in schizophrenia we have further proposed that mental signs and symptoms seen in schizophrenia are produced by NRHypo (as others have also argued) and that these symptoms are a direct result of NRHypo-induced disinhibition of a polysynaptic circuit analogous to the one described in the rat (Olney and Farber, 1995). Consistent with this proposal is the observation that patients with schizophrenia appear to be more sensitive to NRA drugs than healthy humans and following NRA drug exposure can even develop a prolonged recrudescence of the acute psychotic state lasting for days to months (Luby et al., 1959; Ban et al., 1961; Lahti et al., 1995).

Schizophrenia typically has an age of onset in late adolescence or early adulthood. In order to account for growing evidence of an early developmental lesion in patients with schizophrenia, investigators have postulated that unknown maturational events must occur in the brain during adolescence that allows for the expression of

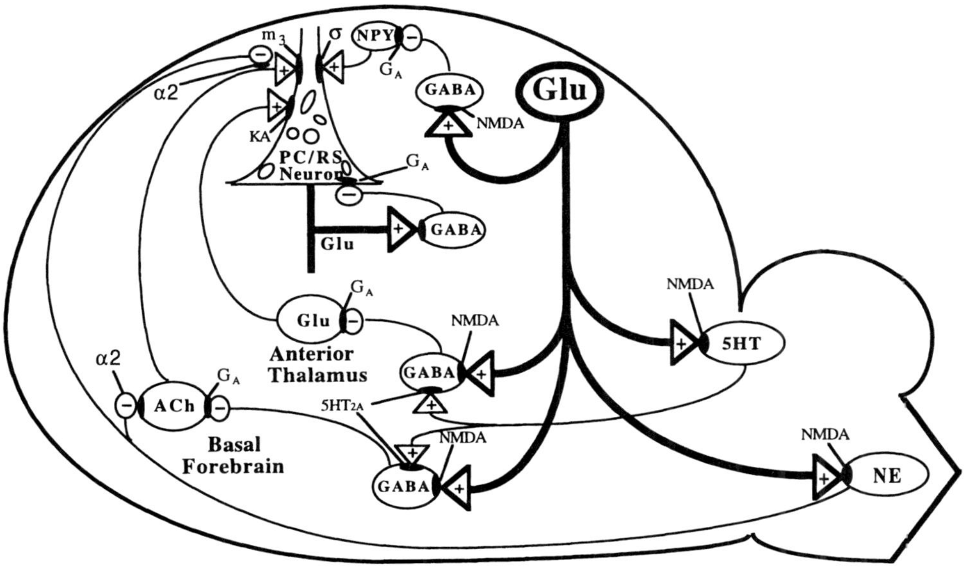


Fig. 3. To explain NRHypo-induced neurotoxicity of PC/RS neurons, we propose that Glu acting through NMDA receptors on GABAergic, serotonergic and noradrenergic neurons maintains tonic inhibitory control over multiple excitatory pathways that convergently innervate PC/RS neurons. Systemic administration of an NMDA receptor antagonist (or NRHypo produced by any mechanism) would simultaneously abolish inhibitory control over multiple excitatory inputs to PC/RS neurons. This would create chaotic disruption among multiple intracellular second messenger systems, thereby causing derangement of cognitive functions subserved by the afflicted neurons, as well as eventual degeneration of these neurons. This circuit diagram focuses exclusively on PC/RS neurons. We hypothesize that a similar disinhibition mechanism and similar but not necessarily identical neural circuits and receptor mechanisms mediate damage induced in other corticolimbic brain regions by sustained NRHypo. (+) = excitatory input; (−) = inhibitory input; ACh = acetylcholine; NE = norepinephrine; NPY = neuropeptide Y; $\sigma$ = sigma receptor; 5HT = serotonin; $\alpha_2$ = $\alpha_2$ subtype of adrenergic receptor; $G_A$ = $GABA_A$ subtype of GABA receptor; $m_3$ = $m_3$ subtype of muscarinic cholinergic receptor; KA = kainic acid subtype of Glu receptor; NMDA = NMDA subtype of Glu receptor. $5HT_{2A}$ = $5HT_{2A}$ subtype of 5HT receptor.

symptoms. As discussed above, NRA drug-induced neurotoxicity and psychosis has a similar age of onset in adolescence, suggesting similar dependence on maturational events. Given this similar age of onset profile, we propose that a common mechanism, namely an NRHypo-induced disinhibited state, underlies the neurotoxic and psychotomimetic effects of NRA drugs as well as the signs and symptoms of schizophrenia, and that similar maturational events allow the onset of both.

*Neurodegeneration and schizophrenia*

The proposal that the same mechanism (NRHypo) underlies the neurotoxic and psychotomimetic action of NMDA antagonists and that this mechanism is also operative in schizophrenia does not imply that NRHypo-induced damage to the adult brain causes the psychosis or even that such damage will necessarily occur in the majority of people with schizophrenia. Rather, we believe that in many cases dysfunction in the NRHypo circuit may lead to mental changes without causing histological damage. Our reasoning is based primarily on the observation that induction of histological damage requires the simultaneous disinhibition of three converging excitatory projections plus the loss of other control elements. This signifies that a patient with schizophrenia would need to have pervasive abnormalities affecting multiple components of the circuit in order to sustain histological damage. In contrast, disruption in only one or two parts of the circuit would be sufficient to cause hyperactivation and excess firing of the pyramidal neuron with consequent mental disturbances.

Regarding the relative absence of scarring in the brains of patients with schizophrenia and the dogma that on-going brain damage in the adult brain should produce glial scarring, it has been observed that when neurons are deleted from the adult brain by the NRHypo mechanism, only a transient and very limited gliotic response occurs (Fix et al., 1995). Thus, NRHypo is an excellent mechanism to explain how brain damage can occur during adulthood in schizophrenia with consequent cognitive deterioration but little or no residual scarring in the brain.

*Abnormal neural development in schizophrenia*

Consistent with evidence for genetic factors and early brain insults in schizophrenia (Lyon et al., 1989), it is proposed that a pathological process leading to NRHypo in certain brain circuits might occur in utero and remain clinically silent until adolescence. As noted above, even though NRA drugs can instill a profound NRHypo state in either the human or rat brain during various stages of development, they do not elicit either a psychotic or neurotoxic response until after puberty. This signifies that the immature brain could harbor a lesion that renders the NMDA receptor system hypofunctional and this defect would not be expressed symptomatically until a critical stage in adolescence when unknown maturational changes make it possible for the NRHypo state to trigger a pathological reaction. We postulate that as the maturation process occurs in each particular brain region, mental functions subserved by that brain region would become dysfunctional. Since each region would mature at different times, one would see the unfolding of the full syndrome over time as each specific clinical sign appears in succession. For example, relatively subtle impairments in cognitive functions such as attention and memory might occur in early adolescence while the more obvious psychotic signs would not become present until later in adolescence. In addition some brain regions and their associated mental functions may be more sensitive to disruption by NRHypo so that different symptoms may present at different times even if brain maturation is relatively homogeneous. In those individuals with a particularly severe and persistent form of NRHypo, secondary structural brain changes accompanied by progressive clinical deterioration would also occur. Thus, two types of structural brain changes are postulated. One would originate in utero as a function of genetic, environmental factors, or both, altering the relevant normal circuit to produce NRHypo. The other would originate in adolescence as the brain matures and becomes vulnerable to the NRHypo state.

There are a number of potential mechanisms that could occur early and produce an NRHypo condition in the pre-adolescent brain. Candidates include: loss of NMDA-receptor bearing GABAergic neurons, dopaminergic hyperactivity, hypofunctional NMDA receptors, and abnormalities in one or more of the other neurotransmitter systems in the NRHypo circuit. In addition abnormal neuronal migration resulting in an abnormal arrangement or "miswiring" of the circuit could lead to an NRHypo-equivalent disinhibited state. Of these candidate mechanisms, the loss of NMDA-receptor bearing GABAergic neurons is of particular interest given that Benes et al. (1991) have reported a loss of GABAergic neurons in postmortem brain tissue of patients with schizophrenia. Because the GABAergic neuron in the feedback loop normally regulates the firing pattern of cerebrocortical pyramidal neurons and aberrant firing of these neurons could cause considerable mental dysfunction, we propose that this population of GABAergic neurons might be differentially abnormal in patients with schizophrenia. Indeed, an excellent formula for psychosis would be loss of GABAergic inhibition simultaneously in this feedback loop and in any one of the three main excitatory inputs to the pyramidal neuron, since this would cause the pyramidal neuron to be persistently hyperactived at the same time that it has lost feedback control over its firing on other neurons.

*Potential inconsistencies of the NRHypo schizophrenia hypothesis*

An apparent inconsistency between the NRA drug model and schizophrenia is that female rats and humans are more sensitive than males to the neurotoxic and psychotomimetic effects, respectively, of NRA drugs, whereas in schizophrenia there are either no gender differences or a slightly earlier age of onset for males. However, the greater sensitivity of females to NRA drugs pertains only to PCP and its structural analogs and is most likely explained by these drugs being metabolized in the liver less efficiently by females than males. In contrast male rats are slightly more sensitive than females to the neurotoxic effects of competitive NMDA antagonists (Olney et al., unpublished observations), presumably because these agents are not PCP analogs and are not subject to the same sex-linked differential in hepatic metabolism. Thus if the hepatic metabolic variable is corrected for, the male/female sensitivity ratio may be quite similar for the NRA drug model and the human disease.

A potential shortcoming of the NRHypo hypothesis as formulated herein is that it does not explain why $D_2$ dopamine receptor antagonists are often therapeutically beneficial in schizophrenia. However, as discussed elsewhere (Olney and Farber, 1995), $D_2$ dopamine receptors may interact with the NRHypo circuitry at points that remain to be identified and such interaction could easily explain the observed therapeutic efficacy of $D_2$ dopamine antagonists in terms of their ability to correct or counteract the NRHypo state.

*Summary of the NRHypo schizophrenia hypothesis*

Given the schizophrenia-like quality of the mental disturbances caused in normal humans by NRA drugs, the exquisite sensitivity of schizophrenic patients to the psychotogenic actions of NRA drugs, the fact that the usual onset of susceptibility to schizophrenia and to the neurotoxic and psychotic effects of NRA drugs has the same age-dependency profile and substantial evidence that the basic mechanism underlying the neuropsychotoxic effects of NRA drugs is an NRHypo-induced disinhibitory state, we hypothesize that in the brains of patients with schizophrenia there is a comparable NRHypo-induced condition that causes a complex network disturbance featuring disinhibition as the core principle. We further postulate that in schizophrenia the NRHypo condition is instilled in the brain early in life but remains quiescent until maturational changes occur in adolescence that permit the disinhibition mechanism to disrupt mental functions and produce the symptoms of schizophrenia. Finally we propose that if the NRHypo state is severe enough it will cause ongoing neurodegenerative changes and progressive cognitive deterioration.

Fundamental features of schizophrenia accommodated by this hypothesis are: (1) the occurrence of structural brain changes during early development that have the potential for producing subsequent clinical manifestations of schizophrenia; (2) a quiescent period in infancy and adolescence before clinical manifestations are expressed; (3) onset in late adolescence of psychotic symptoms; (4) involvement of $D_2$ dopamine receptors in some cases but not others, which would explain why some but not all patients are responsive to typical antipsychotic therapy; (5) ongoing neurodegenerative changes and cognitive deterioration in some patients.

**NRHypo hypothesis of AD**

Over the past decade major strides have been made in discovering important genetic abnormalities in AD. Mutations on four different chromosomes, each of which can promote amyloidopathy, have now been identified as etiologic factors in familial AD and the role of ApoE genotype as a risk factor in sporadic AD has been established. While recent research has elucidated the basic neurochemistry of $\beta$-amyloid and it is clear that abnormal deposition of $\beta$-amyloid in the brain occurs early in AD, it is not at all clear how $\beta$-amyloid deposition contributes to the neurodegenerative events in AD. Based on evidence that a severe degree of NRHypo is present in the human AD brain, that NRHypo can induce an AD-like pattern of neurodegeneration in animal brain and that $\beta$-amyloid deposition may contribute to a worsening of the NRHypo state, we propose NRHypo as a missing link that can help explain major aspects of the pathophysiology of AD that have heretofore remained elusive.

*Similarities between the neuronal degeneration seen in NRHypo and in AD*

There are important similarities between the overall pattern of NRHypo neurodegeneration (Fig. 2) and the pattern that has been described in the AD brain by various researchers. The PC/RS cortex, which is the brain region most vulnerable to NRHypo degeneration, was recently reported to be selectively affected early in the course of AD in a PET study of living patients (Minoshima et al., 1997). The PC/RS cortex has also been shown to be markedly atrophic late in the disease (Braak et al., 1992). In contrast, neurodegeneration in the anterior cingulate cortex is less severe in both the AD brain (Brun and Englund, 1981) and NRHypo animal model (Fig. 2). While it is difficult to make precise anatomical comparisons between the rodent and human brain, the transentorhinal area, considered among the earliest and most severely affected regions in the human AD brain (Braak and Braak, 1995), is roughly homologous to the perirhinal cortex in rat brain, which is second only to the PC/RS cortex in its sensitivity to NRHypo neurodegeneration (Fig. 2). Other brain regions preferentially affected in both the AD brain and NRHypo model include portions of the parietal, temporal, entorhinal, amygdaloid, subicular, hippocampal and insular cortices. A mild but transient microglial and astrocytic response accompanies the neurodegeneration seen with NRHypo. However, consistent with the known pathology of AD, a robust phagocytic response is conspicuously absent (Fix et al., 1995).

The neurons primarily involved in neurofibrillary tangle (NFT) formation in the AD brain are distributed widely throughout cortical and limbic brain regions, but in each region these neurons tend to be pyramidal or multipolar neurons and in certain cortical regions they are distributed in a bilaminar pattern. This fits the description of the subpopulation of neurons primarily affected in the NRHypo neurodegenerative syndrome (Corso et al., 1997). Interneurons in the cerebral cortex are also occasionally involved but the most prominently affected neurons are medium-sized pyramidal or multipolar neurons in each region.

The tortuousity of dendritic processes in the NRHypo model is accompanied by a parallel pattern of tortuosity of the microtubular cytoskeleton within the distorted dendrite. This suggests that changes in the external configuration of the dendrite are due to cytoskeletal changes within the dendrite. The cytoskeleton of the injured neurons appears to be undergoing both degenerative and regenerative processes but the repair effort is not very successful. As described above the mechanism of injury involves simultaneous hyperactivation of the neuron through several excitatory receptors, including a muscarinic ($m_3$) cholinergic receptor and a glutamatergic (kainic acid) receptor (Fig. 3). Second messenger systems associated with the kainic acid receptor are not known at this time, but it is well known that the $m_3$ muscarinic receptor is coupled to a phosphoinositide/$Ca^{2+}$/ protein kinase C second messenger system that mediates protein phosphorylation functions. Thus the erratic pattern of microtubule tortuosity and disarray together with apparent efforts at microtubule regeneration may reflect abnormal hyperactivity of this $m_3$-linked second messenger system and consequent disruption of its protein phosphorylation functions. Since other unknown second messenger systems are also hyperactivated simultaneously in the same neuron, these systems may also contribute to the cytoskeletal disruption pattern.

We are just beginning to study the possible relationship between this cytoskeletal disruption process in the NRHypo model and NFT, its potential counterpart in the AD brain. Thus far, we have not detected ultrastructural evidence of paired helical filaments. However, even if such evidence cannot be found, this might signify species specificity of this particular abnormality without disqualifying the NRHypo degenerative process as a valid model of the mechanism giving rise to NFT in human AD brain. Thus, our findings in the rat are consistent with the conclusion that NRHypo by itself can produce many of the neuropathological features of AD.

*NRHypo is present in aging and to a greater degree in AD*

At least four different laboratories studying three different non-human species (mice, rats, monkeys) have reported that the NMDA receptor transmitter system becomes markedly hypofunctional with advancing age (Gonzales et al., 1991; Wenk et al., 1991; Magnusson and Cotman, 1993; Saransaari and Oja, 1995). In addition it was recently shown in humans that a more severe degree of NRHypo is present in the AD brain than in age-matched normal controls (Ulas and Cotman, 1997). Thus, in the aging human brain the stage may already be set for widespread corticolimbic neurodegeneration to occur. All that is required to explain why it occurs to a more severe degree in the AD brain than in the "normal" aging brain is to identify one or more adjunctive conditions peculiar to the AD brain that may serve as catalysts or promoters of the NRHypo state. Because we have seen no evidence of amyloidosis or amyloid plaque formation in the NRHypo animal model, we propose that genetic or other predisposing factors peculiar to the AD condition are primarily responsible for the amyloidopathy in AD brain and that when amyloidopathy occurs alongside NRHypo the pathological process known as AD develops. How then does amyloidosis interact with NRHypo?

*Amyloidopathy and NRHypo*

Most investigators of amyloidosis have tended to focus exclusively on the potential of $\beta$-amyloid to kill neurons by itself without reference to its potential pathological interaction with NMDA receptors. We think the focus should be redirected. $\beta$-amyloid alone is not toxic except at very high concentrations, whereas at substantially lower concentrations $\beta$-amyloid causes cultured neurons to become hypersensitive to Glu or NMDA excitotoxicity (Koh et al., 1990; Gray and Patel, 1995; Patel, 1995). We propose, therefore, that predisposing factors (e.g., apoE4 genotype in sporadic AD, amyloidogenic mutations in familial AD) promote amyloidosis which, in turn, increases the sensitivity of NMDA receptors so that even normal concentrations of Glu can trigger abnormal currents, which on a chronic low grade excitotoxic basis can destroy NMDA receptor-bearing neurons. Loss of these neurons (and their NMDA receptors) increases the NRHypo burden in the aging brain.

*Other contributing factors*

Other factors such as oxidative stress and disturbances in energy metabolism may contribute to $\beta$-amyloid's augmentation of neuronal sensitivity to Glu. Disturbances in energy metabolism may cause NMDA receptor hypersensitivity by interfering with a voltage dependent mechanism by which $Mg^{2+}$ normally inhibits passage of sodium and calcium currents through the NMDA receptor channel. Because membrane polarization is maintained by energy-dependent mechanisms, impaired energy can trigger partial membrane depolarization which abolishes the $Mg^{2+}$ block and allows normal concentrations of transmitter Glu to drive abnormal currents on a chronic basis. Oxidative stressors may act through a similar mechanism in view of evidence that free radical generation in nitric oxide pathways disrupts glycolytic metabolism, and superoxide radical formation causes hyperactivation of NMDA receptors in cultured neurons. The proposal that oxidative stress may contribute to neurodegeneration in AD is consistent with recent evidence (Sano et al., 1997) that anti-oxidant drugs may retard the progression of cognitive deterioration in AD. Whether impaired energy or oxidative stressors are relatively more active in the aging AD brain than the aging normal brain is not clear at this time. Their presence, even if not more severe than that in the normal aging brain, would augment amyloid's ability to sensitize neurons to Glu's excitotoxic potential. Persistent hyperactivation of NMDA receptors would result in either excitotoxic degeneration of the dendritic spines on which NMDA receptors are located or in degeneration of the entire NMDA receptor-bearing neuron. In either case, NMDA receptors are deleted from the brain and the NMDA receptor system is reduced to a hypofunctional status. NRHypo thus would represent a residual deficit condition caused by NRHyper.

*AD neurodegeneration: A two-stage process*

A major tenet of our proposal is that the NMDA receptor system becomes hypofunctional in either the normal brain or AD brain after having first gone through an early stage of NRHyper. This hypothesis, consistent with the bulk of available data, assumes that the pattern of massive neurodegeneration in AD tends to follow the pattern of NFT formation, and that the neurons that display NFT at the time of autopsy are injured neurons that would be destined to slowly die and leave behind neurofibrillary debris. However, this hypothesis also assumes that there is a less massive pattern of neuronal degeneration that corresponds with the pattern of amyloid deposition. Our hypothesis conceives of the neurodegenerative events in AD occurring in two separate stages, by two separate mechanisms and according to two separate patterns. These have been difficult to tease apart because the two stages have a significant degree of temporal overlap and the two patterns have significant spatial overlap.

*–Stage I*

We propose that the first neurodegenerative stage entails the deposition of low concentrations of amyloid in the brain and interaction of amyloid with certain NMDA receptors in a manner that increases the sensitivity of these receptors to Glu so that the neurons bearing these receptors will be hyperstimulated and destroyed by endogenous Glu. As these neurons degenerate, amyloid plaques may form and incorporate portions of the degenerating neurons and other neural and glial processes in the immediate environment. The pattern of these early neurodegenerative and reactive events will follow the pattern of distribution of the specific neurons vulnerable to this amyloid/NMDA receptor-mediated neuropathological process. We postulate that it may not be a very conspicuous pattern of neuronal loss because it may be restricted to just the NMDA receptor-bearing neurons in our schematic circuit that control the release of transmitters onto the vulnerable pyramidal neuron (Fig. 3). In Stage I, the neurodegenerative process may produce few if any symptoms because it is limited to a small population of neurons. In addition we postulate that the recurrent collateral feedback loop (Fig. 3) remains relatively intact so that pyramidal neurons, as they begin to receive excessive stimulation, will be prevented from firing eratically onto other neurons and thereby prevented from generating florrid symptoms.

*–Stage II*

The second stage commences when the loss of NMDA receptor-bearing neurons is sufficient to substantially unleash the disinhibition syndrome in which many primary cerebrocortical and corticolimbic neurons are pathologically hyperstimulated through several signal transduction pathways at the same time. At this point, cognitive disturbances will become clearly evident. We propose that pyramidal neurons in many cortical and limbic brain regions will be affected and will slowly degenerate and die as the Stage II process progresses. Death and deletion of these neurons will disrupt mental functions just as excessive hyperactivation of these neurons will disrupt these functions.

While these neurons are degenerating, we propose that at least some of them develop NFT on the basis of excessive activation of second messenger pathways associated with muscarinic and/or non-NMDA glutamate receptors. These second messenger systems are coupled to kinases or other possible factors relevant to protein phosphorylation; therefore, hyperactivation of these systems provides a rational explanation for NFT formation which is believed to result from hyperphosphorylation of microtubule-associated proteins. In Stage II, neurodegeneration occurs as a network disturbance. The pattern of degeneration is determined by the pattern of connections within the network, and by the failure of inhibition over certain excitatory pathways within the network, causing specific cortical and limbic neurons innervated by these excitatory pathways to degenerate. This provides a rational explanation for the pattern of degeneration seen in AD.

In summary, we propose that in contrast to schizophrenia, the brain of someone afflicted with AD develops normally and then functions normally throughout most of the adult life. As the brain ages the NMDA receptor system becomes progressively hypofunctional. In those individuals destined to develop AD, other abnormalities (e.g., amyloidopathy and oxidative stress) interact to increase the NRHypo burden and the brain enters into a severe and persistent NRHypo state which leads to widespread neurodegeneration and accompanying cognitive deterioration.

**Comparison of the NRHypo models of schizophrenia and AD**

Since the time of Emil Kraeplin and Alois Alzheimer neuroscientists have delineated the aberrations in mental functioning seen in schizophrenia (dementia praecox) from those seen in Alzheimer's Dementia. It thus seems counterintuitive that a common mechanism (NRHypo) could underlie the pathophysiology of both illnesses. However, each illness has its own set of specific abnormalities and the differences between these specific abnormalities could account for the disparate clinical picture (Table 2). The first difference is that the NRHypo state originates at different times. In schizophrenia the condition is instilled early in life (probably in utero in most cases), whereas in AD NRHypo occurs late in life. Given the maturational processes that occur in the brain during childhood and adolescence, it is likely that introduction of the NRHypo condition early in the life of an individual destined to develop schizophrenia would have a drastically different effect on the functioning of the CNS than it would if it were to occur de novo late in life.

TABLE 2

Differences in the NRHypo models of AD schizophrenia

| | Schizophrenia | AD |
|---|---|---|
| Onset of NRHypo | early in life | late in life |
| Elements of Circuit Involved | a) excitatory limbs and accessory control elements minimally involved<br>b) feedback loop maximally involved | a) excitatory limbs and accessory control elements maximally involved<br>b) feedback loop minimally involved |
| Degree of NRHypo | usually mild and non-progressive | severe and progressive |
| Accessory Abnormalities | hyperdopaminergia<br>abnormal neuronal migration<br>miswiring of the circuit<br>abnormalities in other receptors or neurotransmitters in the circuit | amyloidosis<br>unfavorable ApoE genotype<br>oxidative stressors<br>disturbances in energy metabolism |

Second, the specific elements in each circuit that are abnormal or missing are different. In schizophrenia, the main abnormality is proposed to lie in the feedback control loop with an additional mild abnormality in one or more of the excitatory limbs or accessory control elements. These abnormalities alone would be sufficient to lead to abnormal and uncontrolled output from the vulnerable pyramidal neurons resulting in mental dysfunction characteristic of schizophrenia. However, since the disturbances in many cases are not pervasive or severe enough, the vulnerable pyramidal neurons typically will sustain only limited or no histological damage. In contrast, in AD the main problem is postulated to lie with the excitatory inputs to the pyramidal neurons and the two accessory control elements. The abnormalities are severe, pervasive, and persistent enough that pyramidal neurons develop structural damage and ultimately die. The hyperactivation of these neurons in AD, as they are abnormally stimulated through multiple signal transduction pathways, is proposed to not typically translate into schizophrenia-like symptoms early in the disease because the feedback loop is intact, allowing pyramidal neurons to maintain control over their firing onto other neurons. However, as the disease progresses a point will be reached where the feedback loop is unable to appreciably prevent abnormal output from the pyramidal neuron, resulting in florrid mental disturbances. As a result of excessive excitatory input, the pyramidal neurons may initially lose dendritic spines and accompanying synapses, but eventually these neurons will die.

A third difference is that the extent of dysfunction in each brain region is different between the two disorders. The disseminated pattern that can be seen with NRHypo (Fig. 2) represents the full range of brain areas and neuronal populations susceptible to this process. However, it is likely that the specific neurons predominately involved in schizophrenia, which originates as a developmental aberration, will be different from those predominently involved in AD. These differences, of course, may partially account for the different patterns of mental dysfunction seen in these two illnesses.

The fourth and last major difference is that each illness has its own set of specific transmitter or neurochemical abnormalities that would coexist with the NRHypo state. For example, in AD excessive deposition of amyloid, disturbances in energy metabolism, oxidative stressors, and ApoE genotype are all potential contributors to the pathological process. In schizophrenia, abnormalities such as hyperdopaminergic tone, early loss of GABAergic interneurons, abnormal neuronal migration, miswiring of circuits or abnormal receptors would contribute to the pathogenesis of the illness.

While this list of differences between the two disease processes is not exhaustive, it is offered to illustrate how a similar process can be at the core of two vastly different illnesses. If the hypotheses described herein prove correct, treatment implications may be considerable. An important caveat, however, is that therapeutic strategies developed in animal studies may be less applicable to the treatment of schizophrenia than AD, because in schizophrenia complex plastic changes will occur in reaction to the developmental NRHypo defect and the altered networks may not respond optimally or at all to treatments predicted to be beneficial based on NRA drug studies. For example, administering benzodiazepines to an adult patient with schizophrenia to correct for the loss of GABAergic neurons in fetal life may be largely unsuccessful because of complex changes affecting the GABAergic system during neurodevelopment. Indeed, while benzodiazepines given adjunctively with antipsychotic medications have a defined role in the treatment of schizophrenia, monotherapy with benzodiazepines has minimal if any antipsychotic action. Thus, although the NRHypo model suggests pharmacological methods that could protect the brain against the adverse consequences of disinhibited activity in corticolimbic circuits, these methods were developed in a model based upon the acute induction of NRHypo in adulthood and at least some may not be directly applicable to treatment of a schizophrenia syndrome of developmental origin. However, in AD where the pathological process is initiated in the setting of a normally developed adult CNS, it is more likely that pharmacological methods for preventing NRHypo neurotoxicity could be directly translated into effective treatments.

## Summary

Here we have described a novel excitotoxic process in which hypofunctional NMDA receptors cease driving GABAergic neurons which cease inhibiting excitatory transmitters in the brain. These disinhibited excitatory transmitters then act in concert to slowly hyperstimulate neurons in corticolimbic brain regions. We have discussed how such an abnormality could exist in the brains of individuals with schizophrenia or AD and could account for the clinical stigmata of the two disorders. In addition, we have highlighted how other disorder-specific factors would account for the differences in the clinical presentation of AD and schizophrenia. In an animal model, pharmacological methods have been developed for preventing the overstimulation of these vulnerable corticolimbic pyramidal neurons and at least some of these methods may be applicable for treating AD and schizophrenia.

## Acknowledgements

Supported in part by NIDA Scientist Development Award for Clinicians DA 00290 (NBF), NIMH Independent Scientist Award MH 01510 (JWN), NIMH Research Scientist Award MH 33894 (JWO), MH 53363 (JWN) from NIMH, DA 05072 (JWO) from NIDA and AG 11355 (JWO) from NIA.

## References

Allen, H.L. and Iversen, L.L. (1990) Phencyclidine, dizocilpine, and cerebrocortical neurons. *Science*, 247: 221.

Ban, T.A., Lohrena, J.J. and Lehmann, H.E. (1961) Observations on the action of Sernyl – A new psychotropic drug. *Can. Psychiatr. Assoc. J.*, 6: 150–157.

Benes, F.M., McSparren, J., Bird, E.D., SanGiovanni, J.P. and Vincent, S.L. (1991) Deficits in small interneurons in prefrontal and cingulate cortices of schizophrenic and schizoaffective patients. *Arch. Gen. Psychiatry*, 48: 996–1001.

Braak, H., Braak, E. and Bohl, J. (1992) Retrosplenial region involvement in Alzheimer's Disease. *Neurodegeneration*, 1: 53–57.

Braak, H. and Braak, E. (1995) Staging of Alzheimer's disease-related neurofibrillary changes. *Neurobiology of Aging*, 16: 271–284.

Brun, A. and Englund, E. (1981) Regional pattern of degeneration in Alzheimer's disease: Neuronal loss and histological grading. *Histopathology*, 5: 549–564.

Choi, D.W. (1988) Glutamate neurotoxicity and diseases of the nervous system. *Neuron*, 1: 623–634.

Choi, D.W. (1992) Excitotoxic cell death. *Journal of Neurobiology*, 23: 1261–1276.

Corso, T.D., Sesma, M.A., Tenkova, T.I., Der, T.C., Wozniak, D.F., Farber, N.B. and Olney, J.W. (1997) Multifocal brain damage induced by phencyclidine is augmented by pilocarpine. *Brain Res.*, 752: 1–14.

Domino, E.F. and Luby, E.D. (1981) Abnormal mental states induced by phencyclidine as a model of schizophrenia. In E.F. Domino (Ed.), *PCP (Phencyclidine): Historical and current perspectives*, NPP Books, Ann Arbor, pp. 401–418.

Ellison, G. (1994) Competitive and non-competitive NMDA antagonists induce similar limbic degeneration. *Neuroreport*, 5: 2688–2692.

Farber, N.B., Price, M.T., Labruyere, J., Nemnich, J., St Peter, H., Wozniak, D.F. and Olney, J.W. (1993) Antipsychotic drugs block phencyclidine receptor-mediated neurotoxicity. *Biol. Psychiatry*, 34: 119–121.

Farber, N.B. and Olney, J.W. (1995) $\alpha 2$ adrenergic agonists prevent MK-801 neurotoxicity. *Neuropsychopharmacology*, 12: 347–349.

Farber, N.B., Wozniak, D.F., Price, M.T., Labruyere, J., Huss, J., St Peter, H. and Olney, J.W. (1995) Age specific neurotoxicity in the rat associated with NMDA receptor blockade: potential relevance to schizophrenia? *Biol. Psychiatry*, 38: 788–796.

Farber, N.B., Kim, S.H. and Olney, J.W. (1997) Costimulation of muscarinic and non-NMDA glutamate receptors reproduces NMDA antagonist neurotoxicity. *Soc. Neurosci. Abstr.*, 23: 2308.

Farber, N.B., Hanslick, J., Kirby, C., McWilliams, L. and Olney, J.W. (1998) Serotonergic agents that activate $5HT_{2A}$ receptors prevent NMDA antagonist neurotoxicity. *Neuropsychopharmacology*, 18: 57–62.

Fix, A.S., Wightman, K.A. and O'Callaghan, J.P. (1995) Reactive gliosis induced by MK-801 in the rat posterior cingulate/retrosplenial cortex: GFAP evaluation by sandwich ELISA and immunohistochemistry. *Neurotoxicology*, 16: 229–239.

Gonzales, R.A., Brown, L.M., Jones, T.W., Trent, R.D., Westbrook, S.L. and Leslie, S.W. (1991) *N*-methyl-D-aspartate mediated responses decrease with age in Fischer 344 rat brain. *Neurobiology of Aging*, 12: 219–225.

Gray, C.W. and Patel, A.J. (1995) Neurodegeneration mediated by glutamate and $\beta$-amyloid peptide: A comparison and possible interaction. *Brain Res.*, 691: 169–179.

Grotta, J., Clark, W., Coull, B., Pettigrew, L.C., Mackay, B., Goldstein, L.B., Murphy, D. and LaRue, L. (1995) Safety and tolerability of the glutamate antagonist CGS 19755 (Selfotel) in patients with acute ischemic stroke. Results of a phase IIa randomized trial. *Stroke*, 26: 602–605.

Gue, M., Junien, J.L., Del Rio, C. and Bueno, L. (1992) Neuropeptide Y and $\sigma$ ligand (JO 1784) suppress stress-induced colonic motor disturbances in rats through $\sigma$ and cholecystokinin receptors. *J. Pharmacol. Exp. Ther.*, 261: 850–855.

Hargreaves, R.J., Rigby, M., Smith, D., Hill, R.G. and Iversen, L.L. (1993) Competitive as well as uncompetitive *N*-methyl-D-aspartate receptor antagonists affect cortical neuronal morphology and cerebral glucose metabolism. *Neurochem. Res.*, 18: 1263–1269.

Herrling, P. (1994) D-CPPene (SDZ EAA 494), a competitive NMDA antagonist. Results from animal models and first results in humans. *Neuropsychopharmacology*, 10 (3S/Part 1): 591S.

Horvath, Z.C., Czopf, J. and Buzsaki, G. (1997) MK-801-induced neuronal damage in rats. *Brain Res.*, 753: 181–195.

Ikonomidou, C., Qin, Y.Q., Labruyere, J. and Olney, J.W. (1996) Motor neuron degeneration induced by excitotoxin agonists has features in common with that seen in the SOD-1 transgenic mouse model of amyotrophic lateral sclerosis. *J. Neuropathol. Exp. Neurol.*, 55: 211–224.

Ishimaru, M., Fukamauchi, F. and Olney, J.W. (1995) Halothane prevents MK-801 neurotoxicity in the rat cingulate cortex. *Neurosci. Lett.*, 193: 1–4.

Javitt, D.C. and Zukin, S.R. (1991) Recent advances in the phencyclidine model of schizophrenia. *Am. J. Psychiatry*, 148: 1301–1308.

Jevtovic-Todorovic, V., Kirby, C.O. and Olney, J.W. (1997) Isoflurane and propofol block neurotoxicity caused by MK-801 in the rat posterior cingulate/retrosplenial cortex. *J. Cereb. Blood Flow Metab.*, 17: 168–174.

Kim, S.H., Farber, N.B., Price, M.T. and Olney, J.W. (1997) Clonidine suppresses MK801-induced acetylcholine release in rat posterior cingulate cortex: *in vivo*, microdialysis study. *Soc. Neurosci. Abstr.*, 23: 2308.

Koh, J.Y., Yang, L.L. and Cotman, C.W. (1990) $\beta$-amyloid protein increases the vulnerability of cultured cortical neurons to excitotoxic damage. *Brain Res.*, 533: 315–320.

Kristensen, J.D., Svensson, B. and Gordh, T., Jr. (1992) The NMDA receptor-antagonist CPP abolishes neurogenic 'wind-up pain' after intrathecal administration in humans. *Pain*, 51: 249–253.

Krystal, J.H., Karper, L.P., Seibyl, J.P., Freeman, G.K., Delaney, R., Bremner, J.D., Heninger, G.R., Bowers, M.B.,Jr. and Charney, D.S. (1994) Subanesthetic effects of the noncompetitive NMDA antagonist, ketamine, in humans. Psychotomimetic, perceptual, cognitive, and neuroendocrine responses. *Arch. Gen. Psychiatry*, 51: 199–214.

Lahti, A.C., Holcomb, H.H., Medoff, D.R. and Tamminga, C.A. (1995) Ketamine activates psychosis and alters limbic blood flow in schizophrenia. *Neuroreport*, 6: 869–872.

Luby, E.D., Cohen, B.D., Rosenbaum, G., Gottlieb, J.S. and Kelley, R. (1959) Study of a new schizophrenomimetic drug – Sernyl. *Arch. Neurol. Psychiatry*, 81: 363–369.

Lyon, M., Barr, C.E., Cannon, T.D., Mednick, S.A. and Shore, D. (1989) Fetal neural development and schizophrenia [clinical conference]. *Schizophr. Bull.*, 15: 149–161.

Magbagbeola, J.A. and Thomas, N.A. (1974) Effect of thiopentone on emergence reactions to ketamine anaesthesia. *Can. Anaesth. Soc. J.*, 21: 321–324.

Magnusson, K.R. and Cotman, C.W. (1993) Age-related changes in excitatory amino acid receptors in two mouse strains. *Neurobiology of Aging*, 14: 197–206.

Malhotra, A.K., Pinals, D.A., Weingartner, H., Sirocco, K., Missar, C.D., Pickar, D. and Breier, A. (1996) NMDA receptor function and human cognition: The effects of ketamine in healthy volunteers. *Neuropsychopharmacology*, 14: 301–307.

Marshall, B.E. and Longnecker, D.E. (1990) General anesthetics. In L.S. Goodman, A. Gilman, T.W. Rall, A.S. Nies and P. Taylor (Eds.), *The Pharmacological Basis of Therapeutics*, Pergamon Press, New York, pp. 285–310.

Minoshima, S., Giordani, B., Berebt, S., Frey, K.A., Foster, N.L. and Kuhl, D.E. (1997) Metabolic reduction in the posterior cingulate cortex in very early Alzheimer's disease. *Ann. Neurol.*, 42: 85–94.

Monnet, F.P., Debonnel, G., Junien, J.L. and de Montigny, C. (1990) *N*-methyl-D-aspartate-induced neuronal activation is selectively modulated by sigma receptors. *Eur. J. Pharmacol.*, 179: 441–445.

Monnet, F.P., Debonnel, G. and de Montigny, C. (1992) In vivo electrophysiological evidence for a selective modulation of *N*-methyl-D-aspartate-induced neuronal activation in rat CA3 dorsal hippocampus by sigma ligands. *J. Pharmacol. Exp. Ther.*, 261: 123–130.

Muir, K.W., Grosset, D.G. and Lees, K.R. (1997) Effects of prolonged infusion of the NMDA antagonist aptiganel hydrochloride (CNS 1102) in normal volunteers. *Clin. Neuropharmacol.*, 20: 311–321.

Newcomer, J.W., Farber, N.B., Jevtovic-Todorovic, V., Selke, G., Kelly Melson, A., Hershey, T., Craft, S. and Olney, J.W. (1998) Ketamine-induced NMDA receptor hypofunction as a model of memory impairment in schizophrenia. *Neuropsychopharmacology* (In Press).

Olney, J.W., Labruyere, J. and Price, M.T. (1989) Pathological changes induced in cerebrocortical neurons by phencyclidine and related drugs. *Science*, 244: 1360–1362.

Olney, J.W. (1990) Excitotoxic amino acids and neuropsychiatric disorders. *Annu. Rev. Pharmacol. Toxicol.*, 30: 47–71.

Olney, J.W., Labruyere, J., Wang, G., Wozniak, D.F., Price, M.T. and Sesma, M.A. (1991) NMDA antagonist neurotoxicity: Mechanism and prevention. *Science*, 254: 1515–1518.

Olney, J.W. and Farber, N.B. (1995) Glutamate receptor dysfunction and schizophrenia. *Arch. Gen. Psychiatry*, 52: 998–1007.

Patel, A.J. (1995) Pretreatment with a sublethal concentration of $\beta$-amyloid$_{25\text{-}35}$ potentiates neurodegeneration mediated by glutamate in cultured cortical neurons. *Alzheimer's Research*, 1: 41–44.

Paxinos, G. and Watson, C. (1986): The rat brain in stereotaxic coordinates. 2nd ed. New York, Academic Press.

Price, M.T., Farber, N.B., Labruyere, J., Foster, J. and Olney, J.W. (1994) Tracing the circuitry that mediates NMDA antagonist neurotoxicity. *Soc. Neurosci. Abstr.*, 20: 1532.

Reich, D.L. and Silvay, G. (1989) Ketamine: An update on the first twenty-five years of clinical experience. *Can. J. Anaesth.*, 36: 186–197.

Riviere, P.J., Pascaud, X., Junien, J.L. and Porreca, F.T.I. (1990) Neuropeptide Y and JO 1784, a selective $\sigma$ ligand, alter intestinal ion transport through a common, haloperidol-sensitive site. *Eur. J. Pharmacol.*, 187: 557–559.

Rothstein, J.D., Van Kammen, M., Levey, A.I., Martin, L.J. and Kuncl, R.W. (1995) Selective loss of glial glutamate transporter GLT-1 in amyotropic lateral sclerosis. *Ann. Neurol.*, 38: 73–84.

Sano, M., Ernesto, C., Thomas, R.G., Klauber, M.R., Schafer, K., Grundman, M., Woodbury, P., Growdon, J., Cotman, C.W., Pfeiffer, E., Schneider, L.S. and Thal, L.J. (1997) A controlled trial of selegiline, $\alpha$-tocopherol, or both as treatment for Alzheimer's disease. The Alzheimer's Disease Cooperative Study [see comments]. *N. Engl. J. Med.*, 336: 1216–1222.

Saransaari, P. and Oja, S.S. (1995) Dizocilpine binding to cerebral cortical membranes from developing and ageing mice. *Mechanisms of Ageing & Development*, 85: 171–181.

Sharp, F.R., Jasper, P., Hall, J., Noble, L. and Sagar, S.M. (1991) MK-801 and ketamine induce heat shock protein HSP72 in injured neurons in posterior cingulate and retrosplenial cortex. *Ann. Neurol.*, 30: 801–809.

Sharp, F.R., Butman, M., Wang, S., Koistinaho, J., Graham, S.H., Sagar, S.M., Noble, L., Berger, P. and Longo, F.M. (1992) Haloperidol prevents induction of the hsp70 heat shock gene in neurons injured by phencyclidine (PCP), MK801, and ketamine. *J. Neurosci. Res.*, 33: 605–616.

Sharp, F.R., Butman, M., Koistinaho, J., Aardalen, K., Nakki, R., Massa, S., Swanson, R.A. and Sagar, S.M. (1994) Phencyclidine induction of the hsp70 stress gene in injured pyramidal neurons is mediated via multiple receptors and voltage gated calcium channels. *Neuroscience*, 62: 1079–1092.

Sharp, J.W., Petersen, D.L. and Langford, M.T. (1995) DNQX inhibits phencyclidine (PCP) and ketamine induction of the hsp70 heat shock gene in the rat cingulate and retrosplenial cortex. *Brain Res.*, 687: 114–124.

Ulas, J. and Cotman, C.W. (1997) Decreased expression of *N*-methyl-D-aspartate receptor 1 messenger RNA in select regions of Alzheimer brain. *Neuroscience*, 79: 973–982.

Wenk, G.L., Walker, L.C., Price, D.L. and Cork, L.C. (1991) Loss of NMDA, but not GABA-A, binding in the brains of aged rats and monkeys. *Neurobiology of Aging*, 12: 93–98.

White, P.F., Way, W.L. and Trevor, A.J. (1982) Ketamine – Its pharmacology and therapeutic uses. *Anesthesiology*, 56: 119–136.

SECTION VI

# Perspectives

O.P. Ottersen, I.A. Langmoen and L. Gjerstad (Eds.)
*Progress in Brain Research*, Vol 116


CHAPTER 29

# The glutamate synapse as a therapeutical target: Perspectives for the future

Brian S. Meldrum

*Department of Clinical Neurosciences, Institute of Psychiatry, Kings College, De Crespigny Park, London, SE5 8AF, UK*

## Introduction

The future prospects for research on the glutamate synapse as a therapeutic target can be evaluated either in terms of the potential clinical targets or in terms of the molecular mechanisms that can be modified. This volume provides key data on both approaches. This chapter aims to provide an integrated summary of the current status and future prospects for clinical targets either currently under evaluation or potentially treatable by such an approach. It also provides an overview of the different molecular targets relevant to modifying glutamatergic function, and speculates on how these might be better exploited in the future.

## Clinical targets

The clinical targets have been defined largely on the basis of experiments involving rodent models of neurological and psychiatric disorders although some potential therapeutic applications have derived their strongest support from post mortem and in vitro experiments (e.g., riluzole in motor neurone disease). Table 1 lists the principal clinical targets alongside compounds that have been evaluated in man.

## Epilepsy

The demonstration that NMDA (*N*-methyl-D-aspartate) receptor antagonists block seizures in rodent models of epilepsy identified epilepsy as a therapeutic target fifteen years ago (Croucher et al., 1982). Much evidence has subsequently accumulated showing that activation of both NMDA and AMPA (α-amino-3-hydroxy-5-methyl-4-isoxazole-propionic acid) receptors plays a critical role in the induction of epileptiform discharges in vitro and the spread of seizure activity in vivo (see Chapman, (Ch. 24), this volume). There is also evidence that glutamate receptors may be functionally modified in particular syndromes of epilepsy. Most notably NMDA receptor function is potentiated in kindled limbic seizures in rodents (Köhr & Mody, 1994) and in the hippocampus in complex partial seizures in man (Mody et al., 1997). To date, however, no pure NMDA receptor antagonist has been shown to produce therapeutic benefit in a clinical trial in epilepsy (Sveinbjornsdottir et al., 1994) (see Chapman, this volume). However some drugs recently in clinical trials show NMDA receptor antagonism e.g., remacemide (whose principal metabolite FPL 12495 is potent both as a non-competitive open channel NMDA antagonist and as a blocker of voltage-dependent $Na^+$ channels, Wamil et al., 1996) and felbamate which

TABLE 1

Syndromes in which manipulation of glutamatergic function may be beneficial, with compounds whose efficacy is, or has been, under clinical evaluation

| Syndrome | Therapeutic approach | Clinical trial |
|---|---|---|
| Epilepsy | NMDA antagonists | D-CPPene (SDZ-EAA 494) |
| | | Dizocilpine ( MK-801) |
| | | Dextrorphan |
| Stroke | NMDA antagonists | Eliprodil (SL 82.0715) |
| | | ACEA 1021 |
| | | Cerestat (CNS 1102) |
| | | Selfotel ( CGS 19755) |
| | | GV 150526A |
| Traumatic brain injury | | D-CPPene (SDZ-EAA 494) |
| | | Selfotel (CGS 19755) |
| | | Eliprodil (SL 82.0715) |
| | | Cerestat (CNS 1102) |
| Perinatal asphyxia/ischaemia | NMDA antagonists | |
| Status epilepticus | NMDA antagonists | |
| Motor neurone disease | "Glutamate release" | Riluzole |
| Huntington's chorea | Glutamate release | Lamotrigine |
| | NMDA antagonist | Remacemide |
| Alzheimer's disease | AMPA/NMDA antagonists | |
| | NMDA partial agonist | D-cycloserine |
| Amnesia | AMPAkines | BDP-12 |
| Neuropathic pain | NMDA antagonists | CPP |
| | | Ketamine |
| Anxiety | NMDA antagonists | |
| Addiction/ | NMDA agonist | Acamprosate |
| Alcohol & drug withdrawal | NMDA antagonist | |
| Schizophrenia | NMDA glycine site | Glycine |
| | | Milacemide |
| | glycine partial agonist | D-cycloserine |

is an NMDA receptor antagonist and a GABA potentiating agent (Subramaniam et al., 1996; Rho et al., 1997). Topiramate has AMPA antagonist properties but its GABA potentiating effect may be more significant (White et al., 1997). The way ahead for glutamate receptor antagonists in epilepsy is likely to involve identification of antagonists with selectivity for receptors with particular subunit compositions and matching these to specific syndromes of epilepsy. An important alternative strategy is to attempt to reverse the molecular changes associated with epileptogenesis. The enhanced function of NMDA receptors in acquired epilepsy syndromes may relate to phosphorylation by protein kinase C; manipulating this might reverse epileptogenesis.

## Acute neurodegeneration: Stroke

That NMDA receptor antagonists offered the possibility of protecting against ischaemic brain damage became apparent in 1984 (Simon et al., 1984). Subsequently very many studies have shown that the volume of cortex infarcted following middle cerebral artery occlusion can be substantially reduced by giving NMDA or AMPA receptor antagonists 0–90 minutes after the onset of ischemia (reviews Meldrum, 1990, 1994; Gill

1994). Similar effects are observed with all types of NMDA antagonist (glutamate competitive, glycine competitive, (Bordi et al., 1997) open-channel blockers and polyamine site antagonists for NMDA receptors). Blocking AMPA receptors with either glutamate site competitive antagonists or allosteric site antagonists (see Table III) also partially protects the cortex in MCA occlusion in rats, with about the same efficacy and therapeutic time window as with NMDA receptor antagonists (Gill, 1994; Gill & Lodge 1994; Smith & Meldrum, 1992, 1993; Buchan, 1991; Li & Buchan, 1993).

Clinical trials have been initiated in stroke for the NMDA antagonists listed under stroke in Table I. The trial with Selfotel is not continuing due to unfavourable side effects and outcome. A substantial phase 3 trial with eliprodil failed to show improved outcome. Analysis of data in the trial with cerestat is awaited. The therapeutic time window for glutamate antagonists in human stroke is not known. Possibly it does not extend significantly beyond three hours. The way forward in stroke trials may involve more rapid delivery to emergency care with early assessment and commencement of treatment. Combinations of different protective strategies may also be important. For example early thrombolysis and glutamate antagonism could be combined with strategies that apparently have longer therapeutic time windows such as those relating to reactive oxygen species, neuronal & inducible nitric oxide synthase and neurotrophic factors (see Nakao & Brundin (Ch. 16) this volume).

## Traumatic brain injury

The possible use of glutamate antagonists in traumatic brain injury (TBI) has been most extensively investigated in rodent models of closed head injury, especially the fluid percussion model. Neuroprotection is usually assessed in terms of reduction in pathology (involving the cortex, hippocampus and thalamus) or in terms of neurological or behavioural outcome (see review by McIntosh et al., 1996, and Alessandri & Bullock (Ch. 20), this volume). In rodent models a wide range of NMDA antagonists have been shown to improve behavioural outcome e.g. LY 233053 and LY235959 [3S-(3α,4aα,6β,8aα)]-decahydro-6-phosphonomethyl)-3-isoquinolinecarboxylic acid (competitive antagonists); phencyclidine, MK-801 (dizocilpine), dextromethorphan, HU-211, and ketamine (uncompetitive antagonists); $Mg^{2+}$ salts; eliprodil (polyamine site antagonist). The AMPA antagonist, GYKI-52466, also protects against cognitive impairment in rats. When given intracerebroventricularly 5 min prior to fluid percussion injury the non-selective glutamate metabotropic antagonist α-methyl-4-carboxyphenylglycine improves neurological and cognitive outcome (Gong et al., 1995). That this protection relates to activation of Group I mGluRs is suggested by exacerbation of injury by a Group I agonist and a link to mGluR1 is indicated by the protective action of antisense probes against this receptor (Mukhin et al., 1996). Overall the animal models of TBI have responded less dramatically to glutamate antagonists than have the models of focal ischemia. Nevertheless these studies have lead to a significant number of clinical trials with NMDA receptor antagonists in traumatic brain injury. A major trial with CGS 19755 (Selfotel, CIBA-GEIGY) failed to show clinical benefit. A trial with CNS 1102 (Cerestat, Aptiganel, Cambridge Neuroscience) was terminated early. Results from a trial of D-CPPene (SDZ-EAA-494, D-(-)-(E)-4-(3-phosphonoprop-2-enyl)piperazine-2-carboxylic acid, Sandoz/Novartis) with over 900 patients enrolled are expected to be reported in June 1998. Although the clinical trial results to date are not encouraging, traumatic brain injury remains a significant target for glutamate receptor antagonists.

## Chronic neurodegeneration

There is strong circumstantial evidence implicating glutamate receptor activation in selective patterns of neurodegeneration in Huntington's disease and in amyotrophic lateral sclerosis. Thus the pattern of cell loss in the striatum in Huntington's disease is very precisely mimicked by local application of

NMDA or quinolinate. Various metabolic poisons produce a similar pattern of cell loss and in these models a synergistic interaction between mitochondrial poisoning and NMDA receptor activation in inducing neuronal loss can be demonstrated (see Lancelot & Beal, (Ch. 21), this volume). Thus neuroprotective approaches relating to the chronic administration of NMDA receptor antagonists seem appropriate. It is of course necessary to select antagonists such as memantine and remacemide that are relatively free from motor or cognitive side effects. Phase 2 clinical trial data for remacemide in Huntington's disease have been reported (Kieburtz et al., 1996) and the results of a phase 3 trial currently in progress should be reported in 1998.

In motor neurone disease (amyotrophic lateral sclerosis, ALS) the selective loss of anterior horn cells may be related to activation of AMPA rather than NMDA receptors. In an in vitro model (organotypic spinal cord cultures) of glutamate-mediated motor neuron toxicity it is non-NMDA receptor antagonists (such as GYKI 52466 [1-(4-aminophenyl)-4-methyl-7,8-methylenedioxy-5H-2, 3-benzodiazepine hydrochloride] and PD 144,216) that provide the best protection (Rothstein & Kuncl, 1995). There is a selective deficit in the glial glutamate transporter GLT-1 in amyotrophic lateral sclerosis that correlates with the selective pathology (Rothstein et al., 1995; Medina et al., 1996). This defect in the clearance of glutamate may contribute to an excitotoxic process involving AMPA receptors (Leigh & Meldrum, 1996; Rothstein et al., 1993; Rothstein et al., 1996). Thus selective AMPA receptor antagonists could provide a therapeutic approach. Compounds so far evaluated in acute animal models of neurotoxicity (such as NBQX, 6-nitro-7-sulphamoylbenzo(f) quinoxaline-2,3-dione, and GYKI 52466) have proved to be unsuitable for clinical trial because of their pharmacokinetics and toxicological profile. Attention has therefore been directed to compounds thought to influence glutamate release, such as riluzole and lamotrigine. A double-blind placebo controlled trial of riluzole in 959 ALS patients at three dose levels over 18 months showed no significant effect on functional assessments but a dose related effect on risk of death or tracheostomy (Lacomblez et al., 1996). When AMPA antagonists with suitable pharmacokinetics and side effect profiles become available they will be good candidates for clinical trial in ALS.

The evidence for the involvement of glutamate receptors in neuronal degeneration in Alzheimer's disease is more indirect (see Farber et al., (Ch. 28), this volume). Nevertheless many major pharmaceutical companies have research programmes devoted to developing glutamate antagonists for this objective. A large number of patents for NMDA receptor antagonists propose a clinical indication in Alzheimers disease e.g., from Eli Lilly, LY-233053, LY-235723, LY-274614, LY-235959; from Merck, Sharp & Dohme L-698544, L-701252, L-701273. Clearly clinical trials will only be appropriate if the antagonist can be shown to be relatively devoid of motor and cognitive side effects. This may be possible with glycine site antagonists (Kretschmer et al., 1997) or with drugs with a subtype selective activity profile.

### Anxiolytic actions

The initial observation that NMDA antagonists are anxiolytic (Stephens et al., 1986) has been supported by several more recent studies (Xie & Commissaris, 1992; Anthony & Nevins, 1993; Wiley, 1995; Baron et al., 1997).

Even modest cognitive side effects are however unlikely to be acceptable for this clinical application. Some glycine-site NMDA antagonists seem to show anxiolytic action in the absence of motor side effects.

### Alcohol and opiate addiction: withdrawal signs

Seizures and other behavioural signs associated with alcohol or opiate withdrawal can be suppressed by NMDA receptor antagonists (Morrisett et al., 1990). Chronic alcohol exposure leads to a functional up-regulation of NMDA receptors and a down regulation of $GABA_A$ receptors (Madamba et al., 1996). Acamprosate (calcium acetylhomotaurinate, from LIPHA France) is a compound that was originally thought to be GABAergic. It suppresses the signs of acute alcohol withdrawal in habituated rats and has been extensively tried in patients with alcoholism with some success at preventing relapse in weaned alcoholics (Lhuintre et al., 1985, 1990). Acamprosate was shown to decrease the action of excitatory amino acids on cortical neurons in vitro (Zeise et al., 1993). More recently it has been shown to potentiate NMDA receptor responses to synaptic inputs to pyramidal neurons in hippocampal slices (Madamba et al., 1996). It may therefore be a partial agonist at NMDA receptors.

NMDA receptors are clearly involved in several aspects of addiction behaviours and the use of drugs acting on these receptors merits further investigation.

**Amnesia/Memory enhancement**

Cyclothiazide (a carbonic anhydrase inhibitor used clinically for its diuretic action on the kidney) and aniracetam, a nootropic drug, have been shown to act at a site on the AMPA receptor slowing desensitisation and prolonging channel open time (Wong & Meyer, 1993). A series of benzamide compounds has been developed that mimic this action on AMPA receptors in vitro and show good penetration into the brain. Among such compounds, BDP, 1-(1-,3-benzodioxol-5-ylcarbonyl)piperidine (alternatively perperidine (sic)), has been reported to improve learning and memory in a two-odor discrimination task, in a water maze and in a radial maze (pretreating with the drug and testing 96, 24 and 8 hours later respectively (Staubli et al., 1994) and in a classical conditioning task (Shors et al., 1995). A related compound referred to variously as 1-(quinoxalin-6-ylcarbonyl)piperidine, BDP-12 or CX 516, from Cortex Pharmaceuticals) has been shown to enhance steady state AMPA currents and slow receptor desensitisation in vitro (Arai et al., 1996). This compound also potentiates learning and memory in various rodent models (Granger et al., 1996; Larson et al., 1995; Arai et al., 1996). There is also evidence that delayed recall can be enhanced in aged humans (male volunteers, 65–76 years old) by oral administration of CX-516, 600–900 mg (Lynch et al., 1997).

The concept that allosteric modulators of the AMPA receptor can enhance learning and memory seems to be clearly established. The clinical application of drugs with this property remains to be proven, but it is possible that such drugs could be beneficial in Alzheimer's disease or other dementias and in benign forgetfulness of old age.

There is also a substantial number of studies showing that the NMDA receptor glycine site partial agonist D-cycloserine can enhance memory and learning in animals. Thus cognitive impairment following cerebral ischaemia in the rat is attenuated by D-cycloserine (Myhrer & Paulsen, 1995). Visual recognition memory in rhesus monkeys (Matsuoka, 1996), cognitive performance in traumatic brain injured rats Temple & Hamm, 1996) and spatial memory in aged rats (Baxter et al., 1994) are all enhanced by D-cycloserine. However the cognitive enhancement may be more evident with acute than chronic administration (Quartermain et al., 1994). In post mortem material from Alzheimer brain D-cycloserine potentiated NMDA responses (Chessell et al., 1991). The working memory deficit induced by cholinergic antagonists is reversed by D-cycloserine (Ohno & Watanabe, 1996) supporting the concept that D-cycloserine could be therapeutically active in Alzheimer's disease.

**Hyperalgesia/Neuropathic pain**

That NMDA antagonists block the "wind-up" phenomenon whereby nociceptive responses in the spinal cord are potentiated following high frequency stimulation of C-fibres, was clearly established in 1987 (Davies & Lodge, 1987; Dickenson & Sullivan, 1987; Kristensen et al., 1992a). Subsequent experimental studies are presented in detail by Urban & Gebhart, this volume. Clinical studies have involved the systemic use of ketamine (Maurset et al., 1989; Felsby et al., 1996) or the intrathecal injection of CPP, 3-((RS)-2-carboxypiperazin-4-yl)-propyl-1-phosphonic acid, (Kristensen et al., 1992b; Kristensen, 1997). The future development of NMDA receptor antagonism for hyperalgesia will be dependent on identifying antagonists that lack cognitive and motor side effects but are effective when given systemically.

**Schizophrenia**

The possibility that alterations in glutamatergic transmission may play a role in schizophrenia has been under investigation for more than 15 years. Much significance has been attached to evidence that NMDA receptor antagonists can be psychotomimetic (see Farber et al., (Ch. 28), this volume). In particular phencyclidine (and to a lesser extent other high affinity uncompetitive open-channel NMDA blockers) exacerbate schizophrenic symptoms in patients and induce a syndrome closely mimicking schizophrenia in subjects without a history of psychosis (Luby et al., 1959; Javitt & Zukin, 1991). These and other data have lead to the development of a particular hypothesis about the pathways that might be involved in schizophrenia (see Farber et al., this volume). The role of glutamate in schizophrenia is reviewed by Goff & Wine (1997).

Post-mortem studies of glutamate receptors in schizophrenia initially used autoradiography with ligands showing some specificity for the major types of glutamate receptor. These showed reductions in non-NMDA receptor density in the left hippocampal formation (Kerwin et al., 1988, 1990) but increases in the prefrontal cortex (Deakin et al., 1989). In situ hybridisation studies also indicate that the mRNA for two AMPA receptor subunits (GluR1 & GluR2) is decreased in the hippocampus and parahippocampal gyrus (Eastwood et al., 1995; Eastwood et al., 1997, 1997; Breese et al., 1995). The glutamate content is significantly reduced in prefrontal cortex and hippocampus whereas the N-acetylaspartyl glutamate (NAAG) content is increased in association with a decreased activity of its peptidase, *N*-acetyl-α-linked acidic dipeptidase, NAALADase (Tsai et al., 1995).

High dose glycine has been tried as a therapy in schizophrenia although evidence that it penetrates the blood brain barrier significantly is lacking. In two small scale trials, however, apparent improvement of negative symptoms has been reported with glycine 0.4–0.8 g/kg (Javitt et al., 1994; Heresco-Levy et al., 1996). A glycine prodrug milacemide apparently worsened schizophrenic symptoms but may be acting as an NMDA receptor antagonist.

D-cycloserine was initially given to patients with schizophrenia because of its effects on mood and motivation in patients with tuberculosis (Simeon et al., 1970). More recently D-cycloserine, 50 mg, has been shown to diminish negative symptoms in patients on antipsychotic medication (Goff & Wine, 1997; Leiderman et al., 1996).

**Molecular targets at the glutamate synapse: Enzymes involved in metabolism of glutamate and other agonists**

Most attention in terms of drug synthetic programs, preclinical testing and clinical trials has focussed on postsynaptic ionotropic receptors especially the NMDA receptor. Little attention has been paid to the possibility of modifying the synthesis or further metabolism of glutamate. This is partly because of the complexities of glutamate metabolism (see Kvamme (Ch. 6), this volume). Glutamine synthetase is involved in several metabolic pathways, including that for GABA. The GABA-transaminase inhibitor vigabatrin increases the brain content of GABA which, in turn, appears to inhibit glutamine synthetase and decrease the brain and CSF content of glutamate (Kälviäinen et al., 1993; Waniewski & Martin, 1995). Inhibition of glutaminase by DON (6-diazo-5-oxo L-norleucine) leads to a reduction in immunostaining for glutamate (Conti & Minelli, 1994). Glutaminase inhibitors have been given focally into the brain and shown to produce a modest anticonvulsant action (Chung & Johnson, 1984). A more potent anticonvulsant effect has been reported following the focal (inferior colliculus) injection of L-canaline in Genetically

Epilepsy Prone rats (Faingold et al., 1992) which decreases glutamate synthesis apparently through inhibition of ornithine aminotransferase (Wroblewski et al., 1985).

Other neurotransmitters or endogenous agents acting on glutamate receptors may be a better target in terms of drugs acting on their metabolism. Thus NAAG (N-acetylaspartyl glutamate) is a selective agonist for the mGluR3 receptor and its function can apparently be modified via its synthetic or catabolic enzymes. A specific peptidase, N-acetyl-α-linked acidic dipeptidase, NAALADase, cleaves NAAG into glutamate and N-acetylaspartate. An inhibitor of NAALADase, 2-(phosphonomethyl)pentanedoic acid (PMPA), prevents the rise in extracellular glutamate concentration and reduces cortical infarct volume in a model of transient middle cerebral artery occlusion (Thomas et al., 1997; Lu et al., 1997).

Drugs acting on enzymes of the kynurenine pathway can change the brain content of kynurenate and quinolinic acid and have a variety of functional effects. Kynurenate acts as an NMDA antagonist at the glycine site and is synthesized in astrocytic processes by transamination of kynurenine (Gramsbergen et al., 1997). Kynurenine3-hydroxylase inhibitors such as FCE 28833 can enhance synthesis of kynurenic acid and show a neuroprotective effect in cerebral ischaemia (Speciale et al., 1996).

Thus drugs altering metabolism within the kynurenine pathway may significantly modify neuronal excitability and neurodegenerative processes.

## Molecular targets: Presynaptic modulation of glutamate release

This can be achieved by agonists acting on a wide range of membrane receptor molecules, most of which are linked to G proteins and modify high voltage activated $Ca^{2+}$ channels, and by drugs acting directly on voltage sensitive $Ca^{2+}$ channels of the N or P/Q type (see Table II and reviews by Wu & Saggau, 1997 and by Nicholls (Ch. 2), this volume). Some anticonvulsant compounds acting primarily as use-dependent blockers of voltage-sensitive $Na^+$ channels, such as phenytoin, lamotrigine, riluzole and BW-619C89 (4-amino-2-(4-methyl-1-piperazinyl)-5-(2,3,5-trichlorophenyl) pyrimidine) are sometimes referred to as glutamate release blockers (Taylor & Meldrum, 1995). They undoubtedly can block the increase in extracellular glutamate concentration that occurs during cerebral ischaemia; this effect is predominantly a consequence of their actions on $Na^+$ channels. There may, however, be additional mechanisms involved; in the case of lamotrigine an action on N-type $Ca^{2+}$ channels is likely (Wang et al., 1996; Stefani et al., 1997). A synthetic ω conotoxin, SNX-111 that selectively blocks N-type $Ca^{2+}$ channels protects in rodent models of global and focal cerebral ischemia (Valentino et al., 1993; Takizawa et al., 1995).

Adenosine A1 receptors are expressed presynaptically on neurons in hippocampus and cortex and decrease glutamate release (Rudolphi et al., 1992; Nicholls, this volume). They are active on several types of synapse but their action on glutamatergic synapses appears to be the most significant. Agonists acting on A1 receptors, such as 2-chloroadenosine, CHA (N6-cyclohexyladenosine) and R-PIA (N6-R-phenylisopropyladenosine), are neuroprotective in some in vitro and in vivo models (Fredholm, 1997). The clinical use of such agents is complicated by cardiovascular side effects and by adaptive changes in neuronal receptors with long term administration.

Glutamate metabotropic receptors are a particularly attractive drug target in that each receptor subtype shows distinctive functional properties and a particular distribution both in terms of brain regions and in terms of its location relative to the synaptic cleft (see Page 450 below and reviews by Takumi et al. (Ch. 8) and Bruno et al. (Ch. 14), this volume). Groups II and III are predominantly presynaptic; group II (mGluR2 and mGluR3) being located, however, a significant distance from the synaptic cleft, whereas Group III (mGluR4, mGluR7, mGluR8) are primarily within the synaptic cleft. Their effects are pathway specific; within the hippocampus, for example,

Group II agonists decrease excitatory transmission from mossy fibres to CA3 pyramidal neurons, whereas Group III agonists block transmission from lateral perforant path to dentate granule cells. The role of these presynaptic effects in the anticonvulsant action of Group II and III mGluRs is not clear as the anticonvulsant effect following their focal intracerebral injection can be delayed and prolonged (see Chapman, this volume).

The presynaptic effects of muscarinic, adenosine A1 and NPY receptor activation are mutually occlusive indicating involvement of a common pathway. This may involve activation of the G-protein, Go, to inhibit N-type $Ca^{2+}$ channels (Wu & Saggau, 1997; Qian et al., 1997). Neuropeptide Y receptors are potentially a good target because of the relatively restricted distribution of NPY Y2 receptors.

Kainate receptors occur presynaptically on the mossy fibres and modulate glutamate release. At low concentrations kainate enhances excitatory post synaptic potentials but at high concentrations it inhibits glutamate release (Chittajallu et al., 1996).

Cholecystokinin enhances synaptic release of glutamate in the hippocampus (Breukel et al., 1997) but it has yet to be shown that this can be therapeutically exploited.

**Molecular targets: Postsynaptic glutamate ionotropic receptors**

The subunits that comprise the AMPA, kainate and NMDA receptors in the brain are now relatively well understood in terms of the genes that code for them and the modifications that occur during transcription, (Hollmann & Heinemann, 1994; Nakanishi, 1992; Seeburg, 1993). The influence of subunit composition, splice variants and mRNA editing on the biophysical properties of the receptor complexes is discussed in Ch. 11 by Borges & Dingledine and Ch. 13 by MacDonald et al., this volume. The flip splice variant of the AMPA receptor is more efficient in its response to glutamate than is the flop variant. For the NMDA receptor the NR2 subunit dramatically influences the offset decay time, NR1/NR2A being the fastest, NR1/NR2B and NR1/NR2C intermediate and NR1/NR2D the slowest (Monyer et al., 1994). The relative expression of different subunits of the NMDA or AMPA receptors in particular classes of neurons critically determines their response properties (Laurie & Seeburg, 1994; Geiger et al., 1995) (see also Ch. 10 by Wenthold & Roche and Ch. 12 by Monoghan et al., this volume). Pharmacological tools for differentiating AMPA and kainate receptors and their subtype variants have

TABLE 2

Presynaptic modulation of glutamate release

| Target | Drug |
|---|---|
| Glutamate metabotropic receptors | |
| mGluR2, mGluR3 | LY 354740 |
| | LCCGI (2S,1'S,2'S)-2-(carboxycyclopropyl)glycine |
| mGluR4, mGluR7, mGluR8 | L-AP4 ( L-2-amino-4-phosphonobutyric acid) |
| | L-SOP (L-serine-O-phosphate) |
| Adenosine A1 receptors | 2-chloroadenosine, R-PIA ( R-phenylisopropyladenosine) |
| Kappa opioid receptor | Enadoline, (-)-U-50488. |
| Muscarinic Ach receptors | Carbachol, pilocarpine. |
| GABAB receptors | Baclofen, SKF 97541 |
| Neuropeptide Y2 receptors | Neuropeptide Y(3-36), Peptide YY. |
| CCKB receptors | Cholecystokinin (CCK8) |
| Kainate receptors | Kainate, (2S,4R)-4-methylglutamate |

TABLE 3

**Glutamate ionotropic receptors: agonists and antagonists**

| | AMPA | KAINATE | NMDA |
|---|---|---|---|
| **Subunits** | GluR1 flip,flop | GluR5 | NMDAR1(1a,b,2a,b,3a,b,4a,b) |
| | GluR2 flip,flop | GluR6 | NR2A |
| | GluR3 flip,flop | GluR7 | NR2B |
| | GluR4 flip,flop | KA1 | NR2C |
| | | KA2 | NR2D |
| | | | ? NR3A,B |
| **Agonists** | (S)-AMPA | KAINATE | NMDA |
| | (S)-5-Fluorowillardine | Domoate | 1R,3R-ACPD |
| | L-ODAP | (2S,4R)-4-methylglutamate | Quinolinate |
| | L-Quisqualate | ATPA | (RS)(tetrazol-5-yl)glycine |
| | | | Glycine, D-serine |
| **Antagonists** | **Competitive** | **Competitive** | **Competitive glutamate** |
| | *NBQX, YM 90K* | *NBQX, CNQX* | *D-CPPene, CPP,* |
| | *LY 293558* | *NS102* | *CGS 19755 (Selfotel)* |
| | **Allosteric** | | **Competitive (glycine)** |
| | *GYKI 52466* | | *kynurenate, ACEA 1021,* |
| | *GYKI 53655* | | *GV 150526A, ACPC, L-687,414* |
| | **SYM 2206** | | **Open Channel blockers** |
| | | | *MK180(Diizocilpine),* |
| | | | *CNS 1102 (Cerestat, Aptiganel),* |
| | | | *memantine, remacemide* |
| | | | *FPL 12495 (desglycine* |
| | | | *remacemide),* |
| | | | *HU 211* |
| | | | **Polyamine site/2B selective** |
| | | | *Ro-25–6981,* |
| | | | *Ifenprodil, eliprodil* |
| | | | *CP 101,606* |
| | | | **Steroid modulator** |
| | | | *Pregnenolone sulfate* |

Antagonists in italics have been shown to decrease the volume of cortex infarcted in rodent models of MCAocclusion.

1R,3R-ACPD = (1R,3R)-1-amino-cyclopentane-1,3-dicarboxylate
L-ODAP = $\beta$-oxalylamino-L-alanine
ATPA = (RS)-2-amino-3-(3-hydroxy-5-tert-butylisoxazol-4-yl)propanoic acid
NBQX = 2,3-dihydroxy-6-nitro-7-sulfamoylbenzo[f]quinoxaline
CNQX = 6-cyano-7-nitroquinoxaline-2,3-dione
YM 90K = 6-(1-imidazolyl)-7-nitroquinoxaline-2,3(1H,4H)-dione
NS 102 = 5-nitro-6,7,8,9-tetrahydrobenzo[g]indole-2,3-dione-3-oxime
CPP = 3-(2-carboxypiperazine-4-yl)propyl-1-phosphonic acid
ACPC = 1-aminocyclopropanecarboxylic acid
LY 293558 = (3S,4aR,6R,8aR)-6-[2-1(2)H-Tetrazole-5-yl)ethyl]decahydroisoquinoline-3-carboxylic acid
HU 211 = 7-hydroxy-tetrahydrocannabinol 1,1-bionid
SYM 2206 = (+-)-4-(4-aminophenyl)-1,2-dihydro-1-methyl-2-propycarbamoyl-6,7- ethylenedioxyphthalazine
GYKI 524666 = 1-(aminophenyl)-4-methyl-7,8-methylenedioxy-5H-2,3-benzodiazepine
GYKI 53655 = 1-(4-aminophenyl)-3-methylcarbamyl-4-methyl-7,8-methylenedioxydihydro- 5H-2,3-benzodiazepine

been rather lacking (Bettler & Mulle, 1995) although some antagonists that show subunit selectivity e.g., Evans Blue, have been described (Keller et al., 1993). NS 102 preferentially blocks recombinant GluR6 (Verdoorn et al., 1994). Very recently progress has been made by exploiting the AMPA selectivity of the 2,3-benzodiazepines such as GYKI 53655 and the broader blockade of nonNMDA receptors by quinoxalinediones, such as CNQX and NBQX (Vignes & Collingridge 1997; Castillo et al., 1997). In this way it has been shown that the postsynaptic kainate receptors in hippocampal CA3 pyramidal neurons respond to mossy fibre input with a slow excitatory potential.

A wide variety of NMDA antagonists show partial differential effects according to the NR2 subunit (Laurie & Seeburg, 1994 and see also Monaghan et al. (Ch. 12), this volume). Competitive antagonists of the D-AP7, D-2-amino-7-phosphonoheptanoic acid/D-CPPene type act preferentially on the NR1/NR2A subtype. Polyamine site antagonists, such as eliprodil and Ro 25-6981, act preferentially on NR1/NR2B subunits (Avenet et al., 1997). The eight splice variants of the NR1 subunit (1a,1b,2a,2b,3a,3b,4a,4b) also differ in their biophysical and pharmacological properties (Hollmann et al., 1993).

The use of expression systems for recombinant forms of NMDA, AMPA and kainate receptors to screen novel antagonists should yield compounds with high subtype selectivity which may have distinctive clinical and side effect profiles.

## Molecular targets: Metabotropic glutamate receptors

At least seven distinct glutamate metabotropic receptors are known to be expressed in the brain. Their regional and subcellular distribution is characteristic for each receptor (Shigemoto et al., 1996, 1997; Ohishi et al., 1995; Lujan et al., 1996; Petralia, 1996; see Bruno et al. (Ch. 14), this volume). Both group II (=mGluR2 and mGluR3) and group III (=mGluR4, mGluR7, mGluR8) receptors are found presynaptically on asymmetric (presumed glutamatergic) synapses but the group II receptors are relatively remote from the synaptic cleft whereas the Group III receptors are within it (see Takumi et al., (Ch. 8), this volume. Not only are the individual mGluRs differentially expressed according to the cell type, but so are the splice variants. The post-synaptic neuron also influences the expression of presynaptic mGluRs. mGluRs show widely differing affinities for glutamate.

Much needs to be clarified concerning their functional significance but some broad categorisations can be made. Thus group I agonists such as 3,5 dihdroxyphenylglycine are convulsant and group I antagonists tend to be anticonvulsant. Group II agonists such as LY 354740 are anxiolytic in animal models. A neuroprotective mechanism involving activation of mGluR3 receptors on astrocytes is described by Bruno et al., this volume. One aspect of particular interest concerning metabotropic receptors is their role in longer term changes in neuronal function mediated by neurotrophins and growth factors. The expression of mGluR5 receptors in astrocytes is enhanced by several growth factors (Miller et al., 1995). Thus there are complex interactions between mGluRs and various long term processes likely to be significant for epileptogenesis and chronic neurodegeneration. Thus evaluation of likely applications for mGluR active agents requires the use of specific preclinical models for detecting long term effects.

## Glutamate transporters

Glutamate transporters are found in astrocytic processes and in neuronal membranes and have the primary function of removing synaptically released glutamate from the synaptic cleft and keeping the extracellular glutamate concentration low (see Danbolt et al. (Ch. 3), this volume, and Arriza et al., 1994). The inward transport of glutamate is linked to the inward movement of $Na^+$ and the outward movement of $K^+$ (with a stoichiometry of one glutamate anion cotransported with 2 or 3 $Na^+$ and, 1 $H^+$ and countertransport of 1$K^+$ yielding a net inward current (Zerangue & Kavanaugh, 1996; Billups et al., this volume). Thus

TABLE 4

**Glutamate metabotropic receptors; agonists and antagonists**

| Receptor | Transduction mechanism | Agonists | Antagonists |
|---|---|---|---|
| Group I | Phospholipase C activation | 3,5-DHPG | (S)-4C3HPG |
| (mGlu1a,b,c,d | = PhosphoInositide hydrolysis | quisqualate | AIDA |
| mGlu5a,b) | (DAG & IP3 as 2nd messengers) | trans-ADA | PHCCC |
| | | CHPG [mGlu5] | LY 367385 |
| Group II | Decrease in adenylyl cyclase | L-CCGI | EGLU |
| (mGlu2 | Decrease in $Ca^{2+}$ current | (1S,3S)ACPD | MCCG |
| mGlu3) | Decrease in GABA and | LY 354740 | MTPG |
| | Glutamate release | (1S,3R)ACPD | |
| | | NAAG [mGlu3] | |
| Group III | Decrease in adenylyl cyclase | L-AP4 | CPPG |
| (mGlu4a,b | Decrease in $Ca^{2+}$ currents | L-SOP | MAP4 |
| mGlu6 | Decrease in GABA and | homoAMPA | MPPG |
| mGlu7a,b | Glutamate release | 4Cl-3,5DHPG | MCPA |
| mGlu8) | | | MSOP |
| PLD coupled | Phospholipase D activation | L-CSA | 3,5-DHPG |
| | | L-AP3 | |

IS,,3S ACPD = (1S,3S)-1-Aminocyclopentane-1,3-dicarboxylic acid
trans-ADA = trans-Azetidine-2,4-dicarboxylic acid
AIDA = (RS)-1-Aminoindan-1,5-dicarboxylic acid
L-AP3 = L-(+)-2-amino-3-phosphonopropionic acid
3,5DHPG = 3,5-dihydroxyphenylglycine
CHPG = (RS)-2-Chloro-5-hydroxyphenylglycine
4-Cl-3,5DHPG = 4-chloro-3,5-dihydroxyphenylglycine
CPPG = (RS)-(-cyclopropyl-4-phosphonophenylglycine
L-CSA = L-cysteine sulphinic acid
L-CCG1 = (2S,3S,4S)-2-(carboxycyclopropyl)glycine
(S)-4C3HPG = (S)-4-carboxy-3-hydroxyphenylglycine
EGLU = (2S)-α-Ethylglutamic acid
L-SOP = L-serine-O-phosphate
MCCG = α -methyl-L-CCG1
MTPG = (RS)-α -methyl-4-tetrazoylphenylglycine
MPPG = (RS)-α -methyl-4-phosphonophenylglycine
MAP4 = (S)-2-amino-2-methyl-4-phosphonobutanoic acid
MCPA = (S)-α -methyl-3-carboxyphenylalanine
MSOP = (RS)-α -methylserine-O-phosphate
homoAMPA = 2-amino-4-(3-hydroxy-5-methylisoxazol-4-yl)butyric acid
NAAG = N-acetyl-aspartyl-glutamate
PHCCC = N-phenyl-7-(hydroxyimino)cyclopropa[b]chromen-1a-carboxamide

the $Na^+$ and $K^+$ gradients provide the driving force and the inward movement of glutamate is electrogenic. The neuronal glutamate transporters have an associated
Cl- conductance, which contributes to their capacity to terminate the excitatory effect of synaptically released glutamate (Fairman et al., 1995; Arriza et al., 1997). Studies with specific antibodies and with riboprobes for mRNA (Rothstein et al., 1994; Chaudhry et al., 1995; Torp et al., 1997) show that GLT-1 is expressed in astrocytes throughout the brain but with enhanced expres-

TABLE 5

Glutamate transporters

| RAT<br>HUMAN | GLAST<br>EAAT-1 | GLT-1<br>EAAT-2 | EAAC-1<br>EAAT-3 | <br>EAAT-4 |
|---|---|---|---|---|
| Location | Bergmann glia | Glia | Neurons<br>(intestines) | Purkinje cells |
| Inhibitors | DL-threo-3-hydroxyaspartate<br>L-alpha-aminoadipate<br>Arachidonic acid | Dihydrokainate<br>L-trans-2,4-pyrrolidone-dicarboxylate | DL-threo-3-hydroxyaspartate | |
| Antisense probes | | neurodegeneration | epilepsy | |

sion in processes adjacent to glutamatergic synapses. GLAST is prominent in Bergmann glia. EAAC1 is prominently expressed in particular classes of neurons but is not specific for glutamatergic neurons. GLT-1 accounts for more than 90% of glutamate transport in the rat forebrain. Knockout mice which do not express GLT-1 show spontaneous seizures resembling those induced by NMDA and enhanced acute neurodegeneration (Tanaka et al., 1997).

Various classes of compounds that block glutamate transport are known but their selectivity for the different transporters is limited and seems to depend on the type of cell in which the wild type occurs or the system in which the transporter is experimentally expressed (Swanson et al., 1997). Interestingly arachidonic acid which is liberated in various physiological and pathological processes inhibits human EAAT-1 but potentiates EAAT-2 (the homologue of GLT-1) (Zerangue et al., 1995). The availability of pure expression systems should aid the identification or synthesis of more specific inhibitors. In the meantime information about the role of the different glutamate transporters can best be derived using gene knockdown techniques. Chronic intracerebroventricular administration of antisense probes blocking the synthesis of EAAC1 in the rat leads to epileptic seizures (Rothstein et al., 1996) whereas antisense probes to the glial transporters, GLT-1 and GLAST increase extracellular concentrations of glutamate and produce excitotoxic neurodegeneration. GLT-1 therefore seems to play an important role in preventing glutamate producing excitotoxic damage. Enhancing its expression or function may be therapeutically beneficial. This approach is considered further in the next section.

## Non-pharmacological approaches to synaptic modification

Future targeting of the glutamate synapse may not rely on classical pharmacological tools, such as drugs acting on receptors or enzymes. Rather the means of modifying glutamatergic function may be through altered gene expression. This may sometimes be a result of drug administration. Alternative strategies however include the use of antisense probes to diminish the expression of particular gene products and the use of viral vectors to achieve expression or over-expression of particular gene products in certain cell groups. Alternatively it may be possible to use genetically-engineered fibroblast or stem cells to achieve the local production of specific gene products. These strategies may be intended to substitute for or supplement the function of a gene product that is defective (and contributing to the disease phenotype). Alternatively they may be intended to modify function at a point in the glutamatergic system that is not itself defective. Thus hyperfunction of NMDA receptors in epilepsy might be compensated for by overexpression of presynaptic mGluRs (mGluR2, mGluR3, mGluR4, mGluR7, mGluR8), thereby decreasing glutamate release or by overexpression of neuronal glutamate transporter (thereby shortening postsynaptic excitation).

## Conclusions

Any forecast of future directions for therapies directed at the glutamate synapse must emphasize the probability of designing or discovering more potent agents acting on specific glutamate receptors. Thus many pharmaceutical companies are committing substantial resources to the synthesis of appropriate series of compounds employing combinatorial chemistry and screening in high thoughput systems based on the functional expression of particular receptor molecules (single mGluRs, heteromeric recombinant AMPA, kainate and NMDA receptors with specific subunit composition). Natural products can be similarly screened. In the past displacement of specific ligand binding was a common assay procedure, but the use of ion- or pH-sensitive fluorescent dyes permits a wide range of functional assays which not only provide more convenient assays but may additionally lead to the identification of compounds acting at sites on the receptor complex other than the ligand binding site.

Positive clinical data concerning the use of glutamate antagonists in stroke and traumatic brain injury may be reported in the next year or so. Further clinical data relating to amnesia and Alzheimer's disease will soon be available. In the longer term glutamate metabotropic receptors are a particularly attractive target and appropriately selective potent compounds will become available for preclinical evaluation in the next few years. A full picture of the clinical relevance of drugs acting on the glutamatergic synapse will, however, require a decade or more of studies with molecular biological techniques, with preclinical in vivo models and major clinical trials.

## References


Anthony, E.W. and Nevins M.E. (1993) Anxiolytic-like effects of *N*-methyl-D-aspartate-associated glycine receptor ligands in the rat potentiated startle test. *Eur. J. Pharmacol.*, 250: 317–324.

Arai, A., Kessler, M., Rogers, G. and Lynch, G. (1996) Effects of a memory-enhancing drug on DL-α-amino-3-hydroxy-5-methyl-4-isoxazolepropionic acid receptor currents and synaptic transmission in hippocampus. *J. Pharmacol. Exp. Ther.*, 278: 627–638.

Arriza, J.L., Fairman, W.A., Wadiche, J.I., Murdoch, G.H., Kavanaugh, M.P. and Amara, S.G. (1994) Functional comparisons of three glutamate transporter subtypes cloned from human motor cortex. *J. Neurosci.*, 14: 5559–5569.

Arriza, J.L., Eliasof, S., Kavanaugh, M.P. and Amara, S.G. (1997) Excitatory amino acid transporter 5, a retinal glutamate transporter coupled to a chloride conductance. *Proc. Natl. Acad. Sci., USA*, 94: 4155–4160.

Avenet, P., Leonardon, J., Besnard, F., Graham, D., Depoortere, H. and Scatton, B. (1997) Antagonist properties of eliprodil and other NMDA receptor antagonists at rat NR1A/NR2A and NR1A/NR2B receptors expressed in Xenopus oocytes. *Neurosci. Lett.*, 223: 133–136.

Baron, B.M., Harrison, B.L., Kehne, J.H., Schmidt, C.J., Van Giersbergen, P.L.M., White, H.S., Siegel, B.W., Senyah, Y., McCloskey, T.C., Fadayel, G.M., Taylor, V.L., Murawsky, M.K., Nyce, P. and Salituro, F.G. (1997) Pharmacological characterization of MDL 105,519, an NMDA receptor glycine site antagonist. *Eur. J. Pharmacol.*, 323: 181–192.

Baxter, M.G., Lanthorn, T.H., Frick, K.M., Golski, S., Wan, R.-Q. and Olton, D.S. (1994) D-cycloserine, a novel cognitive enhancer, improves spatial memory in aged rats. *Neurobiol. Aging*, 15: 207–213.

Bettler, B. and Mulle, C. (1995) Review: Neurotransmitter receptors II. AMPA and kainate receptors. *Neuropharmacology*, 34: 123–139.

Bordi, F., Pietra, C., Ziviani, L. and Reggiani, A. (1997) The glycine antagonist GV150526 protects somatosensory evoked potentials and reduces the infarct area in the MCAo model of focal ischemia in the rat. *Exp. Neurol.*, 145: 425–433.

Breese, C.R., Freedman, R. and Leonard, S.S. (1995) Glutamate receptor subtype expression in human postmortem brain tissue from schizophrenics and alcohol abusers. *Brain Res.*, 674: 82–90.

Breukel, A.I.M., Lopes da Silva, F.H. and Ghijsen W.E.J.M (1997) Cholecystokinin (CCK-8) modulates vesicular release of excitatory amino acids in rat hippocampal nerve endings. *Neurosci. Lett.*, 234: 67–70.

Buchan, A.M., Xue, D. Huang, Z., Smith, K.H. and Lesiuk, H. (1991) Delayed AMPA receptor blockade reduces cerebral infarction induced by focal ischemia. *Neuroreport*, 2: 473–476.

Castillo, P.E., Malenka, R.C. and Nicoll, R.A. (1997) Kainate receptors mediate a slow postsynaptic current in hippocampal CA3 neurons. *Nature*, 388: 182–186.

Chaudhry, F.A., Lehre, K.P., Van Lookeren Campagne, M., Ottersen, O.P., Danbolt, N.C. and Storm-Mathisen, J. (1995) Glutamate transporters in glial plasma membranes: Highly differentiated localizations revealed by quantitative ultrastructural immunocytochemistry. *Neuron*, 15: 711–720.

Chessell, I.P., Procter, A.W., Francis, P.T. and Bowen, D.M. (1991) D-Cycloserine, a putative cognitive enhancer, facilitates activation of the *N*-methyl-D-aspartate receptor-ionophore complex in Alzheimer brain. *Brain Res.*, 565: 345–348.

Chung, S.H. and M.S. Johnson (1984) Studies on sound-induced epilepsy in mice. *Proc. R. Soc. Lond. [Biol.]*, 221: 145–168.

Chittajallu, R., Vignes, M., Dev, K.K., Barnes, J.M., Collingridge, G.L. and Henley, J.M. (1996) Regulation of glutamate release by presynaptic kainate receptors in the hippocampus. *Nature*, 379: 78–81.

Conti, F. and Minelli, A. (1994) Glutamate immunoreactivity in rat cerebral cortex is reversibly abolished by 6-diazo-5-oxo-L-norleucine (DON), an inhibitor of phosphate-activated glutaminase. *J. Histochem. Cytochem.*, 42: 717–726.

Croucher, M.J., Collins, J.F. and Meldrum, B.S. (1982) Anticonvulsant action of excitatory amino acid antagonists. *Science*, 216: 899–901.

Davies, S.N. and Lodge, D. (1987) Evidence for involvement of *N*-methyl-aspartate receptors in "wind-up" of class 2 neurones in the dorsal horn of the rat. *Brain Res.*, 424: 402–406.

Deakin, J.F.W., Slater, P., Simpson, M.D.C., Gilchrist, A.C., Skan, W.J., Royston, M.C., Reynolds, G.P. and Cross, A.J. (1989) Frontal cortical and left temporal glutamatergic dysfunction in schizophrenia. *J. Neurochem.*, 52: 1781–1786.

Dickenson, A.H. and Sullivan, A.F. (1987) Evidence for a role of the NMDA receptor in the frequency dependent potentiation of deep rat dorsal horn nociceptive neurones following C fibre stimulation. *Neuropharmacology*, 26: 1235–1238.

Eastwood, S.L., Burnet, P.W.J. and Harrison, P.J (1997) GluR2 glutamate receptor subunit flip and flop isoforms are decreased in the hippocampal formation in schizophrenia: A reverse transcriptase-polymerase chain reaction (RT-PCR) study. *Mol. Brain Res.*, 44: 92–98.

Eastwood, S.L., Kerwin, R.W. and Harrison, P.J. (1997) Immunoautoradiographic evidence for a loss of α-amino-3-hydroxy-5-methyl-4-isoxazole propionate-preferring non-*N*-methyl-D-aspartate glutamate receptors within the medial temporal lobe in schizophrenia. *Biol. Psychiatry*, 41: 636–643.

Eastwood, S.L., McDonald, B., Brunette, P.W.J., Beckwith, J.P., Kerwin, R.W. and Harrison, P.J. (1995) Decreased expression of mRNAs encoding non-NMDA glutamate receptors GluR1 and GluR2 in medial temporal lobe neurons in schizophrenia. *Mol. Brain Res.*, 29: 211–223.

Faingold, C.L., Naritoku, D.K., Copley, C.A., Randall, M.E., Riaz, A., Boersma Anderson, C.A., and Arneric, S.P. (1992) Glutamate in the inferior colliculus plays a critical role in audiogenic seizure initiation. *Epilepsy Res.*, 13: 95–105.

Fairman, W.A., Vandenberg, R.J., Arriza, J.L., Kavanaugh, M.P. and Amara, S.G. (1995) An excitatory amino-acid transporter with properties of a ligand-gated chloride channel. *Nature*, 375: 599–603.

Felsby, S., Nielsen, J., Arendt-Nielsen, L. and Jensen, T.S. (1996) NMDA receptor blockade in chronic neuropathic pain: A comparison of ketamine and magnesium chloride. *Pain*, 64: 283–291.

Fredholm, B.B. (1997) Adenosine and neuroprotection. In: Green, A.R., Cross, A.J., eds. Neuroprotective agents. *San Diego: Academic Press*, pp.259–280.

Geiger, J.R.P., Melcher T., Koh, D.-S., Sakmann, B., Seeburg, P.H., Jonas, P. and Monyer, H. (1995) Relative abundance of subunit mRNAs determines gating and $Ca^{2+}$ permeability of AMPA receptors in principal neurons and interneurons in rat CNS. *Neuron*, 15: 193–204.

Gill, R. (1994) The pharmacology of α-amino-3-hydroxy-5-methyl-4-isoxazole propionate (AMPA)/kainate antagonists and their role in cerebral ischaemia. *Cerebrovasc. Brain Metab. Rev.*, 6: 225–256.

Gill, R. and Lodge, D. (1994) The neuroprotective effects of the decahydroisoquinoline, LY 215490; A novel AMPA antagonist in focal ischaemia. *Neuropharmacology*, 33: 1529–1536.

Gill, R., Nordholm, L. and Lodge, D. (1992) The neuroprotective actions of 2,3-dihydroxy-6-nitro-7-sulfamoyl-benzo(F)quinoxaline (NBQX) in a rat focal ischaemia model. *Brain Res.*, 580: 35–43.

Gong, Q.-Z., Delahunty, T.M., Hamm, R.J. and Lyeth, B.G. (1995) Metabotropic glutamate antagonist, MCPG, treatment of traumatic brain injury in rats. *Brain Res.*, 700: 299–302.

Gramsbergen, J.B.P., Hodgkins, P.S., Rassoulpour, A., Turski, W.A., Guidetti, P. and Schwarcz, R. (1997) Brain-specific modulation of kynurenic acid synthesis in the rat. *J. Neurochem.*, 69: 290–298.

Goff, D.C. and Wine L. (1997) Glutamate in schizophrenia: Clinical and research implications. *Schiz. Res.*, 27: 157–168.

Granger, R., Deadwyler, S., Davis, M., Moskovitz, B., Kessler, M., Rogers, G. and Lynch, G. (1996) Facilitation of glutamate receptors reverses an age-associated memory impairment in rats. *Synapse*, 22: 332–337.

Heresco-Levy, U., Javitt, D.C. and Ermilov, M. (1996) Double-blind, placebo-controlled, crossover trial of glycine adjuvant therapy for treatment-resistant schizophrenia. *Br. J. Psychiatry*, 169: 610–617.

Hollmann, M., Boulter, J., Maron, C., Beasley, L., Sullivan, J., Pecht, G. and Heinemann, S. (1993) Zinc potentiates agonist-induced currents at certain splice variants of the NMDA receptor. *Neuron*, 10: 943–954.

Hollmann, M. and Heinemann, S. (1994) Cloned glutamate receptors. *Annu. Rev. Neurosci.*, 17: 31–108.

Ishimaru, M., Kurumaji, A. and Toru, M. (1996) Glutamate receptors and schizophrenia. *Trends Neurosci.*, 19: 416–417.

Javitt, D.C and Zukin, S.R. (1991) Recent advances in the phencyclidine model of schizophrenia. *Am. J. Psychiatry*, 148: 1301–1308.

Javitt, D.C., Zylberman, L. and Zukin, S.R. (1994) Amelioration of negative symptoms of schizophrenia by glycine. *Am. J. Psychiatry*, 151: 1234–1236.

Kälviäinen, R., Halonen, T., Pitkänen, A. and Riekkinen, P.J. (1993) Amino acid levels in the cerebrospinal fluid of newly diagnosed epileptic patients: Effect of vigabatrin and carbamazepine monotherapies. *J. Neurochem.*, 60: 1244–1250.

Keller, B.U., Blaschke, M., Rivosecchi, R., Hollmann, M., Heinemann, S.F. and Konnerth, A. (1993) Identification of a subunit-specific antagonist of α-amino-3-hydroxy-5-methyl-4-isoxazolepropionate/kainate receptor channels. *Proc. Natl. Acad. Sci. USA*, 90: 605–609.

Kerwin, R.W., Patel, S., Meldrum, B., Czudeck, C. and Reynolds, G.P. (1988) Asymmetrical loss of a glutamate receptor subtype in left hippocampus in schizophrenia. *Lancet*, (i): 583–584.

Kerwin, R., Patel, S. and Meldrum, B. (1990) Quantitative autoradiographic analysis of glutamate binding sites in the hippocampal formation in normal and schizophrenic brain post mortem. *Neuroscience*, 39: 25–32.

Kieburtz, K., Feigin, A., McDermott, M., Como, P., Abwender, D., Zimmerman, C., Hickey, C., Orme, C., Claude, K., Sotack, J., Greenamyre, J.T., Dunn, C. and Shoulson, I. (1996) A controlled trial of remacemide hydrochloride in Huntington's disease. *Mov. Disord.*, 11: 273–277.

Köhr, G. and Mody, I. (1994) Kindling increases *N*-methyl-D-aspartate potency at single *N*-methyl-D-aspartate channels in dentate gyrus granule cells. *Neuroscience*, 62: 975–981.

Kretschmer, B.D., Kratzer, U., Breithecker, K. and Koch, M. (1997) ACEA 1021, a glycine site antagonist with minor psychotomimetic and amnestic effects in rats. *Eur. J. Pharmacol.*, 331: 109–116.

Kristensen, J.D., Karlsten, R., Gordh, T. and Berg, O. (1992) The NMDA antagonist 3-(2-carboxypiperazin-4-yl)propyl-1-phosphonic acid (CPP) has antinociceptive effect after intrathecal injection in the rat. *Pain*, 56: 59–67.

Kristensen, J.D., Svensson, B. and Gordh, T. Jr. (1992) The NMDA-receptor antagonist CPP abolishes neurogenic 'wind-up pain' after intrathecal administration in humans. *Pain*, 51: 249–253.

Kristensen, J.D. (1997) Intrathecal administration of a competitive NMDA receptor antagonist for pain treatment. In P.L. Herrling (ed): Excitatory amino acids: Clinical results with antagonists. San Diego: Academic Press, pp. 23–30.

Lacomblez, L., Bensimon, G., Leigh, P.N., Guillet, P. and Meininger, V. (1996) Dose-ranging study of riluzole in amyotrophic lateral sclerosis. *Lancet*, 347: 1425–1431.

Larson, J., Lieu, T., Petchpradub, V., LeDuc, B., Ngo, H., Rogers, G.A. and Lynch, G. (1995) Facilitation of olfactory learning by a modulator of AMPA receptors. *J. Neurosci.*, 15: 8023–8030.

Laurie, D.J. and Seeburg, P.H. (1994) Ligand affinities at recombinant *N*-methyl-D-aspartate receptors depend on subunit composition. *Eur. J. Pharmacol. Mol. Pharmacol.*, 268: 335–345.

Leiderman, E., Zylberman, I., Zukin, S.R., Cooper, T.B. and Javitt, D.C. (1996) Preliminary investigation of high-dose oral glycine on serum levels and negative symptoms in schizophrenia: *An open-label trial. Biol. Psychiatry*, 39: 213–215.

Lhuintre, J.P., Daoust, M., Moore, N., Chretien, P., Saligaut, C., Tran, G., Boismare, F. and Hillemand, B. (1985) Ability of calcium bis acetyl homotaurine, a GABA agonist, to prevent relapse in weaned alcoholics. *Lancet*, 1: 1014–1016.

Lhuintre, J.P., Moore, N., Tran, G., Steru, L., Lancrenon, S., Daoust, M., Parot, P., Ladure, P., Zarnitsky, C., Boismare, F. and Hillemand, B. (1990) A double-blind randomized multicenter study of calcium acetylhomotaurine in the prevention of alcoholic relapse. *Alcoh. Alcoholism*, 25: 613–622.

Li, H. and Buchan A.M. (1993) Treatment with an AMPA antagonist 12 hours following severe normothermic forebrain ischemia prevents CA1 neuronal injury. *J. Cereb. Blood Flow Metab.*, 13: 933–939.

Ling-Gang, W. and Saggau, P. (1997) Presynaptic inhibition of elicited neurotransmitter release. TINS 20: 204–212.

Lu, X.-C.M., Williams, A., Britton, P., Tortella, F.C., Estrada, J., Jackson, P.F. and Slusher, B.S. (1997) Time course of neuroprotection with NAALADase inhibition following MCAO in rats. *Soc. Neurosci. Abst.*, 23: 2302.

Luby, E.D., Cohen, B.D., Rosenbaum, G, Gottlieb, J.S. and Kelley, R. (1959) Study of a new schizophrenomimetic drug-serenyl. *Arch. Neurol. Psychiat.*, 81: 363–369.

Lujn, R., Nusser, J.Z., Roberts, D.B., Shigemoto, R. and Somogyi, P. (1996) Perisynaptic location of metabotropic glutamate receptors mGluR1 and mGluR5 on dendrites and dendritic spines in the rat hippocampus. *Eur. J. Neurosci.*, 8: 1488–1500.

Lynch, G., Granger, R., Ambros-Ingerson, J, Davis, C.M., Kessler, M. and Schehr, R. (1997) Evidence that a positive modulator of AMPA-type glutamate receptors improves delayed recall in aged humans. *Exp. Neurol.*, 145: 89–92.

Madamba, S.G., Schweitzer, P., Zieglgänsberger, W. and Siggins, G.R. (1996) Acamprosate (calcium acetylhomotaurinate) enhances the *N*-methyl-D-aspartate component of excitatory neurotransmission in rat hippocampal CA1 neurons in vitro. *Alcohol. Clin. Exp. Res.*, 20: 651–658.

Matsuoka, N. and Aigner, T.G. (1996) D-cycloserine, a partial agonist at the glycine site coupled to *N*-methyl-D-aspartate receptors, improves visual recognition memory in rhesus monkeys. *J. Pharmacol. Exp. Ther.*, 278: 891–897.

Maurset, A., Skoglund, L.A., Justveit, O. and Oye, I. (1989) Comparison of ketamine and pethidine in experimental and postoperative pain. *Pain*, 36: 37–41.

McIntosh, T.K., Smith, D.H. and Garde, E. (1996) Therapeutic approaches for the prevention of secondary brain injury. *Europ. J. Anaesthes*, 13: 291–309.

Medina, L., Figueredo-Cardenas, G., Rothstein, J.D. and Reiner, A. (1996) Differential abundance of glutamate transporter subtypes in amyotrophic lateral sclerosis (ALS)-vulnerable versus ALS-resistant brain stem motor cell groups. *Exp. Neurol.*, 142: 287–295.

Meldrum, B.S. (1990) Protection against ischaemic neuronal damage by drugs acting on excitatory neurotransmission. *Cerebrovasc. Brain Metab. Rev.*, 2: 27–57.

Meldrum, B.S. (1994) Neuroprotection by NMDA and non-NMDA glutamate antagonists. In M.G. Palfreyman, I.J. Reynolds, and P. Skolnick (eds): Direct and allosteric control of glutamate receptors. *Boca Raton: CRC Press*, pp. 127–138.

Miller, S., Romano, C. and Cotman, C.W. (1995) Growth factor upregulation of a phosphoinositide-coupled metabotropic glutamate receptor in cortical astrocytes. *J. Neurosci.*, 15: 6103–6109.

Monyer, H., Burnashev, N., Laurie, D.J., Sakmann, B. and Seeburg, P.H. (1994) Developmental and regional expression in the rat brain and functional properties of four NMDA receptors. *Neuron*, 12: 529–540.

Morrisett, R.A., Rezvani, A.H., Overstreet, D., Janowsky, D.S., Wilson, W.A. and Swartzwelder, H.S. (1990) MK-801 potently inhibits alcohol withdrawal seizures in rats. *Eur. J. Pharmacol.*, 176: 103–105.

Mukhin, A., Fan, L. and Faden, A.I. (1996) Activation of metabotropic glutamate receptor subtype mGluR1 contributes to post-traumatic neuronal injury. *J. Neurosci.*, 16: 6012–6020.

Myhrer, T. and Paulsen, R.E. (1995) Memory impairment in rats with glutamatergic temporal systems disrupted is attenuated by D-cycloserine: Effects of postoperative time of injection. *Psychobiology*, 23: 233–239.

Nakanishi, S. (1992) Molecular diversity of glutamate receptors and implications for brain function. *Science*, 258: 597–603.

Ohishi, H., Akazawa, C., Shigemoto, R., Nakanishi, S. and Mizuno, N. (1995) Distributions of the mRNAs for L-2-amino-4-phosphonobutyrate-sensitive metabotropic glutamate receptors, mGluR4 and mGluR7, in the rat brain. *J. Comp. Neurol.*, 360: 555–570.

Ohno, M. and Watanabe, S. (1996) D-cycloserine, a glycine site agonist, reverses working memory failure by hippocampal muscarinic receptor blockade in rats. *Eur. J. Pharmacol.*, 318: 267–271.

Petralia, R.S., Wang, Y.X., Niedzielski, A.S. and Wenthold, R.J. (1996) The metabotropic glutamate receptors, mGluR2 and mGluR3, show unique postsynaptic, presynaptic and glial localizations. *Neuroscience*, 71: 949–976.

Qian, J., Colmers, W.F. and Saggau (1997) Inhibition of synaptic transmission by neuropeptide Y in rat hippocampal area CA1: modulation of presynaptic $Ca^{2+}$ entry. *J. Neurosci.*, 17: 8169–8177.

Quartermain, D., Mower, J., Rafferty, M.F., Herting, R.L. and Lanthorn, T.H. (1994) Acute but not chronic activation of the NMDA-coupled glycine receptor with D-cycloserine facilitates learning and retention. *Eur. J. Pharmacol.*, 257: 7–12.

Rho, J.M., Donevan, S.D. and Rogawski, M.A. (1997) Barbiturate-like actions of the propanediol dicarbamates felbamate and meprobamate. *J. Pharmacol. Exp. Ther.*, 280: 1383–1391.

Rothstein, J.D., Jin, L., Dykes-Hoberg, M. and Kuncl, R.W. (1993) Chronic inhibition of glutamate uptake produces a model of slow neurotoxicity. *Proc. Natl. Acad. Sci., USA*, 90: 6591–6595.

Rothstein, J.D., Dykes-Hoberg, M., Pardo, C.A., Bristol, L.A., Jin, L., Kuncl, R.W., Kanai, Y., Hediger, M.A., Wang, Y.F., Schielke, J.P. and Welty, D.F. (1996) Knockout of glutamate transporters reveals a major role for astroglial transport in excitotoxicity and clearance of glutamate. *Neuron*, 16: 675–686.

Rothstein, J.D. and Kuncl, R.W. (1995) Neuroprotective strategies in a model of chronic glutamate-mediated motor neuron toxicity. *J. Neurochem.*, 65: 643–651.

Rothstein, J.D., Martin, L., Levey, A.I., Dykes-Hoberg, M., Jin, L., Wu, D., Nash, N. And Kuncl, R.W. (1994) Localization of neuronal and glial glutamate transporters. *Neuron*, 13: 713–725.

Rothstein, J.D., Van Kammen, M., Levey, A.I., Martin, L.J. and Kuncl, R.W. (1995) Selective loss of glial glutamate transporter GLT-1 in amyotrophic lateral sclerosis. *Ann. Neurol.*, 38: 73–84.

Rudolphi, K.A., Schubert, P., Parkinson, F.E. and Fredholm, B.B. (1992) Adenosine and brain ischemia. *Cerebrovasc. Brain Metab. Rev.*, 4: 346–369.

Seeburg, P.H. (1993) The TINS/TiPS lecture: The molecular biology of mammalian glutamate receptor channels. TINS 16: 359–365.

Shigemoto, R., Kinoshita, A., Wada, E., Nomura, S., Ohishi, H., Takada, M., Flor, P.J., Neki, A., Abe, T., Nakanishi, S. and Mizuno, N. (1997) Differential presynaptic localization of metabotropic glutamate receptor subtypes in the rat hippocampus. *J. Neurosci.*, 17: 7503–7522.

Shigemoto, R., Kulik, A., Roberts, J.D.B., Ohishi, H., Nusser, Z., Kaneko, T. and Somogyi, P. (1996) Target-cell-specific concentration of a metabotropic glutamate receptor in the presynaptic active zone. *Nature*, 381: 523–525.

Shors, T.J., Servatius, R.J., Thompson, R.F., Rogers, G. and Lynch, G. (1995) Enhanced glutamatergic neurotransmission facilitates classical conditioning in the freely moving rat. *Neurosci. Lett.*, 186: 153–156.

Simeon, J., Fink, M. and Itil, T. (1970) D-cycloserine therapy of psychosis by symptom provocation. *Compr. Psychiatry*, 11: 80–88.

Simon, R.P., Swan, J.H., Griffiths, T. and Meldrum, B.S. (1984) Blockade of *N*-methyl-D-aspartate receptors may protect against ischaemic damage in the brain. *Science*, 226: 850–852.

Smith, S.E. and Meldrum, B.S. (1992) Cerebroprotective effect of a non-*N*-methyl-D-aspartate antagonist, GYKI 52466, after focal ischemia in the rat. *Stroke*, 23: 861–864.

Smith, S.E. and Meldrum, B.S. (1993) Cerebroprotective effect of a non-*N*-methyl-D-aspartate antagonist, NBQX, after focal ischemia in the rat. *Func. Neurol.*, 8: 43–48.

Speciale, C., Cini, M., Wu, H.Q., Salvati, P., Schwarcz, R., Molinari, A., Calabresi, M. and Varasi, M. (1996) Kynurenic acid-enhancing and anti-ischemic effects of the potent kynurenine 3-hydroxylase inhibitor FCE 28833 in rodents. *Adv. Exp. Med. Biol.*, 398: 221–227.

Staubli, U., Rogers, G. and Lynch, G. (1994) Facilitation of glutamate receptors enhances memory. *Proc. Natl. Acad. Sci. USA*, 91: 777–781.

Stefani, A., Spadoni, F. and Bernardi, G. (1997) Differential inhibition by riluzole, lamotrigine, and phenytoin of sodium and calcium currents in cortical neurons: Implications for neuroprotective strategies. *Exp. Neurol.*, 147: 115–122.

Stephens, D.N., Meldrum, B.S., Weidmann, R., Schneider, C. and Grutzner, M. (1986) Does the excitatory amino acid receptor antagonist 2-APH exhibit anxiolytic activity? *Psychopharmacology*, 90: 166–169.

Subramaniam, S., Rho, J.M., Penix, L., Donevan, S.D., Fielding, R.P. and Rogawski, M.A. (1995) Felbamate block of the *N*-methyl-D-aspartate receptor. *J. Pharmacol. Exp. Ther.*, 273: 878–886.

Sveinbjornsdottir, S., Sander, J.W.A.S., Upton, D.,Thompson, P.J., Patsalos, P.N., Hirt, D., Emre, M., Lowe, D. and Duncan, J.S. (1993) The excitatory amino acid antagonist D-CPP-ene (SDZ EAA-494) in patients with epilepsy. *Epilepsy Res.*, 16: 165–174.

Swanson, R.A., Liu, J., Miller, J.W., Rothstein J.D., Farrell, K., Stein, B.A. and Longuemare M.C. (1997) Neuronal regulation of glutamate transporter subtype expression in astrocytes. *J. Neurosci.*, 17: 932–940.

Takaoka, S., Bart, R.D., Pearlstein, R., Brinkhous, A. and Warner, D.S. (1997) Neuroprotective effect of NMDA receptor glycine recognition site antagonism persists when brain temperature is controlled. *J. Cereb. Blood Flow Metab.*, 17: 161–167.

Takizawa, S., Matsushima, K., Fujita, H., Nanri, K., Ogawa, S. and Shinohara,Y. (1995) A selective N-type calcium channel antagonist reduces extracellular glutamate release and infarct volume in focal cerebral ischemia. *J. Cereb. Blood Flow Metab.*, 15: 611–618.

Tanaka, K., Watase, K., Manabe, T., Yamada, K.,Watanabe, M.,Takahashi, K., Iwama, H., Nishikawa T., Ichihara, N., Hori, S. Takimoto, M. and Wada K. (1997) Epilepsy and exacerbation of brain injury in mice lacking the glutamate transporter GLT-1. *Science*, 276: 1699–1702.

Taylor, C.P. and Meldrum, B.S. (1995) $Na^+$ channels as targets for neuroprotective drugs. TIPS 16: 309–316.

Temple, M.D. and Hamm, R.J. (1996) Chronic, post-injury administration of D-cycloserine, an NMDA partial agonist, enhances cognitive performance following experimental brain injury. *Brain Res.*, 741: 246–251.

Thomas A.G., Vornov, J.J., Jackson, P.F. and Slusher, B.S. (1997) Inhibition of NAALADase activity is neuroprotective in cell culture model of ischemia. *Soc. Neurosci. Abst.*, 23: 2302.

Torp, R., Hoover, F., Danbolt, N.C., Storm-Mathisen, J. and Ottersen, O.P. (1997) Differential distribution of the glutamate transporters GLT1 and rEAA1 in rat cerebral cortex and thalamus: An in situ hybridization analysis. *Anat. Embryol.*, 195: 317–326.

Tsai, G., Passani, L.A., Slusher, B.S., Carter, R., Baer, L., Kleinman, J.E. and Coyle, J.T. (1995) Abnormal excitatory neurotransmitter metabolism in schizophrenic brains. *Arch. Gen. Psychiatry*, 52: 829–836.

Valentino, K., Newcomb, R., Gadbois, T., Singh, T., Bowersox, S., Bitner, S., Justice, A., Yamashiro, D., Hoffman, B.B., Ciaranello, R., Miljanich, G. and Ramachandran, J. (1993) A selective N-type calcium channel antagonist protects against neuronal loss after global cerebral ischemia. *Proc. Natl. Acad. Sci. USA*, 90: 7894–7897.

Verdoorn, T.A., Johansen, T.H., Drejer, J. and Nielsen, E.O. (1994) Selective block of recombinant glur6 receptors by NS-102, a novel non-NMDA receptor antagonist. *Eur. J. Pharmacol. Mol. Pharmacol.*, 269: 43–49.

Vignes, M. and Collingridge, G.L. (1997) The synaptic activation of kainate receptors. *Nature*, 388: 179–182.

Wamil, A.W., Cheung, H., Harris, E.W. and McLean, M.J. (1996) Remacemide HCl and its metabolite, FPL 12495AA, limit action potential firing frequency and block NMDA responses of mouse spinal cord neurons in cell culture. *Epilepsy Res.*, 23: 1–14.

Wang, S.J., Huang, C.C., Hsu, K.S., Tsai, J.J. and Gean, P.W. (1996) Inhibition of N-type calcium currents by lamotrigine in rat amygdalar neurones. *Neuroreport*, 7: 3037–3040.

Waniewski, R.A. and Martin, D.L. (1995) Repeated administration of $\gamma$-vinylGABA reduces rat brain glutamine synthetase activity. *J. Neurochem.*, 65: 355–362.

White, H.S., Brown, S.D., Woodhead, J.H., Skeen, G.A. and Wolf, H.H. (1997) Topiramate enhances GABA-mediated chloride flux and GABA-evoked chloride currents in murine brain neurons and increases seizure threshold. *Epilepsy Res.*, 28: 167–179.

Wiley, J.L., Cristello, A.F. and Balster, R.L. (1995) Effects of site-selective NMDA receptor antagonists in an elevated plus-maze model of anxiety in mice. *Eur. J. Pharmacol.*, 294: 101–107.

Wong, L.A. and Mayer, M.L. (1993) Differential modulation by cyclothiazide and concanavalin A of desensitiation at native AMPA- and kainate-preferring glutamate receptors. *Molec. Pharmac.*, 44: 504–510.

Wroblewski, J.T., Blaker, W.D. and and Meek, J.L. (1985) Ornithine as a precursor of neurotransmitter glutamate: Effect of canaline on ornithine aminotransferase activity and glutamate content in the septum of rat brain. *Brain Res.*, 329: 161–168.

Xie, Z. and Commissaris, R.L. (1992) Anxiolytic-like effects of the noncompetitive NMDA antagonist MK 801. *Pharmacol. Biochem. Behav.*, 43: 471–477.

Zeise, M.L., Kasparow, S., Capogna, M. and Zieglgasberger, W. (1993) Acamprosate (calcium acetylhomotaurinate) decreases postsynaptic potentials in the rat neocortex: Possible involvement of excitatory amino acid receptors. *Eur. J. Pharmacol.*, 231: 47–52.

Zerangue, N., Arriza, J.L., Amara, S.G. and Kavanaugh, M.P. (1995) Differential modulation of human glutamate transporter subtypes by arachidonic acid. *J. Biol. Chem.*, 270: 6433–6435.

Zerangue N. and Kavanaugh, M.P. (1996) Flux coupling in a neuronal glutamate transporter. *Nature*, 383: 634–637.

# Subject Index

Edwards Brothers Inc.
Ann Arbor MI. USA
March 28, 2012